T0331017

Local Cohomology

This Second Edition of a successful graduate text provides a careful and detailed algebraic introduction to Grothendieck's local cohomology theory, including in multi-graded situations, and provides many illustrations of the theory in commutative algebra and in the geometry of quasi-affine and quasi-projective varieties. Topics covered include Serre's Affineness Criterion, the Lichtenbaum–Hartshorne Vanishing Theorem, Grothendieck's Finiteness Theorem and Faltings' Annihilator Theorem, local duality and canonical modules, the Fulton–Hansen Connectedness Theorem for projective varieties, and connections between local cohomology and both reductions of ideals and sheaf cohomology.

The book is designed for graduate students who have some experience of basic commutative algebra and homological algebra, and also for experts in commutative algebra and algebraic geometry. Over 300 exercises are interspersed among the text; these range in difficulty from routine to challenging, and hints are provided for some of the more difficult ones.

M. P. Brodmann is Emeritus Professor in the Institute of Mathematics at the University of Zürich.

R. Y. Sharp is Emeritus Professor of Pure Mathematics at the University of Sheffield.

CAMBRIDGE STUDIES IN ADVANCED MATHEMATICS

All the titles listed below can be obtained from good booksellers or from Cambridge University Press.
For a complete series listing visit: http://www.cambridge.org/mathematics.

Local Cohomology

An Algebraic Introduction with Geometric Applications

SECOND EDITION

M. P. BRODMANN
Universität Zürich

R. Y. SHARP
University of Sheffield

 CAMBRIDGE
UNIVERSITY PRESS

CAMBRIDGE
UNIVERSITY PRESS

University Printing House, Cambridge CB2 8BS, United Kingdom

One Liberty Plaza, 20th Floor, New York, NY 10006, USA

477 Williamstown Road, Port Melbourne, VIC 3207, Australia

314-321, 3rd Floor, Plot 3, Splendor Forum, Jasola District Centre, New Delhi - 110025, India

79 Anson Road, #06-04/06, Singapore 079906

Cambridge University Press is part of the University of Cambridge.

It furthers the University's mission by disseminating knowledge in the pursuit of education, learning and research at the highest international levels of excellence.

www.cambridge.org
Information on this title: www.cambridge.org/9780521513630

© Cambridge University Press 1998, 2013

First published 1998
Second Edition 2013

A catalogue record for this publication is available from the British Library

ISBN 978-0-521-51363-0 Hardback

To Alice

from the second author

Contents

Preface to the First Edition

One can take the view that local cohomology is an algebraic child of geometric parents. J.-P. Serre's fundamental paper 'Faisceaux algébriques cohérents' [77] represents a cornerstone of the development of cohomology as a tool in algebraic geometry: it foreshadowed many crucial ideas of modern sheaf cohomology. Serre's paper, published in 1955, also has many hints of themes which are central in local cohomology theory, and yet it was not until 1967 that the publication of R. Hartshorne's 'Local cohomology' Lecture Notes [25] (on A. Grothendieck's 1961 Harvard University seminar) confirmed the effectiveness of local cohomology as a tool in local algebra.

Since the appearance of the Grothendieck–Hartshorne notes, local cohomology has become indispensable for many mathematicians working in the theory of commutative Noetherian rings. But the Grothendieck–Hartshorne notes certainly take a geometric viewpoint at the outset: they begin with the cohomology groups of a topological space X with coefficients in an Abelian sheaf on X and supports in a locally closed subspace.

In the light of this, we feel that there is a need for an algebraic introduction to Grothendieck's local cohomology theory, and this book is intended to meet that need. Our book is designed primarily for graduate students who have some experience of basic commutative algebra and homological algebra; for definiteness, we have assumed that our readers are familiar with many of the basic sections of H. Matsumura's [50] and J. J. Rotman's [71]. Our approach is based on the fundamental 'δ-functor' techniques of homological algebra pioneered by Grothendieck, although we shall use the 'connected sequence' terminology of Rotman (see [71, pp. 212–214]).

However, we have not overlooked the geometric roots of the subject or the significance of the ideas for modern algebraic geometry. Indeed, the book presents several detailed examples designed to illustrate the geometrical significance of aspects of local cohomology; we have chosen examples which require only basic ideas from algebraic geometry. In this spirit, there is one particular example, which we refer to as 'Hartshorne's Example', to which we return several times in order to illustrate various points.

The geometric aspects are, in fact, nearer the surface of our treatment than might initially be realised, because we make much use of ideal transforms and their universal properties, but it is only in the final chapter that we expose the fundamental links, expressed by means of the Deligne Correspondence, between the ideal transform functors and their right derived functors on the one hand, and section functors of sheaves and sheaf cohomology on the other.

We define the local cohomology functors to be the right derived functors of the appropriate torsion functor, although we establish in the first chapter that one can also construct local cohomology modules as direct limits of 'Ext' modules; we also present alternative constructions of local cohomology modules, one via cohomology of Čech complexes, and the other via direct limits of homology modules of Koszul complexes, in Chapter 5. (In fact, we do not use this Koszul complex approach very much at all in this book.)

Chapters 2, 3 and 4 include fundamental ideas concerning ideal transforms and their universal properties, the Mayer–Vietoris Sequence for local cohomology and the Independence and Flat Base Change Theorems: we regard all of these as technical cornerstones of the subject, and we certainly use them over and over again.

The main purpose of Chapters 6 and 7 is the presentation of some of Grothendieck's important vanishing theorems for local cohomology, which relate such vanishing to the concepts of dimension and grade. This work is mainly 'algebraic' in nature. In Chapter 8, we present another vanishing theorem for local cohomology modules, namely the local Lichtenbaum–Hartshorne Vanishing Theorem: this has an 'analytic' flavour, in the sense that it is intimately related with 'formal' methods and techniques, that is, with passage to completions of local rings and with the structure theory for complete local rings. The Lichtenbaum–Hartshorne Theorem has important geometric applications: for example, we show in Chapter 19 how it can be used to obtain major results about the connectivity of algebraic varieties.

Grothendieck's Finiteness Theorem and G. Faltings' Annihilator Theorem for local cohomology are the main subjects of Chapter 9. These two theorems also have major geometric applications, including, for example, in Macaulayfication of schemes. They also have significance for the theory of generalized Cohen–Macaulay modules and Buchsbaum modules, two concepts which feature briefly in the exercises in Chapter 9.

We have delayed the introduction of duality (until Chapters 10 and 11) because quite a lot can be achieved without it, and because, for our discussion of duality, we have had to assume (on account of limitations of space) that the reader is familiar with the Matlis–Gabriel decomposition theory for injective modules over a commutative Noetherian ring (although we have reviewed that theory and provided some detailed proofs). We have not explicitly used dualizing complexes or derived categories, as it seems to us that such technicalities could daunt youthful readers and are not essential for a presentation of the main ideas. After the introduction of local duality in Chapter 11, we show how this duality can be used to derive some results established earlier in the book by different means.

The many recent research papers involving local cohomology of graded rings illustrate the importance of this aspect, and we have made some effort to develop the fundamentals of local cohomology in the graded case carefully in Chapters 12 and 13: various representations of local cohomology modules obtained in the earlier chapters inherit natural gradings when the ring, module and ideal concerned are all graded, and it seems to us that it is important to know that there is really only one sensible way of grading local cohomology modules in such circumstances. Our main aim in Chapter 12 has been to address this point. In Chapter 13, 'graded frills' are added to basic results proved earlier in the book.

The short Chapter 14 establishes some links between graded local cohomology and projective varieties; it has been included to provide a little geometric insight, and in order to motivate the work on Castelnuovo–Mumford regularity in Chapters 15–17, and the connections between ideal transforms and section functors of sheaves presented in Chapter 20.

In Chapter 15, we study the graded local cohomology of a homogeneous positively graded commutative Noetherian ring R with respect to the irrelevant ideal. One of the most important invariants in this context is Castelnuovo–Mumford regularity. This concept has, in addition to fundamental significance in projective algebraic geometry, connections with the degrees of generators of a finitely generated graded R-module M: it turns out that M can be generated by homogeneous elements of degrees not exceeding $\mathrm{reg}(M)$, the Castelnuovo–Mumford regularity of M. In turn, this leads on to connections with the theory of syzygies of finitely generated graded modules over polynomial rings over a field.

In certain circumstances, including when M is the vanishing ideal $I_{\mathbb{P}^r}(V)$ of a projective variety $V \subset \mathbb{P}^r$, the above-mentioned $\mathrm{reg}(M)$ coincides with $\mathrm{reg}^2(M)$, the Castelnuovo–Mumford regularity of M at and above level 2. In Chapters 16 and 17, we present bounds for the invariant $\mathrm{reg}^2(M)$. Chapter 16 contains *a priori* bounds which apply whenever the underlying homogeneous positively graded commutative Noetherian ring R has Artinian 0-th component. Chapter 17 is more specialized, and contains bounds expressed in terms of coefficients of Hilbert polynomials; our development of this theory includes a presentation of basic ideas concerning cohomological Hilbert polynomials. The motivation for our work in Chapter 17 comes from D. Mumford's classical work [54]: Mumford established the existence of bounds of the type we present, but, in the spirit of this book, we have added some precision.

One could view Chapters 18 and 19 as propaganda for the effectiveness of local cohomology as a tool in algebra and geometry. Chapter 18 presents some applications of Castelnuovo–Mumford regularities to reductions of ideals. This is a fast developing area, and we have not attempted to give an encyclopaedic account; instead, we have tried to present the basic ideas and a few recent results to whet the reader's appetite. The highlight of Chapter 18 is a theorem of L. T. Hoa; the statement of this theorem is satisfyingly simple, and makes no mention of local cohomology, and yet Hoa's proof, which we present towards the end of the chapter, makes significant use of graded local cohomology.

Chapter 18 is a good advertisement for local cohomology as a 'hidden tool', and Chapter 19 continues this theme, although here the applications (to the connectivity of algebraic varieties) are more geometrical in nature. The only appearances of local cohomology in Chapter 19 are in just two proofs, where a few central ideas (such as the Mayer–Vietoris Sequence and the Lichtenbaum–Hartshorne Vanishing Theorem) are used in crucial ways. No hypothesis or conclusion of any result in the chapter makes any mention of local cohomology, and yet we are able to show how the two results whose proofs use local cohomology can be developed into a theory which leads to proofs of major results involving connectivity, such as Grothendieck's Connectedness Theorem, the Bertini–Gothendieck Connectivity Theorem, the Connectedness Theorem for Projective Varieties due to W. Barth, to W. Fulton and J. Hansen, and to G. Faltings, and a ring-theoretic version of Zariski's Main Theorem. This chapter is certainly a good advertisement for the power of local cohomology as a tool in algebraic geometry!

Finally, in Chapter 20, we bring the subject 'home to its roots', so to speak, by presenting links between local cohomology and the cohomology of quasi-coherent sheaves over certain Noetherian schemes. (Chapter 20 is the only one for which we have assumed that the reader has some basic knowledge about schemes and sheaves.)

Some parts of our presentation are fairly leisurely: this is deliberate, and has been done with graduate students in mind, because we found several preparatory topics where either we knew of no suitable text-book account, or we felt we had something to add to the existing accounts. Examples are the treatments of Matlis duality in Chapter 10, of *canonical modules in Chapter 13, of reductions of ideals in Chapter 18, and of connectedness dimensions in Chapter 19; also, our presentation in Chapter 5 of some links between Koszul complexes and local cohomology is deliberately slow.

Our philosophy throughout has been to try to give a careful and accessible presentation of basic ideas and some important results, illustrating the ideas with examples, to bring the reader to a level of expertise where he or she can approach with some confidence recent research papers in local cohomology. To help with this, the book contains a large number of exercises, and we have supplied hints for many of the more difficult ones.

We have tried out parts of the book, especially the earlier chapters, on some of our own graduate students, and their comments have influenced the final version. We are particularly grateful to Claudia Albertini, Carlo Matteotti, Francesco Mordasini, Henrike Petzl and Massoud Tousi for acting as 'guinea pigs', so to speak. We should also like to express our gratitude to Peter Gabriel, John Greenlees, Martin Holland and Josef Rung for continual interest and encouragement, and to the Schweizerischer Nationalfonds zur Förderung der wissenschaftlichen Forschung, the Forschungsrat des Instituts für Mathematik der Universität Zürich, and the University of Sheffield Research Fund, for financial support to enable several visits for intense collaboration on the book to take place. Both authors would like to thank Alice Sharp: Markus Brodmann thanks her for kind hospitality during pleasant visits to Sheffield for discussions on the book; and Rodney Sharp thanks her for much sympathetic support through the years during which this book was being written (as well as for many things which have nothing to do with local cohomology). Finally, we are very grateful to David Tranah and Roger Astley of Cambridge University Press for their continual encouragement and assistance over many years, and, not least, for their cooperation over our request that the blue stripe on the cover of the book should match the blue of the Zürich trams!

Markus Brodmann   Rodney Sharp
Zürich   Sheffield

April 1997

Preface to the Second Edition

In the fifteen years since we completed the First Edition of this book, we have had opportunity to reflect on how we could change it in order to enhance its usefulness to the graduate students at whom it is aimed. As a result, this Second Edition shows substantial differences from the First. The main ones are described as follows.

One of the more dramatic changes is the introduction of a complete new chapter, Chapter 12, devoted to the study of canonical modules. The treatment of canonical modules in the First Edition was brief and restricted to the case where the underlying ring is Cohen–Macaulay; we assumed that the reader was familiar with the treatment in this case by W. Bruns and J. Herzog in their book on Cohen–Macaulay rings (see [7]). In our new Chapter 12, we present some of the basic work of Y. Aoyama (see [1] and [2]) and follow M. Hochster and C. Huneke [39] in defining a canonical module over a (not necessarily Cohen–Macaulay) local ring $(R, \mathfrak{m})$ to be a finitely generated R-module whose Matlis dual is isomorphic to the 'top' local cohomology module $H_{\mathfrak{m}}^{\dim R}(R)$. Thus this topic is intimately related to local cohomology.

Canonical modules have connections with the theory of S_2-ifications (here, the 'S_2' refers to Serre's condition), and we realised that the development of S_2-ifications can be facilitated by generalizations of arguments we had used to study ideal transforms in §2.2 of the First Edition. For this reason, instead of dealing just with ideal transforms based on the sequence of powers of a fixed ideal, we treat, in §2.2 of this Second Edition, a generalization based on a set $\mathfrak{B}$ of ideals such that, whenever $\mathfrak{b}, \mathfrak{c} \in \mathfrak{B}$, there exists $\mathfrak{d} \in \mathfrak{B}$ with $\mathfrak{d} \subseteq \mathfrak{bc}$. This represents a significant change to Chapter 2.

Another major change concerns our treatment of graded local cohomology. Our new Chapters 13 and 14 treat local cohomology in the situation where the rings, ideals and modules concerned are graded by $\mathbb{Z}^n$, where n is a positive integer. In the First Edition we dealt only with the case where $n = 1$; in the years since that edition was published, there have been more and more uses of local cohomology in multi-graded situations. It was not difficult for us to adapt the treatment of $\mathbb{Z}$-graded local cohomology from the First Edition to the multi-graded case. The main point of our Chapter 13 is to show that, even though there appear to be various possible approaches, there is really only one sensible way of grading local cohomology. Chapter 14 adds '($\mathbb{Z}^n$-)graded frills' to basic results proved earlier in the book. We illustrate this work with some calculations over polynomial rings and Stanley–Reisner rings. Chapters 13 and 14 are generalizations of Chapters 12 and 13 from the First Edition; however, we found it desirable to present some fundamental results from S. Goto's and K.-i. Watanabe's paper [22] about $\mathbb{Z}^n$-graded rings and modules; those results in the particular case when $n = 1$ are more readily available.

The last two decades have seen a surge in the use of local cohomology as a tool in 'characteristic p' commutative algebra, that is, the study of commutative Noetherian rings of prime characteristic p. The key to this is the fact that, for an ideal $\mathfrak{a}$ of such a ring R, and any integer $i \geq 0$, the i-th local cohomology module $H^i_{\mathfrak{a}}(R)$ of R itself with respect to $\mathfrak{a}$ has a so-called 'Frobenius action'. In the new §5.3, we explain why this Frobenius action exists, and in the new §6.5, we use it to present Hochster's proof of his Monomial Conjecture (in characteristic p), and to give some examples of how local cohomology can be used as an effective tool in tight closure theory.

Another new section is §20.5 about locally free sheaves; here we prove Serre's Cohomological Criterion for Local Freeness, Horrocks' Splitting Criterion and Grothendieck's Splitting Theorem. We have also expanded §20.4 with an additional application to projective schemes: we now include a result of Serre about the global generation of twisted coherent sheaves.

In order to include all this new material, we have had to omit some items that were included in the First Edition but which, we now consider, no longer command sufficiently compelling reasons for inclusion. The main topics that fall under this heading are the old §11.3 containing some applications of local duality (one can take the view that the new Chapter 12, on canonical modules, represents a major application of local duality), and the *a priori* bounds of diagonal type on Castelnuovo–Mumford regularity at and above level 2 that were treated in Chapter 16 of the First Edition (Chapter 17 has been reorganized to smooth over the omission, and expanded by the addition of further bounding results which follow from our generalized version of Mumford's bound on regularity at and above level 2).

There are also many minor changes, designed to improve the presentation or the usefulness of the book. For example, the treatment of Faltings' Annihilator Theorem in Chapter 9 now applies to two arbitrary ideals $\mathfrak{a}$ and $\mathfrak{b}$, whereas in the First Edition we treated only the case where $\mathfrak{b} \subseteq \mathfrak{a}$; the syzygetic characterization of Castelnuovo–Mumford regularity is given a full proof in the Second Edition; and the graded Deligne Isomorphism in §20.2 is presented here for a multi-graded situation.

We should also point out that two comments made about the First Edition in its Preface do not apply to this Second Edition. Firstly, the example studied in 2.3.7, 3.3.5, 4.3.7, ... is not called 'Hartshorne's Example' in this Second Edition (but we do cite Hartshorne's paper [28] when we first consider this example); secondly, there are a few more appearances of local cohomology in Chapter 19 than there were in the First Edition (because we followed a suggestion of M. Varbaro that, in some formulas, the arithmetic rank of an ideal $\mathfrak{a}$ could be replaced by the cohomological dimension of $\mathfrak{a}$). Nevertheless, Chapter 19 still contains several exciting examples of situations which represent 'hidden applications' of local cohomology, in the sense that significant results that do not mention local cohomology in either their hypotheses or their conclusions have proofs in this book that depend on local cohomology. There are other examples of such 'hidden applications' in §6.5 and in Chapter 18.

We would like to add, to the list of people thanked in the Preface to the First Edition, several more of our students, namely Roberto Boldini, Stefan Fumasoli, Simon Kurmann, Nicole Nossem and Fred Rohrer, who all contributed to this Second Edition, either by providing constructive criticism of the First Edition, or by trying out drafts of changed or new sections that we planned to include in the Second Edition. We are grateful to them all.

We thank the Schweizerischer Nationalfonds zur Förderung der wissenschaftlichen Forschung, the Forschungsrat des Instituts für Mathematik der Universität Zürich, and the Department of Pure Mathematics of the University of Sheffield, for financial support for visits for collaboration on this Second Edition. We are also particularly grateful to the Scientific Council of the Centre International de Rencontres Mathématiques (CIRM) at Luminy, Marseille, for their award to us of a two-week 'research in pairs' in Spring 2011 that enabled us, in the excellent environment for mathematical research at CIRM, to produce a complete draft. We again thank Alice Sharp for her continued support and encouragement for our project. It is also a pleasure for us to record our gratitude to Roger Astley and Clare Dennison of Cambridge University Press for their encouragement and support.

Markus Brodmann   Rodney Sharp
Zürich   Sheffield

April 2012

Notation and conventions

All rings considered in this book will have identity elements.

Throughout the book, R will always denote a non-trivial commutative Noetherian ring, and $\mathfrak{a}$ will denote an ideal of R. We shall only assume that R has additional properties (such as being local) when these are explicitly stated; however, the phrase '$(R, \mathfrak{m})$ is a local ring' will mean that R is a commutative Noetherian quasi-local ring with unique maximal ideal $\mathfrak{m}$.

For an ideal $\mathfrak{c}$ of R, we denote $\operatorname{Supp}(R/\mathfrak{c}) = \{\mathfrak{p} \in \operatorname{Spec}(R) : \mathfrak{p} \supseteq \mathfrak{c}\}$ by $\operatorname{Var}(\mathfrak{c})$, and refer to this as *the variety of* $\mathfrak{c}$.

By a *multiplicatively closed subset* of R, we shall mean a subset of R which is closed under multiplication and contains 1. It should be noted (and this comment is relevant for the final chapter) that, if S is a non-empty subset of R which is closed under multiplication, then, even if S does not contain 1, we can form the commutative ring $S^{-1}R$ and, for an R-module M, the $S^{-1}R$-module $S^{-1}M$. In fact, $S^{-1}R \cong (S \cup \{1\})^{-1}R$, and, in $S^{-1}R$, the element sr/s, for $r \in R$ and $s \in S$, is independent of the choice of such s; similar comments apply to $S^{-1}M$.

The symbol $\mathbb{Z}$ will always denote the ring of integers; in addition, $\mathbb{N}$ (respectively $\mathbb{N}_0$) will always denote the set of positive (respectively non-negative) integers. The field of rational (respectively real, complex) numbers will be denoted by $\mathbb{Q}$ (respectively $\mathbb{R}$, $\mathbb{C}$).

The category of all modules and homomorphisms over a commutative ring R' will be denoted by $\mathcal{C}(R')$. When R' is G-graded, where G is a finitely generated, torsion-free Abelian group, the category of all graded R'-modules and homogeneous R'-homomorphisms will be denoted by $^*\mathcal{C}(R')$ (or $^*\mathcal{C}^G(R')$ when it is desirable to indicate the grading group G).

The symbol $\subseteq$ will stand for 'is a subset of'; the symbol $\subset$ will be reserved to denote strict inclusion. Thus, for sets A, B, the expression $A \subset B$ means that $A \subseteq B$ and $A \neq B$.

The identity mapping on a set A will be denoted by Id_A. If $f : A \to C$ is a mapping from the set A to the set C, and $S \subseteq A$, then $f \!\restriction_S\, : S \to C$ will denote the restriction of f to S. Thus $f \!\restriction_S (s) = f(s)$ for all $s \in S$.

Some of the exercises in the book are needed for the main development later in the book, and these exercises are marked with a '♯'; however, exercises which are used later in the book but only in other exercises have not been marked with a '♯'.

1

The local cohomology functors

The main objective of this chapter is to introduce the $\mathfrak{a}$-torsion functor $\Gamma_{\mathfrak{a}}$ (throughout the book, $\mathfrak{a}$ always denotes an ideal in a (non-trivial) commutative Noetherian ring R) and its right derived functors $H^i_{\mathfrak{a}}$ ($i \geq 0$), referred to as the local cohomology functors with respect to $\mathfrak{a}$. We shall see that $\Gamma_{\mathfrak{a}}$ is naturally equivalent to the functor $\varinjlim_{n \in \mathbb{N}} \mathrm{Hom}_R(R/\mathfrak{a}^n, \bullet)$ and, indeed, that $H^i_{\mathfrak{a}}$ is naturally equivalent to the functor $\varinjlim_{n \in \mathbb{N}} \mathrm{Ext}^i_R(R/\mathfrak{a}^n, \bullet)$ for each $i \geq 0$; moreover, as $\Gamma_{\mathfrak{a}}$ turns out to be left exact, the functors $\Gamma_{\mathfrak{a}}$ and $H^0_{\mathfrak{a}}$ are naturally equivalent.

This chapter also serves notice that our approach is based on fundamental techniques of homological commutative algebra, such as ones based on connected sequences of functors (see [71, pp. 212–214]): readers familiar with such ideas, and with the local cohomology functors, might like to just glance through this chapter and to move rapidly on to Chapter 2.

1.1 Torsion functors

1.1.1 Definition. For each R-module M, set $\Gamma_{\mathfrak{a}}(M) = \bigcup_{n \in \mathbb{N}}(0 :_M \mathfrak{a}^n)$, the set of elements of M which are annihilated by some power of $\mathfrak{a}$. Note that $\Gamma_{\mathfrak{a}}(M)$ is a submodule of M. For a homomorphism $f : M \longrightarrow N$ of R-modules, we have $f(\Gamma_{\mathfrak{a}}(M)) \subseteq \Gamma_{\mathfrak{a}}(N)$, and so there is a mapping $\Gamma_{\mathfrak{a}}(f) : \Gamma_{\mathfrak{a}}(M) \longrightarrow \Gamma_{\mathfrak{a}}(N)$ which agrees with f on each element of $\Gamma_{\mathfrak{a}}(M)$.

It is clear that, if $g : M \to N$ and $h : N \to L$ are further homomorphisms of R-modules and $r \in R$, then $\Gamma_{\mathfrak{a}}(h \circ f) = \Gamma_{\mathfrak{a}}(h) \circ \Gamma_{\mathfrak{a}}(f)$, $\Gamma_{\mathfrak{a}}(f + g) = \Gamma_{\mathfrak{a}}(f) + \Gamma_{\mathfrak{a}}(g)$, $\Gamma_{\mathfrak{a}}(rf) = r\Gamma_{\mathfrak{a}}(f)$ and $\Gamma_{\mathfrak{a}}(\mathrm{Id}_M) = \mathrm{Id}_{\Gamma_{\mathfrak{a}}(M)}$. Thus, with these assignments, $\Gamma_{\mathfrak{a}}$ becomes a covariant, R-linear functor from $\mathcal{C}(R)$ to itself. (We say that a functor $T : \mathcal{C}(R) \longrightarrow \mathcal{C}(R)$ is R-linear precisely when it is additive and $T(rf) = rT(f)$ for all $r \in R$ and all homomorphisms f of R-modules.) We call $\Gamma_{\mathfrak{a}}$ the $\mathfrak{a}$-*torsion functor*.

1.1.2 ♯Exercise. Let $\mathfrak{b}$ be a second ideal of R. Show that

$$\Gamma_{\mathfrak{a}}(\Gamma_{\mathfrak{b}}(M)) = \Gamma_{\mathfrak{a}+\mathfrak{b}}(M)$$

for each R-module M.

1.1.3 ♯Exercise. Let $\mathfrak{b}$ be a second ideal of R. Show that $\Gamma_{\mathfrak{a}} = \Gamma_{\mathfrak{b}}$ if and only if $\sqrt{\mathfrak{a}} = \sqrt{\mathfrak{b}}$.

(The notation ♯, attached to some exercises, is explained in the section of 'Notation and conventions' following the Preface to the Second Edition.)

1.1.4 Exercise. Suppose that the ideal $\mathfrak{b}$ of R is a reduction of $\mathfrak{a}$; that is, $\mathfrak{b} \subseteq \mathfrak{a}$ and there exists $s \in \mathbb{N}$ such that $\mathfrak{b}\mathfrak{a}^s = \mathfrak{a}^{s+1}$. Show that $\Gamma_{\mathfrak{a}} = \Gamma_{\mathfrak{b}}$.

1.1.5 Exercise. For a prime number p, find $\Gamma_{p\mathbb{Z}}(\mathbb{Q}/\mathbb{Z})$.

1.1.6 Lemma. *The $\mathfrak{a}$-torsion functor* $\Gamma_{\mathfrak{a}} : \mathcal{C}(R) \longrightarrow \mathcal{C}(R)$ *is left exact.*

Proof. Let $0 \longrightarrow L \stackrel{f}{\longrightarrow} M \stackrel{g}{\longrightarrow} N \longrightarrow 0$ be an exact sequence of R-modules and R-homomorphisms. We must show that

$$0 \longrightarrow \Gamma_{\mathfrak{a}}(L) \stackrel{\Gamma_{\mathfrak{a}}(f)}{\longrightarrow} \Gamma_{\mathfrak{a}}(M) \stackrel{\Gamma_{\mathfrak{a}}(g)}{\longrightarrow} \Gamma_{\mathfrak{a}}(N)$$

is still exact. It is clear that $\Gamma_{\mathfrak{a}}(f)$ is a monomorphism and it follows immediately from 1.1.1 that $\Gamma_{\mathfrak{a}}(g) \circ \Gamma_{\mathfrak{a}}(f) = 0$, so that

$$\operatorname{Im}(\Gamma_{\mathfrak{a}}(f)) \subseteq \operatorname{Ker}(\Gamma_{\mathfrak{a}}(g)).$$

To prove the reverse inclusion, let $m \in \operatorname{Ker}(\Gamma_{\mathfrak{a}}(g))$. Thus $m \in \Gamma_{\mathfrak{a}}(M)$, so that there exists $n \in \mathbb{N}$ such that $\mathfrak{a}^n m = 0$, and $g(m) = 0$. Now there exists $l \in L$ such that $f(l) = m$, and our proof will be complete if we show that $l \in \Gamma_{\mathfrak{a}}(L)$. To achieve this, note that, for each $r \in \mathfrak{a}^n$, we have $f(rl) = rf(l) = rm = 0$, so that $rl = 0$ because f is a monomorphism. Hence $\mathfrak{a}^n l = 0$. □

The result of Lemma 1.1.6 will become transparent to many readers once we have covered a little more theory, and related the $\mathfrak{a}$-torsion functor $\Gamma_{\mathfrak{a}}$ to a functor defined in terms of direct limits of 'Hom' modules. However, before we proceed in that direction, we are going to introduce, at this early stage, the fundamental definition of the local cohomology modules of an R-module M with respect to $\mathfrak{a}$.

1.2 Local cohomology modules

1.2.1 Definitions. For $i \in \mathbb{N}_0$, the i-th right derived functor of $\Gamma_\mathfrak{a}$ is denoted by $H_\mathfrak{a}^i$ and will be referred to as the *i-th local cohomology functor with respect to $\mathfrak{a}$*.

For an R-module M, we shall refer to $H_\mathfrak{a}^i(M)$, that is, the result of applying the functor $H_\mathfrak{a}^i$ to M, as the *i-th local cohomology module of M with respect to $\mathfrak{a}$*, and to $\Gamma_\mathfrak{a}(M)$ as the *$\mathfrak{a}$-torsion submodule of M*. We shall say that M is *$\mathfrak{a}$-torsion-free* precisely when $\Gamma_\mathfrak{a}(M) = 0$, and that M is *$\mathfrak{a}$-torsion* precisely when $\Gamma_\mathfrak{a}(M) = M$, that is, if and only if each element of M is annihilated by some power of $\mathfrak{a}$.

It is probably appropriate for us to stress the implications of the above definition at this point, and list some basic properties of the local cohomology modules.

1.2.2 Properties of local cohomology modules. Let M be an arbitrary R-module.

(i) To calculate $H_\mathfrak{a}^i(M)$, one proceeds as follows. Take an injective resolution

$$I^\bullet : 0 \xrightarrow{d^{-1}} I^0 \xrightarrow{d^0} I^1 \longrightarrow \cdots \longrightarrow I^i \xrightarrow{d^i} I^{i+1} \longrightarrow \cdots$$

of M, so that there is an R-homomorphism $\alpha : M \longrightarrow I^0$ such that the sequence

$$0 \longrightarrow M \xrightarrow{\alpha} I^0 \xrightarrow{d^0} I^1 \longrightarrow \cdots \longrightarrow I^i \xrightarrow{d^i} I^{i+1} \longrightarrow \cdots$$

is exact. Apply the functor $\Gamma_\mathfrak{a}$ to the complex $I^\bullet$ to obtain

$$0 \longrightarrow \Gamma_\mathfrak{a}(I^0) \xrightarrow{\Gamma_\mathfrak{a}(d^0)} \cdots \longrightarrow \Gamma_\mathfrak{a}(I^i) \xrightarrow{\Gamma_\mathfrak{a}(d^i)} \Gamma_\mathfrak{a}(I^{i+1}) \longrightarrow \cdots$$

and take the i-th cohomology module of this complex; the result,

$$\operatorname{Ker}(\Gamma_\mathfrak{a}(d^i)) / \operatorname{Im}(\Gamma_\mathfrak{a}(d^{i-1})),$$

which, by a standard fact of homological algebra, is independent (up to R-isomorphism) of the choice of injective resolution $I^\bullet$ of M, is $H_\mathfrak{a}^i(M)$.

(ii) Since $\Gamma_\mathfrak{a}$ is covariant and R-linear, it is automatic that each local cohomology functor $H_\mathfrak{a}^i$ ($i \in \mathbb{N}_0$) is again covariant and R-linear.

(iii) Since $\Gamma_\mathfrak{a}$ is left exact, $H_\mathfrak{a}^0$ is naturally equivalent to $\Gamma_\mathfrak{a}$. Thus, loosely, we can use this natural equivalence to identify these two functors.

(iv) The reader should be aware of the long exact sequence of local cohomology modules which results from a short exact sequence of R-modules and R-homomorphisms, and so we spell out the details here.

Let $0 \longrightarrow L \xrightarrow{f} M \xrightarrow{g} N \longrightarrow 0$ be an exact sequence of R-modules and R-homomorphisms. Then, for each $i \in \mathbb{N}_0$, there is a connecting homomorphism $H_{\mathfrak{a}}^i(N) \to H_{\mathfrak{a}}^{i+1}(L)$, and these connecting homomorphisms make the resulting long sequence

$$0 \longrightarrow H_{\mathfrak{a}}^0(L) \xrightarrow{H_{\mathfrak{a}}^0(f)} H_{\mathfrak{a}}^0(M) \xrightarrow{H_{\mathfrak{a}}^0(g)} H_{\mathfrak{a}}^0(N)$$

$$\longrightarrow H_{\mathfrak{a}}^1(L) \xrightarrow{H_{\mathfrak{a}}^1(f)} H_{\mathfrak{a}}^1(M) \xrightarrow{H_{\mathfrak{a}}^1(g)} H_{\mathfrak{a}}^1(N)$$

$$\longrightarrow \quad \cdots \qquad\qquad\qquad\qquad \cdots$$

$$\longrightarrow H_{\mathfrak{a}}^i(L) \xrightarrow{H_{\mathfrak{a}}^i(f)} H_{\mathfrak{a}}^i(M) \xrightarrow{H_{\mathfrak{a}}^i(g)} H_{\mathfrak{a}}^i(N)$$

$$\longrightarrow H_{\mathfrak{a}}^{i+1}(L) \longrightarrow \quad \cdots$$

exact. The reader should also be aware of the 'natural' or 'functorial' properties of these long exact sequences: if

$$
\begin{array}{ccccccccc}
0 & \longrightarrow & L & \xrightarrow{f} & M & \xrightarrow{g} & N & \longrightarrow & 0 \\
& & \downarrow{\lambda} & & \downarrow{\mu} & & \downarrow{\nu} & & \\
0 & \longrightarrow & L' & \xrightarrow{f'} & M' & \xrightarrow{g'} & N' & \longrightarrow & 0
\end{array}
$$

is a commutative diagram of R-modules and R-homomorphisms with exact rows, then, for each $i \in \mathbb{N}_0$, we not only have a commutative diagram

$$
\begin{array}{ccccc}
H_{\mathfrak{a}}^i(L) & \xrightarrow{H_{\mathfrak{a}}^i(f)} & H_{\mathfrak{a}}^i(M) & \xrightarrow{H_{\mathfrak{a}}^i(g)} & H_{\mathfrak{a}}^i(N) \\
\downarrow{H_{\mathfrak{a}}^i(\lambda)} & & \downarrow{H_{\mathfrak{a}}^i(\mu)} & & \downarrow{H_{\mathfrak{a}}^i(\nu)} \\
H_{\mathfrak{a}}^i(L') & \xrightarrow{H_{\mathfrak{a}}^i(f')} & H_{\mathfrak{a}}^i(M') & \xrightarrow{H_{\mathfrak{a}}^i(g')} & H_{\mathfrak{a}}^i(N')
\end{array}
$$

(simply because $H_{\mathfrak{a}}^i$ is a functor!), but we also have a commutative diagram

$$
\begin{array}{ccc}
H_{\mathfrak{a}}^i(N) & \longrightarrow & H_{\mathfrak{a}}^{i+1}(L) \\
\downarrow{H_{\mathfrak{a}}^i(\nu)} & & \downarrow{H_{\mathfrak{a}}^{i+1}(\lambda)} \\
H_{\mathfrak{a}}^i(N') & \longrightarrow & H_{\mathfrak{a}}^{i+1}(L')
\end{array}
$$

in which the horizontal maps are the appropriate connecting homomorphisms.

The following remark will be used frequently in applications. It is an easy consequence of Exercise 1.1.3 and the definition of local cohomology functors in 1.2.1.

1.2.3 Remark. Let $\mathfrak{b}$ be a second ideal of R such that $\sqrt{\mathfrak{a}} = \sqrt{\mathfrak{b}}$. Then $H_{\mathfrak{a}}^i = H_{\mathfrak{b}}^i$ for all $i \in \mathbb{N}_0$, so that $H_{\mathfrak{a}}^i(M) = H_{\mathfrak{b}}^i(M)$ for each R-module M and all $i \in \mathbb{N}_0$.

The next four exercises might help the reader to consolidate the properties of local cohomology modules listed in 1.2.2. The first three of these exercises (for which non-trivial results from commutative algebra about injective dimension over the relevant rings are very helpful) give a tiny foretaste of results about the vanishing of local cohomology modules which are central to the subject, and which will feature prominently later in the book.

1.2.4 Exercise. Show that, for every Abelian group (that is, $\mathbb{Z}$-module) G and for every $a \in \mathbb{Z}$, we have $H_{\mathbb{Z}a}^i(G) = 0$ for all $i \geq 2$.

1.2.5 Exercise. Suppose that $(R, \mathfrak{m})$ is a regular local ring of dimension d. Show that, for each R-module M, we have $H_{\mathfrak{a}}^i(M) = 0$ for all $i > d$.

1.2.6 Exercise. Suppose that $(R, \mathfrak{m})$ is a Gorenstein local ring (see, for example, Matsumura [50, p. 142]) of dimension d. Show that, for each finitely generated R-module M of finite projective dimension, we have $H_{\mathfrak{a}}^i(M) = 0$ for all $i > d$. (Here is a hint: use the fact [50, Theorem 18.1] that the injective dimension of R as an R-module is d, and then use induction on the projective dimension of M.)

The next exercise investigates the behaviour of local cohomology modules under fraction formation: its results show that, speaking loosely, the local cohomology functors 'commute' with fraction formation. This is a fundamental fact in the subject; however, we shall actually derive it as an immediate consequence of a more general result in Chapter 4 concerning the behaviour of local cohomology under flat base change (and we shall not make use of it until after Chapter 4). Nevertheless, even at this early stage, its proof should not present much difficulty for a reader familiar with the fact (proved in 10.1.14) that, if I is an injective R-module and S is a multiplicatively closed subset of R, then $S^{-1}I$ is an injective $S^{-1}R$-module.

1.2.7 Exercise. Let M be an R-module and let S be a multiplicatively closed subset of R. Show that $S^{-1}(\Gamma_{\mathfrak{a}}(M)) = \Gamma_{\mathfrak{a}S^{-1}R}(S^{-1}M)$, and that, for all $i \in \mathbb{N}_0$, there is an isomorphism of $S^{-1}R$-modules

$$S^{-1}(H_{\mathfrak{a}}^i(M)) \cong H_{\mathfrak{a}S^{-1}R}^i(S^{-1}M).$$

It is now time for us to relate the $\mathfrak{a}$-torsion functor $\Gamma_{\mathfrak{a}}$ to a functor defined in terms of direct limits of 'Hom' modules. Fundamental to the discussion is the natural isomorphism, for an R-module M and $n \in \mathbb{N}$,

$$\phi := \phi_{\mathfrak{a}^n, M} : \operatorname{Hom}_R(R/\mathfrak{a}^n, M) \overset{\cong}{\longrightarrow} (0 :_M \mathfrak{a}^n)$$

for which $\phi(f) = f(1 + \mathfrak{a}^n)$ for all $f \in \operatorname{Hom}_R(R/\mathfrak{a}^n, M)$. In fact, we are going to put the various $\phi_{\mathfrak{a}^n, M}$ $(n \in \mathbb{N})$ together to obtain a natural isomorphism $\varinjlim_{n \in \mathbb{N}} \operatorname{Hom}_R(R/\mathfrak{a}^n, M) \overset{\cong}{\longrightarrow} \Gamma_{\mathfrak{a}}(M)$, but before we do this it might be helpful to the reader if we give some general considerations about functors and direct limits, as the principles involved will be used numerous times in this book.

1.2.8 Remarks. Let $(\Lambda, \leq)$ be a (non-empty) directed partially ordered set, and suppose that we are given an inverse system of R-modules $(W_\alpha)_{\alpha \in \Lambda}$ over Λ, with constituent R-homomorphisms $h_\beta^\alpha : W_\alpha \to W_\beta$ (for each $(\alpha, \beta) \in \Lambda \times \Lambda$ with $\alpha \geq \beta$). Let $T : \mathcal{C}(R) \times \mathcal{C}(R) \to \mathcal{C}(R)$ be an R-linear functor of two variables which is contravariant in the first variable and covariant in the second. (A functor $U : \mathcal{C}(R) \times \mathcal{C}(R) \longrightarrow \mathcal{C}(R)$ is said to be R-*linear* precisely when it is additive and $U(rf, g) = rU(f, g) = U(f, rg)$ for all $r \in R$ and all homomorphisms f, g of R-modules.) We show now how these data give rise to a covariant, R-linear functor

$$\varinjlim_{\alpha \in \Lambda} T(W_\alpha, \bullet) : \mathcal{C}(R) \longrightarrow \mathcal{C}(R).$$

Let M, N be R-modules and let $f : M \longrightarrow N$ be an R-homomorphism. For $\alpha, \beta \in \Lambda$ with $\alpha \geq \beta$, the homomorphism $h_\beta^\alpha : W_\alpha \longrightarrow W_\beta$ induces an R-homomorphism

$$T(h_\beta^\alpha, M) : T(W_\beta, M) \longrightarrow T(W_\alpha, M),$$

and the fact that T is a functor ensures that the $T(h_\beta^\alpha, M)$ turn the family $(T(W_\alpha, M))_{\alpha \in \Lambda}$ into a direct system of R-modules and R-homomorphisms over Λ. We may therefore form $\varinjlim_{\alpha \in \Lambda} T(W_\alpha, M)$. Moreover, again for $\alpha, \beta \in \Lambda$ with $\alpha \geq \beta$, we have a commutative diagram

$$
\begin{array}{ccc}
T(W_\beta, M) & \xrightarrow{\;T(h_\beta^\alpha, M)\;} & T(W_\alpha, M) \\
\Big\downarrow{\scriptstyle T(W_\beta, f)} & & \Big\downarrow{\scriptstyle T(W_\alpha, f)} \\
T(W_\beta, N) & \xrightarrow{\;T(h_\beta^\alpha, N)\;} & T(W_\alpha, N)
\end{array}
\quad ;
$$

therefore the $T(W_\alpha, f)$ $(\alpha \in \Lambda)$ constitute a morphism of direct systems and so induce an R-homomorphism

$$\varinjlim_{\alpha \in \Lambda} T(W_\alpha, f) : \varinjlim_{\alpha \in \Lambda} T(W_\alpha, M) \longrightarrow \varinjlim_{\alpha \in \Lambda} T(W_\alpha, N).$$

It is now straightforward to check that, in this way, $\varinjlim_{\alpha \in \Lambda} T(W_\alpha, \bullet)$ becomes a covariant, R-linear functor from $\mathcal{C}(R)$ to itself. Observe that, since passage to direct limits preserves exactness, if T is left exact, then so too is this new functor.

1.2.9 Examples. Here we present some examples that are central for our subject.

(i) Probably the most important examples for us of the ideas of 1.2.8 concern the case where we take for Λ the set $\mathbb{N}$ of positive integers with its usual ordering and the inverse system $(R/\mathfrak{a}^n)_{n \in \mathbb{N}}$ of R-modules under the natural homomorphisms $h_m^n : R/\mathfrak{a}^n \to R/\mathfrak{a}^m$ (for $n, m \in \mathbb{N}$ with $n \geq m$) (in such circumstances, $\mathfrak{a}^n \subseteq \mathfrak{a}^m$, of course). In this way, we obtain covariant, R-linear functors

$$\varinjlim_{n \in \mathbb{N}} \operatorname{Hom}_R(R/\mathfrak{a}^n, \bullet) \quad \text{and} \quad \varinjlim_{n \in \mathbb{N}} \operatorname{Ext}_R^i(R/\mathfrak{a}^n, \bullet) \quad (i \in \mathbb{N}_0)$$

from $\mathcal{C}(R)$ to itself. Of course, the natural equivalence between the left exact functors Hom_R and Ext_R^0 leads to a natural equivalence between the left exact functors

$$\varinjlim_{n \in \mathbb{N}} \operatorname{Hom}_R(R/\mathfrak{a}^n, \bullet) \quad \text{and} \quad \varinjlim_{n \in \mathbb{N}} \operatorname{Ext}_R^0(R/\mathfrak{a}^n, \bullet)$$

which we shall use without further comment.

(ii) Very similar considerations, this time based on the inclusion maps $\mathfrak{a}^n \to \mathfrak{a}^m$ (for $n, m \in \mathbb{N}$ with $n \geq m$), lead to functors (which are again covariant and R-linear)

$$\varinjlim_{n \in \mathbb{N}} \operatorname{Hom}_R(\mathfrak{a}^n, \bullet) \quad \text{and} \quad \varinjlim_{n \in \mathbb{N}} \operatorname{Ext}_R^i(\mathfrak{a}^n, \bullet) \quad (i \in \mathbb{N}_0)$$

from $\mathcal{C}(R)$ to itself, and a natural equivalence between the left exact functors

$$\varinjlim_{n \in \mathbb{N}} \operatorname{Hom}_R(\mathfrak{a}^n, \bullet) \quad \text{and} \quad \varinjlim_{n \in \mathbb{N}} \operatorname{Ext}_R^0(\mathfrak{a}^n, \bullet).$$

These functors will be considered in detail in Chapter 2.

It will be convenient for us to consider situations slightly more general than that studied in 1.2.9(i) above.

1.2.10 Definition and Example. Let $(\Lambda, \leq)$ be a (non-empty) directed partially ordered set. By an *inverse family of ideals (of R) over* Λ, we mean a family $(\mathfrak{b}_\alpha)_{\alpha \in \Lambda}$ of ideals of R such that, whenever $(\alpha, \beta) \in \Lambda \times \Lambda$ with $\alpha \geq \beta$, we have $\mathfrak{b}_\alpha \subseteq \mathfrak{b}_\beta$.

For example, if

$$\mathfrak{b}_1 \supseteq \mathfrak{b}_2 \supseteq \cdots \supseteq \mathfrak{b}_n \supseteq \mathfrak{b}_{n+1} \supseteq \cdots$$

is a descending chain of ideals of R, then $(\mathfrak{b}_n)_{n \in \mathbb{N}}$ is an inverse family of ideals over $\mathbb{N}$ (with its usual ordering). In particular, the family $(\mathfrak{a}^n)_{n \in \mathbb{N}}$ is an inverse family of ideals over $\mathbb{N}$.

Let $(\mathfrak{b}_\alpha)_{\alpha \in \Lambda}$ be an inverse family of ideals of R over Λ. Then the natural R-homomorphisms $h^\alpha_\beta : R/\mathfrak{b}_\alpha \to R/\mathfrak{b}_\beta$ (for $\alpha, \beta \in \Lambda$ with $\alpha \geq \beta$) turn $(R/\mathfrak{b}_\alpha)_{\alpha \in \Lambda}$ into an inverse system over Λ, and so we can apply the ideas of 1.2.8 to produce covariant, R-linear functors

$$\varinjlim_{\alpha \in \Lambda} \operatorname{Hom}_R(R/\mathfrak{b}_\alpha, \bullet) \quad \text{and} \quad \varinjlim_{\alpha \in \Lambda} \operatorname{Ext}^i_R(R/\mathfrak{b}_\alpha, \bullet) \ (i \in \mathbb{N}_0)$$

(from $\mathcal{C}(R)$ to itself), the first two of which are left exact and naturally equivalent.

1.2.11 Theorem. *Let $\mathfrak{B} = (\mathfrak{b}_\alpha)_{\alpha \in \Lambda}$ be an inverse family of ideals of R over Λ, as in 1.2.10.*

(i) *There is a covariant, R-linear functor $\Gamma_\mathfrak{B} : \mathcal{C}(R) \to \mathcal{C}(R)$ which is such that, for an R-module M,*

$$\Gamma_\mathfrak{B}(M) = \bigcup_{\alpha \in \Lambda} (0 :_M \mathfrak{b}_\alpha),$$

and, for a homomorphism $f : M \longrightarrow N$ of R-modules, $\Gamma_\mathfrak{B}(f) : \Gamma_\mathfrak{B}(M) \longrightarrow \Gamma_\mathfrak{B}(N)$ is just the restriction of f to $\Gamma_\mathfrak{B}(M)$.

(ii) *There is a natural equivalence*

$$\phi' \ (= \phi'_\mathfrak{B}) : \varinjlim_{\alpha \in \Lambda} \operatorname{Hom}_R(R/\mathfrak{b}_\alpha, \bullet) \xrightarrow{\cong} \Gamma_\mathfrak{B}$$

(of functors from $\mathcal{C}(R)$ to itself) which is such that, for an R-module M and $\alpha \in \Lambda$, the image under ϕ'_M of the natural image of an $h \in \operatorname{Hom}_R(R/\mathfrak{b}_\alpha, M)$ is $h(1 + \mathfrak{b}_\alpha)$. Consequently, $\Gamma_\mathfrak{B}$ is left exact.

(iii) *In particular, there is a natural equivalence*

$$\phi^0 \, (= \phi_\mathfrak{a}^0) : \varinjlim_{n \in \mathbb{N}} \mathrm{Hom}_R(R/\mathfrak{a}^n, \, \bullet\,) \xrightarrow{\cong} \Gamma_\mathfrak{a}$$

which is such that, for an R-module M and $n \in \mathbb{N}$, the image under ϕ_M^0 of the natural image of an $h \in \mathrm{Hom}_R(R/\mathfrak{a}^n, M)$ is $h(1 + \mathfrak{a}^n)$.

Proof. (i) This can be proved by straightforward modification of the ideas of 1.1.1, and so will be left to the reader.

(ii) Let $f : M \longrightarrow N$ be a homomorphism of R-modules. For each $\alpha \in \Lambda$, let $\phi_{\mathfrak{b}_\alpha, M} : \mathrm{Hom}_R(R/\mathfrak{b}_\alpha, M) \longrightarrow (0 :_M \mathfrak{b}_\alpha)$ be the R-isomorphism for which $\phi_{\mathfrak{b}_\alpha, M}(h) = h(1 + \mathfrak{b}_\alpha)$ for all $h \in \mathrm{Hom}_R(R/\mathfrak{b}_\alpha, M)$. Let $\alpha, \beta \in \Lambda$ with $\alpha \geq \beta$, and let $h_\beta^\alpha : R/\mathfrak{b}_\alpha \to R/\mathfrak{b}_\beta$ be as in 1.2.10. Since the diagram

$$
\begin{array}{ccc}
\mathrm{Hom}_R(R/\mathfrak{b}_\beta, M) & \xrightarrow[\cong]{\phi_{\mathfrak{b}_\beta, M}} & (0 :_M \mathfrak{b}_\beta) \\
{\scriptstyle \mathrm{Hom}_R(h_\beta^\alpha, M)} \Big\downarrow & & \Big\downarrow \\
\mathrm{Hom}_R(R/\mathfrak{b}_\alpha, M) & \xrightarrow[\cong]{\phi_{\mathfrak{b}_\alpha, M}} & (0 :_M \mathfrak{b}_\alpha)
\end{array}
$$

(in which the right-hand vertical map is inclusion) commutes, it follows that there is indeed an R-isomorphism

$$\phi_M' : \varinjlim_{\alpha \in \Lambda} \mathrm{Hom}_R(R/\mathfrak{b}_\alpha, M) \xrightarrow{\cong} \Gamma_\mathfrak{B}(M) = \bigcup_{\alpha \in \Lambda} (0 :_M \mathfrak{b}_\alpha)$$

as described in the statement of the theorem. It is easy to check that the diagram

$$
\begin{array}{ccc}
\varinjlim_{\alpha \in \Lambda} \mathrm{Hom}_R(R/\mathfrak{b}_\alpha, M) & \xrightarrow[\cong]{\phi_M'} & \Gamma_\mathfrak{B}(M) \\
{\scriptstyle \varinjlim_{\alpha \in \Lambda} \mathrm{Hom}_R(R/\mathfrak{b}_\alpha, f)} \Big\downarrow & & \Big\downarrow {\scriptstyle \Gamma_\mathfrak{B}(f)} \\
\varinjlim_{\alpha \in \Lambda} \mathrm{Hom}_R(R/\mathfrak{b}_\alpha, N) & \xrightarrow[\cong]{\phi_N'} & \Gamma_\mathfrak{B}(N)
\end{array}
$$

commutes, and the final claim is then immediate from 1.2.10.

(iii) This is immediate from (ii), since when we apply (ii) to the family of ideals $\mathfrak{B} := (\mathfrak{a}^n)_{n \in \mathbb{N}}$, the functor $\Gamma_\mathfrak{B}$ of (i) is just the $\mathfrak{a}$-torsion functor $\Gamma_\mathfrak{a}$. $\square$

We commented earlier that it would in time become transparent that $\Gamma_\mathfrak{a}$ is left exact: we had 1.2.11 in mind when we made that comment.

1.2.12 ♯Exercise. Provide a proof for part (i) of 1.2.11.

1.3 Connected sequences of functors

In this section, we are going to use the concepts of 'connected sequence of functors' and 'strongly connected sequence of functors'. These are explained on p. 212 of Rotman's book [71]. For the reader's convenience, we recall here relevant definitions in the case of negative connected sequences, as we shall be particularly concerned with this case.

1.3.1 Definition. Let R' be a commutative ring.

A sequence $(T^i)_{i \in \mathbb{N}_0}$ of covariant functors from $\mathcal{C}(R)$ to $\mathcal{C}(R')$ is said to be a *negative connected sequence* (respectively, a *negative strongly connected sequence*) if the following conditions are satisfied.

(i) Whenever $0 \longrightarrow L \xrightarrow{f} M \xrightarrow{g} N \longrightarrow 0$ is an exact sequence in $\mathcal{C}(R)$, there are defined connecting R'-homomorphisms

$$T^i(N) \longrightarrow T^{i+1}(L) \quad \text{for all } i \in \mathbb{N}_0$$

such that the long sequence

$$
\begin{array}{llll}
0 \longrightarrow & T^0(L) \xrightarrow{T^0(f)} & T^0(M) \xrightarrow{T^0(g)} & T^0(N) \\
\longrightarrow & T^1(L) \xrightarrow{T^1(f)} & T^1(M) \xrightarrow{T^1(g)} & T^1(N) \\
\longrightarrow & \quad \cdots & & \quad \cdots \\
\longrightarrow & T^i(L) \xrightarrow{T^i(f)} & T^i(M) \xrightarrow{T^i(g)} & T^i(N) \\
\longrightarrow T^{i+1}(L) \longrightarrow & \quad \cdots
\end{array}
$$

is a complex (respectively, is exact).

(ii) Whenever

$$
\begin{array}{ccccccccc}
0 & \longrightarrow & L & \longrightarrow & M & \longrightarrow & N & \longrightarrow & 0 \\
& & \downarrow{\lambda} & & \downarrow{\mu} & & \downarrow{\nu} & & \\
0 & \longrightarrow & L' & \longrightarrow & M' & \longrightarrow & N' & \longrightarrow & 0
\end{array}
$$

is a commutative diagram of R-modules and R-homomorphisms with exact rows, then there is induced, by λ, μ and ν, a chain map of the long complex of (i) for the top row into the corresponding long complex for the bottom row.

It might help if we remind the reader of the convention regarding the raising and lowering of indices in a situation such as that of 1.3.1, under which

T^i would be written as T_{-i}: with this convention, $(T^i)_{i \geq 0}$ can be written as $(T_j)_{j \leq 0}$.

We also point out that, if $T : \mathcal{C}(R) \to \mathcal{C}(R')$ is an additive covariant functor, such as $\Gamma_\mathfrak{a}$, then its sequence of right derived functors $(\mathcal{R}^i T)_{i \in \mathbb{N}_0}$ is a negative strongly connected sequence of covariant functors from $\mathcal{C}(R)$ to $\mathcal{C}(R')$; furthermore, if T is left exact, then $\mathcal{R}^0 T$ is naturally equivalent to T. We shall be concerned so often with left exact, additive, covariant functors that it will considerably simplify the exposition if we adopt now the following convention which will be in force for the rest of the book.

1.3.2 Convention. Whenever R' is a commutative ring and $T : \mathcal{C}(R) \longrightarrow \mathcal{C}(R')$ is a covariant, additive functor which is left exact, then we shall identify T with its 0-th right derived functor $\mathcal{R}^0 T$ in the natural way. Likewise, we shall identify Ext_R^0 with Hom_R in the natural way.

1.3.3 Definition. Let R' be a commutative ring, and let $(T^i)_{i \in \mathbb{N}_0}$, $(U^i)_{i \in \mathbb{N}_0}$ be negative connected sequences of covariant functors from $\mathcal{C}(R)$ to $\mathcal{C}(R')$. A *homomorphism* $\Psi : (T^i)_{i \in \mathbb{N}_0} \longrightarrow (U^i)_{i \in \mathbb{N}_0}$ *of connected sequences* is a family $(\psi^i)_{i \in \mathbb{N}_0}$ where, for each $i \in \mathbb{N}_0$, $\psi^i : T^i \to U^i$ is a natural transformation of functors, and which is such that the following condition is satisfied: whenever $0 \longrightarrow L \longrightarrow M \longrightarrow N \longrightarrow 0$ is an exact sequence of R-modules and R-homomorphisms, then, for each $i \in \mathbb{N}_0$, the diagram

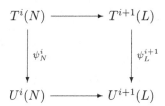

(in which the horizontal maps are the appropriate connecting homomorphisms arising from the connected sequences) commutes.

A homomorphism $\Psi = (\psi^i)_{i \in \mathbb{N}_0} : (T^i)_{i \in \mathbb{N}_0} \longrightarrow (U^i)_{i \in \mathbb{N}_0}$ of connected sequences is said to be an *isomorphism (of connected sequences)* precisely when $\psi^i : T^i \to U^i$ is a natural equivalence of functors for each $i \in \mathbb{N}_0$.

We hope the reader is sufficiently adept at techniques similar to those on pp. 212–214 of [71] to find the following exercise straightforward; if not, he or she might like to study Theorem 10 (and its Corollary) of Section 6.5 of Northcott [60], which together provide a solution.

1.3.4 ♯Exercise. Let R' be a commutative ring, and let $(T^i)_{i \in \mathbb{N}_0}$, $(U^i)_{i \in \mathbb{N}_0}$ be two negative connected sequences of covariant functors from $\mathcal{C}(R)$ to $\mathcal{C}(R')$.

(i) Let $\psi^0 : T^0 \to U^0$ be a natural transformation of functors. Assume that

    (a) the sequence $(T^i)_{i \in \mathbb{N}_0}$ is strongly connected, and

    (b) $T^i(I) = 0$ for all $i \in \mathbb{N}$ and all injective R-modules I.

Show that there exist uniquely determined natural transformations

$$\psi^i : T^i \to U^i \quad (i \in \mathbb{N})$$

such that $(\psi^i)_{i \in \mathbb{N}_0} : (T^i)_{i \in \mathbb{N}_0} \longrightarrow (U^i)_{i \in \mathbb{N}_0}$ is a homomorphism of connected sequences.

(ii) Let $\psi : T^0 \to U^0$ be a natural equivalence of functors. Assume that

    (a) the sequence $(T^i)_{i \in \mathbb{N}_0}$ is strongly connected,

    (b) the sequence $(U^i)_{i \in \mathbb{N}_0}$ is strongly connected, and

    (c) $T^i(I) = U^i(I) = 0$ for all $i \in \mathbb{N}$ and all injective R-modules I.

By part (i), there is a unique homomorphism of connected sequences $\Psi := (\psi^i)_{i \in \mathbb{N}_0} : (T^i)_{i \in \mathbb{N}_0} \longrightarrow (U^i)_{i \in \mathbb{N}_0}$ for which $\psi^0 = \psi$. Show that Ψ is actually an isomorphism of connected sequences.

We shall not state explicitly the analogues of 1.3.1, 1.3.3 and 1.3.4 for positive connected sequences, but we warn the reader now that we shall use such analogues in Chapters 11 and 12.

The following consequence of 1.3.4(ii) essentially provides a characterization of the right derived functors of a left exact, additive, covariant functor from $\mathcal{C}(R)$ to $\mathcal{C}(R')$, where R' is a commutative ring.

1.3.5 Theorem. *Let R' be a commutative ring, and let T be a left exact, additive, covariant functor from $\mathcal{C}(R)$ to $\mathcal{C}(R')$. Let $(T^i)_{i \in \mathbb{N}_0}$ be a negative strongly connected sequence of covariant functors from $\mathcal{C}(R)$ to $\mathcal{C}(R')$ such that there exists a natural equivalence $\psi : T^0 \xrightarrow{\cong} T$ and such that $T^i(I) = 0$ for all $i \in \mathbb{N}$ and all injective R-modules I.*

Then there is a unique isomorphism of connected sequences

$$\Psi = (\psi^i)_{i \in \mathbb{N}_0} : (T^i)_{i \in \mathbb{N}_0} \xrightarrow{\cong} (\mathcal{R}^i T)_{i \in \mathbb{N}_0}$$

(of functors from $\mathcal{C}(R)$ to $\mathcal{C}(R')$) such that $\psi^0 = \psi$. (Of course, we are employing Convention 1.3.2.)

The next exercise strengthens Exercise 1.2.7.

1.3.6 Exercise. Let S be a multiplicatively closed subset of R. Show that

$$\left(S^{-1}(H^i_{\mathfrak{a}}(\,\bullet\,))\right)_{i \in \mathbb{N}_0} \quad \text{and} \quad \left(H^i_{\mathfrak{a}S^{-1}R}(S^{-1}(\,\bullet\,))\right)_{i \in \mathbb{N}_0}$$

are isomorphic connected sequences of functors (from $\mathcal{C}(R)$ to $\mathcal{C}(S^{-1}R)$).

1.3.7 Remarks. Let $\mathfrak{B} = (\mathfrak{b}_\alpha)_{\alpha \in \Lambda}$ be an inverse family of ideals of R over Λ, as in 1.2.10.

Let us temporarily write $U^i := \varinjlim_{\alpha \in \Lambda} \operatorname{Ext}_R^i(R/\mathfrak{b}_\alpha, \bullet)$ for $i \in \mathbb{N}_0$. These functors were introduced in 1.2.10. We are going to show now how they fit together into a negative strongly connected sequence of functors (from $\mathcal{C}(R)$ to itself).

First of all, whenever $0 \longrightarrow L \longrightarrow M \longrightarrow N \longrightarrow 0$ is an exact sequence of R-modules and R-homomorphisms, there are induced, for each $\alpha \in \Lambda$, connecting homomorphisms

$$\operatorname{Ext}_R^i(R/\mathfrak{b}_\alpha, N) \longrightarrow \operatorname{Ext}_R^{i+1}(R/\mathfrak{b}_\alpha, L) \quad (i \in \mathbb{N}_0)$$

which make the induced long sequence

$$0 \longrightarrow \operatorname{Hom}_R(R/\mathfrak{b}_\alpha, L) \longrightarrow \operatorname{Hom}_R(R/\mathfrak{b}_\alpha, M) \longrightarrow \operatorname{Hom}_R(R/\mathfrak{b}_\alpha, N)$$

$$\longrightarrow \operatorname{Ext}_R^1(R/\mathfrak{b}_\alpha, L) \longrightarrow \operatorname{Ext}_R^1(R/\mathfrak{b}_\alpha, M) \longrightarrow \operatorname{Ext}_R^1(R/\mathfrak{b}_\alpha, N)$$

$$\longrightarrow \quad \cdots \qquad\qquad\qquad\qquad\qquad \cdots$$

$$\longrightarrow \operatorname{Ext}_R^i(R/\mathfrak{b}_\alpha, L) \longrightarrow \operatorname{Ext}_R^i(R/\mathfrak{b}_\alpha, M) \longrightarrow \operatorname{Ext}_R^i(R/\mathfrak{b}_\alpha, N)$$

$$\longrightarrow \operatorname{Ext}_R^{i+1}(R/\mathfrak{b}_\alpha, L) \longrightarrow \qquad \cdots$$

exact. Moreover, these connecting homomorphisms are such that, for $\alpha, \beta \in \Lambda$ with $\alpha \geq \beta$, the diagram

$$
\begin{array}{ccc}
\operatorname{Ext}_R^i(R/\mathfrak{b}_\beta, N) & \longrightarrow & \operatorname{Ext}_R^{i+1}(R/\mathfrak{b}_\beta, L) \\
\downarrow{\scriptstyle \operatorname{Ext}_R^i(h_\beta^\alpha, N)} & & \downarrow{\scriptstyle \operatorname{Ext}_R^{i+1}(h_\beta^\alpha, L)} \\
\operatorname{Ext}_R^i(R/\mathfrak{b}_\alpha, N) & \longrightarrow & \operatorname{Ext}_R^{i+1}(R/\mathfrak{b}_\alpha, L)
\end{array}
$$

(in which the horizontal maps are the appropriate connecting homomorphisms and $h_\beta^\alpha : R/\mathfrak{b}_\alpha \to R/\mathfrak{b}_\beta$ is the natural homomorphism) commutes for each $i \in \mathbb{N}_0$. It follows that these diagrams induce 'connecting' R-homomorphisms

$$U^i(N) = \varinjlim_{\alpha \in \Lambda} \operatorname{Ext}_R^i(R/\mathfrak{b}_\alpha, N) \longrightarrow U^{i+1}(L) = \varinjlim_{\alpha \in \Lambda} \operatorname{Ext}_R^{i+1}(R/\mathfrak{b}_\alpha, L)$$

(for $i \in \mathbb{N}_0$); moreover, the fact that passage to direct limits preserves exactness ensures that the resulting long sequence

$$
\begin{array}{ccccc}
0 \longrightarrow & U^0(L) & \longrightarrow & U^0(M) & \longrightarrow & U^0(N) \\
\longrightarrow & U^1(L) & \longrightarrow & U^1(M) & \longrightarrow & U^1(N) \\
\longrightarrow & \cdots & & & & \cdots \\
\longrightarrow & U^i(L) & \longrightarrow & U^i(M) & \longrightarrow & U^i(N) \\
\longrightarrow & U^{i+1}(L) & \longrightarrow & \cdots & &
\end{array}
$$

is exact. Next, standard properties of the extension functors ensure that, whenever

$$
\begin{array}{ccccccccc}
0 & \longrightarrow & L & \longrightarrow & M & \longrightarrow & N & \longrightarrow & 0 \\
& & \downarrow{\scriptstyle \lambda} & & \downarrow{\scriptstyle \mu} & & \downarrow{\scriptstyle \nu} & & \\
0 & \longrightarrow & L' & \longrightarrow & M' & \longrightarrow & N' & \longrightarrow & 0
\end{array}
$$

is a commutative diagram of R-modules and R-homomorphisms with exact rows, then, for all $\alpha \in \Lambda$, the diagram

$$
\begin{array}{ccc}
\operatorname{Ext}_R^i(R/\mathfrak{b}_\alpha, N) & \longrightarrow & \operatorname{Ext}_R^{i+1}(R/\mathfrak{b}_\alpha, L) \\
{\scriptstyle \operatorname{Ext}_R^i(R/\mathfrak{b}_\alpha,\nu)} \downarrow & & \downarrow {\scriptstyle \operatorname{Ext}_R^{i+1}(R/\mathfrak{b}_\alpha,\lambda)} \\
\operatorname{Ext}_R^i(R/\mathfrak{b}_\alpha, N') & \longrightarrow & \operatorname{Ext}_R^{i+1}(R/\mathfrak{b}_\alpha, L')
\end{array}
$$

(in which the horizontal maps are the appropriate connecting homomorphisms) commutes for each $i \in \mathbb{N}_0$. It therefore follows that the diagram

$$
\begin{array}{ccc}
\varinjlim_{\alpha \in \Lambda} \operatorname{Ext}_R^i(R/\mathfrak{b}_\alpha, N) & \longrightarrow & \varinjlim_{\alpha \in \Lambda} \operatorname{Ext}_R^{i+1}(R/\mathfrak{b}_\alpha, L) \\
{\scriptstyle \varinjlim_{\alpha \in \Lambda} \operatorname{Ext}_R^i(R/\mathfrak{b}_\alpha,\nu)} \downarrow & & \downarrow {\scriptstyle \varinjlim_{\alpha \in \Lambda} \operatorname{Ext}_R^{i+1}(R/\mathfrak{b}_\alpha,\lambda)} \\
\varinjlim_{\alpha \in \Lambda} \operatorname{Ext}_R^i(R/\mathfrak{b}_\alpha, N') & \longrightarrow & \varinjlim_{\alpha \in \Lambda} \operatorname{Ext}_R^{i+1}(R/\mathfrak{b}_\alpha, L')
\end{array}
$$

(in which the horizontal maps are again the appropriate connecting homomorphisms) commutes for all $i \in \mathbb{N}_0$.

We have thus made $\left(\varinjlim\limits_{\alpha \in \Lambda} \operatorname{Ext}_R^i(R/\mathfrak{b}_\alpha, \bullet) \right)_{i \in \mathbb{N}_0}$ into a negative strongly connected sequence of covariant functors from $\mathcal{C}(R)$ to $\mathcal{C}(R)$. Since we have $\varinjlim\limits_{\alpha \in \Lambda} \operatorname{Ext}_R^i(R/\mathfrak{b}_\alpha, I) = 0$ for all $i \in \mathbb{N}$ whenever I is an injective R-module, it now follows from 1.3.5 that there is a unique isomorphism of connected sequences

$$\widetilde{\Psi} = \left(\widetilde{\psi}^i \right)_{i \in \mathbb{N}_0} : \left(\varinjlim\limits_{\alpha \in \Lambda} \operatorname{Ext}_R^i(R/\mathfrak{b}_\alpha, \bullet) \right)_{i \in \mathbb{N}_0} \stackrel{\cong}{\longrightarrow} \left(\mathcal{R}^i \Gamma_\mathfrak{B} \right)_{i \in \mathbb{N}_0}$$

for which $\widetilde{\psi}^0$ is the natural equivalence $\phi'_\mathfrak{B}$ of 1.2.11(ii); furthermore, both these connected sequences are isomorphic to the negative (strongly) connected sequence of functors formed by the right derived functors of

$$\varinjlim\limits_{\alpha \in \Lambda} \operatorname{Hom}_R(R/\mathfrak{b}_\alpha, \bullet).$$

A special case of 1.3.7 describes local cohomology modules as direct limits of Ext modules. As this description is of crucial importance for our subject, we state it separately.

1.3.8 Theorem. *There is a unique isomorphism of connected sequences (of functors from $\mathcal{C}(R)$ to $\mathcal{C}(R)$)*

$$\Phi_\mathfrak{a} = \left(\phi_\mathfrak{a}^i \right)_{i \in \mathbb{N}_0} : \left(\varinjlim\limits_{n \in \mathbb{N}} \operatorname{Ext}_R^i(R/\mathfrak{a}^n, \bullet) \right)_{i \in \mathbb{N}_0} \stackrel{\cong}{\longrightarrow} \left(H_\mathfrak{a}^i \right)_{i \in \mathbb{N}_0}$$

which extends the natural equivalence $\phi_\mathfrak{a}^0 : \varinjlim\limits_{n \in \mathbb{N}} \operatorname{Hom}_R(R/\mathfrak{a}^n, \bullet) \stackrel{\cong}{\longrightarrow} \Gamma_\mathfrak{a}$ of 1.2.11(iii). Consequently, for each R-module M and each $i \in \mathbb{N}_0$,

$$H_\mathfrak{a}^i(M) \cong \varinjlim\limits_{n \in \mathbb{N}} \operatorname{Ext}_R^i(R/\mathfrak{a}^n, M).$$

1.3.9 ♯Exercise. Let M be an R-module, not necessarily finitely generated. Let $a_1, \ldots, a_n$ be an M-sequence (of elements of R) (see [50, p. 123]). Let $a_1' \in R$. Show that

(i) if $a_1', a_2, \ldots, a_n$ is also an M-sequence, then so too is $a_1 a_1', a_2, \ldots, a_n$;
(ii) if $h_1, \ldots, h_n \in \mathbb{N}$, then $a_1^{h_1}, \ldots, a_n^{h_n}$ is also an M-sequence; and
(iii) if $a_1, \ldots, a_n$ all belong to $\mathfrak{a}$, then $\operatorname{Ext}_R^i(R/\mathfrak{a}, M) = H_\mathfrak{a}^i(M) = 0$ for all $i = 0, \ldots, n-1$. (This theme will be pursued in Chapter 6.)

2

Torsion modules and ideal transforms

The first section of this chapter contains the essence of a useful reduction technique in the study of local cohomology modules of finitely generated modules. The main points are these: if M is a finitely generated R-module, then it turns out that the R-module $\overline{M} := M/\Gamma_\mathfrak{a}(M)$ is $\mathfrak{a}$-torsion-free, and that $\mathfrak{a}$ contains a non-zerodivisor r on $\overline{M}$; moreover, for $i > 0$, the local cohomology modules $H^i_\mathfrak{a}(M)$ and $H^i_\mathfrak{a}(\overline{M})$ are isomorphic, so that the study of these higher local cohomology modules of M with respect to $\mathfrak{a}$ can be reduced to the study of the corresponding local cohomology modules of $\overline{M}$; the advantage of this is that the exact sequence $0 \longrightarrow \overline{M} \overset{r}{\longrightarrow} \overline{M} \longrightarrow \overline{M}/r\overline{M} \longrightarrow 0$ provides a route to further progress. There are several places later in the book where this strategy is used.

In the second section of this chapter, we develop the basic theory of the functor $D_\mathfrak{a} := \varinjlim_{n \in \mathbb{N}} \mathrm{Hom}_R(\mathfrak{a}^n, \bullet\,)$ which was mentioned in 1.2.9(ii). For an R-module M, the module $D_\mathfrak{a}(M)$ is called the $\mathfrak{a}$-*transform of* M, and we plan to show that such transforms provide a powerful algebraic tool. The use of these ideal transforms is an important part of our approach to local cohomology; we show that they have a certain universal property, and that universal property will help us with many technical details later in the book. In §2.2, we shall actually develop the theory of the *generalized ideal transform functor* with respect to what we call a system of ideals of R; one example of such a system of ideals is the family $(\mathfrak{a}^n)_{n \in \mathbb{N}}$ of powers of $\mathfrak{a}$, and the generalized ideal transform functor with respect to this system is just the $\mathfrak{a}$-transform functor. Our motivation for working in this generality is our wish to apply generalized ideal transforms to the theory of S_2-ifications in Chapter 12.

Towards the end of the chapter, we show that, in certain cases, ideal transforms have geometrical significance: we describe the ring of regular functions on a non-empty open subset of an affine variety V over an algebraically closed field as an ideal transform of $\mathcal{O}(V)$, the ring of regular functions on V. Actually, this is only a brief foretaste of what is to come at the end of the book, in Chapter 20, where we explore the relationship between ideal transforms and sheaf cohomology.

2.1 Torsion modules

We begin with results concerning $\mathfrak{a}$-torsion-free modules and $\mathfrak{a}$-torsion modules. Part (i) of our first lemma is related to Exercise 1.3.9(iv).

2.1.1 Lemma. *Let M be an R-module.*

(i) *If $\mathfrak{a}$ contains a non-zerodivisor on M, then M is $\mathfrak{a}$-torsion-free, that is, $\Gamma_{\mathfrak{a}}(M) = 0$.*

(ii) *Assume now that M is finitely generated. Then M is $\mathfrak{a}$-torsion-free if and only if $\mathfrak{a}$ contains a non-zerodivisor on M.*

Proof. (i) Let $r \in \mathfrak{a}$ be a non-zerodivisor on M, and let $m \in \Gamma_{\mathfrak{a}}(M)$. This means that there exists $n \in \mathbb{N}$ with $\mathfrak{a}^n m = 0$. Thus $r^n m = 0$, from which we deduce that $m = 0$.

(ii) One implication follows from (i), and so we assume that $\mathfrak{a}$ consists entirely of zerodivisors on M. Then $\mathfrak{a} \subseteq \bigcup_{\mathfrak{p} \in \operatorname{Ass} M} \mathfrak{p}$ by [81, Corollary 9.36], and, since M is finitely generated, $\operatorname{Ass} M$ is finite. Hence, by the Prime Avoidance Theorem [81, 3.61], $\mathfrak{a} \subseteq \mathfrak{p}$ for some $\mathfrak{p} \in \operatorname{Ass} M$. Since M has a submodule whose annihilator is exactly $\mathfrak{p}$, it follows that $(0 :_M \mathfrak{a}) \neq 0$, so that $\Gamma_{\mathfrak{a}}(M) \neq 0$. This completes the proof. $\quad\square$

2.1.2 Lemma. *For an R-module M, the module $M/\Gamma_{\mathfrak{a}}(M)$ is $\mathfrak{a}$-torsion-free.*

Proof. Let $m \in M$ be such that the element $m + \Gamma_{\mathfrak{a}}(M)$ of $M/\Gamma_{\mathfrak{a}}(M)$ is annihilated by $\mathfrak{a}^n$, where $n \in \mathbb{N}$. Our aim is to show that $m + \Gamma_{\mathfrak{a}}(M) = 0$, that is, that $m \in \Gamma_{\mathfrak{a}}(M)$.

Now $\mathfrak{a}^n m \subseteq \Gamma_{\mathfrak{a}}(M)$. Since $\mathfrak{a}^n m$ is a finitely generated submodule of $\Gamma_{\mathfrak{a}}(M)$, and each element of $\mathfrak{a}^n m$ is annihilated by some power of $\mathfrak{a}$, it follows that there exists $t \in \mathbb{N}$ such that $\mathfrak{a}^t \mathfrak{a}^n m = 0$. Therefore $m \in (0 :_M \mathfrak{a}^{n+t}) \subseteq \Gamma_{\mathfrak{a}}(M)$. $\quad\square$

2.1.3 Remarks. The following points should be noted.

(i) If M is an $\mathfrak{a}$-torsion R-module, that is, if $M = \Gamma_{\mathfrak{a}}(M)$, then all submodules of M and all R-homomorphic images of M are also $\mathfrak{a}$-torsion.

(ii) Consequently, for each R-module L and each $i \in \mathbb{N}_0$, the i-th local cohomology module $H_\mathfrak{a}^i(L)$ is an $\mathfrak{a}$-torsion R-module. To see this, let

$$I^\bullet : 0 \longrightarrow I^0 \longrightarrow I^1 \longrightarrow \cdots \longrightarrow I^i \longrightarrow I^{i+1} \longrightarrow \cdots$$

be an injective resolution of L; use this in conjunction with 1.2.2(i) to see that $H_\mathfrak{a}^i(L)$ is a homomorphic image of a submodule of the $\mathfrak{a}$-torsion module $\Gamma_\mathfrak{a}(I^i)$; and then appeal to (i).

Our next aim is to show that, if M is an $\mathfrak{a}$-torsion R-module, then $H_\mathfrak{a}^i(M) = 0$ for all $i \in \mathbb{N}$. We approach this by first considering the effect of $\Gamma_\mathfrak{a}$ on an injective R-module.

2.1.4 Proposition. *Let I be an injective R-module. Then $\Gamma_\mathfrak{a}(I)$ is also an injective R-module.*

Proof. Let $\mathfrak{b}$ be an ideal of R, and let $h : \mathfrak{b} \longrightarrow \Gamma_\mathfrak{a}(I)$ be a homomorphism of R-modules. By the Baer Criterion (see [71, Theorem 3.20]), it is enough for us to show that there exists $m \in \Gamma_\mathfrak{a}(I)$ such that $h(r) = rm$ for all $r \in \mathfrak{b}$.

Since I is injective, there exists $w \in I$ such that $h(r) = rw$ for all $r \in \mathfrak{b}$. Since R is Noetherian, $h(\mathfrak{b})$ is a finitely generated submodule of $\Gamma_\mathfrak{a}(I)$, and so there exists $t \in \mathbb{N}$ such that $\mathfrak{a}^t h(\mathfrak{b}) = 0$. Now $h(\mathfrak{b})$ is a submodule of the finitely generated R-module Rw, and so, by the Artin–Rees Lemma [50, Theorem 8.5], there exists $c \in \mathbb{N}$ such that, for all integers $n \geq c$,

$$\mathfrak{a}^n(Rw) \cap h(\mathfrak{b}) = \mathfrak{a}^{n-c}(\mathfrak{a}^c(Rw) \cap h(\mathfrak{b})).$$

Hence $\mathfrak{a}^{t+c}(Rw) \cap h(\mathfrak{b}) \subseteq \mathfrak{a}^t h(\mathfrak{b}) = 0$. Consequently, we can define an R-homomorphism $\tilde{h} : \mathfrak{a}^{t+c} + \mathfrak{b} \longrightarrow \Gamma_\mathfrak{a}(I)$ for which $\tilde{h}(s + r) = rw$ for all $s \in \mathfrak{a}^{t+c}$ and $r \in \mathfrak{b}$: this follows because, if $r_1, r_2 \in \mathfrak{b}$ are such that $r_1 - r_2 \in \mathfrak{a}^{t+c}$, then $r_1 w - r_2 w = (r_1 - r_2)w \in \mathfrak{a}^{t+c}(Rw) \cap h(\mathfrak{b}) = 0$.

Now use again the fact that I is R-injective to see that there exists $m \in I$ such that $\tilde{h}(r) = rm$ for all $r \in \mathfrak{a}^{t+c} + \mathfrak{b}$. Since $\tilde{h}$ extends h, the proof will be complete if we show that $m \in \Gamma_\mathfrak{a}(I)$.

To achieve this, just note that, for all $s \in \mathfrak{a}^{t+c}$,

$$sm = \tilde{h}(s) = \tilde{h}(s + 0) = 0w = 0,$$

by definition of $\tilde{h}$. This completes the proof. $\square$

2.1.5 Corollary. *Let I be an injective R-module. Then the canonical exact sequence $0 \longrightarrow \Gamma_\mathfrak{a}(I) \longrightarrow I \longrightarrow I/\Gamma_\mathfrak{a}(I) \longrightarrow 0$ splits.*

Proof. This is immediate from the fact, established in 2.1.4, that $\Gamma_\mathfrak{a}(I)$ is injective. $\square$

2.1.6 Corollary. *Let M be an $\mathfrak{a}$-torsion R-module. Then there exists an injective resolution of M in which each term is an $\mathfrak{a}$-torsion R-module.*

Proof. First note that, if N is an arbitrary R-module, then there exists an injective R-module I and an R-monomorphism $h : N \to I$. Application of the left exact functor $\Gamma_{\mathfrak{a}}$ yields a monomorphism $\Gamma_{\mathfrak{a}}(h) : \Gamma_{\mathfrak{a}}(N) \to \Gamma_{\mathfrak{a}}(I)$, and $\Gamma_{\mathfrak{a}}(I)$ is injective by 2.1.4.

If we apply the above paragraph to the $\mathfrak{a}$-torsion R-module M, we see that M can be embedded in an $\mathfrak{a}$-torsion injective R-module I^0. Suppose, inductively, that $n \in \mathbb{N}_0$ and we have constructed an exact sequence

$$0 \longrightarrow M \longrightarrow I^0 \longrightarrow \cdots \longrightarrow I^{n-1} \xrightarrow{d^{n-1}} I^n$$

of R-modules and R-homomorphisms in which $I^0, \ldots, I^{n-1}, I^n$ are all $\mathfrak{a}$-torsion injective R-modules. Let $C := \operatorname{Coker} d^{n-1}$, and note that, by 2.1.3(i), C is an $\mathfrak{a}$-torsion module because I^n is. Apply the first paragraph of this proof to C to deduce that there is an $\mathfrak{a}$-torsion injective R-module I^{n+1} and an R-monomorphism $g : C \to I^{n+1}$. Let $d^n : I^n \to I^{n+1}$ be the composition of the natural epimorphism from I^n to C and g.

This completes the inductive step, and the proof. $\qquad\square$

2.1.7 Corollary. *Let M and N be R-modules such that M is $\mathfrak{a}$-torsion. Then*

(i) $H_{\mathfrak{a}}^i(M) = 0$ *for all* $i > 0$;
(ii) $H_{\mathfrak{a}}^i(\Gamma_{\mathfrak{a}}(N)) = 0$ *for all* $i > 0$; *and*
(iii) *the natural map* $\pi : N \to N/\Gamma_{\mathfrak{a}}(N)$ *induces isomorphisms* $H_{\mathfrak{a}}^i(\pi)$: $H_{\mathfrak{a}}^i(N) \xrightarrow{\cong} H_{\mathfrak{a}}^i(N/\Gamma_{\mathfrak{a}}(N))$ *for all* $i > 0$.

Proof. (i) It was remarked in 1.2.2(i) that we can use any injective resolution of M to calculate (up to isomorphism) the $H_{\mathfrak{a}}^i(M)$: by 2.1.6, there is an injective resolution of M in which each term is an $\mathfrak{a}$-torsion R-module, and use of this shows that $H_{\mathfrak{a}}^i(M) = 0$ for all $i > 0$.

(ii) This is immediate from (i) because $\Gamma_{\mathfrak{a}}(N)$ is an $\mathfrak{a}$-torsion R-module.

(iii) This is immediate from (ii) on use of the long exact sequence of local cohomology modules induced by the short exact sequence

$$0 \longrightarrow \Gamma_{\mathfrak{a}}(N) \longrightarrow N \xrightarrow{\pi} N/\Gamma_{\mathfrak{a}}(N) \longrightarrow 0. \qquad\square$$

2.1.8 Exercise. Let M be an $\mathfrak{a}$-torsion R-module, and let

$$I^\bullet : 0 \xrightarrow{d^{-1}} I^0 \xrightarrow{d^0} I^1 \longrightarrow \cdots \longrightarrow I^i \xrightarrow{d^i} I^{i+1} \longrightarrow \cdots$$

be an injective resolution of M. Show that $\Gamma_{\mathfrak{a}}(I^\bullet)$, that is, the complex

$$0 \xrightarrow{\Gamma_{\mathfrak{a}}(d^{-1})} \Gamma_{\mathfrak{a}}(I^0) \longrightarrow \cdots \longrightarrow \Gamma_{\mathfrak{a}}(I^i) \xrightarrow{\Gamma_{\mathfrak{a}}(d^i)} \Gamma_{\mathfrak{a}}(I^{i+1}) \longrightarrow \cdots,$$

is also an injective resolution of M.

2.1.9 ♯Exercise. Let $\mathfrak{b}$ be a second ideal of R, and let M be a $\mathfrak{b}$-torsion R-module. Prove that $H^i_{\mathfrak{a}+\mathfrak{b}}(M) \cong H^i_{\mathfrak{a}}(M)$ for all $i \in \mathbb{N}_0$.

In Chapter 1, we indicated that we shall, at times, find it convenient to consider an inverse family $\mathfrak{B} = (\mathfrak{b}_\alpha)_{\alpha \in \Lambda}$ of ideals of R over a (non-empty) directed partially ordered set Λ, as in 1.2.10. We were able to produce, for such a $\mathfrak{B}$, a functor $\Gamma_\mathfrak{B}$ in 1.2.11, and the right derived functors $\mathcal{R}^i\Gamma_\mathfrak{B}$ $(i \in \mathbb{N}_0)$ of $\Gamma_\mathfrak{B}$ are generalizations of the local cohomology functors $H^i_{\mathfrak{a}}$ $(i \in \mathbb{N}_0)$.

However, we cannot expect the theory of $\Gamma_\mathfrak{B}$ to imitate local cohomology theory completely unless we impose additional conditions on $\mathfrak{B}$. For example, the analogue for $\Gamma_\mathfrak{B}$ of 2.1.4 is not true in general. However, this difficulty does not occur for the concept introduced in the next definition.

2.1.10 Definition. Let $(\Lambda, \leq)$ be a (non-empty) directed partially ordered set. A *system of ideals (of R) over* Λ is an inverse family $\mathfrak{B} = (\mathfrak{b}_\alpha)_{\alpha \in \Lambda}$ of ideals of R over Λ in the sense of 1.2.10 with the additional property that, for all $\alpha, \gamma \in \Lambda$, there exists $\delta \in \Lambda$ such that $\mathfrak{b}_\delta \subseteq \mathfrak{b}_\alpha \mathfrak{b}_\gamma$. (It is clear that the δ in this condition can be chosen so that $\delta \geq \alpha$ and $\delta \geq \gamma$, since $(\Lambda, \leq)$ is a directed set and, by 1.2.10, whenever $(\mu, \nu) \in \Lambda \times \Lambda$ with $\mu \geq \nu$, we have $\mathfrak{b}_\mu \subseteq \mathfrak{b}_\nu$.)

For such a system of ideals $\mathfrak{B}$, we shall denote $\mathcal{R}^i\Gamma_\mathfrak{B}$ by $H^i_\mathfrak{B}$ (for all $i \in \mathbb{N}_0$). The reader should keep in mind that $(\mathfrak{a}^n)_{n \in \mathbb{N}}$ is a fundamental example of a system $\mathfrak{A}$ of ideals (over $\mathbb{N}$); however, we shall continue to use the notation $H^i_{\mathfrak{a}}$ of 1.2.1 (rather than $H^i_{\mathfrak{A}}$).

2.1.11 Examples. Here are some further examples of systems of ideals.

(i) Let $\mathfrak{A}$ be a non-empty set of ideals of R. Then the set of all products of finite families of ideals taken from $\mathfrak{A}$ forms a system of ideals in an obvious way.

  In particular, if $\mathfrak{A}$ is a non-empty multiplicatively closed set of ideals of R, then $\mathfrak{A}$ itself forms a system of ideals.

(ii) Let $n \in \mathbb{N}_0$. The height, ht $\mathfrak{b}$, of a proper ideal $\mathfrak{b}$ of R is defined in [50, p. 31] and [81, 15.6]. Interpret the height of the improper ideal R of R as ∞. Then $\{\mathfrak{b} : \mathfrak{b}$ is an ideal of R and ht $\mathfrak{b} \geq n\}$ forms a system of ideals of R.

2.1.12 ♯Exercise. Let $\mathfrak{B}$ be a system of ideals over Λ in the sense of 2.1.10. Let $\alpha, \beta \in \Lambda$ with $\alpha \geq \beta$. Show that, for each R-module M,

$$\Gamma_{\mathfrak{b}_\beta}(M) \subseteq \Gamma_{\mathfrak{b}_\alpha}(M) \subseteq \Gamma_\mathfrak{B}(M) = \bigcup_{\delta \in \Lambda} \Gamma_{\mathfrak{b}_\delta}(M).$$

2.1.13 ♯Exercise. Let $\mathfrak{B}$ be a system of ideals over Λ in the sense of 2.1.10. Let M be an R-module. We shall say that M is $\mathfrak{B}$-*torsion-free* precisely when $\Gamma_{\mathfrak{B}}(M) = 0$, and that M is $\mathfrak{B}$-*torsion* precisely when $M = \Gamma_{\mathfrak{B}}(M)$.

   (i) Show that $M/\Gamma_{\mathfrak{B}}(M)$ is $\mathfrak{B}$-torsion-free.
   (ii) Show that, if I is an injective R-module, then $\Gamma_{\mathfrak{B}}(I)$ is also an injective R-module. Deduce that, if N is a $\mathfrak{B}$-torsion R-module, then there exists an injective resolution of N in which each term is a $\mathfrak{B}$-torsion R-module, and conclude that $H^i_{\mathfrak{B}}(N) = 0$ for all $i > 0$.
   (iii) Prove that the natural epimorphism $\pi : M \to M/\Gamma_{\mathfrak{B}}(M)$ induces isomorphisms $H^i_{\mathfrak{B}}(\pi) : H^i_{\mathfrak{B}}(M) \xrightarrow{\cong} H^i_{\mathfrak{B}}(M/\Gamma_{\mathfrak{B}}(M))$ for all $i > 0$.

2.1.14 ♯Exercise. Let M be an R-module. Show that the sets $\operatorname{Ass}(\Gamma_{\mathfrak{a}}(M))$ and $\operatorname{Ass}(M/\Gamma_{\mathfrak{a}}(M))$ are disjoint, and that

$$\operatorname{Ass} M = \operatorname{Ass}(\Gamma_{\mathfrak{a}}(M)) \cup \operatorname{Ass}(M/\Gamma_{\mathfrak{a}}(M)).$$

2.2 Ideal transforms and generalized ideal transforms

The principal object of study in this section will be the ideal transform of an R-module M with respect to $\mathfrak{a}$. This is defined as follows.

2.2.1 Definitions. In 1.2.9(ii), we constructed covariant, R-linear functors

$$D_{\mathfrak{a}} := \varinjlim_{n \in \mathbb{N}} \operatorname{Hom}_R(\mathfrak{a}^n, \bullet) \quad \text{and} \quad \varinjlim_{n \in \mathbb{N}} \operatorname{Ext}^i_R(\mathfrak{a}^n, \bullet) \ (i \in \mathbb{N}_0)$$

from $\mathcal{C}(R)$ to itself. We shall refer to $D_{\mathfrak{a}}$ as the $\mathfrak{a}$-*transform functor*; note that, by 1.2.8, this functor is left exact.

For an R-module M, we call $D_{\mathfrak{a}}(M) = \varinjlim_{n \in \mathbb{N}} \operatorname{Hom}_R(\mathfrak{a}^n, M)$ the *ideal transform of M with respect to $\mathfrak{a}$*, or the $\mathfrak{a}$-*transform of M*.

However, instead of working with the powers of a single ideal $\mathfrak{a}$, we are going to work in this section in the more general framework of a system of ideals. This generality is motivated by an application that will be presented in Chapter 12. The reader should keep in mind, throughout this section, that a basic example of a system of ideals is $(\mathfrak{a}^n)_{n \in \mathbb{N}}$.

2.2.2 Notation. Throughout this section, $(\Lambda, \leq)$ will denote a (non-empty) directed partially ordered set, and $\mathfrak{B} = (\mathfrak{b}_\alpha)_{\alpha \in \Lambda}$ will denote a system of ideals of R over Λ in the sense of 2.1.10.

The functors $\Gamma_{\mathfrak{B}}$ and $\varinjlim_{\alpha \in \Lambda} \operatorname{Hom}_R(R/\mathfrak{b}_\alpha, \bullet)$ were introduced in 1.2.11(i) and

1.2.10; for $i \in \mathbb{N}_0$, the functor $\varinjlim_{\alpha \in \Lambda} \operatorname{Ext}_R^i(R/\mathfrak{b}_\alpha, \bullet)$ was discussed in 1.3.7,
where it was shown to be naturally equivalent to $\mathcal{R}^i \Gamma_{\mathfrak{B}}$, the i-th right derived functor of $\Gamma_{\mathfrak{B}}$; in 2.1.10, we agreed to denote the R-linear functor $\mathcal{R}^i \Gamma_{\mathfrak{B}}$ by $H_{\mathfrak{B}}^i$; we shall refer to this as the *i-th generalized local cohomology functor with respect to* $\mathfrak{B}$ (and we shall use natural extensions of this terminology). We denote by

$$\Phi_{\mathfrak{B}} = \left(\phi_{\mathfrak{B}}^i \right)_{i \in \mathbb{N}_0} : \left(\varinjlim_{\alpha \in \Lambda} \operatorname{Ext}_R^i(R/\mathfrak{b}_\alpha, \bullet) \right)_{i \in \mathbb{N}_0} \xrightarrow{\cong} \left(H_{\mathfrak{B}}^i \right)_{i \in \mathbb{N}_0}$$

the unique isomorphism of connected sequences for which $\phi_{\mathfrak{B}}^0$ is the natural equivalence $\phi_{\mathfrak{B}}'$ of 1.2.11(ii): see 1.3.7.

2.2.3 Definitions.   The functors

$$D_{\mathfrak{B}} := \varinjlim_{\alpha \in \Lambda} \operatorname{Hom}_R(\mathfrak{b}_\alpha, \bullet) \quad \text{and} \quad \varinjlim_{\alpha \in \Lambda} \operatorname{Ext}_R^i(\mathfrak{b}_\alpha, \bullet) \, (i \in \mathbb{N}_0)$$

(from $\mathcal{C}(R)$ to itself) can be defined using the ideas of 1.2.8 in conjunction with the inclusion maps $\mathfrak{b}_\alpha \to \mathfrak{b}_\beta$ (for $\alpha, \beta \in \Lambda$ with $\alpha \geq \beta$). We shall refer to $D_{\mathfrak{B}}$ as the *$\mathfrak{B}$-transform functor*; note that, by 1.2.8, this functor is left exact.

For an R-module M, we call $D_{\mathfrak{B}}(M) = \varinjlim_{\alpha \in \Lambda} \operatorname{Hom}_R(\mathfrak{b}_\alpha, M)$ the *generalized ideal transform of M with respect to* $\mathfrak{B}$, or, alternatively, the *$\mathfrak{B}$-transform of M*.

2.2.4 ♯Exercise.   For $i \in \mathbb{N}_0$, we use $\mathcal{R}^i D_{\mathfrak{B}}$ to denote the i-th right derived functor of $D_{\mathfrak{B}}$.

Modify the ideas of 1.3.7 to show that $\left(\varinjlim_{\alpha \in \Lambda} \operatorname{Ext}_R^i(\mathfrak{b}_\alpha, \bullet) \right)_{i \in \mathbb{N}_0}$ is a negative strongly connected sequence of functors from $\mathcal{C}(R)$ to itself. Use 1.3.5 to show that there is a unique isomorphism of connected sequences (of functors from $\mathcal{C}(R)$ to itself)

$$\Psi_{\mathfrak{B}} = \left(\psi_{\mathfrak{B}}^i \right)_{i \in \mathbb{N}_0} : \left(\mathcal{R}^i D_{\mathfrak{B}} \right)_{i \in \mathbb{N}_0} \xrightarrow{\cong} \left(\varinjlim_{\alpha \in \Lambda} \operatorname{Ext}_R^i(\mathfrak{b}_\alpha, \bullet) \right)_{i \in \mathbb{N}_0}$$

which extends the identity natural equivalence from $D_{\mathfrak{B}}$ to itself.

2.2.5 ♯Exercise.   Let M be an R-module. For each $\alpha \in \Lambda$, let

$$\pi_\alpha^M : \operatorname{Hom}_R(\mathfrak{b}_\alpha, M) \to D_{\mathfrak{B}}(M)$$

be the natural homomorphism.

(i) Let $\alpha, \beta \in \Lambda$, and let $f \in \operatorname{Hom}_R(\mathfrak{b}_\alpha, R)$ and $g \in \operatorname{Hom}_R(\mathfrak{b}_\beta, R)$. Since $\mathfrak{B}$ is a system of ideals of R, there exists $\delta \in \Lambda$ such that $\mathfrak{b}_\delta \subseteq \mathfrak{b}_\alpha \mathfrak{b}_\beta$. Observe that $f \upharpoonright_{\mathfrak{b}_\delta}$, the restriction of f to $\mathfrak{b}_\delta$, maps $\mathfrak{b}_\delta$ into $\mathfrak{b}_\beta$. Show that $g \circ (f \upharpoonright_{\mathfrak{b}_\delta}) = f \circ (g \upharpoonright_{\mathfrak{b}_\delta})$.

(ii) Show that there is a binary operation $*$ on $D_{\mathfrak{B}}(R)$ which is such that, for $f \in \operatorname{Hom}_R(\mathfrak{b}_\alpha, R)$ and $g \in \operatorname{Hom}_R(\mathfrak{b}_\beta, R)$,

$$\pi_\alpha^R(f) * \pi_\beta^R(g) = \pi_\delta^R \left(g \circ (f \upharpoonright_{\mathfrak{b}_\delta}) \right)$$

for *any* choice of $\delta \in \Lambda$ with $\mathfrak{b}_\delta \subseteq \mathfrak{b}_\alpha \mathfrak{b}_\beta$; show further that $D_{\mathfrak{B}}(R)$ is a commutative ring with identity with respect to its R-module addition and $*$ as multiplication.

(iii) Show that $D_{\mathfrak{B}}(M)$ has the structure of a $D_{\mathfrak{B}}(R)$-module such that, for $\alpha, \beta \in \Lambda$ and for $f \in \operatorname{Hom}_R(\mathfrak{b}_\alpha, R)$ and $h \in \operatorname{Hom}_R(\mathfrak{b}_\beta, M)$,

$$\pi_\alpha^R(f) \left(\pi_\beta^M(h) \right) = \pi_\delta^M \left(h \circ (f \upharpoonright_{\mathfrak{b}_\delta}) \right)$$

for any choice of $\delta \in \Lambda$ with $\mathfrak{b}_\delta \subseteq \mathfrak{b}_\alpha \mathfrak{b}_\beta$.

(iv) Show that $D_{\mathfrak{B}}$ is an additive, left exact, covariant functor from $\mathcal{C}(R)$ to $\mathcal{C}(D_{\mathfrak{B}}(R))$. Thus all the $\mathcal{R}^i D_{\mathfrak{B}}$ ($i \in \mathbb{N}_0$) can be considered as additive functors from $\mathcal{C}(R)$ to $\mathcal{C}(D_{\mathfrak{B}}(R))$.

2.2.6 Theorem. *Denote the identity functor on the category $\mathcal{C}(R)$ by* Id.

(i) *There are natural transformations of functors (from $\mathcal{C}(R)$ to itself)*

$$\xi \, (= \xi_{\mathfrak{B}}) : \Gamma_{\mathfrak{B}} \longrightarrow \text{Id}, \qquad \eta \, (= \eta_{\mathfrak{B}}) : \text{Id} \longrightarrow D_{\mathfrak{B}}$$

$$\zeta^0 \, (= \zeta_{\mathfrak{B}}^0) : D_{\mathfrak{B}} \longrightarrow H_{\mathfrak{B}}^1$$

such that, for each R-module M,

(a) $\xi_M : \Gamma_{\mathfrak{B}}(M) \longrightarrow M$ *is the inclusion map,*

(b) *for each $g \in M$, $\eta_M(g)$ is the natural image in $D_{\mathfrak{B}}(M)$ of the homomorphism $f_{\alpha,g} \in \operatorname{Hom}_R(\mathfrak{b}_\alpha, M)$ given by $f_{\alpha,g}(r) = rg$ for all $r \in \mathfrak{b}_\alpha$ (for any $\alpha \in \Lambda$), and*

(c) *the sequence*

$$0 \longrightarrow \Gamma_{\mathfrak{B}}(M) \xrightarrow{\xi_M} M \xrightarrow{\eta_M} D_{\mathfrak{B}}(M) \xrightarrow{\zeta_M^0} H_{\mathfrak{B}}^1(M) \longrightarrow 0$$

is exact.

(ii) *Let $i \in \mathbb{N}$, and M be an R-module. For each $\alpha \in \Lambda$, the connecting* *homomorphism* $\beta^i_{\alpha,M} : \operatorname{Ext}^i_R(\mathfrak{b}_\alpha, M) \longrightarrow \operatorname{Ext}^{i+1}_R(R/\mathfrak{b}_\alpha, M)$ *is an isomorphism, and passage to the direct limit yields an R-isomorphism*

$$\beta^i_M : \varinjlim_{\alpha \in \Lambda} \operatorname{Ext}^i_R(\mathfrak{b}_\alpha, M) \xrightarrow{\cong} \varinjlim_{\alpha \in \Lambda} \operatorname{Ext}^{i+1}_R(R/\mathfrak{b}_\alpha, M).$$

Define $\gamma^i_M : \mathcal{R}^i D_\mathfrak{B}(M) \xrightarrow{\cong} H^{i+1}_\mathfrak{B}(M)$ *by*

$$\gamma^i_M := \phi^{i+1}_{\mathfrak{B},M} \circ \beta^i_M \circ \psi^i_{\mathfrak{B},M},$$

where $\phi^{i+1}_\mathfrak{B}$ *and* $\psi^i_\mathfrak{B}$ *are the natural equivalences of 2.2.2 and 2.2.4 respectively. Then, as M varies through the category $\mathcal{C}(R)$, the γ^i_M constitute a natural equivalence of functors* $\gamma^i : \mathcal{R}^i D_\mathfrak{B} \xrightarrow{\cong} H^{i+1}_\mathfrak{B}$.

(iii) *For each $i \in \mathbb{N}$, set $\zeta^i (= \zeta^i_\mathfrak{B}) := (-1)^i \gamma^i$. Then*

$$\left(\zeta^j\right)_{j \in \mathbb{N}_0} : \left(\mathcal{R}^j D_\mathfrak{B}\right)_{j \in \mathbb{N}_0} \longrightarrow \left(H^{j+1}_\mathfrak{B}\right)_{j \in \mathbb{N}_0}$$

is the unique homomorphism of connected sequences which extends the natural transformation $\zeta^0 : D_\mathfrak{B} \longrightarrow H^1_\mathfrak{B}$ of part (i).

Note. When $\mathfrak{B} = (\mathfrak{a}^n)_{n \in \mathbb{N}}$, we shall write $\xi_\mathfrak{a}, \eta_\mathfrak{a}, \zeta^0_\mathfrak{a}$ instead of $\xi_\mathfrak{B}, \eta_\mathfrak{B}, \zeta^0_\mathfrak{B}$.

Proof.   (i), (ii) Let $\alpha, \delta \in \Lambda$ with $\alpha \geq \delta$; let $j^\alpha_\delta : \mathfrak{b}_\alpha \longrightarrow \mathfrak{b}_\delta$ be the inclusion map; and let $h^\alpha_\delta : R/\mathfrak{b}_\alpha \longrightarrow R/\mathfrak{b}_\delta$ be the natural epimorphism. Also, let M, N be R-modules and let $f : M \longrightarrow N$ be an R-homomorphism.

The commutative diagram

$$
\begin{array}{ccccccccc}
0 & \longrightarrow & \mathfrak{b}_\alpha & \longrightarrow & R & \longrightarrow & R/\mathfrak{b}_\alpha & \longrightarrow & 0 \\
 & & \downarrow{\scriptstyle j^\alpha_\delta} & & \| & & \downarrow{\scriptstyle h^\alpha_\delta} & & \\
0 & \longrightarrow & \mathfrak{b}_\delta & \longrightarrow & R & \longrightarrow & R/\mathfrak{b}_\delta & \longrightarrow & 0
\end{array}
$$

(in which the rows are the canonical exact sequences) induces a chain map of the long exact sequence of $\operatorname{Ext}^\bullet_R(\bullet , M)$ modules induced by the top row to that induced by the bottom row. Since R is a projective R-module, and since $\operatorname{Hom}_R(R, M)$ is naturally isomorphic to M, we therefore obtain a commutative diagram

$$
\begin{array}{ccccccccc}
0 \to \operatorname{Hom}_R(R/\mathfrak{b}_\delta, M) & \to & M & \to & \operatorname{Hom}_R(\mathfrak{b}_\delta, M) & \to & \operatorname{Ext}^1_R(R/\mathfrak{b}_\delta, M) & \to & 0 \\
\downarrow & & \| & & \downarrow & & \downarrow & & \\
0 \to \operatorname{Hom}_R(R/\mathfrak{b}_\alpha, M) & \to & M & \to & \operatorname{Hom}_R(\mathfrak{b}_\alpha, M) & \to & \operatorname{Ext}^1_R(R/\mathfrak{b}_\alpha, M) & \to & 0
\end{array}
$$

(in which the rows are exact), and, for each $i \in \mathbb{N}$, a commutative diagram

$$
\begin{array}{ccc}
\operatorname{Ext}_R^i(\mathfrak{b}_\delta, M) & \xrightarrow[\cong]{\beta_{\delta,M}^i} & \operatorname{Ext}_R^{i+1}(R/\mathfrak{b}_\delta, M) \\
{\scriptstyle \operatorname{Ext}_R^i(j_\delta^\alpha, M)} \downarrow & & \downarrow {\scriptstyle \operatorname{Ext}_R^{i+1}(h_\delta^\alpha, M)} \\
\operatorname{Ext}_R^i(\mathfrak{b}_\alpha, M) & \xrightarrow[\cong]{\beta_{\alpha,M}^i} & \operatorname{Ext}_R^{i+1}(R/\mathfrak{b}_\alpha, M) \ .
\end{array}
$$

Now pass to the direct limits, bearing in mind the exactness-preserving properties of this process, and use the natural equivalences $\phi_{\mathfrak{B}}^0$, $\phi_{\mathfrak{B}}^1$ of 2.2.2 to obtain an exact sequence of R-modules and R-homomorphisms

$$
0 \longrightarrow \Gamma_{\mathfrak{B}}(M) \xrightarrow{\xi_M} M \xrightarrow{\eta_M} D_{\mathfrak{B}}(M) \xrightarrow{\zeta_M^0} H_{\mathfrak{B}}^1(M) \longrightarrow 0
$$

(where ξ_M and η_M are as described in (a) and (b) of the statement of the theorem) and, for each $i \in \mathbb{N}$, an isomorphism

$$
\beta_M^i : \varinjlim_{\alpha \in \Lambda} \operatorname{Ext}_R^i(\mathfrak{b}_\alpha, M) \xrightarrow{\cong} \varinjlim_{\alpha \in \Lambda} \operatorname{Ext}_R^{i+1}(R/\mathfrak{b}_\alpha, M).
$$

Moreover, since the diagram

$$
\begin{array}{ccccccccc}
0 & \rightarrow & \operatorname{Hom}_R(R/\mathfrak{b}_\alpha, M) & \rightarrow & M & \rightarrow & \operatorname{Hom}_R(\mathfrak{b}_\alpha, M) & \rightarrow & \operatorname{Ext}_R^1(R/\mathfrak{b}_\alpha, M) & \rightarrow & 0 \\
& & \downarrow & & {\scriptstyle f}\downarrow & & \downarrow & & \downarrow \\
0 & \rightarrow & \operatorname{Hom}_R(R/\mathfrak{b}_\alpha, N) & \rightarrow & N & \rightarrow & \operatorname{Hom}_R(\mathfrak{b}_\alpha, N) & \rightarrow & \operatorname{Ext}_R^1(R/\mathfrak{b}_\alpha, N) & \rightarrow & 0
\end{array}
$$

and, for each $i \in \mathbb{N}$, the diagram

$$
\begin{array}{ccc}
\operatorname{Ext}_R^i(\mathfrak{b}_\alpha, M) & \xrightarrow[\cong]{\beta_{\alpha,M}^i} & \operatorname{Ext}_R^{i+1}(R/\mathfrak{b}_\alpha, M) \\
\downarrow & & \downarrow \\
\operatorname{Ext}_R^i(\mathfrak{b}_\alpha, N) & \xrightarrow[\cong]{\beta_{\alpha,N}^i} & \operatorname{Ext}_R^{i+1}(R/\mathfrak{b}_\alpha, N)
\end{array}
$$

(in all of which, all the unmarked vertical maps are induced by $f : M \longrightarrow N$), are all commutative, it follows that, as M varies through $\mathcal{C}(R)$, the ξ_M, η_M, ζ_M^0 and β_M^i constitute natural transformations of functors ξ, η, ζ^0 and β^i respectively. This completes the proof of parts (i) and (ii).

(iii) Of course $\left(\mathcal{R}^j D_{\mathfrak{B}}\right)_{j \in \mathbb{N}_0}$ is a negative strongly connected sequence of covariant functors from $\mathcal{C}(R)$ to itself; also, $\left(H_{\mathfrak{B}}^{j+1}\right)_{j\in\mathbb{N}_0}$ is a negative connected sequence of covariant functors from $\mathcal{C}(R)$ to itself. It is immediate from 1.3.4(i) that there is a unique homomorphism of the first of these connected sequences to the second which extends the natural transformation ζ^0 : $D_{\mathfrak{B}} \longrightarrow H_{\mathfrak{B}}^1$ of part (i). The fact that this unique homomorphism is actually $\left((-1)^j\gamma^j\right)_{j\in\mathbb{N}_0}$ follows from Rotman [71, Theorem 11.24], which shows that, for each $j \in \mathbb{N}_0$, each exact sequence $0 \longrightarrow L \longrightarrow M \longrightarrow N \longrightarrow 0$ of R-modules and R-homomorphisms and each $\alpha \in \Lambda$, the diagram

$$
\begin{array}{ccc}
\mathrm{Ext}_R^j(\mathfrak{b}_\alpha, N) & \longrightarrow & \mathrm{Ext}_R^{j+1}(\mathfrak{b}_\alpha, L) \\
\downarrow & & \downarrow \\
\mathrm{Ext}_R^{j+1}(R/\mathfrak{b}_\alpha, N) & \longrightarrow & \mathrm{Ext}_R^{j+2}(R/\mathfrak{b}_\alpha, L) \, ,
\end{array}
$$

in which all the homomorphisms are the obvious connecting homomorphisms, is anticommutative. $\square$

2.2.7 Remark. Let the situation be as in 2.2.6. It is immediate from the exact sequence

$$0 \longrightarrow \Gamma_{\mathfrak{B}}(M) \xrightarrow{\xi_M} M \xrightarrow{\eta_M} D_{\mathfrak{B}}(M) \xrightarrow{\zeta_M^0} H_{\mathfrak{B}}^1(M) \longrightarrow 0$$

of 2.2.6(i)(c) that $\eta_M : M \longrightarrow D_{\mathfrak{B}}(M)$ is an isomorphism if and only if $\Gamma_{\mathfrak{B}}(M) = H_{\mathfrak{B}}^1(M) = 0$.

2.2.8 Corollary. *Let M be an R-module, not necessarily finitely generated. Assume that $\mathfrak{a}$ contains an M-sequence of length 2. Then*

$$\eta_M : M \longrightarrow D_{\mathfrak{a}}(M)$$

is an isomorphism.

Proof. By Exercise 1.3.9(iv), we have $\Gamma_{\mathfrak{a}}(M) = H_{\mathfrak{a}}^1(M) = 0$, and so the claim is immediate from 2.2.7. $\square$

2.2.9 Remark. It follows from 2.2.6 that, for each R-module M, there is an R-monomorphism $\theta_M : M/\Gamma_{\mathfrak{B}}(M) \longrightarrow D_{\mathfrak{B}}(M)$, induced by η_M, such that the sequence

$$0 \longrightarrow M/\Gamma_{\mathfrak{B}}(M) \xrightarrow{\theta_M} D_{\mathfrak{B}}(M) \xrightarrow{\zeta_M^0} H_{\mathfrak{B}}^1(M) \longrightarrow 0$$

is exact. As θ_M is induced by η_M, a precise formula for it can be extracted from the statement of 2.2.6. Note also that, as M varies through $\mathcal{C}(R)$, the θ_M

constitute a natural transformation of functors since, whenever $f : M \longrightarrow N$ is a homomorphism of R-modules, the diagram

$$
\begin{array}{ccc}
M/\Gamma_{\mathfrak{B}}(M) & \xrightarrow{\;\theta_M\;} & D_{\mathfrak{B}}(M) \\
\downarrow{\scriptstyle f^*} & & \downarrow{\scriptstyle D_{\mathfrak{B}}(f)} \\
N/\Gamma_{\mathfrak{B}}(N) & \xrightarrow{\;\theta_N\;} & D_{\mathfrak{B}}(N)
\end{array}
$$

(in which f^* denotes the homomorphism induced by f) commutes.

This remark can be particularly helpful when used in conjunction with the fact (see 2.1.13(iii)) that the canonical epimorphism $\pi : M \to M/\Gamma_{\mathfrak{B}}(M)$ induces isomorphisms $H^i_{\mathfrak{B}}(\pi) : H^i_{\mathfrak{B}}(M) \xrightarrow{\cong} H^i_{\mathfrak{B}}(M/\Gamma_{\mathfrak{B}}(M))$ for all $i > 0$. We use this idea in the proof of 2.2.10 below.

The exact sequence

$$
0 \longrightarrow \Gamma_{\mathfrak{B}}(M) \xrightarrow{\xi_M} M \xrightarrow{\eta_M} D_{\mathfrak{B}}(M) \xrightarrow{\zeta^0_M} H^1_{\mathfrak{B}}(M) \longrightarrow 0
$$

of Theorem 2.2.6(i)(c), and the particular case

$$
0 \longrightarrow \Gamma_{\mathfrak{a}}(M) \xrightarrow{\xi_M} M \xrightarrow{\eta_M} D_{\mathfrak{a}}(M) \xrightarrow{\zeta^0_M} H^1_{\mathfrak{a}}(M) \longrightarrow 0
$$

of it, are fundamental. In the next corollary, we present some important applications of such sequences.

2.2.10 Corollary. *Let M be an R-module; we use the notation of* 2.2.6 *and* 2.2.9. *Let $\pi : M \longrightarrow M/\Gamma_{\mathfrak{B}}(M)$ be the canonical epimorphism. Then the following hold:*

(i) $D_{\mathfrak{B}}(\Gamma_{\mathfrak{B}}(M)) = 0;$

(ii) $D_{\mathfrak{B}}(\pi) : D_{\mathfrak{B}}(M) \longrightarrow D_{\mathfrak{B}}(M/\Gamma_{\mathfrak{B}}(M))$ *is an isomorphism;*

(iii) $D_{\mathfrak{B}}(\eta_M) = \eta_{D_{\mathfrak{B}}(M)} : D_{\mathfrak{B}}(M) \longrightarrow D_{\mathfrak{B}}(D_{\mathfrak{B}}(M))$ *is an isomorphism;*

(iv) $\Gamma_{\mathfrak{B}}(D_{\mathfrak{B}}(M)) = 0 = H^1_{\mathfrak{B}}(D_{\mathfrak{B}}(M));$

(v) $H^i_{\mathfrak{B}}(\eta_M) : H^i_{\mathfrak{B}}(M) \longrightarrow H^i_{\mathfrak{B}}(D_{\mathfrak{B}}(M))$ *is an isomorphism for every* $i > 1$.

Proof. (i) Since $\Gamma_{\mathfrak{B}}(M)$ is a $\mathfrak{B}$-torsion R-module, it is enough, in order to prove this part, to show that, if N is a $\mathfrak{B}$-torsion R-module, then $D_{\mathfrak{B}}(N) = 0$. Now, for such an N, we have $H^1_{\mathfrak{B}}(N) = 0$ by 2.1.13(ii), and $\xi_N : \Gamma_{\mathfrak{B}}(N) \to N$ is the identity map. Hence the exact sequence of 2.2.6(i)(c) for N reduces to

$$
0 \longrightarrow \Gamma_{\mathfrak{B}}(N) \xrightarrow{\cong} N \longrightarrow D_{\mathfrak{B}}(N) \longrightarrow 0,
$$

and so $D_\mathfrak{B}(N) = 0$, as required.

(ii) By 2.2.3, the functor $D_\mathfrak{B}$ is left exact. Therefore the canonical exact sequence $0 \longrightarrow \Gamma_\mathfrak{B}(M) \longrightarrow M \xrightarrow{\pi} M/\Gamma_\mathfrak{B}(M) \longrightarrow 0$ induces an exact sequence

$$0 \to D_\mathfrak{B}(\Gamma_\mathfrak{B}(M)) \to D_\mathfrak{B}(M) \xrightarrow{D_\mathfrak{B}(\pi)} D_\mathfrak{B}(M/\Gamma_\mathfrak{B}(M)) \to \mathcal{R}^1 D_\mathfrak{B}(\Gamma_\mathfrak{B}(M)).$$

Now $D_\mathfrak{B}(\Gamma_\mathfrak{B}(M)) = 0$ by (i), while $\mathcal{R}^1 D_\mathfrak{B}(\Gamma_\mathfrak{B}(M)) \cong H^2_\mathfrak{B}(\Gamma_\mathfrak{B}(M))$ by 2.2.6(ii). Since $H^2_\mathfrak{B}(\Gamma_\mathfrak{B}(M)) = 0$ by 2.1.13(ii), it follows that $D_\mathfrak{B}(\pi)$ is an isomorphism.

(iii) It is left as an interesting exercise on direct limits for the reader to show that $D_\mathfrak{B}(\eta_M) = \eta_{D_\mathfrak{B}(M)}$. We show that $D_\mathfrak{B}(\eta_M)$ is an isomorphism.

Since $\eta_M = \theta_M \circ \pi$ (where θ_M is as defined in 2.2.9) and we have already shown in (ii) that $D_\mathfrak{B}(\pi)$ is an isomorphism, it is enough for us to show that $D_\mathfrak{B}(\theta_M)$ is an isomorphism. Since $D_\mathfrak{B}$ is left exact (by 2.2.3), the exact sequence

$$0 \longrightarrow M/\Gamma_\mathfrak{B}(M) \xrightarrow{\theta_M} D_\mathfrak{B}(M) \xrightarrow{\zeta^0_M} H^1_\mathfrak{B}(M) \longrightarrow 0$$

of 2.2.9 yields a further exact sequence

$$0 \longrightarrow D_\mathfrak{B}(M/\Gamma_\mathfrak{B}(M)) \xrightarrow{D_\mathfrak{B}(\theta_M)} D_\mathfrak{B}(D_\mathfrak{B}(M)) \xrightarrow{D_\mathfrak{B}(\zeta^0_M)} D_\mathfrak{B}(H^1_\mathfrak{B}(M)).$$

But $H^1_\mathfrak{B}(M)$ is $\mathfrak{B}$-torsion, and so $D_\mathfrak{B}(H^1_\mathfrak{B}(M)) = 0$ by (i). Hence $D_\mathfrak{B}(\theta_M)$ is an isomorphism, as required.

(iv) This is now immediate from (iii) and 2.2.7.

(v) We again use the fact that $\eta_M = \theta_M \circ \pi$. We already know, from 2.1.13(iii), that $H^i_\mathfrak{B}(\pi)$ is an isomorphism for all $i \in \mathbb{N}$. It is therefore enough, in order to complete the proof, to show that $H^i_\mathfrak{B}(\theta_M)$ is an isomorphism for all $i > 1$.

However, for each $i > 1$, the exact sequence

$$0 \longrightarrow M/\Gamma_\mathfrak{B}(M) \xrightarrow{\theta_M} D_\mathfrak{B}(M) \xrightarrow{\zeta^0_M} H^1_\mathfrak{B}(M) \longrightarrow 0$$

of 2.2.9 induces an exact sequence

$$\cdots \longrightarrow H^{i-1}_\mathfrak{B}(H^1_\mathfrak{B}(M))$$

$$\longrightarrow H^i_\mathfrak{B}(M/\Gamma_\mathfrak{B}(M)) \xrightarrow{H^i_\mathfrak{B}(\theta_M)} H^i_\mathfrak{B}(D_\mathfrak{B}(M)) \xrightarrow{H^i_\mathfrak{B}(\zeta^0_M)} H^i_\mathfrak{B}(H^1_\mathfrak{B}(M))$$

$$\longrightarrow \cdots,$$

and since $H^1_{\mathfrak{B}}(M)$ is a $\mathfrak{B}$-torsion R-module, it follows from 2.1.13(ii) that $H^{i-1}_{\mathfrak{B}}(H^1_{\mathfrak{B}}(M)) = H^i_{\mathfrak{B}}(H^1_{\mathfrak{B}}(M)) = 0$; hence $H^i_{\mathfrak{B}}(\theta_M)$ is an isomorphism, as required. □

2.2.11 ♯Exercise. Complete the proof of 2.2.10(iii). In other words, show that, in the notation of 2.2.10, $D_{\mathfrak{B}}(\eta_M) = \eta_{D_{\mathfrak{B}}(M)}$.

2.2.12 ♯Exercise. Recall that, by Exercise 2.2.5, the $\mathfrak{B}$-transform $D_{\mathfrak{B}}(R)$ of R has the structure of a commutative ring. Show that $\eta_R : R \longrightarrow D_{\mathfrak{B}}(R)$ is a ring homomorphism.

In order to exploit our results on the generalized ideal transform $D_{\mathfrak{B}}(M)$, we are going to obtain a description of $D_{\mathfrak{B}}(M)$ in terms of objects which are perhaps more familiar. This work will, of course, apply to the ideal transform $D_{\mathfrak{a}}(M)$. Towards the end of the section, we shall obtain a particularly simple description of $D_{\mathfrak{a}}(M)$ in the case in which $\mathfrak{a}$ is principal. One approach to these results uses the fact that, for an R-module M, the homomorphism $\eta_M : M \longrightarrow D_{\mathfrak{B}}(M)$ can be viewed as the solution to a universal problem. Our next proposition, which uses ideas similar to ones used by K. Suominen in [85, §1], provides the key to this approach. The influence of the ideas of P. Gabriel [21] should also be acknowledged.

2.2.13 Proposition. (See R. Y. Sharp and M. Tousi [82, Lemma 1.3].) *Let $e : M \longrightarrow M'$ be a homomorphism of R-modules such that $\operatorname{Ker} e$ and $\operatorname{Coker} e$ are both $\mathfrak{B}$-torsion. Let $\psi : M \longrightarrow K$ be a further homomorphism of R-modules.*

(i) *The map $D_{\mathfrak{B}}(e) : D_{\mathfrak{B}}(M) \longrightarrow D_{\mathfrak{B}}(M')$ is an isomorphism.*

(ii) *There is a unique R-homomorphism $\psi' : M' \longrightarrow D_{\mathfrak{B}}(K)$ such that the diagram*

$$
\begin{array}{ccc}
M & \xrightarrow{\;\;e\;\;} & M' \\
{\scriptstyle\psi}\big\downarrow & & \big\downarrow{\scriptstyle\psi'} \\
K & \xrightarrow{\;\eta_K\;} & D_{\mathfrak{B}}(K)
\end{array}
$$

commutes. In fact, $\psi' = D_{\mathfrak{B}}(\psi) \circ D_{\mathfrak{B}}(e)^{-1} \circ \eta_{M'}$.

(iii) *If ψ and $\eta_{M'} : M' \longrightarrow D_{\mathfrak{B}}(M')$ are both isomorphisms, then the homomorphism ψ' of part (ii) is also an isomorphism.*

Proof. (i) We shall use the exact sequences

$$0 \longrightarrow \operatorname{Ker} e \xrightarrow{\;\tau\;} M \xrightarrow{\;\lambda\;} \operatorname{Im} e \longrightarrow 0$$

and

$$0 \longrightarrow \operatorname{Im} e \xrightarrow{\rho} M' \xrightarrow{\sigma} \operatorname{Coker} e \longrightarrow 0,$$

in which the maps are the obvious homomorphisms. Now $e = \rho \circ \lambda$; it is therefore enough for us to show that $D_{\mathfrak{B}}(\rho)$ and $D_{\mathfrak{B}}(\lambda)$ are both isomorphisms. The first of the above exact sequences induces an exact sequence

$$0 \longrightarrow D_{\mathfrak{B}}(\operatorname{Ker} e) \xrightarrow{D_{\mathfrak{B}}(\tau)} D_{\mathfrak{B}}(M) \xrightarrow{D_{\mathfrak{B}}(\lambda)} D_{\mathfrak{B}}(\operatorname{Im} e) \longrightarrow \mathcal{R}^1 D_{\mathfrak{B}}(\operatorname{Ker} e).$$

However, $\mathcal{R}^1 D_{\mathfrak{B}}(\operatorname{Ker} e) \cong H^2_{\mathfrak{B}}(\operatorname{Ker} e)$ by 2.2.6(ii). By hypothesis, $\operatorname{Ker} e$ is $\mathfrak{B}$-torsion. Hence $D_{\mathfrak{B}}(\operatorname{Ker} e) = H^2_{\mathfrak{B}}(\operatorname{Ker} e) = 0$, by 2.2.10(i) and 2.1.13(ii). Therefore $D_{\mathfrak{B}}(\lambda)$ is an isomorphism.

Next, from the exact sequence $0 \longrightarrow \operatorname{Im} e \xrightarrow{\rho} M' \xrightarrow{\sigma} \operatorname{Coker} e \longrightarrow 0$ we obtain an induced exact sequence

$$0 \longrightarrow D_{\mathfrak{B}}(\operatorname{Im} e) \xrightarrow{D_{\mathfrak{B}}(\rho)} D_{\mathfrak{B}}(M') \xrightarrow{D_{\mathfrak{B}}(\sigma)} D_{\mathfrak{B}}(\operatorname{Coker} e).$$

However, by hypothesis, $\operatorname{Coker} e$ is $\mathfrak{B}$-torsion, and so $D_{\mathfrak{B}}(\operatorname{Coker} e) = 0$, by 2.2.10(i). Hence $D_{\mathfrak{B}}(\rho)$ is an isomorphism.

(ii) For this part of the proof, it will be convenient for us to write $K' := D_{\mathfrak{B}}(K)$ and $h := \eta_K : K \longrightarrow D_{\mathfrak{B}}(K) = K'$. Application of the natural transformation $\eta : \operatorname{Id} \longrightarrow D_{\mathfrak{B}}$ to the modules and homomorphisms in the diagram

$$\begin{array}{ccc} M & \xrightarrow{e} & M' \\ \psi \downarrow & & \\ K & \xrightarrow{h} & K' \end{array}$$

yields a commutative diagram

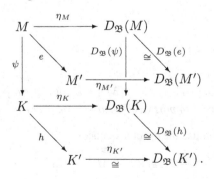

It should be noted that $D_{\mathfrak{B}}(e)$ is an isomorphism, by part (i), that $D_{\mathfrak{B}}(K) = K'$, and that, by 2.2.10(iii),

$$D_{\mathfrak{B}}(h) = D_{\mathfrak{B}}(\eta_K) = \eta_{D_{\mathfrak{B}}(K)} = \eta_{K'}$$

is also an isomorphism.

Thus, if there were an R-homomorphism $\psi' : M' \longrightarrow K'$ such that $\psi' \circ e = h \circ \psi$, then it would satisfy $D_{\mathfrak{B}}(\psi') \circ D_{\mathfrak{B}}(e) = D_{\mathfrak{B}}(h) \circ D_{\mathfrak{B}}(\psi)$ and we would have to have (since $\eta : \mathrm{Id} \longrightarrow D_{\mathfrak{B}}$ is a natural transformation)

$$\psi' = D_{\mathfrak{B}}(h)^{-1} \circ \eta_{K'} \circ \psi' = D_{\mathfrak{B}}(h)^{-1} \circ D_{\mathfrak{B}}(\psi') \circ \eta_{M'}$$
$$= D_{\mathfrak{B}}(\psi) \circ D_{\mathfrak{B}}(e)^{-1} \circ \eta_{M'}.$$

On the other hand, one can easily verify by means of an elementary diagram chase that $D_{\mathfrak{B}}(\psi) \circ D_{\mathfrak{B}}(e)^{-1} \circ \eta_{M'} \circ e = h \circ \psi$.

(iii) This is immediate from part (ii), since, if ψ is an isomorphism, then so too is $D_{\mathfrak{B}}(\psi)$. $\square$

2.2.14 Remark. Let $h : M \longrightarrow N$ be a homomorphism of R-modules. Recall from 2.2.6(i) that $\eta_{\mathfrak{B}} : \mathrm{Id} \longrightarrow D_{\mathfrak{B}}$ is a natural transformation of functors and that $\mathrm{Ker}\,\eta_M$ and $\mathrm{Coker}\,\eta_M$ are both $\mathfrak{B}$-torsion. It therefore follows from 2.2.13 that $D_{\mathfrak{B}}(h) : D_{\mathfrak{B}}(M) \longrightarrow D_{\mathfrak{B}}(N)$ must be the unique R-homomorphism from $D_{\mathfrak{B}}(M)$ to $D_{\mathfrak{B}}(N)$ which makes the diagram

$$
\begin{array}{ccc}
M & \xrightarrow{\ \eta_M\ } & D_{\mathfrak{B}}(M) \\
\downarrow{\scriptstyle h} & & \downarrow \\
N & \xrightarrow{\ \eta_N\ } & D_{\mathfrak{B}}(N)
\end{array}
$$

commute.

2.2.15 Corollary. *Let* $e : M \longrightarrow M'$ *be a homomorphism of R-modules such that* $\mathrm{Ker}\,e$ *and* $\mathrm{Coker}\,e$ *are both* $\mathfrak{B}$-*torsion.*

(i) *The map* $D_{\mathfrak{B}}(e) : D_{\mathfrak{B}}(M) \longrightarrow D_{\mathfrak{B}}(M')$ *is an isomorphism.*

(ii) *There is a unique R-homomorphism* $\psi' : M' \to D_{\mathfrak{B}}(M)$ *such that the diagram*

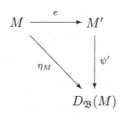

commutes. In fact, $\psi' = D_{\mathfrak{B}}(e)^{-1} \circ \eta_{M'}$.

(iii) *The map ψ' of (ii) is an isomorphism if and only if $\eta_{M'}$ is an isomorphism, and, by 2.2.7, this is the case if and only if $\Gamma_{\mathfrak{B}}(M') = H^1_{\mathfrak{B}}(M') = 0$.*

Proof. Use 2.2.13 with $\mathrm{Id}_M : M \longrightarrow M$ in the rôle of $\psi : M \longrightarrow K$. $\quad\square$

We are now going to examine the special case of 2.2.15 in which $M = R$. Recall from 2.2.5 that $D_{\mathfrak{B}}(R)$ is a commutative ring, and from 2.2.12 that $\eta_R : R \longrightarrow D_{\mathfrak{B}}(R)$ is a ring homomorphism. We shall meet several examples of the situation of 2.2.15 in which $e : R \longrightarrow R'$ is a homomorphism of commutative rings such that, when R' is regarded as an R-module by means of e, both $\mathrm{Ker}\, e$ and $\mathrm{Coker}\, e$ are $\mathfrak{B}$-torsion: 2.2.17 below shows that, in these circumstances, the R-homomorphism $\psi' : R' \longrightarrow D_{\mathfrak{B}}(R)$ given by 2.2.15 is actually a ring homomorphism too.

2.2.16 Proposition. *Let R' be a ring (with identity, but not necessarily commutative), and let $e : R \longrightarrow R'$ be a ring homomorphism such that $\mathrm{Im}\, e$ is contained in the centre of R' and, when R' is regarded as a left R-module by means of e, both $\mathrm{Ker}\, e$ and $\mathrm{Coker}\, e$ are $\mathfrak{B}$-torsion. Assume also that $\Gamma_{\mathfrak{B}}(R') = 0$. Then the ring R' is commutative.*

Proof. By 2.2.15, there is a unique R-homomorphism $\psi' : R' \longrightarrow D_{\mathfrak{B}}(R)$ such that the diagram

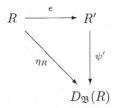

commutes. Let $r'_1, r'_2 \in R'$. Since $\mathrm{Coker}\, e$ is $\mathfrak{B}$-torsion, there exist $\alpha, \delta \in \Lambda$ such that $\mathfrak{b}_\alpha$ (respectively $\mathfrak{b}_\delta$) annihilates the natural image in $\mathrm{Coker}\, e$ of r'_1 (respectively r'_2). Let $b_1 \in \mathfrak{b}_\alpha$ and $b_2 \in \mathfrak{b}_\delta$; then there exist $r_1, r_2 \in R$ such that $b_i r'_i = e(r_i)$, that is $e(b_i)r'_i = e(r_i)$, for $i = 1, 2$. Therefore, since $\mathrm{Im}\, e$ is contained in the centre of R', we have

$$e(b_1)e(b_2)r'_1 r'_2 = e(b_1)r'_1 e(b_2)r'_2 = e(r_1)e(r_2)$$
$$= e(r_2)e(r_1) = e(b_2)r'_2 e(b_1)r'_1 = e(b_1)e(b_2)r'_2 r'_1.$$

Therefore $b_1 b_2 (r'_1 r'_2 - r'_2 r'_1) = 0$. Hence the element $r'_1 r'_2 - r'_2 r'_1$ is annihilated by $\mathfrak{b}_\alpha \mathfrak{b}_\delta$. But $\mathfrak{B}$ is a system of ideals, so that there exists $\mu \in \Lambda$ such that

$\mathfrak{b}_\mu \subseteq \mathfrak{b}_\alpha \mathfrak{b}_\delta$. Since $\Gamma_\mathfrak{B}(R') = 0$, we deduce that $r_1' r_2' - r_2' r_1' = 0$ and R' is commutative. $\qquad\square$

2.2.17 Proposition. *Let R' be a commutative ring (with identity), and let $e : R \longrightarrow R'$ be a ring homomorphism for which the R-modules $\operatorname{Ker} e$ and $\operatorname{Coker} e$ are $\mathfrak{B}$-torsion.*

Then the unique R-homomorphism $\psi' : R' \longrightarrow D_\mathfrak{B}(R)$ such that the diagram

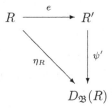

commutes (the existence of which follows from 2.2.15) is a ring homomorphism, and therefore an R-algebra homomorphism.

Proof. Let $r_1', r_2' \in R'$. Since $\operatorname{Coker} e$ is $\mathfrak{B}$-torsion, there exist $\alpha, \delta \in \Lambda$ such that $\mathfrak{b}_\alpha$ (respectively $\mathfrak{b}_\delta$) annihilates the natural image in $\operatorname{Coker} e$ of r_1' (respectively r_2'). Let $b_1 \in \mathfrak{b}_\alpha$ and $b_2 \in \mathfrak{b}_\delta$; then there exist $r_1, r_2 \in R$ such that $b_i r_i' = e(r_i)$, that is $e(b_i) r_i' = e(r_i)$, for $i = 1, 2$.

Note also that, in the commutative ring $D_\mathfrak{B}(R)$, we have

$$b_1 b_2 \psi'(r_1')\psi'(r_2') = \psi'(b_1 r_1')\psi'(b_2 r_2') = \psi'(e(r_1))\psi'(e(r_2))$$
$$= \eta_R(r_1)\eta_R(r_2) \stackrel{.}{=} \eta_R(r_1 r_2) = \psi'(e(r_1 r_2))$$
$$= \psi'(e(r_1)e(r_2)) = \psi'(b_1 r_1' b_2 r_2') = b_1 b_2 \psi'(r_1' r_2'),$$

so that $b_1 b_2 \left(\psi'(r_1')\psi'(r_2') - \psi'(r_1' r_2') \right) = 0$. Hence the element

$$\psi'(r_1')\psi'(r_2') - \psi'(r_1' r_2') \in D_\mathfrak{B}(R)$$

is annihilated by $\mathfrak{b}_\alpha \mathfrak{b}_\delta$. But there exists $\mu \in \Lambda$ such that $\mathfrak{b}_\mu \subseteq \mathfrak{b}_\alpha \mathfrak{b}_\delta$. Since $D_\mathfrak{B}(R)$ is $\mathfrak{B}$-torsion-free by 2.2.10(iv), we have $\psi'(r_1' r_2') = \psi'(r_1')\psi'(r_2')$. Also $\psi'(1_{R'}) = \psi'(e(1_R)) = \eta_R(1_R) = 1_{D_\mathfrak{B}(R)}$. Therefore ψ' is a ring homomorphism. $\qquad\square$

2.2.18 Corollary. *Let M be an R-module, and let S be a multiplicatively closed subset of R which consists entirely of non-zerodivisors on M, and which is such that $S \cap \mathfrak{b}_\alpha \neq \emptyset$ for all $\alpha \in \Lambda$. Then there is a unique R-isomorphism*

$$\psi_M' : \bigcup_{\alpha \in \Lambda} (M :_{S^{-1}M} \mathfrak{b}_\alpha) \longrightarrow D_\mathfrak{B}(M)$$

for which the diagram

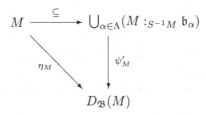

commutes. In the special case in which $M = R$ *(and* S *consists of non-zerodivisors on* R*), the map* ψ'_R *is actually a ring isomorphism.*

Note. Of course, since S consists entirely of non-zerodivisors on M, the canonical R-homomorphism $M \to S^{-1}M$ is injective; we are using this to identify M as an R-submodule of $S^{-1}M$.

Proof. Set $M' := \bigcup_{\alpha \in \Lambda}(M :_{S^{-1}M} \mathfrak{b}_\alpha)$, and let $e : M \longrightarrow M'$ denote the inclusion homomorphism. Since $\operatorname{Coker} e = \Gamma_\mathfrak{B}(S^{-1}M/M)$ is $\mathfrak{B}$-torsion, it is immediate from 2.2.15 that there is a unique R-homomorphism $\psi'_M : M' \longrightarrow D_\mathfrak{B}(M)$ which makes the above diagram commute; it also follows that, in order to show that ψ'_M is an isomorphism, it is sufficient for us to show that $\Gamma_\mathfrak{B}(M') = H^1_\mathfrak{B}(M') = 0$. This we do.

We show first that $H^i_\mathfrak{B}(S^{-1}M) = 0$ for each $i \in \mathbb{N}_0$. Let $y \in H^i_\mathfrak{B}(S^{-1}M)$. Then there exists $\alpha \in \Lambda$ such that $\mathfrak{b}_\alpha y = 0$. By hypothesis, there exists $s \in \mathfrak{b}_\alpha \cap S$, so that $sy = 0$. Since the functor $H^i_\mathfrak{B}$ is R-linear, multiplication by s on $H^i_\mathfrak{B}(S^{-1}M)$ must provide an automorphism; therefore $y = 0$. Therefore $H^i_\mathfrak{B}(S^{-1}M) = 0$, as claimed. Hence $\Gamma_\mathfrak{B}(M') = 0$. It now follows from the exact sequence

$$0 \longrightarrow M' \longrightarrow S^{-1}M \longrightarrow S^{-1}M/M' \longrightarrow 0$$

that $H^1_\mathfrak{B}(M') \cong \Gamma_\mathfrak{B}(S^{-1}M/M')$. However,

$$S^{-1}M/M' \cong (S^{-1}M/M)/(M'/M) = (S^{-1}M/M)/\Gamma_\mathfrak{B}(S^{-1}M/M),$$

and this is $\mathfrak{B}$-torsion-free by 2.1.13(i). Hence $H^1_\mathfrak{B}(M') = 0$. The final claim follows from 2.2.17. $\square$

Of course, all our work so far in this section applies to the particular system of ideals $(\mathfrak{a}^n)_{n\in\mathbb{N}}$, and, indeed, that is a very important example of such a system. As we have now presented, in this section, enough of the theory of generalized ideal transforms for our later needs, we are going to concentrate, for the remainder of this section, on the ordinary $\mathfrak{a}$-transform functor $D_\mathfrak{a}$.

In the case when $\mathfrak{a}$ is principal, there is a result similar to 2.2.18, but under weaker hypotheses, which has important consequences for our work in Chapter 3. We present this result next. Recall that, for an R-module M and $a \in R$, the notation M_a denotes the module of fractions of M with respect to the multiplicatively closed subset $\{a^i : i \in \mathbb{N}_0\}$.

2.2.19 Theorem. *Let $a \in R$. There is a natural equivalence of functors*

$$\omega' : D_{Ra} = \varinjlim_{n \in \mathbb{N}} \operatorname{Hom}_R(Ra^n, \bullet) \longrightarrow (\bullet)_a$$

(from $\mathcal{C}(R)$ to $\mathcal{C}(R)$) such that, for an R-module M, and an $f \in D_{Ra}(M)$ represented by $f_t \in \operatorname{Hom}_R(Ra^t, M)$ (for some $t \in \mathbb{N}$), we have $\omega'_M(f) = f_t(a^t)/a^t$.

Proof. Let M be an R-module. It is immediate from 2.2.15 that there is a unique R-isomorphism $\nu_M : M_a \longrightarrow D_{Ra}(M)$ such that the diagram

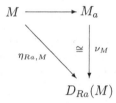

(in which the horizontal homomorphism is the natural one) commutes: note that $H_{Ra}^i(M_a) = 0$ for all $i \in \mathbb{N}_0$ because multiplication by a provides an automorphism on all these local cohomology modules. Define $\omega'_M := \nu_M^{-1}$. It is straightforward to use the commutativity of the above diagram to show that ω'_M satisfies the formula given in the statement of the theorem. Furthermore, it is easy to use that formula to show that, as M varies through the category $\mathcal{C}(R)$, the ω'_M constitute a natural equivalence of functors. $\square$

2.2.20 Remark. With the notation of 2.2.19, consider, for an R-module M, the fundamental exact sequence of 2.2.6(i)(c) in the particular case in which $\mathfrak{B} = (\mathfrak{a}^n)_{n \in \mathbb{N}}$ and $\mathfrak{a} = Ra$: we have

$$0 \longrightarrow \Gamma_{Ra}(M) \xrightarrow{\xi_M} M \xrightarrow{\eta_M} D_{Ra}(M) \xrightarrow{\zeta_M^0} H_{Ra}^1(M) \longrightarrow 0.$$

Note that $\omega'_M \circ \eta_M : M \longrightarrow M_a$ is just the canonical R-homomorphism τ_M. Set $\sigma_M := \zeta_M^0 \circ (\omega'_M)^{-1}$; then there is an exact sequence of R-modules and R-homomorphisms

$$0 \longrightarrow \Gamma_{Ra}(M) \xrightarrow{\xi_M} M \xrightarrow{\tau_M} M_a \xrightarrow{\sigma_M} H_{Ra}^1(M) \longrightarrow 0$$

such that, as M varies through $\mathcal{C}(R)$, the τ_M and σ_M constitute natural transformations of functors

$$\tau : \mathrm{Id} \longrightarrow (\,\bullet\,)_a, \quad \sigma : (\,\bullet\,)_a \longrightarrow H^1_{Ra}.$$

2.2.21 Corollary. *Let M be an R-module and let $a \in R$.*

(i) *The kernel of the natural homomorphism $\tau_M : M \to M_a$ is precisely $\Gamma_{Ra}(M)$, and so, in view of 2.2.20, $M/\Gamma_{Ra}(M)$ can be identified as a submodule of M_a. With this identification,*

$$H^1_{Ra}(M) \cong M_a/(M/\Gamma_{Ra}(M)).$$

(ii) *For all $i \in \mathbb{N}$ with $i > 1$, we have $H^i_{Ra}(M) = 0$.*

Proof. (i) This is immediate from 2.2.20.

(ii) Let $i \in \mathbb{N}$ with $i > 1$. Now $H^i_{Ra}(M) \cong \mathcal{R}^{i-1}D_{Ra}(M)$ by 2.2.6(ii). However, by 2.2.19, the functor D_{Ra} is naturally equivalent to $(\,\bullet\,)_a$; as the latter functor is exact and $i - 1 > 0$, it follows that $\mathcal{R}^{i-1}D_{Ra}(N) = 0$ for all R-modules N. Hence $H^i_{Ra}(M) = 0$. □

The above corollary will prove useful in Chapter 3, as it provides a basis for an argument which uses induction on the number of generators of a, and which relies on the Mayer–Vietoris Sequence for local cohomology for the inductive argument. This Mayer–Vietoris Sequence forms a major part of the subject matter of Chapter 3.

2.2.22 Exercise. Use 2.2.18 to obtain an alternative proof of Corollary 2.2.8.

More precisely, let M be an R-module, not necessarily finitely generated, and assume that a contains an M-sequence x, y of length 2. Use 2.2.18 (with the choice $S := \{x^i : i \in \mathbb{N}_0\}$) and 1.3.9(ii) to show that the map $\eta_M : M \longrightarrow D_a(M)$ of 2.2.6(i) is an isomorphism.

2.2.23 ♯Exercise. Let b be a second ideal of R.

(i) Suppose that $a \subseteq \sqrt{b}$. Show that there is a unique natural transformation of functors (from $\mathcal{C}(R)$ to itself) $\alpha_{b,a} : D_b \longrightarrow D_a$ such that, for each R-module M, the diagram

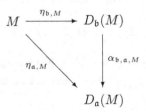

commutes.

(ii) Deduce that, if $\sqrt{\mathfrak{a}} = \sqrt{\mathfrak{b}}$, then $D_{\mathfrak{a}}$ and $D_{\mathfrak{b}}$ are naturally equivalent.

It will be convenient, in our discussion of geometric examples in the next section, for us to have available a result which shows that, in a certain sense, the ideal transform is 'independent of the base ring'. To be more precise, consider a second commutative Noetherian ring R' and a ring homomorphism $f : R \longrightarrow R'$; let M' be an R'-module. At times when we wish to be absolutely precise, we shall use $M' \lceil_R$ to indicate that we are regarding M' as an R-module by means of f. Note that $\lceil_R$ can be regarded as a functor from $\mathcal{C}(R')$ to $\mathcal{C}(R)$. We can form the ideal transform $D_{\mathfrak{a}R'}(M')$ of M' with respect to the extension $\mathfrak{a}R'$ of $\mathfrak{a}$ to R' via f, and then regard this as an R-module by means of f: this is, then, the R-module $D_{\mathfrak{a}R'}(M') \lceil_R$. Alternatively, we can regard M' as the R-module $M' \lceil_R$, and form $D_{\mathfrak{a}}(M' \lceil_R)$. Our next result will show, among other things, that there is an R-isomorphism

$$D_{\mathfrak{a}R'}(M') = D_{\mathfrak{a}R'}(M') \lceil_R \overset{\cong}{\longrightarrow} D_{\mathfrak{a}}(M' \lceil_R) =: D_{\mathfrak{a}}(M'),$$

so that, speaking loosely, it does not matter whether we calculate these ideal transforms over R or R'. The advantage of this in practice is that there is sometimes an obvious choice of an 'R'' over which the calculations are easy.

2.2.24 Theorem. *Let R' be a second commutative Noetherian ring and let $f : R \longrightarrow R'$ be a ring homomorphism. We use the notation introduced above. There is a natural equivalence of functors*

$$\varepsilon : D_{\mathfrak{a}R'}(\,\bullet\,)\lceil_R \longrightarrow D_{\mathfrak{a}}(\,\bullet\,\lceil_R)$$

(from $\mathcal{C}(R')$ to $\mathcal{C}(R)$) which is such that, for each R'-module M', the diagram

$$
\begin{array}{ccc}
M' & \overset{\eta_{\mathfrak{a}R',M'}}{\longrightarrow} & D_{\mathfrak{a}R'}(M') \\
\| & & \downarrow{\scriptstyle \varepsilon_{M'}} \\
M' & \overset{\eta_{\mathfrak{a},M'}}{\longrightarrow} & D_{\mathfrak{a}}(M')
\end{array}
$$

commutes.

Proof. Let M' be an R'-module. By 2.2.6(i)(c), the R'-homomorphism

$$\eta_{\mathfrak{a}R',M'} : M' \longrightarrow D_{\mathfrak{a}R'}(M')$$

has kernel and cokernel which are, respectively, isomorphic to $\Gamma_{\mathfrak{a}R'}(M')$ and $H^1_{\mathfrak{a}R'}(M')$. Hence $\eta_{\mathfrak{a}R',M'} \lceil_R$ has kernel and cokernel which are $\mathfrak{a}$-torsion. It therefore follows from 2.2.15 that there is a unique R-homomorphism $\varepsilon_{M'}$: $D_{\mathfrak{a}R'}(M') \longrightarrow D_{\mathfrak{a}}(M')$ such that the diagram

$$
\begin{array}{ccc}
M' & \xrightarrow{\;\eta_{\mathfrak{a}R',M'}\;} & D_{\mathfrak{a}R'}(M') \\
\| & & \bigg\downarrow{\scriptstyle\varepsilon_{M'}} \\
M' & \xrightarrow{\;\eta_{\mathfrak{a},M'}\;} & D_{\mathfrak{a}}(M')
\end{array}
$$

commutes. In fact, it is easy to use the uniqueness aspect of Proposition 2.2.13 to show that, as M' varies through the category $\mathcal{C}(R')$, the $\varepsilon_{M'}$ constitute a natural transformation of functors, and so it remains only to show that each $\varepsilon_{M'}$ is an isomorphism.

The fact that M' is an (R, R')-bimodule means that $D_{\mathfrak{a}}(M')$ inherits a natural structure as an R'-module (such that $D_{\mathfrak{a}}(r'\operatorname{Id}_{M'})$, for $r' \in R'$, provides multiplication by r'). Furthermore, since ε is a natural transformation of functors, we have

$$
D_{\mathfrak{a}}(r'\operatorname{Id}_{M'}) \circ \varepsilon_{M'} = \varepsilon_{M'} \circ D_{\mathfrak{a}R'}(r'\operatorname{Id}_{M'}) = \varepsilon_{M'} \circ (r'\operatorname{Id}_{D_{\mathfrak{a}R'}(M')})
$$

for all $r' \in R'$, so that $\varepsilon_{M'}$ is an R'-homomorphism. Likewise, $\eta_{\mathfrak{a},M'}$ becomes an R'-homomorphism.

Another use of 2.2.6(i)(c) shows that $\eta_{\mathfrak{a},M'} : M' \longrightarrow D_{\mathfrak{a}}(M')$ has kernel and cokernel which are $\mathfrak{a}$-torsion, so that, when we consider $\eta_{\mathfrak{a},M'}$ as an R'-homomorphism, its kernel and cokernel are $\mathfrak{a}R'$-torsion. It therefore follows from 2.2.15 that there is a unique R'-homomorphism $\lambda_{M'} : D_{\mathfrak{a}}(M') \longrightarrow D_{\mathfrak{a}R'}(M')$ such that the diagram

$$
\begin{array}{ccc}
M' & \xrightarrow{\;\eta_{\mathfrak{a},M'}\;} & D_{\mathfrak{a}}(M') \\
\| & & \bigg\downarrow{\scriptstyle\lambda_{M'}} \\
M' & \xrightarrow{\;\eta_{\mathfrak{a}R',M'}\;} & D_{\mathfrak{a}R'}(M')
\end{array}
$$

commutes. The uniqueness aspect of 2.2.15, together with the facts that $\varepsilon_{M'}$ and $\lambda_{M'}$ are both R- and R'-homomorphisms, now yields that

$$
\lambda_{M'} \circ \varepsilon_{M'} = \operatorname{Id}_{D_{\mathfrak{a}R'}(M')} \quad \text{and} \quad \varepsilon_{M'} \circ \lambda_{M'} = \operatorname{Id}_{D_{\mathfrak{a}}(M')},
$$

so that $\varepsilon_{M'}$ is an isomorphism. $\qquad\square$

2.2.25 ♯Exercise. Let the situation be as in 2.2.24, and consider the natural equivalence ε of that theorem. Show that $\varepsilon_{R'}^{-1} \circ D_\mathfrak{a}(f) : D_\mathfrak{a}(R) \longrightarrow D_{\mathfrak{a}R'}(R')$ is the unique ring homomorphism which makes the diagram

$$
\begin{array}{ccc}
R & \xrightarrow{\;\eta_{\mathfrak{a},R}\;} & D_\mathfrak{a}(R) \\
\downarrow{\scriptstyle f} & & \downarrow \\
R' & \xrightarrow{\;\eta_{\mathfrak{a}R',R'}\;} & D_{\mathfrak{a}R'}(R')
\end{array}
$$

commute. (You might find it helpful to adapt the argument used in the proof of 2.2.17.)

2.2.26 Exercise. Let the situation be as in 2.2.24. Show that there is a natural equivalence of functors

$$
H^1_{\mathfrak{a}R'}(\,\bullet\,)\lceil_R \longrightarrow H^1_\mathfrak{a}(\,\bullet\,\lceil_R)
$$

(from $\mathcal{C}(R')$ to $\mathcal{C}(R)$).

The result of Exercise 2.2.26 is a particular case of a more general (and very important) result concerning 'independence of the base ring' which will be established in Chapter 4.

2.3 Geometrical significance

We are now going to show that the ideal transform $D_\mathfrak{a}(R)$ has an important geometrical significance in the case when R is the ring of regular functions on an (irreducible) affine algebraic variety over an algebraically closed field. Several examples in this book will be concerned with affine algebraic geometry over the field of complex numbers $\mathbb{C}$, and it is convenient for us to introduce some notation which will be consistently used in connection with such examples.

2.3.1 Notation. Let K be an algebraically closed field. For $n \in \mathbb{N}$, we shall use $\mathbb{A}^n(K)$ to denote affine n-space over K, that is, K^n endowed with the Zariski topology; we shall use $\mathbb{A}^n$ to denote complex affine n-space $\mathbb{A}^n(\mathbb{C})$. All unexplained mentions of topological notions, including 'open' and 'closed' subsets, in connection with affine spaces will refer to the Zariski topology.

By an *affine variety* over K we shall mean an irreducible closed subset of $\mathbb{A}^n(K)$ (with the induced topology), and by a *quasi-affine variety* over K we shall mean a non-empty open subset of an affine variety over K (again with the induced topology).

For a quasi-affine variety U over K, we shall use $\mathcal{O}(U)$ to denote the ring of regular functions on U and $K(U)$ to denote the function field of U. Thus, when U is affine, $\mathcal{O}(U)$ is just the coordinate ring of U; this is actually an integral domain (since U is irreducible), and its field of fractions is just $K(U)$.

We shall regard the polynomial ring $K[X_1, \ldots, X_n]$ as the coordinate ring $\mathcal{O}(\mathbb{A}^n(K))$ of $\mathbb{A}^n(K)$ in the obvious way (although we shall tend to use X, Y instead of X_1, X_2 in the case when $n = 2$).

For $f_1, \ldots, f_t \in K[X_1, \ldots, X_n]$, we shall use $V_{\mathbb{A}^n(K)}(f_1, \ldots, f_t)$ to denote the affine algebraic set

$$\{p \in \mathbb{A}^n(K) : f_1(p) = \cdots = f_t(p) = 0\}$$

corresponding to the ideal $(f_1, \ldots, f_t)$ of $K[X_1, \ldots, X_n]$. If V denotes this affine algebraic set, then, for $1 \le i \le n$, the restriction of the coordinate function X_i of $\mathbb{A}^n(K)$ to V will be denoted by $X_i \lceil_V$ or x_i.

For a quasi-affine variety W over K and a function $f \in \mathcal{O}(W)$, we shall use $U_W(f)$ to denote the open subset $\{p \in W : f(p) \ne 0\}$ of W.

We are now going to show that the ring of regular functions on a quasi-affine variety can be expressed in terms of an ideal transform. Before doing so, we remind the reader of the following point. For a non-empty open subset U of an affine variety V over the algebraically closed field K, the ring $\mathcal{O}(U)$ can be identified in a natural way as a subring of $K(V)$, and, when this identification is made, the restriction homomorphism

$$\lceil_U : \mathcal{O}(V) \longrightarrow \mathcal{O}(U)$$

is just the inclusion map.

2.3.2 Theorem. *Let V be an affine variety over the algebraically closed field K. Let $\mathfrak{b}$ be a non-zero ideal of $\mathcal{O}(V)$, let $V(\mathfrak{b})$ denote the closed subset of V determined by $\mathfrak{b}$, and let U be the open subset $V \setminus V(\mathfrak{b})$ of V.*

There is a unique $\mathcal{O}(V)$-isomorphism $\nu_{V,\mathfrak{b}} : \mathcal{O}(U) \xrightarrow{\cong} D_{\mathfrak{b}}(\mathcal{O}(V))$ for which the diagram

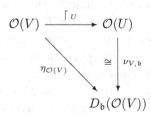

commutes. Furthermore, $\nu_{V,\mathfrak{b}}$ is a ring isomorphism.

Proof. In view of 2.2.18, it is enough for us to show that the submodules $\mathcal{O}(U)$ and $\bigcup_{n\in\mathbb{N}}(\mathcal{O}(V) :_{K(V)} \mathfrak{b}^n)$ of $K(V)$ are equal.

Since the ring $\mathcal{O}(V)$ is Noetherian, $\mathfrak{b}$ is finitely generated, by $h_1,\ldots,h_t$, say. We can, and do, assume that $h_1,\ldots,h_t$ are all non-zero. Note that $U = \bigcup_{i=1}^{t} U_V(h_i)$, and that $\mathcal{O}(U_V(h_i)) = \mathcal{O}(V)_{h_i}$ when these two rings are identified with subrings of $K(V)$ in the natural ways.

Let $f \in \mathcal{O}(U)$. Choose $i \in \mathbb{N}$ with $1 \leq i \leq t$. Since $f\lceil_{U_V(h_i)} \in \mathcal{O}(U_V(h_i)) = \mathcal{O}(V)_{h_i}$, it follows that there exists $n_i \in \mathbb{N}$ such that $h_i^{n_i} f \in \mathcal{O}(V)$. We deduce that there exists $n \in \mathbb{N}$ such that $\mathfrak{b}^n f \subseteq \mathcal{O}(V)$. Hence

$$\mathcal{O}(U) \subseteq \bigcup_{n\in\mathbb{N}}(\mathcal{O}(V) :_{K(V)} \mathfrak{b}^n).$$

Now let $f \in \bigcup_{n\in\mathbb{N}}(\mathcal{O}(V) :_{K(V)} \mathfrak{b}^n)$. Thus there exists $n \in \mathbb{N}$ such that $g_i := h_i^n f \in \mathcal{O}(V)$ for all $i = 1,\ldots,t$. Let $p \in U$, and let $\mathfrak{m}(p)$ be the maximal ideal of $\mathcal{O}(V)$ corresponding to p. Now $p \in U_V(h_i)$ for some i with $1 \leq i \leq t$, and since $h_i(p) \neq 0$, we have $h_i \in \mathcal{O}(V) \setminus \mathfrak{m}(p)$ and

$$f = \frac{g_i}{h_i^n} \in \mathcal{O}(V)_{\mathfrak{m}(p)} = \mathcal{O}_{V,p},$$

the local ring of V at p. It follows that $f \in \bigcap_{p\in U} \mathcal{O}_{V,p} = \mathcal{O}(U)$. Hence

$$\mathcal{O}(U) \supseteq \bigcup_{n\in\mathbb{N}}(\mathcal{O}(V) :_{K(V)} \mathfrak{b}^n),$$

and the proof is complete. $\qquad\Box$

2.3.3 Remark. It follows from 2.3.2 that, in the notation of that proposition, the map $\eta_{\mathcal{O}(V)} : \mathcal{O}(V) \longrightarrow D_{\mathfrak{b}}(\mathcal{O}(V))$ is an epimorphism if and only if the restriction map $\lceil_U : \mathcal{O}(V) \longrightarrow \mathcal{O}(U)$ is surjective, that is, if and only if every regular function on U can be extended to a regular function on V. However, by 2.2.6(i)(c), $\eta_{\mathcal{O}(V)}$ fails to be an epimorphism if and only if $H_{\mathfrak{b}}^1(\mathcal{O}(V)) \neq 0$. Thus we can, in a sense, regard non-zero elements of the local cohomology module $H_{\mathfrak{b}}^1(\mathcal{O}(V))$ as obstructions to the extension of regular functions on U to regular functions on V. We shall exploit this observation later in the chapter in our discussion of the geometric significance of ideal transforms and local cohomology modules in particular examples.

The next exercise establishes a certain 'naturality' property of the isomorphisms given by Theorem 2.3.2.

2.3.4 ♯Exercise. Let $\beta : W \longrightarrow V$ be a morphism of affine varieties over the algebraically closed field K. Let $\mathfrak{b}$ be a non-zero ideal of $\mathcal{O}(V)$, and assume that the extension $\mathfrak{b}\mathcal{O}(W)$ of $\mathfrak{b}$ under the induced K-algebra homomorphism $\beta^* : \mathcal{O}(V) \longrightarrow \mathcal{O}(W)$ is also non-zero.

We use the notation of Theorem 2.3.2. Since $\beta^{-1}(V(\mathfrak{b})) = V(\mathfrak{b}\mathcal{O}(W))$, the restriction of β provides a morphism $\beta\lceil : W \setminus V(\mathfrak{b}\mathcal{O}(W)) \longrightarrow V \setminus V(\mathfrak{b})$ of quasi-affine varieties.

Use Exercise 2.2.25 to show that, with the notation of that exercise, the diagram

$$
\begin{array}{ccc}
\mathcal{O}(V \setminus V(\mathfrak{b})) & \xrightarrow{\ \beta\lceil^* \ } & \mathcal{O}(W \setminus V(\mathfrak{b}\mathcal{O}(W))) \\
\Big\downarrow{\scriptstyle \nu_{V,\mathfrak{b}}} {\scriptstyle \cong} & & {\scriptstyle \cong} \Big\downarrow{\scriptstyle \nu_{W,\mathfrak{b}\mathcal{O}(W)}} \\
D_{\mathfrak{b}}(\mathcal{O}(V)) & \xrightarrow{\ \varepsilon_{\mathcal{O}(W)}^{-1} \circ D_{\mathfrak{b}}(\beta^*) \ } & D_{\mathfrak{b}\mathcal{O}(W)}(\mathcal{O}(W))
\end{array}
$$

of ring homomorphisms commutes.

2.3.5 ♯Exercise. Let $\iota : \mathbb{A}^2 \setminus \{(0,0)\} \to \mathbb{A}^2$ denote the inclusion morphism of varieties. Prove that the induced $\mathbb{C}$-algebra homomorphism

$$
\iota^* : \mathcal{O}(\mathbb{A}^2) \to \mathcal{O}(\mathbb{A}^2 \setminus \{(0,0)\})
$$

is an isomorphism, and deduce that the quasi-affine variety $\mathbb{A}^2 \setminus \{(0,0)\}$ is not affine.

2.3.6 ♯Exercise. Let K be a field, and let $\mathfrak{q}$ be a proper ideal of height 2 in the ring of polynomials $K[X,Y]$ in the indeterminates X, Y. Set $R := K + \mathfrak{q}$, a subring of $K[X,Y]$. Observe that $\mathfrak{q}$ is a maximal ideal of R, and that the simple R-module $R/\mathfrak{q}$ is isomorphic to K, where the R-module structure on K is such that $fa = 0$ for all $f \in \mathfrak{q}$ and all $a \in K$.

(i) Show that the vector space dimension $t := \dim_K(K[X,Y]/\mathfrak{q})$ is finite, and deduce that there is a monic polynomial in $K[X] \cap \mathfrak{q}$ of degree not exceeding t: let p_X be such a polynomial of smallest possible degree. Similarly, let p_Y be a monic polynomial in $K[Y] \cap \mathfrak{q}$ of smallest possible degree. Show that $K[X,Y]$ is a finitely generated R-module.

(ii) Let $c_1, \ldots, c_h$ generate $\mathfrak{q}$ (as an ideal of $K[X,Y]$), let $u := \deg p_X$ and $v := \deg p_Y$. Set

$$
S := \{c_i X^j Y^l : 1 \le i \le h,\ 0 \le j < u,\ 0 \le l < v\} \cup \{p_X, p_Y\} \subset R.
$$

Show that $R = K[S]$, so that R is a finitely generated K-algebra, and therefore Noetherian.

(iii) Show that there is a unique R-isomorphism $\psi' : K[X, Y] \longrightarrow D_q(R)$
such that the diagram

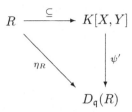

commutes, and that ψ' is a ring isomorphism. Deduce that there is an
exact sequence $0 \longrightarrow R \longrightarrow K[X, Y] \longrightarrow H_q^1(R) \longrightarrow 0$ in $\mathcal{C}(R)$, and
that $\dim_K(H_q^1(R)) = t - 1$.

(iv) In the special case in which $q := XK[X, Y] + Y(Y-1)K[X, Y]$, show
that $R = K[X, XY, Y(Y-1), Y^2(Y-1)]$, that q is the maximal ideal
$(X, XY, Y(Y-1), Y^2(Y-1))$ of R, and that $H_q^1(R) \cong K$.

(v) In the special case in which $q := XK[X, Y] + Y^2K[X, Y]$, show that
$R = K[X, Y^2, XY, Y^3]$, that q is the maximal ideal (X, Y^2, XY, Y^3)
of R, and that $H_q^1(R) \cong K$.

We are now in a position to present one of the geometric examples which
was promised earlier.

2.3.7 Example. (See R. Hartshorne [28, 3.4.2].) With the notation of 2.3.1,
let V be the affine algebraic set in $\mathbb{A}^4$ given by

$$V := V_{\mathbb{A}^4}(X_1 X_4 - X_2 X_3, \ X_1^2 X_3 + X_1 X_2 - X_2^2, \ X_3^3 + X_3 X_4 - X_4^2).$$

It is easy to check that the morphism of varieties $\alpha : \mathbb{A}^2 \to \mathbb{A}^4$ for which
$\alpha((c, d)) = (c, cd, d(d-1), d^2(d-1))$ for all $(c, d) \in \mathbb{A}^2$ satisfies $\operatorname{Im}\alpha \subseteq V$.
In fact, $\operatorname{Im}\alpha = V$, because the map $\beta : V \to \mathbb{A}^2$ defined by

$$\beta((c_1, c_2, c_3, c_4)) = \begin{cases} (c_1, c_2/c_1) & \text{if } c_1 \neq 0, \\ (c_1, c_4/c_3) & \text{if } c_3 \neq 0, \\ (0, 0) & \text{if } c_1 = c_3 = 0 \end{cases}$$

(for all $(c_1, c_2, c_3, c_4) \in V$) satisfies $\alpha \circ \beta = \operatorname{Id}_V$, the identity map on V.

Since $\mathbb{A}^2$ is irreducible, it follows that V is irreducible, too. We propose
to study the coordinate ring, $R := \mathcal{O}(V)$, of this affine variety. Note that the
$\mathbb{C}$-algebra homomorphism

$$\alpha^* : \mathcal{O}(V) \longrightarrow \mathcal{O}(\mathbb{A}^2) = \mathbb{C}[X, Y]$$

induced by α is injective (because α is surjective). Also, if $\iota : V \to \mathbb{A}^4$ denotes the inclusion morphism of varieties, then $\iota^* : \mathbb{C}[X_1, X_2, X_3, X_4] \to \mathcal{O}(V)$ is the natural surjective $\mathbb{C}$-algebra homomorphism given by restriction. Furthermore, $\operatorname{Im} \alpha^* = \mathbb{C}[X, XY, Y(Y-1), Y^2(Y-1)]$, since

$$\alpha^*(\iota^*(X_1)) = X, \quad \alpha^*(\iota^*(X_2)) = XY,$$

$$\alpha^*(\iota^*(X_3)) = Y(Y-1), \quad \alpha^*(\iota^*(X_4)) = Y^2(Y-1).$$

It follows from 2.3.6(iv) that, if we use $\mathfrak{m}$ to denote the maximal ideal of R corresponding to the point $(0,0,0,0)$ of V, then $H^1_{\mathfrak{m}}(R) \cong \mathbb{C}$. Note that the open subset $V \setminus V(\mathfrak{m})$ is just $V \setminus \{(0,0,0,0)\}$. By 2.3.2 and 2.3.3, there must be a regular function on $V \setminus \{(0,0,0,0)\}$ that cannot be extended to a regular function on V. The reader might find it interesting for us to find such a function.

Observe that

$$V \setminus \{(0,0,0,0)\} = U_V(x_1) \cup U_V(x_3)$$

and that, on $U_V(x_1) \cap U_V(x_3)$, the functions x_2/x_1 and x_4/x_3 are both defined and are equal. Thus $\beta_2 : V \setminus \{(0,0,0,0)\} \to \mathbb{C}$ defined by

$$\beta_2((c_1, c_2, c_3, c_4)) = \begin{cases} c_2/c_1 & \text{if } c_1 \neq 0, \\ c_4/c_3 & \text{if } c_3 \neq 0 \end{cases}$$

(for all $(c_1, c_2, c_3, c_4) \in V \setminus \{(0,0,0,0)\}$) is a regular function on the set $V \setminus \{(0,0,0,0)\}$. (In fact, the map from $V \setminus \{(0,0,0,0)\}$ to $\mathbb{A}^2 \setminus \{(0,0), (0,1)\}$ given by $(c_1, c_2, c_3, c_4) \mapsto (c_1, \beta_2((c_1, c_2, c_3, c_4)))$ is actually an isomorphism of (quasi-affine) varieties, because it is inverse to $\alpha\lceil : \mathbb{A}^2 \setminus \{(0,0), (0,1)\} \longrightarrow V \setminus \{(0,0,0,0)\}$.)

Suppose that β_2 can be extended to a regular function $\beta_2' : V \to \mathbb{C}$, and look for a contradiction. Now for $v \in U_V(x_3)$, we have $\beta_2'(v) = x_4(v)/x_3(v)$. Since $x_4(v)^2 = x_3(v)^3 + x_3(v)x_4(v)$, it follows that

$$\beta_2'(v)^2 - \beta_2'(v) = \frac{x_4(v)^2}{x_3(v)^2} - \frac{x_4(v)}{x_3(v)}$$

$$= \frac{x_3(v)^3 + x_3(v)x_4(v)}{x_3(v)^2} - \frac{x_4(v)}{x_3(v)} = x_3(v)$$

for all $v \in U_V(x_3)$. Since $U_V(x_3)$ is a dense open subset of V, we deduce that $\beta_2'^2 - \beta_2' = x_3$. Hence $\beta_2'((0,0,0,0)) = \varepsilon$, where $\varepsilon = 0$ or 1. Now the map $\beta' : V \longrightarrow \mathbb{A}^2 \setminus \{(0, 1-\varepsilon)\}$ given by $(c_1, c_2, c_3, c_4) \longmapsto (c_1, \beta_2'((c_1, c_2, c_3, c_4)))$ is a morphism of varieties. In fact, it is an isomorphism of varieties, because $\alpha\lceil : \mathbb{A}^2 \setminus \{(0, 1-\varepsilon)\} \longrightarrow V$ is an inverse for it. This shows that the quasi-affine variety $\mathbb{A}^2 \setminus \{(0, 1-\varepsilon)\}$ is affine.

However, Exercise 2.3.5 shows that the quasi-affine variety $\mathbb{A}^2 \setminus \{(0,0)\}$ is not affine, and a similar argument will show that $\mathbb{A}^2 \setminus \{(0,1)\}$ is not affine. We have therefore arrived at a contradiction, and this shows that

$$\beta_2 : V \setminus \{(0,0,0,0)\} \to \mathbb{C}$$

is a regular function which cannot be extended to a regular function on V.

2.3.8 Exercise. With the notation of 2.3.1, let V be the affine algebraic set in $\mathbb{A}^4$ given by

$$V := V_{\mathbb{A}^4}(X_1^2 X_2 - X_3^2, \ X_2^3 - X_4^2, \ X_2 X_3 - X_1 X_4, \ X_1 X_2^2 - X_3 X_4).$$

(i) Show that the morphism of varieties $\alpha : \mathbb{A}^2 \to \mathbb{A}^4$ for which

$$\alpha((c, d)) = (c, d^2, cd, d^3) \quad \text{for all } (c, d) \in \mathbb{A}^2$$

is injective and that its image is equal to V. Deduce that V is irreducible, and so is an affine variety.

(ii) Show that

$$\alpha\lceil : \mathbb{A}^2 \setminus \{(0,0)\} \longrightarrow V \setminus \{(0,0,0,0)\}$$

is an isomorphism of (quasi-affine) varieties, with inverse

$$\beta : V \setminus \{(0,0,0,0)\} \longrightarrow \mathbb{A}^2 \setminus \{(0,0)\}$$

given by

$$\beta((c_1, c_2, c_3, c_4)) = \begin{cases} (c_1, c_3/c_1) & \text{if } c_1 \neq 0, \\ (c_1, c_4/c_2) & \text{if } c_2 \neq 0 \end{cases}$$

(for all $(c_1, c_2, c_3, c_4) \in V \setminus \{(0,0,0,0)\}$).

(iii) Let $\mathfrak{m}$ denote the maximal ideal of $\mathcal{O}(V)$ corresponding to the point $(0,0,0,0)$ of V. Show that $H^1_{\mathfrak{m}}(\mathcal{O}(V)) \cong \mathbb{C}$.

(iv) Show that the function $\beta_2 : V \setminus \{(0,0,0,0)\} \to \mathbb{C}$ defined by

$$\beta_2((c_1, c_2, c_3, c_4)) = \begin{cases} c_3/c_1 & \text{if } c_1 \neq 0, \\ c_4/c_2 & \text{if } c_2 \neq 0 \end{cases}$$

(for all $(c_1, c_2, c_3, c_4) \in V \setminus \{(0,0,0,0)\}$) is a regular function on the set $V \setminus \{(0,0,0,0)\}$ that cannot be extended to a regular function on V.

Although Example 2.3.7 and Exercise 2.3.8 look very similar from an algebraic point of view, there are substantial geometric differences between them. This is illustrated by the next two exercises. By the *metric topology* on an affine variety $V \subseteq \mathbb{A}^n$ we mean the (subspace) topology induced on V by the topology defined on $\mathbb{C}^n$ by the standard distance metric of analysis.

2.3.9 Exercise. Let the situation and notation be as in 2.3.7.

Prove that there does not exist a mapping $\beta_2' : V \longrightarrow \mathbb{C}$ which extends β_2 and which is continuous for the metric topology. (Here is a hint: consider points on the path $s : \{t \in \mathbb{R} : 0 \leq t \leq 1\} \to V$ given by $s(t) = \alpha((0,t)) = (0, 0, t(t-1), t^2(t-1))$ (for $0 \leq t \leq 1$) to show that, in the metric topology, $\lim_{v \to (0,0,0,0)} \beta_2(v)$ does not exist.)

2.3.10 Exercise. Let the situation and notation be as in 2.3.8.

Let $\beta' : V \longrightarrow \mathbb{A}^2$ be the map which extends

$$\beta : V \setminus \{(0,0,0,0)\} \longrightarrow \mathbb{A}^2 \setminus \{(0,0)\}$$

and is such that $\beta'((0,0,0,0)) = (0,0)$. Show that β' is actually continuous for the metric topology, and deduce that $\alpha : \mathbb{A}^2 \longrightarrow V$ is a homeomorphism with respect to the metric topology. (Again we offer a hint: show that $\beta_2(v)^2 = x_2(v)$ for all $v \in U_V(x_1) \cup U_V(x_2) = V \setminus \{(0,0,0,0)\}$.)

2.3.11 ♯Exercise. Use the notation of 2.3.1.

(i) Let $0 \neq g \in \mathbb{C}[X, Y] = \mathcal{O}(\mathbb{A}^2)$. Let $U := U_{\mathbb{A}^2}(g)$, a quasi-affine variety over $\mathbb{C}$, and let $p \in U$. Show that each regular function $f : U \setminus \{p\} \to \mathbb{C}$ can be extended to a regular function on U. (Do not forget that U is affine!)

(ii) Let $f \in K(\mathbb{A}^2)$. There is a maximum open subset U' of $\mathbb{A}^2$ on which f is defined: the *poles of* f are precisely the points of $\mathbb{A}^2 \setminus U'$. Use (i) to show that there does not exist an *isolated* pole of f, that is, a pole q of f for which there exists an open subset U'' of $\mathbb{A}^2$ such that $q \in U''$ but q is the only pole of f in U''.

3

The Mayer–Vietoris sequence

Any reader with a basic grounding in algebraic topology will recall the important rôle that the Mayer–Vietoris sequence can play in that subject. There is an analogue of the Mayer–Vietoris sequence in local cohomology theory, and it can play a foundational rôle in this subject. It is our intention in this chapter to present the basic theory of the Mayer–Vietoris sequence in local cohomology, and to prepare for several uses of the idea during the subsequent development.

The Mayer–Vietoris sequence involves two ideals, and so throughout this chapter, $\mathfrak{b}$ will denote a second ideal of R (in addition to $\mathfrak{a}$). Let M be an R-module. The Mayer–Vietoris sequence provides, among other things, a long exact sequence

$$0 \longrightarrow H^0_{\mathfrak{a}+\mathfrak{b}}(M) \longrightarrow H^0_{\mathfrak{a}}(M) \oplus H^0_{\mathfrak{b}}(M) \longrightarrow H^0_{\mathfrak{a}\cap\mathfrak{b}}(M)$$

$$\longrightarrow H^1_{\mathfrak{a}+\mathfrak{b}}(M) \longrightarrow H^1_{\mathfrak{a}}(M) \oplus H^1_{\mathfrak{b}}(M) \longrightarrow H^1_{\mathfrak{a}\cap\mathfrak{b}}(M)$$

$$\longrightarrow \quad \cdots \qquad\qquad\qquad\qquad\qquad \cdots$$

$$\longrightarrow H^i_{\mathfrak{a}+\mathfrak{b}}(M) \longrightarrow H^i_{\mathfrak{a}}(M) \oplus H^i_{\mathfrak{b}}(M) \longrightarrow H^i_{\mathfrak{a}\cap\mathfrak{b}}(M)$$

$$\longrightarrow H^{i+1}_{\mathfrak{a}+\mathfrak{b}}(M) \longrightarrow \qquad \cdots$$

of local cohomology modules. Its potential for use in arguments that employ induction on the number of elements in a generating set for an ideal $\mathfrak{c}$ of R can be explained as follows. Suppose that $\mathfrak{c}$ is generated by n elements $c_1, \ldots, c_n$, where $n > 1$. Set $\mathfrak{a} = Rc_1 + \cdots + Rc_{n-1}$ and $\mathfrak{b} = Rc_n$, so that $\mathfrak{c} = \mathfrak{a} + \mathfrak{b}$. Each of $\mathfrak{a}$ and $\mathfrak{b}$ can be generated by fewer than n elements, but at first sight it seems that the ideal $\mathfrak{a} \cap \mathfrak{b}$, which also appears in the Mayer–Vietoris sequence, could present difficulties. However, $\sqrt{(\mathfrak{a}\cap\mathfrak{b})} = \sqrt{(\mathfrak{a}\mathfrak{b})}$, and so $\Gamma_{\mathfrak{a}\cap\mathfrak{b}} = \Gamma_{\mathfrak{a}\mathfrak{b}}$ by 1.1.3; hence $H^i_{\mathfrak{a}\cap\mathfrak{b}} = H^i_{\mathfrak{a}\mathfrak{b}}$ for all $i \in \mathbb{N}_0$ (see 1.2.3). Moreover, in our situation,

$$\mathfrak{a}\mathfrak{b} = (Rc_1 + \cdots + Rc_{n-1})Rc_n = Rc_1c_n + \cdots + Rc_{n-1}c_n$$

can be generated by $n - 1$ elements. Thus $\mathfrak{a}, \mathfrak{b}$ and $\mathfrak{ab}$ can all be generated by fewer than n elements, and an appropriate inductive hypothesis would apply to all of them. In addition, we have already obtained (in 2.2.21) a certain amount of information about the local cohomology functors H^i_{Ra} ($i \in \mathbb{N}_0$) with respect to a principal ideal Ra of R: this can provide a basis for an inductive argument.

3.1 Comparison of systems of ideals

The result that $H^i_{\mathfrak{a} \cap \mathfrak{b}} = H^i_{\mathfrak{ab}}$ for all $i \in \mathbb{N}_0$ can be viewed as a particular example of a more general phenomenon which concerns a situation where two systems of ideals are 'comparable' in a sense made precise in the proposition below. This comparison result will not only be used to obtain the Mayer–Vietoris sequence; it will also provide an important ingredient, which is the subject of Exercise 3.1.4, in our proof of the local Lichtenbaum–Hartshorne Vanishing Theorem in Chapter 8.

3.1.1 Proposition. *Let* $(\Lambda, \leq)$ *and* $(\Pi, \leq)$ *be (non-empty) directed partially ordered sets, and let* $\mathfrak{B} = (\mathfrak{b}_\alpha)_{\alpha \in \Lambda}$ *be an inverse family of ideals of R over Λ, as in 1.2.10; let* $\mathfrak{C} = (\mathfrak{c}_\beta)_{\beta \in \Pi}$ *be an inverse family of ideals of R over Π.*

Assume that, for all $\alpha \in \Lambda$, *there exists* $\beta \in \Pi$ *such that* $\mathfrak{c}_\beta \subseteq \mathfrak{b}_\alpha$, *and, for all* $\beta' \in \Pi$, *there exists* $\alpha' \in \Lambda$ *such that* $\mathfrak{b}_{\alpha'} \subseteq \mathfrak{c}_{\beta'}$. *Then*

  (i) $\Gamma_\mathfrak{B} = \Gamma_\mathfrak{C}$;
 (ii) *the negative strongly connected sequences of covariant functors*

$$\left(\varinjlim_{\alpha \in \Lambda} \operatorname{Ext}^i_R(R/\mathfrak{b}_\alpha, \, \bullet \,) \right)_{i \in \mathbb{N}_0} \quad and \quad \left(\varinjlim_{\beta \in \Pi} \operatorname{Ext}^i_R(R/\mathfrak{c}_\beta, \, \bullet \,) \right)_{i \in \mathbb{N}_0}$$

  are isomorphic; and
(iii) $\mathfrak{B}$ *is a system of ideals of R over Λ (in the sense of 2.1.10) if and only if* $\mathfrak{C}$ *is a system of ideals of R over Π.*

Note. The functor $\Gamma_\mathfrak{B}$ was defined in 1.2.11, while in 1.3.7 it was explained that the negative strongly connected sequence $(\mathcal{R}^i \Gamma_\mathfrak{B})_{i \in \mathbb{N}_0}$ of its right derived functors is isomorphic to

$$\left(\varinjlim_{\alpha \in \Lambda} \operatorname{Ext}^i_R(R/\mathfrak{b}_\alpha, \, \bullet \,) \right)_{i \in \mathbb{N}_0} ;$$

of course, similar comments apply to $\Gamma_\mathfrak{C}$.

Proof. Part (i) is clear; hence $\mathcal{R}^i\Gamma_{\mathfrak{B}} = \mathcal{R}^i\Gamma_{\mathfrak{C}}$ for all $i \in \mathbb{N}_0$, and the claim in (ii) is an obvious consequence of the above note.

(iii) Suppose that $\mathfrak{B}$ is a system of ideals over Λ, and let $\beta_1, \beta_2 \in \Pi$. By assumption, there exist $\alpha_1, \alpha_2 \in \Lambda$ such that $\mathfrak{b}_{\alpha_1} \subseteq \mathfrak{c}_{\beta_1}$ and $\mathfrak{b}_{\alpha_2} \subseteq \mathfrak{c}_{\beta_2}$. Since $\mathfrak{B}$ is a system of ideals, there exists $\alpha_3 \in \Lambda$ such that $\mathfrak{b}_{\alpha_3} \subseteq \mathfrak{b}_{\alpha_1}\mathfrak{b}_{\alpha_2}$. By assumption, there exists $\beta_3 \in \Pi$ such that $\mathfrak{c}_{\beta_3} \subseteq \mathfrak{b}_{\alpha_3}$. Hence

$$\mathfrak{c}_{\beta_3} \subseteq \mathfrak{b}_{\alpha_3} \subseteq \mathfrak{b}_{\alpha_1}\mathfrak{b}_{\alpha_2} \subseteq \mathfrak{c}_{\beta_1}\mathfrak{c}_{\beta_2}.$$

Hence $\mathfrak{C}$ is a system of ideals over Π.

In view of the symmetry of our hypotheses on $\mathfrak{B}$ and $\mathfrak{C}$, the proof is complete. $\square$

3.1.2 Example. Consider the descending chain $\mathfrak{B} = (\mathfrak{a}^n + \mathfrak{b}^n)_{n \in \mathbb{N}}$ of ideals of R. Since $(\mathfrak{a} + \mathfrak{b})^{2n-1} \subseteq \mathfrak{a}^n + \mathfrak{b}^n \subseteq (\mathfrak{a} + \mathfrak{b})^n$ for all $n \in \mathbb{N}$, it follows from 3.1.1 that $\mathfrak{B}$ is actually a system of ideals of R over $\mathbb{N}$; also, it follows from 1.3.7 and 3.1.1 that the (negative strongly) connected sequences of covariant functors (from $\mathcal{C}(R)$ to $\mathcal{C}(R)$)

$$\left(H^i_{\mathfrak{a}+\mathfrak{b}}\right)_{i \in \mathbb{N}_0}, \quad \left(\varinjlim_{n \in \mathbb{N}} \mathrm{Ext}^i_R(R/(\mathfrak{a}^n + \mathfrak{b}^n), \bullet)\right)_{i \in \mathbb{N}_0} \quad \text{and} \quad \left(H^i_{\mathfrak{B}}\right)_{i \in \mathbb{N}_0}$$

are all isomorphic. In particular, for each $i \in \mathbb{N}_0$, the functors $H^i_{\mathfrak{a}+\mathfrak{b}}$ and $\varinjlim_{n \in \mathbb{N}} \mathrm{Ext}^i_R(R/(\mathfrak{a}^n + \mathfrak{b}^n), \bullet)$ are naturally equivalent.

3.1.3 Exercise. Let $(R, \mathfrak{m})$ be a complete local ring, and let $\mathfrak{B} = (\mathfrak{b}_n)_{n \in \mathbb{N}}$ be a descending chain of $\mathfrak{m}$-primary ideals of R such that $\bigcap_{n \in \mathbb{N}} \mathfrak{b}_n = 0$.

Show that $\mathfrak{B}$ is a system of ideals of R over $\mathbb{N}$, and that the (negative strongly) connected sequences of covariant functors (from $\mathcal{C}(R)$ to $\mathcal{C}(R)$)

$$\left(H^i_{\mathfrak{m}}\right)_{i \in \mathbb{N}_0}, \quad \left(\varinjlim_{n \in \mathbb{N}} \mathrm{Ext}^i_R(R/\mathfrak{b}_n, \bullet)\right)_{i \in \mathbb{N}_0} \quad \text{and} \quad \left(H^i_{\mathfrak{B}}\right)_{i \in \mathbb{N}_0}$$

are all isomorphic. (If you have no idea where to start with this, we suggest that you look up Chevalley's Theorem in, for example, [59, §5.2, Theorem 1] or [89, Chapter VIII, §5, Theorem 13].)

3.1.4 ♯Exercise. Let $(R, \mathfrak{m})$ be a complete local domain, and let $\mathfrak{p}$ be a prime ideal of R of dimension 1, that is, such that $\dim R/\mathfrak{p} = 1$.

(i) Use Chevalley's Theorem ([59, §5.2, Theorem 1] or [89, Chapter VIII, §5, Theorem 13]) to prove that, for each $n \in \mathbb{N}$, there exists $t \in \mathbb{N}$ such that the t-th symbolic power $\mathfrak{p}^{(t)}$ of $\mathfrak{p}$ is contained in $\mathfrak{m}^n$.

(ii) Deduce that $(\mathfrak{p}^{(n)})_{n\in\mathbb{N}}$ is a system of ideals of R over $\mathbb{N}$, and that the (negative strongly) connected sequences of covariant functors

$$\left(H_\mathfrak{p}^i\right)_{i\in\mathbb{N}_0} \quad \text{and} \quad \left(\varinjlim_{n\in\mathbb{N}} \operatorname{Ext}_R^i(R/\mathfrak{p}^{(n)}, \bullet)\right)_{i\in\mathbb{N}_0}$$

(from $\mathcal{C}(R)$ to $\mathcal{C}(R)$) are isomorphic.

3.1.5 Corollary. *The descending chain $(\mathfrak{a}^n \cap \mathfrak{b}^n)_{n\in\mathbb{N}}$ is a system of ideals of R (over $\mathbb{N}$), and the (negative strongly) connected sequences of covariant functors (from $\mathcal{C}(R)$ to $\mathcal{C}(R)$)*

$$\left(\varinjlim_{n\in\mathbb{N}} \operatorname{Ext}_R^i(R/(\mathfrak{a}^n \cap \mathfrak{b}^n), \bullet)\right)_{i\in\mathbb{N}_0} \quad \text{and} \quad \left(H_{\mathfrak{a}\cap\mathfrak{b}}^i\right)_{i\in\mathbb{N}_0}$$

are isomorphic. In particular, for each $i \in \mathbb{N}_0$, the functors $H_{\mathfrak{a}\cap\mathfrak{b}}^i = H_{\mathfrak{a}\mathfrak{b}}^i$ and $\varinjlim_{n\in\mathbb{N}} \operatorname{Ext}_R^i(R/(\mathfrak{a}^n \cap \mathfrak{b}^n), \bullet)$ are naturally equivalent.

Proof. As we mentioned at the beginning of this chapter, it follows from 1.2.3 that $H_{\mathfrak{a}\cap\mathfrak{b}}^i = H_{\mathfrak{a}\mathfrak{b}}^i$ for all $i \in \mathbb{N}_0$. Since $(\mathfrak{a} \cap \mathfrak{b})^n \subseteq \mathfrak{a}^n \cap \mathfrak{b}^n$ for all $n \in \mathbb{N}$, it is enough, in view of 3.1.1, for us to show that, for each $n \in \mathbb{N}$, there exists $q(n) \in \mathbb{N}$ such that $\mathfrak{a}^{q(n)} \cap \mathfrak{b}^{q(n)} \subseteq (\mathfrak{a} \cap \mathfrak{b})^n$. We use the Artin–Rees Lemma [50, Theorem 8.5] to achieve this.

Fix $n \in \mathbb{N}$. By the Artin–Rees Lemma, there is $c \in \mathbb{N}$ such that $\mathfrak{a}^m \cap \mathfrak{b}^n = \mathfrak{a}^{m-c}(\mathfrak{a}^c \cap \mathfrak{b}^n)$ for all integers $m > c$. Hence

$$\mathfrak{a}^{n+c} \cap \mathfrak{b}^{n+c} \subseteq \mathfrak{a}^{n+c} \cap \mathfrak{b}^n = \mathfrak{a}^n(\mathfrak{a}^c \cap \mathfrak{b}^n) \subseteq \mathfrak{a}^n\mathfrak{b}^n \subseteq (\mathfrak{a} \cap \mathfrak{b})^n.$$

The proof is therefore complete.   □

3.1.6 Exercise. For each $h \in \mathbb{N}$, let $\mathfrak{a}^{[h]}$ denote the ideal of R generated by all the h-th powers of elements of $\mathfrak{a}$. Show that, for each $i \in \mathbb{N}_0$, the functor $H_\mathfrak{a}^i$ is naturally equivalent to $\varinjlim_{h\in\mathbb{N}} \operatorname{Ext}_R^i(R/\mathfrak{a}^{[h]}, \bullet)$.

(A variation on this is the result that, if R contains a subfield of characteristic $p > 0$, then, for each $i \in \mathbb{N}_0$, the functor $H_\mathfrak{a}^i$ is naturally equivalent to $\varinjlim_{e\in\mathbb{N}} \operatorname{Ext}_R^i(R/\mathfrak{a}^{[p^e]}, \bullet)$. There are situations where this observation can be very useful: see, for example, C. Peskine and L. Szpiro [66, Chapitre III, Proposition 1.8(3)], and C. L. Huneke and R. Y. Sharp [43, p. 770].)

3.1.7 Exercise. Suppose that the t elements $a_1, \ldots, a_t$ generate $\mathfrak{a}$. Now $\mathbb{N}^t$ is a directed partially ordered set with respect to the ordering $\leq$ defined by, for

$(u_1, \ldots, u_t), (v_1, \ldots, v_t) \in \mathbb{N}^t$,

$(u_1, \ldots, u_t) \leq (v_1, \ldots, v_t)$ if and only if $u_j \leq v_j$ for all $j = 1, \ldots, t$.

Show that, for each $i \in \mathbb{N}_0$, the functor $H_{\mathfrak{a}}^i$ is naturally equivalent to

$$\varinjlim_{(u_1, \ldots, u_t) \in \mathbb{N}^t} \operatorname{Ext}_R^i(R/(a_1^{u_1}, \ldots, a_t^{u_t}), \bullet).$$

3.2 Construction of the sequence

Our first lemma in this section provides a fundamental tool for the construction of the Mayer–Vietoris sequence.

3.2.1 Lemma.   *Let N_1, N_2 be submodules of the R-module M. The sequence of R-modules and R-homomorphisms*

$$0 \longrightarrow M/(N_1 \cap N_2) \overset{\alpha}{\longrightarrow} M/N_1 \oplus M/N_2 \overset{\beta}{\longrightarrow} M/(N_1 + N_2) \longrightarrow 0,$$

in which $\alpha(m + N_1 \cap N_2) = (m + N_1, m + N_2)$ for all $m \in M$ and

$$\beta((x + N_1, y + N_2)) = x - y + (N_1 + N_2) \text{ for all } x, y \in M,$$

is exact.

Proof.   It is clear that α is injective, that β is surjective and that $\beta \circ \alpha = 0$. Let $x, y \in M$ be such that $(x + N_1, y + N_2) \in \operatorname{Ker} \beta$. Then $x - y = n_1 + n_2$ for some $n_1 \in N_1, n_2 \in N_2$, so that $x - n_1 = y + n_2$ and

$$(x + N_1, y + N_2) = (x - n_1 + N_1, y + n_2 + N_2) \in \operatorname{Im} \alpha. \qquad \square$$

Briefly, the general strategy for our construction of the Mayer–Vietoris sequence is to write down, for each $n \in \mathbb{N}$, the long exact sequence of 'Ext' modules which results from application of the functor $\operatorname{Hom}_R(\bullet, M)$ to the exact sequence

$$0 \longrightarrow R/(\mathfrak{a}^n \cap \mathfrak{b}^n) \longrightarrow R/\mathfrak{a}^n \oplus R/\mathfrak{b}^n \longrightarrow R/(\mathfrak{a}^n + \mathfrak{b}^n) \longrightarrow 0$$

resulting from 3.2.1, pass to direct limits, and then appeal to 3.1.2 and 3.1.5 to convert the result into information about local cohomology modules. There is, however, one minor technical point that needs attention first, and that is the identification of $H_{\mathfrak{a}}^i(\bullet) \oplus H_{\mathfrak{b}}^i(\bullet)$ with the functor

$$\varinjlim_{n \in \mathbb{N}} \operatorname{Ext}_R^i(R/\mathfrak{a}^n \oplus R/\mathfrak{b}^n, \bullet)$$

(obtained using the ideas of 1.2.8 in an obvious way). Although perhaps a little tedious, this is not particularly difficult.

3.2.2 ♯Exercise.   In this exercise, we use, for R-modules L and M, the symbolism

$$0 \rightleftarrows L \rightleftarrows L \oplus M \rightleftarrows M \rightleftarrows 0$$

to denote simultaneously the two split exact sequences associated with the direct sum: thus we consider just the arrows pointing to the right to obtain one of these split exact sequences, while the arrows pointing to the left provide the other.

Let $T : \mathcal{C}(R) \times \mathcal{C}(R) \to \mathcal{C}(R)$ be an R-linear functor of two variables which is contravariant in the first variable and covariant in the second. (For example, T could be Hom_R or Ext_R^i for $i \in \mathbb{N}_0$.)

For $n, m \in \mathbb{N}$ with $n \geq m$, let

$$h_m^n : R/\mathfrak{a}^n \longrightarrow R/\mathfrak{a}^m \quad \text{and} \quad k_m^n : R/\mathfrak{b}^n \longrightarrow R/\mathfrak{b}^m$$

denote the natural homomorphisms, and consider the commutative diagrams

$$
\begin{array}{ccccccccc}
0 & \rightleftarrows & R/\mathfrak{a}^n & \rightleftarrows & R/\mathfrak{a}^n \oplus R/\mathfrak{b}^n & \rightleftarrows & R/\mathfrak{b}^n & \rightleftarrows & 0 \\
 & & \downarrow {\scriptstyle h_m^n} & & \downarrow {\scriptstyle h_m^n \oplus k_m^n} & & \downarrow {\scriptstyle k_m^n} & & \\
0 & \rightleftarrows & R/\mathfrak{a}^m & \rightleftarrows & R/\mathfrak{a}^m \oplus R/\mathfrak{b}^m & \rightleftarrows & R/\mathfrak{b}^m & \rightleftarrows & 0 \, ,
\end{array}
$$

in which the rows represent the canonical split exact sequences. Apply the contravariant, additive functor $T(\,\bullet\,, M)$, and pass to direct limits to obtain split exact sequences

$$0 \rightleftarrows \varinjlim_{n \in \mathbb{N}} T(R/\mathfrak{b}^n, M) \rightleftarrows \varinjlim_{n \in \mathbb{N}} T(R/\mathfrak{a}^n \oplus R/\mathfrak{b}^n, M)$$
$$\rightleftarrows \varinjlim_{n \in \mathbb{N}} T(R/\mathfrak{a}^n, M) \rightleftarrows 0.$$

Deduce that there is a natural equivalence between the functors

$$\varinjlim_{n \in \mathbb{N}} T(R/\mathfrak{a}^n \oplus R/\mathfrak{b}^n,\, \bullet\,) \quad \text{and} \quad \varinjlim_{n \in \mathbb{N}} T(R/\mathfrak{a}^n,\, \bullet\,) \oplus \varinjlim_{n \in \mathbb{N}} T(R/\mathfrak{b}^n,\, \bullet\,)$$

(from $\mathcal{C}(R)$ to itself).

Deduce from 1.3.8 that the functors

$$\varinjlim_{n \in \mathbb{N}} \operatorname{Ext}_R^i(R/\mathfrak{a}^n \oplus R/\mathfrak{b}^n,\, \bullet\,) \quad \text{and} \quad H_\mathfrak{a}^i(\,\bullet\,) \oplus H_\mathfrak{b}^i(\,\bullet\,)$$

are naturally equivalent for each $i \in \mathbb{N}_0$.

3.2.3 Theorem: the Mayer–Vietoris sequence. *For each R-module M, there is a long exact sequence (called* the Mayer–Vietoris sequence *for M with respect to $\mathfrak{a}$ and $\mathfrak{b}$)*

$$0 \longrightarrow H^0_{\mathfrak{a}+\mathfrak{b}}(M) \longrightarrow H^0_{\mathfrak{a}}(M) \oplus H^0_{\mathfrak{b}}(M) \longrightarrow H^0_{\mathfrak{a}\cap\mathfrak{b}}(M)$$

$$\longrightarrow H^1_{\mathfrak{a}+\mathfrak{b}}(M) \longrightarrow H^1_{\mathfrak{a}}(M) \oplus H^1_{\mathfrak{b}}(M) \longrightarrow H^1_{\mathfrak{a}\cap\mathfrak{b}}(M)$$

$$\longrightarrow \quad \cdots \qquad\qquad\qquad \cdots$$

$$\longrightarrow H^i_{\mathfrak{a}+\mathfrak{b}}(M) \longrightarrow H^i_{\mathfrak{a}}(M) \oplus H^i_{\mathfrak{b}}(M) \longrightarrow H^i_{\mathfrak{a}\cap\mathfrak{b}}(M)$$

$$\longrightarrow H^{i+1}_{\mathfrak{a}+\mathfrak{b}}(M) \longrightarrow \qquad \cdots$$

such that, whenever $f : M \longrightarrow N$ is a homomorphism of R-modules, the diagram

$$
\begin{array}{ccccccc}
H^i_{\mathfrak{a}+\mathfrak{b}}(M) & \longrightarrow & H^i_{\mathfrak{a}}(M) \oplus H^i_{\mathfrak{b}}(M) & \longrightarrow & H^i_{\mathfrak{a}\cap\mathfrak{b}}(M) & \longrightarrow & H^{i+1}_{\mathfrak{a}+\mathfrak{b}}(M) \\
\downarrow{\scriptstyle H^i_{\mathfrak{a}+\mathfrak{b}}(f)} & & \downarrow{\scriptstyle H^i_{\mathfrak{a}}(f)\oplus H^i_{\mathfrak{b}}(f)} & & \downarrow{\scriptstyle H^i_{\mathfrak{a}\cap\mathfrak{b}}(f)} \quad {\scriptstyle H^{i+1}_{\mathfrak{a}+\mathfrak{b}}(f)}\downarrow & & \\
H^i_{\mathfrak{a}+\mathfrak{b}}(N) & \longrightarrow & H^i_{\mathfrak{a}}(N) \oplus H^i_{\mathfrak{b}}(N) & \longrightarrow & H^i_{\mathfrak{a}\cap\mathfrak{b}}(N) & \longrightarrow & H^{i+1}_{\mathfrak{a}+\mathfrak{b}}(N)
\end{array}
$$

commutes for all $i \in \mathbb{N}_0$.

Proof.   Let $n, m \in \mathbb{N}$ with $n \geq m$, and let

$$h^n_m : R/\mathfrak{a}^n \longrightarrow R/\mathfrak{a}^m \quad \text{and} \quad k^n_m : R/\mathfrak{b}^n \longrightarrow R/\mathfrak{b}^m$$

denote the natural homomorphisms. The diagram

$$
\begin{array}{ccccccc}
0 \longrightarrow & R/(\mathfrak{a}^n \cap \mathfrak{b}^n) & \longrightarrow & R/\mathfrak{a}^n \oplus R/\mathfrak{b}^n & \longrightarrow & R/(\mathfrak{a}^n + \mathfrak{b}^n) & \longrightarrow 0 \\
& \downarrow & & \downarrow{\scriptstyle h^n_m \oplus k^n_m} & & \downarrow & \\
0 \longrightarrow & R/(\mathfrak{a}^m \cap \mathfrak{b}^m) & \longrightarrow & R/\mathfrak{a}^m \oplus R/\mathfrak{b}^m & \longrightarrow & R/(\mathfrak{a}^m + \mathfrak{b}^m) & \longrightarrow 0,
\end{array}
$$

in which the upper row is the exact sequence resulting from application of 3.2.1 to the submodules $\mathfrak{a}^n$ and $\mathfrak{b}^n$ of R, the lower row is the corresponding exact sequence for m instead of n, and the two outer vertical homomorphisms are the natural ones, commutes. Therefore, application of the functor $\operatorname{Hom}_R(\ \bullet\ , M)$

to this yields a long exact sequence

$$0 \longrightarrow \operatorname{Hom}_R(R/(\mathfrak{a}^m + \mathfrak{b}^m), M) \longrightarrow \operatorname{Hom}_R(R/\mathfrak{a}^m \oplus R/\mathfrak{b}^m, M)$$

$$\longrightarrow \operatorname{Hom}_R(R/(\mathfrak{a}^m \cap \mathfrak{b}^m), M) \longrightarrow \operatorname{Ext}_R^1(R/(\mathfrak{a}^m + \mathfrak{b}^m), M)$$

$$\longrightarrow \quad \cdots$$

$$\longrightarrow \operatorname{Ext}_R^i(R/(\mathfrak{a}^m + \mathfrak{b}^m), M) \longrightarrow \operatorname{Ext}_R^i(R/\mathfrak{a}^m \oplus R/\mathfrak{b}^m, M)$$

$$\longrightarrow \operatorname{Ext}_R^i(R/(\mathfrak{a}^m \cap \mathfrak{b}^m), M) \longrightarrow \operatorname{Ext}_R^{i+1}(R/(\mathfrak{a}^m + \mathfrak{b}^m), M)$$

$$\longrightarrow \quad \cdots$$

and a chain map (induced by the vertical homomorphisms in the last commutative diagram) of this long exact sequence into the corresponding one with m replaced by n. Also, a homomorphism $f : M \to N$ of R-modules induces a chain map of the above displayed long exact sequence into the corresponding sequence with M replaced by N. Now pass to direct limits: it follows that there is a long exact sequence

$$0 \longrightarrow \varinjlim_{n \in \mathbb{N}} \operatorname{Hom}_R(R/(\mathfrak{a}^n + \mathfrak{b}^n), M) \longrightarrow \varinjlim_{n \in \mathbb{N}} \operatorname{Hom}_R(R/\mathfrak{a}^n \oplus R/\mathfrak{b}^n, M)$$

$$\longrightarrow \varinjlim_{n \in \mathbb{N}} \operatorname{Hom}_R(R/(\mathfrak{a}^n \cap \mathfrak{b}^n), M) \longrightarrow \varinjlim_{n \in \mathbb{N}} \operatorname{Ext}_R^1(R/(\mathfrak{a}^n + \mathfrak{b}^n), M)$$

$$\longrightarrow \quad \cdots$$

$$\longrightarrow \varinjlim_{n \in \mathbb{N}} \operatorname{Ext}_R^i(R/(\mathfrak{a}^n + \mathfrak{b}^n), M) \longrightarrow \varinjlim_{n \in \mathbb{N}} \operatorname{Ext}_R^i(R/\mathfrak{a}^n \oplus R/\mathfrak{b}^n, M)$$

$$\longrightarrow \varinjlim_{n \in \mathbb{N}} \operatorname{Ext}_R^i(R/(\mathfrak{a}^n \cap \mathfrak{b}^n), M) \longrightarrow \varinjlim_{n \in \mathbb{N}} \operatorname{Ext}_R^{i+1}(R/(\mathfrak{a}^n + \mathfrak{b}^n), M)$$

$$\longrightarrow \quad \cdots$$

and that $f : M \to N$ induces a chain map of this long exact sequence into the corresponding sequence with M replaced by N. The result is now an immediate consequence of the natural equivalences of functors between

$$\varinjlim_{n \in \mathbb{N}} \operatorname{Ext}_R^i(R/(\mathfrak{a}^n + \mathfrak{b}^n), \bullet) \quad \text{and} \quad H_{\mathfrak{a}+\mathfrak{b}}^i,$$

$$\varinjlim_{n \in \mathbb{N}} \operatorname{Ext}_R^i(R/\mathfrak{a}^n \oplus R/\mathfrak{b}^n, \bullet) \quad \text{and} \quad H_{\mathfrak{a}}^i(\bullet) \oplus H_{\mathfrak{b}}^i(\bullet)$$

and

$$\varinjlim_{n \in \mathbb{N}} \operatorname{Ext}_R^i(R/(\mathfrak{a}^n \cap \mathfrak{b}^n), \bullet) \quad \text{and} \quad H_{\mathfrak{a} \cap \mathfrak{b}}^i.$$

(for $i \in \mathbb{N}_0$) established in 3.1.2, 3.2.2 and 3.1.5 respectively. $\qquad\square$

3.2.4 Exercise. Let M be an R-module. Show that the sequence of R-modules and R-homomorphisms

$$0 \longrightarrow \operatorname{Hom}_R(\mathfrak{a} + \mathfrak{b}, M) \overset{\alpha}{\longrightarrow} \operatorname{Hom}_R(\mathfrak{a}, M) \oplus \operatorname{Hom}_R(\mathfrak{b}, M)$$
$$\overset{\beta}{\longrightarrow} \operatorname{Hom}_R(\mathfrak{a} \cap \mathfrak{b}, M)$$

in which $\alpha(f) = (f\lceil_\mathfrak{a}, f\lceil_\mathfrak{b})$ for all $f \in \operatorname{Hom}_R(\mathfrak{a} + \mathfrak{b}, M)$, and $\beta((g, h)) = g\lceil_{\mathfrak{a}\cap\mathfrak{b}} - h\lceil_{\mathfrak{a}\cap\mathfrak{b}}$ for all $(g, h) \in \operatorname{Hom}_R(\mathfrak{a}, M) \oplus \operatorname{Hom}_R(\mathfrak{b}, M)$, is exact.

Show also that, if M is an injective R-module, then β is an epimorphism.

3.2.5 Exercise. Prove, perhaps with the aid of Exercise 3.2.4, that, for each R-module M, there is a long exact sequence

$$0 \longrightarrow D_{\mathfrak{a}+\mathfrak{b}}(M) \longrightarrow D_\mathfrak{a}(M) \oplus D_\mathfrak{b}(M) \longrightarrow D_{\mathfrak{a}\cap\mathfrak{b}}(M)$$
$$\longrightarrow H^2_{\mathfrak{a}+\mathfrak{b}}(M) \longrightarrow H^2_\mathfrak{a}(M) \oplus H^2_\mathfrak{b}(M) \longrightarrow H^2_{\mathfrak{a}\cap\mathfrak{b}}(M)$$
$$\longrightarrow \quad \cdots \qquad\qquad\qquad\qquad\qquad \cdots$$
$$\longrightarrow H^i_{\mathfrak{a}+\mathfrak{b}}(M) \longrightarrow H^i_\mathfrak{a}(M) \oplus H^i_\mathfrak{b}(M) \longrightarrow H^i_{\mathfrak{a}\cap\mathfrak{b}}(M)$$
$$\longrightarrow H^{i+1}_{\mathfrak{a}+\mathfrak{b}}(M) \longrightarrow \qquad \cdots$$

such that each homomorphism $f : M \to N$ of R-modules induces a chain map of the above long exact sequence into the corresponding long exact sequence for N.

As we have already mentioned, we shall make substantial use of the Mayer–Vietoris sequence in this book. We present in the next two sections two illustrations of its use in order to give some idea of its potential to any reader who, wondering why so much effort has been expended in setting up the Mayer–Vietoris sequence, is impatient to see some applications.

3.3 Arithmetic rank

Our first application of the Mayer–Vietoris sequence will be in the proof of a result which relates the number of elements required to generate the ideal $\mathfrak{a}$ (and, more precisely, its so-called 'arithmetic rank') to the vanishing of the local cohomology functors $H^i_\mathfrak{a}$. This vanishing result provides a powerful tool for applications of local cohomology to algebraic geometry, and, in particular, will play a crucial rôle in our presentation of applications to connectivity in Chapter 19.

3.3.1 Theorem. *Suppose that $\mathfrak{a}$ can be generated by t elements. Then, for every R-module M, we have $H_{\mathfrak{a}}^i(M) = 0$ for all $i > t$.*

Proof. This proof is an example of the use of induction outlined in the introduction to this chapter.

When $t = 0$, we have $\mathfrak{a} = 0$ and $\Gamma_{\mathfrak{a}} = \Gamma_{0R}$ is the identity functor, so that $H_{0R}^i(M) = 0$ for all $i > 0$. The result is therefore true when $t = 0$, and it was proved in 2.2.21(ii) in the special case in which $t = 1$.

Now suppose, inductively, that $t > 1$ and the result has been proved for ideals that can be generated by fewer than t elements. Suppose that $\mathfrak{a}$ is generated by t elements $a_1, \ldots, a_t$. Set $\mathfrak{b} = Ra_1 + \cdots + Ra_{t-1}$ and $\mathfrak{c} = Ra_t$, so that $\mathfrak{a} = \mathfrak{b} + \mathfrak{c}$. By the inductive assumption, $H_{\mathfrak{b}}^i(M) = 0$ for all $i > t - 1$ and $H_{\mathfrak{c}}^i(M) = 0$ for all $i > 1$. By the Mayer–Vietoris sequence 3.2.3, we have, for an arbitrary $i > t$, an exact sequence

$$H_{\mathfrak{b} \cap \mathfrak{c}}^{i-1}(M) \longrightarrow H_{\mathfrak{a}}^i(M) \longrightarrow H_{\mathfrak{b}}^i(M) \oplus H_{\mathfrak{c}}^i(M).$$

Now $H_{\mathfrak{b} \cap \mathfrak{c}}^{i-1}(M) = H_{\mathfrak{b}\mathfrak{c}}^{i-1}(M)$ by 3.1.5, and, since

$$\mathfrak{b}\mathfrak{c} = (Ra_1 + \cdots + Ra_{t-1})Ra_t = Ra_1 a_t + \cdots + Ra_{t-1} a_t$$

can be generated by $t - 1$ elements, it follows from the inductive assumption that $H_{\mathfrak{b}\mathfrak{c}}^{i-1}(M) = 0$. Since $H_{\mathfrak{b}}^i(M) \oplus H_{\mathfrak{c}}^i(M) = 0$ also, we have $H_{\mathfrak{a}}^i(M) = 0$. This completes the inductive step. $\quad\square$

We now combine 3.3.1 with 1.2.3 to obtain a result which is often used in geometric applications. To formulate this result, let us recall the following definition from basic commutative algebra.

3.3.2 Definition. The *arithmetic rank of* $\mathfrak{a}$, denoted by $\mathrm{ara}(\mathfrak{a})$, is the least number of elements of R required to generate an ideal which has the same radical as $\mathfrak{a}$. Thus $\mathrm{ara}(\mathfrak{a})$ is equal to the integer

$$\min \left\{ n \in \mathbb{N}_0 : \exists\, b_1, \ldots, b_n \in R \text{ with } \sqrt{(Rb_1 + \cdots + Rb_n)} = \sqrt{\mathfrak{a}} \right\}.$$

Note that $\mathrm{ara}(0R) = 0$.

The next corollary is now immediate from 3.3.1 and 1.2.3.

3.3.3 Corollary. *For every R-module M, we have $H_{\mathfrak{a}}^i(M) = 0$ for all $i > \mathrm{ara}(\mathfrak{a})$.* $\quad\square$

Corollary 3.3.3 leads naturally to the following definition.

3.3.4 Definition. The *cohomological dimension of* $\mathfrak{a}$, denoted $\mathrm{cohd}(\mathfrak{a})$, is defined as the greatest integer i for which there exists an R-module M with $H_{\mathfrak{a}}^i(M) \neq 0$ if any such integers exist, and $-\infty$ otherwise (for example when $\mathfrak{a} = R$). It follows from 3.3.3 that this definition makes sense, and that $\mathrm{cohd}(\mathfrak{a}) \leq \mathrm{ara}\,\mathfrak{a}$.

Theorem 3.3.1 can be used to obtain information about the number of elements required to generate a specified ideal; also, in geometric situations, Corollary 3.3.3 can be used to obtain information about the number of 'equations' needed to define an algebraic variety. The following example illustrates this.

3.3.5 Example. This example concerns the affine variety V in $\mathbb{A}^4$ studied in 2.3.7 and given by

$$V := V_{\mathbb{A}^4}(X_1X_4 - X_2X_3,\ X_1^2X_3 + X_1X_2 - X_2^2,\ X_3^3 + X_3X_4 - X_4^2).$$

By 2.3.7, $V = \{(c, cd, d(d-1), d^2(d-1)) \in \mathbb{A}^4 : c, d \in \mathbb{C}\}$. Thus the lines L, L' in $\mathbb{A}^4$ given by $L := V_{\mathbb{A}^4}(X_2,\ X_3,\ X_4) = \{(c,0,0,0) \in \mathbb{A}^4 : c \in \mathbb{C}\}$ and $L' := V_{\mathbb{A}^4}(X_1 - X_2,\ X_3,\ X_4) = \{(c,c,0,0) \in \mathbb{A}^4 : c \in \mathbb{C}\}$ are both contained in V: our aim here is to use 3.3.3 to show that the subvariety L of V cannot be 'defined by one equation', that is, there does not exist a polynomial $f \in \mathbb{C}[X_1, X_2, X_3, X_4]$ such that

$$L = \{p \in V : f(x_1, x_2, x_3, x_4)(p) = 0\}.$$

Our argument uses the morphism of varieties $\alpha : \mathbb{A}^2 \to V$ of 2.3.7 for which $\alpha((c,d)) = (c, cd, d(d-1), d^2(d-1))$ for all $(c,d) \in \mathbb{A}^2$; recall that $\alpha\lceil :$ $\mathbb{A}^2 \setminus \{(0,0),(0,1)\} \longrightarrow V \setminus \{(0,0,0,0)\}$ is an isomorphism of (quasi-affine) varieties. Set $\overline{L} := \{(c,0) \in \mathbb{A}^2 : c \in \mathbb{C}\}$ and $\overline{L'} := \{(c,1) \in \mathbb{A}^2 : c \in \mathbb{C}\}$. Note that $\alpha(\overline{L}) = L$ and $\alpha(\overline{L'}) = L'$.

Write $\mathcal{O}(\mathbb{A}^2) = \mathbb{C}[X, Y]$; note that $\overline{L} = V_{\mathbb{A}^2}(Y)$ and that $(0,1) \notin \overline{L}$. It therefore follows from 2.3.11 that the restriction map

$$\mathcal{O}(\mathbb{A}^2 \setminus \overline{L}) \longrightarrow \mathcal{O}(\mathbb{A}^2 \setminus (\{(0,1)\} \cup \overline{L}))$$

is surjective. Now $\overline{L'} \setminus \{(0,1)\} \subseteq \mathbb{A}^2 \setminus (\{(0,1)\} \cup \overline{L})$: our immediate aim is to show that the restriction map $\mathcal{O}(\mathbb{A}^2 \setminus (\{(0,1)\} \cup \overline{L})) \longrightarrow \mathcal{O}(\overline{L'} \setminus \{(0,1)\})$ is not surjective.

Set $x := X \lceil_{\overline{L'}}$ and consider the regular function $\gamma : \overline{L'} \setminus \{(0,1)\} \to \mathbb{C}$ defined by $\gamma(v) = x(v)^{-1}$ for all $v \in \overline{L'} \setminus \{(0,1)\} = U_{\overline{L'}}(x)$. If γ could be extended to a regular function on $\mathbb{A}^2 \setminus (\{(0,1)\} \cup \overline{L})$, then it would follow from the consequence of 2.3.11 noted in the previous paragraph that γ could be extended to a regular function $\gamma' : \mathbb{A}^2 \setminus \overline{L} \to \mathbb{C}$. Let γ'' be the restriction of γ' to $\overline{L'}$, a subset of $\mathbb{A}^2 \setminus \overline{L}$. It then follows that, for all $v \in U_{\overline{L'}}(x)$, we have $\gamma''(v)x(v) = 1$. However, $U_{\overline{L'}}(x)$ is a dense open subset of $\overline{L'}$, and so

$\gamma''(v)x(v) = 1$ for all $v \in \overline{L'}$. As this implies that $0 = 1$, we have obtained a contradiction! Thus the restriction map

$$\mathcal{O}(\mathbb{A}^2 \setminus (\{(0,1)\} \cup \overline{L})) \longrightarrow \mathcal{O}(\overline{L'} \setminus \{(0,1)\})$$

is not surjective, as claimed.

Next, $\alpha(\mathbb{A}^2 \setminus (\{(0,1)\} \cup \overline{L})) = V \setminus L$ and

$$\alpha(\overline{L'} \setminus \{(0,1)\}) = L' \setminus \{(0,0,0,0)\}\,.$$

There is therefore a commutative diagram

$$
\begin{array}{ccc}
\mathcal{O}(V \setminus L) & \longrightarrow & \mathcal{O}(L' \setminus \{(0,0,0,0)\}) \\
\cong \downarrow & & \downarrow \cong \\
\mathcal{O}(\mathbb{A}^2 \setminus (\{(0,1)\} \cup \overline{L})) & \longrightarrow & \mathcal{O}(\overline{L'} \setminus \{(0,1)\})
\end{array}
$$

in which the vertical isomorphisms are induced by (the appropriate restrictions of) α. Therefore the map $\sigma : \mathcal{O}(V \setminus L) \longrightarrow \mathcal{O}(L' \setminus \{(0,0,0,0)\})$ given by restriction is not surjective.

Let $\iota : L' \longrightarrow V$ be the inclusion morphism of varieties, and let $\mathfrak{c}$ denote the vanishing ideal $\{f \in \mathcal{O}(V) : f(v) = 0 \text{ for all } v \in L\}$ of L in $\mathcal{O}(V)$. With the notation of 2.3.2 and 2.3.4, we have

$$V(\mathfrak{c}\mathcal{O}(L')) = \iota^{-1}(V(\mathfrak{c})) = \iota^{-1}(L) = L \cap L' = \{(0,0,0,0)\} \subset L',$$

and so $\mathfrak{c}\mathcal{O}(L') \neq 0$. Note also that the restriction map σ of the preceding paragraph is just $\iota\lceil^* : \mathcal{O}(V \setminus V(\mathfrak{c})) \longrightarrow \mathcal{O}(L' \setminus V(\mathfrak{c}\mathcal{O}(L')))$. Therefore $\iota\lceil^*$ is not surjective, and so it follows from 2.3.4 that $D_{\mathfrak{c}}(\iota^*) : D_{\mathfrak{c}}(\mathcal{O}(V)) \longrightarrow D_{\mathfrak{c}}(\mathcal{O}(L'))$ is not surjective. Hence, if $\mathfrak{b} := \operatorname{Ker}(\iota^*)$, we have $\mathcal{R}^1 D_{\mathfrak{c}}(\mathfrak{b}) \neq 0$, so that $H^2_{\mathfrak{c}}(\mathfrak{b}) \neq 0$ by 2.2.6(ii). It therefore follows from 3.3.3 that $\operatorname{ara}(\mathfrak{c}) \geq 2$. Thus the subvariety L of V cannot be 'defined by one equation'.

3.3.6 Exercise. Show that, over the ring $R[X_1, X_2, X_3, X_4, X_5, X_6]$ of polynomials in six indeterminates with coefficients in R,

$$H^4_{(X_1,X_2,X_3)(X_4,X_5,X_6)}(R[X_1, X_2, X_3, X_4, X_5, X_6]) = 0.$$

3.3.7 Exercise. Show that, over the polynomial ring $R[X_1, X_2, X_3]$,

$$H^3_{(X_1^2, X_1 X_2 + X_2^3, X_2^4)}(R[X_1, X_2, X_3]) = 0.$$

3.4 Direct limits

Our second illustration of the use of the Mayer–Vietoris sequence is in our proof that the local cohomology functors 'commute with the formation of direct limits'. This is an important property of local cohomology, which plays a significant rôle in Grothendieck's development in [25]. It should be noted that our approach below avoids the use of an injective resolution of a direct system of R-modules and R-homomorphisms in the category formed by such direct systems. It is possible to approach this theory in a different way, based on the interchange of the order of direct limits; however, we have chosen the approach below because we consider it to be more illuminating.

3.4.1 Terminology. Let $(\Lambda, \leq)$ be a (non-empty) directed partially ordered set, and let $(W_\alpha)_{\alpha \in \Lambda}$ be a direct system of R-modules over Λ, with constituent R-homomorphisms $h_\beta^\alpha : W_\beta \to W_\alpha$ (for each $(\alpha, \beta) \in \Lambda \times \Lambda$ with $\alpha \geq \beta$). Set $W_\infty := \varinjlim_{\alpha \in \Lambda} W_\alpha$, and let $h_\alpha : W_\alpha \to W_\infty$ be the canonical map (for each $\alpha \in \Lambda$).

Let $T : \mathcal{C}(R) \to \mathcal{C}(R)$ be a covariant functor. It is immediate from the definition of functor that the $T(h_\beta^\alpha)$ turn the family $(T(W_\alpha))_{\alpha \in \Lambda}$ into a direct system of R-modules and R-homomorphisms over Λ. Also, whenever $\alpha, \beta \in \Lambda$ with $\alpha \geq \beta$, the commutative diagram

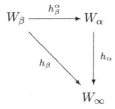

induces the commutative diagram

$$
\begin{array}{ccc}
T(W_\beta) & \xrightarrow{\;T(h_\beta^\alpha)\;} & T(W_\alpha) \\
 & T(h_\beta) \searrow & \big\downarrow T(h_\alpha) \\
 & & T(W_\infty)\,,
\end{array}
$$

and so there is induced an R-homomorphism

$$
\omega_T : \varinjlim_{\alpha \in \Lambda} T(W_\alpha) \longrightarrow T\left(\varinjlim_{\alpha \in \Lambda} W_\alpha \right) = T(W_\infty).
$$

If, for all choices of such a directed set Λ and such a direct system $(W_\alpha)_{\alpha \in \Lambda}$ over Λ, the map ω_T is an isomorphism, then we shall say that T *commutes with the formation of direct limits* or, more loosely, T *commutes with direct limits*.

In the next exercise, we again use the notation M_a, for an R-module M and $a \in R$, to denote the module of fractions of M with respect to the multiplicatively closed subset $\{a^i : i \in \mathbb{N}_0\}$. In fact, we consider the functor $(\bullet)_a : \mathcal{C}(R) \longrightarrow \mathcal{C}(R)$.

3.4.2 ♯Exercise. Let $a \in R$. Show that the functor $(\bullet)_a : \mathcal{C}(R) \longrightarrow \mathcal{C}(R)$ commutes with direct limits.

The next exercise gives another example, again needed later in the book, of a functor which commutes with direct limits.

3.4.3 ♯Exercise. Let $(\Lambda, \leq)$ be a (non-empty) directed partially ordered set, and let $(W_\alpha)_{\alpha \in \Lambda}$ be a direct system of R-modules over Λ, with constituent R-homomorphisms $h_\beta^\alpha : W_\beta \to W_\alpha$ (for each $\alpha, \beta \in \Lambda$ with $\alpha \geq \beta$). Set $W_\infty := \varinjlim_{\alpha \in \Lambda} W_\alpha$.

(i) Show that $\left(\varinjlim_{\alpha \in \Lambda} \mathrm{Ext}_R^i(\bullet, W_\alpha) \right)_{i \in \mathbb{N}_0}$ can be made into a negative strongly connected sequence of contravariant functors from $\mathcal{C}(R)$ to $\mathcal{C}(R)$ in such a way that the natural homomorphisms (of 3.4.1) give rise to a homomorphism

$$\Psi = (\psi^i)_{i \in \mathbb{N}_0} : \left(\varinjlim_{\alpha \in \Lambda} \mathrm{Ext}_R^i(\bullet, W_\alpha) \right)_{i \in \mathbb{N}_0} \longrightarrow \left(\mathrm{Ext}_R^i(\bullet, W_\infty) \right)_{i \in \mathbb{N}_0}$$

of connected sequences.
(ii) Prove that ψ_R^0 is an isomorphism, and deduce that ψ_F^0 is an isomorphism whenever F is a finitely generated free R-module.
(iii) Deduce that ψ_M^0 is an isomorphism whenever M is a finitely generated R-module.
(iv) Prove, by induction on i, that, for each $i \in \mathbb{N}_0$, the homomorphism ψ_M^i is an isomorphism whenever M is a finitely generated R-module.
(v) Deduce that, whenever M is a finitely generated R-module, the functor $\mathrm{Ext}_R^i(M, \bullet)$ commutes with direct limits, for each $i \in \mathbb{N}_0$.

Our main aim during the remainder of this chapter is to show that the local cohomology functors $H_{\mathfrak{a}}^i$ ($i \in \mathbb{N}_0$) commute with direct limits; we show first that the $\mathfrak{a}$-torsion functor commutes with direct limits.

3.4.4 Proposition. *The $\mathfrak{a}$-torsion functor $\Gamma_\mathfrak{a}$ commutes with direct limits.*

Proof. We use the notation of 3.4.1, so that $(\Lambda, \leq)$ is a (non-empty) directed partially ordered set, and $(W_\alpha)_{\alpha \in \Lambda}$ is a direct system of R-modules over Λ, with constituent R-homomorphisms $h_\beta^\alpha : W_\beta \to W_\alpha$ (for each $(\alpha, \beta) \in \Lambda \times \Lambda$ with $\alpha \geq \beta$). We must show that the R-homomorphism

$$\omega_{\Gamma_\mathfrak{a}} : \varinjlim_{\alpha \in \Lambda} \Gamma_\mathfrak{a}(W_\alpha) \longrightarrow \Gamma_\mathfrak{a}\left(\varinjlim_{\alpha \in \Lambda} W_\alpha\right)$$

is an isomorphism. We write $F_\infty := \varinjlim_{\alpha \in \Lambda} \Gamma_\mathfrak{a}(W_\alpha)$ and $W_\infty := \varinjlim_{\alpha \in \Lambda} W_\alpha$, and, for $\alpha \in \Lambda$, we use $f_\alpha : \Gamma_\mathfrak{a}(W_\alpha) \to F_\infty$ to denote the canonical map.

A typical element of $\operatorname{Ker} \omega_{\Gamma_\mathfrak{a}}$ can be expressed as $f_\alpha(w_\alpha)$ for some $\alpha \in \Lambda$ and $w_\alpha \in \Gamma_\mathfrak{a}(W_\alpha)$, where $\Gamma_\mathfrak{a}(h_\alpha)(w_\alpha) = 0$ in $\Gamma_\mathfrak{a}(W_\infty)$. Thus $h_\alpha(w_\alpha) = 0$ in W_∞, and so there exists $\gamma \in \Lambda$ with $\gamma \geq \alpha$ such that $h_\alpha^\gamma(w_\alpha) = 0$. Since $w_\alpha \in \Gamma_\mathfrak{a}(W_\alpha)$, this means that $\Gamma_\mathfrak{a}(h_\alpha^\gamma)(w_\alpha) = 0$ in $\Gamma_\mathfrak{a}(W_\gamma)$, and so $f_\alpha(w_\alpha) = f_\gamma\left(\Gamma_\mathfrak{a}(h_\alpha^\gamma)(w_\alpha)\right) = 0$. Hence $\omega_{\Gamma_\mathfrak{a}}$ is injective.

We now show that $\omega_{\Gamma_\mathfrak{a}}$ is surjective. Let $y \in \Gamma_\mathfrak{a}(W_\infty)$. There exists $\alpha \in \Lambda$ and $w_\alpha \in W_\alpha$ such that $y = h_\alpha(w_\alpha)$. We know that y is annihilated by a power of $\mathfrak{a}$: let $j \in \mathbb{N}$ be such that $\mathfrak{a}^j y = 0$. Let $\mathfrak{a}^j$ be generated by $r_1, \ldots, r_n$. For each $i = 1, \ldots, n$, we have $h_\alpha(r_i w_\alpha) = 0$, so that there exists $\beta_i \in \Lambda$ with $\beta_i \geq \alpha$ such that $h_\alpha^{\beta_i}(r_i w_\alpha) = 0$. Since Λ is directed, there exists $\gamma \in \Lambda$ such that $\gamma \geq \beta_i$ for all $i = 1, \ldots, n$. We now have

$$r_i h_\alpha^\gamma(w_\alpha) = r_i h_{\beta_i}^\gamma(h_\alpha^{\beta_i}(w_\alpha)) = h_{\beta_i}^\gamma(h_\alpha^{\beta_i}(r_i w_\alpha)) = 0$$

for all $i = 1, \ldots, n$, so that $\mathfrak{a}^j h_\alpha^\gamma(w_\alpha) = 0$ and $h_\alpha^\gamma(w_\alpha) \in \Gamma_\mathfrak{a}(W_\gamma)$. But

$$y = h_\alpha(w_\alpha) = h_\gamma(h_\alpha^\gamma(w_\alpha)) = \Gamma_\mathfrak{a}(h_\gamma)(h_\alpha^\gamma(w_\alpha)) \in \operatorname{Im}(\Gamma_\mathfrak{a}(h_\gamma)).$$

Set $z_\gamma := h_\alpha^\gamma(w_\alpha)$, an element of $\Gamma_\mathfrak{a}(W_\gamma)$. Now $f_\gamma(z_\gamma) \in F_\infty$, and

$$\omega_{\Gamma_\mathfrak{a}}(f_\gamma(z_\gamma)) = \Gamma_\mathfrak{a}(h_\gamma)(z_\gamma) = y.$$

Hence $\omega_{\Gamma_\mathfrak{a}}$ is surjective and the proof is complete. $\qquad\square$

3.4.5 ♯Exercise. Let $T : \mathcal{C}(R) \to \mathcal{C}(R)$ be a covariant, additive functor which commutes with direct limits. Let $(L_\theta)_{\theta \in \Omega}$ be a non-empty family of R-modules; for each $\phi \in \Omega$, let $q_\phi : L_\phi \longrightarrow \bigoplus_{\theta \in \Omega} L_\theta$ be the canonical injection.

Prove that $T\left(\bigoplus_{\theta \in \Omega} L_\theta\right)$ is the direct sum of its family of submodules $(\operatorname{Im} T(q_\theta))_{\theta \in \Omega}$, so that

$$T\left(\bigoplus_{\theta \in \Omega} L_\theta\right) = \bigoplus_{\theta \in \Omega} \operatorname{Im} T(q_\theta) \cong \bigoplus_{\theta \in \Omega} T(L_\theta).$$

We shall describe this result by the statement that T *commutes with direct sums.*

To show that the i-th right derived functors of $\Gamma_\mathfrak{a}$ (for $i \in \mathbb{N}$) also commute with direct limits, we plan to argue by induction on the number of elements needed to generate $\mathfrak{a}$, and to use the Mayer–Vietoris sequence to complete the inductive step. The following lemma will be helpful.

3.4.6 Lemma. *Let* Q, S, T, U, V *be covariant functors (from* $\mathcal{C}(R)$ *to* $\mathcal{C}(R)$*) and let* $\sigma : Q \longrightarrow S$, $\tau : S \longrightarrow T$, $\mu : T \longrightarrow U$, $\nu : U \longrightarrow V$ *be natural transformations of functors such that, for each R-module M, the sequence*

$$Q(M) \xrightarrow{\sigma_M} S(M) \xrightarrow{\tau_M} T(M) \xrightarrow{\mu_M} U(M) \xrightarrow{\nu_M} V(M)$$

is exact. Suppose that Q, S, U, V *all commute with direct limits. Then* T *also commutes with direct limits.*

Proof. We again use the notation of 3.4.1. For each $\alpha, \beta \in \Lambda$ with $\alpha \geq \beta$, there is a commutative diagram

$$
\begin{array}{ccccccccc}
Q(W_\beta) & \xrightarrow{\sigma_{W_\beta}} & S(W_\beta) & \xrightarrow{\tau_{W_\beta}} & T(W_\beta) & \xrightarrow{\mu_{W_\beta}} & U(W_\beta) & \xrightarrow{\nu_{W_\beta}} & V(W_\beta) \\
\downarrow{\scriptstyle Q(h_\beta^\alpha)} & & \downarrow{\scriptstyle S(h_\beta^\alpha)} & & \downarrow{\scriptstyle T(h_\beta^\alpha)} & & \downarrow{\scriptstyle U(h_\beta^\alpha)} & & \downarrow{\scriptstyle V(h_\beta^\alpha)} \\
Q(W_\alpha) & \xrightarrow{\sigma_{W_\alpha}} & S(W_\alpha) & \xrightarrow{\tau_{W_\alpha}} & T(W_\alpha) & \xrightarrow{\mu_{W_\alpha}} & U(W_\alpha) & \xrightarrow{\nu_{W_\alpha}} & V(W_\alpha)
\end{array}
$$

with exact rows, and so there is induced an exact sequence

$$Q_\infty \xrightarrow{\sigma_\infty} S_\infty \xrightarrow{\tau_\infty} T_\infty \xrightarrow{\mu_\infty} U_\infty \xrightarrow{\nu_\infty} V_\infty,$$

where $Q_\infty := \varinjlim_{\alpha \in \Lambda} Q(W_\alpha)$, etc., and $\sigma_\infty := \varinjlim_{\alpha \in \Lambda} \sigma_{W_\alpha}$, etc. It is easy to see from the definition of $\omega_Q, \dots, \omega_V$ in 3.4.1 that the diagram

$$
\begin{array}{ccccccccc}
Q_\infty & \xrightarrow{\sigma_\infty} & S_\infty & \xrightarrow{\tau_\infty} & T_\infty & \xrightarrow{\mu_\infty} & U_\infty & \xrightarrow{\nu_\infty} & V_\infty \\
\downarrow{\scriptstyle \omega_Q} & & \downarrow{\scriptstyle \omega_S} & & \downarrow{\scriptstyle \omega_T} & & \downarrow{\scriptstyle \omega_U} & & \downarrow{\scriptstyle \omega_V} \\
Q(W_\infty) & \xrightarrow{\sigma_{W_\infty}} & S(W_\infty) & \xrightarrow{\tau_{W_\infty}} & T(W_\infty) & \xrightarrow{\mu_{W_\infty}} & U(W_\infty) & \xrightarrow{\nu_{W_\infty}} & V(W_\infty)
\end{array}
$$

commutes. Since this diagram has exact rows and $\omega_Q, \omega_S, \omega_U$ and ω_V are isomorphisms by hypothesis, it follows from the Five Lemma that ω_T is an isomorphism too. $\square$

3.4.7 Remark. It is clear that the identity functor Id : $\mathcal{C}(R) \to \mathcal{C}(R)$ and the zero functor $0 : \mathcal{C}(R) \to \mathcal{C}(R)$ (for which $0(M) = 0$ for all R-modules M) commute with direct limits.

3.4.8 Corollary. *Suppose that, for some $i \in \mathbb{N}_0$, the functors $H_{\mathfrak{a}}^i$ and $H_{\mathfrak{b}}^i$ commute with direct limits. Then $H_{\mathfrak{a}}^i(\bullet) \oplus H_{\mathfrak{b}}^i(\bullet)$ commutes with direct limits.*

Proof. This is an easy consequence of 3.4.6 and 3.4.7: for each R-module M, consider the canonical split exact sequences

$$0 \rightleftarrows H_{\mathfrak{a}}^i(M) \rightleftarrows H_{\mathfrak{a}}^i(M) \oplus H_{\mathfrak{b}}^i(M) \rightleftarrows H_{\mathfrak{b}}^i(M) \rightleftarrows 0. \ \square$$

3.4.9 Proposition. *Let $a \in R$. Then the local cohomology functor H_{Ra}^1 commutes with direct limits.*

Proof. By 2.2.20, there are natural transformations of functors

$$\tau : \mathrm{Id} \longrightarrow (\bullet)_a \quad \text{and} \quad \sigma : (\bullet)_a \longrightarrow H_{Ra}^1$$

such that, for all R-modules M, the sequence

$$M \xrightarrow{\tau_M} M_a \xrightarrow{\sigma_M} H_{Ra}^1(M) \longrightarrow 0 \longrightarrow 0$$

is exact. We propose to use 3.4.6, and with this in mind we observe that, by 3.4.2, the functor $(\bullet)_a : \mathcal{C}(R) \longrightarrow \mathcal{C}(R)$ commutes with direct limits. (This will be obvious to any reader familiar with the facts that $(\bullet)_a$ is naturally equivalent to $(\bullet) \otimes_R R_a : \mathcal{C}(R) \longrightarrow \mathcal{C}(R)$ and that, for an arbitrary R-module N, the functor $(\bullet) \otimes_R N : \mathcal{C}(R) \longrightarrow \mathcal{C}(R)$ commutes with direct limits. Alternatively, it is easy to prove directly.)

The claim now follows immediately from 3.4.6 and 3.4.7. $\square$

We are now able to prove that the local cohomology functors $H_{\mathfrak{a}}^i$ ($i \in \mathbb{N}_0$) all commute with direct limits.

3.4.10 Theorem. *For all $i \in \mathbb{N}_0$, the local cohomology functor $H_{\mathfrak{a}}^i$ commutes with direct limits.*

Proof. This proof is another example of the use of induction outlined in the introduction to this chapter.

We suppose that $\mathfrak{a}$ can be generated by t elements and proceed by induction on t. When $t = 0$, we have $\mathfrak{a} = 0$ and $\Gamma_{\mathfrak{a}}$ is the identity functor, so that $H_{\mathfrak{a}}^i = 0$ for all $i \in \mathbb{N}$. The result is therefore immediate from 3.4.7 in this case.

We proved in 3.4.4 that $\Gamma_{\mathfrak{a}}$ commutes with direct limits. When $t = 1$, it follows from 3.3.1 that $H_{\mathfrak{a}}^i = 0$ for all $i > 1$, and from 3.4.9 that $H_{\mathfrak{a}}^1$ commutes with direct limits. The result is therefore also proved in the case when $t = 1$.

Now suppose, inductively, that $t > 1$ and the result has been proved for ideals that can be generated by fewer than t elements. Suppose that $\mathfrak{a}$ is generated by t elements $a_1, \ldots, a_t$. Set $\mathfrak{b} = Ra_1 + \cdots + Ra_{t-1}$ and $\mathfrak{c} = Ra_t$, so that $\mathfrak{a} = \mathfrak{b} + \mathfrak{c}$. Note that

$$\mathfrak{b}\mathfrak{c} = (Ra_1 + \cdots + Ra_{t-1})Ra_t = Ra_1a_t + \cdots + Ra_{t-1}a_t$$

can be generated by $t - 1$ elements, and that $H^i_{\mathfrak{b} \cap \mathfrak{c}} = H^i_{\mathfrak{b}\mathfrak{c}}$ for all $i \in \mathbb{N}_0$ by 3.1.5.

It therefore follows from the inductive hypothesis and what we have already proved that the local cohomology functors $H^i_{\mathfrak{b}}$ ($i \in \mathbb{N}_0$), $H^i_{\mathfrak{c}}$ ($i \in \mathbb{N}_0$) and $H^i_{\mathfrak{b} \cap \mathfrak{c}}$ ($i \in \mathbb{N}_0$) all commute with direct limits. Hence, by 3.4.8, all the functors $H^i_{\mathfrak{a}}(\,\bullet\,) \oplus H^i_{\mathfrak{b}}(\,\bullet\,)$ ($i \in \mathbb{N}_0$) commute with direct limits.

It now follows from 3.4.6 and the Mayer–Vietoris sequence (we use here the full statement of 3.2.3) that the functors $H^i_{\mathfrak{b} + \mathfrak{c}}$ ($i \in \mathbb{N}_0$) all commute with direct limits. Since $\mathfrak{a} = \mathfrak{b} + \mathfrak{c}$, the inductive step is complete. $\square$

3.4.11 Corollary. *The $\mathfrak{a}$-transform functor $D_\mathfrak{a}$ commutes with direct limits.*

Consequently, by 3.4.5, the functor $D_\mathfrak{a}$ commutes with direct sums (in the sense of Exercise 3.4.5).

Proof. By 2.2.6(i), there are natural transformations of functors

$$\xi : \Gamma_\mathfrak{a} \longrightarrow \mathrm{Id}, \quad \eta : \mathrm{Id} \longrightarrow D_\mathfrak{a}, \quad \zeta^0 : D_\mathfrak{a} \longrightarrow H^1_\mathfrak{a}$$

such that, for each R-module M, the sequence

$$\Gamma_\mathfrak{a}(M) \xrightarrow{\xi_M} M \xrightarrow{\eta_M} D_\mathfrak{a}(M) \xrightarrow{\zeta^0_M} H^1_\mathfrak{a}(M) \longrightarrow 0$$

is exact. Since $\Gamma_\mathfrak{a}$ and $H^1_\mathfrak{a}$ commute with direct limits (by 3.4.4 and 3.4.10 respectively), the claim follows from 3.4.6. $\square$

3.4.12 Exercise. Suppose that, for some $i \in \mathbb{N}_0$, we have $H^i_\mathfrak{a}(R) = 0$. Show that $H^i_\mathfrak{a}(P) = 0$ for every projective R-module P.

3.4.13 Exercise. Let K' denote the full ring of fractions of R; that is, $K' := S^{-1}R$, where S is the set of all non-zerodivisors of R. Show that, if, for some integer $i \in \mathbb{N}_0$, we have $H^i_\mathfrak{a}(R) = 0$, then $H^i_\mathfrak{a}(K') = 0$ too.

Thus, in particular, if R is an integral domain with field of fractions K and $H^i_\mathfrak{a}(R) = 0$ for some $i \in \mathbb{N}_0$, then $H^i_\mathfrak{a}(K) = 0$.

4

Change of rings

The main results of this chapter concern a homomorphism of commutative Noetherian rings $f : R \longrightarrow R'$. More precisely, we shall prove two fundamental comparison results for local cohomology modules in this context. The first of these, which we shall call the 'Independence Theorem', compares, for an R'-module M' and an $i \in \mathbb{N}_0$, the local cohomology modules $H_{\mathfrak{a}}^i(M')$ and $H_{\mathfrak{a}R'}^i(M')$: to form the first of these, we consider M' as an R-module by restriction of scalars using f; also, $\mathfrak{a}R'$ denotes the extension of $\mathfrak{a}$ to R' under f. Our second main result, which we shall refer to as the 'Flat Base Change Theorem', compares the local cohomology modules $H_{\mathfrak{a}}^i(M) \otimes_R R'$ and $H_{\mathfrak{a}R'}^i(M \otimes_R R')$ for $i \in \mathbb{N}_0$ and an arbitrary R-module M under the additional assumption that the ring homomorphism f is flat.

Our main results rely on the fact that certain modules are acyclic with respect to torsion functors. We say that an R-module A is $\Gamma_{\mathfrak{a}}$-*acyclic* precisely when $H_{\mathfrak{a}}^i(A) = 0$ for all $i > 0$. As was explained in 1.2.2, the most basic method for calculation, for an R-module M and an $i \in \mathbb{N}_0$, of $H_{\mathfrak{a}}^i(M)$ is to take an injective resolution $I^{\bullet}$ of M, apply $\Gamma_{\mathfrak{a}}$ to $I^{\bullet}$ to obtain the complex $\Gamma_{\mathfrak{a}}(I^{\bullet})$, and take the i-th cohomology module of this complex: we have $H_{\mathfrak{a}}^i(M) = H^i(\Gamma_{\mathfrak{a}}(I^{\bullet}))$. However, it is an easy exercise to show that a resolution of M by $\Gamma_{\mathfrak{a}}$-acyclic R-modules will serve this purpose just as well.

Of course, injective R-modules are $\Gamma_{\mathfrak{a}}$-acyclic, but in general the class of $\Gamma_{\mathfrak{a}}$-acyclic R-modules is larger than the class of injective R-modules. Let $f : R \longrightarrow R'$ be a homomorphism of commutative Noetherian rings. We shall show that, if I' is an injective R'-module, then, when viewed as an R-module by restriction of scalars, I' is $\Gamma_{\mathfrak{a}}$-acyclic. We shall also show that, if the ring homomorphism f is flat, then, for each injective R-module I, the R'-module $I \otimes_R R'$ is $\Gamma_{\mathfrak{a}R'}$-acyclic. These results pave the way for the proofs later in the chapter of the Independence Theorem (4.2.1) and the Flat Base Change Theorem (4.3.2).

4.1 Some acyclic modules

For completeness, we begin with the formal definition.

4.1.1 Definition. We say that an R-module A is $\Gamma_{\mathfrak{a}}$-*acyclic* precisely when $H_{\mathfrak{a}}^i(A) = 0$ for all $i > 0$. If $\mathfrak{B}$ is a system of ideals (of R) over Λ as in 2.1.10, then we say that A is $\Gamma_{\mathfrak{B}}$-*acyclic* if and only if $H_{\mathfrak{B}}^i(A) = 0$ for all $i > 0$.

Our presentation does not depend on the following exercise, but the reader should be aware of the result it contains.

4.1.2 Exercise. Show that, given an R-module M, its local cohomology modules with respect to $\mathfrak{a}$ can be calculated by means of a resolution of M by $\Gamma_{\mathfrak{a}}$-acyclic modules as follows. Let

$$A^{\bullet} : 0 \xrightarrow{d^{-1}} A^0 \xrightarrow{d^0} A^1 \longrightarrow \cdots \longrightarrow A^i \xrightarrow{d^i} A^{i+1} \longrightarrow \cdots$$

be a $\Gamma_{\mathfrak{a}}$-acyclic resolution of M, so that $A^0, A^1, \ldots, A^i, \ldots$ are all $\Gamma_{\mathfrak{a}}$-acyclic R-modules and there is an R-homomorphism $\alpha : M \to A^0$ such that the sequence

$$0 \longrightarrow M \xrightarrow{\alpha} A^0 \xrightarrow{d^0} A^1 \longrightarrow \cdots \longrightarrow A^i \xrightarrow{d^i} A^{i+1} \longrightarrow \cdots$$

is exact. Show that

$$H_{\mathfrak{a}}^i(M) \cong \operatorname{Ker}(\Gamma_{\mathfrak{a}}(d^i)) / \operatorname{Im}(\Gamma_{\mathfrak{a}}(d^{i-1})) \quad \text{for all } i \in \mathbb{N}_0.$$

We shall have two applications of our first proposition in this section.

4.1.3 Proposition. *Let M be an R-module, and let $\mathfrak{C}$ be a set of ideals of R such that*

  (a) $\mathfrak{C}$ *is closed under the formation of finite sums and products,*
  (b) $0R \in \mathfrak{C}$, *and*
  (c) *each ideal in $\mathfrak{C}$ is the sum of finitely many principal ideals which belong to $\mathfrak{C}$.*

Assume that $H_{\mathfrak{c}}^1(M) = 0$ for all $\mathfrak{c} \in \mathfrak{C}$. Then M is $\Gamma_{\mathfrak{c}}$-acyclic for all $\mathfrak{c} \in \mathfrak{C}$.

Remark. Of course, if M is $\Gamma_{\mathfrak{c}}$-acyclic for all $\mathfrak{c} \in \mathfrak{C}$, then it is automatic that $H_{\mathfrak{c}}^1(M) = 0$ for all $\mathfrak{c} \in \mathfrak{C}$. Thus this theorem could be phrased as an 'if and only if' result.

Proof. Since Γ_{0R} is the identity functor on $\mathcal{C}(R)$, it is clear that M is Γ_{0R}-acyclic. Also, it was proved in 2.2.21(ii) that, for $a \in R$, we have $H_{Ra}^i(M) = 0$ for all $i \in \mathbb{N}$ with $i > 1$; hence M is $\Gamma_{\mathfrak{c}}$-acyclic whenever $\mathfrak{c}$ is a principal ideal in $\mathfrak{C}$.

Now suppose, inductively, that $t > 1$ and we have proved that, whenever $\mathfrak{c}' \in \mathfrak{C}$ can be expressed as a sum of fewer than t principal ideals which all belong to $\mathfrak{C}$, then M is $\Gamma_{\mathfrak{c}'}$-acyclic. This is certainly the case when $t = 2$. Let $\mathfrak{c} = Rc_1 + \cdots + Rc_t$, where $c_1, \ldots, c_t \in R$ and $Rc_1, \ldots, Rc_t \in \mathfrak{C}$. We shall use the Mayer–Vietoris sequence to show that M is also $\Gamma_{\mathfrak{c}}$-acyclic.

Set $\mathfrak{a} = Rc_1 + \cdots + Rc_{t-1}$ and $\mathfrak{b} = Rc_t$, so that $\mathfrak{c} = \mathfrak{a} + \mathfrak{b}$. Note that $\mathfrak{b} \in \mathfrak{C}$, and, by hypothesis (a) on $\mathfrak{C}$, each of $\mathfrak{a}, \mathfrak{ab}, Rc_1c_t, \ldots, Rc_{t-1}c_t$ belongs to $\mathfrak{C}$. Since $\mathfrak{b}$ is principal, and $\mathfrak{a}$ and

$$\mathfrak{ab} = (Rc_1 + \cdots + Rc_{t-1})Rc_t = Rc_1c_t + \cdots + Rc_{t-1}c_t$$

are both expressible as sums of $t - 1$ principal ideals belonging to $\mathfrak{C}$, it follows from 2.2.21(ii) and the inductive hypothesis that M is $\Gamma_{\mathfrak{a}}$-acyclic, $\Gamma_{\mathfrak{b}}$-acyclic and $\Gamma_{\mathfrak{ab}}$-acyclic.

By the Mayer–Vietoris sequence 3.2.3, we have, for an arbitrary $i > 1$, an exact sequence $H^{i-1}_{\mathfrak{a} \cap \mathfrak{b}}(M) \longrightarrow H^i_{\mathfrak{c}}(M) \longrightarrow H^i_{\mathfrak{a}}(M) \oplus H^i_{\mathfrak{b}}(M)$. Now $H^{i-1}_{\mathfrak{a} \cap \mathfrak{b}}(M) = H^{i-1}_{\mathfrak{ab}}(M)$ by 3.1.5, and so, by what we have already established, $H^{i-1}_{\mathfrak{a} \cap \mathfrak{b}}(M) = H^i_{\mathfrak{a}}(M) = H^i_{\mathfrak{b}}(M) = 0$. Thus $H^i_{\mathfrak{c}}(M) = 0$ for all $i > 1$. Since $H^1_{\mathfrak{c}}(M) = 0$ by hypothesis (because $\mathfrak{c} \in \mathfrak{C}$), this completes the inductive step.

Since each ideal in $\mathfrak{C}$ can be expressed as the sum of finitely many principal ideals which belong to $\mathfrak{C}$, the proof is complete. $\square$

The two important applications of Proposition 4.1.3 which we have in mind concern a homomorphism of commutative Noetherian rings, and it is appropriate for us to clarify notation at this point.

4.1.4 Notation. Throughout this chapter, R' will denote a second commutative Noetherian ring and $f : R \longrightarrow R'$ will denote a ring homomorphism. As in Chapter 2, for an ideal $\mathfrak{b}$ of R, we shall use $\mathfrak{b}R'$ to denote the extension of $\mathfrak{b}$ to R' under f.

For an R-module M, we shall use $\mathrm{Zdv}_R(M)$ (or $\mathrm{Zdv}(M)$ if there is no ambiguity about the underlying ring involved) to denote the set of elements of R which are zerodivisors on M. Thus, for example, if M' is an R'-module, then $R \setminus \mathrm{Zdv}_R(M')$ denotes the set of elements of R which are non-zerodivisors on M' when the latter is regarded as an R-module by means of f.

4.1.5 Lemma. *Let M' be a finitely generated R'-module that is $\mathfrak{a}R'$-torsion-free. Then $\mathfrak{a} \not\subseteq \mathrm{Zdv}_R(M')$; in other words, $\mathfrak{a}$ contains a non-zerodivisor on M' when the latter is regarded as an R-module by means of f.*

Proof. Suppose that $\mathfrak{a} \cap (R \setminus \mathrm{Zdv}_R(M')) = \emptyset$ and look for a contradiction. Then $f(\mathfrak{a}) \subseteq \mathrm{Zdv}_{R'}(M') = \bigcup_{\mathfrak{P} \in \mathrm{Ass}_{R'} M'} \mathfrak{P}$. Since M' is a finitely generated R'-module, it follows from the Prime Avoidance Theorem in the form given in

[81, 3.61] that $f(\mathfrak{a}) \subseteq \mathfrak{P}$ for some $\mathfrak{P} \in \mathrm{Ass}_{R'}\, M'$. Since there exists $m' \in M'$ with $(0 :_{R'} m') = \mathfrak{P}$, we deduce that $0 \neq m' \in (0 :_{M'} \mathfrak{a}R')$. This is a contradiction. $\qquad\square$

We are now ready for the proof of one of the main results of this section.

4.1.6 Theorem. *Let I' be an injective R'-module. Then I', when viewed as an R-module by means of f, is $\Gamma_{\mathfrak{a}}$-acyclic.*

Proof. The strategy of this proof is to show first that $H_{\mathfrak{a}}^1(I') = 0$, and then appeal to Proposition 4.1.3.

By 2.2.24, there is an R-isomorphism $\varepsilon_{I'} : D_{\mathfrak{a}R'}(I') \longrightarrow D_{\mathfrak{a}}(I')$ such that the diagram

$$
\begin{array}{ccc}
I' & \xrightarrow{\;\eta_{\mathfrak{a}R',I'}\;} & D_{\mathfrak{a}R'}(I') \\
\Big\| & & \cong \Big\downarrow \varepsilon_{I'} \\
I' & \xrightarrow{\;\eta_{\mathfrak{a},I'}\;} & D_{\mathfrak{a}}(I')
\end{array}
$$

commutes. Since $H_{\mathfrak{a}R'}^1(I') = 0$, it follows from 2.2.6(i)(c) that $\eta_{\mathfrak{a}R',I'}$ is surjective. Hence $\eta_{\mathfrak{a},I'}$ is surjective, and so $H_{\mathfrak{a}}^1(I') = 0$, again by 2.2.6(i)(c).

It follows that $H_{\mathfrak{a}}^1(I') = 0$ for an *arbitrary* ideal $\mathfrak{a}$ of R. We can now apply 4.1.3 (with the set of all ideals of R as the set $\mathcal{C}$) to deduce that I' is $\Gamma_{\mathfrak{b}}$-acyclic for every ideal $\mathfrak{b}$ of R. $\qquad\square$

We now move on to consider the situation in which the ring homomorphism $f : R \to R'$ is flat, that is, such that R', when viewed as an R-module via f, is flat. We need two preliminary results for this situation; for the first, we appeal to [50] for a proof.

4.1.7 Lemma. (See [50, Theorem 7.11].) *Assume that the ring homomorphism $f : R \to R'$ is flat. Let M be an R-module. Then there is a natural transformation of functors*

$$
\mu : \mathrm{Hom}_R(\,\bullet\,, M) \otimes_R R' \longrightarrow \mathrm{Hom}_{R'}((\,\bullet\,) \otimes_R R', M \otimes_R R')
$$

(from $\mathcal{C}(R)$ to $\mathcal{C}(R')$) which is such that

(i) *$\mu_N(g \otimes r') = r'(g \otimes \mathrm{Id}_{R'})$ for each R-module N, each $r' \in R'$ and each $g \in \mathrm{Hom}_R(N, M)$, and*

(ii) *μ_N is an isomorphism if N is a finitely generated R-module.* $\qquad\square$

4.1.8 Proposition. *Assume that the ring homomorphism $f : R \to R'$ is flat; let I be an injective R-module. Then the natural R'-homomorphism*

$$\beta : I \otimes_R R' \longrightarrow \mathrm{Hom}_{R'}(\mathfrak{a}R', I \otimes_R R')$$

(for which $\beta(m')(r') = r'm'$ for all $r' \in \mathfrak{a}R'$ and all $m' \in I \otimes_R R'$) is surjective.

Proof. Since f is flat, the natural R'-epimorphism $\gamma : \mathfrak{a} \otimes_R R' \longrightarrow \mathfrak{a}R'$ (for which $\gamma(\sum_{i=1}^{t} a_i \otimes r'_i) = \sum_{i=1}^{t} f(a_i)r'_i$ for all $t \in \mathbb{N}$, $a_1, \ldots, a_t \in \mathfrak{a}$ and $r'_1, \ldots, r'_t \in R'$) is an isomorphism (see [50, Theorem 7.7]).

Since $\mathfrak{a}$ is a finitely generated R-module, the R'-homomorphism

$$\mu_{\mathfrak{a}} : \mathrm{Hom}_R(\mathfrak{a}, I) \otimes_R R' \longrightarrow \mathrm{Hom}_{R'}(\mathfrak{a} \otimes_R R', I \otimes_R R')$$

of 4.1.7 is an isomorphism. Since I is R-injective, the R-homomorphism $\sigma : I \to \mathrm{Hom}_R(\mathfrak{a}, I)$ (for which $\sigma(m)(a) = am$ for all $a \in \mathfrak{a}$ and all $m \in I$) is surjective, so that, since tensor product is right exact,

$$\sigma \otimes \mathrm{Id}_{R'} : I \otimes_R R' \longrightarrow \mathrm{Hom}_R(\mathfrak{a}, I) \otimes_R R'$$

is an R'-epimorphism. It follows that the composition

$$\mathrm{Hom}_{R'}(\gamma^{-1}, \mathrm{Id}_{I \otimes_R R'}) \circ \mu_{\mathfrak{a}} \circ (\sigma \otimes \mathrm{Id}_{R'}) : I \otimes_R R' \longrightarrow \mathrm{Hom}_{R'}(\mathfrak{a}R', I \otimes_R R')$$

is surjective. Since this composition is just β, the proof is complete.   □

We are now in a position to prove a theorem which has some similarities to Theorem 4.1.6.

4.1.9 Theorem. *Assume that the ring homomorphism $f : R \to R'$ is flat, and let I be an injective R-module. Then $I \otimes_R R'$ is $\Gamma_{\mathfrak{a}R'}$-acyclic.*

Proof. We shall employ a strategy similar to that which we used in the proof of 4.1.6: we shall show first that $H^1_{\mathfrak{a}R'}(I \otimes_R R') = 0$, and then we shall appeal to Proposition 4.1.3.

For each $n \in \mathbb{N}$, the natural R'-homomorphism

$$I \otimes_R R' \longrightarrow \mathrm{Hom}_{R'}(\mathfrak{a}^n R', I \otimes_R R')$$

of 4.1.8 is surjective, so that, since $\mathfrak{a}^n R' = (\mathfrak{a}R')^n$, the R'-homomorphism

$$\eta_{I \otimes_R R'} : I \otimes_R R' \longrightarrow D_{\mathfrak{a}R'}(I \otimes_R R') = \varinjlim_{n \in \mathbb{N}} \mathrm{Hom}_{R'}((\mathfrak{a}R')^n, I \otimes_R R')$$

of 2.2.6(i) is surjective. Hence $H^1_{\mathfrak{a}R'}(I \otimes_R R') = 0$, by 2.2.6(i)(c).

We can now apply 4.1.3 (with the set $\mathfrak{C}$ taken as the set of all ideals of R' which are extended from ideals of R) to deduce that $I \otimes_R R'$ is $\Gamma_{\mathfrak{b}R'}$-acyclic for every ideal $\mathfrak{b}$ of R.   □

Theorems 4.1.6 and 4.1.9 pave the way for us to present, in the next two sections, the two main results of this chapter.

4.2 The Independence Theorem

In this section, we prove the Independence Theorem, which shows, loosely speaking, that local cohomology is 'independent of the base ring'. We have already seen results of a similar type for ideal transforms in 2.2.24, and for first local cohomology modules in Exercise 2.2.26. Let us again use $\lceil R : \mathcal{C}(R') \to \mathcal{C}(R)$ to denote the functor obtained from restriction of scalars (using f): thus, if M' is an R'-module and $i \in \mathbb{N}_0$, we can form the R-modules $H^i_{\mathfrak{a}R'}(M')\lceil R$ and $H^i_{\mathfrak{a}}(M'\lceil R)$. For the first, we form the i-th local cohomology module of M' with respect to $\mathfrak{a}R'$ and then regard the resulting R'-module as an R-module by means of f; for the second, we first consider M' as an R-module via f, and then take the i-th local cohomology module with respect to $\mathfrak{a}$. Theorem 4.2.1 below shows, among other things, that there is an R-isomorphism

$$H^i_{\mathfrak{a}R'}(M') = H^i_{\mathfrak{a}R'}(M')\lceil R \xrightarrow{\cong} H^i_{\mathfrak{a}}(M'\lceil R) =: H^i_{\mathfrak{a}}(M'),$$

so that, speaking loosely, it does not matter whether we calculate these local cohomology modules over R or R'.

4.2.1 Independence Theorem. *The functors* $\Gamma_{\mathfrak{a}R'}(\bullet)\lceil R$ *and* $\Gamma_{\mathfrak{a}}(\bullet\lceil R)$ *(from* $\mathcal{C}(R')$ *to* $\mathcal{C}(R)$) *are the same, and there is a unique isomorphism*

$$\Lambda = (\lambda^i)_{i \in \mathbb{N}_0} : \left(H^i_{\mathfrak{a}R'}(\bullet)\lceil R\right)_{i \in \mathbb{N}_0} \xrightarrow{\cong} \left(H^i_{\mathfrak{a}}(\bullet\lceil R)\right)_{i \in \mathbb{N}_0}$$

of negative (strongly) connected sequences of covariant functors (from $\mathcal{C}(R')$ *to* $\mathcal{C}(R)$) *such that* λ^0 *is the identity natural equivalence. In particular, for each* $i \in \mathbb{N}_0$, *the functors* $H^i_{\mathfrak{a}R'}(\bullet)\lceil R$ *and* $H^i_{\mathfrak{a}}(\bullet\lceil R)$ *are naturally equivalent.*

Proof. Since $\lceil R : \mathcal{C}(R') \to \mathcal{C}(R)$ is covariant and exact, it is clear that

$$(H^i_{\mathfrak{a}R'}(\bullet)\lceil R)_{i \in \mathbb{N}_0} \quad \text{and} \quad (H^i_{\mathfrak{a}}(\bullet\lceil R))_{i \in \mathbb{N}_0}$$

are negative strongly connected sequences of covariant functors from $\mathcal{C}(R')$ to $\mathcal{C}(R)$.

Now $\mathfrak{a}^n R' = (\mathfrak{a}R')^n$ for all $n \in \mathbb{N}$, and so $\Gamma_{\mathfrak{a}R'}(\bullet)\lceil R$ and $\Gamma_{\mathfrak{a}}(\bullet\lceil R)$ are the same functor. Furthermore, whenever I' is an injective R'-module, it is, of course, automatic that $H^i_{\mathfrak{a}R'}(I') = 0$ for all $i \in \mathbb{N}$, while it follows from 4.1.6 that $H^i_{\mathfrak{a}}(I') = 0$ for all $i \in \mathbb{N}$. The result now follows immediately on application of 1.3.4(ii). □

Remark. Some readers might prefer to approach the final claim of 4.2.1 in the following way. Observe first that $\Gamma_{\mathfrak{a}R'}(\bullet)\lceil_R$ and $\Gamma_{\mathfrak{a}}(\bullet \lceil_R)$ are the same functor. Let M' be an R'-module, and let

$$I'^{\bullet} : 0 \longrightarrow I'^0 \longrightarrow I'^1 \longrightarrow \cdots \longrightarrow I'^i \longrightarrow I'^{i+1} \longrightarrow \cdots$$

be an injective resolution of M' over R'. By 1.2.2, the i-th cohomology module of the complex $\Gamma_{\mathfrak{a}R'}(I'^{\bullet})$ is isomorphic to $H^i_{\mathfrak{a}R'}(M')$ (for $i \in \mathbb{N}_0$).

By 4.1.6, the complex

$$I'^{\bullet}\lceil_R : 0 \longrightarrow I'^0\lceil_R \longrightarrow I'^1\lceil_R \longrightarrow \cdots \longrightarrow I'^i\lceil_R \longrightarrow I'^{i+1}\lceil_R \longrightarrow \cdots$$

is a resolution of $M'\lceil_R$ by $\Gamma_{\mathfrak{a}}$-acyclic R-modules, and so, by 4.1.2, the i-th cohomology module of the complex $\Gamma_{\mathfrak{a}}(I'^{\bullet}\lceil_R)$ is isomorphic to $H^i_{\mathfrak{a}}(M'\lceil_R)$ (for $i \in \mathbb{N}_0$). Since $\Gamma_{\mathfrak{a}R'}(\bullet)\lceil_R = \Gamma_{\mathfrak{a}}(\bullet \lceil_R)$, this gives a quick and relatively transparent proof that $H^i_{\mathfrak{a}R'}(M')\lceil_R \cong H^i_{\mathfrak{a}}(M'\lceil_R)$ for $i \in \mathbb{N}_0$. Note, however, that our use of 1.3.4 in the above proof of 4.2.1 led rather rapidly to additional information.

4.2.2 Example. Suppose that the R-module M is annihilated by the ideal $\mathfrak{b}$ of R. Then, since M is $\mathfrak{b}$-torsion, it is immediate from Exercise 2.1.9 that $H^i_{\mathfrak{a}}(M) \cong H^i_{\mathfrak{a}+\mathfrak{b}}(M)$ for all $i \in \mathbb{N}_0$. However, the reader might find the following alternative approach, which uses the Independence Theorem 4.2.1, illuminating.

Since M is annihilated by $\mathfrak{b}$, we can regard M as a module over $R/\mathfrak{b}$ in a natural way. Let $i \in \mathbb{N}_0$. By the Independence Theorem 4.2.1, there is an R-isomorphism $H^i_{\mathfrak{a}}(M) \cong H^i_{\mathfrak{a}(R/\mathfrak{b})}(M)$. But $\mathfrak{a}(R/\mathfrak{b}) = (\mathfrak{a}+\mathfrak{b})/\mathfrak{b}$, and another use of the Independence Theorem 4.2.1 provides us with an R-isomorphism $H^i_{(\mathfrak{a}+\mathfrak{b})/\mathfrak{b}}(M) \cong H^i_{\mathfrak{a}+\mathfrak{b}}(M)$, so that $H^i_{\mathfrak{a}}(M) \cong H^i_{\mathfrak{a}+\mathfrak{b}}(M)$, as claimed.

4.2.3 Remark. In 3.3.4 we defined the cohomological dimension $\mathrm{cohd}(\mathfrak{a})$ of $\mathfrak{a}$. It is immediate from the Independence Theorem 4.2.1 that $\mathrm{cohd}(\mathfrak{a}R') \leq \mathrm{cohd}(\mathfrak{a})$.

4.2.4 Exercise. Let $n \in \mathbb{N}$ and consider the ring $R[X_1, \ldots, X_n]$ of polynomials over R.

(i) Show that the $R[X_1]$-module $H^1_{(X_1)}(R[X_1])$ is not finitely generated, and therefore non-zero. (Here is a hint: you might find 2.2.21 helpful.)

(ii) Use 1.3.9(iv) to show that, when $n > 1$,

$$H^{n-1}_{(X_1,\ldots,X_n)}(R[X_1,\ldots,X_n]) = 0.$$

(iii) The ring homomorphism ν : $R[X_1, \ldots, X_n] \to R[X_1, \ldots, X_{n-1}]$ (again for $n > 1$) obtained by evaluation at $X_1, \ldots, X_{n-1}, 0$ allows us to regard $R[X_1, \ldots, X_{n-1}]$ as an $R[X_1, \ldots, X_n]$-module. Show that, when this is done,

$$H^n_{(X_1,\ldots,X_n)}(R[X_1,\ldots,X_{n-1}]) = 0.$$

(iv) Assume again that $n > 1$. Deduce from the exact sequence

$$0 \longrightarrow R[X_1,\ldots,X_n] \xrightarrow{X_n} R[X_1,\ldots,X_n]$$
$$\xrightarrow{\nu} R[X_1,\ldots,X_{n-1}] \longrightarrow 0$$

of $R[X_1,\ldots,X_n]$-modules that $H^n_{(X_1,\ldots,X_n)}(R[X_1,\ldots,X_n])$ has a submodule isomorphic to $H^{n-1}_{(X_1,\ldots,X_{n-1})}(R[X_1,\ldots,X_{n-1}])$ (the latter being regarded as an $R[X_1,\ldots,X_n]$-module by means of ν), and that

$$H^n_{(X_1,\ldots,X_n)}(R[X_1,\ldots,X_n]) = X_n H^n_{(X_1,\ldots,X_n)}(R[X_1,\ldots,X_n]).$$

(v) Prove by induction that, for all $n \in \mathbb{N}$, the $R[X_1,\ldots,X_n]$-module $H^n_{(X_1,\ldots,X_n)}(R[X_1,\ldots,X_n])$ is not finitely generated, and so is non-zero.

4.2.5 Exercise. Use 4.2.4 to show that, over the ring $R[X_1, X_2, X_3, X_4]$ of polynomials in four indeterminates with coefficients in R,

$$H^3_{(X_1,X_2)\cap(X_3,X_4)}(R[X_1, X_2, X_3, X_4]) \neq 0.$$

Deduce that the affine algebraic set W in $\mathbb{A}^4$ defined by

$$W := V_{\mathbb{A}^4}(X_1,\ X_2) \cup V_{\mathbb{A}^4}(X_3,\ X_4)$$

cannot be 'defined by two equations', that is, there do not exist two polynomials $f, g \in \mathbb{C}[X_1, X_2, X_3, X_4]$ such that $W = \{p \in \mathbb{A}^4 : f(p) = g(p) = 0\}$.

We are now going to use the Independence Theorem to prove a proposition which presents a non-vanishing result for certain local cohomology modules. We shall use this proposition in Chapter 6 to prove a more general non-vanishing result. The strategy of the proof of the proposition has some similarities with the strategy used in Exercise 4.2.4.

4.2.6 Proposition. *Suppose that $(R, \mathfrak{m})$ is a regular local ring of dimension $d > 0$. Then the R-module $H^d_{\mathfrak{m}}(R)$ is not finitely generated, and therefore non-zero.*

Proof. We argue by induction on d. When $d = 1$, the maximal ideal $\mathfrak{m}$ is generated by one element, π say. Let K denote the field of fractions of R. Apply 2.2.18 (in the case where $\mathfrak{B} = (R\pi^n)_{n\in\mathbb{N}}$, $M = R$ and S is the set of non-zero elements of R), and note that $\bigcup_{n\in\mathbb{N}}(R :_K \pi^n) = K$: it follows from this and 2.2.6(i)(c) that there is an exact sequence

$$0 \longrightarrow R \longrightarrow K \longrightarrow H^1_{R\pi}(R) \longrightarrow 0.$$

Since R is finitely generated as an R-module but K is not, we deduce that $H^1_{R\pi}(R)$ is not finitely generated.

Now suppose, inductively, that $d > 1$ and that the result has been proved for regular local rings of smaller (positive) dimensions. Let $u_1, \ldots, u_d$ be d elements of $\mathfrak{m}$ which generate this maximal ideal. Let $\overline{R} := R/u_dR$, a regular local ring of dimension $d - 1$ with maximal ideal $\overline{\mathfrak{m}} := \mathfrak{m}/u_dR$. By the inductive hypothesis, the $\overline{R}$-module $H^{d-1}_{\overline{\mathfrak{m}}}(\overline{R})$ is not finitely generated. Hence, by the Independence Theorem 4.2.1, the R-module $H^{d-1}_{\mathfrak{m}}(R/u_dR)$ is not finitely generated. The exact sequence

$$0 \longrightarrow R \xrightarrow{u_d} R \longrightarrow R/u_dR \longrightarrow 0$$

induces an exact sequence $H^{d-1}_{\mathfrak{m}}(R) \longrightarrow H^{d-1}_{\mathfrak{m}}(R/u_dR) \longrightarrow H^d_{\mathfrak{m}}(R)$. But $H^{d-1}_{\mathfrak{m}}(R) = 0$ by 1.3.9(iv) since $u_1, \ldots, u_d$ is an R-sequence. Hence $H^d_{\mathfrak{m}}(R)$ has a submodule isomorphic to $H^{d-1}_{\mathfrak{m}}(R/u_dR)$; since the latter is not finitely generated, neither is $H^d_{\mathfrak{m}}(R)$.   □

4.2.7 Exercise. Assume that R contains a field K as a subring. Suppose that $\mathfrak{a}$ can be generated by t elements $a_1, \ldots, a_t$. The ring homomorphism $f : K[X_1, \ldots, X_t] \longrightarrow R$ obtained by evaluation at $a_1, \ldots, a_t$ allows us to regard each R-module as a module over $K[X_1, \ldots, X_t]$.

Show that, for an R-module M, there is a $K[X_1, \ldots, X_t]$-isomorphism

$$H^i_{(X_1,\ldots,X_t)}(M) \cong H^i_{\mathfrak{a}}(M) \quad \text{for each } i \in \mathbb{N}_0.$$

Use the fact that the ring $K[X_1, \ldots, X_t]$ has global dimension t to show that $H^i_{\mathfrak{a}}(M) = 0$ for all $i > t$. Compare this with Theorem 3.3.1.

4.2.8 Exercise. (In this exercise, there is no assumption about R beyond the standard ones.) Suppose that $\mathfrak{a}$ can be generated by t elements. Use an appropriate ring homomorphism $\mathbb{Z}[X_1, \ldots, X_t] \longrightarrow R$ in conjunction with the Independence Theorem 4.2.1 and the fact that the global dimension of $\mathbb{Z}[X_1, \ldots, X_t]$ is $t + 1$ to show that $H^i_{\mathfrak{a}}(M) = 0$ for all $i > t + 1$. Compare this with Theorem 3.3.1.

The following refinement of the Independence Theorem 4.2.1 is occasionally useful.

4.2.9 Exercise. Let the situation be as in the Independence Theorem 4.2.1. Let M' be an R'-module. Show that, for each $i \in \mathbb{N}_0$, the R-module $H^i_{\mathfrak{a}}(M')$ actually has a natural structure as an R'-module under which

$$r'x = H^i_{\mathfrak{a}}(r' \operatorname{Id}_{M'})(x) \quad \text{for all } r' \in R' \text{ and } x \in H^i_{\mathfrak{a}}(M').$$

Show further that, if $f' : M' \longrightarrow N'$ is a homomorphism of R'-modules, then $H^i_{\mathfrak{a}}(f') : H^i_{\mathfrak{a}}(M') \longrightarrow H^i_{\mathfrak{a}}(N')$ is an R'-homomorphism.

Deduce that $(H^i_{\mathfrak{a}}(\bullet \lceil_R))_{i \in \mathbb{N}_0}$ is a negative strongly connected sequence of covariant functors from $\mathcal{C}(R')$ to *itself*, and, as such, is isomorphic to $(H^i_{\mathfrak{a}R'})_{i \in \mathbb{N}_0}$.

4.3 The Flat Base Change Theorem

The second fundamental result about local cohomology that we propose to derive from the work in §4.1 is the Flat Base Change Theorem. This result is a consequence of Theorem 4.1.9 and concerns the situation in which $f : R \to R'$ is flat: we are going to show that, in this situation, speaking loosely, the formation of local cohomology 'commutes' with 'extension of the base ring from R to R''. More precisely, we shall show that, for each R-module M and each $i \in \mathbb{N}_0$, there is an R'-isomorphism $H^i_{\mathfrak{a}R'}(M \otimes_R R') \overset{\cong}{\longrightarrow} H^i_{\mathfrak{a}}(M) \otimes_R R'$. Although our proof of this result has some similarities with our proof of 4.2.1 above, a preparatory lemma will be helpful in this case.

4.3.1 Lemma. *Assume that the ring homomorphism $f : R \to R'$ is flat. There is a natural equivalence of functors (from $\mathcal{C}(R)$ to $\mathcal{C}(R')$)*

$$\rho^0 : \Gamma_{\mathfrak{a}}(\bullet) \otimes_R R' \longrightarrow \Gamma_{\mathfrak{a}R'}((\bullet) \otimes_R R')$$

which is such that, for each R-module M, we have $\rho^0_M(m \otimes r') = m \otimes r'$ for all $m \in \Gamma_{\mathfrak{a}}(M)$ and all $r' \in R'$.

Proof. Let M be an R-module. Since f is flat, the inclusion map induces an R'-monomorphism $\Gamma_{\mathfrak{a}}(M) \otimes_R R' \to M \otimes_R R'$ whose image is contained in $\Gamma_{\mathfrak{a}R'}(M \otimes_R R')$. It follows easily that there is a natural transformation of functors $\rho^0 : \Gamma_{\mathfrak{a}}(\bullet) \otimes_R R' \longrightarrow \Gamma_{\mathfrak{a}R'}((\bullet) \otimes_R R')$ which is such that, for each R-module M, ρ^0_M is monomorphic and satisfies

$$\rho^0_M \left(\sum_{i=1}^t m_i \otimes r'_i \right) = \sum_{i=1}^t m_i \otimes r'_i$$

for all $t \in \mathbb{N}$, $m_1, \ldots, m_t \in \Gamma_{\mathfrak{a}}(M)$ and $r'_1, \ldots, r'_t \in R'$. It remains to show that ρ^0_M is surjective for each R-module M.

We can deduce this from 4.1.7 as follows. Let $n \in \mathbb{N}$. Since $R/\mathfrak{a}^n$ is a finitely generated R-module, 4.1.7 provides us with an R'-isomorphism

$$\mu_{R/\mathfrak{a}^n} : \operatorname{Hom}_R(R/\mathfrak{a}^n, M) \otimes_R R' \xrightarrow{\cong} \operatorname{Hom}_{R'}((R/\mathfrak{a}^n) \otimes_R R', M \otimes_R R')$$

such that $\mu_{R/\mathfrak{a}^n}(h \otimes r') = r'(h \otimes \operatorname{Id}_{R'})$ for all $h \in \operatorname{Hom}_R(R/\mathfrak{a}^n, M)$ and all $r' \in R'$. We can incorporate $\mu := \mu_{R/\mathfrak{a}^n}$ into the composition

$$(0 :_M \mathfrak{a}^n) \otimes_R R' \xrightarrow{\cong} \operatorname{Hom}_R(R/\mathfrak{a}^n, M) \otimes_R R'$$
$$\xrightarrow{\mu} \operatorname{Hom}_{R'}((R/\mathfrak{a}^n) \otimes_R R', M \otimes_R R')$$
$$\xrightarrow{\cong} \operatorname{Hom}_{R'}(R'/\mathfrak{a}^n R', M \otimes_R R')$$
$$\xrightarrow{\cong} (0 :_{M \otimes_R R'} (\mathfrak{a}R')^n)$$

of R'-isomorphisms, in which the other three isomorphisms are the obvious natural ones. The reader can easily check that this composition ε is such that $\varepsilon(\sum_{i=1}^t m_i \otimes r_i') = \sum_{i=1}^t m_i \otimes r_i'$ (for $t \in \mathbb{N}$, $m_1, \ldots, m_t \in (0 :_M \mathfrak{a}^n)$ and $r_1', \ldots, r_t' \in R'$). It is now clear that ρ_M^0 is surjective. $\square$

4.3.2 Flat Base Change Theorem. *Assume that the ring homomorphism $f :$ $R \longrightarrow R'$ is flat. There is a unique isomorphism*

$$(\rho^i)_{i \in \mathbb{N}_0} : (H_\mathfrak{a}^i(\,\bullet\,) \otimes_R R')_{i \in \mathbb{N}_0} \xrightarrow{\cong} (H_{\mathfrak{a}R'}^i((\,\bullet\,) \otimes_R R'))_{i \in \mathbb{N}_0}$$

of negative (strongly) connected sequences of covariant functors (from $\mathcal{C}(R)$ to $\mathcal{C}(R')$) which extends the natural equivalence ρ^0 of 4.3.1. In particular, for each $i \in \mathbb{N}_0$, the functors $H_\mathfrak{a}^i(\,\bullet\,) \otimes_R R'$ and $H_{\mathfrak{a}R'}^i((\,\bullet\,) \otimes_R R')$ are naturally equivalent.

Proof. As in the proof of 4.2.1, we use 1.3.4. Since $(\,\bullet\,) \otimes_R R' : \mathcal{C}(R) \longrightarrow \mathcal{C}(R')$ is exact, it is clear that the two sequences given in the statement of 4.3.2 are indeed negative strongly connected sequences of covariant functors from $\mathcal{C}(R)$ to $\mathcal{C}(R')$. Furthermore, whenever I is an injective R-module, it is, of course, automatic that $H_\mathfrak{a}^i(I) = 0$ for all $i \in \mathbb{N}$, while it follows from 4.1.9 that $H_{\mathfrak{a}R'}^i(I \otimes_R R') = 0$ for all $i \in \mathbb{N}$. The result is now an immediate consequence of 4.3.1 and 1.3.4(ii). $\square$

Remark. Again, some readers might prefer the following direct approach to the final claim of 4.3.2. Let M be an R-module, and let

$$I^\bullet : 0 \longrightarrow I^0 \longrightarrow I^1 \longrightarrow \cdots \longrightarrow I^i \longrightarrow I^{i+1} \longrightarrow \cdots$$

be an injective resolution of M over R. For $i \in \mathbb{N}_0$, the i-th cohomology module of the complex $\Gamma_{\mathfrak{a}}(I^\bullet)$ is, of course, $H^i_{\mathfrak{a}}(M)$. Since $(\bullet) \otimes_R R'$: $\mathcal{C}(R) \longrightarrow \mathcal{C}(R')$ is additive and exact, we have an R'-isomorphism

$$H^i_{\mathfrak{a}}(M) \otimes_R R' = H^i(\Gamma_{\mathfrak{a}}(I^\bullet)) \otimes_R R' \cong H^i(\Gamma_{\mathfrak{a}}(I^\bullet) \otimes_R R').$$

In view of the natural equivalence of 4.3.1, there is an R'-isomorphism

$$H^i(\Gamma_{\mathfrak{a}}(I^\bullet) \otimes_R R') \cong H^i(\Gamma_{\mathfrak{a}R'}(I^\bullet \otimes_R R')).$$

But, by 4.1.9, the complex $I^\bullet \otimes_R R'$ is a $\Gamma_{\mathfrak{a}R'}$-acyclic resolution of the R'-module $M \otimes_R R'$, and so, by 4.1.2, there is an R'-isomorphism

$$H^i(\Gamma_{\mathfrak{a}R'}(I^\bullet \otimes_R R')) \cong H^i_{\mathfrak{a}R'}(M \otimes_R R').$$

This gives us, fairly directly, an R'-isomorphism

$$H^i_{\mathfrak{a}}(M) \otimes_R R' \cong H^i_{\mathfrak{a}R'}(M \otimes_R R').$$

Note again, however, that our use of 1.3.4 in the above proof of 4.3.2 led very quickly to additional information.

If S is a multiplicatively closed subset of R, then, as is very well known, the natural ring homomorphism $R \to S^{-1}R$ is flat. Thus we can obtain the result of Exercise 1.3.6 as a special case of the Flat Base Change Theorem 4.3.2.

4.3.3 Corollary. *Let S be a multiplicatively closed subset of R. Then*

$$\left(S^{-1}(H^i_{\mathfrak{a}}(\bullet))\right)_{i \in \mathbb{N}_0} \quad and \quad \left(H^i_{\mathfrak{a}S^{-1}R}(S^{-1}(\bullet))\right)_{i \in \mathbb{N}_0}$$

are isomorphic connected sequences of functors (from $\mathcal{C}(R)$ to $\mathcal{C}(S^{-1}R)$). In particular, for every $i \in \mathbb{N}_0$ and every R-module M, there is an $S^{-1}R$-isomorphism $S^{-1}(H^i_{\mathfrak{a}}(M)) \cong H^i_{\mathfrak{a}S^{-1}R}(S^{-1}M)$. $\square$

4.3.4 Exercise. Suppose that R is regular; that is, $R_{\mathfrak{p}}$ is a regular local ring for every $\mathfrak{p} \in \mathrm{Spec}(R)$.

Suppose that $\mathfrak{a}$ is proper and has height $t > 0$. Show that the R-module $H^t_{\mathfrak{a}}(R)$ is not finitely generated, and therefore non-zero.

The first part of the next exercise is for those readers who have reached this point without needing the fact that, for an arbitrary R-module N, the functor $(\bullet) \otimes_R N : \mathcal{C}(R) \longrightarrow \mathcal{C}(R)$ commutes with direct limits!

4.3.5 ‡Exercise. Let N be an R-module.

(i) Show that the functor $(\bullet) \otimes_R N : \mathcal{C}(R) \longrightarrow \mathcal{C}(R)$ commutes with direct limits (see 3.4.1).

(ii) Assume that the ring homomorphism $f : R \to R'$ is flat. Show that there is a natural equivalence of functors

$$\varepsilon' : D_{\mathfrak{a}}(\,\bullet\,) \otimes_R R' \longrightarrow D_{\mathfrak{a}R'}((\,\bullet\,) \otimes_R R')$$

(from $\mathcal{C}(R)$ to $\mathcal{C}(R')$) which is such that, for each R-module M, the diagram

$$\begin{array}{ccc}
M \otimes_R R' & \xrightarrow{\ \eta_{\mathfrak{a},M} \otimes_R R'\ } & D_{\mathfrak{a}}(M) \otimes_R R' \\
\| & & \downarrow{\scriptstyle \varepsilon'_M} \\
M \otimes_R R' & \xrightarrow{\ \eta_{\mathfrak{a}R',M \otimes_R R'}\ } & D_{\mathfrak{a}R'}(M \otimes_R R')
\end{array}$$

commutes. (The notations $\eta_{\mathfrak{a},M}$ and $\eta_{\mathfrak{a}R',M \otimes_R R'}$ are explained in the note following the statement of 2.2.6.)

(iii) Deduce that, for a multiplicatively closed subset S of R, the functors

$$S^{-1}(D_{\mathfrak{a}}(\,\bullet\,)) \quad \text{and} \quad D_{\mathfrak{a}S^{-1}R}(S^{-1}(\,\bullet\,))$$

(from $\mathcal{C}(R)$ to $\mathcal{C}(S^{-1}R)$) are naturally equivalent.

We shall end this chapter with a geometric example and exercise; the following remark will be useful.

4.3.6 Remark. Let $r_1, r_2 \in R$. Then

$$\{\mathfrak{p} \in \operatorname{Spec}(R) \ : \ \mathfrak{p} \supseteq r_1 r_2 R + \mathfrak{a}\}$$
$$= \left\{\mathfrak{p} \in \operatorname{Spec}(R) : \mathfrak{p} \supseteq \sqrt{(r_1 R + \mathfrak{a})} \cap \sqrt{(r_2 R + \mathfrak{a})}\right\},$$

and so $\sqrt{(r_1 r_2 R + \mathfrak{a})} = \sqrt{(r_1 R + \mathfrak{a})} \cap \sqrt{(r_2 R + \mathfrak{a})}$.

4.3.7 Example. This example concerns the affine variety V in $\mathbb{A}^4$ studied in 2.3.7 and 3.3.5 and given by

$$V := V_{\mathbb{A}^4}(X_1 X_4 - X_2 X_3,\ X_1^2 X_3 + X_1 X_2 - X_2^2,\ X_3^3 + X_3 X_4 - X_4^2).$$

In 3.3.5 we showed that the subvariety

$$L := V_{\mathbb{A}^4}(X_2, X_3, X_4) = \left\{(c, 0, 0, 0) \in \mathbb{A}^4 : c \in \mathbb{C}\right\}$$

of V (a line) cannot be 'defined by one equation'; here, we shall show that V itself cannot be 'defined by two equations'.

Write, for convenience, $R := \mathbb{C}[X_1, X_2, X_3, X_4]$, and let $\mathfrak{c}$ be the ideal of R given by

$$\mathfrak{c} = (X_1 X_4 - X_2 X_3,\ X_1^2 X_3 + X_1 X_2 - X_2^2,\ X_3^3 + X_3 X_4 - X_4^2).$$

We wish to show that ara($\mathfrak{c}$) > 2, and we shall achieve this by showing that $H_{\mathfrak{c}}^3(R) \neq 0$ and then appealing to 3.3.3. Let $\widehat{R} := \mathbb{C}[[X_1, X_2, X_3, X_4]]$; since the natural ring homomorphism $R \to \widehat{R}$ is flat, it follows from the Flat Base Change Theorem 4.3.2 that $H_{\mathfrak{c}}^3(R) \otimes_R \widehat{R} \cong H_{\mathfrak{c}\widehat{R}}^3(\widehat{R})$, and so it will be enough, in order to achieve our aim, for us to show that $H_{\mathfrak{c}\widehat{R}}^3(\widehat{R}) \neq 0$. This is what we shall do.

Let $u \in \widehat{R}$ be given by

$$u := 1 + 2X_3 - 2X_3^2 + 4X_3^3 - 10X_3^4 + \cdots + \frac{(-1)^{n-1}2(2n-2)!}{n!(n-1)!}X_3^n + \cdots.$$

In fact, $u^2 = 1 + 4X_3$; note that, since the latter is a unit in $\widehat{R}$ (by [81, 1.43]), so too is u. Write

$$f_1 := X_1X_4 - X_2X_3, \quad f_2 := X_1^2X_3 + X_1X_2 - X_2^2, \quad f_3 := X_3^3 + X_3X_4 - X_4^2,$$

so that $\mathfrak{c}\widehat{R} = (f_1, f_2, f_3)\widehat{R}$. Now

$$\begin{aligned}
4f_2 &= 4X_1^2X_3 + X_1^2 - (4X_2^2 - 4X_1X_2 + X_1^2) \\
&= X_1^2u^2 - (2X_2 - X_1)^2 \\
&= (X_1u + 2X_2 - X_1)(X_1u - 2X_2 + X_1) \\
&= (X_1(u-1) + 2X_2)(X_1(u+1) - 2X_2).
\end{aligned}$$

A similar calculation shows that

$$\begin{aligned}
4f_3 &= 4X_3^3 + X_3^2 - (4X_4^2 - 4X_3X_4 + X_3^2) \\
&= (X_3(u-1) + 2X_4)(X_3(u+1) - 2X_4).
\end{aligned}$$

Set

$$g_1 := X_1(u-1) + 2X_2, \quad g_2 := X_1(u+1) - 2X_2,$$

$$h_1 := X_3(u-1) + 2X_4, \quad h_2 := X_3(u+1) - 2X_4.$$

Note that

$$\begin{aligned}
h_1g_2 - h_2g_1 &= X_1(u+1)2X_4 - X_3(u-1)2X_2 \\
&\quad - 2X_2X_3(u+1) + 2X_4X_1(u-1) \\
&= 4u(X_1X_4 - X_2X_3) = 4uf_1,
\end{aligned}$$

so that $f_1 = u^{-1}(h_1g_2 - h_2g_1)/4$.

Therefore, on use of 4.3.6, we have

$$\sqrt{\mathfrak{c}\widehat{R}} = \sqrt{(f_1, f_2, f_3)\widehat{R}} = \sqrt{(f_1, g_1g_2, h_1h_2)\widehat{R}}$$
$$= \sqrt{(f_1, g_1, h_1h_2)\widehat{R}} \cap \sqrt{(f_1, g_2, h_1h_2)\widehat{R}}$$
$$= \sqrt{\mathfrak{d}_{11}} \cap \sqrt{\mathfrak{d}_{12}} \cap \sqrt{\mathfrak{d}_{21}} \cap \sqrt{\mathfrak{d}_{22}}$$

where $\mathfrak{d}_{ij} = (f_1, g_i, h_j)\widehat{R}$ for $i, j = 1, 2$. We now use the equation $f_1 = u^{-1}(h_1g_2 - h_2g_1)/4$ to see that $\mathfrak{d}_{11} = (f_1, g_1, h_1)\widehat{R} = (g_1, h_1)\widehat{R}$ and $\mathfrak{d}_{22} = (f_1, g_2, h_2)\widehat{R} = (g_2, h_2)\widehat{R}$; the same equation and 4.3.6 show that

$$\sqrt{\mathfrak{d}_{12}} = \sqrt{(f_1, g_1, h_2)\widehat{R}}$$
$$= \sqrt{(h_1g_2, g_1, h_2)\widehat{R}} = \sqrt{(h_1, g_1, h_2)\widehat{R}} \cap \sqrt{(g_2, g_1, h_2)\widehat{R}}$$

and

$$\sqrt{\mathfrak{d}_{21}} = \sqrt{(f_1, g_2, h_1)\widehat{R}} = \sqrt{(h_2, g_2, h_1)\widehat{R}} \cap \sqrt{(g_1, g_2, h_1)\widehat{R}}.$$

Now put all this information together to see that

$$\sqrt{\mathfrak{c}\widehat{R}} = \sqrt{(g_1, h_1)\widehat{R}} \cap \sqrt{(g_2, h_2)\widehat{R}}.$$

Next, note that, since u is a unit of $\widehat{R}$, the four elements g_1, g_2, h_1, h_2 generate the maximal ideal $(X_1, X_2, X_3, X_4)\widehat{R}$ of the 4-dimensional regular local ring $\widehat{R}$, and so, by [81, 15.38] for example, $\mathfrak{p} := (g_1, h_1)\widehat{R}$ and $\mathfrak{q} := (g_2, h_2)\widehat{R}$ are prime ideals of $\widehat{R}$. Thus we have the interesting situation where, although $\sqrt{\mathfrak{c}}$ is, by 2.3.7, a prime ideal of R, the ideal $\sqrt{\mathfrak{c}\widehat{R}}$ of $\widehat{R}$ is the intersection of two distinct prime ideals of height 2.

Note that $\mathfrak{p} + \mathfrak{q} = \widehat{\mathfrak{m}} := (X_1, X_2, X_3, X_4)\widehat{R}$, the maximal ideal of $\widehat{R}$. Since X_1, X_2, X_3, X_4 is an $\widehat{R}$-sequence in $\widehat{\mathfrak{m}}$, it follows from 1.3.9(iv) that $H^3_{\mathfrak{p}+\mathfrak{q}}(\widehat{R}) = 0$. However, Proposition 4.2.6 shows that $H^4_{\mathfrak{p}+\mathfrak{q}}(\widehat{R}) \neq 0$. By 3.3.1, we have $H^i_{\mathfrak{p}}(\widehat{R}) \oplus H^i_{\mathfrak{q}}(\widehat{R}) = 0$ for all $i > 2$. It therefore follows from 4.3.2, 1.2.3 and the Mayer–Vietoris sequence 3.2.3 that

$$H^3_{\mathfrak{c}}(R) \otimes_R \widehat{R} \cong H^3_{\mathfrak{c}\widehat{R}}(\widehat{R}) = H^3_{\mathfrak{p}\cap\mathfrak{q}}(\widehat{R}) \cong H^4_{\widehat{\mathfrak{m}}}(\widehat{R}) \neq 0.$$

Hence $H^3_{\mathfrak{c}}(R) \neq 0$, and so $\operatorname{ara}(\mathfrak{c}) > 2$ by 3.3.3. Therefore V cannot be 'defined by two equations'.

4.3.8 Exercise. Let K be a field of characteristic 0, and let R denote the ring $K[X_1, X_2, X_3, X_4]$; let $\mathfrak{c}$ be the ideal of R given by

$$\mathfrak{c} = (X_1^2 + X_1^3 - X_2^2, \; X_3^2 + X_1X_3^2 - X_4^2, \; X_2X_3 - X_1X_4).$$

Let $\widehat{R} := K[[X_1, X_2, X_3, X_4]]$ and let $u \in \widehat{R}$ be given by

$$u := 1 + \frac{1}{2}X_1 - \frac{1}{8}X_1^2 + \frac{1}{16}X_1^3 - \frac{5}{128}X_1^4 + \cdots + \frac{(-1)^{n-1}2(2n-2)!}{4^n n!(n-1)!}X_1^n + \cdots,$$

so that $u^2 = 1 + X_1$. Let $z_1 := X_1 u - X_2$, $z_2 := X_3 u - X_4$, $z_3 := X_1 u + X_2$ and $z_4 := X_3 u + X_4$.

(i) Show that z_1, z_2, z_3, z_4 generate the maximal ideal of $\widehat{R}$.

(ii) Show that, in $\widehat{R}$,

$$X_2 X_3 - X_1 X_4 = X_1 z_2 - X_3 z_1 = X_3 z_3 - X_1 z_4$$

and $2u(X_2 X_3 - X_1 X_4) = z_2 z_3 - z_1 z_4$. Deduce that

$$\sqrt{\mathfrak{c}\widehat{R}} = (z_1, z_2)\widehat{R} \cap (z_3, z_4)\widehat{R}.$$

(iii) Prove that $H_{\mathfrak{c}}^3(R) \neq 0$.

4.3.9 Exercise. With the notation of 2.3.1, let V be the affine algebraic set in $\mathbb{A}^4$ given by

$$V := V_{\mathbb{A}^4}(X_1^2 + X_1^3 - X_2^2, \ X_3^2 + X_1 X_3^2 - X_4^2, \ X_2 X_3 - X_1 X_4).$$

(i) By considering the morphism of varieties $\alpha : \mathbb{A}^2 \to \mathbb{A}^4$ for which $\alpha((c, d)) = (c^2 - 1, c^3 - c, d, cd)$ for all $(c, d) \in \mathbb{A}^2$ and the mapping

$$\beta : V \setminus \{(0, 0, 0, 0)\} \longrightarrow \mathbb{A}^2 \setminus \{(1, 0), (-1, 0)\}$$

given by

$$\beta((c_1, c_2, c_3, c_4)) = \begin{cases} (c_2/c_1, c_3) & \text{if } c_1 \neq 0, \\ (c_4/c_3, c_3) & \text{if } c_3 \neq 0 \end{cases}$$

(for all $(c_1, c_2, c_3, c_4) \in V \setminus \{(0, 0, 0, 0)\}$), show that V is irreducible, and so is an affine variety.

(ii) Deduce from Exercise 4.3.8 above that V cannot be 'defined by two equations'.

5

Other approaches

Although we have now developed enough of the basic algebraic theory of local cohomology so that we could, if we wished, start right away with serious calculations with local cohomology modules, there are two other approaches to the construction of local cohomology modules which are useful, and popular with many workers in the subject. One approach uses cohomology of Čech complexes, and the other uses direct limits of homology modules of Koszul complexes. Links between local cohomology and Koszul complexes and Čech cohomology are described in A. Grothendieck's foundational lecture notes [25, §2]; related ideas are present in J.-P. Serre's fundamental paper [77, §61]. Among other texts which discuss links between local cohomology and the Čech complex or Koszul complexes are those by W. Bruns and J. Herzog [7, §3.5], D. Eisenbud [10, Appendix 4], M. Herrmann, S. Ikeda and U. Orbanz [32, §35], P. Roberts [70, Chapter 3, §2], J. R. Strooker [83, §4.3] and J. Stückrad and W. Vogel [84, Chapter 0, §1.3].

We shall make very little use in this book of the descriptions of local cohomology modules as direct limits of homology modules of Koszul complexes. However, we will use the approach to local cohomology via cohomology of Čech complexes, and so we present the basic ideas of this approach in this chapter. As this work leads naturally to the connection between local cohomology and direct limits of homology modules of Koszul complexes, we also present some aspects of that connection.

The Čech complex approach is particularly useful for calculations in $H_{\mathfrak{a}}^n(R)$ when $\mathfrak{a}$ can be generated by n elements. We shall illustrate this in §5.3 in the case where R has prime characteristic p. Then each local cohomology module $H_{\mathfrak{a}}^i(R)$ ($i \in \mathbb{N}_0$) of R itself carries a natural 'Frobenius action', that is, an Abelian group homomorphism $F : H_{\mathfrak{a}}^i(R) \longrightarrow H_{\mathfrak{a}}^i(R)$ such that $F(rm) = r^p F(m)$ for all $m \in H_{\mathfrak{a}}^i(R)$ and $r \in R$. The Frobenius action on $H_{\mathfrak{a}}^n(R)$, where $\mathfrak{a}$ can be generated by n elements, can be described very simply using the Čech complex.

Later in the book, we shall often study aspects of $H_{\mathfrak{m}}^d(R)$ where $(R, \mathfrak{m})$ is local and of dimension d. Let $a_1, \ldots, a_d$ be a system of parameters for R, and set $\mathfrak{q} := Ra_1 + \cdots + Ra_d$. Since $\sqrt{\mathfrak{q}} = \mathfrak{m}$, we have $H_{\mathfrak{m}}^d(R) = H_{\mathfrak{q}}^d(R)$, and so, in the case when R has prime characteristic p, the above-mentioned Čech complex approach facilitates calculations with the natural Frobenius action on $H_{\mathfrak{m}}^d(R)$. There is an illustration of this idea in §6.5.

5.1 Use of Čech complexes

Throughout this chapter, $a_1, \ldots, a_n$ (where $n > 0$) will denote n elements which generate $\mathfrak{a}$, and M will denote an arbitrary R-module. Our first task is to define the (extended) Čech complex $C(M)^\bullet$ of M with respect to the sequence $a_1, \ldots, a_n$. This has the form

$$0 \longrightarrow C(M)^0 \xrightarrow{d^0} C(M)^1 \longrightarrow \cdots \longrightarrow C(M)^{n-1} \xrightarrow{d^{n-1}} C(M)^n \longrightarrow 0.$$

The following lemma will be helpful. Recall again that, for $a \in R$, the notations R_a and M_a denote, respectively, the ring and module of fractions of R and M with respect to the multiplicatively closed subset $\{a^i : i \in \mathbb{N}_0\}$ of R.

5.1.1 Lemma. *Let $a, b \in R$. There is an isomorphism of R-modules*

$$\mu : M_{ab} \longrightarrow (M_a)_b$$

for which $\mu(m/(ab)^i) = (m/a^i)/b^i$ for all $m \in M$ and all $i \in \mathbb{N}_0$.

Proof. Suppose that $m, y \in M$ and $i, j \in \mathbb{N}_0$ are such that $m/(ab)^i = y/(ab)^j$ in M_{ab}. Then there exists $k \in \mathbb{N}_0$ such that

$$(ab)^k \left((ab)^j m - (ab)^i y\right) = 0.$$

Thus, in M_a, we have $b^{k+j}m/a^i = b^{k+i}y/a^j$, so that $(m/a^i)/b^i = (y/a^j)/b^j$ in $(M_a)_b$. Therefore there is indeed a mapping $\mu : M_{ab} \longrightarrow (M_a)_b$ given by the formula in the statement of the lemma. It is now a very easy exercise to check that μ is an R-isomorphism. $\square$

5.1.2 ♯Exercise. Complete the proof of Lemma 5.1.1.

5.1.3 Remark. In situations such as that of 5.1.1, we shall use μ to identify the R-modules M_{ab} and $(M_a)_b$ without further comment. Thus, when we speak of the natural R-homomorphism $\omega : M_a \longrightarrow M_{ab}$, we shall be referring to the natural R-homomorphism from M_a to $(M_a)_b$ and employing the above identification, so that $\omega(m/a^i) = b^i m/(ab)^i$ for all $m \in M$ and $i \in \mathbb{N}_0$. These ideas are employed in the construction of the Čech complex of M with respect to $a_1, \ldots, a_n$ in 5.1.5 below.

5.1.4 Notation. For $k \in \mathbb{N}$ with $1 \leq k \leq n$, we shall write

$$\mathcal{I}(k,n) := \left\{ (i(1), \ldots, i(k)) \in \mathbb{N}^k : 1 \leq i(1) < i(2) < \ldots < i(k) \leq n \right\},$$

the set of all strictly increasing sequences of length k of positive integers taken from the set $\{1, \ldots, n\}$. For $i \in \mathcal{I}(k,n)$, we shall, for $1 \leq j \leq k$, denote the j-th component of i by $i(j)$, so that $i = (i(1), \ldots, i(k))$.

Now suppose that $k < n$, and $s \in \mathbb{N}$ with $1 \leq s \leq k+1$. Let $j \in \mathcal{I}(k+1,n)$. Then by

$$j^{\widehat{s}} \quad \text{or} \quad (j(1), \ldots, \widehat{j(s)}, \ldots, j(k+1))$$

we mean the sequence $(j(1), \ldots, j(s-1), j(s+1), \ldots, j(k+1))$ of $\mathcal{I}(k,n)$ obtained by omitting the s-th component of j.

It is perhaps worth pointing out here that, if $t \in \mathbb{N}$ with $1 \leq t < s$, then $(j^{\widehat{s}})^{\widehat{t}} = (j^{\widehat{t}})^{\widehat{s-1}}$.

Again under the assumption that $k < n$, let $i \in \mathcal{I}(k,n)$. By the *n-complement of* i we mean the sequence $j \in \mathcal{I}(n-k,n)$ such that

$$\{1, \ldots, n\} = \{i(1), \ldots, i(k), j(1), \ldots, j(n-k)\}.$$

5.1.5 Proposition and Definition. *Define a sequence $C(M)^\bullet$ of R-modules and R-homomorphisms*

$$0 \longrightarrow C(M)^0 \xrightarrow{d^0} C(M)^1 \longrightarrow \cdots \longrightarrow C(M)^{n-1} \xrightarrow{d^{n-1}} C(M)^n \longrightarrow 0$$

as follows:

(a) $C(M)^0 := M$;

(b) *for $k = 1, \ldots, n$, and with the notation of 5.1.4,*

$$C(M)^k = \bigoplus_{i \in \mathcal{I}(k,n)} M_{a_{i(1)} \ldots a_{i(k)}};$$

(c) *$d^0 : C(M)^0 \longrightarrow C(M)^1$ is to be such that, for each $h = 1, \ldots, n$, the composition of d^0 followed by the canonical projection from $C(M)^1$ to M_{a_h} is just the natural map from M to M_{a_h}; and*

(d) *for $k = 1, \ldots, n-1$, $i \in \mathcal{I}(k,n)$ and $j \in \mathcal{I}(k+1,n)$, the composition*

$$M_{a_{i(1)} \ldots a_{i(k)}} \longrightarrow C(M)^k \xrightarrow{d^k} C(M)^{k+1} \longrightarrow M_{a_{j(1)} \ldots a_{j(k+1)}}$$

(in which the first and third maps are the canonical injection and

canonical projection respectively) is to be the natural map from $M_{a_{i(1)}...a_{i(k)}}$ *to* $M_{a_{i(1)}...a_{i(k)}a_{j(s)}}$ *multiplied by* $(-1)^{s-1}$ *if* $i = j^{\widehat{s}}$ *for an s with* $1 \leq s \leq k+1$, *and is to be* 0 *otherwise.*

Then $C(M)^\bullet$ *is a complex, called* the (extended) Čech complex of M *with respect to* $a_1, \ldots, a_n$. *(Henceforth, we shall omit the word 'extended'.)*
We denote $C(R)^\bullet$ *by*

$$C^\bullet : 0 \longrightarrow C^0 \xrightarrow{d^0} C^1 \longrightarrow \cdots \longrightarrow C^i \xrightarrow{d^i} C^{i+1} \longrightarrow \cdots \longrightarrow C^n \longrightarrow 0.$$

Proof. We have only to prove that $d^{k+1} \circ d^k = 0$ for all $k = 0, \ldots, n-2$.

Let $m \in M$. To show that $d^1 \circ d^0 = 0$, it is enough to show that, for each $i \in \mathcal{I}(2, n)$, the component of $d^1 \circ d^0(m)$ in the direct summand $M_{a_{i(1)}a_{i(2)}}$ of $C(M)^2$ is 0. The only contributions to this component that could conceivably be non-zero must come 'through' the direct summands $M_{a_{i(1)}}$ and $M_{a_{i(2)}}$ of $C(M)^1$. It follows that the component of $d^1 \circ d^0(m)$ in the direct summand $M_{a_{i(1)}a_{i(2)}}$ of $C(M)^2$ is $(-1)(m/1) + (m/1) = 0$. Hence $d^1 \circ d^0 = 0$.

Now consider the case where $1 \leq k \leq n-2$. In order to show $d^{k+1} \circ d^k = 0$, it is enough to show that, for each $i \in \mathcal{I}(k, n)$, the restriction of $d^{k+1} \circ d^k$ to the direct summand $M_{a_{i(1)}...a_{i(k)}}$ of $C(M)^k$ is zero, and this is what we shall do. So let $m \in M, v \in \mathbb{N}_0$ and $j \in \mathcal{I}(k+2, n)$: we calculate the component of

$$d^{k+1} \circ d^k \left(\frac{m}{(a_{i(1)} \cdots a_{i(k)})^v} \right)$$

in the direct summand $M_{a_{j(1)}...a_{j(k+2)}}$ of $C(M)^{k+2}$. This component will be (conceivably) non-zero only if $i = (j^{\widehat{s}})^{\widehat{t}}$ for some integers s, t with $1 \leq t < s \leq k+2$, and, when this is so, will (in view of the penultimate paragraph of 5.1.4) be

$$\frac{(-1)^{t-1}(-1)^{s-2}a_{j(t)}^v a_{j(s)}^v m}{(a_{j(1)} \cdots a_{j(k+2)})^v} + \frac{(-1)^{s-1}(-1)^{t-1}a_{j(s)}^v a_{j(t)}^v m}{(a_{j(1)} \cdots a_{j(k+2)})^v},$$

which is zero. Hence $d^{k+1} \circ d^k = 0$, and the proof is complete. $\qquad\square$

5.1.6 Example. The reader might find it helpful if we write down explicitly the Čech complex $C^\bullet = C(R)^\bullet$ of R with respect to $a_1, \ldots, a_n$ when n has a fairly small value: when $n = 3$, the complex is

$$0 \to R \xrightarrow{d^0} R_{a_1} \oplus R_{a_2} \oplus R_{a_3} \xrightarrow{d^1} R_{a_2 a_3} \oplus R_{a_1 a_3} \oplus R_{a_1 a_2} \xrightarrow{d^2} R_{a_1 a_2 a_3} \to 0$$

where the d^i $(i = 0, 1, 2)$ are described as follows. For $r, r_1, r_2, r_3 \in R$ and

$v_1, v_2, v_3 \in \mathbb{N}_0,$

$$d^0(r) = \left(\frac{r}{1}, \frac{r}{1}, \frac{r}{1} \right),$$

$$d^1\left(\left(\frac{r_1}{a_1^{v_1}}, \frac{r_2}{a_2^{v_2}}, \frac{r_3}{a_3^{v_3}} \right) \right)$$

$$= \left(\frac{a_2^{v_3} r_3}{(a_2 a_3)^{v_3}} - \frac{a_3^{v_2} r_2}{(a_2 a_3)^{v_2}}, \frac{a_1^{v_3} r_3}{(a_1 a_3)^{v_3}} - \frac{a_3^{v_1} r_1}{(a_1 a_3)^{v_1}}, \frac{a_1^{v_2} r_2}{(a_1 a_2)^{v_2}} - \frac{a_2^{v_1} r_1}{(a_1 a_2)^{v_1}} \right)$$

and

$$d^2 \left(\left(\frac{r_1}{(a_2 a_3)^{v_1}}, \frac{r_2}{(a_1 a_3)^{v_2}}, \frac{r_3}{(a_1 a_2)^{v_3}} \right) \right)$$

$$= \frac{a_1^{v_1} r_1}{(a_1 a_2 a_3)^{v_1}} - \frac{a_2^{v_2} r_2}{(a_1 a_2 a_3)^{v_2}} + \frac{a_3^{v_3} r_3}{(a_1 a_2 a_3)^{v_3}}.$$

5.1.7 ♯Exercise. Suppose that $n > 1$ and M is Ra_n-torsion. Show that the Čech complex of M with respect to $a_1, \ldots, a_n$ is the Čech complex of M with respect to $a_1, \ldots, a_{n-1}$.

5.1.8 ♯Exercise. Show that a homomorphism $f : M \to N$ of R-modules induces a chain map of complexes $C(f)^\bullet : C(M)^\bullet \longrightarrow C(N)^\bullet$ such that $C(f)^0 : C(M)^0 \to C(N)^0$ is just $f : M \to N$.

Show further that, with these assignments, $C(\,\bullet\,)^\bullet$ becomes a functor from the category $\mathcal{C}(R)$ to the category of complexes of R-modules (and R-homomorphisms) and chain maps of such complexes.

We remind the reader that we use the notation

$$C^\bullet : 0 \longrightarrow C^0 \xrightarrow{d^0} C^1 \longrightarrow \cdots \longrightarrow C^i \xrightarrow{d^i} C^{i+1} \longrightarrow \cdots \longrightarrow C^n \longrightarrow 0$$

for the Čech complex of R itself with respect to $a_1, \ldots, a_n$. We interpret C^i as 0 for $i \in \mathbb{Z} \setminus \{0, 1, \ldots, n\}$, of course.

It should be clear to the reader that $H^i((\,\bullet\,) \otimes_R C^\bullet)$ is, for each $i \in \mathbb{N}_0$, a covariant R-linear functor from $\mathcal{C}(R)$ to itself.

5.1.9 Lemma. *The sequence* $(H^i((\,\bullet\,) \otimes_R C^\bullet))_{i \in \mathbb{N}_0}$ *is a negative strongly connected sequence of covariant functors from* $\mathcal{C}(R)$ *to itself.*

Proof. Let

$$\begin{array}{ccccccccc}
0 & \longrightarrow & L & \xrightarrow{\alpha} & M & \xrightarrow{\beta} & N & \longrightarrow & 0 \\
 & & \downarrow{\lambda} & & \downarrow{\mu} & & \downarrow{\nu} & & \\
0 & \longrightarrow & L' & \xrightarrow{\alpha'} & M' & \xrightarrow{\beta'} & N' & \longrightarrow & 0
\end{array}$$

be a commutative diagram of R-modules and R-homomorphisms with exact rows. Now for each $i \in \mathbb{N}$, the R-module C^i is flat, since it is a direct sum of finitely many modules of fractions of the form $S^{-1}R$ for suitable choices of the multiplicatively closed subset S of R. It follows that there is a commutative diagram

$$
\begin{array}{ccccccccc}
0 & \longrightarrow & L \otimes_R C^\bullet & \xrightarrow{\alpha \otimes C^\bullet} & M \otimes_R C^\bullet & \xrightarrow{\beta \otimes C^\bullet} & N \otimes_R C^\bullet & \longrightarrow & 0 \\
 & & \downarrow{\lambda \otimes C^\bullet} & & \downarrow{\mu \otimes C^\bullet} & & \downarrow{\nu \otimes C^\bullet} & & \\
0 & \longrightarrow & L' \otimes_R C^\bullet & \xrightarrow{\alpha' \otimes C^\bullet} & M' \otimes_R C^\bullet & \xrightarrow{\beta' \otimes C^\bullet} & N' \otimes_R C^\bullet & \longrightarrow & 0
\end{array}
$$

of complexes of R-modules and chain maps of such complexes such that, for each $i \in \mathbb{N}_0$, the sequence

$$
0 \longrightarrow L \otimes_R C^i \xrightarrow{\alpha \otimes C^i} M \otimes_R C^i \xrightarrow{\beta \otimes C^i} N \otimes_R C^i \longrightarrow 0
$$

is exact, and a similar property holds for the lower row. Thus the above commutative diagram of complexes gives rise to a long exact sequence of cohomology modules of the complexes in the top row, a similar long exact sequence for the bottom row, and a chain map of the first long exact sequence into the second. The claim follows from this. $\square$

The technique used in the above proof will solve the following exercise.

5.1.10 ♯Exercise.   Show that $(H^i(C(\,\bullet\,)^\bullet))_{i \in \mathbb{N}_0}$ is a negative strongly connected sequence of covariant functors from $\mathcal{C}(R)$ to itself.

5.1.11 ♯Exercise.   Use the natural isomorphisms

$$
M \otimes_R R_{a_{i(1)} \ldots a_{i(k)}} \xrightarrow{\;\cong\;} M_{a_{i(1)} \ldots a_{i(k)}} \quad (\text{for } i \in \mathcal{I}(k,n))
$$

(where $1 \le k \le n$) to produce an isomorphism of complexes

$$
\omega_M^\bullet : M \otimes_R C^\bullet \xrightarrow{\;\cong\;} C(M)^\bullet.
$$

Show further that, as M varies through $\mathcal{C}(R)$, the $\omega_M^\bullet$ constitute a natural equivalence of functors $\omega^\bullet : (\,\bullet\,) \otimes_R C^\bullet \longrightarrow C(\,\bullet\,)^\bullet$ (from $\mathcal{C}(R)$ to the category of all complexes of R-modules (and R-homomorphisms) and chain maps of such complexes).

Show also that $\omega^\bullet$ induces an isomorphism

$$
(H^i(\omega^\bullet))_{i \in \mathbb{N}_0} : (H^i((\,\bullet\,) \otimes_R C^\bullet))_{i \in \mathbb{N}_0} \xrightarrow{\;\cong\;} (H^i(C(\,\bullet\,)^\bullet))_{i \in \mathbb{N}_0}
$$

of negative strongly connected sequences of covariant functors (from $\mathcal{C}(R)$ to $\mathcal{C}(R)$).

5.1.12 ♯Exercise. In this and the next few exercises, we shall be concerned with relationships between Čech complexes of M with respect to different sequences of elements of R, and, for such discussions, it will help if we use a slightly more complicated notation. Thus, in this and a few subsequent situations, we shall use $C(a_1, \ldots, a_n; M)^\bullet$ to denote the Čech complex of M with respect to $a_1, \ldots, a_n$.

(i) Suppose that $n \geq 2$, and let $m \in \mathbb{N}$ be such that $1 \leq m < n$. Let $b_1, \ldots, b_n$ denote the sequence

$$a_1, \ldots, a_{m-1}, a_{m+1}, a_m, a_{m+2}, \ldots, a_n$$

obtained from the sequence $a_1, \ldots, a_n$ by interchange of the m-th and $(m+1)$-th terms.

Let $k \in \mathbb{N}$ with $1 \leq k \leq n$. Let $i \in \mathcal{I}(k, n)$.

If neither m nor $m + 1$ appears as a term in i, then

$$M_{a_{i(1)}...a_{i(k)}} = M_{b_{i(1)}...b_{i(k)}}:$$

define $\psi(i) : M_{a_{i(1)}...a_{i(k)}} \longrightarrow M_{b_{i(1)}...b_{i(k)}}$ to be the identity mapping in this case.

If $m + 1$ does not appear as a term in i but $m = i(s)$, then

$$M_{a_{i(1)}...a_{i(k)}} = M_{b_{i(1)}...b_{i(s-1)}b_{i(s)+1}b_{i(s+1)}...b_{i(k)}}:$$

define $\psi(i) : M_{a_{i(1)}...a_{i(k)}} \longrightarrow M_{b_{i(1)}...b_{i(s-1)}b_{i(s)+1}b_{i(s+1)}...b_{i(k)}}$ to be the identity mapping in this case.

If m does not appear as a term in i but $m + 1 = i(s)$, then

$$M_{a_{i(1)}...a_{i(k)}} = M_{b_{i(1)}...b_{i(s-1)}b_{i(s)-1}b_{i(s+1)}...b_{i(k)}}:$$

define $\psi(i) : M_{a_{i(1)}...a_{i(k)}} \longrightarrow M_{b_{i(1)}...b_{i(s-1)}b_{i(s)-1}b_{i(s+1)}...b_{i(k)}}$ to be the identity mapping in this case.

If m and $m + 1$ both appear in i, with, say, $m = i(s)$ (so that $m + 1 = i(s) + 1 = i(s + 1)$), then $M_{a_{i(1)}...a_{i(k)}} = M_{b_{i(1)}...b_{i(k)}}$: in this case, define $\psi(i) : M_{a_{i(1)}...a_{i(k)}} \longrightarrow M_{b_{i(1)}...b_{i(k)}}$ to be the identity mapping multiplied by -1.

Let

$$\psi^k = \bigoplus_{i \in \mathcal{I}(k,n)} \psi(i) : C(a_1, \ldots, a_n; M)^k \longrightarrow C(b_1, \ldots, b_n; M)^k.$$

Also, let $\psi^0 : M \longrightarrow M$ be the identity mapping. Show that

$$\Psi = \left(\psi^k\right)_{0 \le k \le n} : C(a_1, \ldots, a_n; M)^\bullet \longrightarrow C(b_1, \ldots, b_n; M)^\bullet$$

is an isomorphism of complexes.

(ii) Let σ be a permutation of the set $\{1, \ldots, n\}$. Show that there is an isomorphism of complexes

$$C(a_1, \ldots, a_n; M)^\bullet \xrightarrow{\cong} C(a_{\sigma(1)}, \ldots, a_{\sigma(n)}; M)^\bullet.$$

5.1.13 Definition. Let $m \in \mathbb{Z}$, and let

$$Q^\bullet : \quad \cdots \longrightarrow Q^k \xrightarrow{d_{Q^\bullet}^k} Q^{k+1} \xrightarrow{d_{Q^\bullet}^{k+1}} Q^{k+2} \longrightarrow \cdots$$

be a complex of R-modules and R-homomorphisms. We define $\{m\}Q^\bullet$, the result of *shifting* $Q^\bullet$ *by m places*, as follows: $(\{m\}Q^\bullet)^k = Q^{m+k}$ for all $k \in \mathbb{Z}$, and the k-th 'differentiation' homomorphism $d_{\{m\}Q^\bullet}^k$ in $\{m\}Q^\bullet$ is given by $d_{\{m\}Q^\bullet}^k = d_{Q^\bullet}^{m+k} : Q^{m+k} \longrightarrow Q^{m+k+1}$. Thus changing $Q^\bullet$ into $\{1\}Q^\bullet$ (respectively $\{-1\}Q^\bullet$) amounts to shifting it one place to the left (respectively right).

5.1.14 ♯Exercise. In this exercise, we use the extended notation for Čech complexes introduced in Exercise 5.1.12.

Let $b \in R$. Show that there is a sequence of complexes (of R-modules and R-homomorphisms) and chain maps

$$0 \longrightarrow \{-1\}C(a_1, \ldots, a_n; M_b)^\bullet \longrightarrow C(a_1, \ldots, a_n, b; M)^\bullet$$
$$\longrightarrow C(a_1, \ldots, a_n; M)^\bullet \longrightarrow 0$$

which is such that, for each $i \in \mathbb{N}_0$, the sequence

$$0 \longrightarrow (\{-1\}C(a_1, \ldots, a_n; M_b)^\bullet)^i \longrightarrow C(a_1, \ldots, a_n, b; M)^i$$
$$\longrightarrow C(a_1, \ldots, a_n; M)^i \longrightarrow 0$$

is exact.

5.1.15 Exercise. In this exercise, we use the extended notation for Čech complexes introduced in Exercise 5.1.12.

Let $a \in R$. Show that $H^i(C(a; M)^\bullet) \cong H_{Ra}^i(M)$ for all $i \in \mathbb{N}_0$.

5.1.16 ♯Exercise. Show that all the cohomology modules of the Čech complex $C(M)^\bullet$ are $\mathfrak{a}$-torsion. (Here are some hints: try induction on the number of elements needed to generate $\mathfrak{a}$, in conjunction with 5.1.14 and 5.1.12.)

5.1.17 Exercise. In this exercise, we again use the extended notation for Čech complexes introduced in Exercise 5.1.12.

Suppose that $b_1, \ldots, b_m$ also generate $\mathfrak{a}$. Prove that

$$H^i(C(a_1, \ldots, a_n; M)^\bullet) \cong H^i(C(b_1, \ldots, b_m; M)^\bullet) \quad \text{for all } i \in \mathbb{N}_0,$$

that is, the cohomology modules of the Čech complex $C(M)^\bullet$ are, up to R-isomorphism, independent of the choice of sequence of generators for $\mathfrak{a}$. (We suggest that you use 5.1.14 to compare

$$C(a_1, \ldots, a_n, b_1; M)^\bullet \quad \text{and} \quad C(a_1, \ldots, a_n; M)^\bullet.$$

Also, you may find 5.1.16 and 5.1.12 helpful.)

Our first major aim in this chapter is to show that, for each $i \in \mathbb{N}_0$, the local cohomology module $H^i_\mathfrak{a}(M)$ is isomorphic to $H^i(C(M)^\bullet)$, the i-th cohomology module of the Čech complex $C(M)^\bullet$. The next remark is a first step.

5.1.18 Remark. In the Čech complex

$$C(M)^\bullet : 0 \longrightarrow M \xrightarrow{d^0} M_{a_1} \oplus M_{a_2} \oplus \cdots \oplus M_{a_n} \longrightarrow \ldots,$$

we have

$$\begin{aligned}
H^0(C(M)^\bullet) &= \operatorname{Ker} d^0 \\
&= \{m \in M : \text{ for each } i = 1, \ldots, n, \text{ there exists} \\
&\qquad\quad h_i \in \mathbb{N} \text{ with } a_i^{h_i} m = 0 \} \\
&= \Gamma_\mathfrak{a}(M).
\end{aligned}$$

It will probably come as no surprise to the reader to learn that we intend to exploit the above Remark 5.1.18 by use of negative strongly connected sequences of functors and 1.3.5.

5.1.19 Remark. It is immediate from 5.1.18 and 5.1.11 that there is an R-isomorphism $\gamma^0_M : \Gamma_\mathfrak{a}(M) \xrightarrow{\cong} H^0(M \otimes_R C^\bullet)$ for which $\gamma^0_M(m) = m \otimes 1$ for all $m \in \Gamma_\mathfrak{a}(M)$, and it is then clear that, as M varies through $\mathcal{C}(R)$, the γ^0_M constitute a natural equivalence of functors $\gamma^0 : \Gamma_\mathfrak{a} \longrightarrow H^0((\,\bullet\,) \otimes_R C^\bullet)$ from $\mathcal{C}(R)$ to itself.

5.1.20 Theorem. *(Recall that $a_1, \ldots, a_n$ (where $n > 0$) denote n elements which generate $\mathfrak{a}$, and $C^\bullet$ denotes the Čech complex of R with respect to $a_1, \ldots, a_n$.) There is a unique isomorphism*

$$(\delta^i)_{i \in \mathbb{N}_0} : (H^i(C(\,\bullet\,)^\bullet))_{i \in \mathbb{N}_0} \xrightarrow{\cong} (H^i_\mathfrak{a})_{i \in \mathbb{N}_0}$$

of negative (strongly) connected sequences of covariant functors (from $\mathcal{C}(R)$ to $\mathcal{C}(R)$) which extends the identity natural equivalence on $\Gamma_{\mathfrak{a}}$.

Consequently, there is a unique isomorphism

$$(\gamma^i)_{i\in\mathbb{N}_0} : (H_{\mathfrak{a}}^i)_{i\in\mathbb{N}_0} \xrightarrow{\cong} (H^i((\,\bullet\,)\otimes_R C^\bullet))_{i\in\mathbb{N}_0}$$

of negative strongly connected sequences of covariant functors (from $\mathcal{C}(R)$ to $\mathcal{C}(R)$) which extends the natural equivalence γ^0 of 5.1.19.

Proof. Recall from 5.1.18 that $H^0(C(\,\bullet\,)^\bullet) = \Gamma_{\mathfrak{a}}$. We shall be able to use Theorem 1.3.5 to prove the first part provided we can show that $H^i(C(I)^\bullet) = 0$ for all $i \in \mathbb{N}$ whenever I is an injective R-module. We shall achieve this by induction on n, and we shall use the extended notation for Čech complexes introduced in Exercise 5.1.12.

When $n = 1$, the Čech complex $C(a_1; I)^\bullet$ is just $0 \longrightarrow I \xrightarrow{\tau_I} I_{a_1} \longrightarrow 0$, where τ_I is the natural map. But, by 2.2.20, Coker $\tau_I \cong H^1_{Ra_1}(I)$, and the latter is zero because I is injective. Thus $H^i(C(I)^\bullet) = 0$ for all $i \in \mathbb{N}$. We note also that Ker $\tau_I = \Gamma_{Ra_1}(I)$, which is injective by 2.1.4. Hence the exact sequence $0 \longrightarrow \Gamma_{Ra_1}(I) \longrightarrow I \xrightarrow{\tau_I} I_{a_1} \longrightarrow 0$ of 2.2.20 splits, and I_{a_1} is an injective R-module.

Now suppose, inductively, that $n > 1$ and the result has been proved for smaller values of n. By 5.1.14, there is a sequence of complexes (of R-modules and R-homomorphisms) and chain maps

$$0 \longrightarrow \{-1\}C(a_1,\dots,a_{n-1};I_{a_n})^\bullet \longrightarrow C(a_1,\dots,a_n;I)^\bullet$$
$$\longrightarrow C(a_1,\dots,a_{n-1};I)^\bullet \longrightarrow 0$$

which is such that, for each $i \in \mathbb{N}_0$, the sequence

$$0 \longrightarrow (\{-1\}C(a_1,\dots,a_{n-1};I_{a_n})^\bullet)^i \longrightarrow C(a_1,\dots,a_n;I)^i$$
$$\longrightarrow C(a_1,\dots,a_{n-1};I)^i \longrightarrow 0$$

is exact. This sequence of complexes therefore induces a long exact sequence of cohomology modules. Since I_{a_n} is an injective R-module (by the immediately preceding paragraph of this proof), we can deduce from our inductive hypothesis that

$$H^i(\{-1\}C(a_1,\dots,a_{n-1};I_{a_n})^\bullet) = 0 \quad \text{for all } i \geq 2$$

and

$$H^i(C(a_1,\dots,a_{n-1};I)^\bullet) = 0 \quad \text{for all } i \geq 1.$$

Furthermore, an easy check shows that the connecting homomorphism

$$H^0(C(a_1,\dots,a_{n-1};I)^\bullet) \longrightarrow H^1(\{-1\}C(a_1,\dots,a_{n-1};I_{a_n})^\bullet)$$

induced by the above sequence of complexes is just the map

$$\Gamma_{Ra_1+\cdots+Ra_{n-1}}(I) \longrightarrow \Gamma_{Ra_1+\cdots+Ra_{n-1}}(I_{a_n})$$

induced by the natural homomorphism $I \to I_{a_n}$, and this is surjective since the canonical exact sequence

$$0 \longrightarrow \Gamma_{Ra_n}(I) \longrightarrow I \longrightarrow I_{a_n} \longrightarrow 0$$

of 2.2.20 splits. It follows from the long exact sequence of cohomology modules that $H^i(C(a_1, \ldots, a_n; I)^\bullet) = 0$ for all $i \geq 1$. This completes the inductive step. We can now use Theorem 1.3.5 to complete the proof of the first part.

For the second part, we recall the isomorphism

$$(H^i(\omega^\bullet))_{i \in \mathbb{N}_0} : (H^i((\,\bullet\,) \otimes_R C^\bullet))_{i \in \mathbb{N}_0} \overset{\cong}{\longrightarrow} (H^i(C(\,\bullet\,)^\bullet))_{i \in \mathbb{N}_0}$$

of negative strongly connected sequences of 5.1.11, and deduce that

$$\left((H^i(\omega^\bullet))_{i \in \mathbb{N}_0}\right)^{-1} \circ \left((\delta^i)_{i \in \mathbb{N}_0}\right)^{-1} : (H_\mathfrak{a}^i)_{i \in \mathbb{N}_0} \overset{\cong}{\longrightarrow} (H^i((\,\bullet\,) \otimes_R C^\bullet))_{i \in \mathbb{N}_0}$$

is an isomorphism of connected sequences; moreover, $(H^0(\omega^\bullet))^{-1} \circ (\delta^0)^{-1}$ is just the natural equivalence γ^0 of 5.1.19, as is easy to check. The uniqueness in the second part follows from 1.3.4. $\qquad\square$

5.1.21 Remark. The reader should note that it is immediate from Theorem 5.1.20 that, when $\mathfrak{a}$ can be generated by n elements, $H_\mathfrak{a}^i(N) = 0$ for all R-modules N whenever $i > n$: we proved this result by means of the Mayer–Vietoris sequence in Theorem 3.3.1. This is one example of a situation where the use of a different approach to the calculation of local cohomology modules can provide a simpler proof and additional insight. The next exercise provides another example.

5.1.22 Exercise. Prove that $H_\mathfrak{a}^n(R) \neq 0$ if and only if there exists $k \in \mathbb{N}$ such that, for every $t \in \mathbb{N}$, it is the case that

$$(a_1 \ldots a_n)^t \notin Ra_1^{t+k} + \cdots + Ra_n^{t+k}.$$

Deduce that, for $h \in \mathbb{N}$, in the ring $R[X_1, \ldots, X_h]$ of polynomials over R, we have $H_{(X_1,\ldots,X_h)}^h(R[X_1, \ldots, X_h]) \neq 0$.

Compare this approach with that of Exercise 4.2.4.

5.1.23 Proposition. *Let $K^1(M) := \operatorname{Ker} d^1$, where $d^1 : C(M)^1 \longrightarrow C(M)^2$ is the first 'differentiation' map in the Čech complex $C(M)^\bullet$ of M with respect to the sequence $a_1, \ldots, a_n$. (It is clear from 5.1.8 that $K^1(\,\bullet\,)$ can easily be made into a functor from $\mathcal{C}(R)$ to itself.)*

There is a natural equivalence of functors $\tilde{\varepsilon} : K^1(\,\bullet\,) \longrightarrow D_{\mathfrak{a}}$ which is such that, for each R-module M, the diagram

$$
\begin{array}{ccc}
M & \xrightarrow{\;d^0\;} & K^1(M) \\
\Big\| & & \Big\downarrow{\scriptstyle \tilde{\varepsilon}_M} \\
M & \xrightarrow{\;\eta_{\mathfrak{a},M}\;} & D_{\mathfrak{a}}(M)
\end{array}
$$

commutes.

Proof. By 5.1.16, the cohomology modules $H^i(C(M)^\bullet)$ are all $\mathfrak{a}$-torsion, and so both the kernel and cokernel of $d^0 : M \longrightarrow K^1(M)$ are $\mathfrak{a}$-torsion. It therefore follows from 2.2.15 that there is a unique R-homomorphism $\tilde{\varepsilon}_M :$ $K^1(M) \longrightarrow D_{\mathfrak{a}}(M)$ such that the diagram

$$
\begin{array}{ccc}
M & \xrightarrow{\;d^0\;} & K^1(M) \\
\Big\| & & \Big\downarrow{\scriptstyle \tilde{\varepsilon}_M} \\
M & \xrightarrow{\;\eta_{\mathfrak{a},M}\;} & D_{\mathfrak{a}}(M)
\end{array}
$$

commutes, and that $\tilde{\varepsilon}_M = D_{\mathfrak{a}}(d^0)^{-1} \circ \eta_{\mathfrak{a},K^1(M)}$. This formula and 2.2.6(i)(c) show that $\tilde{\varepsilon}_M$ is injective, since $\Gamma_{\mathfrak{a}}(K^1(M)) \subseteq \Gamma_{\mathfrak{a}}(C^1(M)) = 0$. Furthermore, it is easy to use the uniqueness aspect of 2.2.13 to deduce that, as M varies through the category $\mathcal{C}(R)$, the $\tilde{\varepsilon}_M$ constitute a natural transformation of functors.

Let $t \in \mathbb{N}$, and let $h \in \operatorname{Hom}_R(\mathfrak{a}^t, M)$. It is straightforward to check that the element

$$
\left(\frac{h(a_1^t)}{a_1^t}, \frac{h(a_2^t)}{a_2^t}, \ldots, \frac{h(a_n^t)}{a_n^t} \right) \in C(M)^1
$$

actually belongs to $K^1(M)$. (Note that $a_i^t h(a_j^t) = h(a_i^t a_j^t) = a_j^t h(a_i^t)$ for integers i, j with $1 \le i, j \le n$.) Hence there is an R-homomorphism $\gamma_{t,M} :$ $\operatorname{Hom}_R(\mathfrak{a}^t, M) \to K^1(M)$ for which

$$
\gamma_{t,M}(h) = \left(\frac{h(a_1^t)}{a_1^t}, \ldots, \frac{h(a_n^t)}{a_n^t} \right) \quad \text{for all } h \in \operatorname{Hom}_R(\mathfrak{a}^t, M).
$$

Also, it is straightforward to check that, for $t, u \in \mathbb{N}$ with $u \geq t$, the diagram

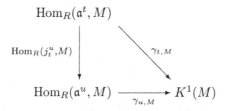

(in which $j_t^u : \mathfrak{a}^u \to \mathfrak{a}^t$ denotes the inclusion map) commutes, and so the $\gamma_{t,M}$ ($t \in \mathbb{N}$) induce an R-homomorphism $\gamma_M : D_\mathfrak{a}(M) \longrightarrow K^1(M)$ for which $\gamma_M \circ \eta_{\mathfrak{a},M} = d^0$. Note that $\tilde{\varepsilon}_M \circ \gamma_M = \operatorname{Id}_{D_\mathfrak{a}(M)}$ by the uniqueness aspect of 2.2.15. Hence $\tilde{\varepsilon}_M$ is surjective, and so is an isomorphism. $\qquad\square$

5.1.24 Exercise. Consider the special case of 5.1.23 in which $M = R$. Note that $C^1 = C(R)^1 = \prod_{i=1}^n R_{a_i}$ has a natural structure as a (commutative Noetherian) ring (with identity), and that $d^0 : R \to C^1$ is a ring homomorphism. Recall also from 2.2.5 that $D_\mathfrak{a}(R)$ has a structure as a commutative ring with identity.

Show that $K^1(R)$ is a subring of C^1, and deduce from 2.2.17 that $\tilde{\varepsilon}_R : K^1(R) \longrightarrow D_\mathfrak{a}(R)$ (where $\tilde{\varepsilon}_R$ is as in 5.1.23) is actually a ring isomorphism.

The next exercise concerns the ring $K^1(R)$ in a geometrical situation.

5.1.25 Exercise. Let V be an affine variety over the algebraically closed field K. Consider C^1 and $K^1(R)$ (of 5.1.24) in the special case in which $R = \mathcal{O}(V)$, and assume that the ideal $\mathfrak{a}$ of $\mathcal{O}(V)$ is non-zero (and generated by $a_1, \ldots, a_n$, all of which are assumed to be non-zero). Let U be the open subset $V \setminus \{p \in V : a_1(p) = \cdots = a_n(p) = 0\}$ of V.

For each $j = 1, \ldots, n$, let U_j denote the open subset $\{p \in V : a_j(p) \neq 0\}$ of U, identify the subring $\mathcal{O}(V)_{a_j}$ of $K(V)$ with the ring $\mathcal{O}(U_j)$ of regular functions on U_j in the natural way, and let $\iota_j : \mathcal{O}(U) \longrightarrow \mathcal{O}(U_j)$ be the restriction homomorphism.

Show that the map $\lambda : \mathcal{O}(U) \longrightarrow \prod_{j=1}^n \mathcal{O}(U_j)$ for which

$$\lambda(g) = (\iota_1(g), \ldots, \iota_n(g)) \quad \text{for all } g \in \mathcal{O}(U)$$

is an injective ring homomorphism with image $K^1(\mathcal{O}(V))$. In this way, we obtain a ring isomorphism $\lambda' : \mathcal{O}(U) \xrightarrow{\cong} K^1(\mathcal{O}(V))$. Show that, with the

notation of 5.1.24, the diagram

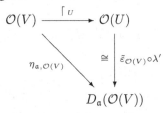

commutes, and deduce from 2.3.2 that $\bar{\varepsilon}_{\mathcal{O}(V)} \circ \lambda'$ is the $\nu_{V,\mathfrak{a}}$ of that theorem.

5.2 Use of Koszul complexes

We revert to the general situation and remind the reader that, throughout this chapter, $a_1, \ldots, a_n$ (where $n > 0$) denote n elements which generate $\mathfrak{a}$, and M denotes an arbitrary R-module.

There is yet another method of calculation of local cohomology modules: this describes them as direct limits of homology modules of Koszul complexes. We present this description here, because it can be derived quickly from our work so far in this chapter on the Čech complex. We need to specify our notation for the Koszul complexes that we shall use.

5.2.1 Notation. For all $u \in \mathbb{N}$, we denote by $K(a^u)_{\bullet}$ (or $K(a_1^u, \ldots, a_n^u)_{\bullet}$) the usual Koszul complex of R with respect to $a_1^u, \ldots, a_n^u$. Thus, if F denotes the free R-module R^n and, for each $i = 1, \ldots, n$, the element

$$(0, \ldots, 0, 1, 0, \ldots, 0) \in R^n$$

which has i-th component 1 and all its other components 0 is denoted by e_i, then $K(a^u)_{\bullet}$ has the form

$$0 \longrightarrow K(a^u)_n \longrightarrow \cdots \longrightarrow K(a^u)_k \xrightarrow{d(a^u)_k} K(a^u)_{k-1} \longrightarrow$$
$$\cdots \longrightarrow K(a^u)_0 \longrightarrow 0,$$

where

$$K(a^u)_k = \bigwedge^k F = \bigwedge^k (R^n) \quad \text{for } k = 0, \ldots, n$$

(so that $K(a^u)_0 = R$) and, for $k = 1, \ldots, n$ and $i \in \mathcal{I}(k, n)$ (with the notation of 5.1.4),

$$d(a^u)_k(e_{i(1)} \wedge \ldots \wedge e_{i(k)}) = \sum_{h=1}^{k} (-1)^{h-1} a_{i(h)}^u e_{i(1)} \wedge \ldots \wedge \widehat{e_{i(h)}} \wedge \ldots \wedge e_{i(k)},$$

where the '$\widehat{e_{i(h)}}$' indicates that $e_{i(h)}$ is omitted.

Of course, we set $K(a^u)_k = 0$ for all $k \in \mathbb{Z} \setminus \{0, 1, \ldots, n\}$.

5.2.2 Lemma. *Let $u, v \in \mathbb{N}$ with $u \le v$. There is a chain map*

$$(\psi_u^v)_\bullet = ((\psi_u^v)_k)_{k \in \mathbb{Z}} : K(a^u)_\bullet \longrightarrow K(a^v)_\bullet$$

of complexes of R-modules and R-homomorphisms such that $(\psi_u^v)_n$ is the identity mapping of $\bigwedge^n F$, such that $(\psi_u^v)_0$ is the endomorphism of R given by multiplication by $(a_1 \ldots a_n)^{v-u}$, and such that, for $k = 1, \ldots, n-1$ and $i \in \mathcal{I}(k, n)$,

$$(\psi_u^v)_k(e_{i(1)} \wedge \ldots \wedge e_{i(k)}) = (a_{j(1)} \ldots a_{j(n-k)})^{v-u} e_{i(1)} \wedge \ldots \wedge e_{i(k)},$$

where $j \in \mathcal{I}(n-k, n)$ is the n-complement of i (see 5.1.4).

Proof. We must check that $d(a^v)_k \circ (\psi_u^v)_k = (\psi_u^v)_{k-1} \circ d(a^u)_k$ for each $k = 1, \ldots, n$. We leave this to the reader in the case in which $k = n$. For $1 \le k < n$ and $i \in \mathcal{I}(k, n), j \in \mathcal{I}(n-k, n)$ as in the statement of the lemma, we have

$$(d(a^v)_k \circ (\psi_u^v)_k)(e_{i(1)} \wedge \ldots \wedge e_{i(k)})$$

$$= \sum_{h=1}^{k} (-1)^{h-1} a_{i(h)}^v (a_{j(1)} \ldots a_{j(n-k)})^{v-u} e_{i(1)} \wedge \ldots \wedge \widehat{e_{i(h)}} \wedge \ldots \wedge e_{i(k)},$$

and

$$((\psi_u^v)_{k-1} \circ d(a^u)_k)(e_{i(1)} \wedge \ldots \wedge e_{i(k)})$$

$$= (\psi_u^v)_{k-1}\left(\sum_{h=1}^{k} (-1)^{h-1} a_{i(h)}^u e_{i(1)} \wedge \ldots \wedge \widehat{e_{i(h)}} \wedge \ldots \wedge e_{i(k)}\right)$$

$$= \sum_{h=1}^{k} (-1)^{h-1} (a_{j(1)} \ldots a_{j(n-k)})^{v-u} a_{i(h)}^v e_{i(1)} \wedge \ldots \wedge \widehat{e_{i(h)}} \wedge \ldots \wedge e_{i(k)}$$

$$= (d(a^v)_k \circ (\psi_u^v)_k)(e_{i(1)} \wedge \ldots \wedge e_{i(k)}).$$

The result follows. $\square$

5.2.3 ♯Exercise. Complete the proof of 5.2.2. In other words, show that, with the notation of the lemma, $d(a^v)_n \circ (\psi_u^v)_n = (\psi_u^v)_{n-1} \circ d(a^u)_n$.

5.2.4 Remark. It is clear that, in the notation of 5.2.2, for $u, v, w \in \mathbb{N}$ with $u \le v \le w$, we have $(\psi_v^w)_\bullet \circ (\psi_u^v)_\bullet = (\psi_u^w)_\bullet$, and that $(\psi_u^u)_\bullet$ is the identity chain map of the complex $K(a^u)_\bullet$ to itself. Thus the $(\psi_u^v)_\bullet$ turn the family

$(K(a^u)_\bullet)_{u \in \mathbb{N}}$ into a direct system of complexes of R-modules and chain maps. Our immediate aim is to show that the direct limit complex, which we denote by $K(a^\infty)_\bullet$, is isomorphic to a shift of the Čech complex $C^\bullet$ of 5.1.5.

5.2.5 Theorem. *There is an isomorphism of complexes*

$$K(a^\infty)_\bullet = \varprojlim_{u \in \mathbb{N}} K(a^u)_\bullet \xrightarrow{\cong} \{n\}C^\bullet$$

between the direct limit complex of Koszul complexes described in 5.2.4 and the shift $\{n\}C^\bullet$ of the Čech complex of R with respect to $a_1, \ldots, a_n$.

Proof. Let $u \in \mathbb{N}$. We first define a chain map of complexes

$$(g_u)_\bullet : K(a^u)_\bullet \longrightarrow \{n\}C^\bullet.$$

Define $(g_u)_n : K(a^u)_n \longrightarrow (\{n\}C^\bullet)_n = C^{-n+n} = C^0 = R$ by requiring that

$$(g_u)_n(re_1 \wedge \ldots \wedge e_n) = (-1)^{1+2+\cdots+(n-1)}r \quad \text{for all } r \in R.$$

(We point out now, to help the reader discern a pattern in what follows, that $\sum_{i=1}^{n-1} i = -n + \sum_{i=1}^{n} i$.) Now let $k \in \{1, \ldots, n-1\}$ and let i be a typical element of $\mathcal{I}(k, n)$ (with the notation of 5.1.4). Define an R-homomorphism

$$(g_u)_k : K(a^u)_k \longrightarrow (\{n\}C^\bullet)_k = C^{n-k}$$

by requiring that, for $i \in \mathcal{I}(k, n)$ as above,

$$(g_u)_k(e_{i(1)} \wedge \ldots \wedge e_{i(k)}) = \frac{(-1)^{i(1)+\cdots+i(k)-k}}{(a_{j(1)} \cdots a_{j(n-k)})^u}$$

in the direct summand $R_{a_{j(1)} \cdots a_{j(n-k)}}$ of C^{n-k}, where $j \in \mathcal{I}(n-k, n)$ is the n-complement of i. Lastly, define an R-homomorphism $(g_u)_0 : K(a^u)_0 = R \longrightarrow (\{n\}C^\bullet)_0 = C^n = R_{a_1 \ldots a_n}$ by

$$(g_u)_0(r) = \frac{r}{(a_1 \ldots a_n)^u} \quad \text{for all } r \in R.$$

In order to show that $((g_u)_k)_{0 \le k \le n}$ gives rise to a chain map of complexes $(g_u)_\bullet$, we must show that (with the notation of 5.1.5 concerning the Čech complex $C^\bullet$ of R)

$$(g_u)_{k-1} \circ d(a^u)_k = d^{n-k} \circ (g_u)_k \quad \text{for all } k = 1, \ldots, n.$$

We leave this to the reader in the case when $k = n$, and deal here with the case when $1 \leq k < n$. Let $i \in \mathcal{I}(k, n)$, and let $j \in \mathcal{I}(n - k, n)$ be the n-complement of i. Then

$$
\begin{aligned}
((g_u)_{k-1} &\circ d(a^u)_k)(e_{i(1)} \wedge \ldots \wedge e_{i(k)}) \\
&= (g_u)_{k-1} \left(\sum_{h=1}^{k} (-1)^{h-1} a_{i(h)}^u e_{i(1)} \wedge \ldots \wedge \widehat{e_{i(h)}} \wedge \ldots \wedge e_{i(k)} \right) \\
&= \sum_{h=1}^{k} (-1)^{h-1} a_{i(h)}^u \frac{(-1)^{i(1)+\cdots+i(h)+\cdots+i(k)-(k-1)}(-1)^{i(h)}}{(a_{j(1)} \cdots a_{j(n-k)} a_{i(h)})^u} \\
&= \sum_{h=1}^{k} (-1)^{i(h)+h} \frac{(-1)^{i(1)+\cdots+i(k)-k} a_{i(h)}^u}{(a_{j(1)} \cdots a_{j(n-k)} a_{i(h)})^u}.
\end{aligned}
$$

On the other hand,

$$
(d^{n-k} \circ (g_u)_k)(e_{i(1)} \wedge \ldots \wedge e_{i(k)}) = d^{n-k} \left(\frac{(-1)^{i(1)+\cdots+i(k)-k}}{(a_{j(1)} \cdots a_{j(n-k)})^u} \right).
$$

We now refer back to 5.1.5: in order to evaluate the right-hand side of the above equation, we need to know, for each $h = 1, \ldots, k$, at which point $i(h)$ should be inserted in the sequence $(j(1), \ldots, j(n - k))$ in order to make an increasing sequence: if it should occupy the s_h-th position in the new sequence, then

$$
\{1, 2, \ldots, i(h) - 1\} = \{i(1), \ldots, i(h-1), j(1), \ldots, j(s_h - 1)\},
$$

so that $i(h) - 1 = h - 1 + s_h - 1$. We thus see that

$$
\begin{aligned}
(d^{n-k} &\circ (g_u)_k)(e_{i(1)} \wedge \ldots \wedge e_{i(k)}) \\
&= \sum_{h=1}^{k} (-1)^{i(h)-h} \frac{(-1)^{i(1)+\cdots+i(k)-k} a_{i(h)}^u}{(a_{j(1)} \cdots a_{j(n-k)} a_{i(h)})^u} \\
&= ((g_u)_{k-1} \circ d(a^u)_k)(e_{i(1)} \wedge \ldots \wedge e_{i(k)}).
\end{aligned}
$$

It follows that we do indeed obtain a chain map of complexes

$$
(g_u)_\bullet : K(a^u)_\bullet \longrightarrow \{n\}C^\bullet.
$$

Next note that, for $u, v \in \mathbb{N}$ with $u \leq v$ and the chain map

$$
(\psi_u^v)_\bullet = ((\psi_u^v)_k)_{k \in \mathbb{Z}} : K(a^u)_\bullet \longrightarrow K(a^v)_\bullet.
$$

of complexes of 5.2.2, we have a commutative diagram

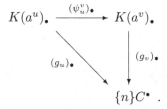

It follows that there is induced a chain map of complexes

$$(g_\infty)_\bullet = ((g_\infty)_k)_{k\in\mathbb{Z}} : K(a^\infty)_\bullet = \varinjlim_{u\in\mathbb{N}} K(a^u)_\bullet \longrightarrow \{n\}C^\bullet$$

which is clearly such that $(g_\infty)_k$ is surjective for all $k \in \mathbb{Z}$. In order to show that $(g_\infty)_\bullet$ is an isomorphism, it is enough to show that, for $u \in \mathbb{N}, k \in \mathbb{N}_0$ with $0 \le k \le n$ and $\beta \in \text{Ker}(g_u)_k$, there exists $v \in \mathbb{N}$ with $u \le v$ such that $\beta \in \text{Ker}(\psi_u^v)_k$. This is very easy (and left to the reader) in the cases when $k = 0$ and $k = n$, and so we suppose that $1 \le k < n$. Let $i \in \mathcal{I}(k, n)$. Now $(g_u)_k$ maps the direct summand $Re_{i(1)} \wedge \ldots \wedge e_{i(k)}$ of $K(a^u)_k$ into the direct summand $R_{a_{j(1)}\ldots a_{j(n-k)}}$ of $C^{n-k} = (\{n\}C^\bullet)_k$, where $j \in \mathcal{I}(n-k, n)$ is the n-complement of i. In view of this and the fact that $\mathcal{I}(k, n)$ is a finite set, we can assume that β has the form $re_{i(1)} \wedge \ldots \wedge e_{i(k)}$ for some $r \in R$. In this case, the fact that $\beta \in \text{Ker}(g_u)_k$ means that there exists $w \in \mathbb{N}_0$ such that $(a_{j(1)} \ldots a_{j(n-k)})^w r = 0$, and then $(\psi_u^{u+w})_k(re_{i(1)} \wedge \ldots \wedge e_{i(k)}) = 0$.

It follows that $(g_\infty)_\bullet$ is an isomorphism of complexes. $\square$

5.2.6 ♯Exercise. Complete the proof of Theorem 5.2.5.

Because the Čech complex $C^\bullet$ can, by 5.1.20, be used to calculate the local cohomology modules $H_\mathfrak{a}^i(M)$ $(i \in \mathbb{N}_0)$, it is a consequence of Theorem 5.2.5 that these local cohomology modules can also be computed in terms of Koszul complexes.

The result of the following exercise will be used in the proof of Theorem 5.2.9.

5.2.7 ♯Exercise. Let N be an R-module. Let $(\Lambda, \le)$ be a (non-empty) directed partially ordered set, and let $((W_\alpha)^\bullet)_{\alpha\in\Lambda}$ be a direct system of complexes of R-modules and chain maps over Λ, with constituent chain maps $(h_\beta^\alpha)^\bullet : (W_\beta)^\bullet \to (W_\alpha)^\bullet$ (for each $(\alpha, \beta) \in \Lambda \times \Lambda$ with $\alpha \ge \beta$). Set

$$(W_\infty)^\bullet := \varinjlim_{\alpha\in\Lambda} (W_\alpha)^\bullet.$$

Show that the chain maps $N \otimes_R (h_\beta^\alpha)^\bullet$ (for $\alpha, \beta \in \Lambda$ with $\alpha \ge \beta$) turn the family $(N \otimes_R (W_\alpha)^\bullet)_{\alpha \in \Lambda}$ into a direct system of complexes of R-modules and R-homomorphisms and chain maps over Λ, and show that there is an isomorphism of complexes

$$\varinjlim_{\alpha \in \Lambda} (N \otimes_R (W_\alpha)^\bullet) \cong N \otimes_R (W_\infty)^\bullet = N \otimes_R \left(\varinjlim_{\alpha \in \Lambda} (W_\alpha)^\bullet \right).$$

5.2.8 ♯Exercise. For each $u \in \mathbb{N}$, let $K(a^u, M)_\bullet$ denote $M \otimes_R K(a^u)_\bullet$, the Koszul complex of M with respect to $a_1^u, \ldots, a_n^u$. Let $\alpha : M \longrightarrow N$ be a homomorphism of R-modules. For $v, w \in \mathbb{N}$ with $v \leq w$, there is a commutative diagram

$$
\begin{array}{ccc}
M \otimes_R K(a^v)_\bullet & \xrightarrow{\alpha \otimes K(a^v)_\bullet} & N \otimes_R K(a^v)_\bullet \\
\downarrow{\scriptstyle M \otimes (\psi_v^w)_\bullet} & & \downarrow{\scriptstyle N \otimes (\psi_v^w)_\bullet} \\
M \otimes_R K(a^w)_\bullet & \xrightarrow{\alpha \otimes K(a^w)_\bullet} & N \otimes_R K(a^w)_\bullet
\end{array}
$$

of complexes and chain maps; use this to show that, for each $j \in \mathbb{Z}$, the family $(H_j(K(a^u, M)_\bullet))_{u \in \mathbb{N}}$ is a direct system, and that $\varinjlim_{u \in \mathbb{N}} H_j(K(a^u, \bullet)_\bullet)$ is a covariant R-linear functor from $\mathcal{C}(R)$ to itself.

Show further that

$$
\left(\varinjlim_{u \in \mathbb{N}} H_{n-i}(K(a^u, \bullet)_\bullet) \right)_{i \in \mathbb{N}_0}
$$

is a negative strongly connected sequence of functors from $\mathcal{C}(R)$ to itself.

5.2.9 Theorem. *In the situation and with the notation of 5.2.8, there is a natural equivalence of functors* $\delta^0 : \varinjlim_{u \in \mathbb{N}} H_n(K(a^u, \bullet)_\bullet) \xrightarrow{\cong} \Gamma_a$ *from* $\mathcal{C}(R)$ *to itself; furthermore, there is a unique isomorphism*

$$
(\delta^i)_{i \in \mathbb{N}_0} : \left(\varinjlim_{u \in \mathbb{N}} H_{n-i}(K(a^u, \bullet)_\bullet) \right)_{i \in \mathbb{N}_0} \xrightarrow{\cong} \left(H_a^i \right)_{i \in \mathbb{N}_0}
$$

of negative (strongly) connected sequences of covariant functors (from $\mathcal{C}(R)$ *to* $\mathcal{C}(R)$*) which extends* δ^0*.*

Consequently, for each $i \in \mathbb{N}_0$ *and each* R*-module* M,

$$
H_a^i(M) \cong \varinjlim_{u \in \mathbb{N}} H_{n-i}(K(a^u, M)_\bullet).
$$

Proof. We shall use Theorem 1.3.5 again.

Let $u \in \mathbb{N}$. Note that (with the notation of 5.2.1) the homomorphism

$$\mathrm{Id}_M \otimes d(a^u)_n : M \otimes_R K(a^u)_n \longrightarrow M \otimes_R K(a^u)_{n-1}$$

has kernel $\{m \otimes (e_1 \wedge \ldots \wedge e_n) : m \in (0 :_M (a_1^u, \ldots, a_n^u))\}$, and that

$$\Gamma_{\mathfrak{a}}(M) = \bigcup_{u \in \mathbb{N}} (0 :_M (a_1^u, \ldots, a_n^u)).$$

Hence, for each $u \in \mathbb{N}$, there is a monomorphism

$$\delta_{u,M} : \mathrm{Ker}(\mathrm{Id}_M \otimes d(a^u)_n) \longrightarrow \Gamma_{\mathfrak{a}}(M)$$

with image $(0 :_M (a_1^u, \ldots, a_n^u))$; furthermore, for $u, v \in \mathbb{N}$ with $u \leq v$, the diagram

$$
\begin{array}{ccc}
\mathrm{Ker}(\mathrm{Id}_M \otimes d(a^u)_n) & & \\
\ \ \Big\downarrow{\subseteq} & \diagdown^{\delta_{u,M}} & \\
\mathrm{Ker}(\mathrm{Id}_M \otimes d(a^v)_n) & \xrightarrow[\delta_{v,M}]{} & \Gamma_{\mathfrak{a}}(M)
\end{array}
$$

commutes. It follows that there is induced an isomorphism

$$\delta_M^0 : \varinjlim_{u \in \mathbb{N}} H_n(K(a^u, M)_{\bullet}) \xrightarrow{\cong} \Gamma_{\mathfrak{a}}(M),$$

and it is easy to check that, as M varies through $\mathcal{C}(R)$, the δ_M^0 constitute a natural equivalence of functors $\delta^0 : \varinjlim_{u \in \mathbb{N}} H_n(K(a^u, \bullet)_{\bullet}) \xrightarrow{\cong} \Gamma_{\mathfrak{a}}$ from $\mathcal{C}(R)$ to itself.

Let I be an injective R-module. The fact that passage to direct limits preserves exactness ensures that

$$\varinjlim_{u \in \mathbb{N}} H_j(K(a^u, I)_{\bullet}) \cong H_j \left(\varinjlim_{u \in \mathbb{N}} (K(a^u, I)_{\bullet}) \right) \quad \text{for all } j \in \mathbb{Z}.$$

It now follows from Exercise 5.2.7 that there is an isomorphism of complexes

$$\varinjlim_{u \in \mathbb{N}} (K(a^u, I)_{\bullet}) = \varinjlim_{u \in \mathbb{N}} (I \otimes_R K(a^u)_{\bullet}) \cong I \otimes_R K(a^\infty)_{\bullet}.$$

But 5.2.5 shows that $K(a^\infty)_{\bullet} \cong \{n\}C^{\bullet}$, and 5.1.20 shows that

$$H^i(I \otimes_R C^{\bullet}) \cong H_{\mathfrak{a}}^i(I) \quad \text{for all } i \in \mathbb{N}_0.$$

Hence $\varinjlim\limits_{u \in \mathbb{N}} H_{n-i}(K(a^u, I)_\bullet) = 0$ for all $i \in \mathbb{N}$.

We can now use Theorem 1.3.5 to complete the proof. □

5.3 Local cohomology in prime characteristic

Since $\mathfrak{a}$ can be generated by n elements, we must have $H_{\mathfrak{a}}^{n+j}(M) = 0$ for all $j \in \mathbb{N}$ (by 3.3.3), so that interest focusses on $H_{\mathfrak{a}}^n(M)$. In this section, we are going to illustrate how the Čech complex approach of this chapter facilitates calculation in $H_{\mathfrak{a}}^n(M)$.

5.3.1 Remark. (Recall that $a_1, \ldots, a_n$ (where $n > 0$) denote n elements which generate $\mathfrak{a}$.) By 5.1.20, the local cohomology module $H_{\mathfrak{a}}^n(M)$ is isomorphic to Coker d^{n-1}, where

$$d^{n-1} : C(M)^{n-1} = \bigoplus_{i=1}^{n} M_{a_1 \ldots a_{i-1} a_{i+1} \ldots a_n} \longrightarrow C(M)^n = M_{a_1 \ldots a_n}$$

is the $(n-1)$-th homomorphism in the Čech complex of M with respect to $a_1, \ldots, a_n$. We use '$[\ \ ]$' to denote natural images of elements of $M_{a_1 \ldots a_n}$ in this cokernel. Thus a typical element of $H_{\mathfrak{a}}^n(M)$ can be represented as $\left[m/(a_1 \ldots a_n)^i\right]$ for some $m \in M$ and $i \in \mathbb{N}_0$. Note that, for $u \in \{1, \ldots, n\}$, we have

$$\left[\frac{a_u^i m}{(a_1 \ldots a_n)^i}\right] = 0 \quad \text{for all } i \in \mathbb{N}_0 \text{ and } m \in M.$$

Thus $[m'/(a_1 \ldots a_n)^i] = 0$ whenever $m' \in (a_1^i, \ldots, a_n^i)M$. It is important for us to know exactly when an element of Coker d^{n-1} is zero, and this is covered in the next lemma.

5.3.2 Lemma. *Denote the product* $a_1 \ldots a_n$ *by* a. *Let* $m, g \in M$ *and* $i, j \in \mathbb{N}_0$. *Then, with the notation of 5.3.1,*

(i) $\left[m/a^i\right] = \left[g/a^j\right]$ *if and only if there exists* $k \in \mathbb{N}_0$ *such that* $k \geq \max\{i, j\}$ *and* $a^{k-i}m - a^{k-j}g \in \sum_{u=1}^{n} a_u^k M$;

(ii) *in particular,* $\left[m/a^i\right] = 0$ *if and only if there exists* $k \in \mathbb{N}_0$ *such that* $k \geq i$ *and* $a^{k-i}m \in \sum_{u=1}^{n} a_u^k M$.

Proof. (i) Since

$$\left[\frac{m}{a^i}\right] - \left[\frac{g}{a^j}\right] = \left[\frac{a^j m}{a^{i+j}}\right] - \left[\frac{a^i g}{a^{i+j}}\right] = \left[\frac{a^j m - a^i g}{a^{i+j}}\right],$$

it is enough for us to prove (ii).

(ii) ($\Leftarrow$) There exist $m_1, \ldots, m_n \in M$ such that $a^{k-i}m = \sum_{u=1}^{n} a_u^k m_u$. Therefore

$$\left[\frac{m}{a^i}\right] = \left[\frac{a^{k-i}m}{a^k}\right] = \left[\frac{\sum_{u=1}^{n} a_u^k m_u}{a^k}\right] = 0$$

because $\left(\sum_{u=1}^{n} a_u^k m_u\right)/a^k \in \operatorname{Im} d^{n-1}$.

($\Rightarrow$) Since $[m/a^i] = 0$, we have $m/a^i \in \operatorname{Im} d^{n-1}$, so that there exist $j_1, \ldots, j_n \in \mathbb{N}_0$ and $m_1, \ldots, m_n \in M$ such that

$$\frac{m}{a^i} = \frac{a_1^{j_1} m_1}{a^{j_1}} + \cdots + \frac{a_n^{j_n} m_n}{a^{j_n}}.$$

Let $j := \max\{j_1, \ldots, j_n\}$; then there exist $m_1', \ldots, m_n' \in M$ such that $m/a^i = (a_1^j m_1' + \cdots + a_n^j m_n')/a^j$, so that there is an $h \in \mathbb{N}_0$ such that

$$a^h(a^j m - a^i(a_1^j m_1' + \cdots + a_n^j m_n')) = 0.$$

Take $k := i + j + h$ to complete the proof. $\square$

5.3.3 Notation for the section. In addition to the standard notation for this chapter, we are going to assume, for the remainder of this section, that R has prime characteristic p. In these circumstances, the map $f : R \longrightarrow R$ for which $f(r) = r^p$ for all $r \in R$ is a ring homomorphism (simply because the binomial coefficient $\binom{p}{i}$ is an integer divisible by p for all $i \in \{1, \ldots, p-1\}$). We call f the *Frobenius homomorphism*.

In this section, we shall use $\lceil_f$ to denote the functor obtained from restriction of scalars using f, rather than the $\lceil_R$ of Chapter 4. We are making this change in the interests of clarity: when the two rings concerned are the same, the notation $\lceil_R$ could be confusing.

Thus $R\lceil_f$ denotes R considered as an R-module via f.

By a *Frobenius action* on the R-module M, we mean an Abelian group homomorphism $F : M \longrightarrow M$ such that $F(rm) = r^p F(m)$ for all $m \in M$ and $r \in R$. For example, the Frobenius homomorphism $f : R \longrightarrow R$ is a Frobenius action on R.

For an ideal $\mathfrak{b}$ of R and $n \in \mathbb{N}_0$, we shall denote by $\mathfrak{b}^{[p^n]}$ the ideal of R generated by all the p^n-th powers of elements of $\mathfrak{b}$. This ideal is called the p^n-th *Frobenius power of* $\mathfrak{b}$. Observe that, if $\mathfrak{b}$ can be generated by $b_1, \ldots, b_t$, then $b_1^{p^n}, \ldots, b_t^{p^n}$ generate $\mathfrak{b}^{[p^n]}$, and that $(\mathfrak{b}^{[p^n]})^{[p]} = \mathfrak{b}^{[p^{n+1}]}$.

Among other things, we aim in this section to show that there is a natural Frobenius action on each local cohomology module $H_{\mathfrak{a}}^i(R)$ of R itself with respect to $\mathfrak{a}$, and to give a detailed description of this action in the case where $i = n$. (Recall our assumption that $\mathfrak{a}$ can be generated by n elements.)

5.3.4 Theorem. *(Recall that R has prime characteristic p, and that $R\lceil_f$ denotes R considered as an R-module via f.) The Frobenius homomorphism $f : R \longrightarrow R\lceil_f$ is a homomorphism of R-modules, and thus induces R-homomorphisms $H^i_{\mathfrak{a}}(f) : H^i_{\mathfrak{a}}(R) \longrightarrow H^i_{\mathfrak{a}}(R\lceil_f)$ for all $i \in \mathbb{N}_0$.*

By the Independence Theorem 4.2.1, there is a unique isomorphism

$$\Lambda = (\lambda^i)_{i \in \mathbb{N}_0} : \left(H^i_{f(\mathfrak{a})R}(\bullet)\lceil_f\right)_{i \in \mathbb{N}_0} \overset{\cong}{\longrightarrow} \left(H^i_{\mathfrak{a}}(\bullet \lceil_f)\right)_{i \in \mathbb{N}_0}$$

of negative strongly connected sequences of covariant functors (from $\mathcal{C}(R)$ to $\mathcal{C}(R)$) such that λ^0 is the identity natural equivalence.

For each $i \in \mathbb{N}_0$, the map $(\lambda^i_R)^{-1} \circ H^i_{\mathfrak{a}}(f)$ is a Frobenius action on $H^i_{\mathfrak{a}}(R)$.

Proof. Note that $f(\mathfrak{a})R = \mathfrak{a}^{[p]}$. Since $\sqrt{\mathfrak{a}} = \sqrt{\mathfrak{a}^{[p]}}$, the local cohomology functor with respect to $\mathfrak{a}$ coincides with the local cohomology functor with respect to $\mathfrak{a}^{[p]}$. Thus $(\lambda^i_R)^{-1}$ is an R-isomorphism from $H^i_{\mathfrak{a}}(R\lceil_f)$ to $H^i_{f(\mathfrak{a})R}(R)\lceil_f = H^i_{\mathfrak{a}^{[p]}}(R)\lceil_f = H^i_{\mathfrak{a}}(R)\lceil_f$. Thus $F := (\lambda^i_R)^{-1} \circ H^i_{\mathfrak{a}}(f) :$ $H^i_{\mathfrak{a}}(R) \longrightarrow H^i_{\mathfrak{a}}(R)\lceil_f$ is an R-homomorphism (and so certainly an Abelian group homomorphism). Since $F(rh) = f(r)F(h) = r^p F(h)$ for all $h \in H^i_{\mathfrak{a}}(R)$ and $r \in R$, we see that F is a Frobenius action on $H^i_{\mathfrak{a}}(R)$. $\square$

5.3.5 Remark. It is important to note that the Frobenius actions defined in 5.3.4 on the $H^i_{\mathfrak{a}}(R)$ ($i \in \mathbb{N}_0$) do not depend on any choice of generators for $\mathfrak{a}$.

However, our next task is to describe the Frobenius action on $H^n_{\mathfrak{a}}(R)$ given by 5.3.4 in terms of our generators $a_1, \ldots, a_n$ for $\mathfrak{a}$.

5.3.6 Theorem. *(Recall that R has prime characteristic p.) The natural Frobenius action F on $H^n_{\mathfrak{a}}(R)$ of 5.3.4 is such that, with the notation of 5.3.1 and when we identify $H^n(C^\bullet)$ with $H^n_{\mathfrak{a}}(R)$ by means of the isomorphism δ^n_R of 5.1.20,*

$$F\left(\left[\frac{r}{(a_1 \ldots a_n)^k}\right]\right) = \left[\frac{r^p}{(a_1 \ldots a_n)^{kp}}\right] \quad \text{for all } r \in R \text{ and } k \in \mathbb{N}_0.$$

Proof. We have $F = (\lambda^n_R)^{-1} \circ H^n_{\mathfrak{a}}(f)$, and so the precise formula that we must establish is that

$$(\delta^n_R)^{-1} \circ (\lambda^n_R)^{-1} \circ H^n_{\mathfrak{a}}(f) \circ \delta^n_R \left(\left[\frac{r}{(a_1 \ldots a_n)^k}\right]\right) = \left[\frac{r^p}{(a_1 \ldots a_n)^{kp}}\right].$$

As in 5.3.4, let $\lceil_f : \mathcal{C}(R) \longrightarrow \mathcal{C}(R)$ denote the functor obtained from restriction of scalars using f. Since δ^n is a natural equivalence of functors, $H^n_{\mathfrak{a}}(f) \circ \delta^n_R = \delta^n_{R\lceil_f} \circ H^n(C(f)^\bullet)$; also

$$H^n(C(f)^\bullet)\left(\left[\frac{r}{(a_1 \ldots a_n)^k}\right]\right) = \left[\frac{r^p}{(a_1 \ldots a_n)^k}\right].$$

It is therefore sufficient for us to show that

$$(\delta_M^n)^{-1} \circ (\lambda_M^n)^{-1} \circ \delta_{M\lceil f}^n \left(\left[\frac{m}{(a_1 \ldots a_n)^k}\right]\right) = \left[\frac{m}{(a_1 \ldots a_n)^{kp}}\right]$$

for all $m \in M$ and $k \in \mathbb{N}_0$.

Let $b \in R$. It is straightforward to check that there is an R-isomorphism $\nu_{M,b} : (M\lceil f)_b \overset{\cong}{\longrightarrow} (M_b)\lceil f$ for which $\nu_{M,b}(m/b^j) = m/b^{pj}$ for all $m \in M$ and $j \in \mathbb{N}_0$. Therefore, for each $k \in \mathbb{N}$ with $1 \le k \le n$, we have an R-isomorphism

$$\tau_M^k := \bigoplus_{i \in \mathcal{I}(k,n)} \nu_{M,a_{i(1)} \ldots a_{i(k)}} : C(M\lceil f)^k \overset{\cong}{\longrightarrow} C(M)^k \lceil f$$

(the notation $\mathcal{I}(k,n)$ was defined in 5.1.4). It is straightforward to check that the τ_M^k ($k \in \{1, \ldots, n\}$), together with the identity map on $M\lceil f$, constitute an isomorphism $\tau_M^\bullet : C(M\lceil f)^\bullet \overset{\cong}{\longrightarrow} C(M)^\bullet \lceil f$ of complexes of R-modules and R-homomorphisms. Note that

$$\tau_M^n(m/(a_1 \ldots a_n)^k) = m/(a_1 \ldots a_n)^{kp} \quad \text{for all } m \in M \text{ and } k \in \mathbb{N}_0.$$

As M varies through $\mathcal{C}(R)$, the $\tau_M^\bullet$ constitute a natural equivalence of functors $\tau^\bullet : C(\,\bullet\,\lceil f)^\bullet \overset{\cong}{\longrightarrow} C(\,\bullet\,)^\bullet \lceil f$ (from $\mathcal{C}(R)$ to the category of all complexes of R-modules (and chain maps of such complexes)), and $\tau^\bullet$ induces an isomorphism

$$(H^i(\tau^\bullet))_{i \in \mathbb{N}_0} : (H^i(C(\,\bullet\,\lceil f)^\bullet))_{i \in \mathbb{N}_0} \overset{\cong}{\longrightarrow} (H^i(C(\,\bullet\,)^\bullet)\lceil f)_{i \in \mathbb{N}_0}$$

of negative strongly connected sequences of covariant functors which extends the identity natural equivalence on $\Gamma_\mathfrak{a}(\,\bullet\,\lceil f) = \Gamma_\mathfrak{a}(\,\bullet\,)\lceil f$.

We now use the isomorphism of connected sequences

$$(\delta^i)_{i \in \mathbb{N}_0} : (H^i(C(\,\bullet\,)^\bullet))_{i \in \mathbb{N}_0} \overset{\cong}{\longrightarrow} (H_\mathfrak{a}^i)_{i \in \mathbb{N}_0}$$

of 5.1.20 to produce further isomorphisms of connected sequences

$$(\delta_{\,\bullet\,\lceil f}^i)_{i \in \mathbb{N}_0} : (H^i(C(\,\bullet\,\lceil f)^\bullet))_{i \in \mathbb{N}_0} \overset{\cong}{\longrightarrow} (H_\mathfrak{a}^i(\,\bullet\,\lceil f))_{i \in \mathbb{N}_0}$$

and

$$(\delta_{\,\bullet\,}^i \lceil f)_{i \in \mathbb{N}_0} : (H^i(C(\,\bullet\,)^\bullet)\lceil f)_{i \in \mathbb{N}_0} \overset{\cong}{\longrightarrow} (H_\mathfrak{a}^i(\,\bullet\,)\lceil f)_{i \in \mathbb{N}_0}$$

which extend the identity natural equivalence on $\Gamma_\mathfrak{a}(\,\bullet\,\lceil f) = \Gamma_\mathfrak{a}(\,\bullet\,)\lceil f$. But then

$$(\delta_{\,\bullet\,\lceil f}^i)_{i \in \mathbb{N}_0} \circ ((H^i(\tau^\bullet))_{i \in \mathbb{N}_0})^{-1} \circ ((\delta_{\,\bullet\,}^i \lceil f)_{i \in \mathbb{N}_0})^{-1} :$$
$$(H_\mathfrak{a}^i(\,\bullet\,)\lceil f)_{i \in \mathbb{N}_0} \overset{\cong}{\longrightarrow} (H_\mathfrak{a}^i(\,\bullet\,\lceil f))_{i \in \mathbb{N}_0}$$

is an isomorphism of connected sequences which extends the identity natural equivalence on $\Gamma_{\mathfrak{a}}(\bullet \lceil_f) = \Gamma_{\mathfrak{a}}(\bullet)\lceil_f$. By the uniqueness aspect of the Independence Theorem 4.2.1, this isomorphism must be the $\Lambda = (\lambda^i)_{i \in \mathbb{N}_0}$ of 5.3.4. Hence $(\lambda^n_M)^{-1} = (\delta^n_M \lceil_f) \circ H^n(\tau^{\bullet}_M) \circ (\delta^n_{M\lceil_f})^{-1}$. Therefore

$$(\delta^n_M)^{-1} \circ (\lambda^n_M)^{-1} \circ \delta^n_{M\lceil_f} \left(\left[\frac{m}{(a_1 \ldots a_n)^k} \right] \right) = H^n(\tau^{\bullet}_M) \left(\left[\frac{m}{(a_1 \ldots a_n)^k} \right] \right)$$

$$= \left[\frac{m}{(a_1 \ldots a_n)^{kp}} \right],$$

as required. $\qquad\square$

5.3.7 Exercise. For $u, v \in \mathbb{N}$ with $u \leq v$, let $h^v_u : R/(a^u_1, \ldots, a^u_n) \longrightarrow R/(a^v_1, \ldots, a^v_n)$ be the R-homomorphism induced by multiplication by a^{v-u}, where $a := a_1 \ldots a_n$. These homomorphisms turn the family

$$(R/(a^u_1, \ldots, a^u_n))_{u \in \mathbb{N}}$$

into a direct system. For each $u \in \mathbb{N}$, let

$$h_u : R/(a^u_1, \ldots, a^u_n) \longrightarrow \varinjlim_{w \in \mathbb{N}} R/(a^w_1, \ldots, a^w_n) =: H$$

be the natural homomorphism. Show that there is an isomorphism $\alpha : H \xrightarrow{\cong} H^n_{\mathfrak{a}}(R)$.

Use the isomorphism α and the Frobenius action of 5.3.4 on $H^n_{\mathfrak{a}}(R)$ to put a Frobenius action F' on H, and show that F' is given by the following rule:

$$F'(h_u(r + (a^u_1, \ldots, a^u_n))) = h_{pu}(r^p + (a^{pu}_1, \ldots, a^{pu}_n)) \text{ for all } u \in \mathbb{N}, r \in R.$$

We plan to exploit the Frobenius action described in 5.3.4 and 5.3.6, but the applications we have in mind will have to be postponed until the final section of the next chapter, by which point we shall have covered the Non-vanishing Theorem and some interactions between local cohomology and regular sequences.

6

Fundamental vanishing theorems

There are many important results concerning the vanishing of local cohomology modules. A few results of this type have already been presented earlier in the book: for example, Exercise 1.3.9(iv) is concerned with the fact that, if an R-module M is such that $\mathfrak{a}$ contains an M-sequence of length n, then $H^i_{\mathfrak{a}}(M) = 0$ for all $i < n$; also, Theorem 3.3.1 shows that, if $\mathfrak{a}$ can be generated by t elements, then, for every R-module M, we have $H^i_{\mathfrak{a}}(M) = 0$ for all $i > t$; and we strengthened the latter result in Corollary 3.3.3, where we showed that, for every R-module M, we have $H^i_{\mathfrak{a}}(M) = 0$ for all $i > \operatorname{ara}(\mathfrak{a})$.

In this chapter, we shall provide a further result of this type: we shall prove Grothendieck's Vanishing Theorem, which states that, if the R-module L (is non-zero and) has (Krull) dimension n, then $H^i_{\mathfrak{a}}(L) = 0$ for all $i > n$. We shall also prove that, when $(R, \mathfrak{m})$ is a local ring and the non-zero, finitely generated R-module M has dimension n, then $H^n_{\mathfrak{m}}(M) \neq 0$, so that, in view of Grothendieck's Vanishing Theorem, $n = \dim M$ is the greatest integer i for which $H^i_{\mathfrak{m}}(M) \neq 0$. Also in this chapter, we shall explore in greater detail the ideas of Exercise 1.3.9, and this investigation will lead to the result that depth M is the least integer i for which $H^i_{\mathfrak{m}}(M) \neq 0$. (Recall that depth M is the common length of all maximal M-sequences.) It will thus follow that, for such an M over such a local ring $(R, \mathfrak{m})$, it is only for integers i satisfying depth $M \leq i \leq \dim M$ that it is possible that $H^i_{\mathfrak{m}}(M)$ could be non-zero, while this local cohomology module is definitely non-zero if i is at either extremity of this range.

In §6.4, we shall exploit our earlier work to obtain a geometrical application: we shall establish a special case of Serre's Affineness Criterion (see [77, §46, Corollaire 1]), concerning the following situation. Let V be an affine variety over the algebraically closed field K. Let $\mathfrak{b}$ be a non-zero ideal of $\mathcal{O}(V)$, let $V(\mathfrak{b})$ denote the closed subset of V determined by $\mathfrak{b}$, and let U be the open subset $V \setminus V(\mathfrak{b})$ of V. Thus U is a quasi-affine variety. It is of fundamental importance in algebraic geometry to be able to determine whether such a quasi-affine variety is itself affine. We shall show that U is affine if and only if $H^i_{\mathfrak{b}}(\mathcal{O}(V)) = 0$ for all $i \geq 2$. Thus this work is well suited to a chapter on 'vanishing theorems'!

In §6.5, we shall present two applications of local cohomology to local algebra. These applications are to results which have no mention of local cohomology in their statements, but which have proofs that make non-trivial use of local cohomology. One concerns the Monomial Conjecture (of M. Hochster [37, Conjecture 1]) that whenever $(a_i)_{i=1}^n$ is a system of parameters for the n-dimensional local ring R, then $(a_1 \ldots a_n)^k \notin (a_1^{k+1}, \ldots, a_n^{k+1})$ for all $k \in \mathbb{N}_0$. We shall present a proof (due to Hochster) of this conjecture in the case where R has prime characteristic p; our proof makes use of the Frobenius action on $H^n_{\mathfrak{m}}(R)$ that was produced in 5.3.4 (and the fundamental fact that $H^n_{\mathfrak{m}}(R) \neq 0$). The other application is also to local rings of characteristic p, and concerns tight closure; it too uses a Frobenius action.

6.1 Grothendieck's Vanishing Theorem

Our first main aim in this chapter is to present a proof of Grothendieck's Vanishing Theorem. We preface this with a reminder about the dimension of an R-module.

6.1.1 Reminder. Let M be a non-zero R-module. The (Krull) *dimension*, $\dim M$ or $\dim_R M$, *of* M is the supremum of lengths of chains of prime ideals in the support of M if this supremum exists, and ∞ otherwise. In the case when M is finitely generated, this is equal to $\dim R/(0 : M)$, the dimension of the ring $R/(0 : M)$, but this need not be the case if M is not finitely generated. We adopt the convention that the dimension of the zero R-module is -1.

6.1.2 Grothendieck's Vanishing Theorem. *Let M be an R-module. Then $H^i_{\mathfrak{a}}(M) = 0$ for all $i > \dim M$.*

Proof. Since, for each $\mathfrak{p} \in \operatorname{Spec}(R)$,

$$\operatorname{Supp}_{R_{\mathfrak{p}}}(M_{\mathfrak{p}}) = \{\mathfrak{q}R_{\mathfrak{p}} : \mathfrak{q} \in \operatorname{Supp} M \text{ and } \mathfrak{q} \subseteq \mathfrak{p}\},$$

it follows from 4.3.3 that it is sufficient for us to prove this result under the additional hypothesis that $(R, \mathfrak{m})$ is local. This is what we shall do.

When $\dim M = -1$, there is nothing to prove, as then $M = 0$. The result is also clear if $\mathfrak{a} = R$, as then $\Gamma_{\mathfrak{a}}$ is the zero functor of 3.4.7. We therefore suppose henceforth in this proof that $M \neq 0$ and $\mathfrak{a} \subseteq \mathfrak{m}$.

We argue by induction on $\dim M$. When $\dim M = 0$, each non-zero element $g \in M$ is annihilated by a power of $\mathfrak{m}$ (and therefore by a power of $\mathfrak{a}$), because the ring $R/(0 : g)$ is of dimension 0 and is therefore Artinian. Thus, in this case, M is $\mathfrak{a}$-torsion, and so it follows from 2.1.7(i) that $H_{\mathfrak{a}}^i(M) = 0$ for all $i > 0 = \dim M$.

Now suppose, inductively, that $\dim M = n > 0$, and the result has been proved for all R-modules of dimensions smaller than n. Since, by 3.4.10, for each $i \in \mathbb{N}_0$, the local cohomology functor $H_{\mathfrak{a}}^i$ commutes with direct limits, and M can be viewed as the direct limit of its finitely generated submodules, it is sufficient for us to prove that $H_{\mathfrak{a}}^i(M') = 0$ for all $i > n$ whenever M' is a finitely generated submodule of M. Since such an M' must have dimension not exceeding n, we can therefore assume, in this inductive step, that M itself is finitely generated. This we do.

By 2.1.7(iii), we have $H_{\mathfrak{a}}^i(M) \cong H_{\mathfrak{a}}^i(M/\Gamma_{\mathfrak{a}}(M))$ for all $i > 0$. Also, $M/\Gamma_{\mathfrak{a}}(M)$ has dimension not exceeding n, and is an $\mathfrak{a}$-torsion-free R-module, by 2.1.2. In view of the inductive hypothesis, we can, and do, assume that M is a (non-zero, finitely generated) $\mathfrak{a}$-torsion-free R-module.

We now use 2.1.1(ii) to deduce that $\mathfrak{a}$ contains an element r which is a non-zerodivisor on M. Let $t, i \in \mathbb{N}$ with $i > n$. The exact sequence

$$0 \longrightarrow M \overset{r^t}{\longrightarrow} M \longrightarrow M/r^t M \longrightarrow 0$$

(in which the second homomorphism is provided by multiplication by r^t) induces an exact sequence $H_{\mathfrak{a}}^{i-1}(M/r^t M) \longrightarrow H_{\mathfrak{a}}^i(M) \overset{r^t}{\longrightarrow} H_{\mathfrak{a}}^i(M)$ of local cohomology modules. (The fact that the second homomorphism is again provided by multiplication by r^t follows from the fact that the functor $H_{\mathfrak{a}}^i$ is R-linear.)

Now $\dim(M/r^t M) < n$ since $r^t \notin \bigcup_{\mathfrak{p} \in \mathrm{Ass}\, M} \mathfrak{p}$ and every minimal member of $\mathrm{Supp}\, M$ belongs to $\mathrm{Ass}\, M$ (see [81, Theorem 9.39]). Hence, by the inductive hypothesis, $H_{\mathfrak{a}}^{i-1}(M/r^t M) = 0$. Thus, for each $t \in \mathbb{N}$, multiplication by r^t provides a monomorphism of $H_{\mathfrak{a}}^i(M)$ into itself. But $r \in \mathfrak{a}$ and $H_{\mathfrak{a}}^i(M)$ is an $\mathfrak{a}$-torsion R-module, so that each element of it is annihilated by some power of r. Therefore $H_{\mathfrak{a}}^i(M) = 0$. This completes the inductive step, and the proof. $\square$

The next theorem can be regarded as a companion to Grothendieck's Vanishing Theorem, because it shows that, in some circumstances, this Vanishing Theorem is best possible. The method of proof employed here uses the powerful technique of reduction to the case where the local ring concerned is a complete local domain. The argument is fairly sophisticated, as it relies on ideas related to the structure theorems for complete local rings. Some readers might like to be informed now that a different proof of the following theorem is provided in Chapter 7. We preface the theorem with an elementary remark.

6.1.3 Remark. Let R' be a second commutative Noetherian ring and let $f :$ $R \to R'$ be a flat ring homomorphism. Let M be a non-zero, finitely generated R-module, generated by $m_1, \ldots, m_t$; set $\mathfrak{c} = (0 :_R M)$.

There is an exact sequence

$$0 \longrightarrow \mathfrak{c} \stackrel{\subseteq}{\longrightarrow} R \stackrel{h}{\longrightarrow} Rm_1 \oplus \cdots \oplus Rm_t$$

of R-modules and R-homomorphisms, where $h(r) = (rm_1, \ldots, rm_t)$ for all $r \in R$. Since R' is a flat R-module, the induced sequence

$$0 \longrightarrow \mathfrak{c} \otimes_R R' \longrightarrow R \otimes_R R' \stackrel{h \otimes_R R'}{\longrightarrow} (Rm_1 \oplus \cdots \oplus Rm_t) \otimes_R R'$$

is again exact, and so it follows easily from the additivity of the tensor product functor that the sequence

$$0 \longrightarrow \mathfrak{c}R' \stackrel{\subseteq}{\longrightarrow} R' \stackrel{h'}{\longrightarrow} (Rm_1 \otimes_R R') \oplus \cdots \oplus (Rm_t \otimes_R R'),$$

where $h'(r') = (m_1 \otimes r', \ldots, m_t \otimes r')$ for all $r' \in R'$, is exact. We thus deduce that $\mathfrak{c}R' = (0 :_{R'} (M \otimes_R R'))$.

It follows that, in particular, if $(R, \mathfrak{m})$ is local with ($\mathfrak{m}$-adic) completion $\widehat{R}$, and M is a non-zero, finitely generated R-module, then

$$\dim_R M = \dim R/\mathfrak{c} = \dim \widehat{R}/\mathfrak{c}\widehat{R} = \dim_{\widehat{R}}(M \otimes_R \widehat{R}).$$

6.1.4 The Non-vanishing Theorem. *Assume that $(R, \mathfrak{m})$ is local, and let M be a non-zero, finitely generated R-module of dimension n. Then $H_{\mathfrak{m}}^n(M) \neq 0$.*

Proof. Let $\widehat{R}$ denote the ($\mathfrak{m}$-adic) completion of R. Since the natural ring homomorphism $R \to \widehat{R}$ is (faithfully) flat, it follows from the Flat Base Change Theorem 4.3.2 that there is an $\widehat{R}$-isomorphism

$$H_{\mathfrak{m}}^n(M) \otimes_R \widehat{R} \cong H_{\mathfrak{m}\widehat{R}}^n(M \otimes_R \widehat{R}),$$

and so it is enough for us to show that $H_{\mathfrak{m}\widehat{R}}^n(M \otimes_R \widehat{R}) \neq 0$. Of course, $\mathfrak{m}\widehat{R}$ is the maximal ideal of the local ring $\widehat{R}$, and $M \otimes_R \widehat{R}$ is a non-zero, finitely generated $\widehat{R}$-module; also 6.1.3 shows that this $\widehat{R}$-module has dimension n. Consequently, we can, and do, assume henceforth in this proof that R is complete.

Let $\mathfrak{p}$ be a minimal member of $\operatorname{Supp} M$ for which $\dim R/\mathfrak{p} = n$. Since $\dim(\mathfrak{p}M) < n + 1$, it follows from Grothendieck's Vanishing Theorem 6.1.2 that the natural epimorphism $M \to M/\mathfrak{p}M$ induces an epimorphism $H_{\mathfrak{m}}^n(M)$ $\longrightarrow H_{\mathfrak{m}}^n(M/\mathfrak{p}M)$, and so it is enough for us to show that $H_{\mathfrak{m}}^n(M/\mathfrak{p}M) \neq 0$.

Now, since $\sqrt{(0 :_R M/\mathfrak{p}M)} = \sqrt{\mathfrak{p} + (0 :_R M)} = \mathfrak{p}$ (by [81, 9.23], for example), we can deduce that $(0 :_R M/\mathfrak{p}M) = \mathfrak{p}$. Hence $\dim_R(M/\mathfrak{p}M) = n$, and so, when we regard $M/\mathfrak{p}M$ as an $R/\mathfrak{p}$-module in the natural way, it is faithful, finitely generated, and of dimension n. By the Independence Theorem 4.2.1, we have $H^n_\mathfrak{m}(M/\mathfrak{p}M) \cong H^n_{\mathfrak{m}/\mathfrak{p}}(M/\mathfrak{p}M)$ as R-modules. Since $R/\mathfrak{p}$ is a complete local domain, it follows that it is enough for us to prove the result under the additional assumptions that R is a complete local domain and that M is a faithful R-module.

At this point, we appeal to Cohen's Structure Theorem for complete local rings: by [50, Theorem 29.4], there exists a complete regular local subring $(R', \mathfrak{m}')$ of R which is such that R is finitely generated as an R'-module. Since R is integral over R', it follows that $\mathfrak{m}$ is the one and only prime ideal of R which has contraction to R' equal to $\mathfrak{m}'$. Hence $\sqrt{(\mathfrak{m}'R)} = \mathfrak{m}$, and so it follows from 1.1.3 and the Independence Theorem 4.2.1 that $H^n_\mathfrak{m}(M) = H^n_{\mathfrak{m}'R}(M) \cong H^n_{\mathfrak{m}'}(M)$ as R'-modules. It is thus sufficient for us to show that $H^n_{\mathfrak{m}'}(M) \neq 0$. Obviously, M is faithful and finitely generated as an R'-module. Moreover, $\dim R' = \dim R = n$ (since R is integral over R'). We can therefore replace R by R' and thus assume that R is a complete regular local ring during the remainder of this proof.

Next, let, for each R-module G,

$$\tau(G) := \{g \in G : \text{ there exists } r \in R \setminus \{0\} \text{ such that } rg = 0\},$$

a submodule of G. Note that $(0 : \tau(M)) \neq 0$, and so $\tau(M)$ has dimension less than n. Another use of Grothendieck's Vanishing Theorem 6.1.2 shows that the natural epimorphism $M \to M/\tau(M)$ induces an isomorphism $H^n_\mathfrak{m}(M) \to H^n_\mathfrak{m}(M/\tau(M))$, and so it is sufficient for us to show that $H^n_\mathfrak{m}(M/\tau(M)) \neq 0$. Note that $\tau(M/\tau(M)) = 0$ and that $M/\tau(M)$ still has dimension n (as its annihilator is 0). Therefore we can, and do, assume that $\tau(M) = 0$.

Let K be the field of fractions of R, and note that, since $\tau(M) = 0$, the natural R-homomorphism $M \to M \otimes_R K$ is injective. Let t denote the (torsion-free) rank of M as an R-module, that is, the vector space dimension $\dim_K(M \otimes_R K)$. There exists $r \in R \setminus \{0\}$ such that M, as an R-module, can be embedded in the submodule

$$R\tfrac{1}{r} \oplus \cdots \oplus R\tfrac{1}{r} \ (t \text{ copies}) \quad \text{of} \quad K \oplus \cdots \oplus K \ (t \text{ copies}).$$

But $R\tfrac{1}{r} \cong R$, and so there is a free R-module F of rank t and an R-monomorphism $h : M \to F$; also, since $h \otimes_R K : M \otimes_R K \to F \otimes_R K$ must be a K-monomorphism between t-dimensional vector spaces over K, and therefore an isomorphism, it follows that $\dim(\operatorname{Coker} h) < n$. It therefore follows from Grothendieck's Vanishing Theorem 6.1.2 that h induces an exact sequence $H^n_{\mathfrak{m}}(M) \longrightarrow H^n_{\mathfrak{m}}(F) \longrightarrow 0$. Now since R is a regular local ring, it follows from 4.2.6 that $H^n_{\mathfrak{m}}(R) \neq 0$; finally, we deduce from the additivity of the functor $H^n_{\mathfrak{m}}$ that $H^n_{\mathfrak{m}}(F) \neq 0$. Hence $H^n_{\mathfrak{m}}(M) \neq 0$ and this completes the proof. □

6.1.5 Exercise. Suppose that M is a finitely generated R-module for which $M \neq \mathfrak{a}M$. Show that there exists $i \in \mathbb{N}_0$ for which $H^i_{\mathfrak{a}}(M) \neq 0$.

In 3.3.4, we introduced the cohomological dimension cohd$(\mathfrak{a})$ of $\mathfrak{a}$.

6.1.6 Lemma. *If $\mathfrak{a}$ is proper, then* ht $\mathfrak{a} \leq$ cohd$(\mathfrak{a})$.

Proof. Let $\mathfrak{p}$ be a minimal prime ideal of $\mathfrak{a}$ such that ht $\mathfrak{p} =$ ht $\mathfrak{a} =: h$. Then, by 1.1.3 and 4.3.3, we have $(H^h_{\mathfrak{a}}(R))_{\mathfrak{p}} \cong H^h_{\mathfrak{a}R_{\mathfrak{p}}}(R_{\mathfrak{p}}) = H^h_{\mathfrak{p}R_{\mathfrak{p}}}(R_{\mathfrak{p}})$, and this is non-zero by the Non-vanishing Theorem 6.1.4. Therefore $H^h_{\mathfrak{a}}(R) \neq 0$, so that ht $\mathfrak{a} = h \leq$ cohd$(\mathfrak{a})$. □

6.1.7 ♯Exercise. Assume that $(R, \mathfrak{m})$ is local, and let M be a non-zero, finitely generated R-module of dimension $n > 0$. Show that $H^n_{\mathfrak{m}}(M)$ is not finitely generated. (Here are some hints: use $M/\Gamma_{\mathfrak{m}}(M)$ to see that one can make the additional assumption that M is an $\mathfrak{m}$-torsion-free R-module; use 2.1.1(ii) to see that, then, $\mathfrak{m}$ contains an element r which is a non-zerodivisor on M; note that, if $H^n_{\mathfrak{m}}(M)$ were finitely generated, then there would exist $t \in \mathbb{N}$ such that $r^t H^n_{\mathfrak{m}}(M) = 0$; and consider the long exact sequence of local cohomology modules induced by the exact sequence $0 \longrightarrow M \xrightarrow{\ r^t\ } M \longrightarrow M/r^t M \longrightarrow 0$ in order to obtain a contradiction.)

6.1.8 Exercise. Provide an example to show that the result of Theorem 6.1.4 is not always true if the hypothesis that M be finitely generated is omitted.

6.1.9 ♯Exercise. Let $T : \mathcal{C}(R) \longrightarrow \mathcal{C}(R)$ be an R-linear covariant functor. For each R-module M, and each $g \in M$, let $\mu_{g,M} : R \to M$ be the R-homomorphism for which $\mu_{g,M}(r) = rg$ for all $r \in R$.

Show that there is a natural transformation of functors

$$\theta : (\bullet) \otimes_R T(R) \longrightarrow T$$

(from $\mathcal{C}(R)$ to $\mathcal{C}(R)$) which is such that, for each R-module M,

$$\theta_M(g \otimes z) = T(\mu_{g,M})(z) \quad \text{for all } g \in M \text{ and } z \in T(R).$$

Show also that θ_F is an isomorphism whenever F is a finitely generated free R-module, and deduce that, if T is right exact, then θ_M is an isomorphism whenever M is a finitely generated R-module.

6.1.10 ♯Exercise. Let n be an integer such that $H_{\mathfrak{a}}^i(M) = 0$ for all $i > n$ and for all R-modules M. (For example, this condition would be satisfied if $n \geq \operatorname{ara}(\mathfrak{a})$ (by 3.3.3), or if $n \geq \dim R$ (by 6.1.2). Use Exercise 6.1.9 to prove that $H_{\mathfrak{a}}^n$ is naturally equivalent to $(\,\bullet\,) \otimes_R H_{\mathfrak{a}}^n(R)$.

We interpret the supremum of the empty set of integers as $-\infty$.

6.1.11 Proposition. *The cohomological dimension* $\operatorname{cohd}(\mathfrak{a})$ *of* $\mathfrak{a}$ *is equal to* $\sup\{i \in \mathbb{N}_0 : H_{\mathfrak{a}}^i(R) \neq 0\}$.

Proof. The result is clear when $\mathfrak{a} = R$, and so we suppose that $\mathfrak{a}$ is proper. Denote by d the greatest integer i such that $H_{\mathfrak{a}}^i(R) \neq 0$. (Such an integer exists, by 3.3.4 and 6.1.6.) Clearly $d \leq \operatorname{cohd}(\mathfrak{a}) =: c$. By 6.1.10, the functors $H_{\mathfrak{a}}^c$ and $(\,\bullet\,) \otimes_R H_{\mathfrak{a}}^c(R)$ are naturally equivalent. Therefore $H_{\mathfrak{a}}^c(R) \neq 0$ and $c \leq d$.   □

6.2 Connections with grade

It is now time for us to explore in detail connections, already hinted at in Exercise 1.3.9, between regular sequences and local cohomology. To set the scene, and clarify precisely what background information the reader will require for this work, we begin by recalling the definition of 'M-sequence' and quoting, without proof, two results from Matsumura [50]; however, we shall use terminology different from his.

6.2.1 Definition. (See [50, p. 123].) Let $a_1, \ldots, a_n \in R$ and let M be an R-module. We say that the sequence $a_1, \ldots, a_n$ is a *poor M-sequence* precisely when

 (i) a_1 is a non-zerodivisor on M and, for each $i = 2, \ldots, n$, the element a_i is a non-zerodivisor on $M / \sum_{j=1}^{i-1} a_j M$.

Furthermore, $a_1, \ldots, a_n$ is said to be an *M-sequence*, or an *M-regular sequence*, precisely when it is a poor M-sequence, that is, it satisfies condition (i) above, and, in addition,

 (ii) $\sum_{j=1}^n a_j M \neq M$.

6.2.2 Theorem. (See [50, Theorem 16.6].) *Let M be a finitely generated R-module such that $\mathfrak{a}M \neq M$; let $n \in \mathbb{N}$. Then the following statements are equivalent:*

 (i) $\operatorname{Ext}_R^i(G, M) = 0$ *for all $i < n$ and each finitely generated R-module G for which $\operatorname{Supp} G \subseteq \operatorname{Var}(\mathfrak{a})$;*

(ii) $\operatorname{Ext}_R^i(R/\mathfrak{a}, M) = 0$ *for all* $i < n$;

(iii) *there is a finitely generated R-module G with* $\operatorname{Supp} G = \operatorname{Var}(\mathfrak{a})$ *for which* $\operatorname{Ext}_R^i(G, M) = 0$ *for all* $i < n$; *and*

(iv) *there exists an M-sequence of length n contained in* $\mathfrak{a}$. $\qquad\square$

Ideas involved in Theorem 6.2.2 can be adapted to prove the result in the following exercise, which is often useful.

6.2.3 ♯Exercise. Let N be an R-module (note that it is not assumed that N is finitely generated) such that there exists a poor N-sequence of length n contained in $\mathfrak{a}$. Show that $\operatorname{Ext}_R^i(G, N) = 0$ for all $i < n$ and each finitely generated R-module G for which $\operatorname{Supp} G \subseteq \operatorname{Var}(\mathfrak{a})$.

Observe that, with the notation of 6.2.2, every M-sequence contained in $\mathfrak{a}$ (even the empty one!) can be extended to a maximal one, since otherwise, as $\mathfrak{a}M \neq M$, there would exist an infinite sequence $(b_i)_{i\in\mathbb{N}}$ of elements of $\mathfrak{a}$ such that $b_1, \ldots, b_n$ is an M-sequence for all $n \in \mathbb{N}$, and this would lead to an infinite strictly ascending chain

$$(b_1) \subset (b_1, b_2) \subset \cdots \subset (b_1, \ldots, b_n) \subset \cdots$$

of ideals of R. We incorporate this observation into the next theorem, which is otherwise taken from Matsumura [50].

6.2.4 Theorem and Definition. (See [50, Theorem 16.7].) *Let M be a finitely generated R-module such that* $\mathfrak{a}M \neq M$.

(i) *There exists an M-sequence contained in* $\mathfrak{a}$ *which cannot be extended to a longer one by the addition of an extra term (such a sequence will henceforth be referred to as a* maximal *M-sequence contained in* $\mathfrak{a}$*).*

(ii) *Every M-sequence contained in* $\mathfrak{a}$ *can be extended to a maximal one.*

(iii) *All maximal M-sequences contained in* $\mathfrak{a}$ *have the same length, namely the least integer i such that* $\operatorname{Ext}_R^i(R/\mathfrak{a}, M) \neq 0$.

We shall refer to the common length of all maximal M-sequences contained in $\mathfrak{a}$ *as the* M-grade *of* $\mathfrak{a}$, *and we shall denote this non-negative integer by* $\operatorname{grade}_M \mathfrak{a}$. $\qquad\square$

6.2.5 Notes. The following points should be noted.

(i) If the ideal $\mathfrak{a}$ is proper, then the R-grade of $\mathfrak{a}$ is defined: we shall follow Rees's original terminology [68] and refer to this simply as the *grade of* $\mathfrak{a}$, and we shall denote it by $\operatorname{grade} \mathfrak{a}$.

(ii) Suppose that $(R, \mathfrak{m})$ is local, and that M is a non-zero finitely generated R-module. Then it follows from Nakayama's Lemma that $\mathfrak{m}M \neq M$, and so $\operatorname{grade}_M \mathfrak{m}$ is defined: this is referred to as the *depth of* M, and denoted by $\operatorname{depth} M$ or $\operatorname{depth}_R M$. Since every M-sequence must be contained in $\mathfrak{m}$, we see that $\operatorname{depth} M$ is equal to the common length of all maximal M-sequences.

6.2.6 Exercise. Let M be an R-module and let $i \in \mathbb{N}_0$. Show that

$$\operatorname{Supp}\left(H_{\mathfrak{a}}^i(M)\right) \subseteq \operatorname{Supp} M \cap \operatorname{Var}(\mathfrak{a}).$$

Deduce that, if M is finitely generated and $\mathfrak{a}M = M$, then $H_{\mathfrak{a}}^j(M) = 0$ for all $j \in \mathbb{N}_0$.

6.2.7 Theorem. *Let M be a finitely generated R-module such that $\mathfrak{a}M \neq M$. Then $\operatorname{grade}_M \mathfrak{a}$ is the least integer i such that $H_{\mathfrak{a}}^i(M) \neq 0$.*

Proof. Let $g := \operatorname{grade}_M \mathfrak{a}$. We use induction on g. When $g = 0$, every element of $\mathfrak{a}$ must be a zerodivisor on M, and so $\Gamma_{\mathfrak{a}}(M) \neq 0$ by 2.1.1(ii). Now suppose that $g > 0$ and that the result has been proved for each finitely generated R-module N with $\mathfrak{a}N \neq N$ and $\operatorname{grade}_N \mathfrak{a} < g$.

There exists $a_1 \in \mathfrak{a}$ such that a_1 is a non-zerodivisor on M. Set $M_1 := M/a_1M$, and observe that ($\mathfrak{a}M_1 \neq M_1$ and) $\operatorname{grade}_{M_1} \mathfrak{a} = g-1$. Therefore, by the inductive hypothesis, $H_{\mathfrak{a}}^i(M_1) = 0$ for all $i < g-1$, while $H_{\mathfrak{a}}^{g-1}(M_1) \neq 0$. The exact sequence $0 \longrightarrow M \xrightarrow{a_1} M \longrightarrow M_1 \longrightarrow 0$ induces, for each $i \in \mathbb{N}$, an exact sequence

$$H_{\mathfrak{a}}^{i-1}(M) \longrightarrow H_{\mathfrak{a}}^{i-1}(M_1) \longrightarrow H_{\mathfrak{a}}^i(M) \xrightarrow{a_1} H_{\mathfrak{a}}^i(M).$$

This shows that, for $i < g$, the element a_1 is a non-zerodivisor on $H_{\mathfrak{a}}^i(M)$, so that, since this module is $\mathfrak{a}$-torsion, it must be zero. We therefore have an exact sequence $0 \longrightarrow H_{\mathfrak{a}}^{g-1}(M_1) \longrightarrow H_{\mathfrak{a}}^g(M)$, and since $H_{\mathfrak{a}}^{g-1}(M_1) \neq 0$, it follows that $H_{\mathfrak{a}}^g(M) \neq 0$. $\square$

6.2.8 Corollary. *Assume that $(R, \mathfrak{m})$ is local, and let M be a non-zero, finitely generated R-module. Then any integer i for which $H_{\mathfrak{m}}^i(M) \neq 0$ must satisfy*

$$\operatorname{depth} M \leq i \leq \dim M,$$

while for i at either extremity of this range we do have $H_{\mathfrak{m}}^i(M) \neq 0$.

Proof. This is immediate from Grothendieck's Vanishing Theorem 6.1.2, the Non-vanishing Theorem 6.1.4 and the above 6.2.7, because $M \neq \mathfrak{m}M$ by Nakayama's Lemma, and $\operatorname{depth} M = \operatorname{grade}_M \mathfrak{m}$. $\square$

6.2.9 Corollary. *Assume that $(R, \mathfrak{m})$ is local, and let M be a non-zero, finitely generated R-module. Then there is exactly one integer i for which $H^i_{\mathfrak{m}}(M) \neq 0$ if and only if $\operatorname{depth} M = \dim M$, that is, if and only if M is a Cohen–Macaulay R-module (see [50, p. 134]).*

In particular, if $(R, \mathfrak{m})$ is a regular local ring of dimension n, then n is the unique integer i for which $H^i_{\mathfrak{m}}(R) \neq 0$. □

6.2.10 Remark. Let M be a finitely generated R-module such that $\mathfrak{a}M \neq M$. It is immediate from Theorems 2.2.6(i)(c) and 6.2.7 that $\eta_M : M \longrightarrow D_{\mathfrak{a}}(M)$ is an isomorphism if and only if $\operatorname{grade}_M \mathfrak{a} \geq 2$.

6.2.11 Exercise. Let $T : \mathcal{C}(R) \longrightarrow \mathcal{C}(R)$ be an R-linear covariant functor with the property that $T(M)$ is $\mathfrak{a}$-torsion for every R-module M.

(i) Show that, for each R-module M, the result $\mathcal{R}^i T(M)$ of applying the i-th right derived functor of T to M is again $\mathfrak{a}$-torsion.

(ii) Assume, in addition, that T is left exact and such that, for each finitely generated R-module M, the ideal $\mathfrak{a}$ contains a non-zerodivisor on M if and only if $T(M) = 0$. Let $r \in \mathbb{N}_0$, and let M be a finitely generated R-module such that $\mathfrak{a}M \neq M$. Prove that $\mathcal{R}^i T(M) = 0$ for all integers $i < r$ if and only if $\mathfrak{a}$ contains an M-sequence of length r.

(iii) Use part (ii) applied to the functor $\operatorname{Hom}_R(R/\mathfrak{a}, \bullet)$ to reprove the equivalence of statements (ii) and (iv) in Theorem 6.2.2.

(iv) Use part (ii) applied to the functor $\Gamma_{\mathfrak{a}}$ to prove the result of 6.2.7.

Corollary 6.2.8 raises the following question: if J is any prescribed finite non-empty set of non-negative integers, does there exist a local ring $(R', \mathfrak{m}')$ having the property that $H^i_{\mathfrak{m}'}(R') \neq 0$ if and only if $i \in J$? An affirmative answer to this question was provided by I. G. Macdonald [46], and the next two exercises sketch the essence of his argument.

6.2.12 ♯Exercise. Let M be an R-module.

(i) Show that the Abelian group $R \oplus M$ is a commutative ring (with identity) with respect to multiplication defined by

$$(r_1, m_1)(r_2, m_2) = (r_1 r_2, r_1 m_2 + r_2 m_1)$$

for all $(r_1, m_1), (r_2, m_2) \in R \oplus M$. This ring is called the *trivial extension of R by M*, and we shall denote it by $R \propto M$.

(ii) Show that the nilradical of $R \propto M$ contains $0 \times M$, and show that $\operatorname{Spec}(R \propto M) = \{\mathfrak{p} \times M : \mathfrak{p} \in \operatorname{Spec}(R)\}$.

(iii) Show that the ring $R \propto M$ is Noetherian if and only if M is a finitely generated R-module.

(iv) Let S be a multiplicatively closed subset of R. Show that $S \times M$ is a multiplicatively closed subset of $R \propto M$ and that there is a ring isomorphism $\phi : (S \times M)^{-1}(R \propto M) \xrightarrow{\cong} S^{-1}R \propto S^{-1}M$ for which $\phi((r,m)/(s,m')) = (r/s, (sm - rm')/s^2)$ for all $r \in R$, $s \in S$ and $m, m' \in M$. In particular, this will show that

$$(R \propto M)_{(\mathfrak{p} \times M)} \cong R_{\mathfrak{p}} \propto M_{\mathfrak{p}} \quad \text{for all } \mathfrak{p} \in \operatorname{Spec}(R).$$

(v) Show that, if R is local and M is finitely generated, then $R \propto M$ is local with maximal ideal $\mathfrak{m} \times M$, and that $\dim(R \propto M) = \dim R$.

6.2.13 Exercise. Let h, n be integers such that $0 \leq h \leq n$, and let J be an arbitrary set of integers such that $\{h, n\} \subseteq J \subseteq \{i \in \mathbb{N}_0 : h \leq i \leq n\}$. Let $(R, \mathfrak{m})$ be a regular local ring of dimension n, and let $u_1, \ldots, u_n$ be n elements which generate the maximal ideal $\mathfrak{m}$. For each $j = 0, \ldots, n-1$, let $\mathfrak{p}_j = Ru_{j+1} + \cdots + Ru_n$. Set

$$M := \bigoplus_{j \in J \setminus \{n\}} R/\mathfrak{p}_j.$$

(i) Show that, for $i \in \mathbb{N}_0$, we have $H^i_{\mathfrak{m}}(M) \neq 0$ if and only if $i \in J \setminus \{n\}$.

(ii) Let $R' := R \propto M$, the trivial extension of R by M of 6.2.12 above. By that exercise, R' is a local ring, with maximal ideal $\mathfrak{m}' := \mathfrak{m} \times M$. The ring homomorphism $\psi : R \propto M \to R$ for which $\psi((r, m)) = r$ for all $(r, m) \in R \propto M$ enables us to regard both R and M as R'-modules. Use the exact sequence $0 \longrightarrow M \xrightarrow{q} R \propto M \xrightarrow{\psi} R \longrightarrow 0$ of R'-modules (in which q is the canonical injection) to show that, for $i \in \mathbb{N}_0$, we have $H^i_{\mathfrak{m}'}(R') \neq 0$ if and only if $i \in J$.

6.2.14 Exercise. Let M be an R-module, and let $a_1, \ldots, a_n \in R$ be such that $M \neq (a_1, \ldots, a_n)M$. Prove that $a_1, \ldots, a_n$ is an M-sequence if and only if

$$H^{i-1}_{(a_1,\ldots,a_i)}(M) = 0 \quad \text{for all } i = 1, \ldots, n.$$

(Here are some hints for the implication '$\Leftarrow$'. Use induction on n: for the inductive step, on the assumption that $n > 1$ and the result has been proved for smaller values of n, use Exercise 1.3.9(iv) (in conjunction with the hypotheses) to help you prove that

$$H^{i-2}_{(a_1,\ldots,a_i)}(M/a_1 M) = 0 \quad \text{for all } i = 2, \ldots, n;$$

then use the Independence Theorem 4.2.1 to deduce that

$$H^{i-2}_{(a_2,\ldots,a_i)}(M/a_1 M) = 0 \quad \text{for all } i = 2, \ldots, n.)$$

6.2.15 Exercise. Let $(R, \mathfrak{m})$ be a regular local ring of dimension $d \geq 2$, and let $u_1, \ldots, u_d$ be d elements which generate $\mathfrak{m}$. Let $t \in \mathbb{N}_0$ with $t < d - 1$, and let $\mathfrak{p} = Ru_1 + \cdots + Ru_t$, a prime ideal of R. Let M denote the R-module $R_{\mathfrak{p}}/R$.

(i) Show that $\mathfrak{p}M \neq M$.
(ii) Show that $u_1, \ldots, u_t$ is a maximal M-sequence contained in $\mathfrak{m}$.
(iii) Show that the least integer i such that $H^i_{\mathfrak{m}}(M) \neq 0$ is $d - 1$.

Contrast this with the situation for finitely generated R-modules!

6.2.16 Exercise. Let $(R, \mathfrak{m})$ be a regular local ring of dimension $d > 2$, let K denote the quotient field of R, and let $N := R \oplus K/R$.

Show that $\mathfrak{m}N \neq N$, that $\mathfrak{m}$ consists entirely of zerodivisors on N, and that $\eta_N : N \to D_{\mathfrak{m}}(N)$ is an isomorphism. Compare this with the situation for finitely generated R-modules described in 6.2.10.

6.3 Exactness of ideal transforms

This section prepares the ground for our presentation, in the next section, of the promised special case of Serre's Affineness Criterion. Central to this preparation are various necessary and sufficient conditions for the exactness of the $\mathfrak{a}$-transform functor $D_{\mathfrak{a}}$; some of these were presented by P. Schenzel in [73].

6.3.1 Lemma. *The following statements are equivalent:*

(i) *the $\mathfrak{a}$-transform functor $D_{\mathfrak{a}}$ is exact;*
(ii) $H^i_{\mathfrak{a}}(R) = 0$ *for all $i \geq 2$;*
(iii) $H^2_{\mathfrak{a}}(M) = 0$ *for each finitely generated R-module M;*
(iv) $H^2_{\mathfrak{a}}(M) = 0$ *for each R-module M.*

Proof. (i) $\Rightarrow$ (ii) By 2.2.6(ii), for each $i \in \mathbb{N}$, the i-th right derived functor $\mathcal{R}^i D_{\mathfrak{a}}$ of $D_{\mathfrak{a}}$ is naturally equivalent to $H^{i+1}_{\mathfrak{a}}$. Now exactness of $D_{\mathfrak{a}}$ implies that $\mathcal{R}^i D_{\mathfrak{a}}(R) = 0$ for all $i \in \mathbb{N}$; therefore $H^i_{\mathfrak{a}}(R) = 0$ for all $i \geq 2$.

(ii) $\Rightarrow$ (iii) Let $\operatorname{ara}(\mathfrak{a}) = t$. By 3.3.3, for every R-module M, we have $H^i_{\mathfrak{a}}(M) = 0$ for all $i > t$. We now argue by descending induction on i. Suppose that $i \in \mathbb{N}$ with $i > 2$, and we have proved that $H^i_{\mathfrak{a}}(M') = 0$ for each finitely generated R-module M'; let M be an arbitrary finitely generated R-module. There exists an exact sequence

$$0 \longrightarrow N \longrightarrow F \longrightarrow M \longrightarrow 0$$

of finitely generated R-modules and R-homomorphisms in which F is free.

This induces an exact sequence $H_\mathfrak{a}^{i-1}(F) \longrightarrow H_\mathfrak{a}^{i-1}(M) \longrightarrow H_\mathfrak{a}^i(N)$. Since $i - 1 \geq 2$, it follows from the additivity of the $(i - 1)$-th local cohomology functor and condition (ii) that $H_\mathfrak{a}^{i-1}(F) = 0$; since $H_\mathfrak{a}^i(N) = 0$ by our inductive assumption, we can deduce that $H_\mathfrak{a}^{i-1}(M) = 0$. This completes the inductive step. Hence statement (iii) is proved by descending induction.

(iii) $\Rightarrow$ (iv) This is immediate from the fact (3.4.10) that $H_\mathfrak{a}^2$ commutes with direct limits, because each R-module can be viewed as the direct limit of its finitely generated submodules.

(iv) $\Rightarrow$ (i) By 2.2.6(ii), the first right derived functor $\mathcal{R}^1 D_\mathfrak{a}$ of $D_\mathfrak{a}$ is naturally equivalent to $H_\mathfrak{a}^2$. It therefore follows from statement (iv) (and the left exactness of $D_\mathfrak{a}$) that $D_\mathfrak{a}$ is exact. $\square$

6.3.2 Exercise. Assume that ht $\mathfrak{p} = 1$ for every minimal prime ideal $\mathfrak{p}$ of $\mathfrak{a}$, and that $R_\mathfrak{q}$ is a UFD for each prime ideal $\mathfrak{q}$ of R which contains $\mathfrak{a}$. Prove that $D_\mathfrak{a} : \mathcal{C}(R) \to \mathcal{C}(R)$ is exact.

6.3.3 Remark. Let R' be a second commutative Noetherian ring and let $f : R \to R'$ be a ring homomorphism. Suppose that the $\mathfrak{a}$-transform functor $D_\mathfrak{a}$ is exact. Then since the restriction functor $\lceil_R : \mathcal{C}(R') \to \mathcal{C}(R)$ is exact and, by 2.2.24, there is a natural equivalence of functors

$$\varepsilon : D_{\mathfrak{a}R'}(\,\bullet\,)\lceil_R \longrightarrow D_\mathfrak{a}(\,\bullet\,\lceil_R),$$

it follows that the $\mathfrak{a}R'$-transform functor $D_{\mathfrak{a}R'} : \mathcal{C}(R') \to \mathcal{C}(R')$ is also exact.

Recall from Exercise 2.2.5 that $D_\mathfrak{a}(R)$ has a natural structure as a commutative ring with identity, and from Exercise 2.2.12 that $\eta_R : R \to D_\mathfrak{a}(R)$ is a ring homomorphism. Thus we can regard $D_\mathfrak{a}(R)$ as an R-algebra by means of η_R. Our next proposition is concerned with this R-algebra structure.

6.3.4 Proposition. *Suppose that $\mathfrak{a}D_\mathfrak{a}(R) = D_\mathfrak{a}(R)$. Then $D_\mathfrak{a}(R)$ is a finitely generated R-algebra.*

Proof. Note that $\Gamma_\mathfrak{a}(R) = R$ if and only if $\mathfrak{a}$ is nilpotent, and that in this case $\Gamma_\mathfrak{a}$ is the identity functor, so that $D_\mathfrak{a}(R) = 0$ by 2.2.6(i)(c) and the claim is clear in this case. Thus we suppose henceforth in this proof that $\Gamma_\mathfrak{a}(R) \neq R$.

Set $\overline{R} := R/\Gamma_\mathfrak{a}(R)$. We mentioned just before the statement of the proposition that $\eta_R : R \longrightarrow D_\mathfrak{a}(R)$ is a ring homomorphism. By 2.2.6(i)(c), we have $\mathrm{Ker}(\eta_R) = \Gamma_\mathfrak{a}(R)$, and so η_R induces an injective ring homomorphism $\theta_R : \overline{R} \longrightarrow D_\mathfrak{a}(R)$. Moreover, as $\mathrm{Coker}\,\eta_R \cong H_\mathfrak{a}^1(R)$ by 2.2.6(i)(c), the $\overline{R}$-module $\mathrm{Coker}\,\theta_R$ is $\overline{\mathfrak{a}}$-torsion, where $\overline{\mathfrak{a}} := \mathfrak{a}\overline{R}$, the extension of $\mathfrak{a}$ to $\overline{R}$.

Therefore, by 2.2.17, the unique $\overline{R}$-homomorphism $\psi : D_\mathfrak{a}(R) \longrightarrow D_{\overline{\mathfrak{a}}}(\overline{R})$

such that the diagram

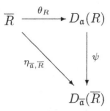

commutes is a ring homomorphism. Furthermore, by the Independence Theorem 4.2.1, we have $H_{\overline{\mathfrak{a}}}^i(D_{\mathfrak{a}}(R)) \cong H_{\mathfrak{a}}^i(D_{\mathfrak{a}}(R))$ for all $i \in \mathbb{N}_0$, and so $H_{\overline{\mathfrak{a}}}^i(D_{\mathfrak{a}}(R)) = 0$ for $i = 0, 1$ by 2.2.10(iv). Hence, in view of 2.2.15(iii), the map ψ is a ring isomorphism. Therefore $\overline{\mathfrak{a}}D_{\overline{\mathfrak{a}}}(\overline{R}) = D_{\overline{\mathfrak{a}}}(\overline{R})$, and, if we can show that $D_{\overline{\mathfrak{a}}}(\overline{R})$ is a finitely generated $\overline{R}$-algebra, then it will follow that $D_{\mathfrak{a}}(R)$ is a finitely generated R-algebra. Since $\Gamma_{\overline{\mathfrak{a}}}(\overline{R}) = 0$ (by 2.1.2), we therefore assume henceforth in this proof that $\Gamma_{\mathfrak{a}}(R) = 0$.

By 2.1.1(ii), this means that $\mathfrak{a}$ contains a non-zerodivisor s on R. Note that $\eta_R : R \to D_{\mathfrak{a}}(R)$ is injective, by 2.2.6(i)(c). We again use R_s to denote the ring of fractions of R with respect to the multiplicatively closed subset $\{s^i : i \in \mathbb{N}_0\}$. By 2.2.18, the subring $D := \bigcup_{n \in \mathbb{N}}(R :_{R_s} \mathfrak{a}^n)$ of R_s satisfies $\mathfrak{a}D = D$, and it will be sufficient for us to show that D is a finitely generated R-algebra.

Let $a_1, \ldots, a_t$ be t elements which generate $\mathfrak{a}$. Since $\mathfrak{a}D = D$, there exist $y_1, \ldots, y_t \in D$ such that $1 = \sum_{i=1}^t a_i y_i$. We aim to show that $D = R[y_1, \ldots, y_t]$. We achieve this by showing, by induction on n, that

$$(R :_{R_s} \mathfrak{a}^n) \subseteq R[y_1, \ldots, y_t] \quad \text{for every } n \in \mathbb{N}_0.$$

This claim is clear for $n = 0$, and so we suppose that $n > 0$ and the claim has been proved for smaller values of n. Let $z \in (R :_{R_s} \mathfrak{a}^n)$. Note that

$$a_i z \in (R :_{R_s} \mathfrak{a}^{n-1}) \subseteq R[y_1, \ldots, y_t] \quad \text{for all } i = 1, \ldots, t.$$

Thus $z = 1z = \sum_{i=1}^t a_i y_i z = \sum_{i=1}^t (a_i z) y_i \in R[y_1, \ldots, y_t]$, and the inductive step is complete.

Therefore $D = R[y_1, \ldots, y_t]$, as claimed, and $D_{\mathfrak{a}}(R)$ is a finitely generated R-algebra. $\qquad\square$

In Lemma 6.3.1, we established several criteria for the exactness of the $\mathfrak{a}$-transform functor $D_{\mathfrak{a}}$. We are now in a position to prove a further such criterion.

6.3.5 Proposition. *The $\mathfrak{a}$-transform functor $D_{\mathfrak{a}}$ is exact if and only if*

$$\mathfrak{a}D_{\mathfrak{a}}(R) = D_{\mathfrak{a}}(R).$$

Proof. ($\Rightarrow$) Assume first that $D_\mathfrak{a}$ is exact. By 6.1.9, we have

$$D_\mathfrak{a}(R/\mathfrak{a}) \cong (R/\mathfrak{a}) \otimes_R D_\mathfrak{a}(R) \cong D_\mathfrak{a}(R)/\mathfrak{a}D_\mathfrak{a}(R).$$

Since $D_\mathfrak{a}(R/\mathfrak{a}) = 0$ by 2.2.10(i), we deduce that $\mathfrak{a}D_\mathfrak{a}(R) = D_\mathfrak{a}(R)$.

($\Leftarrow$) Assume now that $\mathfrak{a}D_\mathfrak{a}(R) = D_\mathfrak{a}(R)$. By Lemma 6.3.1, it is enough for us to show that $H^i_\mathfrak{a}(R) = 0$ for all $i \geq 2$; by Corollary 2.2.10(v), we have $H^i_\mathfrak{a}(R) \cong H^i_\mathfrak{a}(D_\mathfrak{a}(R))$ for all $i \geq 2$, and therefore it is enough for us to show that $H^i_\mathfrak{a}(D_\mathfrak{a}(R)) = 0$ for all $i \geq 2$.

By Proposition 6.3.4, the commutative R-algebra $D_\mathfrak{a}(R)$ is finitely generated, and therefore a Noetherian ring. Therefore, by the Independence Theorem 4.2.1,

$$H^i_\mathfrak{a}(D_\mathfrak{a}(R)) \cong H^i_{\mathfrak{a}D_\mathfrak{a}(R)}(D_\mathfrak{a}(R)) \quad \text{for all } i \in \mathbb{N}_0.$$

However, the assumption that $\mathfrak{a}D_\mathfrak{a}(R) = D_\mathfrak{a}(R)$ means that $\Gamma_{\mathfrak{a}D_\mathfrak{a}(R)}$ is the zero functor, and so $H^i_{\mathfrak{a}D_\mathfrak{a}(R)}(D_\mathfrak{a}(R)) = 0$ for all $i \in \mathbb{N}_0$. This completes the proof. $\qquad\square$

6.3.6 Corollary. *Assume the $\mathfrak{a}$-transform functor $D_\mathfrak{a}$ is exact. Then* $\operatorname{ht} \mathfrak{p} \leq 1$ *for every minimal prime ideal $\mathfrak{p}$ of $\mathfrak{a}$.*

Proof. Suppose that $\mathfrak{p}$ is a minimal prime ideal of $\mathfrak{a}$ with $\operatorname{ht} \mathfrak{p} =: t \geq 2$, and look for a contradiction. Since $D_\mathfrak{a}$ is exact, it follows from Lemma 6.3.1 that $H^t_\mathfrak{a}(R) = 0$. Therefore, by 1.1.3 and 4.3.3,

$$H^t_{\mathfrak{p}R_\mathfrak{p}}(R_\mathfrak{p}) = H^t_{\mathfrak{a}R_\mathfrak{p}}(R_\mathfrak{p}) \cong (H^t_\mathfrak{a}(R))_\mathfrak{p} = 0.$$

However, this contradicts Theorem 6.1.4, since $\dim R_\mathfrak{p} = t$. $\qquad\square$

6.3.7 ‡Exercise. Let $\mathfrak{c}$ be a second ideal of R, and suppose that $\mathfrak{c} \subseteq \mathfrak{a}$. Show that $D_\mathfrak{c}(\eta_{\mathfrak{a},(\bullet)}) : D_\mathfrak{c}(\bullet) \longrightarrow D_\mathfrak{c}(D_\mathfrak{a}(\bullet))$ is a natural equivalence of functors. (Here is a hint: you might find the argument used in the proof of 2.2.10(iii) helpful.)

6.3.8 Lemma. *Let $\mathfrak{b}$ be a second ideal of R and let M be an R-module. Then the R-homomorphism*

$$\alpha_{\mathfrak{a},\mathfrak{a}\mathfrak{b},D_\mathfrak{b}(M)} : D_\mathfrak{a}(D_\mathfrak{b}(M)) \longrightarrow D_{\mathfrak{a}\mathfrak{b}}(D_\mathfrak{b}(M))$$

(which results from application of the natural transformation $\alpha_{\mathfrak{a},\mathfrak{a}\mathfrak{b}}$ of 2.2.23(i) to the R-module $D_\mathfrak{b}(M)$) is an isomorphism.

Proof. The diagram

$$D_b(M) \xrightarrow{\ \eta_{a,D_b(M)}\ } D_a(D_b(M))$$

$$\alpha_{a,ab,D_b(M)}$$

$$D_b(M) \xrightarrow{\ \eta_{ab,D_b(M)}\ } D_{ab}(D_b(M))$$

commutes. By 2.2.10(iv), we have $H_b^i(D_b(M)) = 0$ for $i = 0, 1$. Hence, by the Mayer–Vietoris sequence 3.2.3, for $i = 0, 1$, there is an exact sequence

$$H_a^i(D_b(M)) \longrightarrow H_{ab}^i(D_b(M)) \longrightarrow H_{a+b}^{i+1}(D_b(M)),$$

so that $H_{ab}^i(D_b(M))$ is a-torsion. Hence, by 2.2.6(i)(c) and 2.2.15, there is a unique homomorphism $\theta : D_{ab}(D_b(M)) \longrightarrow D_a(D_b(M))$ such that the diagram

$$D_b(M) \xrightarrow{\ \eta_{ab,D_b(M)}\ } D_{ab}(D_b(M))$$

$$\theta$$

$$D_b(M) \xrightarrow{\ \eta_{a,D_b(M)}\ } D_a(D_b(M))$$

commutes. In fact, it is immediate from the uniqueness aspects of 2.2.15 that

$$\theta \circ \alpha_{a,ab,D_b(M)} = \mathrm{Id}_{D_a(D_b(M))} \quad \text{and} \quad \alpha_{a,ab,D_b(M)} \circ \theta = \mathrm{Id}_{D_{ab}(D_b(M))},$$

and so $\alpha_{a,ab,D_b(M)}$ is an isomorphism. $\qquad\square$

6.3.9 Exercise. Let b be a second ideal of R. Suppose that the functors D_a and D_b are both exact. Let $L \xrightarrow{\ g\ } M \xrightarrow{\ h\ } N$ be an exact sequence of R-modules and R-homomorphisms.

(i) Use 6.3.8 to show that the induced sequence

$$D_{ab}(D_b(L)) \xrightarrow{\ D_{ab}(D_b(g))\ } D_{ab}(D_b(M)) \xrightarrow{\ D_{ab}(D_b(h))\ } D_{ab}(D_b(N))$$

is exact.

(ii) Use 6.3.7 to show that the induced sequence

$$D_{ab}(L) \xrightarrow{\ D_{ab}(g)\ } D_{ab}(M) \xrightarrow{\ D_{ab}(h)\ } D_{ab}(N)$$

is exact, and conclude that the functor D_{ab} is exact.

6.4 An Affineness Criterion due to Serre

Although the main result of this section, 6.4.4, is, strictly speaking, only a special case of Serre's Affineness Criterion (see [77, §46, Corollaire 1]), we shall nevertheless refer to it as 'Serre's Affineness Criterion'.

6.4.1 Notation. In our discussion of Serre's Affineness Criterion, we shall use the following notation. We shall use V to denote an affine variety over the algebraically closed field K, and $\mathfrak{b}$ will denote a non-zero ideal of $\mathcal{O}(V)$; we shall use U to denote the quasi-affine variety $V \setminus V(\mathfrak{b})$, where $V(\mathfrak{b})$ is the closed subset of V determined by $\mathfrak{b}$.

Also, for an affine or quasi-affine variety W over K and a point $q \in W$, we shall frequently denote the maximal ideal $\{f \in \mathcal{O}(W) : f(q) = 0\}$ of $\mathcal{O}(W)$ by $I_W(q)$. The local ring of W at q will be denoted by $\mathcal{O}_{W,q}$.

6.4.2 Reminders. Let the notation be as in 6.4.1.

(i) The quasi-affine variety U is said to be *affine* precisely when there exists an affine variety W over K and an isomorphism of varieties $U \xrightarrow{\cong} W$.

(ii) Let W be an affine variety over K. If $\alpha : U \to W$ is a morphism of varieties, we shall denote by $\alpha^* : \mathcal{O}(W) \to \mathcal{O}(U)$ the homomorphism of K-algebras induced by α. Recall, from [30, Chapter I, Proposition 3.5] for example, that the correspondence $\alpha \mapsto \alpha^*$ provides a bijective map from the set of all morphisms of varieties $U \to W$ to the set of all K-algebra homomorphisms $\mathcal{O}(W) \to \mathcal{O}(U)$.

(iii) Let T be a closed subvariety of V, so that T is a closed, irreducible (and so necessarily non-empty) subset of V and the ideal

$$\mathfrak{p} := \{f \in \mathcal{O}(V) : f(t) = 0 \text{ for all } t \in T\}$$

of $\mathcal{O}(V)$ is prime. Recall that $\operatorname{codim}_V T$, the *codimension of T in V*, is given by

$$\operatorname{codim}_V T = \dim V - \dim T = \dim \mathcal{O}(V) - \dim \mathcal{O}(T) = \operatorname{ht}_{\mathcal{O}(V)} \mathfrak{p}.$$

(iv) Recall that a non-empty closed subset C of V is said to be *of pure codimension r in V* precisely when every irreducible component of C has codimension r in V.

(v) Let $u \in U$. With the natural identifications, we have

$$\mathcal{O}(V) \subseteq \mathcal{O}(U) \subseteq \mathcal{O}_{U,u} = \mathcal{O}(U)_{I_U(u)} \subseteq K(U) = K(V).$$

(vi) Finally, we recall that points on the quasi-affine variety U can be 'separated by regular functions'. More precisely, let $u_1, \ldots, u_r$ be r distinct points of U and let $c_1, \ldots, c_r \in K$. Then there is a function $f \in \mathcal{O}(U)$ such that $f(u_i) = c_i$ for all $i = 1, \ldots, r$. In particular, if $u, u' \in U$ are such that $I_U(u) = I_U(u')$, then $u = u'$.

6.4.3 Lemma. *Let the notation be as in 6.4.1, and let W be an affine variety over K. Let $\alpha : U \to W$ be a surjective morphism of varieties for which $\alpha^* : \mathcal{O}(W) \to \mathcal{O}(U)$ is an isomorphism of K-algebras. Then α is an isomorphism of varieties.*

Proof. Let $w \in W$. Since α is surjective, there exists $u \in U$ with $\alpha(u) = w$. Now, for $f \in \mathcal{O}(W)$, we have $f(w) = f(\alpha(u)) = 0$ if and only if $(\alpha^*(f))(u) = 0$. It follows that $\alpha^*(I_W(w)) = I_U(u)$. This shows that $I_U(u)$ is uniquely determined by w. By 6.4.2(vi), it follows that there is exactly one $u \in U$ for which $\alpha(u) = w$. Hence α is bijective. It thus remains only for us to show that its inverse $\alpha^{-1} : W \to U$ is a morphism of varieties. For this, let U' be a non-empty open subset of U, and let $f \in \mathcal{O}(U')$: it is enough for us to show that $(\alpha^{-1})^{-1}(U') = \alpha(U')$ is an open subset of W and that $f \circ (\alpha^{-1} \lceil_{\alpha(U')}) : \alpha(U') \longrightarrow K$ is regular.

Before establishing these two points, we make one preparatory observation. For each $h \in \mathcal{O}(U)$, the regular function $(\alpha^*)^{-1}(h) \in \mathcal{O}(W)$ has the property that, for each $w \in W$,

$$\left((\alpha^*)^{-1}(h)\right)(w) = \left((\alpha^*)^{-1}(h)\right)\left(\alpha(\alpha^{-1}(w))\right)$$
$$= \alpha^*((\alpha^*)^{-1}(h))(\alpha^{-1}(w)) = h(\alpha^{-1}(w)).$$

We now turn our attention to U'. There is a non-zero ideal $\mathfrak{c}$ of $\mathcal{O}(V)$ such that $U' = V \setminus V(\mathfrak{c})$. Thus, with the notation of 2.3.1, we have $U' = \bigcup_{g \in \mathfrak{c}} U_V(g)$. However, for each $g \in \mathfrak{c}$, we have

$$\alpha(U_V(g)) = \alpha\left(\{v \in V : g(v) \neq 0\}\right)$$
$$= \alpha\left(\{u \in U : (g\lceil_U)(u) \neq 0\}\right)$$
$$= \{w \in W : (g\lceil_U)(\alpha^{-1}(w)) \neq 0\}$$
$$= \{w \in W : ((\alpha^*)^{-1}(g\lceil_U))(w) \neq 0\}$$

by the preceding paragraph; since $(\alpha^*)^{-1}(g\lceil_U)$ is a regular function on W, we see that $\alpha(U_V(g))$ is an open subset of W. Hence $\alpha(U') = \bigcup_{g \in \mathfrak{c}} \alpha(U_V(g))$ is open too.

Next, let $w \in \alpha(U')$. Since $f \in \mathcal{O}(U')$, there exists an open subset U'' of U' with $\alpha^{-1}(w) \in U''$ and regular functions $h, k \in \mathcal{O}(U)$ such that k does not vanish on U'' and $f(p) = h(p)/k(p)$ for all $p \in U''$. It follows from the immediately preceding paragraph that $\alpha(U'')$ is an open subset of W that contains w, and from the paragraph before that that, for all $q \in \alpha(U'')$, we have $\left((\alpha^*)^{-1}(k)\right)(q) = k(\alpha^{-1}(q)) \neq 0$ and

$$\left(f \circ (\alpha^{-1}\lceil_{\alpha(U')})\right)(q) = f(\alpha^{-1}(q)) = \frac{h(\alpha^{-1}(q))}{k(\alpha^{-1}(q))} = \frac{\left((\alpha^*)^{-1}(h)\right)(q)}{\left((\alpha^*)^{-1}(k)\right)(q)}.$$

Since $(\alpha^*)^{-1}(h)$ and $(\alpha^*)^{-1}(k)$ are regular functions on W, it follows that $f \circ (\alpha^{-1}\lceil_{\alpha(U')})$ is a regular function on $\alpha(U')$, as required. This completes the proof that $\alpha^{-1} : W \to U$ is a morphism of varieties. $\qquad\square$

We are now ready to present Serre's Affineness Criterion.

6.4.4 Serre's Affineness Criterion. *Let the notation be as in* 6.4.1. *Then the following conditions are equivalent:*

(i) $U = V \setminus V(\mathfrak{b})$ *is affine;*

(ii) $D_{\mathfrak{b}} : \mathcal{C}(\mathcal{O}(V)) \longrightarrow \mathcal{C}(\mathcal{O}(V))$ *is exact;*

(iii) $H^i_{\mathfrak{b}}(\mathcal{O}(V)) = 0$ *for all* $i \geq 2$;

(iv) $H^2_{\mathfrak{b}}(M) = 0$ *for each finitely generated* $\mathcal{O}(V)$-*module* M;

(v) $H^2_{\mathfrak{b}}(M) = 0$ *for each* $\mathcal{O}(V)$-*module* M;

(vi) $\mathfrak{b}D_{\mathfrak{b}}(\mathcal{O}(V)) = D_{\mathfrak{b}}(\mathcal{O}(V))$.

Proof. The equivalence of the last five conditions (ii) – (vi) was established in 6.3.1 and 6.3.5. It only remains for us to establish that these are also equivalent to statement (i). Let $\iota : U \to V$ denote the inclusion morphism of varieties.

(i) $\Rightarrow$ (vi) Assume that U is affine, so that there is an affine variety W over K and an isomorphism of varieties $\alpha : U \xrightarrow{\cong} W$. Set $\beta := \iota \circ \alpha^{-1} : W \to V$. In view of 2.3.2, we therefore have a commutative diagram

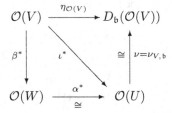

of K-algebra homomorphisms. Let $\mathfrak{n}$ be a maximal ideal of $D_{\mathfrak{b}}(\mathcal{O}(V))$. Then

$$\eta_{\mathcal{O}(V)}^{-1}(\mathfrak{n}) = \iota^{*-1}(\nu^{-1}(\mathfrak{n})) = \beta^{*-1}(\alpha^{*-1}(\nu^{-1}(\mathfrak{n}))).$$

Now $\alpha^{*-1}(\nu^{-1}(\mathfrak{n}))$ is a maximal ideal of the ring $\mathcal{O}(W)$, and so there exists $w \in W$ such that $\alpha^{*-1}(\nu^{-1}(\mathfrak{n})) = I_W(w)$. Hence

$$\eta_{\mathcal{O}(V)}^{-1}(\mathfrak{n}) = \beta^{*-1}(I_W(w)) = I_V(\beta(w)).$$

However, $\beta(w) = \iota(\alpha^{-1}(w)) \in U - V \setminus V(\mathfrak{b})$, and so

$$\mathfrak{b} \not\subseteq I_V(\beta(w)) = \eta^{-1}_{\mathcal{O}(V)}(\mathfrak{n}).$$

This is true for each maximal ideal $\mathfrak{n}$ of $D_\mathfrak{b}(\mathcal{O}(V))$; therefore $\mathfrak{b}D_\mathfrak{b}(\mathcal{O}(V)) = D_\mathfrak{b}(\mathcal{O}(V))$.

(vi) $\Rightarrow$ (i) Assume that $\mathfrak{b}D_\mathfrak{b}(\mathcal{O}(V)) = D_\mathfrak{b}(\mathcal{O}(V))$. Then $D_\mathfrak{b}(\mathcal{O}(V))$ is a finitely generated $\mathcal{O}(V)$-algebra, by 6.3.4. Moreover $D_\mathfrak{b}(\mathcal{O}(V))$ is an integral domain, and so there exists an affine variety W over K for which $\mathcal{O}(W) = D_\mathfrak{b}(\mathcal{O}(V))$. It follows from Theorem 2.3.2 that there is an $\mathcal{O}(V)$-isomorphism $\nu : \mathcal{O}(U) \xrightarrow{\cong} D_\mathfrak{b}(\mathcal{O}(V))$ for which the diagram

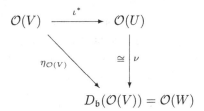

commutes. By 6.4.2(ii), there is a morphism of varieties $\gamma : U \to W$ such that $\gamma^* = \nu^{-1}$, and there is a morphism of varieties $\beta : W \to V$ such that $\beta^* = \eta_{\mathcal{O}(V)}$. Since $(\beta \circ \gamma)^* = \gamma^* \circ \beta^* = \iota^*$, it also follows from 6.4.2(ii) that $\beta \circ \gamma = \iota$. Our strategy is to use Lemma 6.4.3 to show that γ is an isomorphism of varieties, and so our immediate aim is to show that γ is surjective.

Let $w \in W$. Note that $\beta^{*-1}(I_W(w)) = I_V(\beta(w))$. Since

$$(\beta^{*-1}(I_W(w)))D_\mathfrak{b}(\mathcal{O}(V)) \subseteq I_W(w) \subset D_\mathfrak{b}(\mathcal{O}(V)) = \mathfrak{b}D_\mathfrak{b}(\mathcal{O}(V)),$$

it follows that $\mathfrak{b} \not\subseteq \beta^{*-1}(I_W(w)) = I_V(\beta(w))$. Therefore $\beta(w) \in U$. Denote $\beta(w)$ by u. We aim to show that $\gamma(u) = w$.

Suppose that $\gamma(u) \ne w$, and look for a contradiction. By 6.4.2(vi), there exists a function $g \in \mathcal{O}(W)$ such that $g(w) = 0$ and $g(\gamma(u)) \ne 0$. Also, since $\mathfrak{b} \not\subseteq I_V(u)$, there exists $h \in \mathfrak{b} \setminus I_V(u)$. Now $\operatorname{Coker} \eta_{\mathcal{O}(V)} \cong H^1_\mathfrak{b}(\mathcal{O}(V))$, by 2.2.6(i)(c), and so this cokernel is $\mathfrak{b}$-torsion. Therefore, there is $n \in \mathbb{N}$ such that $\beta^*(h^n)g = \eta_{\mathcal{O}(V)}(h^n)g = h^n g \in \eta_{\mathcal{O}(V)}(\mathcal{O}(V)) = \beta^*(\mathcal{O}(V))$. Thus there exists $k \in \mathcal{O}(V)$ such that $\beta^*(h^n)g = \beta^*(k)$. We shall now calculate $k(u)$ in two ways. First of all,

$$k(u) = k(\beta(w)) = (\beta^*(k))(w) = (\beta^*(h^n)g)(w)$$
$$= ((\beta^*(h))(w))^n g(w) = 0.$$

On the other hand,

$$k(u) = k(\iota(u)) = k(\beta \circ \gamma(u)) = k(\beta(\gamma(u))) = (\beta^*(k))(\gamma(u))$$
$$= (\beta^*(h^n)g)(\gamma(u)) = ((\beta^*(h))(\gamma(u)))^n \, g(\gamma(u))$$
$$= (h(\beta \circ \gamma(u)))^n \, g(\gamma(u)) = (h(u))^n g(\gamma(u));$$

this is non-zero by choice of h and g. This contradiction shows that $\gamma(u) = w$.

We have therefore proved that γ is surjective, and so we can now use Lemma 6.4.3 to deduce that γ is an isomorphism of varieties, so that U is affine. $\square$

6.4.5 Corollary. *Let the notation be as in* 6.4.1, *and assume* $U = V \setminus V(\mathfrak{b})$ *is affine. Then* $V \setminus U = V(\mathfrak{b})$ *is of pure codimension 1 in* V.

Proof. By Serre's Affineness Criterion 6.4.4, the functor $D_{\mathfrak{b}}$ is exact. Therefore, by 6.3.6, ht $\mathfrak{p} \leq 1$ for every minimal prime ideal $\mathfrak{p}$ of $\mathfrak{b}$. But $\mathfrak{b} \neq 0$ and $\mathcal{O}(V)$ is an integral domain, so that ht $\mathfrak{p} = 1$ for every minimal prime ideal $\mathfrak{p}$ of $\mathfrak{b}$. Hence every irreducible component of $V(\mathfrak{b})$ has codimension 1 in V. $\square$

Our work on Serre's Affineness Criterion raises the following questions. We again use the notation of 6.4.1. First, if $U = V \setminus V(\mathfrak{b})$ is affine, so that $\mathfrak{b}D_{\mathfrak{b}}(\mathcal{O}(V)) = D_{\mathfrak{b}}(\mathcal{O}(V))$, then it follows from 6.3.4 that $D_{\mathfrak{b}}(\mathcal{O}(V))$ is a finitely generated $\mathcal{O}(V)$-algebra: is the converse statement true, that is, if $D_{\mathfrak{b}}(\mathcal{O}(V))$ is a finitely generated $\mathcal{O}(V)$-algebra, is it necessarily the case that U is affine? Second, is the converse of 6.4.5 true, that is, if $V \setminus U = V(\mathfrak{b})$ is of pure codimension 1 in V, is it necessarily the case that U is affine? Another examination of the example studied in 2.3.7 and 3.3.5 will provide us with negative answers to both questions.

6.4.6 Example. We consider again the affine variety V in $\mathbb{A}^4$ studied in 2.3.7 and 3.3.5 and given by

$$V := V_{\mathbb{A}^4}(X_1X_4 - X_2X_3, \; X_1^2X_3 + X_1X_2 - X_2^2, \; X_3^3 + X_3X_4 - X_4^2).$$

As in 3.3.5, let $L := V_{\mathbb{A}^4}(X_2, \; X_3, \; X_4) = \{(c,0,0,0) \in \mathbb{A}^4 : c \in \mathbb{C}\}$ and $\overline{L} := \{(c,0) \in \mathbb{A}^2 : c \in \mathbb{C}\}$. Our argument uses the morphism of varieties $\alpha : \mathbb{A}^2 \to V$ of 2.3.7 for which $\alpha((c,d)) = (c, cd, d(d-1), d^2(d-1))$ for all $(c,d) \in \mathbb{A}^2$. It was shown in 3.3.5 that the restriction of α provides an isomorphism of (quasi-affine) varieties $\mathbb{A}^2 \setminus (\{(0,1)\} \cup \overline{L}) \xrightarrow{\cong} V \setminus L$. Now L is of pure codimension 1 in V. (As L is actually a subvariety of V, this statement is equivalent to the statement that $\operatorname{codim}_V L = 1$.) If we can show that $U := \mathbb{A}^2 \setminus (\{(0,1)\} \cup \overline{L})$ is not affine and that $\mathcal{O}(U)$ is a finitely generated $\mathbb{C}$-algebra, then it will follow that both questions posed just after 6.4.5 have negative answers. Let $q := (0, 1) \in \mathbb{A}^2$. As $\mathbb{A}^2 \setminus U = \overline{L} \cup \{q\}$ is not of pure codimension 1 in $\mathbb{A}^2$, it follows from 6.4.5 that U is not affine.

Next, $\overline{L} = V_{\mathbb{A}^2}(Y)$, so that $\mathbb{A}^2 \setminus \overline{L}$ is affine, and, by 2.3.2 and 2.2.19, we can identify $\mathcal{O}(\mathbb{A}^2 \setminus \overline{L})$ with the subring $\mathbb{C}[X, Y]_Y = \mathbb{C}[X, Y, Y^{-1}]$ of $\mathbb{C}(X, Y)$.

Note that $I_{\mathbb{A}^2 \setminus \overline{L}}(q) = (X, Y - 1)\mathbb{C}[X, Y, Y^{-1}]$. (Observe that the maximal ideal $(X, Y - 1)$ of $\mathbb{C}[X, Y]$ does not meet $\{Y^i : i \in \mathbb{N}_0\}$.) Since $I_{\mathbb{A}^2 \setminus \overline{L}}(q)$ contains a $\mathbb{C}[X, Y, Y^{-1}]$-sequence $X, Y - 1$ of length 2, it follows from 2.2.8 that

$$\eta_{\mathbb{C}[X,Y,Y^{-1}]} : \mathbb{C}[X, Y, Y^{-1}] \longrightarrow D_{I_{\mathbb{A}^2 \setminus \overline{L}}(q)}(\mathbb{C}[X, Y, Y^{-1}])$$

is an isomorphism. But, by 2.3.2, and as $U = (\mathbb{A}^2 \setminus \overline{L}) \setminus \{q\}$, there is an isomorphism

$$D_{I_{\mathbb{A}^2 \setminus \overline{L}}(q)}(\mathbb{C}[X, Y, Y^{-1}]) \cong \mathcal{O}((\mathbb{A}^2 \setminus \overline{L}) \setminus \{q\}) = \mathcal{O}(U)$$

of $\mathcal{O}(\mathbb{A}^2 \setminus \overline{L})$-algebras. Hence $\mathcal{O}(U)$ is a finitely generated $\mathbb{C}$-algebra. Also, the fact that the map $\eta_{\mathbb{C}[X,Y,Y^{-1}]}$ is an isomorphism shows that

$$(X, Y - 1)D_{I_{\mathbb{A}^2 \setminus \overline{L}}(q)}(\mathbb{C}[X, Y, Y^{-1}]) \neq D_{I_{\mathbb{A}^2 \setminus \overline{L}}(q)}(\mathbb{C}[X, Y, Y^{-1}]),$$

and so we see again (this time from Serre's Affineness Criterion 6.4.4) that the quasi-affine variety U is not affine.

Thus both questions posed just after 6.4.5 have negative answers.

6.4.7 Exercise. Let the notation be as in 6.4.1. Also, let $\mathfrak{b}'$ be a second non-zero ideal of $\mathcal{O}(V)$ and let U' denote the quasi-affine variety $V \setminus V(\mathfrak{b}')$. Deduce from Exercise 6.3.9 and Serre's Affineness Criterion 6.4.4 that, if U and U' are affine, then $U \cap U'$ is also affine.

6.5 Applications to local algebra in prime characteristic

In this section, we present some applications of local cohomology to the study of algebra over local rings of prime characteristic. These applications concern results that do not involve local cohomology in their statements, but which have proofs that make non-trivial use of local cohomology.

In this work, we shall use some techniques that assist calculation with regular sequences, and our first exercises in the section are concerned with these. The reader should recall the definition of poor M-sequence (where M is an R-module) given in 6.2.1

6.5.1 ♯Exercise. Let M be an R-module, and let $r_1, \ldots, r_n$ be a poor M-sequence, where $n \geq 2$. Show that r_1 is a non-zerodivisor on $M/(r_2, \ldots, r_n)M$.

6.5.2 ♯Exercise. Let M be an R-module, let $r_2, \ldots, r_n, a, b \in R$, where $n \geq 2$, and suppose that $a, r_2, \ldots, r_n$ is a poor M-sequence.

(i) Assume that

$$abm_1 + r_2 m_2 + \cdots + r_n m_n = a m_1' + r_2 m_2' + \cdots + r_n m_n',$$

where $m_1, \ldots, m_n, m_1', \ldots, m_n' \in M$. Show that

$$m_1' \in (b, r_2, \ldots, r_n)M.$$

(ii) Deduce that, if $b, r_2, \ldots, r_n$ is also a poor M-sequence, then $ab, r_2, \ldots, r_n$ is a poor M-sequence.

6.5.3 ♯Exercise. Let M be an R-module, and $r_1, \ldots, r_n \in R$, where $n \in \mathbb{N}$.

(i) Let $i \in \{1, \ldots, n\}$, and suppose that r_i can be written as $r_i = ab$, where $a, b \in R$. Show that $r_1, \ldots, r_{i-1}, r_i, r_{i+1}, \ldots, r_n$ is a poor M-sequence if and only if

$$r_1, \ldots, r_{i-1}, a, r_{i+1}, \ldots, r_n \quad \text{and} \quad r_1, \ldots, r_{i-1}, b, r_{i+1}, \ldots, r_n$$

are poor M-sequences.

(ii) Let $t_1, \ldots, t_n$ be arbitrary positive integers. Show that $r_1, \ldots, r_n$ is a poor M-sequence if and only if $r_1^{t_1}, \ldots, r_n^{t_n}$ is a poor M-sequence.

6.5.4 Notation for the section. Throughout the section, M will denote an R-module, n will denote a positive integer, and $L_n(R)$ will denote the set of $n \times n$ lower triangular matrices with entries in R. For $H \in L_n(R)$, we shall use $|H|$ to denote the determinant of H, that is, the product of the diagonal entries of H. We shall use ^{T} to denote matrix transpose; displayed matrices will be shown between rectangular brackets.

Let $d_1, \ldots, d_n \in R$. We shall use $\mathrm{diag}(d_1, \overset{\cdot}{\ldots}, d_n)$ to denote the diagonal matrix in $L_n(R)$ whose (i, i)-th entry is d_i (for each $i = 1, \ldots, n$).

6.5.5 Remark. Suppose that $x_1, \ldots, x_n, y_1, \ldots, y_n \in R$ and $H = [h_{ij}] \in L_n(R)$ are such that $[y_1 \cdots y_n]^T = H [x_1 \cdots x_n]^T$. The fact that the adjoint matrix $\mathrm{Adj}\, H$ satisfies $(\mathrm{Adj}\, H)H = |H|I_n$ ensures that

$$|H|(x_1, \ldots, x_n)M \subseteq (y_1, \ldots, y_n)M.$$

Multiplication by $|H|$ therefore induces an R-homomorphism

$$M/(x_1, \ldots, x_n)M \longrightarrow M/(y_1, \ldots, y_n)M.$$

Let $k \in \{1, \ldots, n\}$. Since H is lower triangular, the $k \times k$ submatrix H_k of H obtained by deleting the $(k+1)$-th,..., n-th rows and columns of H satisfies $[y_1 \cdots y_k]^T = H_k [x_1 \cdots x_k]^T$. We deduce from the above paragraph that $|H_k|(x_1, \ldots, x_k)M \subseteq (y_1, \ldots, y_k)M$, so that

$$|H|(x_1, \ldots, x_k)M \subseteq (y_1, \ldots, y_k)M$$

because $|H_k|$ is a factor of $|H|$.

6.5.6 Theorem. (L. O'Carroll [64, Theorem 3.2]) *Suppose that $x_1, \ldots, x_n$, $y_1, \ldots, y_n \in R$ and $H \in L_n(R)$ are such that*

(i) $[y_1 \cdots y_n]^T = H [x_1 \cdots x_n]^T$, *and*
(ii) $y_1, \ldots, y_n$ *is a poor M-sequence.*

Then the R-homomorphism $\alpha : M/(x_1, \ldots, x_n)M \longrightarrow M/(y_1, \ldots, y_n)M$ induced by multiplication by $|H|$ is a monomorphism, and $x_1, \ldots, x_n$ is also a poor M-sequence.

Proof. We use induction on n. When $n = 1$, the result is immediate from 6.5.3(i). We therefore suppose that $n > 1$ and that the result has been proved for sequences of length smaller than n. Let h_{ij} denote the (i, j)-th entry of H (for all $i, j = 1, \ldots, n$).

Now $[x_1 \, y_2 \cdots y_n]^T = H' [x_1 \, x_2 \cdots x_n]^T$, where $H' = [h'_{ij}] \in L_n(R)$ is specified as follows:

$$h'_{ij} = \begin{cases} 1 & \text{if } i = j = 1, \\ h_{ij} & \text{otherwise.} \end{cases}$$

Let $\beta : M/(x_1, x_2, \ldots, x_n)M \longrightarrow M/(x_1, y_2, \ldots, y_n)M$ be the R-homomorphism induced by multiplication by $|H'|$. Note also that

$$[y_1 \, y_2 \cdots y_n]^T = D [x_1 \, y_2 \cdots y_n]^T \quad \text{where } D := \text{diag}(h_{11}, 1, \ldots, 1).$$

Let $\gamma : M/(x_1, y_2, \ldots, y_n)M \longrightarrow M/(y_1, y_2, \ldots, y_n)M$ be the R-homomorphism induced by multiplication by $|D| = h_{11}$. Since $|H| = |D||H'|$, we have $\gamma \circ \beta = \alpha$. In order to show that α is a monomorphism, it is therefore sufficient for us to show that both γ and β are monomorphisms.

Since $y_1 = h_{11}x_1$, we see by 6.5.3(i) that $h_{11}, y_2, \ldots, y_n$ and $x_1, y_2, \ldots, y_n$ are poor M-sequences. Let $m \in M$ be such that $h_{11}m \in (y_1, y_2, \ldots, y_n)M$; thus $h_{11}m = y_1m_1 + \cdots + y_nm_n$ for some $m_1, \ldots, m_n \in M$. Since $y_1 = h_{11}x_1$, we obtain $h_{11}(m - x_1m_1) \in (y_2, \ldots, y_n)M$. Since $h_{11}, y_2, \ldots, y_n$ is a poor M-sequence, we can use 6.5.1 to see that h_{11} is a non-zerodivisor on $M/(y_2, \ldots, y_n)M$; hence $m - x_1m_1 \in (y_2, \ldots, y_n)M$. It follows that γ is a monomorphism.

We now turn our attention to β. Set $\overline{R} = R/x_1 R$ and $\overline{M} = M/x_1 M$; for $r \in R$, denote the natural image of r in $\overline{R}$ by $\bar{r}$. We noted in the last paragraph that $x_1, y_2, \ldots, y_n$ is a poor M-sequence. Therefore $\overline{y_2}, \ldots, \overline{y_n}$ is a poor $\overline{M}$-sequence in $\overline{R}$. Moreover $[\overline{y_2} \cdots \overline{y_n}]^T = G [\overline{x_2} \cdots \overline{x_n}]^T$, where $G = [g_{ij}] \in L_{n-1}(\overline{R})$ is given by $g_{ij} = \overline{h}_{i+1,j+1}$ for all $i, j \in \{1, \ldots, n-1\}$.

Let $\bar{\beta} : \overline{M}/(\overline{x_2}, \ldots, \overline{x_n})\overline{M} \longrightarrow \overline{M}/(\overline{y_2}, \ldots, \overline{y_n})\overline{M}$ denote the $\overline{R}$-homomorphism induced by multiplication by $|G| = \overline{|H'|}$. By the inductive hypothesis, $\bar{\beta}$ is a monomorphism. An easy calculation then shows that β is a monomorphism. Also, the facts that $y_1 = h_{11}x_1$ and y_1 is a non-zerodivisor on M ensure that x_1 is a non-zerodivisor on M. The inductive hypothesis yields that $\overline{x_2}, \ldots, \overline{x_n}$ is a poor $\overline{M}$-sequence, and so we can conclude that $x_1, \ldots, x_n$ is a poor M-sequence.

This completes the inductive step, and the proof. $\square$

6.5.7 Corollary. *Suppose that* $a_1, \ldots, a_n$ *are* n *elements of* R *that generate* $\mathfrak{a}$, *and let* M *be an* R-*module. Set* $a := a_1 \ldots a_n$. *Use the notation of 5.3.1 to denote natural images of elements of* M_a *in the* n-*th cohomology module of the Čech complex of* M *with respect to* $a_1, \ldots, a_n$.

Suppose that $a_1, \ldots, a_n$ *is a poor* M-*sequence, and that* $m \in M$ *and* $i \in \mathbb{N}$ *are such that* $[m/a^i] = 0$. *Then* $m \in (a_1^i, \ldots, a_n^i)M$.

Proof. By 5.3.2, there is $k \in \mathbb{N}_0$ such that $k \geq i$ and $a^{k-i}m \in \sum_{u=1}^{n} a_u^k M$. Now $\operatorname{diag}(a_1^{k-i}, \ldots, a_n^{k-i}) [a_1^i \cdots a_n^i]^T = [a_1^k \cdots a_n^k]^T$ and $a_1^k, \ldots, a_n^k$ is a poor M-sequence, by 6.5.3(ii). Since

$$| \operatorname{diag}(a_1^{k-i}, \ldots, a_n^{k-i})|m = a^{k-i}m \in (a_1^k, \ldots, a_n^k)M,$$

it follows from O'Carroll's Theorem 6.5.6 that $m \in (a_1^i, \ldots, a_n^i)M$. $\square$

6.5.8 Remark. Suppose that $(R, \mathfrak{m})$ is local with $\dim R = n > 0$. Recall that a *system of parameters for* R is a sequence $(r_i)_{i=1}^{n}$ of n elements of $\mathfrak{m}$ such that the ideal $(r_1, \ldots, r_n)$ of R generated by the terms of the sequence is $\mathfrak{m}$-primary. By a *subsystem of parameters for* R we mean a sequence $(r_i')_{i=1}^{t}$ of t elements of $\mathfrak{m}$, with $t \leq n$, which can be extended to a system of parameters for R by the addition of $n - t$ extra terms.

Let $(a_i)_{i=1}^{n}$ be a system of parameters for R.

When R is Cohen–Macaulay, so that $a_1, \ldots, a_n$ is an R-sequence, we must have $(a_1 \ldots a_n)^k \notin (a_1^{k+1}, \ldots, a_n^{k+1})$ for all $k \in \mathbb{N}_0$, as we now show. If $(a_1 \ldots a_n)^k \in (a_1^{k+1}, \ldots, a_n^{k+1})$, then, since $a_1^{k+1}, \ldots, a_n^{k+1}$ is an R-sequence by 6.5.3(ii), we can use O'Carroll's Theorem 6.5.6 in conjunction with the equation

$$\operatorname{diag}(a_1^k, \ldots, a_n^k) [a_1 \cdots a_n]^T = [a_1^{k+1} \cdots a_n^{k+1}]^T$$

to deduce that $1 \in (a_1, \ldots, a_n)$, a contradiction.

This observation leads to the following famous conjecture.

6.5.9 The Monomial Conjecture. (See M. Hochster [37, Conjecture 1].) Suppose that $(R, \mathfrak{m})$ is local (but not necessarily Cohen–Macaulay) and that $\dim R = n > 0$. The conjecture that $(a_1 \ldots a_n)^k \notin (a_1^{k+1}, \ldots, a_n^{k+1})$ for all $k \in \mathbb{N}_0$ and all systems of parameters $(a_i)_{i=1}^n$ for R is known as *the Monomial Conjecture*.

6.5.10 Theorem. (See M. Hochster [37, pp. 33–34].) *Suppose that $(R, \mathfrak{m})$ is local with $\dim R = n > 0$, and has prime characteristic p. Then the conclusion of the Monomial Conjecture is true in R.*

Proof. Let $(a_i)_{i=1}^n$ be a system of parameters for R. Suppose there exists $k \in \mathbb{N}_0$ such that $(a_1 \ldots a_n)^k \in (a_1^{k+1}, \ldots, a_n^{k+1})$, and seek a contradiction.

Set $a := a_1 \ldots a_n$, and represent elements of $H_{\mathfrak{m}}^n(R)$ using $a_1, \ldots, a_n$ and a in the manner described in 5.3.1. Let F denote the Frobenius action on $H_{\mathfrak{m}}^n(R)$ of 5.3.4 and 5.3.6. Let $y := [1/a] \in H_{\mathfrak{m}}^n(R)$; then the supposition that $a^k \in (a_1^{k+1}, \ldots, a_n^{k+1})$ means that there exist $r_1, \ldots, r_n \in R$ such that $a^k = a_1^{k+1} r_1 + \cdots + a_n^{k+1} r_n$, so that, by the comments in 5.3.1,

$$y = \left[\frac{1}{a} \right] = \left[\frac{a^k}{a^{k+1}} \right] = \left[\frac{a_1^{k+1} r_1 + \cdots + a_n^{k+1} r_n}{a^{k+1}} \right] = 0.$$

But an arbitrary element z of $H_{\mathfrak{m}}^n(R)$ can be expressed as $z = [r/a^t]$ for some $r \in R$ and $t \in \mathbb{N}_0$. Choose $h \in \mathbb{N}$ such that $p^h \geq t$. On use of 5.3.6 we now see that

$$z = \left[\frac{r}{a^t} \right] = \left[\frac{a^{p^h - t} r}{a^{p^h}} \right] = a^{p^h - t} r \left[\frac{1}{a^{p^h}} \right] = a^{p^h - t} r F^h(y) = 0.$$

We have therefore shown that $H_{\mathfrak{m}}^n(R) = 0$, contrary to the Non-vanishing Theorem 6.1.4. $\qquad\square$

Hochster's theorem above provides one example of an important result in local algebra whose statement makes no mention of local cohomology but for which local cohomology provides a proof. Below we present another such result, from the theory of tight closure. Some definitions are needed.

6.5.11 Definitions. Suppose that R has prime characteristic p. We use R° to denote the complement in R of the union of the minimal prime ideals of R. An element $r \in R$ belongs to the *tight closure* $\mathfrak{a}^*$ *of* $\mathfrak{a}$ if and only if there exists $c \in R^\circ$ such that $cr^{p^n} \in \mathfrak{a}^{[p^n]}$ for all $n \gg 0$. We say that $\mathfrak{a}$ is *tightly closed* precisely when $\mathfrak{a}^* = \mathfrak{a}$. The theory of tight closure was invented by M.

Hochster and C. Huneke [38], and many applications have been found for the theory: see [41] and [42], for example.

The next exercise establishes some basic properties of tight closure.

6.5.12 ♯Exercise.   Suppose that R has prime characteristic p.

(i) Show that the tight closure $\mathfrak{a}^*$ of $\mathfrak{a}$ is an ideal of R.

(ii) Let $\mathfrak{b}$ be a second ideal of R with $\mathfrak{a} \subseteq \mathfrak{b}$. Show that $\mathfrak{a}^* \subseteq \mathfrak{b}^*$.

(iii) Show that $(\mathfrak{a}^*)^* = \mathfrak{a}^*$, so that tight closure really is a 'closure operation'.

6.5.13 Exercise.   Suppose that R has prime characteristic p, and let $\mathfrak{a}$, $\mathfrak{b}$ be ideals of R.

(i) Show that $(\mathfrak{a} + \mathfrak{b})^* = (\mathfrak{a}^* + \mathfrak{b}^*)^*$ and $(\mathfrak{a}\mathfrak{b})^* = (\mathfrak{a}^*\mathfrak{b}^*)^*$.

(ii) Show that $0^* = \sqrt{0}$, and conclude that $\sqrt{0} \subseteq \mathfrak{c}^*$ for every ideal $\mathfrak{c}$ of R.

(iii) Let $\pi : R \longrightarrow R/\sqrt{0}$ be the natural ring homomorphism. Show that
$$\mathfrak{a}^* = \pi^{-1}\left(\left(\pi(\mathfrak{a})(R/\sqrt{0})\right)^*\right).$$

(iv) Show that, if $\mathfrak{a}$ is tightly closed, then so too is $(\mathfrak{a} : \mathfrak{b})$.

(v) Show that every maximal ideal of R is tightly closed.

6.5.14 Exercise.   Let K be a field of prime characteristic p, and let R be the subring of the ring of formal power series $K[[X]]$ given by

$$R := \left\{\textstyle\sum_{i=0}^{\infty} a_i X^i \in K[[X]] : a_1 = 0, a_0, a_2, a_3, \ldots \in K\right\}.$$

In R, calculate $(X^2R)^*$, $(X^3R)^*$, $(X^2R)^* \cap (X^3R)^*$, $(X^2R \cap X^3R)^*$, $((X^2R)^* : (X^3R)^*)$ and $(X^2R : X^3R)^*$.

6.5.15 Proposition.   *Suppose that $(R, \mathfrak{m})$ is local, of prime characteristic p, and Cohen–Macaulay with $\dim R = n > 0$. Let $(a_i)_{i=1}^n$ be a system of parameters for R, and suppose that the ideal $\mathfrak{q} := (a_1, \ldots, a_n)$ is tightly closed. Then the ideal $(a_1^{t_1}, \ldots, a_n^{t_n})$ is tightly closed for all $t_1, \ldots, t_n \in \mathbb{N}$.*

Proof.   We argue by induction on $t := \sum_{i=1}^n t_i$. Suppose that $t > n$ and that the desired result has been proved for smaller values of t.

Without loss of generality, we can, and do, assume that $t_1 > 1$. By our inductive hypothesis, $(a_1^{t_1-1}, a_2^{t_2}, \ldots, a_n^{t_n})$ is tightly closed. Let $r \in (a_1^{t_1}, \ldots, a_n^{t_n})^*$. Then $r \in (a_1^{t_1-1}, a_2^{t_2}, \ldots, a_n^{t_n})^* = (a_1^{t_1-1}, a_2^{t_2}, \ldots, a_n^{t_n})$, and so there exist $r_1, \ldots, r_n \in R$ such that

$$r = r_1 a_1^{t_1-1} + r_2 a_2^{t_2} + \cdots + r_n a_n^{t_n}.$$

Now $r_2 a_2^{t_2} + \cdots + r_n a_n^{t_n} \in (a_1^{t_1}, \ldots, a_n^{t_n}) \subseteq (a_1^{t_1}, \ldots, a_n^{t_n})^*$, and so

$$r_1 a_1^{t_1-1} = r - (r_2 a_2^{t_2} + \cdots + r_n a_n^{t_n}) \in (a_1^{t_1}, \ldots, a_n^{t_n})^*.$$

Therefore, there exists $c \in R^\circ$ such that

$$c\left(r_1 a_1^{t_1-1}\right)^{p^j} \in \left(a_1^{t_1 p^j}, \ldots, a_n^{t_n p^j}\right) \quad \text{for all } j \gg 0.$$

Now $a_1^{t_1 p^j}, \ldots, a_n^{t_n p^j}$ (for $j \in \mathbb{N}$) is an R-sequence, and the diagonal matrix

$$D := \text{diag}(a_1^{(t_1-1)p^j}, 1, \ldots, 1) \in L_n(R)$$

satisfies $D \left[a_1^{p^j} \, a_2^{t_2 p^j} \, \cdots \, a_n^{t_n p^j}\right]^T = \left[a_1^{t_1 p^j} \, a_2^{t_2 p^j} \, \cdots \, a_n^{t_n p^j}\right]^T$. We can therefore use O'Carroll's Theorem 6.5.6 to deduce that

$$cr_1^{p^j} \in \left(a_1^{p^j}, a_2^{t_2 p^j}, \ldots, a_n^{t_n p^j}\right) \quad \text{for all } j \gg 0.$$

Therefore $r_1 \in (a_1, a_2^{t_2}, \ldots, a_n^{t_n})^*$. By the inductive hypothesis, the ideal $(a_1, a_2^{t_2}, \ldots, a_n^{t_n})$ is tightly closed, so that $r_1 \in (a_1, a_2^{t_2}, \ldots, a_n^{t_n})$. Therefore

$$r = r_1 a_1^{t_1-1} + r_2 a_2^{t_2} + \cdots + r_n a_n^{t_n} \in (a_1^{t_1}, a_2^{t_2}, \ldots, a_n^{t_n}).$$

Thus $(a_1^{t_1}, a_2^{t_2}, \ldots, a_n^{t_n})$ is tightly closed. This completes the inductive step, and the proof. $\square$

6.5.16 Theorem. (See R. Fedder and K.-i. Watanabe [17, Proposition 2.2].) *Suppose that $(R, \mathfrak{m})$ is local, of prime characteristic p, and Cohen–Macaulay with $\dim R = n > 0$. Suppose that there is one system of parameters $(a_i)_{i=1}^n$ for R which generates a tightly closed ideal. Then the ideal of R generated by each system of parameters for R is tightly closed, as is the ideal generated by each subsystem of parameters for R.*

Proof. First of all, by 6.5.15, the ideal $(a_1^{t_1}, \ldots, a_n^{t_n})$ is tightly closed for all $t_1, \ldots, t_n \in \mathbb{N}$. Let F denote the Frobenius action on $H^n_{\mathfrak{m}}(R)$ of 5.3.4 and 5.3.6. We plan to prove that, if $y \in H^n_{\mathfrak{m}}(R)$ is such that there exists $c \in R^\circ$ with $cF^j(y) = 0$ for all $j \gg 0$, then $y = 0$.

Represent elements of $H^n_{\mathfrak{m}}(R) = H^n_{(a_1, \ldots, a_n)}(R)$ using $a_1, \ldots, a_n$ and $a := a_1 \ldots a_n$ in the manner described in 5.3.1. Thus $y \in H^n_{\mathfrak{m}}(R)$ can be written as $y = [r/a^t]$ for some $r \in R$ and $t \in \mathbb{N}_0$. Suppose that there exists $c \in R^\circ$ with $cF^j(y) = 0$ for all $j \gg 0$. By 5.3.6, this means that $[cr^{p^j}/a^{tp^j}] = 0$ for all $j \gg 0$. Since $a_1, \ldots, a_n$ is an R-sequence, we can deduce from 6.5.7 that $cr^{p^j} \in (a_1^{tp^j}, \ldots, a_n^{tp^j})$ for all $j \gg 0$. Therefore $r \in (a_1^t, \ldots, a_n^t)^*$. Since $(a_1^t, \ldots, a_n^t)$ is tightly closed, we see that $r \in (a_1^t, \ldots, a_n^t)$, so that $y = [r/a^t] = 0$ by the comments in 5.3.1.

Let $(b_i)_{i=1}^n$ be an arbitrary system of parameters for R. We show now that the ideal $(b_1, \ldots, b_n)$ is tightly closed. Let $s \in (b_1, \ldots, b_n)^*$, so that there exists $c' \in R^\circ$ such that $c's^{p^j} \in (b_1^{p^j}, \ldots, b_n^{p^j})$ for all $j \gg 0$. Represent elements of $H_{\mathfrak{m}}^n(R) = H_{(b_1,\ldots,b_n)}^n(R)$ using $b_1,\ldots,b_n$ and $b := b_1 \ldots b_n$ in the manner described in 5.3.1, and consider $z := [s/b] \in H_{\mathfrak{m}}^n(R)$. At this point, it is important to recall from 5.3.5 that the Frobenius action F on $H_{\mathfrak{m}}^n(R)$ does not depend on any choice of generators for any $\mathfrak{m}$-primary ideal of R. Therefore, using 5.3.6 and 5.3.1 again, we can deduce that

$$c'F^j(z) = \left[\frac{c's^{p^j}}{b^{p^j}}\right] = 0 \quad \text{for all } j \gg 0$$

because $c's^{p^j} \in (b_1^{p^j},\ldots,b_n^{p^j})$ for all $j \gg 0$. The claim proved in the immediately preceding paragraph therefore shows that $z = [s/b] = 0$, so that $s \in (b_1,\ldots,b_n)$ by 6.5.7 because $b_1,\ldots,b_n$ is an R-sequence. Therefore $(b_1,\ldots,b_n)$ is tightly closed.

To complete the proof, let $i \in \{0,1,\ldots,n-1\}$ and set $\mathfrak{c} := (b_1,\ldots,b_i)$; we have to show that $\mathfrak{c}$ is tightly closed. (Interpret $\mathfrak{c}$ as 0 when $i = 0$.) By the immediately preceding paragraph, the ideal $(b_1,\ldots,b_i,b_{i+1}^t,\ldots,b_n^t)$ is tightly closed for all $t \in \mathbb{N}$. Let $v \in \mathfrak{c}^*$; then v belongs to the tight closure of every ideal of R that contains $\mathfrak{c}$ (by 6.5.12(ii)). Hence

$$v \in \bigcap_{t\in\mathbb{N}}(b_1,\ldots,b_i,b_{i+1}^t,\ldots,b_n^t)^* = \bigcap_{t\in\mathbb{N}}(b_1,\ldots,b_i,b_{i+1}^t,\ldots,b_n^t)$$

$$\subseteq \bigcap_{t\in\mathbb{N}}((b_1,\ldots,b_i) + \mathfrak{m}^t) = \bigcap_{t\in\mathbb{N}}(\mathfrak{c} + \mathfrak{m}^t) = \mathfrak{c}$$

by Krull's Intersection Theorem. Therefore $\mathfrak{c}$ is tightly closed, and the proof is complete. $\qquad\qquad\Box$

6.5.17 Remark. Suppose that $(R,\mathfrak{m})$ is local and of prime characteristic. We say that R is *F-rational* if and only if every proper ideal $\mathfrak{c}$ of R which can be generated by ht $\mathfrak{c}$ elements is tightly closed. The Theorem 6.5.16 of Fedder and Watanabe shows that a Cohen–Macaulay local ring of prime characteristic is F-rational if one single system of parameters generates a tightly closed ideal. It can be proved that, if R is excellent, or a homomorphic image of a Cohen–Macaulay local ring, and is F-rational, then it must be Cohen–Macaulay; however, that result is beyond the scope of this book. We have included 6.5.16 because, firstly, it gives another example of a significant result in local algebra whose statement makes no mention of local cohomology, but for which local cohomology provides a proof, and, secondly, because it gives some hints about the important rôle that local cohomology can play in tight closure theory.

7

Artinian local cohomology modules

In this chapter, we shall show that certain local cohomology modules are Artinian, that is, satisfy the descending chain condition on submodules, and we shall use our results to provide a different proof of a theorem in Chapter 6.

Suppose, temporarily, that $(R, \mathfrak{m})$ is local. In the Non-vanishing Theorem 6.1.4, we proved that, if M is a non-zero, finitely generated R-module of dimension n, then $H_{\mathfrak{m}}^n(M) \neq 0$. One consequence of our work in this chapter is that we can give an alternative proof of this result, and, at the same time, obtain more information than we deduced in Chapter 6. Our approach in this chapter exploits the facts that the R-module $H_{\mathfrak{m}}^n(M)$ is actually Artinian, and that, for Artinian modules over commutative rings, there is available a theory of secondary representation that is, in several respects, dual to the theory of primary decomposition of Noetherian modules over commutative rings.

7.1 Artinian modules

We begin with some arguments due to L. Melkersson ([51], [52]) which will enable us to show that, for a non-zero, finitely generated module M of dimension n over the local ring $(R, \mathfrak{m})$, the n-th local cohomology module $H_{\mathfrak{a}}^n(M)$ of M with respect to $\mathfrak{a}$ is Artinian, and all the local cohomology modules $H_{\mathfrak{m}}^i(M)$ ($i \in \mathbb{N}_0$) of M with respect to $\mathfrak{m}$ are Artinian.

We revert to our standard hypotheses concerning R and $\mathfrak{a}$ (although it is worth pointing out that the first two results below (due to Melkersson) actually hold under weaker hypotheses).

7.1.1 Lemma. (L. Melkersson [51, Lemma 2.1]). *Let $a \in R$ and let M be an aR-torsion R-module. Suppose that N, N' are submodules of M such that $N' \subseteq N$ and $a^i(0 :_N a^{i+1}) = a^i(0 :_{N'} a^{i+1})$ for all $i \in \mathbb{N}_0$. Then $N = N'$.*

Proof. Since each element of N is annihilated by some power of a, it is enough for us to show that $(0 :_N a^i) \subseteq N'$ for all $i \in \mathbb{N}$. We prove this by induction on i. By hypothesis, $(0 :_N a) = (0 :_{N'} a) \subseteq N'$, and so we assume that, for $i \in \mathbb{N}$, we have proved that $(0 :_N a^i) \subseteq N'$.

Let $z \in (0 :_N a^{i+1})$. Then $a^i z \in a^i(0 :_N a^{i+1}) = a^i(0 :_{N'} a^{i+1})$, and so there exists $z' \in (0 :_{N'} a^{i+1})$ such that $a^i z = a^i z'$. Hence $z - z' \in (0 :_N a^i)$, and so $z - z' \in N'$ by the inductive assumption. Therefore $z = (z - z') + z' \in N'$, and the inductive step is complete. $\qquad\square$

7.1.2 Theorem. (L. Melkersson [51, Theorem 1.3]). *Assume that M is an $\mathfrak{a}$-torsion R-module for which $(0 :_M \mathfrak{a})$ is Artinian. Then M is Artinian.*

Proof. We suppose that $\mathfrak{a}$ can be generated by t elements and proceed by induction on t. When $t = 0$, we have $\mathfrak{a} = 0$ and $(0 :_M \mathfrak{a}) = M$, so that there is nothing to prove in this case.

Now suppose that $t = 1$ and $\mathfrak{a} = Ra$ for $a \in R$. Let

$$L_1 \supseteq L_2 \supseteq \cdots \supseteq L_n \supseteq L_{n+1} \supseteq \cdots$$

be a descending chain of submodules of M. Observe that, for each $i \in \mathbb{N}_0$ and each submodule L of M, we have $a^i(0 :_L a^{i+1}) \subseteq (0 :_M a)$. In fact, for each $n \in \mathbb{N}$,

$$(0 :_{L_n} a) \supseteq \cdots \supseteq a^i(0 :_{L_n} a^{i+1}) \supseteq a^{i+1}(0 :_{L_n} a^{i+2}) \supseteq \cdots$$

is a descending chain of submodules of the Artinian R-module $(0 :_M a)$, and so is eventually stationary: let E_n denote its eventual stationary value, so that there is $k_n \in \mathbb{N}$ such that $a^i(0 :_{L_n} a^{i+1}) = E_n$ for all $i \geq k_n$.

Since

$$\begin{aligned}
E_n &= a^{k_n + k_{n+1}}(0 :_{L_n} a^{k_n + k_{n+1} + 1}) \\
&\supseteq a^{k_n + k_{n+1}}(0 :_{L_{n+1}} a^{k_n + k_{n+1} + 1}) = E_{n+1},
\end{aligned}$$

we see that $E_1 \supseteq E_2 \supseteq \cdots \supseteq E_n \supseteq E_{n+1} \supseteq \cdots$ is a descending chain of submodules of the Artinian R-module $(0 :_M a)$, and so there exists $t \in \mathbb{N}$ such that $E_n = E_t$ for all $n \geq t$. Now for all $i, n \in \mathbb{N}$ with $n \geq t$ and $i \geq k_t$, we have

$$E_t = a^{k_t}(0 :_{L_t} a^{k_t + 1}) \supseteq a^i(0 :_{L_t} a^{i+1}) \supseteq a^i(0 :_{L_n} a^{i+1}) \supseteq E_n = E_t.$$

Thus $a^i(0 :_{L_n} a^{i+1}) = E_t$ for all $n \geq t$ and $i \geq k_t$.

We plan to use Lemma 7.1.1. With this in mind, we consider, for each integer $i = 0, 1, 2, \ldots, k_t - 1$, the descending chain

$$a^i(0 :_{L_1} a^{i+1}) \supseteq \cdots \supseteq a^i(0 :_{L_n} a^{i+1}) \supseteq a^i(0 :_{L_{n+1}} a^{i+1}) \supseteq \cdots$$

of submodules of the Artinian R-module $(0 :_M a)$: there exists $u \in \mathbb{N}$ with $u \geq t$ such that $a^i(0 :_{L_n} a^{i+1}) = a^i(0 :_{L_u} a^{i+1})$ for all $n \geq u$ and all $i = 0, 1, 2, \ldots, k_t - 1$.

We now have that, for each integer $n \geq u$,

$$a^i(0 :_{L_n} a^{i+1}) = a^i(0 :_{L_{n+1}} a^{i+1}) \quad \text{for all } i \in \mathbb{N}_0:$$

this equation is true for all $i = 0, 1, 2, \ldots, k_t - 1$ by the last paragraph and for all $i \geq k_t$ by the paragraph before that. Therefore, by Lemma 7.1.1, we have $L_n = L_{n+1}$ for all $n \geq u$. It follows that M is Artinian.

Now suppose, inductively, that $t > 1$ and the result has been proved for ideals that can be generated by fewer than t elements. Suppose that a is generated by t elements $a_1, \ldots, a_t$. Set $b = Ra_1 + \cdots + Ra_{t-1}$ and $N = (0 :_M b)$. Then N is Ra_t-torsion and $(0 :_N a_t) = (0 :_M a)$. It therefore follows from what we have already proved in the case where $t = 1$ that $N = (0 :_M b)$ is Artinian. Since M is, of course, b-torsion and b can be generated by $t - 1$ elements, we can apply the inductive hypothesis to deduce that M is Artinian. The inductive step is complete. $\square$

We immediately exploit Theorem 7.1.2 to prove that certain local cohomology modules are Artinian. The proofs presented below of the next two theorems are due to L. Melkersson (see [52, Theorems 2.1 and 2.2]), although the results themselves are somewhat older.

7.1.3 Theorem. *Assume that $(R, \mathfrak{m})$ is local, and let M be a finitely generated R-module. Then the R-module $H^i_{\mathfrak{m}}(M)$ is Artinian for all $i \in \mathbb{N}_0$.*

Proof. We use induction on i. First, $\Gamma_{\mathfrak{m}}(M)$ is a finitely generated R-module annihilated by a power of $\mathfrak{m}$; hence $H^0_{\mathfrak{m}}(M)$ has finite length.

Now suppose, inductively, that $i > 0$ and we have shown that $H^{i-1}_{\mathfrak{m}}(M')$ is Artinian for every finitely generated R-module M'. Now

$$H^i_{\mathfrak{m}}(M) \cong H^i_{\mathfrak{m}}(M/\Gamma_{\mathfrak{m}}(M))$$

by 2.1.7(iii). Also, 2.1.2 shows that $M/\Gamma_{\mathfrak{m}}(M)$ is $\mathfrak{m}$-torsion-free. We therefore assume in addition that M is an $\mathfrak{m}$-torsion-free R-module.

We now use 2.1.1(ii) to deduce that $\mathfrak{m}$ contains an element r which is a non-zerodivisor on M. The exact sequence

$$0 \longrightarrow M \overset{r}{\longrightarrow} M \longrightarrow M/rM \longrightarrow 0$$

induces an exact sequence $H^{i-1}_{\mathfrak{m}}(M/rM) \longrightarrow H^i_{\mathfrak{m}}(M) \overset{r}{\longrightarrow} H^i_{\mathfrak{m}}(M)$ of local cohomology modules. Since M/rM is a finitely generated R-module, it follows from the inductive hypothesis that $H^{i-1}_{\mathfrak{m}}(M/rM)$ is Artinian, so that, in view of the above exact sequence, the R-module $(0 :_{H_{\mathfrak{m}}^{i}(M)} r)$ is Artinian. Since $H_{\mathfrak{m}}^{i}(M)$ is $\mathfrak{m}$-torsion and therefore Rr-torsion, it follows from Theorem 7.1.2 that $H_{\mathfrak{m}}^{i}(M)$ is Artinian. The inductive step is complete. $\square$

The following exercise generalizes Theorem 7.1.3 to non-local situations.

7.1.4 ♯Exercise. Assume that $R/\mathfrak{a}$ is Artinian, and let M be a finitely generated R-module. Show that $H_{\mathfrak{a}}^{i}(M)$ is Artinian for all $i \in \mathbb{N}_0$.

7.1.5 Exercise. Suppose that $\mathfrak{m}$ is a maximal ideal of R, and that $\mathfrak{q}$ is an $\mathfrak{m}$-primary ideal of R. Let M be an R-module, and let $i \in \mathbb{N}_0$.

(i) Let $s \in R \setminus \mathfrak{m}$. Show that multiplication by s provides an automorphism of $H_{\mathfrak{q}}^{i}(M)$, so that this local cohomology module has a natural structure as an $R_{\mathfrak{m}}$-module.

(ii) Show that a subset of $H_{\mathfrak{q}}^{i}(M)$ is an $R_{\mathfrak{m}}$-submodule if and only if it is an R-submodule.

(iii) Deduce from (i), (ii) and Theorem 7.1.3 that, when M is finitely generated, $H_{\mathfrak{q}}^{i}(M)$ is an Artinian R-module.

7.1.6 Theorem. *Assume $(R, \mathfrak{m})$ is local, and let M be a non-zero, finitely generated R-module of dimension n. Then the R-module $H_{\mathfrak{a}}^{n}(M)$ is Artinian.*

Proof. We use induction on n. If $n = 0$, then $\sqrt{(0 : M)} = \mathfrak{m}$, so that M is annihilated by a power of $\mathfrak{m}$ and so has finite length; therefore its submodule $\Gamma_{\mathfrak{a}}(M)$ also has finite length.

Now suppose, inductively, that $n > 0$ and we have established the result for (non-zero, finitely generated) R-modules of dimension smaller than n. Now $H_{\mathfrak{a}}^{n}(M) \cong H_{\mathfrak{a}}^{n}(M/\Gamma_{\mathfrak{a}}(M))$ by 2.1.7(iii). If $\dim(M/\Gamma_{\mathfrak{a}}(M)) < n$, then $H_{\mathfrak{a}}^{n}(M/\Gamma_{\mathfrak{a}}(M)) = 0$ by Grothendieck's Vanishing Theorem 6.1.2, and there is nothing to prove. Since $M/\Gamma_{\mathfrak{a}}(M)$ is $\mathfrak{a}$-torsion-free (by 2.1.2), we therefore make the additional assumption that M is an $\mathfrak{a}$-torsion-free R-module.

The argument now proceeds like that used in the proof of Theorem 7.1.3 above. By 2.1.1(ii), the ideal $\mathfrak{a}$ contains an element r which is a non-zerodivisor on M. The exact sequence $0 \longrightarrow M \xrightarrow{r} M \longrightarrow M/rM \longrightarrow 0$ induces an exact sequence $H_{\mathfrak{a}}^{n-1}(M/rM) \longrightarrow H_{\mathfrak{a}}^{n}(M) \xrightarrow{r} H_{\mathfrak{a}}^{n}(M)$. Since r is not in any minimal member of Supp M, we have $\dim(M/rM) \leq n - 1$, so that, by the inductive hypothesis (or Grothendieck's Vanishing Theorem 6.1.2), the R-module $H_{\mathfrak{a}}^{n-1}(M/rM)$ is Artinian. Therefore, in view of the above exact sequence, $(0 :_{H_{\mathfrak{a}}^{n}(M)} r)$ is Artinian. Since $H_{\mathfrak{a}}^{n}(M)$ is Rr-torsion, it follows from Theorem 7.1.2 that $H_{\mathfrak{a}}^{n}(M)$ is Artinian. The inductive step is complete. $\square$

It is perhaps worth pointing out that, in our applications of Melkersson's Theorem 7.1.2 in the proofs of 7.1.3 and 7.1.6, we have only used the special case of 7.1.2 in which $\mathfrak{a}$ is principal. However, this is perhaps not surprising, as most of our proof of 7.1.2 was devoted to the case when $\mathfrak{a}$ is principal. The following exercise generalizes Theorem 7.1.6 to non-local situations.

7.1.7 Exercise. Let M be a non-zero, finitely generated R-module of finite dimension n. Show that the R-module $H_{\mathfrak{a}}^n(M)$ is Artinian.

7.2 Secondary representation

We intend to exploit Theorem 7.1.3 to provide another proof of the Non-vanishing Theorem 6.1.4, which states that, if M is a non-zero, finitely generated module of dimension n over the local ring $(R, \mathfrak{m})$, then $H_{\mathfrak{m}}^n(M) \neq 0$. We are going to use the theory of secondary representations of Artinian modules. As this theory, although mentioned in Matsumura [50, Section 6, Appendix], is not as well known as the theory of primary decomposition, we shall guide the reader through the main points by means of a series of exercises. Although we shall maintain our standard hypothesis that R is Noetherian, the reader might be interested to learn that this condition is not strictly necessary for the development of a worthwhile theory: see D. Kirby [44], I. G. Macdonald [45] or D. G. Northcott [62].

7.2.1 Definitions and ♯Exercise. Let S be an R-module. We say that S is *secondary* precisely when $S \neq 0$ and, for each $r \in R$, either $rS = S$ or there exists $n \in \mathbb{N}$ such that $r^n S = 0$. Show that, when this is the case, $\mathfrak{p} := \sqrt{(0 :_R S)}$ is a prime ideal of R: in these circumstances, we say that S is a $\mathfrak{p}$-*secondary R-module.*

Show that a non-zero homomorphic image of a $\mathfrak{p}$-secondary R-module is again $\mathfrak{p}$-secondary.

Show that, if $S_1, \ldots, S_n$ (where $n \in \mathbb{N}$) are $\mathfrak{p}$-secondary submodules of an R-module M, then so too is $\sum_{i=1}^n S_i$.

7.2.2 Definitions and ♯Exercise. Let M be an R-module. A *secondary representation of M* is an expression for M as a sum of finitely many secondary submodules of M. Such a secondary representation

$$M = S_1 + \cdots + S_n \quad \text{with } S_i \ \mathfrak{p}_i\text{-secondary } (1 \leq i \leq n)$$

of M is said to be *minimal* precisely when

(i) $\mathfrak{p}_1, \ldots, \mathfrak{p}_n$ are n different prime ideals of R; and

(ii) for all $j = 1, \ldots, n$, we have $S_j \nsubseteq \sum\limits_{\substack{i=1 \\ i \neq j}}^{n} S_i$.

We say that M is a *representable R-module* precisely when it has a secondary representation. As the sum of the empty family of submodules of an R-module is zero, we shall regard a zero R-module as representable.

Show that a representable R-module has a minimal secondary representation.

7.2.3 ♯Exercise. The First Uniqueness Theorem. Let M be a representable R-module and let

$$M = S_1 + \cdots + S_n \quad \text{with } S_i \ \mathfrak{p}_i\text{-secondary } (1 \leq i \leq n)$$

and

$$M = S_1' + \cdots + S_{n'}' \quad \text{with } S_i' \ \mathfrak{p}_i'\text{-secondary } (1 \leq i \leq n')$$

be two minimal secondary representations of M. Prove that $n = n'$ and

$$\{\mathfrak{p}_1, \ldots, \mathfrak{p}_n\} = \{\mathfrak{p}_1', \ldots, \mathfrak{p}_n'\}.$$

(Here is hint: show that, for $\mathfrak{p} \in \operatorname{Spec}(R)$, it is the case that $\mathfrak{p}$ is one of $\mathfrak{p}_1, \ldots, \mathfrak{p}_n$ if and only if there is a homomorphic image of M which is $\mathfrak{p}$-secondary.)

7.2.4 Definition. Let M be a representable R-module and let

$$M = S_1 + \cdots + S_n \quad \text{with } S_i \ \mathfrak{p}_i\text{-secondary } (1 \leq i \leq n)$$

be a minimal secondary representation of M. Then the n-element set

$$\{\mathfrak{p}_1, \ldots, \mathfrak{p}_n\},$$

which is independent of the choice of minimal secondary representation of M by 7.2.3, is called *the set of attached prime ideals of M* and denoted by $\operatorname{Att} M$ or $\operatorname{Att}_R M$. The members of $\operatorname{Att} M$ are referred to as the *attached prime ideals* or the *attached primes* of M.

7.2.5 ♯Exercise. Suppose that the R-module M is representable, and let $\mathfrak{p} \in \operatorname{Spec}(R)$. Use the Noetherian property of R to show that $\mathfrak{p} \in \operatorname{Att} M$ if and only if there is a homomorphic image of M which has annihilator equal to $\mathfrak{p}$.

7.2.6 Exercise. Let $0 \longrightarrow L \longrightarrow M \longrightarrow N \longrightarrow 0$ be an exact sequence of representable R-modules and R-homomorphisms. Prove that

$$\operatorname{Att} N \subseteq \operatorname{Att} M \subseteq \operatorname{Att} L \cup \operatorname{Att} N.$$

7.2.7 ♯Exercise. The Second Uniqueness Theorem. Let M be a represent-able R-module and let

$$M = S_1 + \cdots + S_n \quad \text{with } S_i \ \mathfrak{p}_i\text{-secondary } (1 \le i \le n)$$

be a minimal secondary representation of M.

Suppose that $\mathfrak{p}_j$ is a minimal member of $\{\mathfrak{p}_1, \ldots, \mathfrak{p}_n\}$ with respect to inclusion. Prove that $S_j = \bigcap_{r \in R \setminus \mathfrak{p}_j} rM$.

In the light of the First Uniqueness Theorem 7.2.3, this means that, in a minimal secondary representation of M, each secondary term corresponding to a minimal member of Att M is uniquely determined by M and independent of the choice of minimal secondary representation.

In order to make use of this theory of secondary representation, we shall need the fact that every Artinian R-module is representable. This fact is the subject of the next two exercises.

7.2.8 Definition and ♯Exercise. We say that an R-module N is *sum-irreducible* precisely when it is non-zero and cannot be expressed as the sum of two proper submodules of itself.

Prove that an Artinian sum-irreducible R-module is secondary.

7.2.9 ♯Exercise. Let A be an Artinian R-module. Show that A can be expressed as a sum of finitely many sum-irreducible submodules, and deduce from Exercise 7.2.8 above that A is representable.

We can therefore form the finite set Att A. Note that Att $A = \emptyset$ if and only if $A = 0$.

Actually, the class of R-modules which possess secondary representations is, in general, larger than the class of Artinian R-modules: this is illustrated by the next exercise.

7.2.10 Exercise. Let E be an injective R-module.

(i) Suppose that Q is an R-module with the property that its zero submodule is a $\mathfrak{p}$-primary submodule of Q. Prove that $\operatorname{Hom}_R(Q, E)$, if non-zero, is $\mathfrak{p}$-secondary.

(ii) Let M be a finitely generated R-module. Prove that $\operatorname{Hom}_R(M, E)$ is representable, and that $\operatorname{Att}_R(\operatorname{Hom}_R(M, E)) \subseteq \operatorname{Ass}_R M$.

(iii) Deduce that E is representable and that $\operatorname{Att}_R E \subseteq \operatorname{Ass} R = \operatorname{ass} 0$.

(iv) The injective R-module E is said to be an *injective cogenerator* for R precisely when $\operatorname{Hom}_R(N, E) \ne 0$ for every non-zero R-module N. Prove that, when this is the case, $\operatorname{Att}_R(\operatorname{Hom}_R(M, E)) = \operatorname{Ass}_R M$ for every finitely generated R-module M.

7.2.11 Proposition. *Let A be an Artinian R-module, and let $r \in R$. Then*

(i) *$rA = A$ if and only if $r \in R \setminus \bigcup_{\mathfrak{p} \in \operatorname{Att} A} \mathfrak{p}$; and*

(ii) *$\sqrt{(0 : A)} = \bigcap_{\mathfrak{p} \in \operatorname{Att} A} \mathfrak{p}$.*

Proof. Clearly, we can assume that $A \neq 0$, since $\operatorname{Att} 0 = \emptyset$. Let

$$A = S_1 + \cdots + S_n \quad \text{with } S_i \text{ } \mathfrak{p}_i\text{-secondary } (1 \leq i \leq n)$$

be a minimal secondary representation of M.

(i) Suppose that $r \in R \setminus \bigcup_{\mathfrak{p} \in \operatorname{Att} A} \mathfrak{p}$; then $rS_i = S_i$ for all $i = 1, \ldots, n$, and so $rA = A$. On the other hand, if $r \in \mathfrak{p}_j$ for some j with $1 \leq j \leq n$, then $r^h S_j = 0$ for a sufficiently large integer h, and so

$$r^h A = r^h S_1 + \cdots + r^h S_n \subseteq \sum_{\substack{i=1 \\ i \neq j}}^{n} S_i \subset A.$$

(ii) To prove this, just note that $\sqrt{(0 : A)} = \bigcap_{i=1}^{n} \sqrt{(0 : S_i)} = \bigcap_{i=1}^{n} \mathfrak{p}_i$. $\quad\square$

Part (i) of 7.2.11 provides an Artinian analogue of the well-known fact that, if N is a Noetherian R-module and $r \in R$, then r is a non-zerodivisor on N if and only if r lies outside all the associated prime ideals of N.

There is more to the theory of secondary representation than the brief outline presented above: the interested reader should consult the references cited earlier for this topic, especially [45]. We have presented little more than the part of the theory we shall need to use. Its relevance for us lies in the fact that, when $(R, \mathfrak{m})$ is local, and M is a finitely generated R-module, then, for each $i \in \mathbb{N}_0$, the local cohomology module $H^i_{\mathfrak{m}}(M)$ is, by 7.1.3, Artinian, and so we can form the finite set of prime ideals $\operatorname{Att}(H^i_{\mathfrak{m}}(M))$ and use the theory of secondary representation for these modules. In the subsequent work, we shall be interested in whether, for particular j, the local cohomology module $H^j_{\mathfrak{m}}(M)$ is finitely generated (and so, since it is in any case Artinian, of finite length). With this in mind, we present here one more result about secondary representation before going on to apply the theory to local cohomology modules.

7.2.12 Corollary. *Let $(R, \mathfrak{m})$ be local and let A be an Artinian R-module. Then A is finitely generated, and so of finite length, if and only if $\operatorname{Att} A \subseteq \{\mathfrak{m}\}$.*

Proof. ($\Rightarrow$) When A is of finite length, there exists $h \in \mathbb{N}$ such that $\mathfrak{m}^h A = 0$, so that A is either 0 or $\mathfrak{m}$-secondary.

($\Leftarrow$) If $\operatorname{Att} A \subseteq \{\mathfrak{m}\}$, then, by 7.2.11(ii), there exists $h \in \mathbb{N}$ such that $\mathfrak{m}^h A = 0$, and it then follows from, for example, [81, 7.30] that A has finite length. $\quad\square$

7.3 The Non-vanishing Theorem again

Our first lemma of this section is in preparation for the main theorem of this chapter. Recall our convention that the dimension of the zero R-module is -1.

7.3.1 Lemma. *Let $(R, \mathfrak{m})$ be local and let M be a non-zero, finitely generated R-module of dimension n. Then the set*

$$\Sigma := \{N' : N' \text{ is a submodule of } M \text{ and } \dim N' < n\}$$

has a largest element with respect to inclusion, N say. Set $G := M/N$. Then

(i) $\dim G = n$;
(ii) *G has no non-zero submodule of dimension less than n;*
(iii) $\operatorname{Ass} G = \{\mathfrak{p} \in \operatorname{Ass} M : \dim R/\mathfrak{p} = n\}$; *and*
(iv) $H_{\mathfrak{m}}^n(G) \cong H_{\mathfrak{m}}^n(M)$.

Proof. Since M is a Noetherian R-module, the set Σ has a maximal member, N say. Since the sum of any two members of Σ is again in Σ, it follows that N contains every member of Σ, and so is the largest element of Σ.

(i) It follows from the canonical exact sequence

$$0 \longrightarrow N \longrightarrow M \longrightarrow G \longrightarrow 0$$

that $\operatorname{Supp} G \subseteq \operatorname{Supp} M$, and that any $\mathfrak{p} \in \operatorname{Ass} M$ with $\dim R/\mathfrak{p} = n$ must belong to $\operatorname{Supp} G$ since it cannot belong to $\operatorname{Supp} N$. Hence $\dim G = n$ and $\{\mathfrak{p} \in \operatorname{Ass} M : \dim R/\mathfrak{p} = n\} \subseteq \operatorname{Ass} G$.

(ii) Suppose that L is a submodule of M such that $N \subseteq L \subseteq M$ and $\dim(L/N) < n$. Consideration of the canonical exact sequence

$$0 \longrightarrow N \longrightarrow L \longrightarrow L/N \longrightarrow 0$$

shows that $\dim L < n$; hence $L \subseteq N$ and $L/N = 0$.

(iii) Let $\mathfrak{p} \in \operatorname{Ass} G$. By (ii), $\dim R/\mathfrak{p} = n$, so that, since $\operatorname{Supp} G \subseteq \operatorname{Supp} M$, we must have $\mathfrak{p} \in \operatorname{Ass} M$. Thus

$$\operatorname{Ass} G \subseteq \{\mathfrak{p} \in \operatorname{Ass} M : \dim R/\mathfrak{p} = n\} .$$

Since the reverse inclusion was established in our proof of (i) above, we have completed the proof of (iii).

(iv) Since $\dim N < n$, it follows from Grothendieck's Vanishing Theorem 6.1.2 that $H_{\mathfrak{m}}^n(N) = H_{\mathfrak{m}}^{n+1}(N) = 0$. The claim therefore follows from the long exact sequence of local cohomology modules (with respect to $\mathfrak{m}$) that results from the exact sequence $0 \longrightarrow N \longrightarrow M \longrightarrow G \longrightarrow 0$. $\square$

The next theorem is the main result of this chapter. With it, we not only offer an alternative proof of the result of the Non-vanishing Theorem 6.1.4 (as promised just before 6.1.3), but we also provide more information than was given in Theorem 6.1.4. (We should perhaps point out that our proof below in 7.3.2 does quote from Theorem 6.2.7 the fact that, if M is a non-zero, finitely generated module of depth d over the local ring $(R, \mathfrak{m})$, then $H_{\mathfrak{m}}^d(M) \neq 0$; however, no use was made of 6.1.4 in our proof of 6.2.7, and, indeed, the argument given in the proof of 6.2.7 is elementary.)

7.3.2 Theorem. *Assume* $(R, \mathfrak{m})$ *is local, and let* M *be a non-zero, finitely generated* R-*module of dimension* n. *Then* $H_{\mathfrak{m}}^n(M) \neq 0$ *and*

$$\operatorname{Att}(H_{\mathfrak{m}}^n(M)) = \{\mathfrak{p} \in \operatorname{Ass} M : \dim R/\mathfrak{p} = n\}.$$

Proof. (This proof is due to I. G. Macdonald and R. Y. Sharp [47, Theorem 2.2].) Throughout this proof, we make tacit use of the fact, proved in Theorem 7.1.3, that, for each $i \in \mathbb{N}_0$, the module $H_{\mathfrak{m}}^i(M)$ is Artinian, so that, by 7.2.9, it has a secondary representation and we can form the set $\operatorname{Att}(H_{\mathfrak{m}}^i(M))$.

We use induction on n. When $n = 0$, the module M has finite length and so is annihilated by some power of $\mathfrak{m}$. Hence $H_{\mathfrak{m}}^0(M) \cong \Gamma_{\mathfrak{m}}(M) = M \neq 0$. By 7.2.9 and 7.2.12,

$$\operatorname{Att}(H_{\mathfrak{m}}^0(M)) = \operatorname{Att} M = \{\mathfrak{m}\} = \operatorname{Ass} M = \{\mathfrak{p} \in \operatorname{Ass} M : \dim R/\mathfrak{p} = 0\}.$$

The result has been proved in this case.

Assume, inductively, that $n > 0$ and that the result has been proved for non-zero, finitely generated R-modules of dimension $n - 1$. By 7.3.1, we can, after factoring out, if necessary, the largest submodule of M of dimension smaller than n, assume that M has no non-zero submodule of dimension smaller than n. We shall make this assumption for the remainder of the inductive step, and with this assumption our aim is to show that $\operatorname{Att}(H_{\mathfrak{m}}^n(M)) = \operatorname{Ass} M$.

Since $n > 0$, we have $\mathfrak{m} \notin \operatorname{Ass} M$, and so there exists $r \in \mathfrak{m}$ which is a non-zerodivisor on M. We suppose that $H_{\mathfrak{m}}^n(M) = 0$, and look for a contradiction. If $n = 1$, we have $1 \leq \operatorname{grade}_M \mathfrak{m} = \operatorname{depth} M \leq \dim M = 1$, so that $\operatorname{grade}_M \mathfrak{m} = 1$ and we have a contradiction to 6.2.7. Thus, in our search for a contradiction, we can, and do, assume that $n > 1$.

Now, for each $r \in \mathfrak{m}$ which is a non-zerodivisor on M, the module M/rM (is non-zero and finitely generated and) has $\dim(M/rM) = n - 1$, and the exact sequence $0 \longrightarrow M \xrightarrow{r} M \longrightarrow M/rM \longrightarrow 0$ induces a long exact sequence of local cohomology modules

$$H_{\mathfrak{m}}^{n-1}(M) \xrightarrow{r} H_{\mathfrak{m}}^{n-1}(M) \longrightarrow H_{\mathfrak{m}}^{n-1}(M/rM) \longrightarrow 0$$

in view of our supposition that $H_{\mathfrak{m}}^n(M) = 0$. Thus, for each $r \in \mathfrak{m}$ which is a non-zerodivisor on M, we have $H_{\mathfrak{m}}^{n-1}(M)/rH_{\mathfrak{m}}^{n-1}(M) \cong H_{\mathfrak{m}}^{n-1}(M/rM)$, and this is non-zero by the inductive hypothesis. Therefore $H_{\mathfrak{m}}^{n-1}(M) \neq 0$.

Our next step is to prove that $\mathfrak{m} \in \mathrm{Att}(H_{\mathfrak{m}}^{n-1}(M))$. We suppose that $\mathfrak{m} \notin \mathrm{Att}(H_{\mathfrak{m}}^{n-1}(M))$ and look for a contradiction. Then, by the Prime Avoidance Theorem,

$$\mathfrak{m} \not\subseteq \left(\bigcup_{\mathfrak{p} \in \mathrm{Att}(H_{\mathfrak{m}}^{n-1}(M))} \mathfrak{p} \right) \cup \left(\bigcup_{\mathfrak{q} \in \mathrm{Ass}\, M} \mathfrak{q} \right),$$

so that, in the light of 7.2.11(i), there exists $r_1 \in \mathfrak{m}$ which is a non-zerodivisor on M and such that $H_{\mathfrak{m}}^{n-1}(M) = r_1 H_{\mathfrak{m}}^{n-1}(M)$; this contradicts the fact that $H_{\mathfrak{m}}^{n-1}(M/r_1 M) \neq 0$.

Thus $\mathfrak{m} \in \mathrm{Att}(H_{\mathfrak{m}}^{n-1}(M))$: let $\mathfrak{p}_1, \ldots, \mathfrak{p}_t$ be the remaining members of $\mathrm{Att}(H_{\mathfrak{m}}^{n-1}(M))$. Again by the Prime Avoidance Theorem, there exists

$$r_2 \in \mathfrak{m} \setminus \left(\bigcup_{i=1}^{t} \mathfrak{p}_i \right) \cup \left(\bigcup_{\mathfrak{q} \in \mathrm{Ass}\, M} \mathfrak{q} \right).$$

Since $r_2 \in \mathfrak{m}$ and r_2 is a non-zerodivisor on M, we again have

$$H_{\mathfrak{m}}^{n-1}(M)/r_2 H_{\mathfrak{m}}^{n-1}(M) \cong H_{\mathfrak{m}}^{n-1}(M/r_2 M),$$

and, by the inductive hypothesis, $H_{\mathfrak{m}}^{n-1}(M/r_2 M) \neq 0$ and

$$\mathrm{Att}(H_{\mathfrak{m}}^{n-1}(M/r_2 M)) \subseteq \{\mathfrak{p} \in \mathrm{Ass}(M/r_2 M) : \dim R/\mathfrak{p} = n - 1\}.$$

But, by 7.2.5,

$$\mathrm{Att}(H_{\mathfrak{m}}^{n-1}(M)/r_2 H_{\mathfrak{m}}^{n-1}(M)) \subseteq \{\mathfrak{p} \in \mathrm{Att}(H_{\mathfrak{m}}^{n-1}(M)) : r_2 \in \mathfrak{p}\},$$

and $\mathfrak{m}$ is the only member of the latter set. Since $n > 1$, we have obtained a contradiction.

Thus we have proved that $H_{\mathfrak{m}}^n(M) \neq 0$. To complete the inductive step, since M now has no non-zero submodule of dimension smaller than n, it remains for us to prove that $\mathrm{Att}(H_{\mathfrak{m}}^n(M)) = \mathrm{Ass}\, M$.

We know that $\mathrm{grade}_M \mathfrak{m} \geq 1$; also, for each $r \in \mathfrak{m}$ which is a non-zerodivisor on M, we have $\dim(M/rM) = n - 1$, so that $H_{\mathfrak{m}}^n(M/rM) = 0$ by Grothendieck's Vanishing Theorem 6.1.2, and the exact sequence of local cohomology modules induced by the exact sequence

$$0 \longrightarrow M \overset{r}{\longrightarrow} M \longrightarrow M/rM \longrightarrow 0$$

yields that $H_{\mathfrak{m}}^n(M) = rH_{\mathfrak{m}}^n(M)$. It therefore follows from 7.2.11(i) that

$$\mathfrak{m} \setminus \bigcup_{\mathfrak{p} \in \mathrm{Ass}\, M} \mathfrak{p} \subseteq \mathfrak{m} \setminus \bigcup_{\mathfrak{q} \in \mathrm{Att}(H_{\mathfrak{m}}^n(M))} \mathfrak{q}.$$

Let $\mathfrak{q} \in \mathrm{Att}(H_{\mathfrak{m}}^n(M))$: it follows from the above inclusion relation and the

Prime Avoidance Theorem that $\mathfrak{q} \subseteq \mathfrak{p}$ for some $\mathfrak{p} \in \operatorname{Ass} M$. Since $H_{\mathfrak{m}}^n$ is an R-linear functor, it follows that

$$(0 : M) \subseteq (0 : H_{\mathfrak{m}}^n(M)) \subseteq \mathfrak{q} \subseteq \mathfrak{p}.$$

As $n = \dim R/(0 : M) = \dim R/\mathfrak{p}$, it follows that $\mathfrak{q} = \mathfrak{p}$. Hence

$$\operatorname{Att}(H_{\mathfrak{m}}^n(M)) \subseteq \operatorname{Ass} M.$$

To establish the reverse inclusion, let $\mathfrak{p} \in \operatorname{Ass} M$, so that $\dim R/\mathfrak{p} = n$. By the theory of primary decomposition, there exists a $\mathfrak{p}$-primary submodule Q of M; thus M/Q is a non-zero finitely generated R-module with $\operatorname{Ass}(M/Q) = \{\mathfrak{p}\}$. Note that M/Q cannot have any non-zero submodule of dimension less than n (or else it would have an associated prime ideal other than $\mathfrak{p}$). Thus, by the work in the preceding six paragraphs applied to M/Q rather than M, we have $H_{\mathfrak{m}}^n(M/Q) \neq 0$ and

$$\emptyset \neq \operatorname{Att}(H_{\mathfrak{m}}^n(M/Q)) \subseteq \operatorname{Ass}(M/Q) = \{\mathfrak{p}\}.$$

Hence $\operatorname{Att}(H_{\mathfrak{m}}^n(M/Q)) = \{\mathfrak{p}\}$. Since $\dim Q < n + 1$, Grothendieck's Vanishing Theorem 6.1.2 tells us that $H_{\mathfrak{m}}^{n+1}(Q) = 0$; therefore, the canonical exact sequence $0 \longrightarrow Q \longrightarrow M \longrightarrow M/Q \longrightarrow 0$ induces an epimorphism $H_{\mathfrak{m}}^n(M) \longrightarrow H_{\mathfrak{m}}^n(M/Q)$. It now follows from 7.2.5 that

$$\{\mathfrak{p}\} = \operatorname{Att}(H_{\mathfrak{m}}^n(M/Q)) \subseteq \operatorname{Att}(H_{\mathfrak{m}}^n(M)).$$

Hence $\operatorname{Ass} M \subseteq \operatorname{Att}(H_{\mathfrak{m}}^n(M))$, and therefore $\operatorname{Ass} M = \operatorname{Att}(H_{\mathfrak{m}}^n(M))$. This completes the inductive step. $\qquad\square$

We show next how to deduce the result of Exercise 6.1.7 very quickly from Theorem 7.3.2.

7.3.3 Corollary. *Assume $(R, \mathfrak{m})$ is local, and let M be a non-zero, finitely generated R-module of dimension $n > 0$. Then $H_{\mathfrak{m}}^n(M)$ is not finitely generated.*

Proof. By 7.1.3 and 7.3.2, the local cohomology module $H_{\mathfrak{m}}^n(M)$ is a non-zero Artinian module which has a non-maximal attached prime ideal. Hence $H_{\mathfrak{m}}^n(M)$ is not finitely generated, by 7.2.12. $\qquad\square$

7.3.4 ♯Exercise. Let M be a finitely generated R-module for which the ideal $\mathfrak{a} + (0 : M)$ is proper. Let $\mathfrak{p}$ be a minimal prime ideal of $\mathfrak{a} + (0 : M)$, and let $t := \dim_{R_{\mathfrak{p}}} M_{\mathfrak{p}}$. Prove that $H_{\mathfrak{a}}^t(M) \neq 0$, and that, if $t > 0$, then $H_{\mathfrak{a}}^t(M)$ is not finitely generated.

8

The Lichtenbaum–Hartshorne Theorem

In this chapter, we take up again the main theme of Chapter 6, and establish another vanishing theorem for local cohomology modules, namely the local Lichtenbaum–Hartshorne Vanishing Theorem (see R. Hartshorne [29, Theorem 3.1], and C. Peskine and L. Szpiro [66, chapitre III, théorème 3.1]). While two important vanishing theorems in Chapter 6, namely Grothendieck's Vanishing Theorem 6.1.2, and Theorem 6.2.7, which shows that, for a finitely generated R-module M such that $\mathfrak{a}M \neq M$, we have $H^i_\mathfrak{a}(M) = 0$ for all $i < \operatorname{grade}_M \mathfrak{a}$, can be regarded as 'algebraic' in nature, the Lichtenbaum–Hartshorne Theorem is of 'analytic' nature, in the sense that it is intimately related with 'formal' methods and techniques, that is, with passage to completions of local rings and with the structure theory for complete local rings.

Results of a 'formal' nature in algebraic geometry sometimes provide powerful tools: the Lichtenbaum–Hartshorne Theorem is one such example, for we shall see in Chapter 19 that it can play a crucial rôle in the study of connectivity in algebraic varieties.

The local Lichtenbaum–Hartshorne Vanishing Theorem gives necessary and sufficient conditions, over an n-dimensional local ring $(R, \mathfrak{m})$, for the vanishing of n-th local cohomology modules with respect to $\mathfrak{a}$, where $\mathfrak{a}$ is proper. The sufficiency of these conditions is rather harder to prove than the necessity, and so we shall just discuss the sufficiency in this introduction. Let $\widehat{R}$ denote the completion of R. It turns out that $H^n_\mathfrak{a}(M) = 0$ for every R-module M if, for each $\mathfrak{P} \in \operatorname{Spec}(\widehat{R})$ satisfying $\dim \widehat{R}/\mathfrak{P} = n$, we have $\dim \widehat{R}/(\mathfrak{a}\widehat{R} + \mathfrak{P}) > 0$.

In the case when R is an n-dimensional complete local domain, the statement simplifies: then, $H^n_\mathfrak{a}(M) = 0$ for every R-module M if $\dim R/\mathfrak{a} > 0$. The general statement can be deduced from this special case by means of standard reductions, and we now outline our strategy for proof of the special case. This strategy is similar to that used by M. Brodmann and C. Huneke in [4].

Of course, the fact (3.4.10) that the local cohomology functor $H^i_\mathfrak{a}$ commutes with direct limits means that it is enough to prove that $H_{\mathfrak{a}}^n(M') = 0$ for each finitely generated R-module M' in order to obtain the above result; in fact, Lemma 8.1.7 below will show that it is enough to prove merely that $H_{\mathfrak{a}}^n(R) = 0$. Then we shall use the Noetherian property of R to reduce to the case where $\mathfrak{a}$ is a prime ideal $\mathfrak{p}$ with $\dim R/\mathfrak{p} = 1$. As in our proof in Chapter 6 of Theorem 6.1.4, we again use ideas related to the structure theorems for complete local rings to reduce to the case where R is a Gorenstein ring.

The arguments outlined in the last paragraph will reduce our proof to the following: given a complete Gorenstein local domain R of dimension n and $\mathfrak{p} \in \mathrm{Spec}(R)$ with $\dim R/\mathfrak{p} = 1$, how can we show that $H_{\mathfrak{p}}^n(R) = 0$? Our approach to this will use the ideas concerning the symbolic powers of $\mathfrak{p}$ of Exercise 3.1.4 (for which we shall provide a solution!) to see that

$$H_{\mathfrak{p}}^n(R) \cong \varinjlim_{j \in \mathbb{N}} \mathrm{Ext}_R^n(R/\mathfrak{p}^{(j)}, R);$$

we shall then use the facts that R is Gorenstein and $\mathrm{depth}\, R/\mathfrak{p}^{(j)} > 0$ to see that $\mathrm{Ext}_R^n(R/\mathfrak{p}^{(j)}, R) = 0$ for all $j \in \mathbb{N}$.

8.1 Preparatory lemmas

Our intention is to use Noetherian induction to reduce part of the proof of the Lichtenbaum–Hartshorne Theorem to a case where (R is a complete local domain and) $\mathfrak{a}$ is a prime ideal $\mathfrak{p}$ of R such that $\dim R/\mathfrak{p} = 1$. To achieve this, we wish to make our first application of a very useful proposition (8.1.2 below) that relates local cohomology with respect to $\mathfrak{a} + Rb$, where b is an element of R, to local cohomology with respect to $\mathfrak{a}$. Our proof of 8.1.2 uses the following lemma about injective modules.

8.1.1 Lemma. *Let $b \in R$, and let I be an injective R-module. Then the sequence of R-modules and R-homomorphisms*

$$0 \longrightarrow \Gamma_{Rb}(I) \xrightarrow{\xi_I} I \xrightarrow{\tau_I} I_b \longrightarrow 0,$$

in which ξ_I is the inclusion map and τ_I is the natural homomorphism, is split exact. Consequently, when I_b is regarded as an R-module in the natural way, it is injective; furthermore, application of the additive functor $\Gamma_{\mathfrak{a}}$ to the above split exact sequence yields a further split exact sequence of injective R-modules

$$0 \longrightarrow \Gamma_{\mathfrak{a}+Rb}(I) \longrightarrow \Gamma_{\mathfrak{a}}(I) \longrightarrow \Gamma_{\mathfrak{a}}(I_b) \longrightarrow 0.$$

Proof. Since I is injective, $H^1_{Rb}(I) = 0$; therefore, the exactness of the first sequence in the statement of the lemma is immediate from 2.2.20. By 2.1.4, the R-module $\Gamma_{Rb}(I)$ is injective, and so this exact sequence splits, and I_b, as R-module, is injective. The second part is then immediate once one recalls from 1.1.2 that $\Gamma_{\mathfrak{a}}(\Gamma_{Rb}(I)) = \Gamma_{\mathfrak{a}+Rb}(I)$, and from 2.1.4 that $\Gamma_{\mathfrak{a}}(I')$ is injective whenever I' is an injective R-module. $\qquad\square$

8.1.2 Proposition. *Let $b \in R$ and let $f : M \longrightarrow N$ be a homomorphism of R-modules.*

(i) *There is a long exact sequence of R-modules and R-homomorphisms*

$$0 \longrightarrow H^0_{\mathfrak{a}+Rb}(M) \longrightarrow H^0_{\mathfrak{a}}(M) \longrightarrow H^0_{\mathfrak{a}}(M_b)$$
$$\longrightarrow H^1_{\mathfrak{a}+Rb}(M) \longrightarrow H^1_{\mathfrak{a}}(M) \longrightarrow H^1_{\mathfrak{a}}(M_b)$$
$$\longrightarrow \quad \cdots \qquad\qquad\qquad \cdots$$
$$\longrightarrow H^i_{\mathfrak{a}+Rb}(M) \longrightarrow H^i_{\mathfrak{a}}(M) \longrightarrow H^i_{\mathfrak{a}}(M_b)$$
$$\longrightarrow H^{i+1}_{\mathfrak{a}+Rb}(M) \longrightarrow \quad \cdots$$

such that the diagram

$$
\begin{array}{ccccccc}
H^i_{\mathfrak{a}+Rb}(M) & \longrightarrow & H^i_{\mathfrak{a}}(M) & \longrightarrow & H^i_{\mathfrak{a}}(M_b) & \longrightarrow & H^{i+1}_{\mathfrak{a}+Rb}(M) \\
\downarrow{\scriptstyle H^i_{\mathfrak{a}+Rb}(f)} & & \downarrow{\scriptstyle H^i_{\mathfrak{a}}(f)} & {\scriptstyle H^i_{\mathfrak{a}}(f_b)}\downarrow & & \downarrow{\scriptstyle H^{i+1}_{\mathfrak{a}+Rb}(f)} & \\
H^i_{\mathfrak{a}+Rb}(N) & \longrightarrow & H^i_{\mathfrak{a}}(N) & \longrightarrow & H^i_{\mathfrak{a}}(N_b) & \longrightarrow & H^{i+1}_{\mathfrak{a}+Rb}(N)
\end{array}
$$

commutes for all $i \in \mathbb{N}_0$.

(ii) *For each $i \in \mathbb{N}_0$, there is a commutative diagram*

$$
\begin{array}{ccccccccc}
0 & \longrightarrow & H^1_{Rb}(H^i_{\mathfrak{a}}(M)) & \longrightarrow & H^{i+1}_{\mathfrak{a}+Rb}(M) & \longrightarrow & \Gamma_{Rb}(H^{i+1}_{\mathfrak{a}}(M)) & \longrightarrow & 0 \\
& & \downarrow{\scriptstyle H^1_{Rb}(H^i_{\mathfrak{a}}(f))} & & \downarrow{\scriptstyle H^{i+1}_{\mathfrak{a}+Rb}(f)} & & \downarrow{\scriptstyle \Gamma_{Rb}(H^{i+1}_{\mathfrak{a}}(f))} & & \\
0 & \longrightarrow & H^1_{Rb}(H^i_{\mathfrak{a}}(N)) & \longrightarrow & H^{i+1}_{\mathfrak{a}+Rb}(N) & \longrightarrow & \Gamma_{Rb}(H^{i+1}_{\mathfrak{a}}(N)) & \longrightarrow & 0
\end{array}
$$

with exact rows. The top row is referred to as the comparison exact sequence *for M.*

Proof. Let

$$I^{\bullet} : 0 \xrightarrow{d^{-1}} I^0 \xrightarrow{d^0} I^1 \longrightarrow \cdots \longrightarrow I^i \xrightarrow{d^i} I^{i+1} \longrightarrow \cdots$$

be an injective resolution of M, so that there is an R-homomorphism $\alpha : M \to I^0$ such that the sequence

$$0 \longrightarrow M \xrightarrow{\ \alpha\ } I^0 \xrightarrow{\ d^0\ } I^1 \longrightarrow \cdots \longrightarrow I^i \xrightarrow{\ d^i\ } I^{i+1} \longrightarrow \cdots$$

is exact. Similarly, let the exact sequence

$$0 \longrightarrow N \xrightarrow{\ \beta\ } J^0 \xrightarrow{\ e^0\ } J^1 \longrightarrow \cdots \longrightarrow J^i \xrightarrow{\ e^i\ } J^{i+1} \longrightarrow \cdots$$

provide an injective resolution for N. By the (dual of) the Comparison Theorem [71, 6.9], there exists a chain map $\phi^\bullet = (\phi^i)_{i \in \mathbb{N}_0} : I^\bullet \longrightarrow J^\bullet$ for which the diagram

$$
\begin{array}{ccc}
M & \xrightarrow{\ \alpha\ } & I^0 \\
\downarrow{\scriptstyle f} & & \downarrow{\scriptstyle \phi^0} \\
N & \xrightarrow{\ \beta\ } & J^0
\end{array}
$$

commutes.

It is immediate from Lemma 8.1.1 that there is a commutative diagram

$$
\begin{array}{ccccccccc}
0 & \longrightarrow & \Gamma_{\mathfrak{a}+Rb}(I^\bullet) & \longrightarrow & \Gamma_{\mathfrak{a}}(I^\bullet) & \longrightarrow & \Gamma_{\mathfrak{a}}((I^\bullet)_b) & \longrightarrow & 0 \\
& & \downarrow{\scriptstyle \Gamma_{\mathfrak{a}+Rb}(\phi^\bullet)} & & \downarrow{\scriptstyle \Gamma_{\mathfrak{a}}(\phi^\bullet)} & & \downarrow{\scriptstyle \Gamma_{\mathfrak{a}}((\phi^\bullet)_b)} & & \\
0 & \longrightarrow & \Gamma_{\mathfrak{a}+Rb}(J^\bullet) & \longrightarrow & \Gamma_{\mathfrak{a}}(J^\bullet) & \longrightarrow & \Gamma_{\mathfrak{a}}((J^\bullet)_b) & \longrightarrow & 0
\end{array}
$$

of complexes of R-modules and chain maps of such complexes such that, for each $i \in \mathbb{N}_0$, the sequence $0 \to \Gamma_{\mathfrak{a}+Rb}(I^i) \to \Gamma_{\mathfrak{a}}(I^i) \to \Gamma_{\mathfrak{a}}((I^i)_b) \to 0$ is exact, and a similar property holds for the lower row. Thus the above commutative diagram of complexes gives rise to a long exact sequence of cohomology modules of the complexes in the top row, a similar long exact sequence for the bottom row, and a chain map of the first long exact sequence into the second. Since, by 8.1.1, $(I^\bullet)_b$ provides an injective resolution for M_b as R-module and $(J^\bullet)_b$ provides an injective resolution for N_b as R-module, all the claims in (i) follow.

(ii) For each R-module K, let $\tau_K : K \longrightarrow K_b$ be the natural map.

For an R-submodule L of I^i, we use the R_b-monomorphism $L_b \longrightarrow (I^i)_b$ induced by inclusion to identify L_b as an R_b-submodule of $(I^i)_b$. With this convention, it is routine to check that

$$\operatorname{Ker}(\Gamma_{\mathfrak{a}}((d^i)_b)) = (\operatorname{Ker}(\Gamma_{\mathfrak{a}}(d^i)))_b, \ \ \operatorname{Im}(\Gamma_{\mathfrak{a}}((d^{i-1})_b)) = (\operatorname{Im}(\Gamma_{\mathfrak{a}}(d^{i-1})))_b.$$

Hence

$$
\begin{aligned}
(H_{\mathfrak{a}}^i(M))_b &= \big(\mathrm{Ker}(\Gamma_{\mathfrak{a}}(d^i))/\mathrm{Im}(\Gamma_{\mathfrak{a}}(d^{i-1}))\big)_b \\
&\cong (\mathrm{Ker}(\Gamma_{\mathfrak{a}}(d^i)))_b/(\mathrm{Im}(\Gamma_{\mathfrak{a}}(d^{i-1})))_b \\
&= \mathrm{Ker}(\Gamma_{\mathfrak{a}}((d^i)_b))/\mathrm{Im}(\Gamma_{\mathfrak{a}}((d^{i-1})_b)) = H_{\mathfrak{a}}^i(M_b).
\end{aligned}
$$

This natural isomorphism $(H_{\mathfrak{a}}^i(M))_b \cong H_{\mathfrak{a}}^i(M_b)$, and the corresponding one for N, enable us to deduce from (i) that there is a commutative diagram

$$
\begin{array}{ccccccc}
H_{\mathfrak{a}+Rb}^i(M) & \longrightarrow & H_{\mathfrak{a}}^i(M) & \xrightarrow{\ \tau_{H_{\mathfrak{a}}^i(M)}\ } & (H_{\mathfrak{a}}^i(M))_b & \longrightarrow & H_{\mathfrak{a}+Rb}^{i+1}(M) \\
\Big\downarrow{\scriptstyle H_{\mathfrak{a}+Rb}^i(f)} & & \Big\downarrow{\scriptstyle H_{\mathfrak{a}}^i(f)} & & \Big\downarrow{\scriptstyle (H_{\mathfrak{a}}^i(f))_b} & & \Big\downarrow{\scriptstyle H_{\mathfrak{a}+Rb}^{i+1}(f)} \\
H_{\mathfrak{a}+Rb}^i(N) & \longrightarrow & H_{\mathfrak{a}}^i(N) & \xrightarrow{\ \tau_{H_{\mathfrak{a}}^i(N)}\ } & (H_{\mathfrak{a}}^i(N))_b & \longrightarrow & H_{\mathfrak{a}+Rb}^{i+1}(N)
\end{array}
$$

with exact rows. The result now follows from 2.2.20, which shows that

$$
\mathrm{Ker}\,\tau_{H_{\mathfrak{a}}^{i+1}(M)} = \Gamma_{Rb}(H_{\mathfrak{a}}^{i+1}(M)),
$$

that $\mathrm{Ker}\,\tau_{H_{\mathfrak{a}}^{i+1}(N)} = \Gamma_{Rb}(H_{\mathfrak{a}}^{i+1}(N))$ and that there is a commutative diagram

$$
\begin{array}{ccccccc}
H_{\mathfrak{a}}^i(M) & \xrightarrow{\ \tau_{H_{\mathfrak{a}}^i(M)}\ } & (H_{\mathfrak{a}}^i(M))_b & \longrightarrow & H_{Rb}^1(H_{\mathfrak{a}}^i(M)) & \longrightarrow & 0 \\
\Big\downarrow{\scriptstyle H_{\mathfrak{a}}^i(f)} & & \Big\downarrow{\scriptstyle (H_{\mathfrak{a}}^i(f))_b} & & \Big\downarrow{\scriptstyle H_{Rb}^1(H_{\mathfrak{a}}^i(f))} & & \\
H_{\mathfrak{a}}^i(N) & \xrightarrow{\ \tau_{H_{\mathfrak{a}}^i(N)}\ } & (H_{\mathfrak{a}}^i(N))_b & \longrightarrow & H_{Rb}^1(H_{\mathfrak{a}}^i(N)) & \longrightarrow & 0
\end{array}
$$

with exact rows. $\qquad\square$

8.1.3 Remark. In 3.3.4 we introduced $\mathrm{cohd}(\mathfrak{a})$, the cohomological dimension of $\mathfrak{a}$. We deduce from 8.1.2(i) that, if $b \in R$, then

$$
\mathrm{cohd}(\mathfrak{a} + Rb) \le \mathrm{cohd}(\mathfrak{a}) + 1.
$$

8.1.4 Exercise. Let $\mathfrak{b}$ be a second ideal of R and let $i \in \mathbb{N}_0$. Show that the restrictions of the functors $H_{\mathfrak{a}+\mathfrak{b}}^i$ and $H_{\mathfrak{a}}^i$ to the full subcategory of $\mathcal{C}(R)$ whose objects are the $\mathfrak{b}$-torsion R-modules are naturally equivalent.

8.1.5 Exercise. Let $f : M \longrightarrow N$ be a homomorphism of R-modules and let $b \in R$. Use 8.1.2(ii) and 8.1.4 to show that there is a commutative diagram

$$
\begin{CD}
0 @>>> H^1_{\mathfrak{a}+Rb}(H^i_\mathfrak{a}(M)) @>>> H^{i+1}_{\mathfrak{a}+Rb}(M) @>>> \Gamma_{\mathfrak{a}+Rb}(H^{i+1}_\mathfrak{a}(M)) @>>> 0 \\
@. @VV{H^1_{\mathfrak{a}+Rb}(H^i_\mathfrak{a}(f))}V @VV{H^{i+1}_{\mathfrak{a}+Rb}(f)}V @VV{\Gamma_{\mathfrak{a}+Rb}(H^{i+1}_\mathfrak{a}(f))}V @. \\
0 @>>> H^1_{\mathfrak{a}+Rb}(H^i_\mathfrak{a}(N)) @>>> H^{i+1}_{\mathfrak{a}+Rb}(N) @>>> \Gamma_{\mathfrak{a}+Rb}(H^{i+1}_\mathfrak{a}(N)) @>>> 0
\end{CD}
$$

with exact rows.

We make use of Proposition 8.1.2 in the following lemma.

8.1.6 Lemma. *Let $(R, \mathfrak{m})$ be a local integral domain of dimension n; suppose that the set*

$$
\mathcal{B} := \{\mathfrak{b} : \mathfrak{b} \text{ is an ideal of } R, \ H^n_\mathfrak{b}(R) \neq 0 \text{ and } \dim R/\mathfrak{b} > 0\}
$$

is non-empty. Let $\mathfrak{p}$ be a maximal member of $\mathcal{B}$. (The fact that R is Noetherian ensures that such a $\mathfrak{p}$ exists.)

Then $\mathfrak{p}$ is prime and $\dim R/\mathfrak{p} = 1$.

Proof. Suppose that the statement '$\mathfrak{p}$ is prime and $\dim R/\mathfrak{p} = 1$' is false. This means that there exists $\mathfrak{q} \in \operatorname{Spec}(R)$ such that $\dim R/\mathfrak{q} = 1$ and $\mathfrak{p} \subset \mathfrak{q}$. Let $b \in \mathfrak{q} \setminus \mathfrak{p}$. Then, by 8.1.2(i), there is an exact sequence

$$
H^n_{\mathfrak{p}+Rb}(R) \longrightarrow H^n_\mathfrak{p}(R) \longrightarrow H^n_\mathfrak{p}(R_b).
$$

By the Independence Theorem 4.2.1, there is an isomorphism of R-modules $H^n_\mathfrak{p}(R_b) \cong H^n_{\mathfrak{p}R_b}(R_b)$. Since $\dim R_b < n$ (as $b \in \mathfrak{m}$), we can deduce from Grothendieck's Vanishing Theorem 6.1.2 that $H^n_{\mathfrak{p}R_b}(R_b) = 0$. It therefore follows from the above exact sequence that $H^n_{\mathfrak{p}+Rb}(R) \neq 0$, and since $\mathfrak{p} \subset \mathfrak{p}+Rb$ and $\dim R/(\mathfrak{p} + Rb) > 0$, this contradicts the maximality of $\mathfrak{p}$. □

Our next lemma will come as no surprise to any reader who has solved Exercises 6.1.9 and 6.1.10. However, we provide a more direct argument, which avoids use of 6.1.9.

8.1.7 Lemma. *Assume that $\dim R = n$ and $H^n_\mathfrak{a}(R) = 0$. Then $H^n_\mathfrak{a}(M) = 0$ for every R-module M.*

Proof. Since $H^i_\mathfrak{a}$ commutes with direct limits (by 3.4.10) and each R-module is the direct limit of its finitely generated submodules, it is enough for us to prove that $H^n_\mathfrak{a}(M) = 0$ whenever M is finitely generated. However, in that case, there exists an exact sequence $0 \longrightarrow N \longrightarrow F \longrightarrow M \longrightarrow 0$ of finitely generated R-modules and R-homomorphisms in which F is free. This induces an exact sequence $H_\mathfrak{a}^n(F) \longrightarrow H_\mathfrak{a}^n(M) \longrightarrow H_\mathfrak{a}^{n+1}(N)$. It follows from the additivity of the local cohomology functor $H_\mathfrak{a}^n$ and the assumption that $H_\mathfrak{a}^n(R) = 0$ that $H_\mathfrak{a}^n(F) = 0$; since $H_\mathfrak{a}^{n+1}(N) = 0$ by Grothendieck's Vanishing Theorem 6.1.2, we can deduce that $H_\mathfrak{a}^n(M) = 0$, as required. $\square$

8.1.8 Exercise. Assume that $\dim R = n$ and that $\mathfrak{a}$ is proper, having minimal prime ideals $\mathfrak{p}_1, \dots, \mathfrak{p}_t$. Let M be an R-module for which $H_{\mathfrak{p}_i}^n(M) = 0$ for all $i = 1, \dots, t$. Show that $H_\mathfrak{a}^n(M) = 0$.

In the course of our proof of the local Lichtenbaum–Hartshorne Vanishing Theorem, we shall reduce to the case where R is a complete local domain and then appeal to Cohen's Structure Theorem for complete local rings: we shall again quote [50, Theorem 29.4] to deduce that there exists a complete regular local subring R' of R which is such that R is finitely generated as an R'-module. The next lemma will be applied to this situation.

8.1.9 Lemma. *Suppose that $(R, \mathfrak{m})$ is local, and that $(R', \mathfrak{m}')$ is a local subring of R which is such that R is finitely generated as an R'-module. Let $\mathfrak{p} \in \operatorname{Spec}(R)$ be such that $\dim R/\mathfrak{p} = 1$. There exists $b \in R$ such that, if B denotes the subring $R'[b]$ of R, then $\sqrt{(\mathfrak{p} \cap B)R} = \mathfrak{p}$.*

Proof. Of course, R is integral over R'. By the Incomparability Theorem (see [81, 13.33], for example), $\mathfrak{p}$ must be a minimal prime ideal of $(\mathfrak{p} \cap R')R$: let $\mathfrak{p}_2, \dots, \mathfrak{p}_t$ be the other minimal prime ideals of $(\mathfrak{p} \cap R')R$ (there may be none). By the Prime Avoidance Theorem, there exists $b \in \mathfrak{p} \setminus \bigcup_{i=2}^t \mathfrak{p}_i$. Then $B := R'[b]$, necessarily a Noetherian ring, is finitely generated as an R'-module, and so is integral over R'.

To complete the proof, it is enough for us to show that $\mathfrak{p}$ is the one and only minimal prime ideal of $(\mathfrak{p} \cap B)R$. It certainly is one, since R is integral over B so that we can appeal to the Incomparability Theorem again, as we did at the beginning of this proof. Let us suppose that $\mathfrak{q}$ is another minimal prime ideal of $(\mathfrak{p} \cap B)R$, different from $\mathfrak{p}$.

Obviously $\mathfrak{q} \neq \mathfrak{m}$. Since R is integral over R', it follows that $\mathfrak{q} \cap R' \neq \mathfrak{m}'$ (by [81, 13.31], for example). Since $\mathfrak{p} \cap R' \subseteq \mathfrak{p} \cap B \subseteq \mathfrak{q}$, it follows that $\mathfrak{p} \cap R' \subseteq \mathfrak{q} \cap R' \subset \mathfrak{m}'$. But $R/\mathfrak{p}$ is an integral extension of $R'/(\mathfrak{p} \cap R')$, and so $\dim R'/(\mathfrak{p} \cap R') = \dim R/\mathfrak{p} = 1$. We can therefore deduce that $\mathfrak{p} \cap R' = \mathfrak{q} \cap R'$. Another use of the Incomparability Theorem as before shows that $\mathfrak{q}$ must be a minimal prime ideal of $(\mathfrak{p} \cap R')R$, and so must be one of $\mathfrak{p}_2, \dots, \mathfrak{p}_t$. But $b \in \mathfrak{p} \cap R'[b] \subseteq (\mathfrak{p} \cap B)R \subseteq \mathfrak{q}$, and this is a contradiction because b was chosen outside $\mathfrak{p}_2 \cup \dots \cup \mathfrak{p}_t$. $\square$

In our application of Lemma 8.1.9 in which R' is a complete regular local ring, it will be important for us to know that the ring $B = R'[b]$ is a Gorenstein ring. This is the motivation behind the next lemma. In fact, the first part of this lemma is actually a special case of [81, Proposition 13.40], but we include a proof for the convenience of the reader.

8.1.10 Lemma. *Let R be a subring of the integral domain S, and suppose that R is integrally closed and $S = R[b]$, where $b \in S$ is integral over R. Let K be the field of fractions of R. Then b is algebraic over K and its minimal polynomial f over K belongs to $R[X]$. Also, the surjective ring homomorphism $\phi : R[X] \to R[b]$ given by evaluation at b has kernel $fR[X]$, so that ϕ induces an isomorphism of R-algebras*

$$\overline{\phi} : R[X]/fR[X] \xrightarrow{\cong} R[b] = S.$$

Proof. Since b is integral over R, it is certainly algebraic over K. Let its minimal polynomial over K be

$$f = X^h + a_{h-1}X^{h-1} + \cdots + a_1 X + a_0 \in K[X].$$

We aim to show that $a_0, \ldots, a_{h-1} \in R$.

There exists a field extension L of the field of fractions of S such that f splits into linear factors in $L[X]$: let $s = s_1, \ldots, s_h \in L$ be such that $f = (X - s_1)(X - s_2) \ldots (X - s_h)$ in $L[X]$. Equate coefficients to see that each of $a_0, \ldots, a_{h-1}$ can be written as a 'homogeneous polynomial' (in fact, a 'symmetric function') in $s_1, \ldots, s_h$ with coefficients ± 1; in particular, $a_0, \ldots, a_{h-1} \in R[s_1, \ldots, s_h]$.

Since b is integral over R, there exists

$$g = X^n + r_{n-1}X^{n-1} + \cdots + r_1 X + r_0 \in R[X]$$

such that $g(b) = 0$. Therefore $g \in \operatorname{Ker}\phi \subseteq fK[X]$, and so there exists $g_1 \in K[X]$ such that $g = fg_1$ in $K[X]$. For each $i = 1, \ldots, h$, evaluate both sides of this equation at $s_i \in L$ to see that $g(s_i) = 0$. Thus all the s_i $(i = 1, \ldots, h)$ are integral over R, and so, by [81, 13.21] for example, the ring $R[s_1, \ldots, s_h]$ is a finitely generated R-module; it therefore follows that $a_0, \ldots, a_{h-1}$ are all integral over R, since they belong to $R[s_1, \ldots, s_h]$. But $a_0, \ldots, a_{h-1} \in K$ and R is integrally closed; hence $a_0, \ldots, a_{h-1} \in R$.

Finally, $\operatorname{Ker}\phi = fK[X] \cap R[X]$, and it is elementary to deduce from the facts that f is monic and has all its coefficients in R that $fK[X] \cap R[X] = fR[X]$. $\square$

8.1.11 Lemma. *Suppose that $(R, \mathfrak{m})$ is local and of dimension n, and also that R is a Gorenstein ring. Let M be a non-zero, finitely generated R-module. Then $\operatorname{depth} M > 0$ if and only if $\operatorname{Ext}_R^n(M, R) = 0$.*

Proof. Recall from [50, Theorem 18.1] (or from [3]) that R has injective dimension n as a module over itself.

($\Rightarrow$) There exists $r \in \mathfrak{m}$ such that r is a non-zerodivisor on M. The exact sequence $0 \longrightarrow M \xrightarrow{\ r\ } M \longrightarrow M/rM \longrightarrow 0$ induces an exact sequence

$$\mathrm{Ext}^n_R(M/rM, R) \longrightarrow \mathrm{Ext}^n_R(M, R) \xrightarrow{\ r\ } \mathrm{Ext}^n_R(M, R) \longrightarrow 0$$

since $\mathrm{Ext}^{n+1}_R(M/rM, R) = 0$. Application of Nakayama's Lemma to the finitely generated R-module $\mathrm{Ext}^n_R(M, R)$ now shows that this module is zero.

($\Leftarrow$) By [50, Theorem 18.1], we have $\mathrm{Ext}^n_R(R/\mathfrak{m}, R) \neq 0$. It follows that $\mathfrak{m} \notin \mathrm{Ass}\, M$, for otherwise there would be an exact sequence

$$0 \longrightarrow R/\mathfrak{m} \longrightarrow M \longrightarrow C \longrightarrow 0$$

of R-modules, and the induced exact sequence

$$\mathrm{Ext}^n_R(M, R) \longrightarrow \mathrm{Ext}^n_R(R/\mathfrak{m}, R) \longrightarrow \mathrm{Ext}^{n+1}_R(C, R) = 0$$

would provide a contradiction. $\qquad\Box$

The next exercise extends Lemma 8.1.11 in a precise manner.

8.1.12 Exercise. Suppose that $(R, \mathfrak{m})$ is a Gorenstein local ring of dimension n. Let M be a non-zero, finitely generated R-module. Show that $\mathrm{depth}\, M$ is the least integer i such that $\mathrm{Ext}^{n-i}_R(M, R) \neq 0$.

The next lemma provides the promised solution for Exercise 3.1.4.

8.1.13 Lemma. *Let $(R, \mathfrak{m})$ be a complete local domain, and let $\mathfrak{p}$ be a prime ideal of R of dimension 1, that is, such that $\dim R/\mathfrak{p} = 1$.*

(i) *For each $n \in \mathbb{N}$, there exists $t \in \mathbb{N}$ such that $\mathfrak{p}^{(t)} \subseteq \mathfrak{m}^n$.*

(ii) *The family $(\mathfrak{p}^{(n)})_{n\in\mathbb{N}}$ is a system of ideals of R over $\mathbb{N}$ in the sense of 2.1.10, and the (negative strongly) connected sequences of covariant functors (from $\mathcal{C}(R)$ to $\mathcal{C}(R)$)*

$$\left(H^i_{\mathfrak{p}} \right)_{i\in\mathbb{N}_0} \quad \text{and} \quad \left(\varinjlim_{j\in\mathbb{N}} \mathrm{Ext}^i_R(R/\mathfrak{p}^{(j)}, \bullet) \right)_{i\in\mathbb{N}_0}$$

are isomorphic.

Proof. Let $\psi : R \to R_{\mathfrak{p}}$ denote the natural homomorphism.

(i) Recall that $\mathfrak{p}^{(n)} := \psi^{-1}(\mathfrak{p}^n R_{\mathfrak{p}}) = \psi^{-1}((\mathfrak{p}R_{\mathfrak{p}})^n)$. Hence

$$\bigcap_{n\in\mathbb{N}} \mathfrak{p}^{(n)} = \psi^{-1}\left(\bigcap_{n\in\mathbb{N}} (\mathfrak{p}R_{\mathfrak{p}})^n \right) = \psi^{-1}(0) = \mathrm{Ker}\, \psi,$$

by Krull's Intersection Theorem [50, Theroem 8.9]. Since R is a domain, it follows that $\bigcap_{n\in\mathbb{N}} \mathfrak{p}^{(n)} = 0$.

We can now use Chevalley's Theorem ([59, Section 5.2, Theorem 1] or [89, Chapter VIII, Section 5, Theorem 13]) to deduce that, for each $n \in \mathbb{N}$, there exists $t \in \mathbb{N}$ such that $\mathfrak{p}^{(t)} \subseteq \mathfrak{m}^n$.

(ii) Let $n \in \mathbb{N}$. Note that, by [81, 5.47(i)] for example, $\mathfrak{p}$ is the unique minimal prime ideal of $\mathfrak{p}^n$, and $\mathfrak{p}^{(n)}$ is the (uniquely determined) $\mathfrak{p}$-primary term in any minimal primary decomposition of $\mathfrak{p}^n$. Since $\dim R/\mathfrak{p} = 1$, the only possible associated prime ideal of $\mathfrak{p}^n$ in addition to $\mathfrak{p}$ is $\mathfrak{m}$. Thus either $\mathfrak{p}^n = \mathfrak{p}^{(n)}$, or there exists $h \in \mathbb{N}$ such that $\mathfrak{p}^{(n)} \cap \mathfrak{m}^h \subseteq \mathfrak{p}^n$. It now follows from part (i) that there exists $t \in \mathbb{N}$ such that $\mathfrak{p}^{(t)} \subseteq \mathfrak{p}^n$.

Of course, $\mathfrak{p}^n \subseteq \mathfrak{p}^{(n)}$ for all $n \in \mathbb{N}$. Since it is clear that $(\mathfrak{p}^{(n)})_{n \in \mathbb{N}}$ is an inverse family of ideals of R over $\mathbb{N}$ in the sense of 1.2.10, it now follows from Proposition 3.1.1(iii) that $(\mathfrak{p}^{(n)})_{n \in \mathbb{N}}$ is actually a system of ideals of R over $\mathbb{N}$, and from part (ii) of the same proposition that the negative strongly connected sequences of covariant functors

$$\left(\varinjlim_{j \in \mathbb{N}} \mathrm{Ext}^i_R(R/\mathfrak{p}^j, \bullet) \right)_{i \in \mathbb{N}_0} \quad \text{and} \quad \left(\varinjlim_{j \in \mathbb{N}} \mathrm{Ext}^i_R(R/\mathfrak{p}^{(j)}, \bullet) \right)_{i \in \mathbb{N}_0}$$

are isomorphic. The proof can now be completed by an appeal to 1.3.8.   □

8.2 The main theorem

We are now in a position to put all our various lemmas of §8.1 together to produce a proof of the main theorem of this chapter.

8.2.1 The Lichtenbaum–Hartshorne Vanishing Theorem. (See R. Hartshorne [29, Theorem 3.1].) *Suppose that $(R, \mathfrak{m})$ is local of dimension n, and also that $\mathfrak{a}$ is proper. Then the following statements are equivalent:*

  (i) $H^n_\mathfrak{a}(R) = 0;$
  (ii) *for each (necessarily minimal) prime ideal $\mathfrak{P}$ of $\widehat{R}$, the completion of R, satisfying $\dim \widehat{R}/\mathfrak{P} = n$, we have $\dim \widehat{R}/(\mathfrak{a}\widehat{R} + \mathfrak{P}) > 0$.*

Proof. (i) $\Rightarrow$ (ii) Assume that $H^n_\mathfrak{a}(R) = 0$ and that there exists a prime ideal $\mathfrak{P}$ of $\widehat{R}$ such that $\dim \widehat{R}/\mathfrak{P} = n$ but $\dim \widehat{R}/(\mathfrak{a}\widehat{R} + \mathfrak{P}) = 0$. Since the natural ring homomorphism $R \to \widehat{R}$ is flat, it follows from the Flat Base Change Theorem 4.3.2 that there is an $\widehat{R}$-isomorphism $H^n_\mathfrak{a}(R) \otimes_R \widehat{R} \cong H^n_{\mathfrak{a}\widehat{R}}(\widehat{R})$, and so $H^n_{\mathfrak{a}\widehat{R}}(\widehat{R}) = 0$.

Now $\mathfrak{m}\widehat{R}$ is the maximal ideal of the local ring $\widehat{R}$, and our assumptions mean that $(\widehat{R}/\mathfrak{P}, \mathfrak{m}\widehat{R}/\mathfrak{P})$ is an n-dimensional local ring and $(\mathfrak{a}\widehat{R} + \mathfrak{P})/\mathfrak{P}$ is an $(\mathfrak{m}\widehat{R}/\mathfrak{P})$-primary ideal of this ring. It therefore follows from 1.1.3 and Theorem 6.1.4 (or Theorem 7.3.2) that $H^n_{(\mathfrak{a}\widehat{R}+\mathfrak{P})/\mathfrak{P}}(\widehat{R}/\mathfrak{P}) \neq 0$. We now deduce from the Independence Theorem 4.2.1 that $H^n_{\mathfrak{a}\widehat{R}}(\widehat{R}/\mathfrak{P}) \neq 0$. Therefore, by 8.1.7, we must have $H^n_{\mathfrak{a}\widehat{R}}(\widehat{R}) \neq 0$, and this is a contradiction.

(ii) $\Rightarrow$ (i) Assume that $\dim \widehat{R}/(\mathfrak{a}\widehat{R} + \mathfrak{P}) > 0$ for each $\mathfrak{P} \in \operatorname{Spec}(\widehat{R})$ for which $\dim \widehat{R}/\mathfrak{P} = n$. We suppose also that $H^n_{\mathfrak{a}}(R) \neq 0$, and we again look for a contradiction. We again use the Flat Base Change Theorem 4.3.2 to see that there is an $\widehat{R}$-isomorphism $H^n_{\mathfrak{a}}(R) \otimes_R \widehat{R} \cong H^n_{\mathfrak{a}\widehat{R}}(\widehat{R})$, from which we deduce that $H^n_{\mathfrak{a}\widehat{R}}(\widehat{R}) \neq 0$ (because the natural ring homomorphism $R \to \widehat{R}$ is faithfully flat). We can therefore assume, in our search for a contradiction, that R is complete: we make this assumption in what follows.

There is an ascending chain $0 = \mathfrak{b}_0 \subset \mathfrak{b}_1 \subset \cdots \subset \mathfrak{b}_{h-1} \subset \mathfrak{b}_h = R$ of ideals of R such that, for each $i = 1, \ldots, h$, there exists $\mathfrak{p}_i \in \operatorname{Spec}(R)$ with $\mathfrak{b}_i/\mathfrak{b}_{i-1} \cong R/\mathfrak{p}_i$. It follows from the half-exactness of $H^n_{\mathfrak{a}}$ that $H^n_{\mathfrak{a}}(R/\mathfrak{p}_i) \neq 0$ for at least one i between 1 and h. The Independence Theorem 4.2.1 now enables us to deduce that $H^n_{(\mathfrak{a}+\mathfrak{p}_i)/\mathfrak{p}_i}(R/\mathfrak{p}_i) \neq 0$. Note that, by Grothendieck's Vanishing Theorem 6.1.2, the complete local domain $R/\mathfrak{p}_i$ has dimension n. Therefore, our assumptions imply that $\dim(R/\mathfrak{p}_i)/((\mathfrak{a} + \mathfrak{p}_i)/\mathfrak{p}_i) > 0$. Therefore we can, and do, assume henceforth in our search for a contradiction that R is an n-dimensional complete local domain and the proper ideal $\mathfrak{a}$ satisfies $H^n_{\mathfrak{a}}(R) \neq 0$ and $\dim R/\mathfrak{a} > 0$. We shall show that such a situation is impossible.

By Lemma 8.1.6, there exists $\mathfrak{p} \in \operatorname{Spec}(R)$ such that $H^n_{\mathfrak{p}}(R) \neq 0$ and $\dim R/\mathfrak{p} = 1$. We now appeal to Cohen's Structure Theorem for complete local rings: by [50, Theorem 29.4], there exists a complete regular local subring $(R', \mathfrak{m}')$ of R which is such that R is finitely generated as an R'-module. By Lemma 8.1.9, there exists $b \in R$ such that, if B denotes the subring $R'[b]$ of R, then $\sqrt{(\mathfrak{p} \cap B)R} = \mathfrak{p}$. Note that, since R is integral over B, the latter is also local and of dimension n, with unique maximal ideal $\mathfrak{m} \cap B$. Note also that, since B is a finitely generated R'-module, it is a complete local ring.

Since R' is integrally closed, it follows from Lemma 8.1.10 that there is a monic polynomial $f \in R'[X]$ and an isomorphism of R'-algebras

$$R'[X]/fR'[X] \cong R'[b] = B.$$

By [50, Theorem 19.5], the ring $R'[X]$ is regular. As f is monic, it is a nonzerodivisor in $R'[X]$. We now deduce that B is a Gorenstein local ring as follows. There exists exactly one maximal ideal $\mathfrak{Q}$ of $R'[X]$ that contains f;

we have

$$B \cong (R'[X]/fR'[X])_{\mathfrak{Q}/fR'[X]} \cong R'[X]_{\mathfrak{Q}}/(f/1)R'[X]_{\mathfrak{Q}};$$

and the latter is a Gorenstein local ring by [50, Exercise 18.1].

Since $\sqrt{(\mathfrak{p} \cap B)R} = \mathfrak{p}$, we can deduce from 1.1.3 that $H^n_{(\mathfrak{p} \cap B)R}(R) \neq 0$, and thence by the Independence Theorem 4.2.1 that $H^n_{(\mathfrak{p} \cap B)}(R) \neq 0$. Therefore $H^n_{(\mathfrak{p} \cap B)}(B) \neq 0$ by Lemma 8.1.7. Let $\mathfrak{q} := \mathfrak{p} \cap B$; observe that, since R is integral over B, we have $\dim B/\mathfrak{q} = \dim R/\mathfrak{p} = 1$.

We have nearly arrived at a contradiction: B is a complete n-dimensional Gorenstein local domain, $\mathfrak{q} \in \mathrm{Spec}(B)$ has $\dim B/\mathfrak{q} = 1$, and $H^n_{\mathfrak{q}}(B) \neq 0$. This is, in fact, impossible, because, by Lemma 8.1.13(ii),

$$H^n_{\mathfrak{q}}(B) \cong \varinjlim_{j \in \mathbb{N}} \mathrm{Ext}^n_B(B/\mathfrak{q}^{(j)}, B)$$

and the latter module is zero by Lemma 8.1.11, since $\mathrm{depth}\, B/\mathfrak{q}^{(j)} > 0$ for all $j \in \mathbb{N}$ because each $\mathfrak{q}^{(j)}$ is a $\mathfrak{q}$-primary ideal. $\qquad \square$

8.2.2 Corollary. *Suppose that* $(R, \mathfrak{m})$ *is local and has dimension* n*, and that* $\dim \widehat{R}/(\mathfrak{a}\widehat{R} + \mathfrak{P}) > 0$ *for every prime ideal* $\mathfrak{P}$ *of* $\widehat{R}$ *such that* $\dim \widehat{R}/\mathfrak{P} = n$*. Then* $H^i_{\mathfrak{a}}(M) = 0$ *for all* $i \geq n$ *and for every* R*-module* M*.*

Proof. This follows from Grothendieck's Vanishing Theorem 6.1.2, Lemma 8.1.7, and the Lichtenbaum–Hartshorne Vanishing Theorem 8.2.1. $\qquad \square$

8.2.3 Exercise. Suppose that $(R, \mathfrak{m})$ is local and that M is a non-zero finitely generated R-module of dimension n. Assume that

$$\dim \widehat{R}/(\mathfrak{a}\widehat{R} + \mathfrak{P}) > 0$$

for every prime ideal $\mathfrak{P}$ of $\mathrm{Supp}_{\widehat{R}}(M \otimes_R \widehat{R})$ such that $\dim \widehat{R}/\mathfrak{P} = n$. Prove that $H^i_{\mathfrak{a}}(M) = 0$ for all $i \geq n$.

8.2.4 ♯Exercise. Suppose that $(R, \mathfrak{m})$ is local and that A is an Artinian R-module. Show that, for each $a \in A$, there exists $t \in \mathbb{N}$ such that $\mathfrak{m}^t a = 0$. Deduce that A has a natural structure as a module over $\widehat{R}$, the completion of R, that a subset of A is an R-submodule if and only if it is an $\widehat{R}$-submodule, and that the map $\phi : A \to A \otimes_R \widehat{R}$ defined by $\phi(a) = a \otimes 1$ for all $a \in A$ is an isomorphism of $\widehat{R}$-modules.

The next two exercises involve the theory of secondary representation and attached primes of Artinian modules, discussed in §7.2.

8.2.5 Exercise. Let R' denote a second commutative Noetherian ring and let $f : R \to R'$ be a ring homomorphism. Suppose that the R'-module M' has a secondary representation (7.2.2). Show that, when M' is viewed as an R-module by means of f, it has a secondary representation as an R-module, and that

$$\mathrm{Att}_R \, M' = \left\{ f^{-1}(\mathfrak{p}') : \mathfrak{p}' \in \mathrm{Att}_{R'} \, M' \right\}.$$

Is it possible for M' to have fewer attached prime ideals as an R-module than it has as an R'-module? Justify your response.

8.2.6 Exercise. Suppose that $(R, \mathfrak{m})$ is local and has dimension n, and also that $\mathfrak{a}$ is proper. By Theorem 7.1.6, the R-module $H_\mathfrak{a}^n(R)$ is Artinian.

(i) Use the Lichtenbaum–Hartshorne Vanishing Theorem 8.2.1 to prove that

$$\mathrm{Att}_R(H_\mathfrak{a}^n(R)) = \{\mathfrak{P} \cap R : \mathfrak{P} \in \mathrm{Spec}(\widehat{R}), \ \dim \widehat{R}/\mathfrak{P} = n,$$

$$\text{and } \dim \widehat{R}/(\mathfrak{a}\widehat{R} + \mathfrak{P}) = 0 \}.$$

(It is perhaps worth pointing out that the result of this exercise (which is due to R. Y. Sharp [80, Corollary 3.5]) amounts, in effect, to a refinement of the statement of the local Lichtenbaum–Hartshorne Vanishing Theorem, because the set on the right-hand side of the above display is empty if and only if $H_\mathfrak{a}^n(R) = 0$.)

(ii) Deduce that, if $n > 0$, then $H_\mathfrak{a}^n(R)$, if non-zero, is not finitely generated.

8.2.7 Remark. Let $n \in \mathbb{N}$ and let K be a field. This comment is concerned with the ring $K[X_1, \ldots, X_n]$ of polynomials over K.

Each $f \in K[X_1, \ldots, X_n] \setminus (X_1, \ldots, X_n)$ is a unit in $K[[X_1, \ldots, X_n]]$, and so there is a natural injective ring homomorphism

$$\phi : K[X_1, \ldots, X_n]_{(X_1, \ldots, X_n)} \longrightarrow K[[X_1, \ldots, X_n]].$$

It is not difficult to see that we can use ϕ to regard $K[[X_1, \ldots, X_n]]$ as the completion of the regular local ring $K[X_1, \ldots, X_n]_{(X_1, \ldots, X_n)}$. Let $\mathfrak{m}$ denote the maximal ideal $(X_1, \ldots, X_n)$ of $K[X_1, \ldots, X_n]$.

It follows that, if $\mathfrak{b}$ is an ideal of $K[X_1, \ldots, X_n]$ contained in $\mathfrak{m}$, and we use x_i to denote the natural image of X_i in $K[X_1, \ldots, X_n]/\mathfrak{b}$ (for each $i = 1, \ldots, n$), then the injective ring homomorphism

$$K[x_1, \ldots, x_n]_{(x_1, \ldots, x_n)} \longrightarrow K[[X_1, \ldots, X_n]]/\mathfrak{b}K[[X_1, \ldots, X_n]]$$

obtained by composing the natural ring isomorphism

$$K[x_1, \ldots, x_n]_{(x_1, \ldots, x_n)} = (K[X_1, \ldots, X_n]/\mathfrak{b})_{\mathfrak{m}/\mathfrak{b}}$$

$$\xrightarrow{\cong} K[X_1, \ldots, X_n]_\mathfrak{m}/\mathfrak{b}K[X_1, \ldots, X_n]_\mathfrak{m}$$

with the injective ring homomorphism

$$K[X_1, \ldots, X_n]_{\mathfrak{m}}/\mathfrak{b}K[X_1, \ldots, X_n]_{\mathfrak{m}}$$
$$\longrightarrow K[[X_1, \ldots, X_n]]/\mathfrak{b}K[[X_1, \ldots, X_n]]$$

induced by ϕ provides the completion of $K[x_1, \ldots, x_n]_{(x_1, \ldots, x_n)}$.

8.2.8 Exercise. Consider the special case of the situation of Remark 8.2.7 in which $n = 3$ and $\mathfrak{b} = (X_1 X_2)$. With these choices, let

$$R = K[x_1, x_2, x_3]_{(x_1, x_2, x_3)} = (K[X_1, X_2, X_3]/(X_1 X_2))_{(X_1, X_2, X_3)/(X_1 X_2)}.$$

Let $\mathfrak{p}$ be the prime ideal $(x_1/1, x_3/1)$ of R. Use the exact sequence

$$0 \longrightarrow R \longrightarrow R/(x_1/1) \oplus R/(x_2/1) \longrightarrow R/(x_1/1, x_2/1) \longrightarrow 0$$

of Lemma 3.2.1 to show that $H_{\mathfrak{p}}^2(R)$ is not finitely generated. Use the description of the completion $\widehat{R}$ of R provided by 8.2.7 to find a prime ideal $\mathfrak{P}$ of $\widehat{R}$ such that $\dim \widehat{R}/\mathfrak{P} = 2$ but $\dim \widehat{R}/(\mathfrak{p}\widehat{R} + \mathfrak{P}) = 0$.

8.2.9 Exercise. Consider the special case of the situation of Remark 8.2.7 in which K has characteristic 0, $n = 3$ and $\mathfrak{b} = (X_2^2 - X_1^2 - X_1^3)$. Set $R' := K[X_1, X_2, X_3]$ and $\mathfrak{m}' := (X_1, X_2, X_3)$. Let

$$\mathfrak{c} := (X_1 + X_2 - X_2 X_3)R' + ((X_3 - 1)^2(X_1 + 1) - 1)R'.$$

(i) Show that $\mathfrak{b} \in \operatorname{Spec}(R')$, that $\mathfrak{b} \subseteq \mathfrak{c}$, and also that $\mathfrak{c}R'_{\mathfrak{m}'} \in \operatorname{Spec}(R'_{\mathfrak{m}'})$. (Here is a hint: do not forget that $R'_{\mathfrak{m}'}$ is a regular local ring.)

(ii) Follow the procedure of 8.2.7, and set

$$R := (R'/\mathfrak{b})_{\mathfrak{m}'/\mathfrak{b}} \quad \text{and} \quad \widehat{R} = K[[X_1, X_2, X_3]]/\mathfrak{b}K[[X_1, X_2, X_3]];$$

regard $\widehat{R}$ as the completion of R in the manner described in 8.2.7. It will be convenient to use x_i to denote the natural image of X_i in $R'/\mathfrak{b}$ and in $\widehat{R}$ (for $i = 1, 2, 3$). Show that R is a 2-dimensional local domain, and that

$$\mathfrak{p} := (x_1 + x_2 - x_2 x_3)R + ((x_3 - 1)^2(x_1 + 1) - 1)R$$

is a prime ideal of R with $\dim R/\mathfrak{p} = 1$.

(iii) Let $u \in \widehat{R}$ be given by

$$u := 1 + \frac{1}{2}x_1 - \frac{1}{8}x_1^2 + \frac{1}{16}x_1^3 - \frac{5}{128}x_1^4 + \cdots + \frac{(-1)^{n-1}2(2n-2)!}{4^n n!(n-1)!}x_1^n + \cdots,$$

so that u is a unit of $\widehat{R}$ and $u^2 = 1 + x_1$. Show that

$$\mathfrak{p}\widehat{R} = (x_1 + x_2 - x_2 x_3)\widehat{R} + (x_3 - 1 + u^{-1})\widehat{R},$$

and that this is a prime ideal of $\widehat{R}$.

(iv) Use the exact sequence

$$0 \to \widehat{R} \longrightarrow \widehat{R}/(x_2 - x_1 u) \oplus \widehat{R}/(x_2 + x_1 u) \longrightarrow \widehat{R}/(x_2 - x_1 u, x_2 + x_1 u) \to 0$$

of Lemma 3.2.1 to show that $H^2_{p\widehat{R}}(\widehat{R})$ is not finitely generated, and deduce that $H^2_p(R)$ is not finitely generated.

(v) Find a prime ideal $\mathfrak{P}$ of $\widehat{R}$ such that $\dim \widehat{R}/\mathfrak{P} = 2$ but

$$\dim \widehat{R}/(p\widehat{R} + \mathfrak{P}) = 0.$$

8.2.10 Definition and ♯Exercise. Suppose that $(R, \mathfrak{m})$ is local. We say that R is *analytically irreducible* precisely when $\widehat{R}$, the completion of R, is an integral domain.

Note that an analytically irreducible local ring must itself be a domain.

Suppose that $(R, \mathfrak{m})$ is analytically irreducible, and that $\mathfrak{a}$ is proper. Let $\dim R = n$. Show that $H^n_{\mathfrak{a}}(R) = 0$ if and only if $\dim R/\mathfrak{a} > 0$, and that, when this is the case, $H^i_{\mathfrak{a}}(M) = 0$ for all $i \geq n$ and for every R-module M.

8.2.11 Exercise. Suppose $(R, \mathfrak{m})$ is an analytically irreducible (see 8.2.10 above) local domain of dimension 2, and that $\dim R/\mathfrak{a} > 0$. Show that the $\mathfrak{a}$-transform functor $D_{\mathfrak{a}}$ is exact.

8.2.12 Exercise. Show by means of an example that, if the phrase 'analytically irreducible' is omitted from the hypotheses in Exercise 8.2.11 above, then the resulting statement is no longer always true.

8.2.13 Exercise. For this exercise, we recommend that the reader refers to the example studied in 2.3.7, 3.3.5 and 4.3.7, and applies the ideas of Remark 8.2.7 in the special case in which $n = 3$ and

$$\mathfrak{b} = \sqrt{(X_1 X_4 - X_2 X_3,\ X_1^2 X_3 + X_1 X_2 - X_2^2,\ X_3^3 + X_3 X_4 - X_4^2)}.$$

Let S' denote the subring $\mathbb{C}[X, XY, Y(Y - 1), Y^2(Y - 1)]$ of $\mathbb{C}[X, Y]$. Consider the ideals

$$\mathfrak{n} := (X, XY, Y(Y-1), Y^2(Y-1)) \quad \text{and} \quad \mathfrak{r} := (XY, Y(Y-1), Y^2(Y-1))$$

of S'.

(i) Show that $\mathfrak{r}S'_{\mathfrak{q}}$ is a principal ideal of $S'_{\mathfrak{q}}$ for every maximal ideal $\mathfrak{q}$ of S' different from $\mathfrak{n}$.

(ii) Show that $S := S'_{\mathfrak{n}}$ is a local domain of dimension 2, and that $\mathfrak{p} := \mathfrak{r}S'_{\mathfrak{n}}$ is a prime ideal of S such that $\dim S/\mathfrak{p} = 1$.

(iii) Use a calculation made at the end of 3.3.5 to show that $H^2_{\mathfrak{p}}(S) \neq 0$.

(iv) Use the description of the completion $(\widehat{S}, \widehat{\mathfrak{M}})$ of S provided by Remark 8.2.7, together with calculations made in 4.3.7, to see that $\widehat{S}$ has exactly two minimal prime ideals, $\mathfrak{P}_1$ and $\mathfrak{P}_2$ say, and that

$$\mathfrak{P}_1 + \mathfrak{P}_2 = \widehat{\mathfrak{M}} \quad \text{and} \quad \mathfrak{P}_i + \mathfrak{p}\widehat{S} = \widehat{\mathfrak{M}} \text{ for } i = 1 \text{ or } 2.$$

Show that $H^2_{\mathfrak{p}\widehat{S}}(\widehat{S})$ is not finitely generated, and deduce that $H^2_{\mathfrak{p}}(S)$ is not finitely generated.

8.2.14 Exercise. Let R be the localization $\mathbb{C}[X, Y^2, XY, Y^3]_{(X,Y^2,XY,Y^3)}$ of the subring $\mathbb{C}[X, Y^2, XY, Y^3]$ of $\mathbb{C}[X, Y]$. Show that $H^2_{\mathfrak{a}}(R) = 0$ for every ideal $\mathfrak{a}$ of R which is not primary to the maximal ideal.

8.2.15 Definition and Exercise. Let V be a quasi-affine variety over an algebraically closed field K, and let $p \in V$.

We say that V is *analytically irreducible at p* precisely when the completion $\widehat{\mathcal{O}_{V,p}}$ of the local ring $\mathcal{O}_{V,p}$ of V at p is an integral domain, that is, if and only if $\mathcal{O}_{V,p}$ is analytically irreducible (see 8.2.10); otherwise, we say that V is *analytically reducible at p*.

(i) Now let V be an affine surface over K (that is, an affine algebraic variety over K of dimension 2), and let $C \subset V$ be a (not necessarily irreducible) curve (that is, a non-empty closed subset of pure codimension 1 in V (see 6.4.2(iv))) which avoids all the points of V at which V is analytically reducible. Use Serre's Affineness Criterion 6.4.4 to show that the quasi-affine variety $V \setminus C$ is, in fact, affine.

(ii) By considering the example studied in 2.3.7, 3.3.5, 4.3.7 and (especially) 6.4.6, show that the condition that C 'avoids all the points of V at which V is analytically reducible' cannot be omitted from the hypotheses of the result established in part (i) above.

8.2.16 Exercise. Let V be an affine surface over an algebraically closed field K; let $C \subset V$ be a (not necessarily irreducible) curve (see 8.2.15(i)). Let $\mathfrak{b}$ be the ideal of C in $\mathcal{O}(V)$ and let $p \in C$. We say that C is *fully branched in V at p* precisely when $\dim \widehat{\mathcal{O}_{V,p}}/(\mathfrak{b}\widehat{\mathcal{O}_{V,p}} + \mathfrak{P}) > 0$ for all minimal primes $\mathfrak{P}$ of $\widehat{\mathcal{O}_{V,p}}$ satisfying $\dim \widehat{\mathcal{O}_{V,p}}/\mathfrak{P} = 2$.

(i) Show that C is fully branched in V at p if and only if

$$H^2_{\mathfrak{b}\mathcal{O}_{V,p}}(\mathcal{O}_{V,p}) = 0.$$

(ii) Show that $V \setminus C$ is affine if and only if C is fully branched in V at all its points.

8.2.17 Exercise. This exercise is concerned with the situation of Exercise 8.2.9, and so we use the notation of that exercise. It was not necessary in 8.2.9 for the reader to know that the ideal $\mathfrak{c}$ is actually prime, but, in fact, it is and the first three parts of this exercise sketch a route to a proof of this fact.

(i) Note that $K[X_1, X_2, X_3] = K[\widetilde{X}_1, X_2, \widetilde{X}_3]$, where $\widetilde{X}_1 = X_1 + 1$ and $\widetilde{X}_3 = X_3 - 1$. Note also that $\widetilde{X}_1, X_2, \widetilde{X}_3$ are algebraically independent over K.

(ii) Set $A := K[\widetilde{X}_1, X_2, \widetilde{X}_3]/(\widetilde{X}_3^2 \widetilde{X}_1 - 1)$; consider $A_0 := K[\widetilde{X}_3]_{\widetilde{X}_3}$ and $K[X_2, \widetilde{X}_3]_{\widetilde{X}_3}$ as subrings of $K(X_2, \widetilde{X}_3)$ in the natural ways. Show that

$$A \cong K[X_2, \widetilde{X}_3]_{\widetilde{X}_3} = A_0[X_2].$$

(iii) Show that

$$K[\widetilde{X}_1, X_2, \widetilde{X}_3]/(\widetilde{X}_1 - X_2\widetilde{X}_3 - 1, \widetilde{X}_3^2\widetilde{X}_1 - 1)$$
$$\cong K[X_2, \widetilde{X}_3]_{\widetilde{X}_3}/(X_2 + \widetilde{X}_3^{-1} - \widetilde{X}_3^{-3}) \cong A_0,$$

and deduce that $\mathfrak{c}$ is a prime ideal of $K[X_1, X_2, X_3]$.

(iv) Now assume that K is algebraically closed.

Let $C := V_{\mathbb{A}^3(K)}(X_1 + X_2 - X_2X_3, \ (X_3 - 1)^2(X_1 + 1) - 1)$ and $V := V_{\mathbb{A}^3(K)}(X_2^2 - X_1^2 - X_1^3)$. Show that $C \subseteq V$ and (with or without the aid of Exercise 8.2.16 above) that $V \setminus C$ is not affine.

9

The Annihilator and Finiteness Theorems

There have been several examples earlier in this book of non-finitely generated local cohomology modules of finitely generated modules: for example, in 6.1.7, and also in 7.3.3, we saw that, if $(R, \mathfrak{m})$ is local, and N is a non-zero, finitely generated R-module of dimension $n > 0$, then $H^n_{\mathfrak{m}}(N)$ is not finitely generated. Since $H^0_{\mathfrak{m}}(N)$, being isomorphic to a submodule of N, is certainly finitely generated, it becomes of interest to identify the least integer i for which $H^i_{\mathfrak{m}}(N)$ is not finitely generated. This integer is referred to as the finiteness dimension $f_{\mathfrak{m}}(N)$ of N relative to $\mathfrak{m}$. Our work in this chapter on Grothendieck's Finiteness Theorem will enable us to see that, in this situation, and under mild restrictions on R,

$$f_{\mathfrak{m}}(N) = \min \left\{ \operatorname{depth}_{R_{\mathfrak{p}}} N_{\mathfrak{p}} + \dim R/\mathfrak{p} : \mathfrak{p} \in \operatorname{Supp} N \setminus \{\mathfrak{m}\} \right\}.$$

However, our approach will not restrict attention to the case where the ideal with respect to which local cohomology is calculated is the maximal ideal of a local ring; also, the approach we shall take will show that questions about such finiteness dimensions are intimately related to questions about precisely which ideals annihilate local cohomology modules. We shall, in fact, establish Grothendieck's Finiteness Theorem as a special case of Faltings' Annihilator Theorem, which is concerned with the following question: given a second ideal $\mathfrak{b}$ of R, and given a finitely generated R-module M, is there a greatest integer i for which $H^i_{\mathfrak{a}}(M)$ is annihilated by some power of $\mathfrak{b}$, and, if so, what is it?

9.1 Finiteness dimensions

Our first two results will provide motivation for the formal introduction of the concept of finiteness dimension. Let M be a finitely generated R-module. It is clear that, if, for some $i \in \mathbb{N}$, the local cohomology module $H_{\mathfrak{a}}^i(M)$ is finitely generated, then $\mathfrak{a}^u H_{\mathfrak{a}}^i(M) = 0$ for some $u \in \mathbb{N}$, and so

$$\mathfrak{a} \subseteq \sqrt{(0 : H_{\mathfrak{a}}^i(M))}.$$

It is not quite so clear that there is a sort of converse to this result: we show below, in our first proposition of this chapter, that, if, for some $t \in \mathbb{N}$, it is the case that $\mathfrak{a} \subseteq \sqrt{(0 : H_{\mathfrak{a}}^i(M))}$ for all $i < t$, then it follows that $H_{\mathfrak{a}}^i(M)$ is finitely generated for all $i < t$.

9.1.1 Lemma. *Let $L \xrightarrow{f} M \xrightarrow{g} N$ be an exact sequence of R-modules and R-homomorphisms, and suppose that $\mathfrak{a} \subseteq \sqrt{(0 : L)}$ and $\mathfrak{a} \subseteq \sqrt{(0 : N)}$. Then $\mathfrak{a} \subseteq \sqrt{(0 : M)}$ also.*

Proof. Let $r \in \mathfrak{a}$. Then, for some $t \in \mathbb{N}$, we have $r^t g(M) = 0$, and so $r^t M \subseteq \operatorname{Ker} g = \operatorname{Im} f$. However, there exists $u \in \mathbb{N}$ such that $r^u L = 0$, and so $r^u(\operatorname{Im} f) = 0$. Hence $r^{u+t} M = 0$. □

9.1.2 Proposition. *Let M be a finitely generated R-module, and let $t \in \mathbb{N}$. Then the following statements are equivalent:*

(i) $H_{\mathfrak{a}}^i(M)$ *is finitely generated for all $i < t$;*

(ii) $\mathfrak{a} \subseteq \sqrt{(0 : H_{\mathfrak{a}}^i(M))}$ *for all $i < t$.*

Proof. (i) $\Rightarrow$ (ii) As we remarked immediately before the statement of 9.1.1, this implication is clear.

(ii) $\Rightarrow$ (i) We use induction on t. When $t = 1$, there is nothing to prove, since $H_{\mathfrak{a}}^0(M) = \Gamma_{\mathfrak{a}}(M)$ is a submodule of M, and so is finitely generated. So suppose that $t > 1$ and that the result has been proved for smaller values of t. By this assumption, $H_{\mathfrak{a}}^i(M)$ is finitely generated for $i = 0, 1, \ldots, t-2$, and it only remains for us to prove that $H_{\mathfrak{a}}^{t-1}(M)$ is finitely generated.

It follows from 2.1.7(iii) that $H_{\mathfrak{a}}^i(M) \cong H_{\mathfrak{a}}^i(M/\Gamma_{\mathfrak{a}}(M))$ for all $i > 0$. Also, $M/\Gamma_{\mathfrak{a}}(M)$ is an $\mathfrak{a}$-torsion-free R-module, by 2.1.2. Hence we can, and do, assume that M is an $\mathfrak{a}$-torsion-free R-module.

We now use 2.1.1(ii) to deduce that $\mathfrak{a}$ contains an element r which is a non-zerodivisor on M. Since $\mathfrak{a} \subseteq \sqrt{(0 : H_{\mathfrak{a}}^{t-1}(M))}$, there exists $u \in \mathbb{N}$ such that $r^u H_{\mathfrak{a}}^{t-1}(M) = 0$. The exact sequence

$$0 \longrightarrow M \xrightarrow{r^u} M \longrightarrow M/r^u M \longrightarrow 0$$

induces a long exact sequence

$$0 \longrightarrow H^0_{\mathfrak{a}}(M) \xrightarrow{\;r^u\;} H^0_{\mathfrak{a}}(M) \longrightarrow H^0_{\mathfrak{a}}(M/r^u M)$$
$$\longrightarrow H^1_{\mathfrak{a}}(M) \xrightarrow{\;r^u\;} H^1_{\mathfrak{a}}(M) \longrightarrow H^1_{\mathfrak{a}}(M/r^u M)$$
$$\longrightarrow \quad \cdots \qquad\qquad\qquad\qquad \cdots$$
$$\longrightarrow H^i_{\mathfrak{a}}(M) \xrightarrow{\;r^u\;} H^i_{\mathfrak{a}}(M) \longrightarrow H^i_{\mathfrak{a}}(M/r^u M)$$
$$\longrightarrow H^{i+1}_{\mathfrak{a}}(M) \longrightarrow \quad \cdots .$$

It follows from this long exact sequence and 9.1.1 that

$$\mathfrak{a} \subseteq \sqrt{(0 : H^i_{\mathfrak{a}}(M/r^u M))} \quad \text{for all } i < t - 1,$$

so that, by the inductive hypothesis, $H^i_{\mathfrak{a}}(M/r^u M)$ is finitely generated for all $i < t - 1$. In particular, we see that $H^{t-2}_{\mathfrak{a}}(M/r^u M)$ is finitely generated; since $r^u H^{t-1}_{\mathfrak{a}}(M) = 0$, it follows from the above long exact sequence that $H^{t-1}_{\mathfrak{a}}(M)$ is a homomorphic image of $H^{t-2}_{\mathfrak{a}}(M/r^u M)$, and so is finitely generated. This completes the inductive step.   □

Proposition 9.1.2 provides some motivation for the following definition. Here, and throughout the book, we adopt the convention that the infimum of the empty set of integers is to be taken as ∞.

9.1.3 Definition.   Let M be a finitely generated R-module. In the light of Proposition 9.1.2, we define the *finiteness dimension* $f_{\mathfrak{a}}(M)$ *of M relative to* $\mathfrak{a}$ by

$$f_{\mathfrak{a}}(M) = \inf \left\{ i \in \mathbb{N} : H^i_{\mathfrak{a}}(M) \text{ is not finitely generated} \right\}$$
$$= \inf \left\{ i \in \mathbb{N} : \mathfrak{a} \not\subseteq \sqrt{(0 : H^i_{\mathfrak{a}}(M))} \right\}.$$

Note that $f_{\mathfrak{a}}(M)$ is either a positive integer or ∞, and that, since $H^0_{\mathfrak{a}}(M)$ is finitely generated,

$$f_{\mathfrak{a}}(M) = \inf \left\{ i \in \mathbb{N}_0 : H^i_{\mathfrak{a}}(M) \text{ is not finitely generated} \right\}.$$

9.1.4 Exercise.   For a finitely generated R-module M, show that $f_{\mathfrak{a}}(M) > 1$ if and only if the $\mathfrak{a}$-transform $D_{\mathfrak{a}}(M)$ is finitely generated.

In the situation of 9.1.3, it is reasonable to regard the condition that $\mathfrak{a} \subseteq \sqrt{(0 : H^i_{\mathfrak{a}}(M))}$ as asserting that $H^i_{\mathfrak{a}}(M)$ is 'small' in a sense, because if this condition holds for all i less than some positive integer t, then $H^i_{\mathfrak{a}}(M)$ is finitely generated for all $i < t$ (by 9.1.2). However, sometimes it is more realistic to hope for a weaker condition than '$\mathfrak{a} \subseteq \sqrt{(0 : H_{\mathfrak{a}}^i(M))}$': we introduce a second ideal $\mathfrak{b}$ of R, and, when $\mathfrak{b} \subseteq \mathfrak{a}$, think of $H_{\mathfrak{a}}^i(M)$ as being 'small' relative to $\mathfrak{b}$ if $\mathfrak{b} \subseteq \sqrt{(0 : H_{\mathfrak{a}}^i(M))}$.

9.1.5 Definition. Let M be a finitely generated R-module and let $\mathfrak{b}$ be a second ideal of R. We define the $\mathfrak{b}$-*finiteness dimension* $f_{\mathfrak{a}}^{\mathfrak{b}}(M)$ *of M relative to* $\mathfrak{a}$ by

$$f_{\mathfrak{a}}^{\mathfrak{b}}(M) := \inf \left\{ i \in \mathbb{N}_0 : \mathfrak{b} \not\subseteq \sqrt{(0 : H_{\mathfrak{a}}^i(M))} \right\}.$$

Note that $f_{\mathfrak{a}}^{\mathfrak{b}}(M)$ is either ∞ or a non-negative integer not exceeding $\dim M$ and that $f_{\mathfrak{a}}^{\mathfrak{a}}(M) = f_{\mathfrak{a}}(M)$ because $\mathfrak{a} \subseteq \sqrt{(0 : \Gamma_{\mathfrak{a}}(M))}$. Note that, if $\mathfrak{b} \subseteq \sqrt{\mathfrak{a}}$, then $\mathfrak{b} \subseteq \sqrt{(0 : \Gamma_{\mathfrak{a}}(M))}$, so that we can then write

$$f_{\mathfrak{a}}^{\mathfrak{b}}(M) := \inf \left\{ i \in \mathbb{N} : \mathfrak{b} \not\subseteq \sqrt{(0 : H_{\mathfrak{a}}^i(M))} \right\}.$$

However, in general we shall not assume that $\mathfrak{b} \subseteq \sqrt{\mathfrak{a}}$.

9.1.6 ♯Exercise. Let the situation be as in 9.1.5, and let S be a multiplicatively closed subset of R. Show that

$$f_{\mathfrak{a}}^{\mathfrak{b}}(M) \leq f_{S^{-1}\mathfrak{a}}^{S^{-1}\mathfrak{b}}(S^{-1}M).$$

9.1.7 ♯Exercise. Let R' be a second commutative Noetherian ring and let $f : R \to R'$ be a ring homomorphism; assume that R', when regarded as an R-module by means of f, is finitely generated.

Let M' be a finitely generated R'-module and let $\mathfrak{b}$ be a second ideal of R. Prove that

$$f_{\mathfrak{a}}^{\mathfrak{b}}(M') = f_{\mathfrak{a}R'}^{\mathfrak{b}R'}(M').$$

9.1.8 Lemma. *Let $\mathfrak{b}$, $\mathfrak{c}$ be further ideals of R, and let M be a finitely generated R-module.*

(i) *If $\mathfrak{b} \subseteq \mathfrak{c}$, then $f_{\mathfrak{a}}^{\mathfrak{b}}(M) = f_{\mathfrak{a}}^{\mathfrak{b}}(M/\Gamma_{\mathfrak{c}}(M))$.*
(ii) *If $f_{\mathfrak{a}}^{\mathfrak{b}}(M) > 0$, then $f_{\mathfrak{a}}^{\mathfrak{b}}(M) = f_{\mathfrak{a}}^{\mathfrak{b}}(M/\Gamma_{\mathfrak{a}}(M))$.*

Proof. Suppose that there exists $n \in \mathbb{N}$ such that $\mathfrak{b}^n \Gamma_{\mathfrak{c}}(M) = 0$. Then $\mathfrak{b}^n H_{\mathfrak{a}}^j(\Gamma_{\mathfrak{c}}(M)) = 0$ for all $j \in \mathbb{N}_0$. Now the exact sequence

$$0 \longrightarrow \Gamma_{\mathfrak{c}}(M) \longrightarrow M \longrightarrow M/\Gamma_{\mathfrak{c}}(M) \longrightarrow 0$$

induces, for each $i \in \mathbb{N}_0$, an exact sequence

$$H_{\mathfrak{a}}^i(\Gamma_{\mathfrak{c}}(M)) \longrightarrow H_{\mathfrak{a}}^i(M) \longrightarrow H_{\mathfrak{a}}^i(M/\Gamma_{\mathfrak{c}}(M)) \longrightarrow H_{\mathfrak{a}}^{i+1}(\Gamma_{\mathfrak{c}}(M)),$$

and, since $\mathfrak{b} \subseteq \sqrt{(0 : H_{\mathfrak{a}}^j(\Gamma_{\mathfrak{c}}(M)))}$ for all $j \in \mathbb{N}_0$, it follows from 9.1.1 that

$$\mathfrak{b} \subseteq \sqrt{(0 : H_{\mathfrak{a}}^i(M))} \quad \text{if and only if} \quad \mathfrak{b} \subseteq \sqrt{(0 : H_{\mathfrak{a}}^i(M/\Gamma_{\mathfrak{c}}(M)))},$$

so that $f_{\mathfrak{a}}^{\mathfrak{b}}(M) = f_{\mathfrak{a}}^{\mathfrak{b}}(M/\Gamma_{\mathfrak{c}}(M))$.

(i) There exists $n \in \mathbb{N}$ such that $\mathfrak{c}^n\Gamma_{\mathfrak{c}}(M) = 0$. If $\mathfrak{b} \subseteq \mathfrak{c}$, then $\mathfrak{b}^n\Gamma_{\mathfrak{c}}(M) = 0$, and the desired result follows from the above.

(ii) Assume that $f_{\mathfrak{a}}^{\mathfrak{b}}(M) > 0$. This means that $\mathfrak{b}^n\Gamma_{\mathfrak{a}}(M) = 0$ for some $n \in \mathbb{N}_0$. Now use the first paragraph of this proof with $\mathfrak{a}$ taken as $\mathfrak{c}$. $\square$

9.1.9 Exercise. Let M be a finitely generated R-module for which $M \neq \mathfrak{a}M$. Show that $f_{\mathfrak{a}}^R(M) = \text{grade}_M \, \mathfrak{a}$.

9.2 Adjusted depths

The $\mathfrak{b}$-finiteness dimension $f_{\mathfrak{a}}^{\mathfrak{b}}(M)$ of 9.1.5 is one of the invariants used in Faltings' Annihilator Theorem. We now motivate the introduction of another of the ingredients. This motivation concerns the situation of Exercise 7.3.4, and we provide now a solution to the second part of that exercise.

Let M be a finitely generated R-module for which the ideal $\mathfrak{a} + (0 : M)$ is proper. Let $\mathfrak{p}$ be a minimal prime of $\mathfrak{a} + (0 : M)$, and suppose that $t := \dim_{R_{\mathfrak{p}}} M_{\mathfrak{p}} > 0$. Then, by 4.3.3, 4.2.2, and 1.1.3 used in conjunction with the fact that $\mathfrak{p}$ is a minimal prime of $\mathfrak{a} + (0 : M)$, we have

$$(H_{\mathfrak{a}}^t(M))_{\mathfrak{p}} \cong H_{\mathfrak{a}R_{\mathfrak{p}}}^t(M_{\mathfrak{p}}) \cong H_{\mathfrak{a}R_{\mathfrak{p}}+(0:_{R_{\mathfrak{p}}}M_{\mathfrak{p}})}^t(M_{\mathfrak{p}})$$

$$= H_{(\mathfrak{a}+(0:M))R_{\mathfrak{p}}}^t(M_{\mathfrak{p}}) = H_{\mathfrak{p}R_{\mathfrak{p}}}^t(M_{\mathfrak{p}}).$$

By 7.3.3, this $R_{\mathfrak{p}}$-module is not finitely generated, since $t := \dim_{R_{\mathfrak{p}}} M_{\mathfrak{p}} > 0$; hence $H_{\mathfrak{a}}^t(M)$ is not finitely generated. We can therefore write $f_{\mathfrak{a}}(M) \leq t$. There exists a minimal prime $\mathfrak{q}$ of $(0 : M)$ such that $\mathfrak{q} \subset \mathfrak{p}$ and $\text{ht}\,\mathfrak{p}/\mathfrak{q} = t$. Note that $\text{depth}_{R_{\mathfrak{q}}} M_{\mathfrak{q}} = 0$, that $\mathfrak{a} \not\subseteq \mathfrak{q}$, and that $\mathfrak{p}$ is a minimal prime of $\mathfrak{a} + \mathfrak{q}$; furthermore, we can write $f_{\mathfrak{a}}(M) \leq \text{depth}_{R_{\mathfrak{q}}} M_{\mathfrak{q}} + \text{ht}\,\mathfrak{p}/\mathfrak{q}$.

Next suppose that $r \in R$ is a non-zerodivisor on M and that there is a minimal prime $\mathfrak{p}'$ of $\mathfrak{a} + (0 : M/rM)$ such that $t' := \dim_{R_{\mathfrak{p}'}} (M/rM)_{\mathfrak{p}'} > 0$. Another application of Exercise 7.3.4 yields that $H_{\mathfrak{a}}^{t'}(M/rM)$ is not finitely generated. It therefore follows from the long exact sequence of local cohomology modules induced by the exact sequence

$$0 \longrightarrow M \overset{r}{\longrightarrow} M \longrightarrow M/rM \longrightarrow 0$$

that either $H_{\mathfrak{a}}^{t'}(M)$ or $H_{\mathfrak{a}}^{t'+1}(M)$ is not finitely generated. The reasoning in the preceding paragraph applied to M/rM (rather than M) shows that there exists a minimal prime $\mathfrak{q}'$ of $(0 : M/rM)$ such that $\mathfrak{q}' \subset \mathfrak{p}'$ and $\text{ht}\,\mathfrak{p}'/\mathfrak{q}' = t'$. Again, note that $\text{depth}_{R_{\mathfrak{q}'}} M_{\mathfrak{q}'} = 1$, that $\mathfrak{a} \not\subseteq \mathfrak{q}'$, that $\mathfrak{p}'$ is a minimal prime of

$\mathfrak{a} + \mathfrak{q}'$, and that

$$f_\mathfrak{a}(M) \leq 1 + t' = \operatorname{depth}_{R_{\mathfrak{q}'}} M_{\mathfrak{q}'} + \operatorname{ht} \mathfrak{p}'/\mathfrak{q}'.$$

It is hoped that, after this little discussion, the reader will not be too dismayed to find that we now begin to consider the values of the expressions

$$\operatorname{depth}_{R_\mathfrak{s}} M_\mathfrak{s} + \operatorname{ht}(\mathfrak{a} + \mathfrak{s})/\mathfrak{s}$$

for prime ideals $\mathfrak{s} \in \operatorname{Supp} M$. (Here, $\operatorname{ht} R/\mathfrak{s}$ is to be interpreted as ∞.) Note that, for such an $\mathfrak{s}$, we have $\operatorname{ht}(\mathfrak{a} + \mathfrak{s})/\mathfrak{s} > 0$ if and only if $(\mathfrak{a} + \mathfrak{s})/\mathfrak{s}$ is a non-zero ideal of the integral domain $R/\mathfrak{s}$, that is, if and only if $\mathfrak{s} \notin \operatorname{Var}(\mathfrak{a})$. We can think, loosely, of $\operatorname{ht}(\mathfrak{a} + \mathfrak{s})/\mathfrak{s}$ as the 'distance' from $\mathfrak{s}$ to $\operatorname{Var}(\mathfrak{a})$ in $\operatorname{Spec}(R)$, as it is, at least in the case when R is catenary (see [50, p. 31]), the minimum length of a saturated ascending chain of prime ideals starting with $\mathfrak{s}$ (as its smallest term) and ending in $\operatorname{Var}(\mathfrak{a})$.

9.2.1 Notation and Conventions. Let M be a finitely generated R-module. For $\mathfrak{p} \in \operatorname{Supp} M$ we shall abbreviate $\operatorname{depth}_{R_\mathfrak{p}} M_\mathfrak{p}$ by $\operatorname{depth} M_\mathfrak{p}$; we shall adopt the convention that the depth of a zero module over a local ring is ∞, and accordingly, for $\mathfrak{p} \in \operatorname{Spec}(R) \setminus \operatorname{Supp} M$, we shall write $\operatorname{depth} M_\mathfrak{p} = \infty$.

We interpret the height of the improper ideal R of R as ∞; accordingly, for a proper ideal $\mathfrak{d}$ of R, we write $\operatorname{ht} R/\mathfrak{d} = \infty$.

9.2.2 Definitions. For a $\mathfrak{p} \in \operatorname{Spec}(R)$ and a finitely generated R-module M, we define the $\mathfrak{a}$-*adjusted depth of* M *at* $\mathfrak{p}$, denoted $\operatorname{adj}_\mathfrak{a} \operatorname{depth} M_\mathfrak{p}$, by

$$\operatorname{adj}_\mathfrak{a} \operatorname{depth} M_\mathfrak{p} := \operatorname{depth} M_\mathfrak{p} + \operatorname{ht}(\mathfrak{a} + \mathfrak{p})/\mathfrak{p}.$$

Note that this is ∞ unless $\mathfrak{p} \in \operatorname{Supp} M$ and $\mathfrak{a} + \mathfrak{p} \subset R$, and then it is a non-negative integer not exceeding $\dim M$.

Let $\mathfrak{b}$ be a second ideal of R. We define the $\mathfrak{b}$-*minimum* $\mathfrak{a}$-*adjusted depth of* M, denoted by $\lambda_\mathfrak{a}^\mathfrak{b}(M)$, by

$$\lambda_\mathfrak{a}^\mathfrak{b}(M) = \inf \left\{ \operatorname{adj}_\mathfrak{a} \operatorname{depth} M_\mathfrak{p} : \mathfrak{p} \in \operatorname{Spec}(R) \setminus \operatorname{Var}(\mathfrak{b}) \right\}$$
$$= \inf \left\{ \operatorname{depth} M_\mathfrak{p} + \operatorname{ht}(\mathfrak{a} + \mathfrak{p})/\mathfrak{p} : \mathfrak{p} \in \operatorname{Spec}(R) \setminus \operatorname{Var}(\mathfrak{b}) \right\}.$$

Thus $\lambda_\mathfrak{a}^\mathfrak{b}(M)$ is either ∞ or a non-negative integer not exceeding $\dim M$.

Faltings' Annihilator Theorem, the main result of this chapter, asserts that, under mild restrictions on R, the invariants $\lambda_\mathfrak{a}^\mathfrak{b}(M)$ and $f_\mathfrak{a}^\mathfrak{b}(M)$ (where M is a finitely generated R-module and $\mathfrak{b}$ is a second ideal of R) are equal. Faltings' original proof in [14] is rather different from the one we shall present below.

The special case of Faltings' Annihilator Theorem in which $\mathfrak{b} = \mathfrak{a}$ reduces to Grothendieck's Finiteness Theorem (see [26, Exposé VIII, Corollaire 2.3]), which is another fundamental result of local cohomology. This theorem provides information about the finiteness dimension $f_\mathfrak{a}(M)$ of M relative to $\mathfrak{a}$ (again under mild restrictions on R).

Our approach to the proof of Faltings' Annihilator Theorem begins with an investigation of some of the properties of the invariant $\lambda_\mathfrak{a}^\mathfrak{b}(M)$ introduced in 9.2.2.

9.2.3 Remarks. Let $\mathfrak{b}$, $\mathfrak{c}$ and $\mathfrak{d}$ be further ideals of R such that $\mathfrak{c} \subseteq \mathfrak{b}$ and $\mathfrak{a} \subseteq \mathfrak{d}$. Let M be a finitely generated R-module. Then

(i) $\lambda_\mathfrak{a}^\mathfrak{b}(M) = \lambda_{\sqrt{\mathfrak{a}}}^\mathfrak{b}(M) = \lambda_{\sqrt{\mathfrak{a}}}^{\sqrt{\mathfrak{b}}}(M)$;

(ii) $\lambda_\mathfrak{a}^\mathfrak{b}(M) \leq \lambda_\mathfrak{d}^\mathfrak{b}(M)$; and

(iii) $\lambda_\mathfrak{a}^\mathfrak{b}(M) \leq \lambda_\mathfrak{a}^\mathfrak{c}(M)$.

9.2.4 Lemma. *Let $\mathfrak{b}$, $\mathfrak{c}$ be further ideals of R, and let M be a finitely generated·R-module.*

(i) *If $\mathfrak{b} \subseteq \mathfrak{c}$, then $\lambda_\mathfrak{a}^\mathfrak{b}(M) = \lambda_\mathfrak{a}^\mathfrak{b}(M/\Gamma_\mathfrak{c}(M))$.*

(ii) *If $\lambda_\mathfrak{a}^\mathfrak{b}(M) > 0$, then $\lambda_\mathfrak{a}^\mathfrak{b}(M) = \lambda_\mathfrak{a}^\mathfrak{b}(M/\Gamma_\mathfrak{a}(M))$.*

Proof. Suppose that, for each $\mathfrak{p} \in \operatorname{Spec}(R) \setminus \operatorname{Var}(\mathfrak{b})$, we have $(\Gamma_\mathfrak{c}(M))_\mathfrak{p} = 0$. Then $(M/\Gamma_\mathfrak{c}(M))_\mathfrak{p} \cong M_\mathfrak{p}$ and

$$\operatorname{depth}(M/\Gamma_\mathfrak{c}(M))_\mathfrak{p} + \operatorname{ht}(\mathfrak{a} + \mathfrak{p})/\mathfrak{p} = \operatorname{depth} M_\mathfrak{p} + \operatorname{ht}(\mathfrak{a} + \mathfrak{p})/\mathfrak{p},$$

so that $\lambda_\mathfrak{a}^\mathfrak{b}(M) = \lambda_\mathfrak{a}^\mathfrak{b}(M/\Gamma_\mathfrak{c}(M))$.

(i) Suppose that $\mathfrak{b} \subseteq \mathfrak{c}$. Then, for each $\mathfrak{p} \in \operatorname{Spec}(R) \setminus \operatorname{Var}(\mathfrak{b})$ we have $\mathfrak{c} \not\subseteq \mathfrak{p}$, so that $(\Gamma_\mathfrak{c}(M))_\mathfrak{p} = 0$. The claim therefore follows from the above paragraph.

(ii) Assume that $\lambda_\mathfrak{a}^\mathfrak{b}(M) > 0$. Let $\mathfrak{p} \in \operatorname{Spec}(R) \setminus \operatorname{Var}(\mathfrak{b})$. We claim that $(\Gamma_\mathfrak{a}(M))_\mathfrak{p} = 0$. This is clearly the case if $\mathfrak{a} \not\subseteq \mathfrak{p}$, so suppose that $\mathfrak{a} \subseteq \mathfrak{p}$. If $(\Gamma_\mathfrak{a}(M))_\mathfrak{p}$ were not zero, then there would be an associated prime ideal $\mathfrak{q}$ of $\Gamma_\mathfrak{a}(M)$ with $\mathfrak{q} \subseteq \mathfrak{p}$, and, necessarily, $\mathfrak{a} \subseteq \mathfrak{q}$. But then we would have $\mathfrak{q} \in \operatorname{Spec}(R) \setminus \operatorname{Var}(\mathfrak{b})$ with $\operatorname{adj}_\mathfrak{a} \operatorname{depth} M_\mathfrak{q} = \operatorname{depth} M_\mathfrak{q} + \operatorname{ht}(\mathfrak{a} + \mathfrak{q})/\mathfrak{q} = 0$, contrary to the assumption that $\lambda_\mathfrak{a}^\mathfrak{b}(M) > 0$. Thus $(\Gamma_\mathfrak{a}(M))_\mathfrak{p} = 0$ in all cases.

The desired result now follows from the first paragraph of this proof with $\mathfrak{a}$ taken as $\mathfrak{c}$.   □

9.2.5 Lemma. *Let $\mathfrak{b}$ be a second ideal of R, let M be a finitely generated R-module, and let S be a multiplicatively closed subset of R. Then*

$$\lambda_\mathfrak{a}^\mathfrak{b}(M) \leq \lambda_{S^{-1}\mathfrak{a}}^{S^{-1}\mathfrak{b}}(S^{-1}M).$$

Proof. Let $\mathfrak{P} \in \operatorname{Spec}(S^{-1}R) \setminus \operatorname{Var}(S^{-1}\mathfrak{b})$. Then there exists $\mathfrak{p} \in \operatorname{Spec}(R)$ such that $\mathfrak{p} \cap S = \emptyset$ and $\mathfrak{P} = S^{-1}\mathfrak{p}$. Now $\mathfrak{p} \notin \operatorname{Var}(\mathfrak{b})$, and so

$$\lambda_{\mathfrak{a}}^{\mathfrak{b}}(M) \leq \operatorname{depth} M_{\mathfrak{p}} + \operatorname{ht}(\mathfrak{a} + \mathfrak{p})/\mathfrak{p}.$$

Now $(S^{-1}M)_{\mathfrak{P}} = (S^{-1}M)_{S^{-1}\mathfrak{p}} \cong M_{\mathfrak{p}}$ as $R_{\mathfrak{p}}$-modules (when $(S^{-1}M)_{S^{-1}\mathfrak{p}}$ is regarded as an $R_{\mathfrak{p}}$-module by means of the natural isomorphism $R_{\mathfrak{p}} \longrightarrow (S^{-1}R)_{S^{-1}\mathfrak{p}}$). Also

$$\operatorname{ht}(\mathfrak{a} + \mathfrak{p})/\mathfrak{p} \leq \operatorname{ht}(S^{-1}\mathfrak{a} + S^{-1}\mathfrak{p})/S^{-1}\mathfrak{p},$$

and so $\lambda_{\mathfrak{a}}^{\mathfrak{b}}(M) \leq \operatorname{depth}(S^{-1}M)_{\mathfrak{P}} + \operatorname{ht}(S^{-1}\mathfrak{a} + \mathfrak{P})/\mathfrak{P}$. The result follows. $\square$

The next lemma shows that $\lambda_{\mathfrak{a}}^{\mathfrak{b}}(M)$ is, in a certain sense, independent of the base ring.

9.2.6 Lemma. *Let $\mathfrak{b}$ be a second ideal of R, let M be a finitely generated R-module, and let $\mathfrak{c}$ be an ideal of R such that $\mathfrak{c} \subseteq (0 : M)$. Then*

$$\lambda_{\mathfrak{a}}^{\mathfrak{b}}(M) = \lambda_{(\mathfrak{a}+\mathfrak{c})/\mathfrak{c}}^{(\mathfrak{b}+\mathfrak{c})/\mathfrak{c}}(M).$$

Proof. Let $\mathfrak{p} \in \operatorname{Spec}(R) \setminus \operatorname{Var}(\mathfrak{b})$. Either $\mathfrak{p} \not\supseteq \mathfrak{c}$ or $\mathfrak{p} \supseteq \mathfrak{c}$. If $\mathfrak{p} \not\supseteq \mathfrak{c}$, then $M_{\mathfrak{p}} = 0$ and $\operatorname{depth} M_{\mathfrak{p}} + \operatorname{ht}(\mathfrak{a} + \mathfrak{p})/\mathfrak{p} = \infty$. This means that

$$\lambda_{\mathfrak{a}}^{\mathfrak{b}}(M) = \inf\left\{\operatorname{depth} M_{\mathfrak{p}'} + \operatorname{ht}(\mathfrak{a} + \mathfrak{p}')/\mathfrak{p}' : \mathfrak{p}' \in \operatorname{Var}(\mathfrak{c}) \setminus \operatorname{Var}(\mathfrak{b})\right\}.$$

However, if $\mathfrak{p} \in \operatorname{Var}(\mathfrak{c}) \setminus \operatorname{Var}(\mathfrak{b})$, then $\mathfrak{p}/\mathfrak{c} \in \operatorname{Spec}(R/\mathfrak{c}) \setminus \operatorname{Var}((\mathfrak{b} + \mathfrak{c})/\mathfrak{c})$, and it is an elementary exercise to check that $\operatorname{depth} M_{\mathfrak{p}/\mathfrak{c}} = \operatorname{depth} M_{\mathfrak{p}}$ and

$$\operatorname{ht}(((\mathfrak{a} + \mathfrak{c})/\mathfrak{c}) + (\mathfrak{p}/\mathfrak{c}))/(\mathfrak{p}/\mathfrak{c}) = \operatorname{ht}(\mathfrak{a} + \mathfrak{p})/\mathfrak{p}.$$

The claim follows from these observations. $\square$

9.3 The first inequality

Our proof of Faltings' Annihilator Theorem consists of demonstrations that, with the notation of 9.1.5 and 9.2.2, we have $f_{\mathfrak{a}}^{\mathfrak{b}}(M) \leq \lambda_{\mathfrak{a}}^{\mathfrak{b}}(M)$ and, under mild restrictions on R, $\lambda_{\mathfrak{a}}^{\mathfrak{b}}(M) \leq f_{\mathfrak{a}}^{\mathfrak{b}}(M)$. We start with the first inequality. We prepare the ground with several lemmas.

9.3.1 Lemma. *Let $\mathfrak{b}$ be a second ideal of R, and let M be a finitely generated R-module. Suppose that $\mathfrak{p} \in \operatorname{Ass} M \setminus \operatorname{Var}(\mathfrak{b})$ has $\operatorname{ht}(\mathfrak{a} + \mathfrak{p})/\mathfrak{p} = 1$. Then $\mathfrak{b} \not\subseteq \sqrt{(0 : H_{\mathfrak{a}}^1(M))}$.*

Proof. It suffices to show that $\mathfrak{a} \cap \mathfrak{b} \not\subseteq \sqrt{(0 : H_{\mathfrak{a}}^1(M))}$. Since $\mathrm{ht}(\mathfrak{a}+\mathfrak{p})/\mathfrak{p} = 1$, we must have that $\mathfrak{a} \not\subseteq \mathfrak{p}$, so that $\mathfrak{p} \notin \mathrm{Var}(\mathfrak{a} \cap \mathfrak{b})$. Therefore we may replace $\mathfrak{b}$ by $\mathfrak{a} \cap \mathfrak{b}$; we therefore assume henceforth in this proof that $\mathfrak{b} \subseteq \mathfrak{a}$.

As $\mathfrak{p} \in \mathrm{Ass}\, M$, there is an exact sequence

$$0 \longrightarrow R/\mathfrak{p} \longrightarrow M \longrightarrow N \longrightarrow 0$$

of R-modules and R-homomorphisms; this induces an exact sequence

$$H_{\mathfrak{a}}^0(N) \longrightarrow H_{\mathfrak{a}}^1(R/\mathfrak{p}) \longrightarrow H_{\mathfrak{a}}^1(M).$$

Since N is finitely generated, $\mathfrak{b} \subseteq \mathfrak{a} \subseteq \sqrt{(0 : H_{\mathfrak{a}}^0(N))}$. Therefore, by 9.1.1, it is enough for us to show that $\mathfrak{b} \not\subseteq \sqrt{(0 : H_{\mathfrak{a}}^1(R/\mathfrak{p}))}$. By the Independence Theorem 4.2.1, there is an R-isomorphism $H_{\mathfrak{a}}^1(R/\mathfrak{p}) \cong H_{(\mathfrak{a}+\mathfrak{p})/\mathfrak{p}}^1(R/\mathfrak{p})$, and so it is enough for us to show that

$$(\mathfrak{b} + \mathfrak{p})/\mathfrak{p} \not\subseteq \sqrt{\left(0 : H_{(\mathfrak{a}+\mathfrak{p})/\mathfrak{p}}^1(R/\mathfrak{p})\right)}.$$

It is therefore enough for us to prove that, if R is a domain, $\mathrm{ht}\,\mathfrak{a} = 1$ and $0 \neq \mathfrak{b} \subseteq \mathfrak{a}$, then $\mathfrak{b} \not\subseteq \sqrt{(0 : H_{\mathfrak{a}}^1(R))}$. We shall achieve this by showing that, in these circumstances, $rH_{\mathfrak{a}}^1(R) \neq 0$ for every $r \in \mathfrak{a} \setminus \{0\}$.

Suppose that, on the contrary, $rH_{\mathfrak{a}}^1(R) = 0$ for some $r \in \mathfrak{a} \setminus \{0\}$. Let $\mathfrak{q}$ be a minimal prime of $\mathfrak{a}$ with $\mathrm{ht}\,\mathfrak{q} = 1$. Let $R' := R_{\mathfrak{q}}$ and $\mathfrak{m}' := \mathfrak{q}R_{\mathfrak{q}}$, so that $(R', \mathfrak{m}')$ is a 1-dimensional local domain. We deduce from 1.1.3 and 4.3.3 that

$$rH_{\mathfrak{m}'}^1(R') = rH_{\mathfrak{a}R_{\mathfrak{q}}}^1(R_{\mathfrak{q}}) \cong r(H_{\mathfrak{a}}^1(R))_{\mathfrak{q}} = 0.$$

Of course, $r \in \mathfrak{m}' \setminus \{0\}$. Since $(R', \mathfrak{m}')$ is a 1-dimensional local domain, we have $\mathfrak{m}' = \sqrt{rR'}$. Therefore, by 1.1.3 and 2.2.21(i),

$$H_{\mathfrak{m}'}^1(R') = H_{rR'}^1(R') \cong R_r'/R'.$$

Thus $rR_r' \subseteq R'$, so that $r(1/r^2) \in R'$. Hence r is a unit of R'. This contradiction completes the proof. $\qquad\square$

9.3.2 Lemma. *Let $\mathfrak{b}$ be a second ideal of R, and let M be a finitely generated R-module. Suppose that $\mathfrak{p} \in \mathrm{Supp}_R M \cap \mathrm{Var}(\mathfrak{a}) \setminus \mathrm{Var}(\mathfrak{b})$ and let $t := \mathrm{grade}_{M_{\mathfrak{p}}}(\mathfrak{a}R_{\mathfrak{p}})$. Then $\mathfrak{b} \not\subseteq \sqrt{(0 : H_{\mathfrak{a}}^t(M))}$.*

Proof. The hypotheses ensure that $M_{\mathfrak{p}} \neq \mathfrak{a}R_{\mathfrak{p}}M_{\mathfrak{p}}$. In view of 4.3.3 and 6.2.7, we have $(H_{\mathfrak{a}}^t(M))_{\mathfrak{p}} \cong H_{\mathfrak{a}R_{\mathfrak{p}}}^t(M_{\mathfrak{p}}) \neq 0$. By hypothesis, there exists $b \in \mathfrak{b} \setminus \mathfrak{p}$. It follows that $(H_{\mathfrak{a}}^t(M))_b \neq 0$, so that $b \notin \sqrt{(0 : H_{\mathfrak{a}}^t(M))}$. $\qquad\square$

9.3.3 Exercise. Show that, for a finitely generated R-module M for which $M \neq \mathfrak{a}M$, we have $\lambda_{\mathfrak{a}}^R(M) = \mathrm{grade}_M\,\mathfrak{a}$. (Here is a hint: observe that

$$\mathrm{grade}_M\,\mathfrak{a} = \inf\{\mathrm{depth}\, M_{\mathfrak{p}} : \mathfrak{p} \in \mathrm{Var}(\mathfrak{a})\}.)$$

9.3.4 Lemma. *Let M be a finitely generated R-module, and $\mathfrak{p}, \mathfrak{s} \in \operatorname{Spec}(R)$ with $\mathfrak{p} \subseteq \mathfrak{s}$. Then*

$$\operatorname{depth} M_{\mathfrak{s}} \leq \operatorname{depth} M_{\mathfrak{p}} + \operatorname{ht} \mathfrak{s}/\mathfrak{p}.$$

Proof. We use induction on $h := \operatorname{ht} \mathfrak{s}/\mathfrak{p}$, there being nothing to prove when $h = 0$.

Consider the case in which $h = 1$. Clearly, we can assume $\mathfrak{p} \in \operatorname{Supp} M$. Now $M_{\mathfrak{p}}$ is $R_{\mathfrak{p}}$-isomorphic to the localization of $M_{\mathfrak{s}}$ at the prime ideal $\mathfrak{p} R_{\mathfrak{s}}$ (when that localization is regarded as an $R_{\mathfrak{p}}$-module by means of the natural isomorphism $R_{\mathfrak{p}} \xrightarrow{\cong} (R_{\mathfrak{s}})_{\mathfrak{p} R_{\mathfrak{s}}}$). In order to establish the desired result when $h = 1$, it is therefore enough for us to show that, if $(R, \mathfrak{m})$ is local, M is a finitely generated R-module, and $\mathfrak{p} \in \operatorname{Supp} M$ is such that $\dim R/\mathfrak{p} = 1$, then $\operatorname{depth} M \leq \operatorname{depth} M_{\mathfrak{p}} + 1$. This we do.

Let $\operatorname{grade}_M \mathfrak{p} = t$ (see 6.2.4), and let $a_1, \ldots, a_t$ be a maximal M-sequence contained in $\mathfrak{p}$. Then $a_1/1, \ldots, a_t/1$ is an $M_{\mathfrak{p}}$-sequence contained in $\mathfrak{p} R_{\mathfrak{p}}$. Let $N := M/\sum_{j=1}^{t} a_j M$. Then

$$\operatorname{depth} N = \operatorname{depth} M - t, \quad \operatorname{depth} N_{\mathfrak{p}} = \operatorname{depth} M_{\mathfrak{p}} - t$$

and $\operatorname{grade}_N \mathfrak{p} = 0$. Thus we can achieve our aim by showing that $\operatorname{depth} N \leq \operatorname{depth} N_{\mathfrak{p}} + 1$. We can assume that $\operatorname{depth} N > 0$, since otherwise there is nothing to prove. This assumption means that $\mathfrak{m} \notin \operatorname{Ass} N$. But $\mathfrak{p}$ consists entirely of zerodivisors on N, and so $\mathfrak{p} \subseteq \bigcup_{\mathfrak{p}' \in \operatorname{Ass} N} \mathfrak{p}'$. It therefore follows from the Prime Avoidance Theorem and the fact that $\dim R/\mathfrak{p} = 1$ that $\mathfrak{p} \in \operatorname{Ass} N$, and so $\operatorname{depth} N \leq \dim R/\mathfrak{p} = 1$ by [50, Theorem 17.2].

The claim in the lemma has therefore been proved when $h = 1$; suppose now that $h > 1$ and the claim has been proved for smaller values of h. Since $\operatorname{ht} \mathfrak{s}/\mathfrak{p} = h$, there exists $\mathfrak{q} \in \operatorname{Spec}(R)$ such that $\mathfrak{p} \subset \mathfrak{q} \subset \mathfrak{s}$, $\operatorname{ht} \mathfrak{q}/\mathfrak{p} = 1$ and $\operatorname{ht} \mathfrak{s}/\mathfrak{q} = h - 1$. We can now use the inductive hypothesis and the (already established) truth of the claim in the case when $h = 1$ to see that

$$\operatorname{depth} M_{\mathfrak{s}} \leq \operatorname{depth} M_{\mathfrak{q}} + h - 1$$
$$\leq \operatorname{depth} M_{\mathfrak{p}} + 1 + h - 1 = \operatorname{depth} M_{\mathfrak{p}} + h.$$

This completes the inductive step, and the proof. $\square$

9.3.5 Lemma. *Let M be a finitely generated R-module. Suppose that $\mathfrak{p}, \mathfrak{q} \in \operatorname{Spec}(R)$ are such that $\mathfrak{q}$ is a minimal prime of $\mathfrak{a} + \mathfrak{p}$ and $\operatorname{ht} \mathfrak{q}/\mathfrak{p} = \operatorname{ht}(\mathfrak{a} + \mathfrak{p})/\mathfrak{p}$. Then*

$$\operatorname{adj}_{\mathfrak{a}} \operatorname{depth} M_{\mathfrak{s}} \leq \operatorname{adj}_{\mathfrak{a}} \operatorname{depth} M_{\mathfrak{p}}$$

for all $\mathfrak{s} \in \operatorname{Spec}(R)$ with $\mathfrak{p} \subseteq \mathfrak{s} \subseteq \mathfrak{q}$.

Proof. It is clear that $\mathfrak{q} \supseteq \mathfrak{a} + \mathfrak{s}$, and so $\operatorname{ht}\mathfrak{q}/\mathfrak{s} \geq \operatorname{ht}(\mathfrak{a} + \mathfrak{s})/\mathfrak{s}$. Hence, on use of 9.3.4, we have

$$
\begin{aligned}
\operatorname{adj}_{\mathfrak{a}} \operatorname{depth} M_{\mathfrak{s}} &= \operatorname{depth} M_{\mathfrak{s}} + \operatorname{ht}(\mathfrak{a} + \mathfrak{s})/\mathfrak{s} \\
&\leq \operatorname{depth} M_{\mathfrak{s}} + \operatorname{ht}\mathfrak{q}/\mathfrak{s} \\
&\leq \operatorname{depth} M_{\mathfrak{p}} + \operatorname{ht}\mathfrak{s}/\mathfrak{p} + \operatorname{ht}\mathfrak{q}/\mathfrak{s} \\
&\leq \operatorname{depth} M_{\mathfrak{p}} + \operatorname{ht}\mathfrak{q}/\mathfrak{p} \\
&= \operatorname{depth} M_{\mathfrak{p}} + \operatorname{ht}(\mathfrak{a} + \mathfrak{p})/\mathfrak{p} = \operatorname{adj}_{\mathfrak{a}} \operatorname{depth} M_{\mathfrak{p}},
\end{aligned}
$$

as required. $\qquad\square$

9.3.6 Lemma. *Let* $\mathfrak{s}, \mathfrak{q} \in \operatorname{Spec}(R)$ *with* $\mathfrak{s} \subset \mathfrak{q}$ *be such that* $\operatorname{ht}\mathfrak{q}/\mathfrak{s} > 1$. *Let* $a \in \mathfrak{q} \setminus \mathfrak{s}$. *Then there exists* $\mathfrak{t} \in \operatorname{Spec}(R)$ *with* $\mathfrak{s} \subset \mathfrak{t} \subset \mathfrak{q}$ *such that* $a \notin \mathfrak{t}$.

Proof. Let $\overline{R} := R/\mathfrak{s}$; for $c \in R$, denote the natural image of c in $\overline{R}$ by $\overline{c}$.

Let the minimal primes of the proper principal ideal $\overline{R}\overline{a}$ be $\mathfrak{p}_1/\mathfrak{s}, \ldots, \mathfrak{p}_h/\mathfrak{s}$, where $\mathfrak{p}_1, \ldots, \mathfrak{p}_h \in \operatorname{Spec}(R)$. Now $\operatorname{ht}\mathfrak{p}_i/\mathfrak{s} = 1$ for all $i = 1, \ldots, h$, by Krull's Principal Ideal Theorem. Therefore, since $\operatorname{ht}\mathfrak{q}/\mathfrak{s} > 1$, it follows from the Prime Avoidance Theorem that $\mathfrak{q}/\mathfrak{s} \not\subseteq \bigcup_{i=1}^{h} \mathfrak{p}_i/\mathfrak{s}$. Hence, there exists $b \in R$ such that

$$
\overline{b} \in \mathfrak{q}/\mathfrak{s} \setminus \bigcup_{i=1}^{h} \mathfrak{p}_i/\mathfrak{s}.
$$

Since $\overline{R}\overline{b} \subseteq \mathfrak{q}/\mathfrak{s}$, there exists $\mathfrak{t} \in \operatorname{Spec}(R)$ with $\mathfrak{s} \subset \mathfrak{t} \subseteq \mathfrak{q}$ such that $\mathfrak{t}/\mathfrak{s}$ is a minimal prime of $\overline{R}\overline{b}$. Since $\operatorname{ht}\mathfrak{q}/\mathfrak{s} > 1$ and $\operatorname{ht}\mathfrak{t}/\mathfrak{s} = 1$, it follows that $\mathfrak{t} \subset \mathfrak{q}$. Since $b \in \mathfrak{t} \setminus \bigcup_{i=1}^{h} \mathfrak{p}_i$, we can deduce that $\mathfrak{t}$ is different from all of $\mathfrak{p}_1, \ldots, \mathfrak{p}_h$. Hence $a \notin \mathfrak{t}$. $\qquad\square$

9.3.7 Theorem. *Let* $\mathfrak{b}$ *be a second ideal of* R, *and let* M *be a finitely generated* R-*module. Then*

$$
f_{\mathfrak{a}}^{\mathfrak{b}}(M) \leq \lambda_{\mathfrak{a}}^{\mathfrak{b}}(M).
$$

Proof. Set $\lambda := \lambda_{\mathfrak{a}}^{\mathfrak{b}}(M)$. If $\lambda = \infty$, then there is nothing to prove; we therefore suppose that λ is finite, and argue by induction on λ.

When $\lambda = 0$, there exists $\mathfrak{p} \in \operatorname{Spec}(R) \setminus \operatorname{Var}(\mathfrak{b})$ with

$$
\operatorname{adj}_{\mathfrak{a}} \operatorname{depth} M_{\mathfrak{p}} = \operatorname{depth} M_{\mathfrak{p}} + \operatorname{ht}(\mathfrak{a} + \mathfrak{p})/\mathfrak{p} = 0.
$$

This means that $\operatorname{depth} M_{\mathfrak{p}} = 0$ and $\mathfrak{p} \in \operatorname{Var}(\mathfrak{a})$, so that

$$
\mathfrak{p}R_{\mathfrak{p}} \in \operatorname{Ass}_{R_{\mathfrak{p}}} M_{\mathfrak{p}} \cap \operatorname{Var}(\mathfrak{a}R_{\mathfrak{p}})
$$

and $\operatorname{grade}_{M_{\mathfrak{p}}}(\mathfrak{a}R_{\mathfrak{p}}) = 0$. Therefore $\mathfrak{b} \not\subseteq \sqrt{(0 : H_{\mathfrak{a}}^0(M))}$ by 9.3.2, so that $f_{\mathfrak{a}}^{\mathfrak{b}}(M) = 0$.

Now suppose that $\lambda > 0$ and assume, inductively, that the desired inequality has been proved for smaller values of λ. We can assume that $f_{\mathfrak{a}}^{\mathfrak{b}}(M) > 0$. By 9.1.8(ii) and 9.2.4(ii), we have $f_{\mathfrak{a}}^{\mathfrak{b}}(M) = f_{\mathfrak{a}}^{\mathfrak{b}}(M/\Gamma_{\mathfrak{a}}(M))$ and $\lambda_{\mathfrak{a}}^{\mathfrak{b}}(M) = \lambda_{\mathfrak{a}}^{\mathfrak{b}}(M/\Gamma_{\mathfrak{a}}(M))$. We may therefore replace M by $M/\Gamma_{\mathfrak{a}}(M)$, and, in view of 2.1.2, assume that M is $\mathfrak{a}$-torsion-free for the remainder of this proof.

Choose $\mathfrak{p} \in \operatorname{Spec}(R) \setminus \operatorname{Var}(\mathfrak{b})$ with

$$\operatorname{adj}_{\mathfrak{a}} \operatorname{depth} M_{\mathfrak{p}} = \operatorname{depth} M_{\mathfrak{p}} + \operatorname{ht}(\mathfrak{a} + \mathfrak{p})/\mathfrak{p} = \lambda.$$

Assume first that $\mathfrak{p} \in \operatorname{Var}(\mathfrak{a})$. Then $t := \operatorname{grade}_{M_{\mathfrak{p}}} \mathfrak{a} R_{\mathfrak{p}} \leq \operatorname{depth} M_{\mathfrak{p}} = \lambda$, so that $f_{\mathfrak{a}}^{\mathfrak{b}}(M) \leq t \leq \lambda$ by 9.3.2.

Assume now that $\mathfrak{p} \notin \operatorname{Var}(\mathfrak{a})$. If $\lambda = 1$, then

$$\operatorname{adj}_{\mathfrak{a}} \operatorname{depth} M_{\mathfrak{p}} = \operatorname{depth} M_{\mathfrak{p}} + \operatorname{ht}(\mathfrak{a} + \mathfrak{p})/\mathfrak{p} = 1.$$

These conditions mean that $\operatorname{ht}(\mathfrak{a} + \mathfrak{p})/\mathfrak{p} = 1$ and $\operatorname{depth} M_{\mathfrak{p}} = 0$. Hence $\mathfrak{p} \in \operatorname{Ass} M$, and so it follows from Lemma 9.3.1 that $\mathfrak{b} \not\subseteq \sqrt{(0 : H_{\mathfrak{a}}^{1}(M))}$. Hence $f_{\mathfrak{a}}^{\mathfrak{b}}(M) \leq 1 = \lambda$.

So we may assume that $\lambda > 1$ (and $\mathfrak{p} \notin \operatorname{Var}(\mathfrak{a})$). Since $\mathfrak{p} \notin \operatorname{Var}(\mathfrak{a} \cap \mathfrak{b})$, there exists $a \in \mathfrak{a} \cap \mathfrak{b} \setminus \mathfrak{p}$. Let $\mathfrak{q}$ be a minimal prime of $\mathfrak{a} + \mathfrak{p}$ such that $\operatorname{ht} \mathfrak{q}/\mathfrak{p} = \operatorname{ht}(\mathfrak{a} + \mathfrak{p})/\mathfrak{p}$. Then $\mathfrak{p}$ belongs to the set

$$\Sigma := \left\{ \mathfrak{s}' \in \operatorname{Spec}(R) : \mathfrak{p} \subseteq \mathfrak{s}' \subseteq \mathfrak{q} \text{ and } a \notin \mathfrak{s}' \right\};$$

let $\mathfrak{s}$ be a maximal member of Σ. Now $a \in \mathfrak{a} \cap \mathfrak{b} \subseteq \mathfrak{a} \subseteq \mathfrak{q}$, and so $\mathfrak{s} \subset \mathfrak{q}$. Lemma 9.3.6 shows that $\operatorname{ht} \mathfrak{q}/\mathfrak{s} = 1$. Note that the fact that $a \notin \mathfrak{s}$ ensures that $\mathfrak{s} \in \operatorname{Spec}(R) \setminus \operatorname{Var}(\mathfrak{b})$.

We can now deduce from 9.3.5 and the definition of λ that

$$\lambda \leq \operatorname{adj}_{\mathfrak{a}} \operatorname{depth} M_{\mathfrak{s}} \leq \operatorname{adj}_{\mathfrak{a}} \operatorname{depth} M_{\mathfrak{p}} = \lambda.$$

Therefore $\operatorname{adj}_{\mathfrak{a}} \operatorname{depth} M_{\mathfrak{s}} = \lambda$. Since $a \in \mathfrak{a} \setminus \mathfrak{s}$, we have $\mathfrak{a} \not\subseteq \mathfrak{s}$. It follows that $\mathfrak{q}$ is a minimal prime of $\mathfrak{a} + \mathfrak{s}$ such that $\operatorname{ht} \mathfrak{q}/\mathfrak{s} = \operatorname{ht}(\mathfrak{a} + \mathfrak{s})/\mathfrak{s} = 1$. Thus we can replace $\mathfrak{p}$ by $\mathfrak{s}$ and so make the additional assumption that $\operatorname{ht} \mathfrak{q}/\mathfrak{p} = \operatorname{ht}(\mathfrak{a} + \mathfrak{p})/\mathfrak{p} = 1$.

We now propose to localize at $\mathfrak{q}$. Note that $\mathfrak{p} R_{\mathfrak{q}} \in \operatorname{Spec}(R_{\mathfrak{q}}) \setminus \operatorname{Var}(\mathfrak{b} R_{\mathfrak{q}})$, that $\mathfrak{q} R_{\mathfrak{q}}$ is a minimal prime of $\mathfrak{a} R_{\mathfrak{q}} + \mathfrak{p} R_{\mathfrak{q}}$, that

$$\operatorname{ht} \mathfrak{q} R_{\mathfrak{q}}/\mathfrak{p} R_{\mathfrak{q}} = \operatorname{ht}(\mathfrak{a} R_{\mathfrak{q}} + \mathfrak{p} R_{\mathfrak{q}})/\mathfrak{p} R_{\mathfrak{q}} = 1,$$

that $M_{\mathfrak{q}}$ is a finitely generated $\mathfrak{a} R_{\mathfrak{q}}$-torsion-free $R_{\mathfrak{q}}$-module, and that

$$\operatorname{adj}_{\mathfrak{a} R_{\mathfrak{q}}} \operatorname{depth}(M_{\mathfrak{q}})_{\mathfrak{p} R_{\mathfrak{q}}} = \operatorname{depth}(M_{\mathfrak{q}})_{\mathfrak{p} R_{\mathfrak{q}}} + \operatorname{ht}(\mathfrak{a} R_{\mathfrak{q}} + \mathfrak{p} R_{\mathfrak{q}})/\mathfrak{p} R_{\mathfrak{q}}$$
$$= \operatorname{depth} M_{\mathfrak{p}} + 1 = \lambda.$$

We can therefore deduce from 9.2.5 that $\lambda_{\mathfrak{a}R_{\mathfrak{q}}}^{\mathfrak{b}R_{\mathfrak{q}}}(M_{\mathfrak{q}}) = \lambda$. Also, by 9.1.6, we have $f_{\mathfrak{a}}^{\mathfrak{b}}(M) \leq f_{\mathfrak{a}R_{\mathfrak{q}}}^{\mathfrak{b}R_{\mathfrak{q}}}(M_{\mathfrak{q}})$. These considerations mean that it is enough for us to prove the desired result under the additional assumption that $(R, \mathfrak{q})$ is local, and we make this assumption in what follows.

Our next aim is to show that $\mathfrak{p}$ contains a non-zerodivisor on M. Suppose that this is not the case, so that $\mathfrak{p} \subseteq \mathfrak{s}$ for some $\mathfrak{s} \in \operatorname{Ass} M$. Since M is $\mathfrak{a}$-torsion-free, $\mathfrak{a}$ contains a non-zerodivisor on M, by 2.1.1(ii), so that, as $\mathfrak{q} \supseteq \mathfrak{a}$, we see that $\mathfrak{s} \subset \mathfrak{q}$. As $\operatorname{ht} \mathfrak{q}/\mathfrak{p} = 1$, it follows that $\mathfrak{p} = \mathfrak{s} \in \operatorname{Ass} M$. This means that $\operatorname{depth} M_{\mathfrak{p}} = 0$, and $\lambda = \operatorname{depth} M_{\mathfrak{p}} + \operatorname{ht}(\mathfrak{a}+\mathfrak{p})/\mathfrak{p} = 1$. This contradiction shows that there exists $r \in \mathfrak{p}$ which is a non-zerodivisor on M. Now

$$\lambda_{\mathfrak{a}}^{\mathfrak{b}}(M/rM) \leq \operatorname{adj}_{\mathfrak{a}} \operatorname{depth}(M/rM)_{\mathfrak{p}} = \operatorname{depth}(M/rM)_{\mathfrak{p}} + 1$$
$$= \operatorname{depth} M_{\mathfrak{p}} - 1 + 1 < \lambda.$$

Therefore, by the induction hypothesis, $f_{\mathfrak{a}}^{\mathfrak{b}}(M/rM) \leq \lambda_{\mathfrak{a}}^{\mathfrak{b}}(M/rM) < \lambda$. Set $u := f_{\mathfrak{a}}^{\mathfrak{b}}(M/rM)$, so that $\mathfrak{b} \not\subseteq \sqrt{(0 : H_{\mathfrak{a}}^{u}(M/rM))}$. But the exact sequence

$$0 \longrightarrow M \overset{r}{\longrightarrow} M \longrightarrow M/rM \longrightarrow 0$$

induces an exact sequence

$$H_{\mathfrak{a}}^{u}(M) \longrightarrow H_{\mathfrak{a}}^{u}(M/rM) \longrightarrow H_{\mathfrak{a}}^{u+1}(M),$$

and so it follows from 9.1.1 that $\mathfrak{b} \not\subseteq \sqrt{(0 : H_{\mathfrak{a}}^{i}(M))}$ for $i = u$ or $i = u + 1$. Thus

$$f_{\mathfrak{a}}^{\mathfrak{b}}(M) \leq u + 1 = f_{\mathfrak{a}}^{\mathfrak{b}}(M/rM) + 1 \leq \lambda.$$

This completes the inductive step, and the proof. $\qquad\square$

9.4 The second inequality

We now embark on the more difficult part of our proof of Faltings' Annihilator Theorem, namely the proof that, when R is a homomorphic image of a regular ring, then, with the notation of 9.1.5 and 9.2.2, we have $\lambda_{\mathfrak{a}}^{\mathfrak{b}}(M) \leq f_{\mathfrak{a}}^{\mathfrak{b}}(M)$. (Recall (see [50, p. 157]) that a commutative Noetherian ring is said to be *regular* precisely when its localizations at all of its prime ideals are regular local rings.)

9.4.1 Lemma. *Let M be a finitely generated R-module, and let*

$$\mathfrak{p} \in \operatorname{Spec}(R) \setminus \operatorname{Supp} M.$$

Then there exists $s \in R \setminus \mathfrak{p}$ such that, for every ideal $\mathfrak{b}$ of R, we have

$$sH_{\mathfrak{b}}^i(M) = 0 \quad \textit{for all } i \in \mathbb{N}_0.$$

Proof. Since M is finitely generated, there exists $s \in (0 : M) \setminus \mathfrak{p}$, and this s has the desired properties because the functors $H_{\mathfrak{b}}^i$ ($i \in \mathbb{N}_0$) are R-linear. $\square$

9.4.2 Lemma. *Let M be a finitely generated R-module, and let $\mathfrak{p} \in \operatorname{Spec}(R)$ be such that $M_{\mathfrak{p}}$ is a non-zero free $R_{\mathfrak{p}}$-module. Then there exist $t \in \mathbb{N}$ and an R-homomorphism $\pi : R^t \longrightarrow M$ such that $(\operatorname{Ker} \pi)_{\mathfrak{p}} = (\operatorname{Coker} \pi)_{\mathfrak{p}} = 0$.*

Proof. Set $t := \operatorname{rank}_{R_{\mathfrak{p}}} M_{\mathfrak{p}}$. There exist $m_1, \ldots, m_t \in M$ such that $m_1/1$, $\ldots, m_t/1$ form a base for the free $R_{\mathfrak{p}}$-module $M_{\mathfrak{p}}$. Let $\pi : R^t \longrightarrow M$ be the R-homomorphism for which $\pi((r_1, \ldots, r_t)) = \sum_{i=1}^t r_i m_i$ for all $(r_1, \ldots, r_t) \in R^t$. Then the localization of π at $\mathfrak{p}$ is an isomorphism, and so $(\operatorname{Ker} \pi)_{\mathfrak{p}} = (\operatorname{Coker} \pi)_{\mathfrak{p}} = 0$. $\square$

9.4.3 Lemma. *Let M be a finitely generated R-module, and let $\mathfrak{p} \in \operatorname{Spec}(R)$ be such that $M_{\mathfrak{p}}$ is a non-zero free $R_{\mathfrak{p}}$-module. Then there exists $s \in R \setminus \mathfrak{p}$ such that, for every proper ideal $\mathfrak{b}$ of R, we have*

$$sH_{\mathfrak{b}}^i(M) = 0 \quad \textit{for all } i < \operatorname{grade} \mathfrak{b}.$$

Proof. By 9.4.2, there exist $t \in \mathbb{N}$ and an R-homomorphism $\pi : R^t \longrightarrow M$ such that $(\operatorname{Ker} \pi)_{\mathfrak{p}} = (\operatorname{Coker} \pi)_{\mathfrak{p}} = 0$. By 9.4.1, there exist $u, v \in R \setminus \mathfrak{p}$ such that, for every ideal $\mathfrak{b}$ of R, we have

$$uH_{\mathfrak{b}}^i(\operatorname{Ker} \pi) = vH_{\mathfrak{b}}^i(\operatorname{Coker} \pi) = 0 \quad \text{for all } i \in \mathbb{N}_0.$$

Set $s := uv$. Let $\mathfrak{b}$ be a proper ideal of R and let $i \in \mathbb{N}_0$ with $i < \operatorname{grade} \mathfrak{b}$.

Now the exact sequence $0 \longrightarrow \operatorname{Ker} \pi \longrightarrow R^t \longrightarrow \operatorname{Im} \pi \longrightarrow 0$ induces an exact sequence $H_{\mathfrak{b}}^i(R^t) \longrightarrow H_{\mathfrak{b}}^i(\operatorname{Im} \pi) \longrightarrow H_{\mathfrak{b}}^{i+1}(\operatorname{Ker} \pi)$. Since $i < \operatorname{grade} \mathfrak{b}$ we have $H_{\mathfrak{b}}^i(R) = 0$, by 6.2.7, so that $H_{\mathfrak{b}}^i(R^t) = 0$ in view of the additivity of the functor $H_{\mathfrak{b}}^i$. By the immediately preceding paragraph, $uH_{\mathfrak{b}}^{i+1}(\operatorname{Ker} \pi) = 0$. Therefore $uH_{\mathfrak{b}}^i(\operatorname{Im} \pi) = 0$.

Next, the exact sequence $0 \longrightarrow \operatorname{Im} \pi \longrightarrow M \longrightarrow \operatorname{Coker} \pi \longrightarrow 0$ induces an exact sequence $H_{\mathfrak{b}}^i(\operatorname{Im} \pi) \longrightarrow H_{\mathfrak{b}}^i(M) \longrightarrow H_{\mathfrak{b}}^i(\operatorname{Coker} \pi)$. As $uH_{\mathfrak{b}}^i(\operatorname{Im} \pi) = 0$ and $vH_{\mathfrak{b}}^i(\operatorname{Coker} \pi) = 0$, it follows that $sH_{\mathfrak{b}}^i(M) = uvH_{\mathfrak{b}}^i(M) = 0$. $\square$

9.4.4 Conventions. Let M be an R-module. We shall denote the projective dimension of M by $\operatorname{proj\,dim} M$ or, occasionally, by $\operatorname{proj\,dim}_R M$ if it is essential to specify the underlying ring concerned. In particular, the reader is warned that, when S is a multiplicatively closed subset of R, we shall always write $\operatorname{proj\,dim} S^{-1}M$ rather than $\operatorname{proj\,dim}_{S^{-1}R} S^{-1}M$. We adopt the convention that a zero module has projective dimension $-\infty$.

9.4.5 Lemma.   *Let M be a finitely generated R-module, and let $\mathfrak{p} \in \operatorname{Spec}(R)$ be such that* $\operatorname{proj dim} M_\mathfrak{p} < \infty$. *Then there exists $s \in R \backslash \mathfrak{p}$ such that, for every proper ideal $\mathfrak{b}$ of R, we have*

$$sH_\mathfrak{b}^i(M) = 0 \quad \text{for all } i < \operatorname{grade} \mathfrak{b} - \operatorname{proj dim} M_\mathfrak{p}.$$

Note.   In the case when $M_\mathfrak{p} = 0$, one should interpret the '$- - \infty$' in the above statement as '∞'.

Proof.   Set $h := \operatorname{proj dim} M_\mathfrak{p}$. We use induction on h. If $h = -\infty$, then $M_\mathfrak{p} = 0$ and the result is clear from Lemma 9.4.1. When $h = 0$, the desired result follows from Lemma 9.4.3, since then $M_\mathfrak{p}$ is a non-zero free $R_\mathfrak{p}$-module.

We therefore assume, inductively, that $h > 0$ and the result has been proved for smaller values of h. There is a non-zero, finitely generated free R-module F and an exact sequence $0 \longrightarrow N \longrightarrow F \longrightarrow M \longrightarrow 0$ of R-modules and homomorphisms. Localization yields an exact sequence

$$0 \longrightarrow N_\mathfrak{p} \longrightarrow F_\mathfrak{p} \longrightarrow M_\mathfrak{p} \longrightarrow 0.$$

Therefore $\operatorname{proj dim} N_\mathfrak{p} = h - 1$ (since $h > 0$), and so, by the inductive hypothesis, there exists $s \in R \backslash \mathfrak{p}$ such that, for every proper ideal $\mathfrak{b}$ of R, we have

$$sH_\mathfrak{b}^i(N) = 0 \quad \text{for all } i < \operatorname{grade} \mathfrak{b} - h + 1.$$

Thus $sH_\mathfrak{b}^{i+1}(N) = 0$ for all $i < \operatorname{grade} \mathfrak{b} - h$. Let $\mathfrak{b}$ be a proper ideal of R and let $i \in \mathbb{N}_0$ with $i < \operatorname{grade} \mathfrak{b} - h$.

The exact sequence $0 \longrightarrow N \longrightarrow F \longrightarrow M \longrightarrow 0$ induces a further exact sequence $H_\mathfrak{b}^i(F) \longrightarrow H_\mathfrak{b}^i(M) \longrightarrow H_\mathfrak{b}^{i+1}(N)$. Since $i < \operatorname{grade} \mathfrak{b}$ we have $H_\mathfrak{b}^i(F) = 0$, by 6.2.7 and the additivity of the functor $H_\mathfrak{b}^i$. By the immediately preceding paragraph, $sH_\mathfrak{b}^{i+1}(N) = 0$. Therefore $sH_\mathfrak{b}^i(M) = 0$. This completes the inductive step.   □

9.4.6 Lemma.   *Let M be a finitely generated R-module, and let $\mathfrak{p} \in \operatorname{Spec}(R)$ be such that* $\operatorname{proj dim} M_\mathfrak{p} < \infty$. *Then there exists $s \in R \backslash \mathfrak{p}$ such that* $\operatorname{proj dim} M_s = \operatorname{proj dim} M_\mathfrak{p}$.

Proof.   Set $h := \operatorname{proj dim} M_\mathfrak{p}$. We use induction on h. If $h = -\infty$, then $M_\mathfrak{p} = 0$ and there exists $s \in R \backslash \mathfrak{p}$ such that $sM = 0$. Hence $M_s = 0$ and $\operatorname{proj dim} M_s = -\infty = h$.

When $h = 0$, then, by Lemma 9.4.2, there exist $t \in \mathbb{N}$ and an R-homomorphism $\pi : R^t \longrightarrow M$ such that $(\operatorname{Ker} \pi)_\mathfrak{p} = (\operatorname{Coker} \pi)_\mathfrak{p} = 0$. Then there exists $s \in R \backslash \mathfrak{p}$ such that $s \operatorname{Ker} \pi = s \operatorname{Coker} \pi = 0$. It follows that $(\operatorname{Ker} \pi)_s = (\operatorname{Coker} \pi)_s = 0$ and $\pi_s : (R^t)_s \longrightarrow M_s$ is an isomorphism. Since $M_s \neq 0$, we have $\operatorname{proj dim} M_s = 0 = h$.

We therefore assume, inductively, that $h > 0$ and the result has been proved for smaller values of h. There is a non-zero, finitely generated free R-module F and an exact sequence $0 \longrightarrow N \longrightarrow F \longrightarrow M \longrightarrow 0$ of R-modules and homomorphisms. Localization yields an exact sequence

$$0 \longrightarrow N_{\mathfrak{p}} \longrightarrow F_{\mathfrak{p}} \longrightarrow M_{\mathfrak{p}} \longrightarrow 0.$$

Note that $\operatorname{proj\,dim} N_{\mathfrak{p}} = h - 1$, and so, by the inductive hypothesis, there exists $s \in R \setminus \mathfrak{p}$ such that $\operatorname{proj\,dim} N_s = \operatorname{proj\,dim} N_{\mathfrak{p}} = h - 1$. But then the exact sequence

$$0 \longrightarrow N_s \longrightarrow F_s \longrightarrow M_s \longrightarrow 0$$

shows that $\operatorname{proj\,dim} M_s \le h$. Since $\mathfrak{p} \cap \{s^n : n \in \mathbb{N}_0\} = \emptyset$, we see that $M_{\mathfrak{p}} \cong (M_s)_{\mathfrak{p} R_s}$ (when the latter is regarded as an $R_{\mathfrak{p}}$-module by means of the natural isomorphism $R_{\mathfrak{p}} \longrightarrow (R_s)_{\mathfrak{p} R_s}$). Hence $h = \operatorname{proj\,dim} M_{\mathfrak{p}} \le \operatorname{proj\,dim} M_s \le h$, so that $\operatorname{proj\,dim} M_s = \operatorname{proj\,dim} M_{\mathfrak{p}}$. This completes the inductive step. $\square$

9.4.7 Corollary. *Let M be a finitely generated R-module. Then for each $t \in \mathbb{N}_0 \cup \{-\infty\}$, the set*

$$U_t(M) := \{\mathfrak{p} \in \operatorname{Spec}(R) : \operatorname{proj\,dim} M_{\mathfrak{p}} \le t\}$$

is an open subset of $\operatorname{Spec}(R)$ *(in the Zariski topology).*

Proof. Let $\mathfrak{p} \in U_t(M)$, so that $h := \operatorname{proj\,dim} M_{\mathfrak{p}} \le t$. By 9.4.6, there exists $s \in R \setminus \mathfrak{p}$ such that $\operatorname{proj\,dim} M_s = h$. Therefore, for each $\mathfrak{q}$ in the open neighbourhood $\operatorname{Spec}(R) \setminus \operatorname{Var}(sR)$ of $\mathfrak{p}$, we have $\operatorname{proj\,dim} M_{\mathfrak{q}} \le \operatorname{proj\,dim} M_s \le t$ (because $M_{\mathfrak{q}} \cong (M_s)_{\mathfrak{q} R_s}$ when the latter is regarded as an $R_{\mathfrak{q}}$-module by means of the natural isomorphism $R_{\mathfrak{q}} \longrightarrow (R_s)_{\mathfrak{q} R_s}$), so that $\mathfrak{q} \in U_t(M)$. $\square$

9.4.8 Notation and Remarks. Let M be a finitely generated R-module. For each $t \in \mathbb{N}_0 \cup \{-\infty\}$, let

$$U_t(M) := \{\mathfrak{p} \in \operatorname{Spec}(R) : \operatorname{proj\,dim} M_{\mathfrak{p}} \le t\},$$

let $C_t(M) = \operatorname{Spec}(R) \setminus U_t(M)$, and let $\mathfrak{c}_t(M) = \bigcap_{\mathfrak{p} \in C_t(M)} \mathfrak{p}$. Then

(i) $U_t(M)$ is an open, and so $C_t(M)$ is a closed, subset of $\operatorname{Spec}(R)$, for all $t \in \mathbb{N}_0$ (by 9.4.7);

(ii) $\operatorname{Spec}(R) \setminus \operatorname{Supp} M = U_{-\infty}(M)$ and

$$U_{-\infty}(M) \subseteq U_0(M) \subseteq U_1(M) \subseteq \cdots \subseteq U_t(M) \subseteq \cdots;$$

(iii) $\operatorname{Supp} M = C_{-\infty}(M) \supseteq C_0(M) \supseteq C_1(M) \supseteq \cdots \supseteq C_t(M) \supseteq \cdots;$

(iv) $\sqrt{(0 : M)} = \mathfrak{c}_{-\infty}(M) \subseteq \mathfrak{c}_0(M) \subseteq \mathfrak{c}_1(M) \subseteq \cdots \subseteq \mathfrak{c}_t(M) \subseteq \cdots;$

(v) for each $t \in \mathbb{N}_0 \cup \{-\infty\}$, the ideal $\mathfrak{c}_t(M)$ is radical (that is, is equal to its own radical), and $\mathrm{Var}(\mathfrak{c}_t(M)) = C_t(M)$; and

(vi) if R is regular, so that, for all $\mathfrak{p} \in \mathrm{Spec}(R)$, the localization $R_\mathfrak{p}$ has finite global dimension and $\mathrm{proj\,dim}\, M_\mathfrak{p}$ is finite, then

$$\bigcup_{t \in \mathbb{N}_0 \cup \{-\infty\}} U_t(M) = \mathrm{Spec}(R) \quad \text{and} \quad \bigcap_{t \in \mathbb{N}_0 \cup \{-\infty\}} C_t(M) = \emptyset.$$

9.4.9 ♯Exercise. Let M be a finitely generated R-module and let S be a multiplicatively closed subset of R. Show that, with the notation of 9.4.8,

$$\mathfrak{c}_t(S^{-1}M) = S^{-1}(\mathfrak{c}_t(M)) \quad \text{for all } t \in \mathbb{N}_0 \cup \{-\infty\}.$$

9.4.10 Proposition. *Let M be a finitely generated R-module, and let $t \in \mathbb{N}_0 \cup \{-\infty\}$. We use the notation of 9.4.8. There exists $n \in \mathbb{N}$ such that, for every proper ideal $\mathfrak{d}$ of R, we have*

$$\mathfrak{c}_t(M)^n H^i_\mathfrak{d}(M) = 0 \quad \text{for all } i < \mathrm{grade}\,\mathfrak{d} - t.$$

Note. In the case when $t = -\infty$, one should interpret the '$- - \infty$' in the above statement as '∞'.

Proof. Let $\mathfrak{p} \in U_t(M)$, so that $\mathrm{proj\,dim}\, M_\mathfrak{p} \leq t$. By 9.4.5, there exists $s_\mathfrak{p} \in R \setminus \mathfrak{p}$ such that, for every proper ideal $\mathfrak{d}$ of R, we have $s_\mathfrak{p} H^i_\mathfrak{d}(M) = 0$ for all $i < \mathrm{grade}\,\mathfrak{d} - \mathrm{proj\,dim}\, M_\mathfrak{p}$, and therefore for all $i < \mathrm{grade}\,\mathfrak{d} - t$.

Set $\mathfrak{g} := \sum_{\mathfrak{p}' \in U_t(M)} s_{\mathfrak{p}'} R$, and observe that, for every proper ideal $\mathfrak{d}$ of R, we have $\mathfrak{g} H^i_\mathfrak{d}(M) = 0$ for all $i < \mathrm{grade}\,\mathfrak{d} - t$. Note that no prime ideal in $U_t(M)$ contains $\mathfrak{g}$. The latter statement implies that $\mathrm{Var}(\mathfrak{g}) \subseteq C_t(M)$, so that $\mathfrak{c}_t(M) \subseteq \sqrt{\mathfrak{g}}$. Hence there exists $n \in \mathbb{N}$ such that $\mathfrak{c}_t(M)^n \subseteq \mathfrak{g}$, and the result follows from this. □

9.4.11 Exercise. Assume that R is a homomorphic image of a regular (commutative Noetherian) ring, and let M be a finitely generated R-module with the property that $\mathrm{Supp}\, M$ has exactly one minimal member. Assume that $\mathfrak{c}$ is an ideal of R such that $M_\mathfrak{p}$ is a Cohen–Macaulay $R_\mathfrak{p}$-module for all $\mathfrak{p} \in \mathrm{Spec}(R) \setminus \mathrm{Var}(\mathfrak{c})$. Prove that there exists $n \in \mathbb{N}$ such that

$$\mathfrak{c}^n H^i_\mathfrak{q}(M) = 0 \quad \text{for all } \mathfrak{q} \in \mathrm{Var}(\mathfrak{c}) \text{ and all } i < \dim_{R_\mathfrak{q}} M_\mathfrak{q}.$$

9.4.12 Exercise. Let the situation be as in Proposition 9.4.10. Prove that there exists $n \in \mathbb{N}$ such that, for every choice of flat ring homomorphism $f : R \to R'$ of commutative Noetherian rings and every choice of proper ideal $\mathfrak{B}'$ of R', we have

$$\mathfrak{c}_t(M)^n H^i_{\mathfrak{B}'}(M \otimes_R R') = 0 \quad \text{for all } i < \mathrm{grade}\,\mathfrak{B}' - t.$$

(Here is a hint: make appropriate modifications to the preparatory results, such as 9.4.1, 9.4.3 and 9.4.5, that were used in our approach to the proof of Proposition 9.4.10.)

9.4.13 Exercise. (The result of this exercise is due to C. L. Huneke.) Assume that R is a homomorphic image of a regular (commutative Noetherian) ring for which there exists a non-zerodivisor $c \in R$ such that the ring R_c is Cohen–Macaulay. Assume that either (a) R is an integral domain, or (b) R is an equidimensional (see [50, p. 250]) local ring.

Use Exercise 9.4.12 above to prove that there exists $n \in \mathbb{N}$ such that, for all choices of $r \in \mathbb{N}_0$ and all choices of an ideal $\mathfrak{B}'$ of the polynomial ring $R[X_1, \ldots, X_r]$ (interpret this as R in the case when $r = 0$), we have

$$c^n H^i_{\mathfrak{B}'}(R[X_1, \ldots, X_r]) = 0 \quad \text{for all } i < \operatorname{ht} \mathfrak{B}'.$$

9.4.14 Lemma. *Let $(R, \mathfrak{m})$ be a regular local ring of dimension d, let M be a finitely generated R-module, and let $\mathfrak{b}$ be an ideal of R. Then, with the notation of 9.4.8, we have $\mathfrak{b} \subseteq \mathfrak{c}_{d - \lambda^{\mathfrak{b}}_{\mathfrak{m}}(M)}(M)$.*

Proof. Set $t := d - \lambda^{\mathfrak{b}}_{\mathfrak{m}}(M)$. It is enough to prove that, for each $\mathfrak{p} \in C_t(M)$, we must have $\mathfrak{b} \subseteq \mathfrak{p}$. Suppose that, for one such $\mathfrak{p}$, we have $\mathfrak{b} \nsubseteq \mathfrak{p}$, and look for a contradiction. Then ($\mathfrak{p} \in \operatorname{Supp} M$ and)

$$\infty > \operatorname{adj}_{\mathfrak{m}} \operatorname{depth} M_{\mathfrak{p}} = \operatorname{depth} M_{\mathfrak{p}} + \operatorname{ht}(\mathfrak{m} + \mathfrak{p})/\mathfrak{p} \geq \lambda^{\mathfrak{b}}_{\mathfrak{m}}(M).$$

Since R is a catenary domain, $\operatorname{ht}(\mathfrak{m} + \mathfrak{p})/\mathfrak{p} = \operatorname{ht} \mathfrak{m}/\mathfrak{p} = d - \operatorname{ht} \mathfrak{p}$. As $R_{\mathfrak{p}}$ is a regular local ring, it follows from the Auslander–Buchsbaum–Serre Theorem that $\operatorname{depth} M_{\mathfrak{p}} + \operatorname{proj dim} M_{\mathfrak{p}} = \dim R_{\mathfrak{p}} = \operatorname{ht} \mathfrak{p}$. Therefore

$$\operatorname{ht} \mathfrak{p} - \operatorname{proj dim} M_{\mathfrak{p}} + d - \operatorname{ht} \mathfrak{p} \geq \lambda^{\mathfrak{b}}_{\mathfrak{m}}(M),$$

so that $\operatorname{proj dim} M_{\mathfrak{p}} \leq t$. This contradiction completes the proof. $\square$

We remind the reader that our present major aim is a proof of the fact that, when R is a homomorphic image of a regular ring, then, with the notation of 9.1.5 and 9.2.2, we have $\lambda^{\mathfrak{b}}_{\mathfrak{a}}(M) \leq f^{\mathfrak{b}}_{\mathfrak{a}}(M)$. In view of 9.1.7 and 9.2.6, it will be enough to do this in the case when R itself is regular. In the light of this, the next result already proves our desired inequality in a special case.

9.4.15 Proposition. *Assume that R is regular, and that $\dim R/\mathfrak{a} = 0$. Let $\mathfrak{b}$ be a second ideal of R, and let M be a finitely generated R-module. Then*

$$\lambda^{\mathfrak{b}}_{\mathfrak{a}}(M) \leq f^{\mathfrak{b}}_{\mathfrak{a}}(M).$$

Proof. Since $\dim R/\mathfrak{a} = 0$, we have $\operatorname{Var}(\mathfrak{a}) = \operatorname{ass} \mathfrak{a}$ is a finite set of maximal ideals of R: let its members be $\mathfrak{m}_1, \ldots, \mathfrak{m}_h$. Consider an integer j with $1 \leq$

$j \leq h$. Set $t_j := \operatorname{ht} \mathfrak{m}_j - \lambda^{\mathfrak{b}R_{\mathfrak{m}_j}}_{\mathfrak{m}_j R_{\mathfrak{m}_j}}(M_{\mathfrak{m}_j})$. Since $\operatorname{ht} \mathfrak{m}_j = \operatorname{grade} \mathfrak{m}_j R_{\mathfrak{m}_j}$, we see that $\operatorname{grade} \mathfrak{m}_j R_{\mathfrak{m}_j} - t_j = \lambda^{\mathfrak{b}R_{\mathfrak{m}_j}}_{\mathfrak{m}_j R_{\mathfrak{m}_j}}(M_{\mathfrak{m}_j})$. By 9.4.10, there exists $n_j \in \mathbb{N}$ such that

$$\mathfrak{c}_{t_j}(M_{\mathfrak{m}_j})^{n_j} H^i_{\mathfrak{m}_j R_{\mathfrak{m}_j}}(M_{\mathfrak{m}_j}) = 0 \quad \text{for all } i < \lambda^{\mathfrak{b}R_{\mathfrak{m}_j}}_{\mathfrak{m}_j R_{\mathfrak{m}_j}}(M_{\mathfrak{m}_j}).$$

By 9.4.14, we have $\mathfrak{b}R_{\mathfrak{m}_j} \subseteq \mathfrak{c}_{t_j}(M_{\mathfrak{m}_j})$.

Set $n := \max\{n_1, \ldots, n_h\}$. We can use 1.1.3 and 4.3.3 to see that

$$H^i_{\mathfrak{m}_j R_{\mathfrak{m}_j}}(M_{\mathfrak{m}_j}) = H^i_{\mathfrak{a}R_{\mathfrak{m}_j}}(M_{\mathfrak{m}_j}) \cong (H^i_{\mathfrak{a}}(M))_{\mathfrak{m}_j} \quad \text{for all } i \in \mathbb{N}_0.$$

Also

$$\lambda^{\mathfrak{b}R_{\mathfrak{m}_j}}_{\mathfrak{m}_j R_{\mathfrak{m}_j}}(M_{\mathfrak{m}_j}) = \lambda^{\mathfrak{b}R_{\mathfrak{m}_j}}_{\mathfrak{a}R_{\mathfrak{m}_j}}(M_{\mathfrak{m}_j}) \geq \lambda^{\mathfrak{b}}_{\mathfrak{a}}(M),$$

by 9.2.3(i) and 9.2.5. It therefore follows that

$$(\mathfrak{b}^n H^i_{\mathfrak{a}}(M))_{\mathfrak{m}_j} \cong (\mathfrak{b}R_{\mathfrak{m}_j})^n H^i_{\mathfrak{m}_j R_{\mathfrak{m}_j}}(M_{\mathfrak{m}_j}) = 0 \text{ for } 1 \leq j \leq h, \ i < \lambda^{\mathfrak{b}}_{\mathfrak{a}}(M).$$

Since a local cohomology module with respect to $\mathfrak{a}$ is $\mathfrak{a}$-torsion, the support of such a local cohomology module must be contained in $\operatorname{Var}(\mathfrak{a}) = \{\mathfrak{m}_1, \ldots, \mathfrak{m}_h\}$. Hence

$$\operatorname{Supp}\left(\mathfrak{b}^n H^i_{\mathfrak{a}}(M)\right) = \emptyset \quad \text{for all } i < \lambda^{\mathfrak{b}}_{\mathfrak{a}}(M).$$

In view of the definition of $f^{\mathfrak{b}}_{\mathfrak{a}}(M)$ (see 9.1.5), this completes the proof. $\qquad\square$

9.4.16 Theorem. *Assume that R is a homomorphic image of a regular (commutative Noetherian) ring. Let $\mathfrak{b}$ be a second ideal of R, and let M be a finitely generated R-module. Then*

$$\lambda^{\mathfrak{b}}_{\mathfrak{a}}(M) \leq f^{\mathfrak{b}}_{\mathfrak{a}}(M).$$

Proof. In view of 9.1.7 and 9.2.6, we can, and do, assume that R itself is regular.

We suppose that $\lambda^{\mathfrak{b}}_{\mathfrak{a}}(M) > f^{\mathfrak{b}}_{\mathfrak{a}}(M)$ and look for a contradiction. Let $\mathcal{I}(R)$ denote the set of all ideals of R. Since R is Noetherian, we can, and do, assume that $\mathfrak{a}$ is a maximal member of the set

$$\{\mathfrak{a}' \in \mathcal{I}(R) : \lambda^{\mathfrak{b}}_{\mathfrak{a}'}(M) > f^{\mathfrak{b}}_{\mathfrak{a}'}(M)\}.$$

Note that $\dim R/\mathfrak{a} > 0$, by 9.4.15. Set $\lambda := \lambda^{\mathfrak{b}}_{\mathfrak{a}}(M)$, and let $\mathfrak{q}$ be a minimal prime of $\mathfrak{a}$ such that $\operatorname{ht} \mathfrak{q} = \operatorname{ht} \mathfrak{a}$. By 9.4.14, 9.4.9, 9.2.5, 9.2.3(i) and 9.4.8(iv),

$$\mathfrak{b}R_{\mathfrak{q}} \subseteq \mathfrak{c}_{\operatorname{ht}\mathfrak{a}-\lambda^{\mathfrak{b}R_{\mathfrak{q}}}_{\mathfrak{q}R_{\mathfrak{q}}}(M_{\mathfrak{q}})}(M_{\mathfrak{q}}) = \left(\mathfrak{c}_{\operatorname{ht}\mathfrak{a}-\lambda^{\mathfrak{b}R_{\mathfrak{q}}}_{\mathfrak{a}R_{\mathfrak{q}}}(M_{\mathfrak{q}})}(M)\right)_{\mathfrak{q}} \subseteq (\mathfrak{c}_{\operatorname{ht}\mathfrak{a}-\lambda}(M))_{\mathfrak{q}}.$$

Therefore, there exists $u \in R \setminus \mathfrak{q}$ such that $\mathfrak{b}R_u \subseteq (\mathfrak{c}_{\operatorname{ht}\mathfrak{a}-\lambda}(M))R_u$. Since

$\dim R/\mathfrak{a} > 0$, there exists $\mathfrak{s} \in \mathrm{Var}(\mathfrak{a})$ which is not a minimal prime of $\mathfrak{a}$: let $v \in \mathfrak{s} \setminus \mathfrak{q}$ and put $s := uv$. Note that (by 9.4.9 again)

$$\mathfrak{a} \subset \mathfrak{a} + Rs \subset R \quad \text{and} \quad \mathfrak{b}R_s \subseteq (\mathfrak{c}_{\mathrm{ht}\,\mathfrak{a}-\lambda}(M))_s = \mathfrak{c}_{\mathrm{ht}(\mathfrak{a}R_s)-\lambda}(M_s).$$

Since $\mathrm{ht}\,\mathfrak{a}R_s = \mathrm{grade}\,\mathfrak{a}R_s$, it follows from 9.4.10 that there exists $n \in \mathbb{N}$ such that

$$\left(\mathfrak{c}_{\mathrm{ht}(\mathfrak{a}R_s)-\lambda}(M_s)\right)^n H^i_{\mathfrak{a}R_s}(M_s) = 0 \quad \text{for all } i < \lambda.$$

Hence $\mathfrak{b}^n H^i_{\mathfrak{a}R_s}(M_s) = 0$ for all $i < \lambda$ (when $H^i_{\mathfrak{a}R_s}(M_s)$ is regarded as an R-module in the natural way). Therefore, by the Independence Theorem 4.2.1, we have $\mathfrak{b}^n H^i_{\mathfrak{a}}(M_s) = 0$ for all $i < \lambda$.

Since $\mathfrak{a} \subset \mathfrak{a} + Rs$, it follows from the 'maximality' assumption about $\mathfrak{a}$ made in the second paragraph of this proof that $\lambda^{\mathfrak{b}}_{\mathfrak{a}+Rs}(M) \leq f^{\mathfrak{b}}_{\mathfrak{a}+Rs}(M)$. Now $\lambda = \lambda^{\mathfrak{b}}_{\mathfrak{a}}(M) \leq \lambda^{\mathfrak{b}}_{\mathfrak{a}+Rs}(M)$, by 9.2.3(ii), and so there exists $n' \in \mathbb{N}$ such that $\mathfrak{b}^{n'} H^i_{\mathfrak{a}+Rs}(M) = 0$ for all $i < \lambda$.

By 8.1.2(i), there is, for each $i < \lambda$, an exact sequence

$$H^i_{\mathfrak{a}+Rs}(M) \longrightarrow H^i_{\mathfrak{a}}(M) \longrightarrow H^i_{\mathfrak{a}}(M_s).$$

It now follows from 9.1.1 that $\mathfrak{b} \subseteq \sqrt{(0 : H^i_{\mathfrak{a}}(M))}$ for all $i < \lambda$. This shows that $\lambda = \lambda^{\mathfrak{b}}_{\mathfrak{a}}(M) \leq f^{\mathfrak{b}}_{\mathfrak{a}}(M)$, and this contradiction completes the proof. $\square$

9.5 The main theorems

We can now put together Theorems 9.3.7 and 9.4.16 to prove Faltings' Annihilator Theorem.

9.5.1 Faltings' Annihilator Theorem. (See G. Faltings [14].) *Assume that R is a homomorphic image of a regular (commutative Noetherian) ring. Let $\mathfrak{b}$ be a second ideal of R, and let M be a finitely generated R-module. Then*

$$\lambda^{\mathfrak{b}}_{\mathfrak{a}}(M) = f^{\mathfrak{b}}_{\mathfrak{a}}(M).$$

Proof. This is now immediate from Theorems 9.3.7 and 9.4.16. $\square$

The special case of Faltings' Annihilator Theorem in which $\mathfrak{a} = \mathfrak{b}$ is Grothendieck's Finiteness Theorem.

9.5.2 Grothendieck's Finiteness Theorem. (See A. Grothendieck [26, Exposé VIII, Corollaire 2.3].) *Assume that R is a homomorphic image of a regular (commutative Noetherian) ring, and let M be a finitely generated R-module. Then*

$$\lambda^{\mathfrak{a}}_{\mathfrak{a}}(M) = f_{\mathfrak{a}}(M).$$

In other words, there exists $i \in \mathbb{N}$ such that $H_\mathfrak{a}^i(M)$ is not finitely generated if and only if $\operatorname{Supp} M \not\subseteq \operatorname{Var}(\mathfrak{a})$; moreover, when this is the case, the least $i \in \mathbb{N}$ such that $H_\mathfrak{a}^i(M)$ is not finitely generated is equal to

$$\min\{\operatorname{depth} M_\mathfrak{p} + \operatorname{ht}(\mathfrak{a}+\mathfrak{p})/\mathfrak{p} : \mathfrak{p} \in \operatorname{Supp} M \setminus \operatorname{Var}(\mathfrak{a})\}.$$

Proof. Put $\mathfrak{a} = \mathfrak{b}$ in Faltings' Annihilator Theorem 9.5.1, and use 9.1.2 for the last part. □

9.5.3 Example. Let K be a field, and let $R = K[X, Y^2, XY, Y^3]$, a sub-ring of the ring of polynomials $K[X,Y]$. Let $\mathfrak{m}$ denote the maximal ideal (X, Y^2, XY, Y^3) of R.

In the ring of polynomials $R[Z]$, let $\mathfrak{N} = \mathfrak{m}R[Z] + ZR[Z]$. We first calculate the invariant $\lambda_\mathfrak{N}^\mathfrak{N}(R[Z])$. Let $\mathfrak{P} \in \operatorname{Spec}(R[Z]) \setminus \operatorname{Var}(\mathfrak{N})$: we calculate $\operatorname{adj}_\mathfrak{N} \operatorname{depth} R[Z]_\mathfrak{P}$. Two cases arise: if $\mathfrak{P} = 0$, then the fact that $\operatorname{ht} \mathfrak{N} = 3$ implies that $\operatorname{adj}_\mathfrak{N} \operatorname{depth} R[Z]_0 = 3$; if $\mathfrak{P} \neq 0$, then the fact that $R[Z]_\mathfrak{P}$ is a local domain which is not a field ensures that

$$\operatorname{adj}_\mathfrak{N} \operatorname{depth} R[Z]_\mathfrak{P} = \operatorname{depth} R[Z]_\mathfrak{P} + \operatorname{ht}(\mathfrak{N}+\mathfrak{P})/\mathfrak{P} \geq 1 + 1 = 2.$$

Thus $\lambda_\mathfrak{N}^\mathfrak{N}(R[Z]) \geq 2$. To see that $\lambda_\mathfrak{N}^\mathfrak{N}(R[Z])$ is exactly 2, note that $\mathfrak{m}R[Z] \in \operatorname{Spec}(R[Z]) \setminus \operatorname{Var}(\mathfrak{N})$. Now $\operatorname{ht} \mathfrak{N}/\mathfrak{m}R[Z] = 1$ since the ideal $\mathfrak{N}/\mathfrak{m}R[Z]$ is principal. Moreover, it follows from 2.3.6(v) that $H_\mathfrak{m}^1(R) \neq 0$; hence, since R is a domain, we can deduce from 6.2.7 that $\operatorname{grade} \mathfrak{m} = 1$. Now X is a non-zerodivisor on R; the fact that $\mathfrak{m}$ is a maximal ideal of R ensures that $\mathfrak{m} \in \operatorname{ass}_R RX$; and it is now easy to deduce that $\operatorname{depth} R[Z]_{\mathfrak{m}R[Z]} = 1$. Hence

$$\operatorname{adj}_\mathfrak{N} \operatorname{depth} R[Z]_{\mathfrak{m}R[Z]} = \operatorname{depth} R[Z]_{\mathfrak{m}R[Z]} + \operatorname{ht} \mathfrak{N}/\mathfrak{m}R[Z] = 1 + 1 = 2.$$

Therefore $\lambda_\mathfrak{N}^\mathfrak{N}(R[Z]) = 2$.

Grothendieck's Finiteness Theorem 9.5.2 thus tells us that the finiteness dimension $f_\mathfrak{N}(R[Z])$ is 2. Let us show this directly.

Since Z, X is an $R[Z]$-sequence contained in $\mathfrak{N}$, it follows from 6.2.7 that $H_\mathfrak{N}^0(R[Z]) = H_\mathfrak{N}^1(R[Z]) = 0$. Further, for each $t \in \mathbb{N}$, the exact sequence

$$0 \longrightarrow R[Z] \xrightarrow{Z^t} R[Z] \longrightarrow R[Z]/Z^t R[Z] \longrightarrow 0$$

induces an exact sequence

$$0 \longrightarrow H_\mathfrak{N}^1(R[Z]/Z^t R[Z]) \longrightarrow H_\mathfrak{N}^2(R[Z]) \xrightarrow{Z^t} H_\mathfrak{N}^2(R[Z]);$$

hence $(0 :_{H_\mathfrak{N}^2(R[Z])} Z^t) \cong H_\mathfrak{N}^1(R[Z]/Z^t R[Z])$. But, by 1.1.3 and 4.2.2,

$$H_\mathfrak{N}^1(R[Z]/Z^t R[Z]) = H_{\mathfrak{m}R[Z]+Z^t R[Z]}^1(R[Z]/Z^t R[Z])$$
$$\cong H_{\mathfrak{m}R[Z]}^1(R[Z]/Z^t R[Z]).$$

The Independence Theorem 4.2.1 shows that, as R-modules,

$$H^1_{\mathfrak{m}R[Z]}(R[Z]/Z^t R[Z]) \cong H^1_{\mathfrak{m}}(R[Z]/Z^t R[Z]) \cong H^1_{\mathfrak{m}}(R^t);$$

this is isomorphic to the R-module K^t, by 2.3.6(v). Therefore

$$\left(0 :_{H^2_{\mathfrak{N}}(R[Z])} Z^t\right) \subset \left(0 :_{H^2_{\mathfrak{N}}(R[Z])} Z^{t+1}\right) \quad \text{for all } t \in \mathbb{N},$$

and so $H^2_{\mathfrak{N}}(R[Z])$ is not finitely generated. Thus we have shown directly that the finiteness dimension $f_{\mathfrak{N}}(R[Z])$ is 2.

9.5.4 Exercise. Assume that the local ring $(R, \mathfrak{m})$ is a homomorphic image of a regular local ring, and let M be a finitely generated R-module.

(i) Show that

$$f_{\mathfrak{m}}(M) = \inf \left\{ \operatorname{depth} M_{\mathfrak{p}} + 1 : \mathfrak{p} \in \operatorname{Spec}(R) \text{ and } \dim R/\mathfrak{p} = 1 \right\}.$$

(ii) Suppose that $n := \dim M > 0$. Prove that $f_{\mathfrak{m}}(M) = n$ if and only if $M_{\mathfrak{p}}$ is a Cohen–Macaulay $R_{\mathfrak{p}}$-module of dimension $n - \dim R/\mathfrak{p}$ for all $\mathfrak{p} \in \operatorname{Supp} M \setminus \{\mathfrak{m}\}$.

9.5.5 Exercise. Assume that $(R, \mathfrak{m})$ is a local domain which is a homomorphic image of a regular local ring; assume also that $d := \dim R > 0$, and that $R_{\mathfrak{p}}$ is a Cohen–Macaulay ring for all $\mathfrak{p} \in \operatorname{Spec}(R) \setminus \{\mathfrak{m}\}$.

Show that, if $H^i_{\mathfrak{m}}(R) = 0$ for all integers i such that $1 < i < d$, then there exists a Cohen–Macaulay subring of the quotient field of R which contains R and is a finitely generated R-module. (Here is a hint: think of $D_{\mathfrak{m}}(R)$.)

9.5.6 Exercise. Assume that $(R, \mathfrak{m})$ is a local domain which is a homomorphic image of a regular local ring, and that the ideal $\mathfrak{a}$ is proper. Show that $f_{\mathfrak{a}}(R) > 1$ if and only if $\operatorname{ht} \mathfrak{a} \neq 1$.

9.5.7 Exercise. Assume that $(R, \mathfrak{m})$ is local, and let M be a non-zero finitely generated R-module of dimension $n > 0$. We say that M is a *generalized Cohen–Macaulay R-module* precisely when $H^i_{\mathfrak{m}}(M)$ is finitely generated for all $i \neq n$. (Such modules were called 'quasi-Cohen–Macaulay modules' by P. Schenzel in [72, p. 238]; these modules were investigated by Schenzel, N. V. Trung and N. T. Cuong in [76], and, since the publication of that paper, the terminology 'generalized Cohen–Macaulay module' seems to have become more commonplace than 'quasi-Cohen–Macaulay module'.)

(i) Show that, if M is a generalized Cohen–Macaulay R-module, then we have $\dim R/\mathfrak{p} = n$ for all $\mathfrak{p} \in \operatorname{Ass} M \setminus \{\mathfrak{m}\}$ and $M_{\mathfrak{q}}$ is a Cohen–Macaulay $R_{\mathfrak{q}}$-module for all $\mathfrak{q} \in \operatorname{Supp} M \setminus \{\mathfrak{m}\}$.

(ii) Show conversely, that, if

(a) R is a homomorphic image of a regular ring,

(b) $\dim R/\mathfrak{p} = n$ whenever $\mathfrak{p}$ is a minimal member of $\operatorname{Supp} M$, and

(c) $M_{\mathfrak{q}}$ is a Cohen–Macaulay $R_{\mathfrak{q}}$-module for all $\mathfrak{q} \in \operatorname{Supp} M \setminus \{\mathfrak{m}\}$,

then M is a generalized Cohen–Macaulay R-module.

9.5.8 Exercise. Assume that $(R, \mathfrak{m})$ is local, and that the non-zero finitely generated R-module M of dimension $n > 0$ is a generalized Cohen–Macaulay R-module (see 9.5.7 above). Note that $\mathfrak{m} \notin \operatorname{Ass}(M/\Gamma_{\mathfrak{m}}(M))$ and $M_{\mathfrak{p}} \cong (M/\Gamma_{\mathfrak{m}}(M))_{\mathfrak{p}}$ for all $\mathfrak{p} \in \operatorname{Spec}(R) \setminus \{\mathfrak{m}\}$.

   (i) Let $r \in \mathfrak{m}$ be such that $\dim M/rM = n - 1$. Show that r is a non-zerodivisor on $M/\Gamma_{\mathfrak{m}}(M)$ and that, if $n > 1$, then M/rM is a generalized Cohen–Macaulay R-module.

  (ii) Use part (i) and induction on n to show that every saturated chain of prime ideals from a minimal member of $\operatorname{Supp} M$ (as smallest term) to $\mathfrak{m}$ (as largest term) has length n.

 (iii) Deduce from part (ii) that the ring $R/(0 : M)$ is catenary.

9.5.9 Definitions. Assume that $(R, \mathfrak{m})$ is local, and let M be a non-zero finitely generated R-module of dimension $n > 0$.

   (i) By a *system of parameters for M* we mean a sequence $(r_i)_{i=1}^{n}$ of n elements of $\mathfrak{m}$ such that $M/\sum_{i=1}^{n} r_i M$ has finite length. We say that $r_1, \ldots, r_n$ *form a system of parameters for M* precisely when $(r_i)_{i=1}^{n}$ is a system of parameters for M. By *a parameter for M*, we mean a member of a system of parameters for M.

  (ii) Let $(r_i)_{i=1}^{n}$ be a system of parameters for M; let $\mathfrak{q} := \sum_{i=1}^{n} r_i R$. We say that $(r_i)_{i=1}^{n}$ is a *standard system of parameters for M* precisely when

$$\mathfrak{q} H_{\mathfrak{m}}^{j}\left(M/\sum_{i=1}^{k} r_i M\right) = 0 \quad \text{for all } j, k \in \mathbb{N}_0 \text{ with } j + k < n.$$

(See Stückrad and Vogel [84, p. 261].)

9.5.10 Exercise. Assume $(R, \mathfrak{m})$ is local, and let M be a non-zero finitely generated R-module of dimension $n > 1$. Let $(v_i)_{i=1}^{n}$ be a system of parameters for M, let $t \in \mathbb{N}$ with $t < n$, and let $M' := M/\sum_{i=1}^{t} v_i M$. Let $r_1, \ldots, r_{n-t} \in \mathfrak{m}$. Show that

   (i) $\dim M' = n - t$;

  (ii) $v_1, \ldots, v_t, r_1, \ldots, r_{n-t}$ form a system of parameters for M if and only if $(r_i)_{i=1}^{n-t}$ is a system of parameters for M';

 (iii) if $(r_i)_{i=1}^{n-t}$ is a system of parameters for M' and $h \in \mathbb{N}_0$ is such that $h < n - t$, then $M'/\sum_{i=1}^{h} r_i M' \cong M/(\sum_{i=1}^{t} v_i M + \sum_{i=1}^{h} r_i M)$;

(iv) if every system of parameters for M is standard, then every system of parameters for M' is standard.

9.5.11 Exercise. Assume that $(R, \mathfrak{m})$ is local, and that M is a generalized Cohen–Macaulay R-module of dimension $n > 1$. Let r be a parameter for M, and write $\overline{M} = M/\Gamma_{\mathfrak{m}}(M)$. Show that

(i) $(0 : H_{\mathfrak{m}}^i(\overline{M}))(0 : H_{\mathfrak{m}}^{i+1}(\overline{M})) \subseteq (0 : H_{\mathfrak{m}}^i(\overline{M}/r\overline{M}))$ for all $i = 1, \ldots,$ $n - 2$;

(ii) $(0 : H_{\mathfrak{m}}^0(M))(0 : H_{\mathfrak{m}}^1(M/(0 :_M r))) \subseteq (0 : H_{\mathfrak{m}}^0(M/rM))$;

(iii) $(0 :_M r)$ is $\mathfrak{m}$-torsion; and

(iv) $(0 : H_{\mathfrak{m}}^i(M))(0 : H_{\mathfrak{m}}^{i+1}(M)) \subseteq (0 : H_{\mathfrak{m}}^i(M/rM))$ for all $i = 0, 1, \ldots,$ $n - 2$.

9.5.12 Exercise. Assume that $(R, \mathfrak{m})$ is local, and that M is a generalized Cohen–Macaulay R-module of dimension $n > 0$. Let

$$\mathfrak{q}_M := \left(0 : H_{\mathfrak{m}}^0(M)\right)^{\binom{n-1}{0}} \left(0 : H_{\mathfrak{m}}^1(M)\right)^{\binom{n-1}{1}} \ldots \left(0 : H_{\mathfrak{m}}^{n-1}(M)\right)^{\binom{n-1}{n-1}}.$$

Prove that every system of parameters for M composed of elements in $\mathfrak{q}_M$ is standard (see 9.5.9). (Here is a hint: use Exercises 9.5.11 and 9.5.8, together with induction.)

9.5.13 Exercise. Assume $(R, \mathfrak{m})$ is local, and let M be a non-zero finitely generated R-module of dimension $n > 0$. Assume that, for every system of parameters $(r_i)_{i=1}^n$ for M,

$$\mathfrak{m}\Gamma_{r_{k+1}R}\left(M/\textstyle\sum_{i=1}^k r_i M\right) = 0 \quad \text{for all } k \in \mathbb{N}_0 \text{ with } k < n.$$

(i) Show that $(0 :_M r) = \Gamma_{\mathfrak{m}}(M)$ for each parameter r for M.

(ii) Use part (i) and induction to prove that $\mathfrak{m}H_{\mathfrak{m}}^j(M) = 0$ for all $j = 0, \ldots, n - 1$.

9.5.14 Definitions and Exercise. Assume that $(R, \mathfrak{m})$ is local, and let M be a non-zero finitely generated R-module of dimension $n > 0$.

Let $a_1, \ldots, a_h \in \mathfrak{m}$. We say that $a_1, \ldots, a_h$ is a *weak M-sequence* precisely when

$$\left(\textstyle\sum_{j=1}^{i-1} a_j M :_M a_i\right) = \left(\textstyle\sum_{j=1}^{i-1} a_j M :_M \mathfrak{m}\right) \quad \text{for all } i = 1, \ldots, h.$$

We say that M is a *Buchsbaum R-module* precisely when every sequence $r_1, \ldots, r_n$ forming a system of parameters for M is a weak M-sequence; we say that R is a *Buchsbaum ring* if and only if it is a Buchsbaum module over itself.

Show that M is a Buchsbaum R-module if and only if, for every system of parameters $(r_i)_{i=1}^n$ for M,

$$\mathfrak{m}\Gamma_{r_{k+1}R}\left(M/\sum_{i=1}^k r_i M\right) = 0 \quad \text{for all } k \in \mathbb{N}_0 \text{ with } k < n.$$

There is an extensive theory of Buchsbaum rings and modules: see Stückrad and Vogel [84]. The following final exercise in this section provides some alternative characterizations of Buchsbaum modules.

9.5.15 Exercise.   Assume $(R, \mathfrak{m})$ is local, and let M be a non-zero finitely generated R-module of dimension $n > 0$.

Prove that the following conditions on M are equivalent:

(i) for every system of parameters $(r_i)_{i=1}^n$ for M,

$$\mathfrak{m}H_{\mathfrak{m}}^j\left(M/\sum_{i=1}^k r_i M\right) = 0 \quad \text{for all } j, k \in \mathbb{N}_0 \text{ with } j + k < n;$$

(ii) every system of parameters for M is standard (see 9.5.9);
(iii) M is a Buchsbaum R-module (see 9.5.14).

(It is perhaps appropriate to give some hints. For the implication '(ii) $\Rightarrow$ (iii)', note that $\mathfrak{m}$ can be generated by parameters for M, and show that, if $r_1, \ldots, r_n$ form a system of parameters for M, then condition (ii) implies that $M/\sum_{i=1}^k r_i M$ is a generalized Cohen–Macaulay R-module (see Exercise 9.5.7) for all $k = 0, \ldots, n - 1$. For the implication '(iii) $\Rightarrow$ (i)', use 9.5.14 and 9.5.13.)

9.6 Extensions

The remaining exercises in this chapter are directed at those readers who are familiar with some fairly advanced concepts in commutative algebra, including formal fibres of local rings, universally catenary rings, and L. J. Ratliff's Theorem [50, Theorem 31.7] that a local ring $(R, \mathfrak{m})$ is universally catenary if and only if, for every $\mathfrak{p} \in \operatorname{Spec}(R)$ and for every minimal prime $\mathfrak{P}$ of the ideal $\mathfrak{p}\widehat{R}$ of $\widehat{R}$, we have $\dim \widehat{R}/\mathfrak{P} = \dim R/\mathfrak{p}$.

So far in this chapter, we have established the main results, that is, the Annihilator and Finiteness Theorems, under the hypothesis that the underlying ring is a homomorphic image of a regular ring. The remaining exercises in this chapter, in conjunction with the Local-global Principle 9.6.2 for Finiteness Dimensions, show that the result of the Finiteness Theorem holds under slightly weaker hypotheses, namely under the assumption that the underlying ring is universally catenary and has the property that all the formal fibres of all its localizations are Cohen–Macaulay rings. We consider the Local-global Principle, which is a consequence of the theorem of Faltings presented in 9.6.1 below, to be of considerable independent interest.

9.6.1 Theorem. (See G. Faltings [16, Satz 1].) *Let M be a finitely generated R-module, and let $t \in \mathbb{N}$. Then the following statements are equivalent:*

(i) *$H^i_{\mathfrak{a}}(M)$ is finitely generated for all $i < t$;*

(ii) *$H^i_{\mathfrak{a}R_{\mathfrak{p}}}(M_{\mathfrak{p}})$ is a finitely generated $R_{\mathfrak{p}}$-module for all $i < t$ and all $\mathfrak{p} \in \mathrm{Spec}(R)$.*

Proof. (i) $\Rightarrow$ (ii) Since 4.3.3 shows that $H^i_{\mathfrak{a}R_{\mathfrak{p}}}(M_{\mathfrak{p}}) \cong (H^i_{\mathfrak{a}}(M))_{\mathfrak{p}}$ for all $i \in \mathbb{N}_0$ and all $\mathfrak{p} \in \mathrm{Spec}(R)$, this implication is clear.

(ii) $\Rightarrow$ (i) We use induction on t. When $t = 1$, there is nothing to prove, since $H^0_{\mathfrak{a}}(M) = \Gamma_{\mathfrak{a}}(M)$ is a submodule of M, and so is finitely generated. So suppose that $t > 1$ and that the result has been proved for smaller values of t. By this assumption, $H^i_{\mathfrak{a}}(M)$ is finitely generated for $i = 0, 1, \ldots, t - 2$, and it only remains for us to prove that $H^{t-1}_{\mathfrak{a}}(M)$ is finitely generated.

Set $\overline{M} := M/\Gamma_{\mathfrak{a}}(M)$. Then, for each $\mathfrak{p} \in \mathrm{Spec}(R)$,

$$\overline{M}_{\mathfrak{p}} \cong M_{\mathfrak{p}}/(\Gamma_{\mathfrak{a}}(M))_{\mathfrak{p}} = M_{\mathfrak{p}}/\Gamma_{\mathfrak{a}R_{\mathfrak{p}}}(M_{\mathfrak{p}}),$$

and so it follows from 2.1.7(iii) that $H^i_{\mathfrak{a}R_{\mathfrak{p}}}(M_{\mathfrak{p}}) \cong H^i_{\mathfrak{a}R_{\mathfrak{p}}}(\overline{M}_{\mathfrak{p}})$ for all $i \in \mathbb{N}$ and all $\mathfrak{p} \in \mathrm{Spec}(R)$. It follows that $\overline{M}$ also satisfies condition (ii) in the statement of the theorem. Moreover, it would be sufficient for us to prove that $H^i_{\mathfrak{a}}(\overline{M})$ is finitely generated for all $i < t$, since $H^i_{\mathfrak{a}}(M) \cong H^i_{\mathfrak{a}}(\overline{M})$ for all $i \in \mathbb{N}$ (by 2.1.7(iii)) and it is automatic that $H^0_{\mathfrak{a}}(M)$ is finitely generated. Now $\overline{M}$ is an $\mathfrak{a}$-torsion-free R-module, by 2.1.2. Hence we can, and do, assume that M is an $\mathfrak{a}$-torsion-free R-module.

We now use 2.1.1(ii) to deduce that $\mathfrak{a}$ contains an element r which is a non-zerodivisor on M. Let $n \in \mathbb{N}$ and let $\mathfrak{p} \in \mathrm{Spec}(R)$. The localization of the exact sequence $0 \longrightarrow M \xrightarrow{r^n} M \longrightarrow M/r^n M \longrightarrow 0$ at $\mathfrak{p}$ induces, for each $i \in \mathbb{N}_0$, an exact sequence

$$H^{i-1}_{\mathfrak{a}R_{\mathfrak{p}}}(M_{\mathfrak{p}}) \longrightarrow H^{i-1}_{\mathfrak{a}R_{\mathfrak{p}}}((M/r^n M)_{\mathfrak{p}}) \longrightarrow H^i_{\mathfrak{a}R_{\mathfrak{p}}}(M_{\mathfrak{p}}).$$

When $i < t$, the two outer modules in this last exact sequence are finitely generated, and therefore so also is the middle one. Thus $H^{i-1}_{\mathfrak{a}R_{\mathfrak{p}}}((M/r^n M)_{\mathfrak{p}})$ is a finitely generated $R_{\mathfrak{p}}$-module for all $i < t$ and all $\mathfrak{p} \in \mathrm{Spec}(R)$. Therefore, by the inductive hypothesis, $H^{i-1}_{\mathfrak{a}}(M/r^n M)$ is finitely generated for all $i < t$. Since the exact sequence $0 \longrightarrow M \xrightarrow{r^n} M \longrightarrow M/r^n M \longrightarrow 0$ induces an exact sequence

$$H_{\mathfrak{a}}^{t-2}(M/r^n M) \longrightarrow H_{\mathfrak{a}}^{t-1}(M) \xrightarrow{r^n} H_{\mathfrak{a}}^{t-1}(M),$$

it therefore follows that $(0 :_{H_{\mathfrak{a}}^{t-1}(M)} r^n)$ is finitely generated.

Set $H := H_{\mathfrak{a}}^{t-1}(M)$, and $H_n := (0 :_H r^n)$ for all $n \in \mathbb{N}$. Recall that our aim is to prove that H is finitely generated; we have just proved that H_n is finitely generated for all $n \in \mathbb{N}$. Note that

$$H_1 \subseteq H_2 \subseteq \cdots \subseteq H_n \subseteq H_{n+1} \subseteq \cdots,$$

and that, since $r \in \mathfrak{a}$ and H is an $\mathfrak{a}$-torsion R-module, $H = \bigcup_{n \in \mathbb{N}} H_n$. We shall achieve our aim by showing that there exists $k \in \mathbb{N}$ such that $H_k = H_{k+i}$ for all $i \in \mathbb{N}$: it will then follow that $H = H_k$, and the latter module is finitely generated.

For each $n \in \mathbb{N}$, let U_n denote the open subset $\mathrm{Spec}(R) \setminus \mathrm{Supp}(H_{n+1}/H_n)$ of $\mathrm{Spec}(R)$. The next two stages in our proof aim to show that $\mathrm{Spec}(R) = \bigcup_{n \in \mathbb{N}} U_n$ and

$$U_1 \subseteq U_2 \subseteq \cdots \subseteq U_i \subseteq U_{i+1} \subseteq \cdots.$$

Let $\mathfrak{p} \in \mathrm{Spec}(R)$. Then $H_{\mathfrak{p}} = (H_{\mathfrak{a}}^{t-1}(M))_{\mathfrak{p}} \cong H_{\mathfrak{a}R_{\mathfrak{p}}}^{t-1}(M_{\mathfrak{p}})$, and this is a finitely generated, $\mathfrak{a}R_{\mathfrak{p}}$-torsion, $R_{\mathfrak{p}}$-module. Therefore, there exists $n \in \mathbb{N}$ such that $H_{\mathfrak{p}} = (0 :_{H_{\mathfrak{p}}} r^n/1)$. Since

$$(0 :_{H_{\mathfrak{p}}} r^n/1) \subseteq (0 :_{H_{\mathfrak{p}}} r^{n+1}/1) \subseteq H_{\mathfrak{p}},$$

it follows that $(0 :_{H_{\mathfrak{p}}} r^n/1) = (0 :_{H_{\mathfrak{p}}} r^{n+1}/1)$. But $(0 :_{H_{\mathfrak{p}}} r^i/1) = (H_i)_{\mathfrak{p}}$ for all $i \in \mathbb{N}$. Therefore $(H_{n+1}/H_n)_{\mathfrak{p}} = 0$, and so $\mathfrak{p} \in U_n$. It follows that $\mathrm{Spec}(R) = \bigcup_{n \in \mathbb{N}} U_n$.

Next, let $i \in \mathbb{N}$ and $\mathfrak{p} \in U_i$. Then $(0 :_{H_{\mathfrak{p}}} r^i/1) = (0 :_{H_{\mathfrak{p}}} r^{i+1}/1)$, and so

$$(0 :_{H_{\mathfrak{p}}} r^{i+2}/1) = \big((0 :_{H_{\mathfrak{p}}} r^{i+1}/1) :_{H_{\mathfrak{p}}} r/1\big)$$
$$= \big((0 :_{H_{\mathfrak{p}}} r^i/1) :_{H_{\mathfrak{p}}} r/1\big) = (0 :_{H_{\mathfrak{p}}} r^{i+1}/1).$$

Thus $(H_{i+2}/H_{i+1})_{\mathfrak{p}} = 0$, and so $\mathfrak{p} \in U_{i+1}$. Hence $U_i \subseteq U_{i+1}$ for all $i \in \mathbb{N}$.

Thus the sets U_i $(i \in \mathbb{N})$ form an ascending open cover of the quasi-compact topological space $\mathrm{Spec}(R)$. Therefore there exists $k \in \mathbb{N}$ such that $U_k = U_{k+i} = \mathrm{Spec}(R)$ for all $i \in \mathbb{N}$. Thus

$$\mathrm{Supp}(H_{k+1}/H_k) = \mathrm{Supp}(H_{k+i+1}/H_{k+i}) = \emptyset \quad \text{for all } i \in \mathbb{N},$$

so that $H_k = H_{k+i}$ for all $i \in \mathbb{N}$. Thus $H = H_k$, and so is finitely generated. This completes the inductive step. $\qquad\qquad\qquad\qquad\qquad\qquad\qquad\qquad\square$

9.6.2 Local-global Principle for Finiteness Dimensions. *Let M be a finitely generated R-module. Then*

$$f_{\mathfrak{a}}(M) = \inf\left\{ f_{\mathfrak{a}R_{\mathfrak{p}}}(M_{\mathfrak{p}}) : \mathfrak{p} \in \operatorname{Spec}(R) \right\}.$$

Proof. This is now immediate from Theorem 9.6.1 above and the definition of finiteness dimension in 9.1.3. □

9.6.3 Exercise. Assume that $(R, \mathfrak{m})$ is local. Let $\mathfrak{b}$ be a second ideal of R, and let M be a finitely generated R-module. Prove that

$$f_{\mathfrak{a}\widehat{R}}^{\mathfrak{b}\widehat{R}}(M \otimes_R \widehat{R}) = f_{\mathfrak{a}}^{\mathfrak{b}}(M).$$

9.6.4 Exercise. Assume that $(R, \mathfrak{m})$ is local, that M is a finitely generated R-module, and that $\mathfrak{p} \in \operatorname{Spec}(R) \setminus \operatorname{Var}(\mathfrak{b})$, where $\mathfrak{b}$ is a second ideal of R. Let $h := \operatorname{ht}(\mathfrak{a} + \mathfrak{p})/\mathfrak{p}$.

(i) Prove that $\operatorname{ht}(\mathfrak{a}\widehat{R} + \mathfrak{p}\widehat{R})/\mathfrak{p}\widehat{R} = h$.
(ii) Let $\mathfrak{Q} \in \operatorname{Spec}(\widehat{R})$ be a minimal prime of $\mathfrak{a}\widehat{R} + \mathfrak{p}\widehat{R}$ with $\operatorname{ht} \mathfrak{Q}/\mathfrak{p}\widehat{R} = h$, and let $\mathfrak{P}$ be a minimal prime of $\mathfrak{p}\widehat{R}$ such that $\mathfrak{P} \subseteq \mathfrak{Q}$ and $\operatorname{ht} \mathfrak{Q}/\mathfrak{P} = h$. Show that

    (a) $\operatorname{ht}(\mathfrak{a}\widehat{R} + \mathfrak{P})/\mathfrak{P} \leq h$,
    (b) $\mathfrak{P} \cap R = \mathfrak{p}$ and $\mathfrak{P} \notin \operatorname{Var}(\mathfrak{b}\widehat{R})$,
    (c) $\operatorname{depth}(M \otimes_R \widehat{R})_{\mathfrak{P}} = \operatorname{depth} M_{\mathfrak{p}}$, and
    (d) $\operatorname{adj}_{\mathfrak{a}\widehat{R}} \operatorname{depth}(M \otimes_R \widehat{R})_{\mathfrak{P}} \leq \operatorname{adj}_{\mathfrak{a}} \operatorname{depth} M_{\mathfrak{p}}$.

(iii) Deduce that $\lambda_{\mathfrak{a}\widehat{R}}^{\mathfrak{b}\widehat{R}}(M \otimes_R \widehat{R}) \leq \lambda_{\mathfrak{a}}^{\mathfrak{b}}(M)$.

9.6.5 Exercise. Let the situation be as in Exercise 9.6.4 above, and assume in addition that R is universally catenary and that all its formal fibres are Cohen–Macaulay rings.

(i) Let $\mathfrak{Q}' \in \operatorname{Spec}(\widehat{R}) \setminus \operatorname{Var}(\mathfrak{b}\widehat{R})$ and set $\mathfrak{q}' := \mathfrak{Q}' \cap R$. Show that $\mathfrak{q}' \in \operatorname{Spec}(R) \setminus \operatorname{Var}(\mathfrak{b})$ and that

$$\operatorname{adj}_{\mathfrak{a}} \operatorname{depth} M_{\mathfrak{q}'} \leq \operatorname{adj}_{\mathfrak{a}\widehat{R}} \operatorname{depth}(M \otimes_R \widehat{R})_{\mathfrak{Q}'}.$$

(You might find Ratliff's Theorem [50, Theorem 31.7] helpful.)
(ii) Deduce from part (i) and Exercise 9.6.4(iii) above that

$$\lambda_{\mathfrak{a}\widehat{R}}^{\mathfrak{b}\widehat{R}}(M \otimes_R \widehat{R}) = \lambda_{\mathfrak{a}}^{\mathfrak{b}}(M).$$

9.6.6 Exercise. Assume that $(R, \mathfrak{m})$ is a universally catenary local ring all of whose formal fibres are Cohen–Macaulay rings. Prove that the conclusion of the Annihilator Theorem holds over R. In other words, prove that, if M is a finitely generated R-module and $\mathfrak{b}$ is a second ideal of R, then $\lambda_{\mathfrak{a}}^{\mathfrak{b}}(M) = f_{\mathfrak{a}}^{\mathfrak{b}}(M)$.

9.6.7 Exercise.   Assume that R is universally catenary and all the formal fibres of all its localizations are Cohen–Macaulay rings. Deduce from Exercise 9.6.6 above and the Local-global Principle 9.6.2 that the conclusion of the Finiteness Theorem holds over R. In other words, prove that, if M is a finitely generated R-module, then $\lambda_{\mathfrak{a}}^{\mathfrak{a}}(M) = f_{\mathfrak{a}}(M)$.

9.6.8 Exercise.   Assume that $(R, \mathfrak{m})$ is a universally catenary local ring all of whose formal fibres are Cohen–Macaulay rings. Let M be a non-zero finitely generated R-module of dimension $n > 0$ such that

  (a) $\dim R/\mathfrak{p} = n$ whenever $\mathfrak{p}$ is a minimal member of $\operatorname{Supp} M$, and
  (b) $M_{\mathfrak{q}}$ is a Cohen–Macaulay $R_{\mathfrak{q}}$-module for all $\mathfrak{q} \in \operatorname{Supp} M \setminus \{\mathfrak{m}\}$.

Prove that M is a generalized Cohen–Macaulay R-module (see 9.5.7).

9.6.9 Exercise.   (This exercise is related to Exercise 9.4.13.) Assume that R is a universally catenary semi-local integral domain with the property that all the formal fibres of all its localizations are Cohen–Macaulay rings. Assume that there exists $0 \neq c \in R$ such that the ring R_c is Cohen–Macaulay. Prove that there exists $n \in \mathbb{N}$ such that, for all choices of $r \in \mathbb{N}_0$ and all choices of an ideal $\mathfrak{B}'$ of the polynomial ring $R[X_1, \ldots, X_r]$, we have

$$c^n H_{\mathfrak{B}'}^i(R[X_1, \ldots, X_r]) = 0 \quad \text{for all } i < \operatorname{ht} \mathfrak{B}'.$$

10

Matlis duality

Prior to this point in the book, we have not made use of the decomposition theory (due to E. Matlis [49]) for injective modules over our (Noetherian) ring R. However, our work in the next Chapter 11 on local duality will involve use of the structure of the terms in the minimal injective resolution of a Gorenstein local ring, and so we can postpone no longer use of the decomposition theory for injective modules. Our discussion of local duality in Chapter 11 will also involve Matlis duality.

Our purpose in this chapter is to prepare the ground for Chapter 11 by reviewing, sometimes in detail, those parts of Matlis's theories that we shall need later in the book. Sometimes we simply refer to [50, Section 18] for proofs; in other cases, we provide alternative proofs for the reader's convenience. An experienced reader who is familiar with this work of Matlis should omit this chapter and progress straight to the discussion of local duality in Chapter 11: for one thing, there is no local cohomology theory in this chapter! However, graduate students might find this chapter helpful.

10.1 Indecomposable injective modules

10.1.1 Reminders. Let M be a submodule of the R-module L.

(i) We say that L is an *essential extension* of M precisely when $B \cap M \neq 0$ for every non-zero submodule B of L.

(ii) We say that L is an *injective envelope* (or *injective hull*) of M precisely when L is an injective R-module that is an essential extension of M.

(iii) (See [50, Theorem B4, p. 281].) The R-module M is injective if and only if the only essential extension of M is M itself.

(iv) If L is an injective envelope of M, and $g : M \longrightarrow K$ is an R-monomorphism from M into an injective R-module K, then there is a homomorphism $g' : L \to K$ such that the diagram

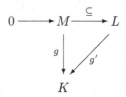

commutes. Since $\operatorname{Ker} g' \cap M = \operatorname{Ker} g = 0$, it follows from the fact that L is an essential extension of M that $\operatorname{Ker} g' = 0$ and g' is a monomorphism. Hence L is isomorphic to a direct summand of K.

(v) (See [50, p. 281].) Each R-module has an injective envelope which is uniquely determined up to isomorphism. In fact, if E and E' are both injective envelopes of M (so that each of E and E' is an injective R-module which contains M as a submodule, and each of E and E' is an essential extension of its submodule M), then there is an isomorphism $f : E \to E'$ such that the diagram

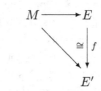

(in which the unnamed homomorphisms are the inclusion maps) commutes.

(vi) We denote by $E(M)$ (or $E_R(M)$ if it is necessary to specify the underlying ring) one choice of injective envelope of M.

10.1.2 Example and Warning. The reader should be warned that the isomorphism $f : E \to E'$ of 10.1.1(v) need not be uniquely determined; for this reason, we cannot regard $E(\bullet)$ as a functor from $\mathcal{C}(R)$ to itself.

To illustrate this point, consider the submodule

$$C_2(\infty) := \left\{ \alpha \in \mathbb{Q}/\mathbb{Z} : \alpha = \frac{r}{2^n} + \mathbb{Z} \text{ for some } r \in \mathbb{Z} \text{ and } n \in \mathbb{N}_0 \right\}$$

of the $\mathbb{Z}$-module $\mathbb{Q}/\mathbb{Z}$. Since $C_2(\infty)$ is a divisible Abelian group, it is an injective $\mathbb{Z}$-module, by [71, Theorem 3.24] for example. Also $C_2(\infty)$ is an essential extension of its $\mathbb{Z}$-submodule $C_2 := \mathbb{Z}(\frac{1}{2} + \mathbb{Z})$. Now '$f = \operatorname{Id}_{C_2(\infty)}$' and '$f =$

$-\operatorname{Id}_{C_2(\infty)}$' are two different choices of automorphism $f : C_2(\infty) \to C_2(\infty)$ for which the diagram

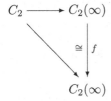

(in which the unnamed homomorphisms are the inclusion maps) commutes.

Injective envelopes play an essential rôle in the very satisfactory decomposition theory for injective R-modules: for one thing, the fundamental 'building blocks' on which the whole theory is based are described using the concept of injective envelope.

10.1.3 Reminders. We shall assume that the reader is familiar with the following facts about injective R-modules.

(i) Let $(M_\iota)_{\iota \in \Lambda}$ be a non-empty family of R-modules. It is immediate from the definition of injective module that the direct product $\prod_{\iota \in \Lambda} M_\iota$ is an injective R-module if and only if M_ι is injective for all $\iota \in \Lambda$. However, since our ring R is Noetherian, it is also the case that the direct sum $\bigoplus_{\iota \in \Lambda} M_\iota$ is injective if and only if M_ι is injective for all $\iota \in \Lambda$: see [50, Theorem 18.5(i)].

(ii) An R-module is said to be *indecomposable* precisely when it is non-zero and cannot be written as the direct sum of two proper submodules. For each $\mathfrak{p} \in \operatorname{Spec}(R)$, the injective R-module $E(R/\mathfrak{p})$ is indecomposable (see [50, Theorem 18.4(i)]); moreover, each indecomposable injective R-module is isomorphic to $E(R/\mathfrak{q})$ for some $\mathfrak{q} \in \operatorname{Spec}(R)$ (see [50, Theorem 18.4(ii)]).

(iii) Let $\mathfrak{p} \in \operatorname{Spec}(R)$ and let $r \in R \setminus \mathfrak{p}$. Then multiplication by r provides an automorphism of $E(R/\mathfrak{p})$ (see [50, Theorem 18.4(iii)]); moreover, each element of $E(R/\mathfrak{p})$ is annihilated by some power of $\mathfrak{p}$, that is, $E(R/\mathfrak{p})$ is $\mathfrak{p}$-torsion (see [50, Theorem 18.4(v)]).

(iv) Let $\mathfrak{p}, \mathfrak{q} \in \operatorname{Spec}(R)$. Then $E(R/\mathfrak{p}) \cong E(R/\mathfrak{q})$ if and only if $\mathfrak{p} = \mathfrak{q}$ (see [50, Theorem 18.4(iv)]).

(v) The results recounted in parts (ii) and (iv) above can be reformulated as follows: there is a *set* of isomorphism classes of indecomposable injective R-modules, and there is a bijective correspondence between this set and $\mathrm{Spec}(R)$, under which a $\mathfrak{p} \in \mathrm{Spec}(R)$ corresponds to the isomorphism class of $E(R/\mathfrak{p})$.

10.1.4 Exercise. Let $M_1, \ldots, M_n$ be R-modules. Show that the map

$$M_1 \oplus \cdots \oplus M_n \longrightarrow E(M_1) \oplus \cdots \oplus E(M_n)$$

(obtained by taking the direct sum of the inclusion maps) provides an injective envelope of $\bigoplus_{i=1}^{n} M_i$.

10.1.5 Exercise. Let M be a non-zero R-module. Show that the following statements are equivalent:

(i) $E(M)$ is indecomposable;
(ii) $E(M)$ is an injective envelope of every non-zero submodule of itself;
(iii) the zero submodule of M cannot be expressed as the intersection of two non-zero submodules of M.

10.1.6 Exercise. Let $\mathfrak{a}$ be a proper ideal of R. Show that $E(R/\mathfrak{a})$ is indecomposable if and only if $\mathfrak{a}$ is irreducible.

Let $\mathfrak{q}$ be a $\mathfrak{p}$-primary ideal of R. Prove that $E(R/\mathfrak{q})$ is isomorphic to a direct sum of finitely many copies of $E(R/\mathfrak{p})$. What can you say about the number of copies? (Here is a hint: recall that $\mathfrak{q}$ can be expressed as an intersection $\mathfrak{q} = \bigcap_{i=1}^{n} \mathfrak{j}_i$, where each $\mathfrak{j}_i$ (for $1 \le i \le n$) is irreducible and irredundant in the intersection. If you still find this exercise difficult, you might like to consult [81, Exercise 8.30].)

10.1.7 Exercise. Let $\mathfrak{p} \in \mathrm{Spec}(R)$.

(i) Let $0 \ne x \in E(R/\mathfrak{p})$. Show that $(0 : x)$ is an irreducible ideal of R.
(ii) In 10.1.3(iii), we saw that $E(R/\mathfrak{p}) = \bigcup_{n \in \mathbb{N}} (0 :_{E(R/\mathfrak{p})} \mathfrak{p}^n)$. Show that $E(R/\mathfrak{p}) = \bigcup_{n \in \mathbb{N}} (0 :_{E(R/\mathfrak{p})} \mathfrak{p}^{(n)})$.

10.1.8 ♯Exercise. Let I be a non-zero injective R-module. Show that I is a direct sum of indecomposable injective submodules, perhaps by means of the following intermediate steps.

(i) Apply Zorn's Lemma to the set of all sets of indecomposable injective submodules of I whose sum is direct, in order to find a maximal member $\mathcal{M}$ of this set.
(ii) Let $J := \bigoplus_{D \in \mathcal{M}} D$. Suppose that $J \subset I$ and seek a contradiction. Use the injectivity of J to find a submodule K of I such that $I = J \oplus K$.
(iii) Let $\mathfrak{q} \in \mathrm{Ass}\, K$. Show that K has a submodule isomorphic to $E_R(R/\mathfrak{q})$.
(iv) Use 10.1.3(ii) to find a contradiction to the maximality of $\mathcal{M}$, and deduce that $J = I$.

In 10.1.3, we reviewed fundamental facts about indecomposable injective R-modules. One of the reasons for the importance of these modules is provided by the following.

10.1.9 Reminder. (See [50, Theorem 18.5(ii)].) By 10.1.8, each injective R-module I is a direct sum of indecomposable injective submodules. Therefore, by 10.1.3(ii), there is a family $(\mathfrak{p}_\alpha)_{\alpha \in \Lambda}$ of prime ideals of R for which $I \cong \bigoplus_{\alpha \in \Lambda} E(R/\mathfrak{p}_\alpha)$.

10.1.10 Remark. Let the situation be as in 10.1.9. Then $\mathfrak{p}$ is equal to $\mathfrak{p}_\alpha$ for some $\alpha \in \Lambda$ if and only if $\mathfrak{p} \in \operatorname{Ass} I$.

10.1.11 Exercise. Let I be an injective R-module, so that, by 10.1.9, there is a family $(\mathfrak{p}_\alpha)_{\alpha \in \Lambda}$ of prime ideals of R for which $I \cong \bigoplus_{\alpha \in \Lambda} E(R/\mathfrak{p}_\alpha)$. Prove that

$$\Gamma_\mathfrak{a}(I) \cong \bigoplus_{\substack{\alpha \in \Lambda \\ \mathfrak{a} \subseteq \mathfrak{p}_\alpha}} E(R/\mathfrak{p}_\alpha).$$

(This exercise provides another proof of the result of Proposition 2.1.4 that $\Gamma_\mathfrak{a}(I)$ is an injective R-module.)

In fact, the direct decompositions described in 10.1.9 have uniqueness properties. The route taken by Matsumura in [50, §18] to these uniqueness properties involves facts about the behaviour of indecomposable injective R-modules under localization. We review this behaviour next.

10.1.12 Lemma. *Let S be a multiplicatively closed subset of R, and let G be an $S^{-1}R$-module. Then G is R-injective (that is, injective when viewed as an R-module by means of the natural homomorphism $R \to S^{-1}R$) if and only if G is $S^{-1}R$-injective.*

Proof. ($\Rightarrow$) Let H be an $S^{-1}R$-submodule of the $S^{-1}R$-module J, and let $h : H \to G$ be an $S^{-1}R$-homomorphism. Then H is an R-submodule of the R-module J, and h is an R-homomorphism. Since G is R-injective, there exists an R-homomorphism $j : J \to G$ which extends h. It is easy to check that j must be an $S^{-1}R$-homomorphism.

($\Leftarrow$) Let M be an R-submodule of the R-module N, and let $\lambda : M \to G$ be an R-homomorphism. As G is an $S^{-1}R$-module, the natural map $\psi : G \to S^{-1}G$ is not only an R-isomorphism but also an $S^{-1}R$-isomorphism. Thus $S^{-1}M$ is an $S^{-1}R$-submodule of the $S^{-1}R$-module $S^{-1}N$, and the composition $\psi^{-1} \circ S^{-1}\lambda : S^{-1}M \to G$ is an $S^{-1}R$-homomorphism.

Since G is $S^{-1}R$-injective, there is an $S^{-1}R$-homomorphism $\mu' : S^{-1}N \to$

G which extends $\psi^{-1} \circ S^{-1}\lambda$. Then $\mu' \circ \theta : N \to G$, where $\theta : N \to S^{-1}N$ is the natural map, is an R-homomorphism which extends λ.   □

10.1.13 Lemma.   *Let S be a multiplicatively closed subset of R, and let $\mathfrak{p} \in$ $\mathrm{Spec}(R)$ be such that $\mathfrak{p} \cap S = \emptyset$. By 10.1.3(iii), the indecomposable injective R-module $E_R(R/\mathfrak{p})$ has a natural structure as an $S^{-1}R$-module.*

As $S^{-1}R$-module, $E_R(R/\mathfrak{p})$ is isomorphic to $E_{S^{-1}R}(S^{-1}R/S^{-1}\mathfrak{p})$.

Furthermore, $E_{S^{-1}R}(S^{-1}R/S^{-1}\mathfrak{p})$, when considered as an R-module by means of the natural homomorphism $R \to S^{-1}R$, is isomorphic to $E_R(R/\mathfrak{p})$.

Proof.   It follows from Lemma 10.1.12 that the $S^{-1}R$-module $E_R(R/\mathfrak{p})$ is $S^{-1}R$-injective. Since an $S^{-1}R$-submodule of $E_R(R/\mathfrak{p})$ is automatically an R-submodule, it is immediate from 10.1.3(ii) that $E_R(R/\mathfrak{p})$ is indecomposable as $S^{-1}R$-module; we can therefore deduce from the same result that $E_R(R/\mathfrak{p})$ is $S^{-1}R$-isomorphic to $E_{S^{-1}R}(S^{-1}R/\mathfrak{P})$ for some $\mathfrak{P} \in \mathrm{Spec}(S^{-1}R)$. By 10.1.3(iii),

$$\mathfrak{P} = \left\{ \alpha \in S^{-1}R : \alpha \, \mathrm{Id}_{E_R(R/\mathfrak{p})} \text{ is not an isomorphism} \right\}.$$

Hence the contraction of $\mathfrak{P}$ to R is just

$$\left\{ r \in R : r \, \mathrm{Id}_{E_R(R/\mathfrak{p})} \text{ is not an isomorphism} \right\} = \mathfrak{p}.$$

Therefore $\mathfrak{P} = S^{-1}\mathfrak{p}$.

The final claim is now immediate.   □

10.1.14 Proposition.   *Let S be a multiplicatively closed subset of R.*

(i) *Let $\mathfrak{p} \in \mathrm{Spec}(R)$. Then*

$$S^{-1}(E_R(R/\mathfrak{p})) \begin{cases} = 0 & \text{if } \mathfrak{p} \cap S \neq \emptyset, \\ \cong E_{S^{-1}R}(S^{-1}R/S^{-1}\mathfrak{p}) & \text{if } \mathfrak{p} \cap S = \emptyset. \end{cases}$$

(ii) *Let I be an injective R-module. Then $S^{-1}I$ is both $S^{-1}R$-injective and R-injective.*

Proof.   (i) Suppose that $s \in \mathfrak{p} \cap S$. By 10.1.3(iii), each $x \in E(R/\mathfrak{p})$ is annihilated by some power of s, and so $S^{-1}(E_R(R/\mathfrak{p})) = 0$.

If, on the other hand, $\mathfrak{p} \cap S = \emptyset$, then, by 10.1.3(iii) again, $E(R/\mathfrak{p})$ has a natural structure as an $S^{-1}R$-module, so that $S^{-1}(E(R/\mathfrak{p})) \cong E(R/\mathfrak{p})$ both as R-modules and $S^{-1}R$-modules. We can now use 10.1.13 to complete the proof of part (i).

(ii) By 10.1.9, there is a family $(\mathfrak{p}_\alpha)_{\alpha \in \Lambda}$ of prime ideals of R for which $I \cong$ $\bigoplus_{\alpha \in \Lambda} E(R/\mathfrak{p}_\alpha)$. Then $S^{-1}I \cong \bigoplus_{\alpha \in \Lambda} S^{-1}(E(R/\mathfrak{p}_\alpha))$ (as $S^{-1}R$-modules); since, by (i), each $S^{-1}(E(R/\mathfrak{p}_\alpha))$ in the above display is $S^{-1}R$-injective, it follows from 10.1.3(i) that $S^{-1}I$ is $S^{-1}R$-injective. An appeal to 10.1.12 now completes the proof. □

The behaviour of injective R-modules under fraction formation is used in Matsumura's proof of the following fundamental uniqueness property of the direct sum decompositions described in 10.1.9.

10.1.15 Reminder. (See [50, Theorem 18.5(iii)].) Let I be an injective R-module. By 10.1.9, there is a family $(\mathfrak{p}_\alpha)_{\alpha \in \Lambda}$ of prime ideals of R for which $I \cong \bigoplus_{\alpha \in \Lambda} E(R/\mathfrak{p}_\alpha)$.

Let $\mathfrak{p} \in \mathrm{Spec}(R)$, and let $k(\mathfrak{p}) = R_\mathfrak{p}/\mathfrak{p}R_\mathfrak{p}$, the residue field of the local ring $R_\mathfrak{p}$. Then the cardinality of the set $\{\alpha \in \Lambda : \mathfrak{p}_\alpha = \mathfrak{p}\}$ depends only on I and $\mathfrak{p}$ and not on the particular decomposition of I (as a direct sum of indecomposable injective submodules) chosen; in fact, this cardinality is equal to the vector space dimension $\dim_{k(\mathfrak{p})} \mathrm{Hom}_{R_\mathfrak{p}}(k(\mathfrak{p}), I_\mathfrak{p})$.

10.1.16 Lemma. *Suppose that the R-module M is annihilated by the ideal $\mathfrak{b}$ of R. We can regard M and $(0 :_{E_R(M)} \mathfrak{b})$ as modules over $R/\mathfrak{b}$ in natural ways: when this is done, $(0 :_{E_R(M)} \mathfrak{b}) \cong E_{R/\mathfrak{b}}(M)$.*

Proof. Let $\iota : (0 :_{E_R(M)} \mathfrak{b}) \to E_R(M)$ denote the inclusion map. Let

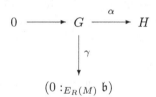

be a diagram of $R/\mathfrak{b}$-modules and $R/\mathfrak{b}$-homomorphisms in which the row is exact. We can regard these modules as R-modules by means of the natural homomorphism $R \to R/\mathfrak{b}$, and then α, γ and $\iota \circ \gamma$ become R-homomorphisms. Since $E_R(M)$ is an injective R-module, there exists an R-homomorphism $\beta : H \longrightarrow E_R(M)$ such that $\beta \circ \alpha = \iota \circ \gamma$. However, since $\mathfrak{b}H = 0$, it follows that $\mathrm{Im}\,\beta \subseteq (0 :_{E_R(M)} \mathfrak{b})$. We can regard β as a map from H to $(0 :_{E_R(M)} \mathfrak{b})$, and, when we do that, β is an $R/\mathfrak{b}$-homomorphism such that $\beta \circ \alpha = \gamma$.

Thus $(0 :_{E_R(M)} \mathfrak{b})$ is injective as an $R/\mathfrak{b}$-module. Obviously $(0 :_{E_R(M)} \mathfrak{b})$ is an essential extension of its $R/\mathfrak{b}$-submodule M. □

10.2 Matlis duality

Although there is a treatment of Matlis duality in [50, Theorem 18.6], we are going to present, for the reader's convenience, a different approach to this theory. Our approach makes use of Melkersson's Theorem [51, Theorem 1.3]: we presented this theorem in 7.1.2, and applied it to local cohomology modules in Chapter 7.

We begin by specifying some notation which we shall frequently use during our discussion of Matlis duality.

10.2.1 Notation and Remarks.   Suppose $(R, \mathfrak{m})$ is local. Set $E = E(R/\mathfrak{m})$, the injective envelope of the simple R-module $R/\mathfrak{m}$. As usual, we shall use $\widehat{R}$ to denote the $\mathfrak{m}$-adic completion $\varprojlim_{n \in \mathbb{N}} R/\mathfrak{m}^n$ of R. We shall use D to denote the exact, contravariant, R-linear functor $\operatorname{Hom}_R(\, \bullet \, , E)$ from $\mathcal{C}(R)$ to itself. For each R-module G, we shall refer to $D(G)$ as the *Matlis dual* of G. Note that $D(R)$ is naturally isomorphic to E, and that $D(E) = \operatorname{Hom}_R(E, E)$ is just the R-endomorphism ring of E considered as an R-module in the natural way.

For each R-module G, let

$$\mu_G : G \longrightarrow DD(G) = \operatorname{Hom}_R(\operatorname{Hom}_R(G, E), E)$$

be the natural R-homomorphism for which $(\mu_G(x))(f) = f(x)$ for all $x \in G$ and $f \in \operatorname{Hom}_R(G, E)$.

Note that, as G varies through $\mathcal{C}(R)$, the μ_G constitute a natural transformation μ from the identity functor to the functor DD.

If an R-module M has finite length, then we shall denote that length by $\ell(M)$.

10.2.2 Remarks.   Suppose that $(R, \mathfrak{m})$ is local, and use the notation of 10.2.1. Let G be an R-module.

(i) The R-homomorphism μ_G is injective, since if $0 \neq x \in G$, there is an R-homomorphism $f' : Rx \to R/\mathfrak{m}$ for which $f'(rx) = r + \mathfrak{m}$ for all $r \in R$, and the composition of f' and the inclusion map $R/\mathfrak{m} \to E$ can be extended to an $f \in \operatorname{Hom}_R(G, E)$ for which $f(x) \neq 0$.

(ii) The annihilators of G and its Matlis dual $D(G)$ are equal, because the fact that D is an R-linear functor ensures that

$$(0 : G) \subseteq (0 : D(G)) \subseteq (0 : DD(G)),$$

while the injectivity of $\mu_G : G \longrightarrow DD(G)$ (proved in (i) above) ensures that $(0 : DD(G)) \subseteq (0 : G)$.

First we analyse the case when $(R, \mathfrak{m})$ is an Artinian local ring. Recall that, then, an R-module G is Artinian if and only if it is Noetherian, and this is the case if and only if G is finitely generated.

10.2.3 Proposition. *Suppose that* $(R, \mathfrak{m})$ *is local and Artinian, and use the notation of* 10.2.1. *Then*

(i) $D(R/\mathfrak{m}) \cong R/\mathfrak{m}$ *(as R-modules);*

(ii) *for each finitely generated R-module G (that is, for each R-module G of finite length), the Matlis dual $D(G)$ is also finitely generated and $\ell(D(G)) = \ell(G)$;*

(iii) *E is finitely generated (and so Artinian), and $\ell(E) = \ell(R)$;*

(iv) *for each finitely generated R-module G (that is, for each Artinian R-module G), the map $\mu_G : G \longrightarrow DD(G)$ is an isomorphism; and*

(v) *for each $f \in \operatorname{Hom}_R(E, E)$, there is a unique $r_f \in R$ such that $f(x) = r_f x$ for all $x \in E$.*

Note. Condition (v) is equivalent to the statement that the homomorphism $\theta : R \longrightarrow \operatorname{Hom}_R(E, E)$ for which $\theta(r) = r \operatorname{Id}_E$, for all $r \in R$, is an isomorphism.

Proof. (i) Let k denote the residue field of R. There is an isomorphism of k-modules

$$D(R/\mathfrak{m}) = \operatorname{Hom}_R(R/\mathfrak{m}, E) \cong (0 :_E \mathfrak{m}).$$

But $(0 :_E \mathfrak{m}) \cong E_k(k)$ by 10.1.16; since every k-module, that is, every vector space over k, is injective, $E_k(k) = k$. The claim follows from this.

(ii) Use induction on length: remember that D is an exact functor.

(iii) This is now immediate from (ii), since $E \cong D(R)$.

(iv) Let G be a finitely generated R-module. Two uses of part (ii) show that $DD(G)$ is also finitely generated and has $\ell(DD(G)) = \ell(G)$; since μ_G is injective by 10.2.2(i), it follows that μ_G must be an isomorphism.

(v) By part (iv), the map $\mu_R : R \to \operatorname{Hom}_R(\operatorname{Hom}_R(R, E), E)$ is an isomorphism. The claim therefore follows from the fact that the composition

$$\theta : R \xrightarrow{\mu_R} \operatorname{Hom}_R(\operatorname{Hom}_R(R, E), E) \xrightarrow{\cong} \operatorname{Hom}_R(E, E),$$

in which the second isomorphism is the obvious natural one, is such that $\theta(r)$, for $r \in R$, is the endomorphism of E given by multiplication by r. $\square$

10.2.4 Exercise. Suppose that $(R, \mathfrak{m})$ is local and Artinian, and set $E := E(R/\mathfrak{m})$. Let L be a submodule of E. Show that L is faithful if and only if $L = E$. (Here is a hint: apply the functor $\operatorname{Hom}_R(\bullet, E)$ to the canonical exact sequence $0 \longrightarrow L \longrightarrow E \longrightarrow E/L \longrightarrow 0$.)

Proposition 10.2.3 provides the main ingredients of Matlis duality for the (very special!) case of an Artinian local ring. However, as we shall show below, this special case can be used to build up to the general case quickly. We proved in 10.2.3(iii) that, if $(R, \mathfrak{m})$ is local and Artinian, then $E(R/\mathfrak{m})$ is Artinian. We now use Melkersson's Theorem [51, Theorem 1.3] (see 7.1.2) to obtain a much more general statement.

10.2.5 Theorem. *Let $\mathfrak{m}$ be a maximal ideal of R. Then $E(R/\mathfrak{m})$ is an Artinian injective R-module.*

Proof. By 10.1.16, we have $(0 :_{E_R(R/\mathfrak{m})} \mathfrak{m}) \cong E_{R/\mathfrak{m}}(R/\mathfrak{m})$. Since $R/\mathfrak{m}$ is a field, $E_{R/\mathfrak{m}}(R/\mathfrak{m}) = R/\mathfrak{m}$. Hence $(0 :_{E_R(R/\mathfrak{m})} \mathfrak{m})$ is an Artinian R-module. Now $E_R(R/\mathfrak{m})$ is $\mathfrak{m}$-torsion, by 10.1.3(iii), and so it follows from Melkersson's Theorem 7.1.2 that $E_R(R/\mathfrak{m})$ is Artinian. $\qquad\square$

10.2.6 Definition. Let M be an R-module. The *socle* $\mathrm{Soc}(M)$ *of* M is defined to be the sum of all simple submodules of M.

10.2.7 Remark. Let A be a non-zero Artinian R-module. Then A is an essential extension of its own socle, because each non-zero submodule B of A must contain a simple submodule (since a minimal member of the set of non-zero submodules of B must be simple).

10.2.8 Corollary. *Suppose that $(R, \mathfrak{m})$ is local, and set $E := E(R/\mathfrak{m})$. Let M be an R-module. Then M is Artinian if and only if M is isomorphic to a submodule of E^t, the direct sum of t copies of E, for some $t \in \mathbb{N}$.*

Proof. ($\Leftarrow$) It is clear from 10.2.5 that, for every $t \in \mathbb{N}$, every submodule of E^t is Artinian.

($\Rightarrow$) Assume that M is Artinian. Clearly, we can assume that $M \neq 0$. Now M is an essential extension of $\mathrm{Soc}(M)$, by 10.2.7, and $\mathrm{Soc}(M)$ is an Artinian module annihilated by $\mathfrak{m}$. Thus $\mathrm{Soc}(M)$ has a natural structure as a (finite-dimensional) vector space over $R/\mathfrak{m}$, and so there exists $t \in \mathbb{N}$ such that $\mathrm{Soc}(M) \cong (R/\mathfrak{m})^t$. Now compose this isomorphism with the direct sum of the inclusion maps to obtain an R-monomorphism $f : \mathrm{Soc}(M) \to E^t$; since E^t is injective (by 10.1.3(i)), this f can be extended to an R-homomorphism $f' : M \to E^t$; and f' must also be a monomorphism since M is an essential extension of $\mathrm{Soc}(M)$, by 10.2.7. $\qquad\square$

Key to the theory of Matlis duality are the facts that, when $(R, \mathfrak{m})$ is local, $E := E(R/\mathfrak{m})$ has a natural structure as an $\widehat{R}$-module, and, for each $f \in \mathrm{Hom}_R(E, E)$, there is a unique $\widehat{r_f} \in \widehat{R}$ such that $f(x) = \widehat{r_f}x$ for all $x \in E$. We approach these facts next, and, in doing so, we touch on ideas mentioned in Exercise 8.2.4.

10.2.9 Remark. Suppose that $(R, \mathfrak{m})$ is local and that A is an Artinian R-module.

Let $a \in A$. It is immediate from 10.2.8 and 10.1.3(iii) that there exists $t \in \mathbb{N}$ such that $\mathfrak{m}^t a = 0$. (In fact, one can prove this in a much more elementary manner: if $a \neq 0$, then, since the R-module Ra is Artinian, $R/(0 : a)$ is an Artinian local ring, and so its maximal ideal is nilpotent.) Let, with an obvious notation,

$$\widehat{r} = (r_n + \mathfrak{m}^n)_{n \in \mathbb{N}} = (r'_n + \mathfrak{m}^n)_{n \in \mathbb{N}} \in \varprojlim_{n \in \mathbb{N}} R/\mathfrak{m}^n = \widehat{R},$$

so that $r_n - r'_n \in \mathfrak{m}^n$ for all $n \in \mathbb{N}$ and $r_{n+h} - r_n \in \mathfrak{m}^n$ for all $n, h \in \mathbb{N}$. Then $r_{t+h} a = r_t a = r'_t a = r'_{t+h} a$ for all $h \in \mathbb{N}$. It is straightforward to check that A can be given the structure of an $\widehat{R}$-module such that, with the above notation, $\widehat{r} a = r_t a$. Note that, if we regard this $\widehat{R}$-module as an R-module by means of the natural ring homomorphism $R \to \widehat{R}$, then we recover the original R-module structure on A; note also that a subset of A is an R-submodule if and only if it is an $\widehat{R}$-submodule.

10.2.10 ♯Exercise. Suppose that $(R, \mathfrak{m})$ is local, and set $E := E(R/\mathfrak{m})$. By 10.2.5, the R-module E is Artinian; therefore, by 10.2.9, it has a natural structure as an $\widehat{R}$-module. Prove that there is an $\widehat{R}$-isomorphism $E(R/\mathfrak{m}) \cong E_{\widehat{R}}(\widehat{R}/\widehat{\mathfrak{m}})$, where $\widehat{\mathfrak{m}}$ denotes the maximal ideal of $\widehat{R}$.

We are now able to prove the key result for Matlis duality that we mentioned just before 10.2.9.

10.2.11 Theorem. *Suppose that* $(R, \mathfrak{m})$ *is local, and set* $E := E(R/\mathfrak{m})$. *By* 10.2.5, *the R-module E is Artinian; therefore, by* 10.2.9, *it has a natural structure as an $\widehat{R}$-module.*

The natural R-homomorphism $\theta : \widehat{R} \longrightarrow \operatorname{Hom}_R(E, E)$ *for which* $\theta(\widehat{r}) = \widehat{r} \operatorname{Id}_E$ *for all* $\widehat{r} \in \widehat{R}$ *is an isomorphism. Thus, for each* $f \in \operatorname{Hom}_R(E, E)$, *there is a unique* $\widehat{r}_f \in \widehat{R}$ *such that* $f(x) = \widehat{r}_f x$ *for all* $x \in E$.

Proof. For each $n \in \mathbb{N}$, set $E_n := (0 :_E \mathfrak{m}^n)$. Let $f \in \operatorname{Hom}_R(E, E)$; let $t \in \mathbb{N}$. Of course, $R/\mathfrak{m}$ is annihilated by $\mathfrak{m}^t$. By 10.1.16, there is an isomorphism of $R/\mathfrak{m}^t$-modules $E_t \cong E_{R/\mathfrak{m}^t}((R/\mathfrak{m}^t)/(\mathfrak{m}/\mathfrak{m}^t))$. Now $f(E_t) \subseteq E_t$, and, of course, $R/\mathfrak{m}^t$ is an Artinian local ring. Therefore, by 10.2.3(v), there exists $r_t \in R$ such that $f(e) = r_t e$ for all $e \in E_t$, and, moreover, if r'_t is any other element of R such that $f(e) = r'_t e$ for all $e \in E_t$, then the uniqueness aspect of 10.2.3(v) ensures that $r_t + \mathfrak{m}^t = r'_t + \mathfrak{m}^t$, that is, $r_t - r'_t \in \mathfrak{m}^t$.

We can proceed as above for each $t \in \mathbb{N}$, and so construct a uniquely determined sequence $(r_n + \mathfrak{m}^n)_{n \in \mathbb{N}} \in \prod_{n \in \mathbb{N}} R/\mathfrak{m}^n$ with the property that, for every

$n \in \mathbb{N}$, we have $f(e) = r_n e$ for all $e \in E_n$. Furthermore, for $n, h \in \mathbb{N}$, we have $E_n \subseteq E_{n+h}$, and so, since $f(e) = r_n e = r_{n+h} e$ for all $e \in E_n$, it follows from the immediately preceding paragraph in this proof that $r_n - r_{n+h} \in \mathfrak{m}^n$.

We have therefore found a uniquely determined sequence

$$(r_n + \mathfrak{m}^n)_{n \in \mathbb{N}} \in \varprojlim_{n \in \mathbb{N}} R/\mathfrak{m}^n = \widehat{R}$$

such that, for every $n \in \mathbb{N}$, we have $f(e) = r_n e$ for all $e \in E_n$. Since $E = \bigcup_{n \in \mathbb{N}} E_n$ by 10.1.3(iii), it follows that there is exactly one $\widehat{r} \in \widehat{R}$ such that $f(x) = \widehat{r}x$ for all $x \in E$.   □

We are now able to present the main aspects of Matlis duality over a complete local ring.

10.2.12 Matlis Duality Theorem. *Suppose that $(R, \mathfrak{m})$ is local and complete, and use the notation of* 10.2.1. *(Thus E denotes the injective envelope $E(R/\mathfrak{m})$ of the simple R-module and $D := \operatorname{Hom}_R(\bullet , E)$.)*

(i) *For each $f \in \operatorname{Hom}_R(E, E)$, there is a unique $r_f \in R$ such that $f(x) = r_f x$ for all $x \in E$.*

(ii) *Whenever N is a finitely generated R-module, the natural homomorphism $\mu_N : N \longrightarrow DD(N)$ is an isomorphism and $D(N)$ is Artinian.*

(iii) *Whenever A is an Artinian R-module, the natural homomorphism $\mu_A : A \longrightarrow DD(A)$ is an isomorphism and $D(A)$ is Noetherian.*

Proof. (i) This is immediate from Theorem 10.2.11.

(ii) The composition

$$\theta : R \xrightarrow{\mu_R} \operatorname{Hom}_R(\operatorname{Hom}_R(R, E), E) \xrightarrow{\cong} \operatorname{Hom}_R(E, E),$$

in which the second map is the obvious natural isomorphism, is such that $\theta(r) = r \operatorname{Id}_E$ for all $r \in R$. We have seen in part (i) that θ is an isomorphism; therefore μ_R is an isomorphism.

The identity functor and DD are both additive, and the result of application of an additive functor to a split short exact sequence is again a split short exact sequence; also, μ is a natural transformation of functors. We can therefore deduce, by induction on rank, that μ_F is an isomorphism whenever F is a finitely generated free R-module.

Let N be an arbitrary finitely generated R-module. Then N can be included in an exact sequence $F_1 \longrightarrow F_0 \longrightarrow N \longrightarrow 0$ in which F_1 and F_0 are finitely generated free R-modules. Since the functor DD is additive and exact, and

μ is a natural transformation of functors, the above exact sequence induces a commutative diagram

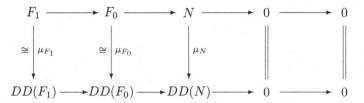

with exact rows. It therefore follows from the Five Lemma that μ_N is an isomorphism.

Finally, application of the contravariant, exact functor D to the exact sequence $F_0 \longrightarrow N \longrightarrow 0$ shows that $D(N)$ is isomorphic to a submodule of $D(F_0)$; since $D(F_0)$ is isomorphic to a direct sum of finitely many copies of $D(R)$ and $D(R) \cong E$, it follows from 10.2.8 that $D(N)$ is Artinian.

(iii) The composition

$$E \xrightarrow{\;\mu_E\;} \operatorname{Hom}_R(\operatorname{Hom}_R(E,E),E) \xrightarrow[\cong]{\operatorname{Hom}_R(\theta,\operatorname{Id}_E)} \operatorname{Hom}_R(R,E) \xrightarrow{\;\cong\;} E,$$

where θ is the isomorphism used in the proof of part (ii) above, is just the identity map; therefore μ_E is an isomorphism. We can now use the additivity of DD, and the natural transformation μ, to deduce, by induction on t, that μ_{E^t} is an isomorphism for all $t \in \mathbb{N}$.

Let A be an arbitrary Artinian R-module. Two uses of Corollary 10.2.8 show that there is an exact sequence $0 \longrightarrow A \longrightarrow E^{n_0} \longrightarrow E^{n_1}$ for suitable positive integers n_0 and n_1. We can now use the exactness of the functor DD, together with the natural transformation μ, as we did in the above proof of part (ii), to obtain from this exact sequence a commutative diagram with exact rows, and another application of the Five Lemma will yield the desired conclusion that μ_A is an isomorphism.

Finally, application of the contravariant, exact functor D to the exact sequence $0 \longrightarrow A \longrightarrow E^{n_0}$ shows that $D(A)$ is a homomorphic image of $D(E^{n_0}) \cong (D(E))^{n_0}$; since $\theta : R \to D(E)$ is an isomorphism, $D(A)$ is a homomorphic image of a finitely generated free R-module, and so is Noetherian. $\square$

10.2.13 ♯Exercise. Suppose that $(R, \mathfrak{m})$ is local and complete, and use the notation of 10.2.1. Let G be an R-module of finite length, so that, by the Matlis Duality Theorem 10.2.12, the Matlis dual $D(G)$ also has finite length. Prove that $\ell(D(G)) = \ell(G)$.

10.2.14 Remark. Suppose that $(R, \mathfrak{m})$ is local and complete, and use the notation of 10.2.1.

The Matlis Duality Theorem 10.2.12 allows statements about Noetherian R-modules to be translated into 'dual' statements about Artinian R-modules, and *vice versa*. For example, it shows that every Noetherian R-module is isomorphic to the Matlis dual of an Artinian R-module, and that every Artinian R-module is isomorphic to the Matlis dual of a Noetherian R-module.

To give a sample of this type of 'translation', let N be a Noetherian R-module, so that $D(N)$ is an Artinian R-module: we can use the Matlis Duality Theorem 10.2.12 to show quickly that $\operatorname{Att} D(N) = \operatorname{Ass} N$, as follows. Let $\mathfrak{p} \in \operatorname{Spec}(R)$. Then $\mathfrak{p} \in \operatorname{Ass} N$ if and only if N has a submodule with annihilator equal to $\mathfrak{p}$. Now D is an exact, contravariant functor and $DD(N) \cong N$; also, by 10.2.2(ii), the annihilators of an R-module M and its Matlis dual $D(M)$ are equal. Therefore N has a submodule with annihilator equal to $\mathfrak{p}$ if and only if $D(N)$ has a homomorphic image with annihilator equal to $\mathfrak{p}$; and, by 7.2.5, this is the case if and only if $\mathfrak{p} \in \operatorname{Att} D(N)$.

10.2.15 Exercise. Suppose that $(R, \mathfrak{m})$ is local and complete, and use the notation of 10.2.1.

Let N be a Noetherian R-module and let A be an Artinian R-module; let $n \in \mathbb{N}_0$ and $h \in \mathbb{N}$.

(i) Prove that $D(\mathfrak{a}^n N/\mathfrak{a}^{n+h} N) \cong (0 :_{D(N)} \mathfrak{a}^{n+h})/(0 :_{D(N)} \mathfrak{a}^n)$.

(ii) Prove that $D\left((0 :_A \mathfrak{a}^{n+h})/(0 :_A \mathfrak{a}^n)\right) \cong \mathfrak{a}^n D(A)/\mathfrak{a}^{n+h} D(A)$.

(iii) Prove that $\operatorname{Att}(0 :_A \mathfrak{a}^h) = \operatorname{Ass}\left(D(A)/\mathfrak{a}^h D(A)\right)$.

So far, we have restricted our account of Matlis duality to situations where the underlying local ring is complete. However, it is possible to obtain from Theorem 10.2.11 a satisfactory partial result which is valid over any local ring. We include this because we shall find a corollary of it useful in our applications of the local duality theory developed in Chapter 11. We shall need the following technical lemma.

10.2.16 Lemma. *Let M, I, J be R-modules.*

(i) *There exists a (unique) R-homomorphism*

$$\xi_{M,I,J} : M \otimes_R \operatorname{Hom}_R(I, J) \longrightarrow \operatorname{Hom}_R(\operatorname{Hom}_R(M, I), J)$$

such that, for $m \in M$, $f \in \operatorname{Hom}_R(I, J)$ and $g \in \operatorname{Hom}_R(M, I)$, we have $(\xi_{M,I,J}(m \otimes f))(g) = f(g(m))$. Furthermore, as M, I, J vary *through the category* $\mathcal{C}(R)$*, the* $\xi_{M,I,J}$ *constitute a natural transformation of functors*

$$\xi_{\bullet,\bullet,\bullet} : (\bullet) \otimes_R \operatorname{Hom}_R(\bullet,\bullet) \longrightarrow \operatorname{Hom}_R(\operatorname{Hom}_R(\bullet,\bullet),\bullet)$$

(from $\mathcal{C}(R) \times \mathcal{C}(R) \times \mathcal{C}(R)$ *to* $\mathcal{C}(R)$*).*

(ii) *If* J *is injective, then* $\xi_{M,I,J}$ *is an isomorphism whenever* M *is finitely generated.*

Proof. (i) This is straightforward and left to the reader.

(ii) It will be convenient to use V to denote $(\bullet) \otimes_R \operatorname{Hom}_R(I, J)$, to use W to denote the functor $\operatorname{Hom}_R(\operatorname{Hom}_R(\bullet, I), J)$, and to use $\zeta_{\bullet}$ to denote $\xi_{\bullet, I, J}$. The composition $\operatorname{Hom}_R(I, J) \xrightarrow{\cong} V(R) \xrightarrow{\zeta_R} W(R) \xrightarrow{\cong} \operatorname{Hom}_R(I, J)$, in which the first and last isomorphisms are the obvious natural ones, is just the identity map; hence $\zeta_R = \xi_{R,I,J}$ is an isomorphism.

The functors V and W are both additive, and ζ is a natural transformation of functors. We can therefore deduce, by induction on rank, that ζ_F is an isomorphism whenever F is a finitely generated free R-module.

Let M be an arbitrary finitely generated R-module. Then M can be included in an exact sequence $F_1 \longrightarrow F_0 \longrightarrow M \longrightarrow 0$ in which F_1 and F_0 are finitely generated free R-modules. Of course the functor V is right exact; since J is injective, the functor W is also right exact; therefore, since ζ is a natural transformation of functors, the above exact sequence induces a commutative diagram

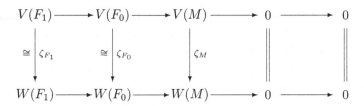

with exact rows. It thus follows from the Five Lemma that ζ_M is an isomorphism. $\qquad\square$

10.2.17 ♯Exercise. Supply a proof for part (i) of Theorem 10.2.16.

10.2.18 Remarks. Suppose that $(R, \mathfrak{m})$ is local (but not necessarily complete), and use the notation of 10.2.1. (Thus E denotes the injective envelope $E(R/\mathfrak{m})$ of the simple R-module and $D := \operatorname{Hom}_R(\bullet, E)$.) Let $\widehat{\mathfrak{m}}$ denote the maximal ideal of $\widehat{R}$.

(i) By 10.2.9, an Artinian R-module A can be given a natural structure as an $\widehat{R}$-module such that

   (a) for each $a \in A$ and each $\widehat{r} \in \widehat{R}$, there exists $r \in R$ with $\widehat{r}a = ra$, and

   (b) if we regard this $\widehat{R}$-module as an R-module by means of the natural ring homomorphism $R \to \widehat{R}$, then we recover the original R-module structure on A.

   It follows that A is an Artinian $\widehat{R}$-module.

(ii) Let A' be an Artinian $\widehat{R}$-module. Let $a' \in A'$. By 10.2.9, there exists $t \in \mathbb{N}$ such that a' is annihilated by $\widehat{\mathfrak{m}}^t = \mathfrak{m}^t \widehat{R}$. Now let $\widehat{r} \in \widehat{R}$. Then there exists $r \in R$ such that $\widehat{r} - r \in \mathfrak{m}^t \widehat{R}$. Therefore $\widehat{r}a' = ra'$, so that, when A' is regarded as an R-module by means of the natural ring homomorphism $R \to \widehat{R}$, every R-submodule of A' is automatically an $\widehat{R}$-submodule, and therefore A' is an Artinian R-module. If we then regard the Artinian R-module A' as an $\widehat{R}$-module using the method of 10.2.9, we recover the original $\widehat{R}$-module structure on A'.

(iii) Thus, by parts (i) and (ii), the category of all Artinian R-modules and all R-homomorphisms between them is equivalent to the category of all Artinian $\widehat{R}$-modules and all $\widehat{R}$-homomorphisms between them; in fact, we can regard these two categories as essentially the same.

(iv) Let ν be the natural transformation from the identity functor to the functor $(\, \bullet \,) \otimes_R \operatorname{Hom}_R(E, E)$ (from $\mathcal{C}(R)$ to itself) given by the formula $\nu_M(m) = m \otimes \operatorname{Id}_E$ for each R-module M and each $m \in M$.

   Let N be a finitely generated R-module. Then, since N is a homomorphic image of R^h for some $h \in \mathbb{N}$, the Matlis dual $D(N)$ is isomorphic to a submodule of $D(R^h)$; since $D(R^h) \cong D(R)^h \cong E^h$, we deduce from 10.2.8 that $D(N)$ is Artinian. Furthermore, it is straightforward to check that the diagram

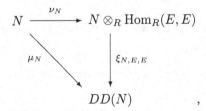

$$
\begin{array}{ccc}
N & \xrightarrow{\ \nu_N\ } & N \otimes_R \operatorname{Hom}_R(E, E) \\
& \mu_N \searrow & \downarrow \xi_{N,E,E} \\
& & DD(N)
\end{array}
\qquad ,
$$

where $\xi_{N,E,E}$ is the isomorphism given by 10.2.16, commutes. Now, by 10.2.2(i), the natural map μ_N is injective; by 10.2.11, $\operatorname{Hom}_R(E, E) \cong \widehat{R}$ under an isomorphism which maps Id_E to $1_{\widehat{R}}$, and so, speaking loosely, we can regard the duality map $\mu_N : N \longrightarrow DD(N)$ as the embedding of N into its completion.

(v) Let A be an Artinian R-module. Part (i) shows that A can be regarded as an Artinian $\widehat{R}$-module in a natural way; by 10.2.5, the R-module E is Artinian, and so it too can be regarded as an Artinian $\widehat{R}$-module in a natural way; by 10.2.10, there is an $\widehat{R}$-isomorphism $E(R/\mathfrak{m}) \cong E_{\widehat{R}}(\widehat{R}/\widehat{\mathfrak{m}})$. Also, $D(A) = \operatorname{Hom}_R(A, E) = \operatorname{Hom}_{\widehat{R}}(A, E)$, by part (i) (and 10.2.9), and so $D(A)$ has a natural (unambiguous) $\widehat{R}$-module structure. It follows from these observations that, as $\widehat{R}$-module, $D(A) \cong \operatorname{Hom}_{\widehat{R}}(A, E_{\widehat{R}}(\widehat{R}/\widehat{\mathfrak{m}}))$; the Matlis Duality Theorem 10.2.12 shows that this is a Noetherian $\widehat{R}$-module.

We now summarize some of the main points of 10.2.18 in a Partial Matlis Duality Theorem.

10.2.19 Partial Matlis Duality Theorem. *Suppose $(R, \mathfrak{m})$ is local (but not necessarily complete), and use the notation of* 10.2.1. *(Thus $E := E(R/\mathfrak{m})$ and $D := \operatorname{Hom}_R(\,\bullet\,, E)$.) By* 10.2.5 *and* 10.2.9, *the R-module E has a natural structure as an $\widehat{R}$-module. Let $\widehat{\mathfrak{m}}$ denote the maximal ideal of $\widehat{R}$.*

(i) *By* 10.2.11, *for each $f \in \operatorname{Hom}_R(E, E)$, there is a unique $\widehat{r}_f \in \widehat{R}$ such that $f(x) = \widehat{r}_f x$ for all $x \in E$.*

(ii) *Whenever N is a finitely generated R-module, the natural homomorphism $\mu_N : N \longrightarrow DD(N)$ is injective, $D(N)$ is Artinian, and there is a commutative diagram*

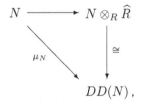

in which the horizontal homomorphism is the canonical one.

(iii) *Whenever A is an Artinian R-module, it has a natural structure as an $\widehat{R}$-module, it is Artinian as such, and $D(A)$ is a Noetherian $\widehat{R}$-module $\widehat{R}$-isomorphic to the Matlis dual of A over $\widehat{R}$.*

Proof. All the claims were explained in 10.2.18. □

We show how the Partial Matlis Duality Theorem 10.2.19 can be used to extend 10.2.14 to the case where the underlying local ring is not necessarily complete.

10.2.20 Corollary. *Suppose that* $(R, \mathfrak{m})$ *is local (but not necessarily complete), and use the notation of* 10.2.1.

Let N *be a Noetherian* R*-module, so that* $D(N)$ *is an Artinian* R*-module by* 10.2.19(ii), *and we can form the set* $\operatorname{Att} D(N)$. *Then* $\operatorname{Att} D(N) = \operatorname{Ass} N$.

Proof. Let $\mathfrak{p} \in \operatorname{Ass} N$, so that N has a submodule B with annihilator equal to $\mathfrak{p}$. Now D is an exact, contravariant functor; also, by 10.2.2(ii), the annihilators of an R-module M and its Matlis dual $D(M)$ are equal. Therefore $D(N)$ has a homomorphic image $D(B)$ with annihilator equal to $\mathfrak{p}$, and so $\mathfrak{p} \in \operatorname{Att} D(N)$ by 7.2.5.

Conversely, let $\mathfrak{p} \in \operatorname{Att}_R D(N)$, so that $D(N)$ has an R-homomorphic image A with annihilator equal to $\mathfrak{p}$, by 7.2.5. Application of D and another use of 10.2.2(ii) now show that $DD(N)$ has an R-submodule with annihilator equal to $\mathfrak{p}$. By 10.2.19(ii), there is an R-isomorphism $DD(N) \cong N \otimes_R \widehat{R}$. Therefore there exists an R-submodule C of $N \otimes_R \widehat{R}$ with $(0 :_R C) = \mathfrak{p}$.

Let $C\widehat{R}$ be the $\widehat{R}$-submodule of $N \otimes_R \widehat{R}$ generated by C. Note that $C\widehat{R}$ is a finitely generated $\widehat{R}$-module, and that

$$(0 :_{\widehat{R}} C\widehat{R}) \cap R = (0 :_R C\widehat{R}) = (0 :_R C) = \mathfrak{p}.$$

It follows that there exists $\mathfrak{P} \in \operatorname{Ass}_{\widehat{R}}(C\widehat{R})$ such that $\mathfrak{P} \cap R = \mathfrak{p}$. Of course, $\mathfrak{P} \in \operatorname{Ass}_{\widehat{R}}(N \otimes_R \widehat{R})$. We can now use [50, Theorem 23.2] to deduce that $\mathfrak{p} \in \operatorname{Ass}_R N$. $\quad\square$

10.2.21 Exercise. Suppose that $(R, \mathfrak{m})$ is local and complete, and use the notation of 10.2.1. Assume that $\dim R > 0$. Prove that there exists an R-module G which is neither Noetherian nor Artinian but for which $\mu_G : G \longrightarrow DD(G)$ is an isomorphism.

11

Local duality

Suppose, temporarily, that $(R, \mathfrak{m})$ is local, and that M is a finitely generated R-module. In Theorem 7.1.3, we showed that $H_{\mathfrak{m}}^i(M)$ is Artinian for all $i \in \mathbb{N}_0$. When R is complete, Matlis duality (see 10.2.12) provides a very satisfactory correspondence between the category of Artinian R-modules and the category of Noetherian R-modules, and so it is natural to ask, in this situation, which Noetherian R-modules correspond to the local cohomology modules $H_{\mathfrak{m}}^i(M)$ $(i \in \mathbb{N}_0)$. Local duality provides a really useful answer to this question.

In fact, the Local Duality Theorem 11.2.6 concerns the situation where R is a (not necessarily complete) homomorphic image of a Gorenstein local ring $(R', \mathfrak{m}')$ of dimension n'. (Cohen's Structure Theorem for complete local rings (see [50, Theorem 29.4(ii)], for example) ensures that this hypothesis would be satisfied if R were complete.) The Local Duality Theorem tells us that, if M is a finitely generated R-module, then, for each $i \in \mathbb{N}_0$, the local cohomology module $H_{\mathfrak{m}}^i(M)$ is isomorphic to the Matlis dual of the finitely generated R-module $\mathrm{Ext}_{R'}^{n'-i}(M, R')$, and, as R' is Gorenstein, quite a lot is known about these 'Ext' modules. The Local Duality Theorem provides a fundamental tool for the study of local cohomology modules with respect to the maximal ideal of a local ring. Although it only applies to local rings which can be expressed as homomorphic images of Gorenstein local rings, this is not a great restriction, because the class of such local rings includes the local rings of points on affine and quasi-affine varieties, and, as mentioned above, all complete local rings.

At the end of the chapter, some exercises are given to show that a local ring which cannot be expressed as a homomorphic image of a Gorenstein local ring need not have all the good properties flowing from the Local Duality Theorem.

11.1 Minimal injective resolutions

The work in this chapter will involve knowledge of the structure of a minimal injective resolution of a Gorenstein local ring; we begin by reviewing the concept of minimal injective resolution.

11.1.1 Definition. Let M be an R-module. A *minimal injective resolution of* M is an injective resolution

$$I^{\bullet} : 0 \longrightarrow I^0 \xrightarrow{d^0} I^1 \longrightarrow \cdots \longrightarrow I^i \xrightarrow{d^i} I^{i+1} \longrightarrow \cdots$$

of M such that I^n is an essential extension of $\operatorname{Ker} d^n$ for every $n \in \mathbb{N}_0$.

11.1.2 ♯Exercise. Let M be an R-module. Prove that M has a minimal injective resolution. (Start with the monomorphism $M \longrightarrow E(M)$, and, successively, take the injective envelope of the cokernel of the last map you have constructed.)

Suppose that, with the notation of 11.1.1, the injective resolution $I^{\bullet}$ of M is minimal. There is an augmentation R-homomorphism $\alpha : M \to I^0$ such that the sequence

$$0 \longrightarrow M \xrightarrow{\alpha} I^0 \xrightarrow{d^0} I^1 \longrightarrow \cdots \longrightarrow I^i \xrightarrow{d^i} I^{i+1} \longrightarrow \cdots$$

is exact. Since I^0 is an essential extension of $\operatorname{Ker} d^0$ and $\operatorname{Ker} d^0 = \operatorname{Im} \alpha \cong M$, it follows that $I^0 \cong E(M)$, the injective envelope of M, and so is uniquely determined up to isomorphism by M (see 10.1.1(v)). Actually, all the terms in a minimal injective resolution of M are uniquely determined up to isomorphism: this is the subject of the next exercise.

11.1.3 ♯Exercise. Let M be an R-module. Let

$$I^{\bullet} : 0 \longrightarrow I^0 \xrightarrow{d^0} I^1 \longrightarrow \cdots \longrightarrow I^i \xrightarrow{d^i} I^{i+1} \longrightarrow \cdots$$

and

$$J^{\bullet} : 0 \longrightarrow J^0 \xrightarrow{\delta^0} J^1 \longrightarrow \cdots \longrightarrow J^i \xrightarrow{\delta^i} J^{i+1} \longrightarrow \cdots$$

be minimal injective resolutions of M, so that there are R-homomorphisms $\alpha : M \to I^0$ and $\beta : M \to J^0$ such that the sequences

$$0 \longrightarrow M \xrightarrow{\alpha} I^0 \xrightarrow{d^0} I^1 \longrightarrow \cdots \longrightarrow I^i \xrightarrow{d^i} I^{i+1} \longrightarrow \cdots$$

and

$$0 \longrightarrow M \xrightarrow{\beta} J^0 \xrightarrow{\delta^0} J^1 \longrightarrow \cdots \longrightarrow J^i \xrightarrow{\delta^i} J^{i+1} \longrightarrow \cdots$$

are both exact. Show that there is an isomorphism of complexes

$$\phi^{\bullet} = \left(\phi^i\right)_{i \in \mathbb{N}_0} : I^{\bullet} \longrightarrow J^{\bullet}$$

such that the diagram

commutes.

11.1.4 Notation and Definition. Let M be an R-module. By 11.1.3, for each $i \in \mathbb{N}_0$, the i-th term in a minimal injective resolution of M is uniquely determined, up to isomorphism, by M, and is independent of the choice of minimal injective resolution of M. We denote this i-th term by $E^i(M)$, or by $E^i_R(M)$ when it is desirable to emphasize the underlying ring.

Observe that $E^0(M) \cong E(M)$, the injective envelope of M.

For each $i \in \mathbb{N}_0$ and each $\mathfrak{p} \in \mathrm{Spec}(R)$, we define the *i-th Bass number of M with respect to* $\mathfrak{p}$ as follows. By 10.1.9, there is a family $(\mathfrak{p}_\alpha)_{\alpha \in \Lambda}$ of prime ideals of R for which $E^i(M) \cong \bigoplus_{\alpha \in \Lambda} E(R/\mathfrak{p}_\alpha)$. By 10.1.15 (see [50, Theorem 18.5(iii)]), the cardinality of the set $\{\alpha \in \Lambda : \mathfrak{p}_\alpha = \mathfrak{p}\}$ depends only on $E^i(M)$ and $\mathfrak{p}$ (and therefore only on M and $\mathfrak{p}$) and not on the particular decomposition of $E^i(M)$ (as a direct sum of indecomposable injective submodules) chosen. This cardinality is denoted by $\mu^i(\mathfrak{p}, M)$, and is referred to as the i-th Bass number of M with respect to $\mathfrak{p}$. Symbolically, we write

$$E^i(M) \cong \bigoplus_{\mathfrak{p} \in \mathrm{Spec}(R)} \mu^i(\mathfrak{p}, M) E(R/\mathfrak{p}),$$

where $\oplus \mu E$ denotes the direct sum of μ copies of E.

In the notation of 11.1.4, we can already give a description of $\mu^0(\mathfrak{p}, M)$, since it follows from 10.1.15 (and the fact that $E^0(M) \cong E(M)$) that

$$\mu^0(\mathfrak{p}, M) = \dim_{k(\mathfrak{p})} \mathrm{Hom}_{R_\mathfrak{p}}(k(\mathfrak{p}), E(M)_\mathfrak{p}),$$

where $k(\mathfrak{p}) = R_\mathfrak{p}/\mathfrak{p}R_\mathfrak{p}$, the residue field of the local ring $R_\mathfrak{p}$. In fact, a refinement of this, and analogues for the higher Bass numbers, are available: we shall refer to Matsumura [50, Theorem 18.7] for these results, although we first give, for the reader's convenience, a slight refinement of one of Matsumura's lemmas.

11.1.5 Lemma. (See [50, §18, Lemma 6].) *Let $f : L \to M$ be a homomorphism of R-modules such that M is an essential extension of $\operatorname{Im} f$. Let S be a multiplicatively closed subset of R. Then $S^{-1}M$ is an essential extension of its submodule $\operatorname{Im}(S^{-1}f)$ (where $S^{-1}f : S^{-1}L \to S^{-1}M$ is the $S^{-1}R$-homomorphism induced by f).*

Proof. Suppose that x/s, where $x \in M$ and $s \in S$, is a non-zero element of $S^{-1}M$. Then $S^{-1}(Rx)$ is a non-zero submodule of $S^{-1}M$, and so there exists $\mathfrak{p} \in \operatorname{Ass}(Rx)$ such that $\mathfrak{p} \cap S = \emptyset$. Also, there exists $r \in R$ such that $(0 : rx) = \mathfrak{p}$.

Since M is an essential extension of $\operatorname{Im} f$, there exists $r' \in R$ such that $0 \neq r'rx = f(y)$ for some $y \in L$. Now $(0 : r'rx) = \mathfrak{p} \subseteq R \setminus S$, and so

$$0 \neq \frac{r'r}{1}\frac{x}{s} = \frac{r'rx}{s} = \frac{f(y)}{s} = (S^{-1}f)\left(\frac{y}{s}\right) \in \operatorname{Im}(S^{-1}f). \qquad \square$$

11.1.6 Corollary. *Let M be an R-module, and let*

$$I^\bullet : 0 \longrightarrow I^0 \xrightarrow{d^0} I^1 \longrightarrow \cdots \longrightarrow I^i \xrightarrow{d^i} I^{i+1} \longrightarrow \cdots$$

be a minimal injective resolution of M. Let S be a multiplicatively closed subset of R. Then

$$0 \longrightarrow S^{-1}(I^0) \longrightarrow \cdots \longrightarrow S^{-1}(I^i) \xrightarrow{S^{-1}(d^i)} S^{-1}(I^{i+1}) \longrightarrow \cdots$$

is a minimal injective resolution of the $S^{-1}R$-module $S^{-1}M$.

Thus, for all $i \in \mathbb{N}_0$,

(i) *$S^{-1}(E^i_R(M)) \cong E^i_{S^{-1}R}(S^{-1}M)$ (as $S^{-1}R$-modules); and*

(ii) *$\mu^i(S^{-1}\mathfrak{p}, S^{-1}M) = \mu^i(\mathfrak{p}, M)$ for all $\mathfrak{p} \in \operatorname{Spec}(R)$ with $\mathfrak{p} \cap S = \emptyset$.*

Proof. All the claims follow easily from 10.1.14 and 11.1.5. $\qquad \square$

11.1.7 Corollary. *Let the situation and notation be as in 11.1.6. Then, for all $\mathfrak{p} \in \operatorname{Spec}(R)$ and $i \in \mathbb{N}_0$, the induced homomorphism $\operatorname{Hom}_{R_\mathfrak{p}}(R_\mathfrak{p}/\mathfrak{p}R_\mathfrak{p}, d^i_\mathfrak{p})$ is zero.*

Proof. In view of 11.1.6, we can, and do, assume that $(R, \mathfrak{m})$ is local and that $\mathfrak{p} = \mathfrak{m}$, and then we have to show that $\operatorname{Hom}_R(R/\mathfrak{m}, d^i) = 0$. To achieve this, it is sufficient for us to show that $(0 :_{I^i} \mathfrak{m}) \subseteq \operatorname{Ker} d^i$. However, $(0 :_{I^i} \mathfrak{m})$ is a direct sum of simple R-modules, and each simple R-submodule of I^i has non-trivial intersection with $\operatorname{Ker} d^i$ and so is contained in $\operatorname{Ker} d^i$. $\qquad \square$

11.1.8 Theorem. (See [50, Theorem 18.7].) *Let M be an R-module. For each $i \in \mathbb{N}_0$ and each $\mathfrak{p} \in \operatorname{Spec}(R)$, the i-th Bass number $\mu^i(\mathfrak{p}, M)$ is given by*

$$\mu^i(\mathfrak{p}, M) = \dim_{k(\mathfrak{p})} \operatorname{Ext}^i_{R_\mathfrak{p}}(k(\mathfrak{p}), M_\mathfrak{p}) = \dim_{k(\mathfrak{p})}\left(\operatorname{Ext}^i_R(R/\mathfrak{p}, M)\right)_\mathfrak{p},$$

where $k(\mathfrak{p}) = R_{\mathfrak{p}}/\mathfrak{p}R_{\mathfrak{p}}$, the residue field of the local ring $R_{\mathfrak{p}}$. □

We have not explained the use of the word 'minimal' in the phrase 'minimal injective resolution'. We do this next.

11.1.9 Conventions. The injective dimension of an R-module M will be denoted by inj dim M or, occasionally, by inj $\dim_R M$ if it is desirable to specify the underlying ring concerned. We adopt the convention that a zero module has injective dimension $-\infty$.

11.1.10 Remark. Let M be an R-module, and let $n \in \mathbb{N}_0$.

(i) If $E^n(M) = 0$, then $E^{n+i}(M) = 0$ for all $i \in \mathbb{N}$, and, consequently, inj dim $M < n$.
(ii) Conversely, if inj dim $M < n$, then, for all $\mathfrak{p} \in \mathrm{Spec}(R)$, we have $\mathrm{Ext}_R^n(R/\mathfrak{p}, M) = 0$, so that $\mu^n(\mathfrak{p}, M) = 0$ by 11.1.8; it follows that $E^n(M) = 0$.

11.1.11 Exercise. Let M be an R-module, and let

$$I^{\bullet} : 0 \longrightarrow I^0 \xrightarrow{d^0} I^1 \longrightarrow \cdots \longrightarrow I^i \xrightarrow{d^i} I^{i+1} \longrightarrow \cdots$$

be a minimal injective resolution of M and

$$J^{\bullet} : 0 \longrightarrow J^0 \xrightarrow{\delta^0} J^1 \longrightarrow \cdots \longrightarrow J^i \xrightarrow{\delta^i} J^{i+1} \longrightarrow \cdots$$

be an arbitrary injective resolution of M. Prove that there is a chain map of complexes

$$\phi^{\bullet} = \left(\phi^i\right)_{i \in \mathbb{N}_0} : I^{\bullet} \longrightarrow J^{\bullet}$$

such that the diagram

$$
\begin{array}{ccc}
M & \xrightarrow{\alpha} & I^0 \\
\| & & \downarrow{\phi^0} \\
M & \xrightarrow{\beta} & J^0
\end{array}
$$

(in which the horizontal maps are the appropriate augmentation homomorphisms) commutes, and which is such that ϕ^i is a monomorphism for all $i \in \mathbb{N}_0$. Deduce that, for each $i \in \mathbb{N}_0$, the i-th term J^i has a direct summand isomorphic to $E^i(M)$.

11.2 Local Duality Theorems

We shall again refer to Matsumura's treatment of Gorenstein rings in [50, pp. 139–145]. We remind the reader of the following necessary and sufficient condition, in terms of the Bass numbers $\mu^i(\mathfrak{p}, R)$, for R to be a Gorenstein ring.

11.2.1 Reminder. (See [50, Theorems 18.1 and 18.8].) The ring R is a Gorenstein ring if and only if, for all $i \in \mathbb{N}_0$ and all $\mathfrak{p} \in \operatorname{Spec}(R)$,

$$\mu^i(\mathfrak{p}, R) = \begin{cases} 0 & \text{if } i \neq \operatorname{ht} \mathfrak{p}, \\ 1 & \text{if } i = \operatorname{ht} \mathfrak{p}, \end{cases}$$

that is, if and only if $E^i(R) \cong \bigoplus_{\substack{\mathfrak{p} \in \operatorname{Spec}(R) \\ \operatorname{ht} \mathfrak{p} = i}} E(R/\mathfrak{p})$ for all $i \in \mathbb{N}_0$.

11.2.2 Exercise. Suppose that $(R, \mathfrak{m})$ is a Gorenstein local ring of dimension n, and let

$$I^\bullet : 0 \longrightarrow I^0 \xrightarrow{d^0} I^1 \longrightarrow \cdots \longrightarrow I^{n-1} \xrightarrow{d^{n-1}} I^n \longrightarrow 0$$

be a minimal injective resolution of R. (We are using 11.2.1 when we 'terminate' this resolution after the n-th term.) Set $E := I^n$, so that $E \cong E(R/\mathfrak{m})$ by 11.2.1. Let $V := \{\mathfrak{p} \in \operatorname{Spec}(R) : \operatorname{ht} \mathfrak{p} = n - 1 \text{ and } \mathfrak{p} \supseteq \mathfrak{a}\}$, and, for each $\mathfrak{p} \in V$, let $\mathfrak{c}(\mathfrak{p})$ denote the kernel of the natural ring homomorphism $R \longrightarrow R_\mathfrak{p}$.

(i) Prove that $\operatorname{Im} \Gamma_\mathfrak{a}(d^{n-1}) = \sum_{\mathfrak{p} \in V} \left(\bigcup_{i \in \mathbb{N}} (0 :_E \mathfrak{p}^{(i)}) \right)$.

(ii) Assume now that R is complete. Prove that

$$H_\mathfrak{a}^n(R) \cong \operatorname{Hom}_R \left(\bigcap_{\mathfrak{p} \in V} \mathfrak{c}(\mathfrak{p}), E \right).$$

(The results of this exercise provided the key steps in a short proof of the local Lichtenbaum–Hartshorne Vanishing Theorem (see 8.2.1) given by F. W. Call and R. Y. Sharp in [8]: a reader who finds this exercise difficult is referred to that paper for details.)

We now start our approach to the Local Duality Theorem.

11.2.3 Lemma. *Let* $(R, \mathfrak{m})$ *be a Gorenstein local ring of dimension n. Then* $H_\mathfrak{m}^n(R) \cong E(R/\mathfrak{m})$, *the injective envelope of the simple R-module $R/\mathfrak{m}$.*

Proof. Let

$$I^\bullet : 0 \longrightarrow I^0 \xrightarrow{d^0} I^1 \longrightarrow \cdots \longrightarrow I^{n-1} \xrightarrow{d^{n-1}} I^n \longrightarrow 0$$

be a minimal injective resolution of R. We calculate $H_{\mathfrak{m}}^n(R)$ by working out the n-th cohomology module of the complex $\Gamma_{\mathfrak{m}}(I^{\bullet})$.

By 11.2.1,

$$I^i \cong \bigoplus_{\substack{\mathfrak{p} \in \text{Spec}(R) \\ \text{ht } \mathfrak{p} = i}} E(R/\mathfrak{p}) \quad \text{for all } i \in \mathbb{N}_0.$$

Let $\mathfrak{p} \in \text{Spec}(R)$ with $\mathfrak{p} \neq \mathfrak{m}$. Then there exists $r \in \mathfrak{m} \setminus \mathfrak{p}$. Now, by 10.1.3(iii), multiplication by r provides an automorphism of $E(R/\mathfrak{p})$, and so

$$\Gamma_{\mathfrak{m}}(E(R/\mathfrak{p})) = 0.$$

Hence $\Gamma_{\mathfrak{m}}(I^i) = 0$ for all $i \in \mathbb{N}_0$ with $i \neq n$.

On the other hand, $E(R/\mathfrak{m})$ is $\mathfrak{m}$-torsion, again by 10.1.3(iii). Thus all the terms other than the n-th of the complex $\Gamma_{\mathfrak{m}}(I^{\bullet})$ are zero, while its n-th term is isomorphic to $E(R/\mathfrak{m})$. Hence $H_{\mathfrak{m}}^n(R) \cong E(R/\mathfrak{m})$. $\qquad\square$

11.2.4 ♯Exercise. This exercise is central to our treatment of local duality.

(i) Let $T : \mathcal{C}(R) \longrightarrow \mathcal{C}(R)$ be a covariant R-linear functor, and let B be an R-module.

Show that there is a natural transformation of functors

$$\phi : T \longrightarrow \text{Hom}_R(\text{Hom}_R(\,\bullet\,, B), T(B))$$

(from $\mathcal{C}(R)$ to itself) which, for an R-module M, is such that

$$(\phi_M(y))(f) = T(f)(y) \quad \text{for all } y \in T(M) \text{ and all } f \in \text{Hom}_R(M, B).$$

Show also that, if each endomorphism of B can be realized as multiplication by precisely one element of R, then ϕ_B is an isomorphism.

(ii) Now suppose that $(R, \mathfrak{m})$ is a Gorenstein local ring of dimension n; set $E := E(R/\mathfrak{m})$ and $D := \text{Hom}_R(\,\bullet\,, E)$.

Use part (i) and 11.2.3 to show that there is a natural transformation of functors

$$\phi_0 : H_{\mathfrak{m}}^n \longrightarrow D(\text{Hom}_R(\,\bullet\,, R))$$

(from $\mathcal{C}(R)$ to itself) which is such that $\phi_{0\,R}$ is an isomorphism.

Let M be a finitely generated R-module. Use Grothendieck's Vanishing Theorem 6.1.2, together with the fact that M can be included in an exact sequence $F_1 \longrightarrow F_0 \longrightarrow M \longrightarrow 0$ in which F_1 and F_0 are finitely generated free R-modules, to show that $\phi_{0\,M}$ is an isomorphism.

We can now present the Local Duality Theorem for a Gorenstein local ring.

11.2.5 Local duality for a Gorenstein local ring. *Let* $(R, \mathfrak{m})$ *be a local Gorenstein ring of dimension* n*; set* $E := E(R/\mathfrak{m})$ *and* $D := \operatorname{Hom}_R(\,\bullet\,, E)$*. By 11.2.4, there is a natural transformation of functors*

$$\phi_0 : H_{\mathfrak{m}}^n \longrightarrow D(\operatorname{Hom}_R(\,\bullet\,, R))$$

(from $\mathcal{C}(R)$ *to itself) which is such that* $\phi_{0\,M}$ *is an isomorphism for every finitely generated* R*-module* M*.*

There is a unique extension of ϕ_0 *to a homomorphism*

$$\Phi := (\phi_i)_{i \in \mathbb{N}_0} : \left(H_{\mathfrak{m}}^{n-i} \right)_{i \in \mathbb{N}_0} \longrightarrow \left(D(\operatorname{Ext}_R^i(\,\bullet\,, R)) \right)_{i \in \mathbb{N}_0}$$

of (positive strongly) connected sequences of covariant functors from $\mathcal{C}(R)$ *to* $\mathcal{C}(R)$*. Furthermore,* $\phi_{i\,M}$ *is an isomorphism for all* $i \in \mathbb{N}_0$ *whenever* M *is a finitely generated* R*-module.*

In particular, for each finitely generated R*-module* M*,*

$$H_{\mathfrak{m}}^{n-i}(M) \cong D(\operatorname{Ext}_R^i(M, R)) \quad \text{for all } i \in \mathbb{Z}.$$

Proof. By Grothendieck's Vanishing Theorem 6.1.2, the functor $H_{\mathfrak{m}}^n$ is right exact; it is an easy consequence of this that $\left(H_{\mathfrak{m}}^{n-i} \right)_{i \in \mathbb{N}_0}$ is a positive strongly connected sequence of covariant functors.

Since depth $R = n$, it follows from 6.2.8 and 3.4.10 that $H_{\mathfrak{m}}^{n-i}(P) = 0$ for all $i \in \mathbb{N}$ and all projective R-modules P; also $D(\operatorname{Ext}_R^i(P, R)) = 0$ for all $i \in \mathbb{N}$ and all projective R-modules P. It follows from the analogue of 1.3.4 for positive connected sequences that ϕ_0 can be incorporated into a (uniquely determined) homomorphism

$$\Phi := (\phi_i)_{i \in \mathbb{N}_0} : \left(H_{\mathfrak{m}}^{n-i} \right)_{i \in \mathbb{N}_0} \longrightarrow \left(D(\operatorname{Ext}_R^i(\,\bullet\,, R)) \right)_{i \in \mathbb{N}_0}$$

of connected sequences. Furthermore, it is easy to prove by induction that, for each $i \in \mathbb{N}$, the homomorphism $\phi_{i\,M}$ is an isomorphism whenever M is a finitely generated R-module: use the fact that such an M can be included in an exact sequence $0 \longrightarrow K \longrightarrow F \longrightarrow M \longrightarrow 0$ in which F is a finitely generated free R-module.   □

11.2.6 Local Duality Theorem. *Suppose that* $(R, \mathfrak{m})$ *is a local ring which is a homomorphic image of a Gorenstein local ring* $(R', \mathfrak{m}')$ *of dimension* n'*: let* $f : R' \to R$ *be a surjective ring homomorphism. Set* $E := E(R/\mathfrak{m})$ *and* $D := \operatorname{Hom}_R(\,\bullet\,, E)$*.*

An R*-module* M *can be regarded as an* R'*-module by means of* f*: then* M*, and, for each* $j \in \mathbb{N}_0$*, the modules* $\operatorname{Ext}_{R'}^j(M, R')$ *and* $D(\operatorname{Ext}_{R'}^j(M, R'))$*, inherit natural* (R, R')*-bimodule structures.*

There is a homomorphism

$$\Phi := (\phi^i)_{i \in \mathbb{N}_0} : \left(H_{\mathfrak{m}}^i\right)_{i \in \mathbb{N}_0} \longrightarrow \left(D(\mathrm{Ext}_{R'}^{n'-i}(\,\bullet\, , R'))\right)_{i \in \mathbb{N}_0}$$

of (negative strongly) connected sequences of covariant functors from $\mathcal{C}(R)$ to $\mathcal{C}(R)$ which is such that ϕ_M^i is an isomorphism for all $i \in \mathbb{N}_0$ whenever M is a finitely generated R-module.

Consequently (since $\mathrm{inj}\dim_{R'} R' = n'$*), for each finitely generated R-module M,*

$$H_{\mathfrak{m}}^i(M) \cong D(\mathrm{Ext}_{R'}^{n'-i}(M, R')) = \mathrm{Hom}_R(\mathrm{Ext}_{R'}^{n'-i}(M, R'), E)$$

for all $i \in \mathbb{Z}$.

Proof.   Let $E' := E_{R'}(R'/\mathfrak{m}')$, and let $D' := \mathrm{Hom}_{R'}(\,\bullet\, , E')$.

By 11.2.5, there is a homomorphism

$$\Psi := (\psi_i)_{i \in \mathbb{N}_0} : \left(H_{\mathfrak{m}'}^{n'-i}\right)_{i \in \mathbb{N}_0} \longrightarrow \left(D'(\mathrm{Ext}_{R'}^i(\,\bullet\, , R'))\right)_{i \in \mathbb{N}_0}$$

of (positive strongly) connected sequences of covariant functors from $\mathcal{C}(R')$ to $\mathcal{C}(R')$ which is such that $\psi_{i\,M'}$ is an isomorphism for all $i \in \mathbb{N}_0$ whenever M' is a finitely generated R'-module. We can interpret

$$(\psi_{n'-i})_{i \in \mathbb{N}_0} : \left(H_{\mathfrak{m}'}^i\right)_{i \in \mathbb{N}_0} \longrightarrow \left(D'(\mathrm{Ext}_{R'}^{n'-i}(\,\bullet\, , R'))\right)_{i \in \mathbb{N}_0}$$

as a homomorphism of negative connected sequences.

Let $\lceil_{R'} : \mathcal{C}(R) \to \mathcal{C}(R')$ denote the functor obtained from restriction of scalars (using f). If we precede each $\psi_{n'-i}$ by $\lceil_{R'}$, we can deduce from the Independence Theorem 4.2.1 that there is a homomorphism

$$(\widetilde{\psi}^i)_{i \in \mathbb{N}_0} : \left(H_{\mathfrak{m}}^i\right)_{i \in \mathbb{N}_0} \longrightarrow \left(D'(\mathrm{Ext}_{R'}^{n'-i}(\,\bullet\, , R'))\right)_{i \in \mathbb{N}_0}$$

of (negative strongly) connected sequences of covariant functors from $\mathcal{C}(R)$ to $\mathcal{C}(R')$ which is such that $\widetilde{\psi}_M^i$ is an isomorphism for all $i \in \mathbb{N}_0$ whenever M is a finitely generated R-module.

Let $\mathfrak{c} := \mathrm{Ker}\, f$. Let M be an arbitrary R-module. Then $(0 :_{E'} \mathfrak{c})$, and, for each $j \in \mathbb{N}_0$, the modules $\mathrm{Ext}_{R'}^j(M, R')$ and $D'(\mathrm{Ext}_{R'}^j(M, R'))$ all inherit natural R-module structures: for each of these modules, and for $H_{\mathfrak{m}}^j(M)$, we have $f(r')y = r'y$ for all $r' \in R'$ and y in the module. It follows that

$$D'(\mathrm{Ext}_{R'}^j(M, R')) = \mathrm{Hom}_{R'}(\mathrm{Ext}_{R'}^j(M, R'), E')$$
$$= \mathrm{Hom}_{R'}(\mathrm{Ext}_{R'}^j(M, R'), (0 :_{E'} \mathfrak{c}))$$
$$= \mathrm{Hom}_R(\mathrm{Ext}_{R'}^j(M, R'), (0 :_{E'} \mathfrak{c})).$$

Thus, for each $i \in \mathbb{N}_0$, we can regard

$$\widetilde{\psi}_M^i : H_{\mathfrak{m}}^i(M) \longrightarrow \operatorname{Hom}_R(\operatorname{Ext}_{R'}^{n'-i}(M, R'), (0 :_{E'} \mathfrak{c}))$$

as an R-homomorphism. All that remains in order to complete the proof is to note that, by 10.1.16, there are R-isomorphisms

$$(0 :_{E'} \mathfrak{c}) \cong E_R(R'/\mathfrak{m}') \cong E_R(R/\mathfrak{m}) = E. \qquad \square$$

The following exercise will be useful in applications of the Local Duality Theorem.

11.2.7 ♯Exercise. Suppose that $(R, \mathfrak{m})$ is a local ring which is a homomorphic image of a local ring $(R', \mathfrak{m}')$: let $f : R' \to R$ be a surjective ring homomorphism. Let $\mathfrak{p} \in \operatorname{Spec}(R)$, and let $\mathfrak{p}' = f^{-1}(\mathfrak{p}), \in \operatorname{Spec}(R')$. Of course, $R'/\mathfrak{p}' \cong R/\mathfrak{p}$, so that these rings have equal dimensions. Moreover, observe that f induces a surjective ring homomorphism $f' : R'_{\mathfrak{p}'} \longrightarrow R_{\mathfrak{p}}$ which is such that $f'(r'/s') = f(r')/f(s')$ for all $r' \in R'$ and $s' \in R' \setminus \mathfrak{p}'$.

Let M be a finitely generated R-module, and let $j \in \mathbb{N}_0$. We can form $\operatorname{Ext}_{R'_{\mathfrak{p}'}}^j(M_{\mathfrak{p}}, R'_{\mathfrak{p}'})$, which has a natural $R_{\mathfrak{p}}$-module structure. We can also localize the R-module $\operatorname{Ext}_{R'}^j(M, R')$ at $\mathfrak{p}$. Show that

$$\operatorname{Ext}_{R'_{\mathfrak{p}'}}^j(M_{\mathfrak{p}}, R'_{\mathfrak{p}'}) \cong \left(\operatorname{Ext}_{R'}^j(M, R') \right)_{\mathfrak{p}} \qquad \text{as } R_{\mathfrak{p}}\text{-modules.}$$

The next exercise illustrates how the Local Duality Theorem can be used to obtain some of the results of 7.3.2 and 7.3.3 in the case where the local ring concerned is a homomorphic image of a Gorenstein local ring.

11.2.8 Exercise. Suppose that $(R, \mathfrak{m})$ is a local ring which is a homomorphic image of a Gorenstein local ring. Let M be a finitely generated R-module. Let $\mathfrak{p} \in \operatorname{Spec}(R)$, and let $\dim R/\mathfrak{p} = t$.

(i) Let $\mathfrak{q} \in \operatorname{Spec}(R)$ be such that $\mathfrak{q} \subseteq \mathfrak{p}$, and let $i \in \mathbb{Z}$. By 7.1.3, the $R_{\mathfrak{p}}$-module $H_{\mathfrak{p}R_{\mathfrak{p}}}^i(M_{\mathfrak{p}})$ is Artinian, and $H_{\mathfrak{m}}^{i+t}(M)$ is an Artinian R-module. Show that

$$\mathfrak{q}R_{\mathfrak{p}} \in \operatorname{Att}_{R_{\mathfrak{p}}}\left(H_{\mathfrak{p}R_{\mathfrak{p}}}^i(M_{\mathfrak{p}}) \right) \quad \text{if and only if} \quad \mathfrak{q} \in \operatorname{Att}_R(H_{\mathfrak{m}}^{i+t}(M)).$$

(ii) Suppose $\mathfrak{p} \in \operatorname{Ass} M$. Show that $H_{\mathfrak{m}}^t(M) \neq 0$ and $\mathfrak{p} \in \operatorname{Att}(H_{\mathfrak{m}}^t(M))$. Show further that, if $t > 0$, then $H_{\mathfrak{m}}^t(M)$ is not finitely generated.

(iii) Deduce from (ii) that, if M is non-zero and of dimension n, then $H_{\mathfrak{m}}^n(M) \neq 0$ and $\{\mathfrak{p} \in \operatorname{Ass} M : \dim R/\mathfrak{p} = n\} \subseteq \operatorname{Att}(H_{\mathfrak{m}}^n(M))$; deduce also that, if $n > 0$, then $H_{\mathfrak{m}}^n(M)$ is not finitely generated. Compare 7.3.2 and 7.3.3.

11.2.9 Exercise. Suppose that $(R, \mathfrak{m})$ is a local ring which is a homomorphic image of a Gorenstein local ring. Let M be a finitely generated R-module, and let j be an integer for which $H_{\mathfrak{m}}^j(M) \neq 0$. Show that, for each $\mathfrak{p} \in \mathrm{Att}_R(H_{\mathfrak{m}}^j(M))$, we have $\dim R/\mathfrak{p} \leq j$.

Among the hypotheses for 11.2.8 and 11.2.9 was the assumption that the underlying local ring is a homomorphic image of a Gorenstein local ring. Unfortunately, the results of those exercises are not all true for general local rings, as can be seen from Exercise 11.2.13 below.

The next two exercises show that one of the implications in 11.2.8(i) is true in general. These two exercises are directed at those readers who are experienced at working with the fibres of flat homomorphisms of commutative Noetherian rings.

11.2.10 Exercise. Suppose that $(R, \mathfrak{m})$ is a local ring, that $(R', \mathfrak{m}')$ is a second local ring, and that $f : R \longrightarrow R'$ is a flat local ring homomorphism such that the extension $\mathfrak{m}R'$ of $\mathfrak{m}$ to R' under f is $\mathfrak{m}'$-primary. Let $i \in \mathbb{N}_0$, and let M be a non-zero, finitely generated R-module with the property that $H_{\mathfrak{m}}^i(M) \neq 0$.

(i) Use the Flat Base Change Theorem 4.3.2 to see that there is an R'-isomorphism $H_{\mathfrak{m}'}^i(M \otimes_R R') \cong H_{\mathfrak{m}}^i(M) \otimes_R R'$.

(ii) Let

$$H_{\mathfrak{m}}^i(M) = S_1 + \cdots + S_h \quad \text{with } S_j \ \mathfrak{p}_j\text{-secondary } (1 \leq j \leq h)$$

be a minimal secondary representation of $H_{\mathfrak{m}}^i(M)$. For all $j = 1, \ldots, h$, let $u_j : S_j \to H_{\mathfrak{m}}^i(M)$ be the inclusion map, and let

$$T_j := (u_j \otimes \mathrm{Id}_{R'})(S_j \otimes_R R').$$

Show that $H_{\mathfrak{m}}^i(M) \otimes_R R' = T_1 + \cdots + T_h$ and that

$$\{f^{-1}(\mathfrak{P}) : \mathfrak{P} \in \mathrm{Att}_{R'}(T_j)\} = \{\mathfrak{p}_j\} \quad \text{for all } j = 1, \ldots, h.$$

(iii) Show that

$$\mathrm{Att}_R(H_{\mathfrak{m}}^i(M)) = \{f^{-1}(\mathfrak{P}) : \mathfrak{P} \in \mathrm{Att}_{R'}(H_{\mathfrak{m}'}^i(M \otimes_R R'))\} .$$

11.2.11 Exercise. Suppose that $(R, \mathfrak{m})$ is a local ring. Let M be a finitely generated R-module. Let $\mathfrak{p} \in \mathrm{Spec}(R)$ and let $\dim R/\mathfrak{p} = t$. Let $i \in \mathbb{Z}$ and $\mathfrak{q} \in \mathrm{Spec}(R)$ be such that $\mathfrak{q} \subseteq \mathfrak{p}$ and $\mathfrak{q}R_{\mathfrak{p}} \in \mathrm{Att}_{R_{\mathfrak{p}}}\left(H_{\mathfrak{p}R_{\mathfrak{p}}}^i(M_{\mathfrak{p}})\right)$. Prove that $\mathfrak{q} \in \mathrm{Att}_R(H_{\mathfrak{m}}^{i+t}(M))$.

(Here are some hints: use Exercise 11.2.10, and Cohen's Structure Theorem that a complete local ring is a homomorphic image of a regular local ring, in conjunction with 11.2.8(i).)

11.2.12 Exercise.   Use 11.2.11 to extend the result of 11.2.8(ii) to all local rings.

Thus, in detail, suppose that $(R, \mathfrak{m})$ is a local ring. Let M be a non-zero, finitely generated R-module, and let $\mathfrak{p} \in \operatorname{Ass} M$. Suppose that $\dim R/\mathfrak{p} = t$. Prove that $H_{\mathfrak{m}}^t(M) \neq 0$ and $\mathfrak{p} \in \operatorname{Att}(H_{\mathfrak{m}}^t(M))$.

11.2.13 Exercise.   This exercise shows that, if, in 11.2.9, the hypothesis that the local ring $(R, \mathfrak{m})$ be a homomorphic image of a Gorenstein local ring is dropped, then the corresponding statement is no longer always true. At the same time, it shows that the result of 11.2.8(i) is not true for every local ring.

For a counterexample, suppose that $(R, \mathfrak{m})$ is a 2-dimensional local domain whose completion $(\widehat{R}, \widehat{\mathfrak{m}})$ possesses an embedded prime ideal $\mathfrak{P}$ (associated to its zero ideal). Such an example has been constructed by D. Ferrand and M. Raynaud in [18]. Use 11.2.12 and 11.2.10 to prove that $0 \in \operatorname{Att}(H_{\mathfrak{m}}^1(R))$, and deduce that the result of 11.2.9 is not true for every local ring.

Show also that, for each $\mathfrak{p} \in \operatorname{Spec}(R)$ having $\dim R/\mathfrak{p} = 1$,

$$0R_{\mathfrak{p}} \notin \operatorname{Att}_{R_{\mathfrak{p}}}\left(H_{\mathfrak{p}R_{\mathfrak{p}}}^0(R_{\mathfrak{p}})\right).$$

Deduce that the result of 11.2.8(i) is not true for every local ring.

12

Canonical modules

Suppose that $(R, \mathfrak{m})$ is local of dimension $n > 0$. We have seen earlier in the book some illustrations of the importance of the so-called 'top' local cohomology module $H_{\mathfrak{m}}^n(R)$ of R. (It is referred to as the 'top' local cohomology module because $H_{\mathfrak{m}}^i(R) = 0$ for all $i > n$ by Grothendieck's Vanishing Theorem 6.1.2, whereas $H_{\mathfrak{m}}^n(R) \neq 0$ by the Non-vanishing Theorem 6.1.4.) Now $H_{\mathfrak{m}}^n(R)$ is an Artinian R-module (by 7.1.3) and is not finitely generated; commutative algebraists tend to be brought up to work with finitely generated modules, and Artinian modules are perhaps a little less familiar. The philosophy behind the concept of canonical module is the following: wouldn't it be nice if we could, in some sense, replace $H_{\mathfrak{m}}^n(R)$ by some finitely generated R-module that we could work with effectively to achieve our desired results?

The theory of canonical modules for Cohen–Macaulay local rings is developed by Bruns and Herzog in [7, Chapter 3]; their account is partly based on the lecture notes of Herzog and E. Kunz [33]. However, here we are going to work in the more general setting of an arbitrary local ring $(R, \mathfrak{m})$, and define a canonical module for R to be a finitely generated R-module K whose Matlis dual $\operatorname{Hom}_R(K, E_R(R/\mathfrak{m}))$ is isomorphic to $H_{\mathfrak{m}}^n(R)$, where $n := \dim R$. In the special case where R is Cohen–Macaulay, this condition turns out to be equivalent to Bruns' and Herzog's definition, namely that a canonical module for R is a Cohen–Macaulay R-module K of dimension n for which $\operatorname{inj\,dim}_R K < \infty$ and $\mu^n(\mathfrak{m}, K) = 1$. The more general approach that we shall take is based on work of Y. Aoyama [1], [2]; the definition we use is that employed by Hochster and Huneke in [39].

If R is a homomorphic image of a Gorenstein local ring R' of dimension n' then the Local Duality Theorem 11.2.6 shows that the R-module

$$\operatorname{Ext}_{R'}^{n'-n}(R, R')$$

is a canonical module for R. Thus the assumption that our local ring is a homomorphic image of a Gorenstein local ring again appears in hypotheses. In fact, there is a result, proved independently by H.-B. Foxby [19] and I. Reiten [69], to the effect that the Cohen–Macaulay local ring R admits a canonical module if and only if R is a homomorphic image of a Gorenstein local ring.

It turns out that canonical modules are intimately related to Serre's condition S_2. Recall that a non-zero finitely generated R-module M (here R is not assumed to be local) is said to satisfy Serre's condition S_2 if and only if $\text{depth}_{R_\mathfrak{p}} M_\mathfrak{p} \geq \min\{2, \dim_{R_\mathfrak{p}} M_\mathfrak{p}\}$ for all $\mathfrak{p} \in \text{Supp } M$. When R is a normal domain, R itself satisfies S_2. It turns out that, if K is a canonical module for the local ring R, then $\text{depth } K \geq \min\{2, \dim K\}$, and this is the key to the connection with the condition S_2. Furthermore, the endomorphism ring $\text{Hom}_R(K, K)$ is a semi-local commutative Noetherian ring that, under mild conditions on R, acts as the so-called 'S_2-ification' of R. (We did not invent that name!) There are links between this concept and the generalized ideal transforms we studied in Chapter 2.

The concept of canonical module is of fundamental importance in the study of Cohen–Macaulay local rings, and therefore we shall reconcile our approach with that of Bruns and Herzog. In connection with the fact that a canonical module over a Cohen–Macaulay local ring has finite injective dimension, it is perhaps worth noting that the present second author proved in [79, Corollary 2.3], that, over a Cohen–Macaulay homomorphic image R of a Gorenstein local ring, the finitely generated R-modules of finite injective dimension are precisely those R-modules M that can be included in an exact sequence

$$0 \longrightarrow G_s \longrightarrow G_{s-1} \longrightarrow \cdots \longrightarrow G_1 \longrightarrow G_0 \longrightarrow M \longrightarrow 0$$

in which each of $G_0, \ldots, G_s$ is a direct sum of finitely many copies of the canonical R-module K. However, the proof of that result is beyond the scope of this book.

12.1 Definition and basic properties

Many of the results in this and the next section are due to Y. Aoyama [1], [2]. Here, we shall adopt the definition of canonical module used by M. Hochster and C. Huneke in [39].

12.1.1 Notation. Throughout this section, we shall assume $(R, \mathfrak{m})$ is local, and we shall use n to denote $\dim R$ and $(\widehat{R}, \widehat{\mathfrak{m}})$ to denote the completion of R.

12.1.2 Definition. A *canonical module for* R is a finitely generated R-module K such that $\text{Hom}_R(K, E_R(R/\mathfrak{m})) \cong H^n_\mathfrak{m}(R)$.

12.1.3 Remarks. We can make the following observations right away.

(i) Recall that the R-module $H^n_{\mathfrak{m}}(R)$ is Artinian (by 7.1.3), and that the Matlis dual of a finitely generated R-module is (by 10.2.19(ii)) also Artinian.

(ii) Suppose that R has a canonical module K. There are $\widehat{R}$-isomorphisms

$$E_{\widehat{R}}(\widehat{R}/\widehat{\mathfrak{m}}) \cong E_R(R/\mathfrak{m}) \otimes_R \widehat{R}$$

(by 8.2.4, 10.2.9 and 10.2.10) and $H^n_{\widehat{\mathfrak{m}}}(\widehat{R}) \cong H^n_{\mathfrak{m}}(R) \otimes_R \widehat{R}$ (by the Flat Base Change Theorem 4.3.2). Therefore, by [50, Theorem 7.11], there are $\widehat{R}$-isomorphisms

$$H^n_{\widehat{\mathfrak{m}}}(\widehat{R}) \cong H^n_{\mathfrak{m}}(R) \otimes_R \widehat{R} \cong \operatorname{Hom}_R(K, E_R(R/\mathfrak{m})) \otimes_R \widehat{R}$$
$$\cong \operatorname{Hom}_{\widehat{R}}(K \otimes_R \widehat{R}, E_R(R/\mathfrak{m}) \otimes_R \widehat{R})$$
$$\cong \operatorname{Hom}_{\widehat{R}}(K \otimes_R \widehat{R}, E_{\widehat{R}}(\widehat{R}/\widehat{\mathfrak{m}})).$$

Therefore $K \otimes_R \widehat{R}$ is a canonical module for $\widehat{R}$.

(iii) It is immediate from the Local Duality Theorem 11.2.6 that, if R is a homomorphic image of an n'-dimensional Gorenstein local ring R', then R has a canonical module, namely $\operatorname{Ext}^{n'-n}_{R'}(R, R')$.

(iv) In particular, if R itself is Gorenstein, then R is a canonical module for itself.

(v) In particular, when R is complete, so that it is a homomorphic image of a regular local ring by Cohen's Structure Theorem, R has a canonical module; in fact, in that case, it follows from the Matlis Duality Theorem 10.2.12 that each canonical module for R is isomorphic to $\operatorname{Hom}_R(H^n_{\mathfrak{m}}(R), E_R(R/\mathfrak{m}))$; thus, when R is complete, there is, up to isomorphism, exactly one canonical module for R.

(vi) Suppose, in the case where the local ring $(R, \mathfrak{m})$ is not necessarily complete, that K' is an R-module for which there is a $\widehat{R}$-isomorphism

$$K' \otimes_R \widehat{R} \cong \operatorname{Hom}_{\widehat{R}}(H^n_{\widehat{\mathfrak{m}}}(\widehat{R}), E_{\widehat{R}}(\widehat{R}/\widehat{\mathfrak{m}})).$$

Then, by the Matlis Duality Theorem 10.2.12 and the faithful flatness of $\widehat{R}$ over R, it follows that K' is a finitely generated R-module such that $K' \otimes_R \widehat{R}$ is a canonical module for $\widehat{R}$. One can use [50, Theorem 7.11], 8.2.4, 10.2.10 and 4.3.2 again to see that $\operatorname{Hom}_R(K', E_R(R/\mathfrak{m})) \otimes_R \widehat{R} \cong H^n_{\mathfrak{m}}(R) \otimes_R \widehat{R}$ as $\widehat{R}$-modules. Since $\operatorname{Hom}_R(K', E_R(R/\mathfrak{m}))$ and $H^n_{\mathfrak{m}}(R)$ are Artinian R-modules (by 10.2.19(ii) and 7.1.6 respectively), it follows from 8.2.4 that there is an R-isomorphism $\operatorname{Hom}_R(K', E_R(R/\mathfrak{m})) \cong H^n_{\mathfrak{m}}(R)$. Thus K' is a canonical module for R.

This and part (ii) reconcile the above Definition 12.1.2 with that used by Herzog and Kunz in [33, Definition 5.6].

Our next aim is to extend 12.1.3(v) by showing that, if R has a canonical module, then any two canonical modules for R are isomorphic.

12.1.4 Lemma. *Let K and L be two finitely generated R-modules. Then the set $\mathrm{Iso}_R(K, L)$ of all R-isomorphisms from K to L is an open subset of $\mathrm{Hom}_R(K, L)$ in the $\mathfrak{m}$-adic topology. In fact, if $\alpha \in \mathrm{Iso}_R(K, L)$, then*

$$\alpha + \mathfrak{m} \mathrm{Hom}_R(K, L) \subseteq \mathrm{Iso}_R(K, L).$$

Proof. If K and L are not isomorphic, then $\mathrm{Iso}_R(K, L) = \emptyset$, an open subset of $\mathrm{Hom}_R(K, L)$. So suppose that $\alpha : K \xrightarrow{\cong} L$ is an isomorphism. We shall show that $\alpha + \mathfrak{m} \mathrm{Hom}_R(K, L) \subseteq \mathrm{Iso}_R(K, L)$, and this will be enough to complete the proof.

So let $\beta \in \alpha + \mathfrak{m} \mathrm{Hom}_R(K, L)$. Then there exist $r_1, \ldots, r_t \in \mathfrak{m}$ and $\lambda_1, \ldots, \lambda_t \in \mathrm{Hom}_R(K, L)$ such that $\beta = \alpha + \sum_{i=1}^{t} r_i \lambda_i$. To show that $\beta \in \mathrm{Iso}_R(K, L)$, it is enough for us to show that $\alpha^{-1} \circ \beta$ is an automorphism of K, and, since K is finitely generated, it is therefore enough for us to show that $\alpha^{-1} \circ \beta$ is surjective (by [81, Exercise 7.2]). This we do. Now

$$\alpha^{-1} \circ \beta = \alpha^{-1} \circ \left(\alpha + \sum_{i=1}^{t} r_i \lambda_i\right) = \mathrm{Id}_K + \sum_{i=1}^{t} r_i \alpha^{-1} \circ \lambda_i.$$

Let $g \in K$. Then

$$g = \mathrm{Id}_K(g) = \left(\alpha^{-1} \circ \beta - \sum_{i=1}^{t} r_i \alpha^{-1} \circ \lambda_i\right)(g) \in \mathrm{Im}(\alpha^{-1} \circ \beta) + \mathfrak{m}K.$$

Therefore $K = \mathrm{Im}(\alpha^{-1} \circ \beta) + \mathfrak{m}K$, and so it follows from Nakayama's Lemma that $\alpha^{-1} \circ \beta$ is surjective, as required. $\quad\square$

12.1.5 Lemma. *Let R' and R'' be two commutative Noetherian rings and let $T : \mathcal{C}(R') \longrightarrow \mathcal{C}(R'')$ be an exact additive functor which is faithful in the sense that $T(M) = 0$, for an R'-module M, implies that $M = 0$. Let $f : M \longrightarrow N$ be an R'-homomorphism.*

(i) *If T is covariant and $T(f)$ is a monomorphism (respectively, an epimorphism), then f is a monomorphism (respectively, an epimorphism).*

(ii) *If T is contravariant and $T(f)$ is a monomorphism (respectively, an epimorphism), then f is an epimorphism (respectively, a monomorphism).*

Proof. These are all proved similarly, and so we just prove the first part of (ii). Let $C := \mathrm{Coker} f$. Apply T to the exact sequence

$$M \xrightarrow{f} N \longrightarrow C \longrightarrow 0$$

to obtain an exact sequence $0 \longrightarrow T(C) \longrightarrow T(N) \xrightarrow{T(f)} T(M)$. Since $T(f)$ is a monomorphism, we must have $T(C) = 0$. Since T is faithful, $C = 0$, so that f is an epimorphism. $\qquad\square$

12.1.6 Theorem. *Any two canonical modules K and K' for R are isomorphic.*

Proof. By 12.1.3(ii), both $K \otimes_R \widehat{R}$ and $K' \otimes_R \widehat{R}$ are canonical modules for $\widehat{R}$, and so, by 12.1.3(v), there is an $\widehat{R}$-isomorphism $\alpha : K \otimes_R \widehat{R} \xrightarrow{\cong} K' \otimes_R \widehat{R}$.

Since K and K' are finitely generated R-modules, $\mathrm{Hom}_R(K, K')$ is finitely generated, and so the canonical injective map

$$\mathrm{Hom}_R(K, K') \longrightarrow \mathrm{Hom}_R(K, K') \otimes_R \widehat{R}$$

provides the completion of $\mathrm{Hom}_R(K, K')$. Now, by [50, Theorem 7.11], there is an $\widehat{R}$-isomorphism

$$\lambda : \mathrm{Hom}_R(K, K') \otimes_R \widehat{R} \xrightarrow{\cong} \mathrm{Hom}_{\widehat{R}}(K \otimes_R \widehat{R}, K' \otimes_R \widehat{R})$$

which is such that $\lambda(f \otimes \widehat{r}) = \widehat{r}(f \otimes \mathrm{Id}_{\widehat{R}})$ for all $f \in \mathrm{Hom}_R(K, K')$ and $\widehat{r} \in \widehat{R}$.

Since the image of $\mathrm{Hom}_R(K, K')$ is dense in $\mathrm{Hom}_R(K, K') \otimes_R \widehat{R}$, there exists $g \in \mathrm{Hom}_R(K, K')$ such that

$$g \otimes 1 \in \lambda^{-1}(\alpha) + \widehat{\mathfrak{m}} \left(\mathrm{Hom}_R(K, K') \otimes_R \widehat{R} \right).$$

Apply λ to deduce that $g \otimes \mathrm{Id}_{\widehat{R}} \in \alpha + \widehat{\mathfrak{m}} \, \mathrm{Hom}_{\widehat{R}}(K \otimes_R \widehat{R}, K' \otimes_R \widehat{R})$. But, by 12.1.4, the coset $\alpha + \widehat{\mathfrak{m}} \, \mathrm{Hom}_{\widehat{R}}(K \otimes_R \widehat{R}, K' \otimes_R \widehat{R})$ consists entirely of isomorphisms; therefore $g \otimes \mathrm{Id}_{\widehat{R}}$ is an isomorphism from $K \otimes_R \widehat{R}$ to $K' \otimes_R \widehat{R}$, so that $g : K \longrightarrow K'$ is an isomorphism by 12.1.5, because $\widehat{R}$ is faithfully flat over R. $\qquad\square$

12.1.7 Remark. Suppose that there exists a canonical module for R. By 12.1.6, any two canonical modules for R are isomorphic. We shall denote by ω_R one choice of canonical module for R. We shall sometimes use the clause 'ω_R exists' as an abbreviation for 'there exists a canonical module for R'.

Note that, in particular, if R is a homomorphic image of an n'-dimensional Gorenstein local ring R', and also a homomorphic image of an n''-dimensional Gorenstein local ring R'', then, in view of 12.1.3(iii), there is an isomorphism $\mathrm{Ext}_{R'}^{n'-n}(R, R') \cong \mathrm{Ext}_{R''}^{n''-n}(R, R'')$ of R-modules.

12.1.8 Exercise. Let M be a non-zero finitely generated R-module, and suppose that R is a homomorphic image of a Gorenstein local ring R'. For $i \in \mathbb{N}_0$, let $K^i(M)$ denote the R-module $\mathrm{Ext}_{R'}^{\dim R' - i}(M, R')$.

(i) Show that $K^i(M) = 0$ for $i < \operatorname{depth}_R M$ and for $i > \dim M$, and that $K^i(M) \neq 0$ for $i = \operatorname{depth}_R M$ and for $i = \dim M$.

(ii) Show that, if R is complete, then, for a fixed $i \in \mathbb{N}_0$, the R-module $K^i(M)$ is, up to R-isomorphism, independent of the choice of the local Gorenstein ring R' (of which R is a homomorphic image).

(iii) Prove the result of (ii) without the assumption that R is complete. In other words, assume only that R is a homomorphic image of a local Gorenstein ring R' and prove that $K^i(M)$ is, up to R-isomorphism, independent of the choice of R'. (Here is a hint: use 12.1.4 in a manner similar to the way it was used in 12.1.6.)

The module $K^i(M)$, for $i \neq \dim M$, is called the *i-th module of deficiency of* M. In a sense, these modules give, if M is not Cohen–Macaulay, an indication of the extent of the failure of M to be Cohen–Macaulay. They have been studied by P. Schenzel in [74] and [75].

We shall see during the chapter that the existence of a canonical module for a local ring R imposes some restrictions on R.

12.1.9 Proposition. *Suppose that* $\dim R = n > 0$ *and that* ω_R *exists. Let* $a_1, \ldots, a_n$ *be a system of parameters for* R. *Then*

(i) $\operatorname{Ass}_R \omega_R = \{\mathfrak{p} \in \operatorname{Spec}(R) : \dim R/\mathfrak{p} = n\}$; *and*

(ii) a_1 *is a non-zerodivisor on* ω_R *and, if* $n \geq 2$, *then* a_1, a_2 *is an* ω_R*-sequence.*

Proof. Let D denote the Matlis duality functor $\operatorname{Hom}_R(\bullet, E_R(R/\mathfrak{m}))$.

(i) By definition, $D(\omega_R) \cong H^n_{\mathfrak{m}}(R)$; also,

$$\operatorname{Att}(H^n_{\mathfrak{m}}(R)) = \{\mathfrak{p} \in \operatorname{Spec}(R) : \dim R/\mathfrak{p} = n\}$$

by 7.3.2; therefore, by 10.2.20,

$$\operatorname{Ass} \omega_R = \operatorname{Att}(D(\omega_R)) = \{\mathfrak{p} \in \operatorname{Spec}(R) : \dim R/\mathfrak{p} = n\}.$$

(ii) Since $\dim R/Ra_1 = n - 1$, it follows from part (i) that a_1 does not belong to any associated prime of ω_R. Therefore a_1 is an ω_R-sequence.

Now suppose that $n \geq 2$. Let $\alpha : R/(0 :_R a_1) \longrightarrow R$ be the R-monomorphism induced by multiplication by a_1. The exact sequence

$$0 \longrightarrow R/(0 :_R a_1) \overset{\alpha}{\longrightarrow} R \longrightarrow R/Ra_1 \longrightarrow 0$$

yields an exact sequence

$$H^{n-1}_{\mathfrak{m}}(R/a_1 R) \longrightarrow H^n_{\mathfrak{m}}(R/(0 :_R a_1)) \overset{H^n_{\mathfrak{m}}(\alpha)}{\longrightarrow} H^n_{\mathfrak{m}}(R) \longrightarrow 0$$

because $\dim R/a_1R < n$. Let $\pi : R \longrightarrow R/(0 :_R a_1)$ be the natural epimorphism. The exact sequence

$$0 \longrightarrow (0 :_R a_1) \longrightarrow R \xrightarrow{\pi} R/(0 :_R a_1) \longrightarrow 0$$

yields the isomorphism $H_{\mathfrak{m}}^n(\pi) : H_{\mathfrak{m}}^n(R) \xrightarrow{\cong} H_{\mathfrak{m}}^n(R/(0 :_R a_1))$ because $\dim(0 :_R a_1) < n$. It follows that there is an exact sequence

$$H_{\mathfrak{m}}^{n-1}(R/a_1R) \longrightarrow H_{\mathfrak{m}}^n(R) \xrightarrow{a_1} H_{\mathfrak{m}}^n(R) \longrightarrow 0.$$

But the exact sequence $0 \longrightarrow \omega_R \xrightarrow{a_1} \omega_R \longrightarrow \omega_R/a_1\omega_R \longrightarrow 0$ yields the exact sequence

$$0 \longrightarrow D(\omega_R/a_1\omega_R) \longrightarrow D(\omega_R) \xrightarrow{a_1} D(\omega_R) \longrightarrow 0.$$

Since $H_{\mathfrak{m}}^n(R) \cong D(\omega_R)$, it follows that $D(\omega_R/a_1\omega_R)$ is a homomorphic image of $H_{\mathfrak{m}}^{n-1}(R/a_1R)$.

The natural images of $a_2, \ldots, a_n$ in R/a_1R are a system of parameters for that $(n-1)$-dimensional local ring, and, by the Independence Theorem 4.2.1, we have $H_{\mathfrak{m}}^{n-1}(R/a_1R) \cong H_{\mathfrak{m}/a_1R}^{n-1}(R/a_1R)$. Now it follows from the above proof of part (i) that the natural image of a_2 is not in any attached prime ideal of $H_{\mathfrak{m}/a_1R}^{n-1}(R/a_1R)$, and therefore $a_2 H_{\mathfrak{m}}^{n-1}(R/a_1R) = H_{\mathfrak{m}}^{n-1}(R/a_1R)$. Since $D(\omega_R/a_1\omega_R)$ is a homomorphic image of $H_{\mathfrak{m}}^{n-1}(R/a_1R)$, it follows that

$$a_2 D(\omega_R/a_1\omega_R) = D(\omega_R/a_1\omega_R).$$

Therefore, as D is a contravariant faithful exact R-linear additive functor from $\mathcal{C}(R)$ to itself, we can use 12.1.5(ii) to see that a_2 is a non-zerodivisor on $\omega_R/a_1\omega_R$, so that a_1, a_2 is an ω_R-sequence. $\square$

12.1.10 ♯Exercise. Let the situation and notation be as in 12.1.9, and let M be a finitely generated R-module for which $\operatorname{Hom}_R(M, \omega_R) \neq 0$. Show that a_1 is a non-zerodivisor on $\operatorname{Hom}_R(M, \omega_R)$ and, if $n \geq 2$, then a_1, a_2 is an $\operatorname{Hom}_R(M, \omega_R)$-sequence.

Our next aim is to identify, in the case where ω_R exists, the annihilator of ω_R. By Matlis Duality, this is the same as the annihilator of $H_{\mathfrak{m}}^{\dim R}(R)$: see 10.2.2(ii).

12.1.11 ♯Exercise. Let $\mathfrak{c}$ be the intersection of the (uniquely determined) primary components $\mathfrak{q}$ of the zero ideal of R for which $\dim R/\mathfrak{q} = n$, and let

$$\mathfrak{g} := \bigcap_{\substack{\mathfrak{p} \in \operatorname{Ass} R \\ \dim R/\mathfrak{p} < n}} \mathfrak{p}.$$

(i) Show that

$$\mathfrak{c} = \Gamma_{\mathfrak{g}}(R)$$
$$= \{r \in R : \mathfrak{b}r = 0 \text{ for some ideal } \mathfrak{b} \subset R \text{ with } \dim R/\mathfrak{b} < n\}.$$

(ii) Let $\mathcal{J}(R)$ denote the set of all ideals of R. Set

$$\mathcal{L}(R) = \{\mathfrak{b} \in \mathcal{J}(R) : \dim \mathfrak{b} < n\}$$
$$= \{\mathfrak{b} \in \mathcal{J}(R) : \mathfrak{b}R_{\mathfrak{p}} = 0 \; \forall \, \mathfrak{p} \in \mathrm{Spec}(R) \text{ with } \dim R/\mathfrak{p} = n\}.$$

Recall from 7.3.1 that there is a maximum member $\mathfrak{d}$ of the set $\mathcal{L}(R)$. Show that $\mathfrak{d} = \mathfrak{c}$.

12.1.12 Notation. Bearing in mind 12.1.11, we shall denote the maximum member of the set

$$\{\mathfrak{b} : \mathfrak{b} \text{ is an ideal of } R \text{ with } \dim \mathfrak{b} < n\}$$

by $\mathfrak{u}_R(0)$. (Note that we really are discussing $\dim \mathfrak{b}$ for an ideal $\mathfrak{b}$, as opposed to $\dim R/\mathfrak{b}$.)

We shall use some of the properties of $\mathfrak{u}_R(0)$ established in 12.1.11.

12.1.13 Lemma. *A canonical module for R is annihilated by $\mathfrak{u}_R(0)$. Moreover, a finitely generated R-module K that is annihilated by $\mathfrak{u}_R(0)$ (so that it can be regarded as an $R/\mathfrak{u}_R(0)$-module in the natural way) is a canonical module for R if and only if it is a canonical module for $R/\mathfrak{u}_R(0)$.*

Proof. Since $\dim \mathfrak{u}_R(0) < n$, it follows from Grothendieck's Vanishing Theorem 6.1.2 and the exact sequence

$$0 \longrightarrow \mathfrak{u}_R(0) \longrightarrow R \longrightarrow R/\mathfrak{u}_R(0) \longrightarrow 0$$

that there is an isomorphism $H_{\mathfrak{m}}^n(R) \stackrel{\cong}{\longrightarrow} H_{\mathfrak{m}}^n(R/\mathfrak{u}_R(0))$. Therefore, if ω_R is a canonical module for R, then

$$\mathrm{Hom}_R(\omega_R, E_R(R/\mathfrak{m})) \cong H_{\mathfrak{m}}^n(R) \cong H_{\mathfrak{m}}^n(R/\mathfrak{u}_R(0)).$$

Since $\mathfrak{u}_R(0)$ annihilates $H_{\mathfrak{m}}^n(R/\mathfrak{u}_R(0))$, it follows that $\mathfrak{u}_R(0)$ annihilates

$$\mathrm{Hom}_R(\omega_R, E_R(R/\mathfrak{m})).$$

As an R-module and its Matlis dual have the same annihilator (by 10.2.2(ii)), it follows that $\mathfrak{u}_R(0)$ annihilates ω_R.

Now let K be a finitely generated R-module that is annihilated by $\mathfrak{u}_R(0)$, and denote $E_R(R/\mathfrak{m})$ by E. The second displayed isomorphism in the last paragraph, together with the Independence Theorem 4.2.1, show that there is an R-isomorphism $H_{\mathfrak{m}}^n(R) \cong H_{\mathfrak{m}/\mathfrak{u}_R(0)}^n(R/\mathfrak{u}_R(0))$. Now, in view of 10.1.16, we have

$$\mathrm{Hom}_R(K, E) = \mathrm{Hom}_R(K, (0 :_E \mathfrak{u}_R(0))) = \mathrm{Hom}_{R/\mathfrak{u}_R(0)}(K, (0 :_E \mathfrak{u}_R(0)))$$
$$\cong \mathrm{Hom}_{R/\mathfrak{u}_R(0)}\left(K, E_{R/\mathfrak{u}_R(0)}\left((R/\mathfrak{u}_R(0))/(\mathfrak{m}/\mathfrak{u}_R(0))\right)\right).$$

Therefore there is an R-isomorphism $\mathrm{Hom}_R(K, E) \cong H_{\mathfrak{m}}^n(R)$ if and only if there is an $R/\mathfrak{u}_R(0)$-isomorphism

$$\mathrm{Hom}_{R/\mathfrak{u}_R(0)}(K, E_{R/\mathfrak{u}_R(0)}((R/\mathfrak{u}_R(0))/(\mathfrak{m}/\mathfrak{u}_R(0)))) \cong H_{\mathfrak{m}/\mathfrak{u}_R(0)}^n(R/\mathfrak{u}_R(0));$$

thus K is a canonical module for R if and only if it is a canonical module for $R/\mathfrak{u}_R(0)$. $\qquad\square$

12.1.14 Remark. Often in this chapter, including in the proof of the next theorem, we shall want to work under the hypothesis that R is a homomorphic image of an n'-dimensional Gorenstein local ring $(R', \mathfrak{m}')$, so that there is a surjective ring homomorphism $g : R' \longrightarrow R$. We explain here why it is possible to assume that $n' = n$.

Let $\mathfrak{c} = \mathrm{Ker}\, g$. Now $\mathrm{ht}\, \mathfrak{c} = n' - n$, and, since a Gorenstein ring is Cohen–Macaulay, there exists an R'-sequence $r_1', \ldots, r_{n'-n}'$ contained in $\mathfrak{c}$ of length $n' - n$. Now $R'/(r_1', \ldots, r_{n'-n}')$ is a Gorenstein local ring of dimension n (see [50, Exercise 18.1]), and so we can assume that $n' = n$.

12.1.15 Theorem. *Suppose that ω_R exists. Then*

$$(0 :_R \omega_R) = (0 :_R H_{\mathfrak{m}}^n(R)) = \mathfrak{u}_R(0).$$

Proof. It follows from 10.2.2(ii) that $(0 :_R \omega_R) = (0 :_R H_{\mathfrak{m}}^n(R))$, and we proved in 12.1.13 that $\mathfrak{u}_R(0) \subseteq (0 :_R \omega_R)$. It is therefore sufficient for us to show that $(0 :_R \omega_R) \subseteq \mathfrak{u}_R(0)$.

We first consider the case where R is a homomorphic image of an n'-dimensional Gorenstein local ring $(R', \mathfrak{m}')$, so that there is a surjective ring homomorphism $g : R' \longrightarrow R$. In view of 12.1.14, we assume that $n' = n$. By 12.1.3(iii), we have $\omega_R \cong \mathrm{Hom}_{R'}(R, R')$. By 12.1.9(i), $\mathrm{Ass}\, \omega_R = \{\mathfrak{p} \in \mathrm{Spec}(R) : \dim R/\mathfrak{p} = n\}$. Thus, in order to deal with this case, it is enough for us to show that, if $\mathfrak{p} \in \mathrm{Spec}(R)$ has $\dim R/\mathfrak{p} = n$, then the $R_{\mathfrak{p}}$-module $(\omega_R)_{\mathfrak{p}}$ has zero annihilator, for that would show that $(0 :_R \omega_R)R_{\mathfrak{p}} = (0 :_{R_{\mathfrak{p}}} (\omega_R)_{\mathfrak{p}}) = 0$. See 12.1.11 and 12.1.12.

To achieve this, let $\mathfrak{p}' := g^{-1}(\mathfrak{p}), \in \mathrm{Spec}(R')$, and use 11.2.7 to see that there are $R_{\mathfrak{p}}$-isomorphisms

$$(\omega_R)_{\mathfrak{p}} \cong (\mathrm{Hom}_{R'}(R, R'))_{\mathfrak{p}} \cong \mathrm{Hom}_{R'_{\mathfrak{p}'}}(R_{\mathfrak{p}}, R'_{\mathfrak{p}'}).$$

However, $R'_{\mathfrak{p}'}$ is a 0-dimensional Gorenstein local ring, and so

$$R'_{\mathfrak{p}'} \cong E_{R'_{\mathfrak{p}'}}(R'_{\mathfrak{p}'}/\mathfrak{p}'R'_{\mathfrak{p}'}).$$

Therefore $\left(0 :_{R'_{\mathfrak{p}'}} \operatorname{Hom}_{R'_{\mathfrak{p}'}}(R_{\mathfrak{p}}, R'_{\mathfrak{p}'})\right) = (0 :_{R'_{\mathfrak{p}'}} R_{\mathfrak{p}})$ by 10.2.2(ii), so that $(0 :_{R_{\mathfrak{p}}} (\omega_R)_{\mathfrak{p}}) = 0$, as required.

We have therefore proved the theorem under the additional assumption that R is a homomorphic image of a Gorenstein local ring. We now deal with the general case. We use 12.1.3(ii) to see that $\omega_{\widehat{R}}$ exists and there is an $\widehat{R}$-isomorphism $\omega_R \otimes_R \widehat{R} \cong \omega_{\widehat{R}}$. Now $\widehat{R}$ is a homomorphic image of a regular local ring by Cohen's Structure Theorem, and so it follows from the first two paragraphs of this proof that $(0 :_{\widehat{R}} \omega_R \otimes_R \widehat{R}) = \mathfrak{u}_{\widehat{R}}(0)$. Note that

$$(0 :_R \omega_R) = (0 :_{\widehat{R}} \omega_R \otimes_R \widehat{R}) \cap R = \mathfrak{u}_{\widehat{R}}(0) \cap R,$$

and so our proof will be complete if we can show that it is not possible for there to be a $\mathfrak{P} \in \operatorname{ass}_{\widehat{R}} 0$ with $\dim \widehat{R}/\mathfrak{P} < n$ and $\dim R/(\mathfrak{P} \cap R) = n$. So we suppose that such a $\mathfrak{P}$ exists and seek a contradiction. Denote $\mathfrak{P} \cap R$ by $\mathfrak{p}$.

The inclusion homomorphism $R \longrightarrow \widehat{R}$ induces a flat local homomorphism $f : R_{\mathfrak{p}} \longrightarrow \widehat{R}_{\mathfrak{P}}$. Note that $\operatorname{depth}_{\widehat{R}_{\mathfrak{P}}} \widehat{R}_{\mathfrak{P}} = 0$, so that the depth of the fibre ring of f over the maximal ideal of $R_{\mathfrak{p}}$ is 0 (by [50, Theorem 23.3]). Since $\mathfrak{p} \in \operatorname{Ass} \omega_R$ by 12.1.9(i), we have $\operatorname{depth}_{R_{\mathfrak{p}}} (\omega_R)_{\mathfrak{p}} = 0$, so that another use of [50, Theorem 23.3] shows that

$$\operatorname{depth}_{\widehat{R}_{\mathfrak{P}}}\left((\omega_R)_{\mathfrak{p}} \otimes_{R_{\mathfrak{p}}} \widehat{R}_{\mathfrak{P}}\right) = 0.$$

But $(\omega_R)_{\mathfrak{p}} \otimes_{R_{\mathfrak{p}}} \widehat{R}_{\mathfrak{P}}$ is $\widehat{R}_{\mathfrak{P}}$-isomorphic to $(\omega_R \otimes_R \widehat{R})_{\mathfrak{P}}$ and $\omega_R \otimes_R \widehat{R} \cong \omega_{\widehat{R}}$. Since $\dim \widehat{R}/\mathfrak{P} < n$, there exists $\widehat{a} \in \mathfrak{P}$ that is a parameter for $\widehat{R}$. By 12.1.9(ii), the element $\widehat{a}$ is an $(\omega_R \otimes_R \widehat{R})$-sequence, so that

$$\operatorname{depth}_{\widehat{R}_{\mathfrak{P}}}\left((\omega_R)_{\mathfrak{p}} \otimes_{R_{\mathfrak{p}}} \widehat{R}_{\mathfrak{P}}\right) = \operatorname{depth}_{\widehat{R}_{\mathfrak{P}}}(\omega_R \otimes_R \widehat{R})_{\mathfrak{P}} \geq 1.$$

This is a contradiction, and the proof is complete. $\square$

We now remind the reader of Serre's conditions S_i ($i \in \mathbb{N}$). Recall our convention whereby the depth of the zero module over a local ring is interpreted as ∞, and also our occasional abbreviation of $\operatorname{depth}_{R_{\mathfrak{p}}} M_{\mathfrak{p}}$ by $\operatorname{depth} M_{\mathfrak{p}}$ (for a finitely generated R-module M and $\mathfrak{p} \in \operatorname{Spec}(R)$).

12.1.16 Definition. Let R' be a commutative Noetherian ring. Let $k \in \mathbb{N}$ and let M be a faithful finitely generated R'-module. We say that M *satisfies Serre's condition* S_k, or, more loosely, that M *is* S_k, precisely when $\operatorname{depth} M_{\mathfrak{p}} \geq \min\{k, \operatorname{ht} \mathfrak{p}\}$ for all $\mathfrak{p} \in \operatorname{Spec}(R')$.

More generally, we say that a non-zero finitely generated R'-module N satisfies *Serre's condition* S_k precisely when it is S_k as an $R'/(0 :_{R'} N)$-module, that is, if and only if $\operatorname{depth}_{R'_\mathfrak{p}} N_\mathfrak{p} \geq \min\{k, \dim_{R'_\mathfrak{p}} N_\mathfrak{p}\}$ for all $\mathfrak{p} \in \operatorname{Supp} N$, or, equivalently (in view of our convention concerning the depth of zero modules), if and only if $\operatorname{depth}_{R'_\mathfrak{p}} N_\mathfrak{p} \geq \min\{k, \dim_{R'_\mathfrak{p}} N_\mathfrak{p}\}$ for all $\mathfrak{p} \in \operatorname{Spec}(R')$.

12.1.17 ♯Exercise. Let R' be a commutative Noetherian ring, and let N be a non-zero finitely generated R'-module. Show that N is S_2 if and only if

(i) each associated prime of N is a minimal member of $\operatorname{Supp} N$, and
(ii) for every $a \in R'$ which is a non-zerodivisor on N, each associated prime of N/aN is a minimal member of $\operatorname{Supp}_{R'}(N/aN)$.

It is important for us to identify the support of a canonical R-module, if one exists.

12.1.18 Theorem. *Suppose that ω_R exists.*

(i) *We have*

$$\operatorname{Supp} \omega_R = \{\mathfrak{p} \in \operatorname{Spec}(R) : \operatorname{ht} \mathfrak{p} + \dim R/\mathfrak{p} = \dim R\};$$

also the R-modules ω_R and $\operatorname{Hom}_R(\omega_R, \omega_R)$ satisfy the condition S_2.
(ii) *If R is a homomorphic image of a Gorenstein local ring, then, for each $\mathfrak{p} \in \operatorname{Supp} \omega_R$, the localization $(\omega_R)_\mathfrak{p}$ is a canonical module for $R_\mathfrak{p}$.*

Note. Some readers may be surprised by the additional hypothesis in part (ii). In fact, the statement is true without the hypothesis that R be a homomorphic image of a Gorenstein local ring, but the proof is beyond the scope of this book. Interested readers are referred to Aoyama [2, Corollary 4.3].

Proof. We deal first with the case where R is a homomorphic image of an n'-dimensional Gorenstein local ring $(R', \mathfrak{m}')$ via a surjective ring homomorphism $g : R' \longrightarrow R$. In view of 12.1.14, we assume that $n' = n$, so that we can take $\omega_R = \operatorname{Hom}_{R'}(R, R')$.

Let $\mathfrak{p} \in \operatorname{Supp} \omega_R$, so that there exists $\mathfrak{q} \in \operatorname{Ass} \omega_R$ with $\mathfrak{p} \supseteq \mathfrak{q}$. By 12.1.9, $\dim R/\mathfrak{q} = n$. Since R is catenary, $\operatorname{ht} \mathfrak{p}/\mathfrak{q} + \dim R/\mathfrak{p} = \dim R/\mathfrak{q}$, so that $\operatorname{ht} \mathfrak{p} + \dim R/\mathfrak{p} = n$.

Now let $\mathfrak{p} \in \operatorname{Spec}(R)$ be such that $\operatorname{ht} \mathfrak{p} + \dim R/\mathfrak{p} = n$. Let $\mathfrak{p}' := g^{-1}(\mathfrak{p})$, $\in \operatorname{Spec}(R')$, and use 11.2.7 to see that there are $R_\mathfrak{p}$-isomorphisms

$$(\omega_R)_\mathfrak{p} \cong (\operatorname{Hom}_{R'}(R, R'))_\mathfrak{p} \cong \operatorname{Hom}_{R'_{\mathfrak{p}'}}(R_\mathfrak{p}, R'_{\mathfrak{p}'}).$$

We have $\dim R'/\mathfrak{p}' = \dim R/\mathfrak{p}$ and $\operatorname{ht} \mathfrak{p}' \geq \operatorname{ht} \mathfrak{p}$. Therefore we must have $\operatorname{ht} \mathfrak{p}' + \dim R'/\mathfrak{p}' = n$ and $\operatorname{ht} \mathfrak{p}' = \operatorname{ht} \mathfrak{p}$. Therefore $\operatorname{Hom}_{R'_{\mathfrak{p}'}}(R_\mathfrak{p}, R'_{\mathfrak{p}'})$ is a canonical module for $R_{\mathfrak{p}}$, so that there is an $R_{\mathfrak{p}}$-isomorphism $(\omega_R)_{\mathfrak{p}} \cong \omega_{R_{\mathfrak{p}}}$. In particular, $\mathfrak{p} \in \operatorname{Supp} \omega_R$. Also, it follows from 12.1.9 that $\operatorname{depth} \omega_{R_{\mathfrak{p}}} \geq \min\{2, \dim R_{\mathfrak{p}}\}$, and from 12.1.10 that

$$\operatorname{depth} \operatorname{Hom}_{R_{\mathfrak{p}}}(\omega_{R_{\mathfrak{p}}}, \omega_{R_{\mathfrak{p}}}) \geq \min\{2, \dim R_{\mathfrak{p}}\}.$$

We can therefore conclude that both ω_R and $\operatorname{Hom}_R(\omega_R, \omega_R)$ are S_2.

All parts of the theorem have now been proved in the case where R is a homomorphic image of a Gorenstein local ring. We now deal with the general case. We use 12.1.3(ii) to see that $\omega_{\widehat{R}}$ exists and there is an $\widehat{R}$-isomorphism $\omega_R \otimes_R \widehat{R} \cong \omega_{\widehat{R}}$. Now $\widehat{R}$ is a homomorphic image of a regular local ring by Cohen's Structure Theorem, and so we know that the claims of part (i) hold true for the $\widehat{R}$-canonical module $\omega_R \otimes_R \widehat{R}$. In particular, the $\widehat{R}$-modules $\omega_R \otimes_R \widehat{R}$ and $\operatorname{Hom}_{\widehat{R}}(\omega_R \otimes_R \widehat{R}, \omega_R \otimes_R \widehat{R})$ are S_2. An argument similar to one in the proof of [50, Theorem 23.9] shows that the R-modules ω_R and $\operatorname{Hom}_R(\omega_R, \omega_R)$ are both S_2.

We now introduce some notation for a general prime ideal $\mathfrak{p}$ of R. Let $\mathfrak{P} \in \operatorname{Spec}(\widehat{R})$ be a minimal prime ideal of $\mathfrak{p}\widehat{R}$ for which $\dim \widehat{R}/\mathfrak{P} = \dim \widehat{R}/\mathfrak{p}\widehat{R} = \dim R/\mathfrak{p}$. Then $\mathfrak{P} \cap R = \mathfrak{p}$, and the inclusion homomorphism $R \longrightarrow \widehat{R}$ induces a flat local homomorphism $f : R_{\mathfrak{p}} \longrightarrow \widehat{R}_{\mathfrak{P}}$ with the property that its fibre ring over the maximal ideal of $R_{\mathfrak{p}}$ is Artinian. Hence $\operatorname{ht} \mathfrak{P} = \operatorname{ht} \mathfrak{p}$, and we have $\operatorname{ht} \mathfrak{p} + \dim R/\mathfrak{p} = \operatorname{ht} \mathfrak{P} + \dim \widehat{R}/\mathfrak{P} \leq n$. Note also that there is a $\widehat{R}_{\mathfrak{P}}$-isomorphism

$$(\omega_R \otimes_R \widehat{R}) \otimes_{\widehat{R}} \widehat{R}_{\mathfrak{P}} \cong (\omega_R \otimes_R R_{\mathfrak{p}}) \otimes_{R_{\mathfrak{p}}} \widehat{R}_{\mathfrak{P}}.$$

Since f is faithfully flat, it follows that $\mathfrak{p} \in \operatorname{Supp} \omega_R$ if and only if $\mathfrak{P} \in \operatorname{Supp}_{\widehat{R}}(\omega_R \otimes_R \widehat{R})$.

Now suppose that $\mathfrak{p} \in \operatorname{Supp} \omega_R$. Then $\mathfrak{P} \in \operatorname{Supp}_{\widehat{R}}(\omega_R \otimes_R \widehat{R})$, so that $\operatorname{ht} \mathfrak{P} + \dim \widehat{R}/\mathfrak{P} = n$ because $\widehat{R}$ is a homomorphic image of a Gorenstein local ring. Therefore $\operatorname{ht} \mathfrak{p} + \dim R/\mathfrak{p} = n$.

Conversely, suppose that $\operatorname{ht} \mathfrak{p} + \dim R/\mathfrak{p} = n$. Then $\operatorname{ht} \mathfrak{P} + \dim \widehat{R}/\mathfrak{P} = n$, so that $\mathfrak{P} \in \operatorname{Supp}_{\widehat{R}}(\omega_R \otimes_R \widehat{R})$ because $\widehat{R}$ is a homomorphic image of a Gorenstein local ring; therefore $\mathfrak{p} \in \operatorname{Supp} \omega_R$.

All parts of the theorem have now been proved. □

12.1.19 Exercise. Suppose R is a homomorphic image of a Gorenstein local ring and that $\mathfrak{p} \in \operatorname{Supp} \omega_R$. Show that $\mathfrak{u}_{R_{\mathfrak{p}}}(0) = \mathfrak{u}_R(0)R_{\mathfrak{p}}$.

The following variant of the Local Duality Theorem will be useful.

12.1.20 Theorem. *Suppose that ω_R exists. Set $E := E(R/\mathfrak{m})$, and let $D := \operatorname{Hom}_R(\bullet, E)$.*

(i) *There is a natural transformation of functors*

$$\phi_0 : H_{\mathfrak{m}}^n \longrightarrow D(\mathrm{Hom}_R(\,\bullet\,,\omega_R))$$

which is such that $\phi_{0\,M}$ is an isomorphism whenever M is a finitely generated R-module.

(ii) *Now suppose that R is Cohen–Macaulay. Then there is a unique extension of ϕ_0 to a homomorphism*

$$\Phi := (\phi_i)_{i\in\mathbb{N}_0} : \left(H_{\mathfrak{m}}^{n-i}\right)_{i\in\mathbb{N}_0} \longrightarrow \left(D(\mathrm{Ext}_R^i(\,\bullet\,,\omega_R))\right)_{i\in\mathbb{N}_0}$$

of (positive strongly) connected sequences of covariant functors from $\mathcal{C}(R)$ to $\mathcal{C}(R)$. Furthermore, $\phi_{i\,M}$ is an isomorphism for all $i \in \mathbb{N}_0$ whenever M is a finitely generated R-module.

In particular, for each finitely generated R-module M,

$$H_{\mathfrak{m}}^{n-i}(M) \cong D(\mathrm{Ext}_R^i(M,\omega_R)) \quad \text{for all } i \in \mathbb{Z}.$$

Proof. (i) By definition, $H_{\mathfrak{m}}^n(R) \cong D(\omega_R)$. By 6.1.10, the functor $H_{\mathfrak{m}}^n$ is naturally equivalent to $(\,\bullet\,) \otimes_R H_{\mathfrak{m}}^n(R)$. Therefore $H_{\mathfrak{m}}^n$ is naturally equivalent to $(\,\bullet\,) \otimes_R \mathrm{Hom}_R(\omega_R, E)$. We can now make use of 10.2.16 in order to arrive at a natural transformation of functors

$$\phi_0 : H_{\mathfrak{m}}^n \longrightarrow D(\mathrm{Hom}_R(\,\bullet\,,\omega_R))$$

which is such that $\phi_{0\,M}$ is an isomorphism whenever M is a finitely generated R-module.

(ii) Now suppose that R is Cohen–Macaulay. Here, we reason as in the proof of 11.2.5. Since $\mathrm{depth}\,R = n$, it follows from 6.2.8 and 3.4.10 that $H_{\mathfrak{m}}^{n-i}(P) = 0$ for all $i \in \mathbb{N}$ and all projective R-modules P; also

$$D(\mathrm{Ext}_R^i(P,\omega_R)) = 0$$

for all $i \in \mathbb{N}$ and all projective R-modules P. It follows from the analogue of 1.3.4 for positive connected sequences that ϕ_0 can be incorporated into a (uniquely determined) homomorphism

$$\Phi := (\phi_i)_{i\in\mathbb{N}_0} : \left(H_{\mathfrak{m}}^{n-i}\right)_{i\in\mathbb{N}_0} \longrightarrow \left(D(\mathrm{Ext}_R^i(\,\bullet\,,\omega_R))\right)_{i\in\mathbb{N}_0}$$

of connected sequences, and it is easy to prove by induction that, for each $i \in \mathbb{N}$, the homomorphism $\phi_{i\,M}$ is an isomorphism whenever M is a finitely generated R-module. For $i < 0$, we have $H_{\mathfrak{m}}^{n-i}(M) = \mathrm{Ext}_R^i(M,\omega_R) = 0$ for each R-module M. $\square$

12.1.21 Corollary. *Suppose that the local ring R is Cohen–Macaulay, and that ω_R exists. Then ω_R has finite injective dimension equal to n, and is a Cohen–Macaulay R-module of dimension n.*

Proof. By 12.1.20 and the fact that $E_R(R/\mathfrak{m})$ is an injective cogenerator for R, we have $\text{Ext}_R^i(M, \omega_R) = 0$ for all $i > n$ and all finitely generated R-modules M, so that, in particular, $\text{Ext}_R^i(R/\mathfrak{p}, \omega_R) = 0$ for all $i > n$ and all $\mathfrak{p} \in \text{Spec}(R)$. Thus all the Bass numbers $\mu^i(\mathfrak{p}, \omega_R)$ are zero for all $i > n$ and all $\mathfrak{p} \in \text{Spec}(R)$ (by 11.1.8). Also $\text{Ext}_R^n(R/\mathfrak{m}, \omega_R) \neq 0$ because $D(\text{Ext}_R^n(R/\mathfrak{m}, \omega_R)) \cong \Gamma_\mathfrak{m}(R/\mathfrak{m}) = R/\mathfrak{m} \neq 0$. Therefore ω_R has finite injective dimension equal to n.

It also follows similarly from 12.1.20 that

$$D(\text{Ext}_R^i(R/\mathfrak{m}, \omega_R)) \cong H_\mathfrak{m}^{n-i}(R/\mathfrak{m}) = 0 \quad \text{for all } i \in \{0, \ldots, n-1\},$$

so that $\text{depth}\, \omega_R = n$ and ω_R is a Cohen–Macaulay R-module. $\quad\square$

12.1.22 Corollary. *Suppose that the local ring R is Cohen–Macaulay, and that ω_R exists. Then R is Gorenstein if and only if $\omega_R \cong R$, that is, if and only if R is a canonical module for itself.*

Proof. We observed in 12.1.3(iv) that a Gorenstein local ring is a canonical module for itself. Conversely, suppose that R is a Cohen–Macaulay local ring and that ω_R exists and is isomorphic to R. Then $\text{inj}\dim_R R < \infty$, by 12.1.21, and so R is Gorenstein. $\quad\square$

12.1.23 ♯Exercise. Suppose that the local ring R is Cohen–Macaulay and that ω_R exists. Show that $\mu^i(\mathfrak{m}, \omega_R) = \delta_{i,\text{ht}\,\mathfrak{m}}$ (the Kronecker delta) for all $i \in \mathbb{N}_0$. (Here is a hint: use the method of proof in 12.1.21.)

12.1.24 ♯Exercise. Suppose that the local ring R is Cohen–Macaulay and a homomorphic image of a Gorenstein local ring, so that R has a canonical module ω_R, by 12.1.3(iii). Show that, for all $i \in \mathbb{N}_0$ and all $\mathfrak{p} \in \text{Spec}(R)$,

$$\mu^i(\mathfrak{p}, \omega_R) = \begin{cases} 0 & \text{if } i \neq \text{ht}\,\mathfrak{p}, \\ 1 & \text{if } i = \text{ht}\,\mathfrak{p}, \end{cases}$$

that is, $\mu^i(\mathfrak{p}, \omega_R)$ is equal to the Kronecker delta $\delta_{i,\text{ht}\,\mathfrak{p}}$. (Here is a hint: use 12.1.18 and 12.1.23.)

12.1.25 ♯Exercise. Suppose that the local ring R is Cohen–Macaulay. Suppose that K is a finitely generated R-module such that $\mu^i(\mathfrak{m}, K) = \delta_{i,\text{ht}\,\mathfrak{m}}$ for all $i \in \mathbb{N}_0$. The purposes of this exercise are to show that K is actually a canonical module for R, and to prove a result of I. Reiten [69] and H.-B. Foxby [19] that R is a homomorphic image of a Gorenstein local ring. Recall that n denotes $\dim R$.

(i) Show that there exists an R-sequence $r_1, \ldots, r_n$ which is also a K-sequence.

(ii) Denote the local ring $R/(r_1, \ldots, r_n)$ by $(\overline{R}, \overline{\mathfrak{m}})$ and use $\overline{K}$ to denote $K/(r_1, \ldots, r_n)K$. Use [7, 3.1.16] to calculate the $\mu_{\overline{R}}^j(\overline{\mathfrak{m}}, \overline{K})$ $(j \in \mathbb{N}_0)$, and conclude that there is an $\overline{R}$-isomorphism $\overline{K} \cong E_{\overline{R}}(\overline{R}/\overline{\mathfrak{m}})$. Deduce that

$$(0 :_R K/(r_1, \ldots, r_n)K) = (r_1, \ldots, r_n).$$

(iii) By considering $r_1^t, \ldots, r_n^t$ for $t \in \mathbb{N}$, show that $(0 :_R K) = 0$.

(iv) Let R' be the trivial extension of R by K, introduced in 6.2.12. It follows from that exercise that R' is an n-dimensional local ring. Show that $(r_1, 0), \ldots, (r_n, 0)$ is an R'-sequence, and deduce that R' is Cohen–Macaulay.

(v) Show that $R'/((r_1, 0), \ldots, (r_n, 0))$ is isomorphic to the trivial extension of $\overline{R}$ by $\overline{K}$, and calculate the socle of the latter ring. Deduce that R' is Gorenstein.

(vi) Conclude that R is a homomorphic image of a Gorenstein local ring, so that ω_R exists, by 12.1.3(iii). Prove that $K \cong \omega_R$. (Here is a hint: recall from 12.1.3(iii) that $\omega_R \cong \mathrm{Hom}_{R'}(R, R')$.)

12.1.26 Remark. Suppose that the local ring R is Cohen–Macaulay. It follows from 12.1.3(iii), 12.1.23 and 12.1.25 that R has a canonical module if and only if R is a homomorphic image of a Gorenstein local ring, and that, when this is the case, an R-module K is a canonical module for R if and only if it is finitely generated and $\mu^i(\mathfrak{m}, K) = \delta_{i,\mathrm{ht}\,\mathfrak{m}}$ for all $i \in \mathbb{N}_0$.

In their book on Cohen–Macaulay rings [7], Bruns and Herzog (essentially) defined a canonical module for a Cohen–Macaulay local ring $(R, \mathfrak{m})$ to be a finitely generated R-module K for which $\mu^i(\mathfrak{m}, K) = \delta_{i,\mathrm{ht}\,\mathfrak{m}}$ for all $i \in \mathbb{N}_0$. Thus 12.1.23 and 12.1.25 reconcile their approach with the one we have taken in this chapter.

12.1.27 Remark. Suppose $(R, \mathfrak{m})$ is a Cohen–Macaulay local ring. It follows from 12.1.25, 12.1.26 and 12.1.24 that a finitely generated R-module C is a canonical module for R if and only if $\mu^i(\mathfrak{m}, C) = \delta_{i,\mathrm{ht}\,\mathfrak{m}}$ (Kronecker delta) for all $i \in \mathbb{N}_0$, and that, when this is the case, R is a homomorphic image of a Gorenstein local ring and $C_\mathfrak{p}$ is a canonical module for $R_\mathfrak{p}$ for all $\mathfrak{p} \in \mathrm{Spec}(R)$.

Bruns and Herzog also extended the definition of canonical module to the case where the underlying ring is not necessarily local (but is Cohen–Macaulay).

12.1.28 Definition. Let R' be a (not necessarily local) Cohen–Macaulay (commutative Noetherian) ring. Let C be a finitely generated R'-module. We say that C is a *canonical module for* R' precisely when $C_\mathfrak{m}$ is a canonical module for $R'_\mathfrak{m}$ for all maximal ideals $\mathfrak{m}$ of R'.

In that case, $C_\mathfrak{p}$ is a canonical module for $R'_\mathfrak{p}$ for all $\mathfrak{p} \in \mathrm{Spec}(R')$, by 12.1.27.

12.1.29 Remark. Let R' be a (not necessarily local) Cohen–Macaulay (commutative Noetherian) ring. Let C be a finitely generated R'-module. It follows from 12.1.27 that C is a canonical module for R' if and only if $\mu^i(\mathfrak{p}, C) = \delta_{i,\mathrm{ht}\,\mathfrak{p}}$ for all $\mathfrak{p} \in \mathrm{Spec}(R')$ and all $i \in \mathbb{N}_0$.

12.1.30 Proposition. *Let R' be a (not necessarily local) Cohen–Macaulay (commutative Noetherian) ring, and assume that R' has a canonical module C. Then the trivial extension $R' \propto C$ of R' by C (see 6.2.12) is a Gorenstein ring, so that R' is a homomorphic image of a Gorenstein ring.*

Proof. By 6.2.12(ii), a general prime ideal $\mathfrak{P}$ of $R' \propto C$ has the form $\mathfrak{p} \times C$ for a prime ideal $\mathfrak{p}$ of R'; also, 6.2.12(iv) shows that $(R' \propto C)_\mathfrak{P} \cong R'_\mathfrak{p} \propto C_\mathfrak{p}$, which is Gorenstein by 12.1.25. Therefore $R' \propto C$ is Gorenstein. The ring homomorphism $\phi : R' \propto C \longrightarrow R'$ for which $\phi((r, c)) = r$ for all $r \in R'$ and $c \in C$ is surjective.   □

12.2 The endomorphism ring

When $(R, \mathfrak{m})$ is local and ω_R exists, it turns out that $\mathrm{Hom}_R(\omega_R, \omega_R)$ has some very interesting and useful properties, some of which are relevant to the theory of S_2-ifications that we shall develop in the final section of this chapter. In this section, we shall concentrate on the R-module structure of $\mathrm{Hom}_R(\omega_R, \omega_R)$; we shall consider its ring structure later in the chapter.

12.2.1 Notation. Throughout this section also, we shall assume that $(R, \mathfrak{m})$ is local, we shall use n to denote $\dim R$, and we shall denote the completion of R by $(\widehat{R}, \widehat{\mathfrak{m}})$.

In order to present the nice properties of $\mathrm{Hom}_R(\omega_R, \omega_R)$, we are going to concentrate first on the case where $(R, \mathfrak{m})$ is a Cohen–Macaulay local ring (and ω_R exists): by 12.1.15, we have $(0 :_R \omega_R) = \mathfrak{u}_R(0)$, and this is zero when R is Cohen–Macaulay; in fact, we shall show, in that case, that each R-endomorphism of ω_R is given by multiplication by a uniquely determined element of R.

12.2.2 Proposition. *Suppose the local ring R is Cohen–Macaulay, and ω_R exists. Let $a_1, \ldots, a_j \in \mathfrak{m}$. Then $a_1, \ldots, a_j$ is an R-sequence if and only if it* *is an ω_R-sequence; moreover, when this is the case, $\omega_{R/(a_1,\ldots,a_j)}$ exists and $\omega_R/(a_1,\ldots,a_j)\omega_R \cong \omega_{R/(a_1,\ldots,a_j)}$.*

Proof. Let D denote the Matlis duality functor $\operatorname{Hom}_R(\ \bullet\ , E_R(R/\mathfrak{m}))$. Note that there is nothing to prove if $n = 0$, and so we suppose that $n > 0$.

We first deal with the case in which $j = 1$. Since R is Cohen–Macaulay, $\operatorname{Ass} R = \{\mathfrak{p} \in \operatorname{Spec}(R) : \dim R/\mathfrak{p} = n\}$, and this is equal to $\operatorname{Ass}\omega_R$, by 12.1.9(i). Therefore a_1 is a non-zerodivisor on R if and only if it is a non-zerodivisor on ω_R. Suppose that this is the case. Since $\operatorname{grade}\mathfrak{m} = n$, we have $H^{n-1}_\mathfrak{m}(R) = 0$, by 6.2.7. Therefore the exact sequence

$$0 \longrightarrow R \xrightarrow{\ a_1\ } R \longrightarrow R/a_1R \longrightarrow 0$$

yields an exact sequence

$$0 \longrightarrow H^{n-1}_\mathfrak{m}(R/a_1R) \longrightarrow H^n_\mathfrak{m}(R) \xrightarrow{\ a_1\ } H^n_\mathfrak{m}(R) \longrightarrow 0.$$

Also, the exact sequence $0 \longrightarrow \omega_R \xrightarrow{\ a_1\ } \omega_R \longrightarrow \omega_R/a_1\omega_R \longrightarrow 0$ induces an exact sequence $0 \longrightarrow D(\omega_R/a_1\omega_R) \longrightarrow D(\omega_R) \xrightarrow{\ a_1\ } D(\omega_R) \longrightarrow 0$. By the definition of canonical module, we have $D(\omega_R) \cong H^n_\mathfrak{m}(R)$. It follows that there is an R-isomorphism $D(\omega_R/a_1\omega_R) \cong H^{n-1}_\mathfrak{m}(R/a_1R)$.

Now $H^{n-1}_\mathfrak{m}(R/a_1R) \cong H^{n-1}_{\mathfrak{m}/a_1R}(R/a_1R)$ by the Independence Theorem 4.2.1. Also

$$D(\omega_R/a_1\omega_R) = \operatorname{Hom}_R(\omega_R/a_1\omega_R, E_R(R/\mathfrak{m}))$$
$$= \operatorname{Hom}_R(\omega_R/a_1\omega_R, (0 :_{E_R(R/\mathfrak{m})} a_1)),$$

and 10.1.16 shows that there is an R/a_1R-isomorphism

$$(0 :_{E_R(R/\mathfrak{m})} a_1) \cong E_{R/a_1R}((R/a_1R)/(\mathfrak{m}/a_1R)).$$

We can therefore conclude that there is an R/a_1R-isomorphism

$$\operatorname{Hom}_{R/a_1R}(\omega_R/a_1\omega_R, E_{R/a_1R}((R/a_1R)/(\mathfrak{m}/a_1R))) \cong H^{n-1}_{\mathfrak{m}/a_1R}(R/a_1R).$$

Since the local ring R/a_1R has dimension $n - 1$, this shows that $\omega_R/a_1\omega_R$ is a canonical module for it.

To complete the proof, proceed by induction on j. □

12.2.3 Lemma. *Suppose that the local ring R is Cohen–Macaulay, and that ω_R exists. Let $a \in \mathfrak{m}$ be a non-zerodivisor on R. Then the R-homomorphism*

$$\xi : \operatorname{Hom}_R(\omega_R, \omega_R) \longrightarrow \operatorname{Hom}_R(\omega_R/a\omega_R, \omega_R/a\omega_R)$$

for which $\xi(g)(x + a\omega_R) = g(x) + a\omega_R$, for each endomorphism g of ω_R and each $x \in \omega_R$, is surjective, and $\operatorname{Ker}\xi = a\operatorname{Hom}_R(\omega_R, \omega_R)$.

Proof. Denote $\omega_R / a\omega_R$ by $\overline{\omega_R}$, and let $\pi : \omega_R \longrightarrow \overline{\omega_R}$ denote the natural epimorphism.

Proposition 12.1.9(ii) shows that a is an ω_R-sequence, and 12.2.2 shows that $\operatorname{depth}_R \omega_R = n$. Therefore $H_{\mathfrak{m}}^{n-1}(\omega_R) = 0$, and it follows from 12.1.20 that $\operatorname{Ext}_R^1(\omega_R, \omega_R) = 0$.

Therefore the exact sequence $0 \longrightarrow \omega_R \overset{a}{\longrightarrow} \omega_R \longrightarrow \overline{\omega_R} \longrightarrow 0$ yields the exact sequence

$$0 \longrightarrow \operatorname{Hom}_R(\omega_R, \omega_R) \overset{a}{\longrightarrow} \operatorname{Hom}_R(\omega_R, \omega_R) \longrightarrow \operatorname{Hom}_R(\omega_R, \overline{\omega_R}) \longrightarrow 0.$$

It also yields the exact sequence

$$0 \longrightarrow \operatorname{Hom}_R(\overline{\omega_R}, \overline{\omega_R}) \longrightarrow \operatorname{Hom}_R(\omega_R, \overline{\omega_R}) \overset{a}{\longrightarrow} \operatorname{Hom}_R(\omega_R, \overline{\omega_R}).$$

But a annihilates $\operatorname{Hom}_R(\omega_R, \overline{\omega_R})$, and so the latter exact sequence shows that $\operatorname{Hom}(\pi, \overline{\omega_R}) : \operatorname{Hom}_R(\overline{\omega_R}, \overline{\omega_R}) \longrightarrow \operatorname{Hom}_R(\omega_R, \overline{\omega_R})$ is an isomorphism. It is straightforward to check that the diagram

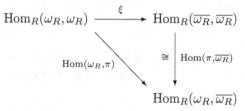

commutes. The sequence

$$0 \longrightarrow \operatorname{Hom}_R(\omega_R, \omega_R) \overset{a}{\longrightarrow} \operatorname{Hom}_R(\omega_R, \omega_R) \overset{\xi}{\longrightarrow} \operatorname{Hom}_R(\overline{\omega_R}, \overline{\omega_R}) \longrightarrow 0$$

is therefore exact, and this proves all the claims of the lemma. $\square$

12.2.4 Notation.   Suppose that ω_R exists. We shall use

$$h_R : R \longrightarrow \operatorname{Hom}_R(\omega_R, \omega_R)$$

to denote the natural homomorphism for which $h_R(r) = r \operatorname{Id}_{\omega_R}$ for all $r \in R$.

Note that h_R is both an R-module homomorphism and a ring homomorphism, and that $\operatorname{Im} h_R$ is contained in the centre of the ring $\operatorname{Hom}_R(\omega_R, \omega_R)$. We are going to be interested in the kernel and cokernel of h_R, and, in particular, we shall show that h is an isomorphism when R is Cohen–Macaulay.

12.2.5 Remark.   Suppose that ω_R exists and that $n > 0$. By 12.1.11 and 12.1.15,

$$\operatorname{Ker} h_r = \mathfrak{u}_R(0)$$
$$= \{r \in R : \mathfrak{b}r = 0 \text{ for some ideal } \mathfrak{b} \subset R \text{ with } \dim R/\mathfrak{b} < n\}.$$

12.2.6 Theorem. *Suppose that the local ring R is Cohen–Macaulay, and that ω_R exists. Then $h_R : R \longrightarrow \operatorname{Hom}_R(\omega_R, \omega_R)$ is an isomorphism.*

Proof. We argue by induction on n. When $n = 0$, the ring R is an Artinian local ring, and therefore complete; also $\Gamma_{\mathfrak{m}}(R) = R$; thus 12.1.3(v) shows that

$$\omega_R \cong \operatorname{Hom}_R(H^0_{\mathfrak{m}}(R), E_R(R/\mathfrak{m})) \cong \operatorname{Hom}_R(R, E_R(R/\mathfrak{m})) \cong E_R(R/\mathfrak{m}).$$

By 10.2.3(v), for each endomorphism f of $E_R(R/\mathfrak{m})$, there is a unique $r_f \in R$ such that $f(x) = r_f x$ for all $x \in E_R(R/\mathfrak{m})$. It follows easily that, for our endomorphism g of ω_R, there is a unique $r_g \in R$ such that $g(x) = r_g x$ for all $x \in \omega_R$.

Now suppose that $n > 0$ and assume, inductively, that the result has been proved for all Cohen–Macaulay local rings of dimension $n - 1$ that possess canonical modules. We already know that h_R is injective, because $\operatorname{Ker} h_R = (0 :_R \omega_R) = \mathfrak{u}_R(0)$ by 12.1.15, and this ideal is zero because R is a Cohen–Macaulay local ring. So let $C := \operatorname{Coker} \phi_R$.

Let $a \in \mathfrak{m}$ be a non-zerodivisor on R, and let $\overline{R}$ denote the $(n-1)$-dimensional Cohen–Macaulay local ring R/Ra. Now $\overline{R}$ has a canonical module and we can identify $\omega_{R/Ra} = \omega_R/a\omega_R$ (by 12.2.2). The exact sequence $0 \longrightarrow R \xrightarrow{h_R} \operatorname{Hom}_R(\omega_R, \omega_R) \longrightarrow C \longrightarrow 0$ induces the exact sequence

$$R \otimes_R \overline{R} \xrightarrow{h_R \otimes \overline{R}} \operatorname{Hom}_R(\omega_R, \omega_R) \otimes_R \overline{R} \longrightarrow C \otimes_R \overline{R} \longrightarrow 0.$$

But it follows from 12.2.3 that there is an R-isomorphism

$$\overline{\beta} : \operatorname{Hom}_R(\omega_R, \omega_R) \otimes_R \overline{R} \xrightarrow{\cong} \operatorname{Hom}_R(\omega_R/a\omega_R, \omega_R/a\omega_R)$$

such that $\overline{\beta} \circ (h_R \otimes \overline{R})$ maps $1 \otimes (r + Ra)$, for $r \in R$, to $h_{\overline{R}}(r + Ra)$. Since $h_{\overline{R}}$ is an isomorphism by the inductive hypothesis, it follows that $h_R \otimes \overline{R}$ is an isomorphism, so that $C/aC = 0$. Since C is finitely generated because ω_R is, we can use Nakayama's Lemma to deduce that $C = 0$. Therefore h_R is surjective, and so is an isomorphism.

This completes the inductive step, and the proof. $\square$

12.2.7 Theorem. *Suppose that ω_R exists. Then*

$$h_R : R \longrightarrow \operatorname{Hom}_R(\omega_R, \omega_R)$$

has $\operatorname{Supp}(\operatorname{Coker} h_R) \subseteq \{\mathfrak{p} \in \operatorname{Spec}(R) : \operatorname{ht} \mathfrak{p} \geq 2\}$. Thus each element of $\operatorname{Coker} h_R$ has annihilator of height at least 2.

Proof. Let $C := \operatorname{Coker} h_R$. Let $\mathfrak{p} \in \operatorname{Spec}(R)$ have $\operatorname{ht} \mathfrak{p} \leq 1$; it is enough for us to show that $C_{\mathfrak{p}} = 0$. Since $\operatorname{Supp} C \subseteq \operatorname{Supp}(\operatorname{Hom}_R(\omega_R, \omega_R)) \subseteq \operatorname{Supp} \omega_R$, we can, and do, assume that $\mathfrak{p} \in \operatorname{Supp} \omega_R$. Therefore $\operatorname{ht} \mathfrak{p} + \dim R/\mathfrak{p} = n$, by 12.1.18.

Now argue as in that part of the proof of 12.1.18 where the general case is treated to find $\mathfrak{P} \in \operatorname{Spec}(\widehat{R})$ such that $\mathfrak{P} \cap R = \mathfrak{p}$, $\dim \widehat{R}/\mathfrak{P} = \dim R/\mathfrak{p}$ and $\operatorname{ht} \mathfrak{P} = \operatorname{ht} \mathfrak{p}$. Then the inclusion homomorphism $R \longrightarrow \widehat{R}$ induces a flat local homomorphism $f : R_{\mathfrak{p}} \longrightarrow \widehat{R}_{\mathfrak{P}}$, and so it is enough for us to show that $C_{\mathfrak{p}} \otimes_{R_{\mathfrak{p}}} \widehat{R}_{\mathfrak{P}} = 0$. Hence it is enough for us to show that $(C \otimes_R \widehat{R}) \otimes_{\widehat{R}} \widehat{R}_{\mathfrak{P}} = 0$.

Now, by 12.1.3(ii), we have $\omega_R \otimes_R \widehat{R} \cong \omega_{\widehat{R}}$. In view of the natural $\widehat{R}$-isomorphism $\operatorname{Hom}_R(\omega_R, \omega_R) \otimes_R \widehat{R} \xrightarrow{\cong} \operatorname{Hom}_{\widehat{R}}(\omega_R \otimes_R \widehat{R}, \omega_R \otimes_R \widehat{R})$ (see [50, Theorem 7.11]), we can therefore assume for the remainder of this proof that R is complete. Recall that $\mathfrak{p} \in \operatorname{Supp} \omega_R$, $\operatorname{ht} \mathfrak{p} \leq 1$ and our aim is to show that $C_{\mathfrak{p}} = 0$.

By 12.1.15 and 12.1.13, we have $(0 :_R \omega_R) = \mathfrak{u}_R(0)$ and ω_R is a canonical module for $R/\mathfrak{u}_R(0)$. Note that $\mathfrak{p} \supseteq \mathfrak{u}_R(0)$ and $\operatorname{ht} \mathfrak{p} = \operatorname{ht} \mathfrak{p}/\mathfrak{u}_R(0)$, in view of 12.1.15 and 12.1.18(i). Therefore, we can, and do, assume henceforth in this proof not only that R is complete, but also that $\mathfrak{u}_R(0) = 0$. Since $\operatorname{ht} \mathfrak{p} \leq 1$ and $\mathfrak{u}_R(0) = 0$, the localization $R_{\mathfrak{p}}$ is Cohen–Macaulay. By 12.1.18(ii), we have $(\omega_R)_{\mathfrak{p}} = \omega_{R_{\mathfrak{p}}}$. It therefore follows from 12.2.6 that $h_{R_{\mathfrak{p}}}$ is an isomorphism. As the standard $R_{\mathfrak{p}}$-isomorphism

$$\psi : (\operatorname{Hom}_R(\omega_R, \omega_R))_{\mathfrak{p}} \xrightarrow{\cong} \operatorname{Hom}_{R_{\mathfrak{p}}}((\omega_R)_{\mathfrak{p}}, (\omega_R)_{\mathfrak{p}})$$

is such that $\psi \circ (h_R)_{\mathfrak{p}} = h_{R_{\mathfrak{p}}}$, it follows that $C_{\mathfrak{p}} = 0$. This completes the proof. $\qquad \square$

The following proposition, due to Hartshorne and Grothendieck [28, Proposition 2.1 and Remark 2.4.1] will be helpful in the determination of a necessary and sufficient condition (in the case where ω_R exists) for $h_R : R \longrightarrow \operatorname{Hom}_R(\omega_R, \omega_R)$ to be an isomorphism.

12.2.8 Proposition. *Suppose $(R, \mathfrak{m})$ is local and S_2, and let $0 = \bigcap_{i=1}^{t+u} \mathfrak{q}_i$ be a minimal primary decomposition of the zero ideal of R, where both t and u are positive integers. Then*

(i) *every minimal prime ideal of $\bigcap_{i=1}^{t} \mathfrak{q}_i + \bigcap_{j=t+1}^{t+u} \mathfrak{q}_j$ has height 1;*

(ii) *if, in addition, R is catenary, then $\dim R/\mathfrak{p}' = \dim R$ for all $\mathfrak{p}' \in \operatorname{ass} 0$.*

Proof. Write $\mathfrak{a} := \bigcap_{i=1}^{t} \mathfrak{q}_i$ and $\mathfrak{b} = \bigcap_{i=t+1}^{t+u} \mathfrak{q}_i$, and let $\sqrt{\mathfrak{q}_i} = \mathfrak{p}_i$ for all $i = 1, \ldots, t + u$. Observe that $\mathfrak{a} + \mathfrak{b}$ is a proper ideal.

(i) Let $\mathfrak{p}$ be a minimal prime of $\mathfrak{a} + \mathfrak{b}$. There exists $i \in \{1, \ldots, t\}$ such that $\mathfrak{p} \supseteq \mathfrak{p}_i$, and there exists $j \in \{t+1, \ldots, t+u\}$ such that $\mathfrak{p} \supseteq \mathfrak{p}_j$. Localize at $\mathfrak{p}$:

we obtain, in $R_{\mathfrak{p}}$, that

$$\mathfrak{a}R_{\mathfrak{p}} = \bigcap_{\substack{i=1 \\ \mathfrak{p}_i \subseteq \mathfrak{p}}}^{t} \mathfrak{q}_i R_{\mathfrak{p}}, \quad \mathfrak{b}R_{\mathfrak{p}} = \bigcap_{\substack{j=t+1 \\ \mathfrak{p}_j \subseteq \mathfrak{p}}}^{t+u} \mathfrak{q}_j R_{\mathfrak{p}} \quad \text{and} \quad 0 = \bigcap_{\substack{i=1 \\ \mathfrak{p}_i \subseteq \mathfrak{p}}}^{t+u} \mathfrak{q}_i R_{\mathfrak{p}}$$

are all minimal primary decompositions, that $0 = \mathfrak{a}R_{\mathfrak{p}} \cap \mathfrak{b}R_{\mathfrak{p}}$ and that $\mathfrak{p}R_{\mathfrak{p}}$ is the only associated prime ideal of $\mathfrak{a}R_{\mathfrak{p}} + \mathfrak{b}R_{\mathfrak{p}}$, and it is enough, in order for us to complete the proof of this part, to show that $\operatorname{ht}\mathfrak{p}R_{\mathfrak{p}} = 1$. It is thus enough for us to prove the claim under the additional assumption that $\mathfrak{p} = \mathfrak{m}$, and we now assume this.

As R is S_2, all associated primes of 0 are minimal; since there is an $i \in \{1,\ldots,t\}$ such that $\mathfrak{m} \supseteq \mathfrak{p}_i$, and there is a $j \in \{t+1,\ldots,t+u\}$ such that $\mathfrak{m} \supseteq \mathfrak{p}_j$, we have $\operatorname{ht}\mathfrak{m} \geq 1$. Thus $\mathfrak{m} \notin \operatorname{ass}0$, $\mathfrak{m} \notin \operatorname{ass}\mathfrak{a}$ and $\mathfrak{m} \notin \operatorname{ass}\mathfrak{b}$; on the other hand, $\mathfrak{m} \in \operatorname{ass}(\mathfrak{a}+\mathfrak{b})$. Thus $\Gamma_{\mathfrak{m}}(R) = \Gamma_{\mathfrak{m}}(R/\mathfrak{a}) = \Gamma_{\mathfrak{m}}(R/\mathfrak{b}) = 0$, whereas $\Gamma_{\mathfrak{m}}(R/(\mathfrak{a}+\mathfrak{b})) \neq 0$. By 3.2.1, there is an exact sequence

$$0 \longrightarrow R \longrightarrow (R/\mathfrak{a}) \oplus (R/\mathfrak{b}) \longrightarrow R/(\mathfrak{a}+\mathfrak{b}) \longrightarrow 0.$$

Application of local cohomology (with respect to $\mathfrak{m}$) to this sequence therefore yields that $H^1_{\mathfrak{m}}(R) \neq 0$, so that $\operatorname{depth}R = 1$. Since R is S_2, we must have $\dim R = \operatorname{ht}\mathfrak{m} = 1$.

(ii) Now suppose, in addition, that R is catenary. Suppose that there exists $\mathfrak{p}' \in \operatorname{ass}0$ such that $\dim R/\mathfrak{p}' < n$, and seek a contradiction. By this assumption, we may suppose that the numbering, used above, of the primary components of the zero ideal is such that $\dim R/\mathfrak{p}_i = n$ for all $i = 1,\ldots,t$ and $\dim R/\mathfrak{p}_j < n$ for all $j = t+1,\ldots,t+u$. Then, with the above notation, $\mathfrak{a}+\mathfrak{b}$ will have a minimal prime ideal $\mathfrak{p}$, and by part (i), $\operatorname{ht}\mathfrak{p} = 1$. Now $\mathfrak{p} \supseteq \mathfrak{p}_i$ for some $i \in \{1,\ldots,t\}$ and $\mathfrak{p} \supseteq \mathfrak{p}_j$ for some $j \in \{t+1,\ldots,t+u\}$. Then, by the catenarity, we have

$$n = \dim R/\mathfrak{p}_i = \dim R/\mathfrak{p} + \operatorname{ht}\mathfrak{p}/\mathfrak{p}_i = \dim R/\mathfrak{p} + 1$$
$$= \dim R/\mathfrak{p} + \operatorname{ht}\mathfrak{p}/\mathfrak{p}_j = \dim R/\mathfrak{p}_j,$$

and this is a contradiction. $\square$

12.2.9 ♯Exercise. Suppose R is a homomorphic image of a Gorenstein local ring, so that ω_R exists. Show that $h_R : R \longrightarrow \operatorname{Hom}_R(\omega_R, \omega_R)$ is an isomorphism if and only if R satisfies the condition S_2.

Here are some hints for the implication '($\Leftarrow$)'.

(a) Argue by induction on $\dim R$. Use 12.2.6 to establish the claim in the cases where $\dim R \leq 2$.

(b) When $\dim R > 2$, use 12.2.8(ii) and 12.2.5 to show that h_R is monomorphic, and consider the exact sequence

$$0 \longrightarrow R \xrightarrow{h_R} \operatorname{Hom}_R(\omega_R, \omega_R) \longrightarrow \operatorname{Coker} h_R \longrightarrow 0.$$

12.2.10 Lemma. *Suppose that R is a homomorphic image of a Gorenstein local ring, and that $\mathfrak{u}_R(0) = 0$. Then $\{\mathfrak{p} \in \operatorname{Spec}(R) : R_\mathfrak{p}$ is $S_2\}$, which we refer to as the S_2-locus of R, is an open subset of $\operatorname{Spec}(R)$ in the Zariski topology.*

Proof. Consider $h_R : R \longrightarrow \operatorname{Hom}_R(\omega_R, \omega_R)$, and let $C := \operatorname{Coker} h_R$, a finitely generated R-module. Since $\mathfrak{u}_R(0) = 0$, we see that h_R is monomorphic and $\operatorname{Supp} \omega_R = \operatorname{Spec}(R)$, by 12.1.15. Let $\mathfrak{p} \in \operatorname{Spec}(R)$. On use of the natural $R_\mathfrak{p}$-isomorphism between $(\operatorname{Hom}_R(\omega_R, \omega_R))_\mathfrak{p}$ and $\operatorname{Hom}_{R_\mathfrak{p}}((\omega_R)_\mathfrak{p}, (\omega_R)_\mathfrak{p})$, and the fact that $(\omega_R)_\mathfrak{p} \cong \omega_{R_\mathfrak{p}}$ (by 12.1.18(ii)), we see from 12.2.9 that $R_\mathfrak{p}$ is not S_2 if and only if $\mathfrak{p} \in \operatorname{Supp} C$. $\qquad\square$

The next exercise, which can be solved by use of 12.2.9, establishes a result due to Y. Aoyama.

12.2.11 Exercise (Y. Aoyama [1, Proposition 2]). Suppose that ω_R exists. Show that $h_R : R \longrightarrow \operatorname{Hom}_R(\omega_R, \omega_R)$ is an isomorphism if and only if the completion $\widehat{R}$ satisfies the condition S_2.

12.2.12 Exercise. Suppose that ω_R exists, that R is S_2 and that all the formal fibres of R are Cohen–Macaulay. Show that h_R is an isomorphism.

12.2.13 Exercise. Suppose that R is a homomorphic image of a Gorenstein local ring and that R is S_2. Show that, for all $\mathfrak{p} \in \operatorname{Spec}(R)$, the localization $(\omega_R)_\mathfrak{p}$ is a canonical module for $R_\mathfrak{p}$.

The following exercise has important significance for algebraic geometry.

12.2.14 Exercise. Suppose that $(R, \mathfrak{m})$ is a Cohen–Macaulay local domain of dimension n that is not Gorenstein but admits a canonical module $\omega := \omega_R$.

(i) Show that ω is isomorphic to an ideal of R. (Here is a hint: consider $\omega \otimes_R Q$, where Q is the quotient field of R.)
(ii) In the light of (i), regard ω as an ideal of R. Show that
  (a) $\operatorname{ht} \omega = 1$ and ω is unmixed;
  (b) R/ω is a Cohen–Macaulay ring of dimension $n - 1$; and
  (c) R/ω is Gorenstein.

(Here are some more hints: for (ii)(a) and (ii)(b), apply local cohomology to the exact sequence $0 \longrightarrow \omega \longrightarrow R \longrightarrow R/\omega \longrightarrow 0$; for (c), use 12.1.20 to see that the R/ω-module $\operatorname{Ext}_R^1(R/\omega, \omega)$ is a canonical module for R/ω, and then apply the functor $\operatorname{Hom}_R(\bullet, \omega)$ to the above-mentioned exact sequence in order to deduce that R/ω is a canonical module for itself. If you still find the exercise difficult, you might like to consult [7, Proposition 3.3.18].)

12.2.15 Exercise. Prove the following result of M. P. Murthy [55]: a Cohen–Macaulay UFD that is a homomorphic image of a Gorenstein local ring must itself be Gorenstein.

12.3 S_2-ifications

The purpose of this section is to relate the theory of canonical modules to the concept of S_2-ification of a local ring discussed by Hochster and Huneke in [39, Discussion (2.3)]. Our starting point is, however, a little more general, and concerns the theory of generalized ideal transforms, discussed in §2.2.

12.3.1 Notation. In this section, we shall only assume that R is local when that is explicitly stated. Throughout this section, $(\Lambda, \leq)$ will denote a (non-empty) directed partially ordered set, and $\mathfrak{B} = (\mathfrak{b}_\alpha)_{\alpha \in \Lambda}$ will denote a system of ideals of R over Λ in the sense of 2.1.10.

Later in the section, we shall take $\mathfrak{B}$ to be a particular system of ideals relevant to the condition S_2.

By a *subsystem of* $\mathfrak{B}$ we shall mean a system of ideals $\mathfrak{C}$ of R such that each ideal in the family $\mathfrak{C}$ is a member of $\mathfrak{B}$ and $\mathfrak{C}$ can be written as $(\mathfrak{b}_\alpha)_{\alpha \in \Theta}$ for some directed subset Θ of the indexing set Λ.

12.3.2 Proposition. *Let M be a finitely generated R-module whose support is equal to the whole of $\operatorname{Spec}(R)$. Then $H_{\mathfrak{B}}^0(M) = H_{\mathfrak{B}}^1(M) = 0$ if and only if $H_{\mathfrak{b}_\alpha}^0(M) = H_{\mathfrak{b}_\alpha}^1(M) = 0$ for all $\alpha \in \Lambda$.*

Proof. Since
$$\Gamma_{\mathfrak{b}_\alpha}(M) \subseteq \Gamma_{\mathfrak{B}}(M) \subseteq \bigcup_{\beta \in \Lambda} \Gamma_{\mathfrak{b}_\beta}(M),$$
we see that $H_{\mathfrak{B}}^0(M) = 0$ if and only if $H_{\mathfrak{b}_\alpha}^0(M) = 0$ for all $\alpha \in \Lambda$. Suppose that this is so.

Let S be the set of non-zerodivisors on M. Then, by 2.1.1(ii), $S \cap \mathfrak{b}_\alpha \neq \emptyset$ for all $\alpha \in \Lambda$. It follows from 2.2.6(i) that $H_{\mathfrak{B}}^1(M) = 0$ if and only if $\eta_{\mathfrak{B}, M}$ is an isomorphism, and $H_{\mathfrak{b}_\alpha}^1(M) = 0$ (for a given $\alpha \in \Lambda$) if and only if $\eta_{\mathfrak{b}_\alpha, M}$

is an isomorphism. Moreover, by 2.2.18, the map $\eta_{\mathfrak{B},M}$ is an isomorphism if and only if the inclusion monomorphism $M \xrightarrow{\subseteq} \bigcup_{\alpha \in \Lambda}(M :_{S^{-1}M} \mathfrak{b}_\alpha)$ is an isomorphism, and, for an $\alpha \in \Lambda$, the map $\eta_{\mathfrak{b}_\alpha,M}$ is an isomorphism if and only if the inclusion monomorphism $M \xrightarrow{\subseteq} \bigcup_{n \in \mathbb{N}}(M :_{S^{-1}M} \mathfrak{b}_\alpha^n)$ is an isomorphism. The proof can now be completed easily. □

12.3.3 Reminders. We remind the reader of some properties of the natural ring homomorphism $\eta_{\mathfrak{B},R} : R \longrightarrow D_{\mathfrak{B}}(R)$ that were established in 2.2.6(i), 2.2.15, 2.2.16 and 2.2.17.

(i) Both the kernel and cokernel of $\eta_{\mathfrak{B},R}$ are $\mathfrak{B}$-torsion.

(ii) Suppose that R' is a ring (with identity, but not necessarily commutative), and let $e : R \longrightarrow R'$ be a ring homomorphism such that $\mathrm{Im}\, e$ is contained in the centre of R' and, when R' is regarded as a left R-module by means of e, both $\mathrm{Ker}\, e$ and $\mathrm{Coker}\, e$ are $\mathfrak{B}$-torsion. If $\Gamma_{\mathfrak{B}}(R') = 0$, then the ring R' is commutative.

(iii) If now R' is a commutative ring (with identity) and $e : R \longrightarrow R'$ is a ring homomorphism such that the R-modules $\mathrm{Ker}\, e$ and $\mathrm{Coker}\, e$ are $\mathfrak{B}$-torsion, then the unique R-homomorphism $\psi' : R' \longrightarrow D_{\mathfrak{B}}(R)$ such that the diagram

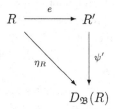

commutes is actually a ring homomorphism. The fact that this diagram commutes can simply be recorded by the statement that ψ' is an R-algebra homomorphism.

(iv) In the situation of part (iii), it follows from the formula for ψ' in 2.2.15 that ψ' is injective if and only if $\eta_{\mathfrak{B},R'}$ is injective, and this is the case if and only if $\Gamma_{\mathfrak{B}}(R') = 0$. Furthermore, ψ' is an R-algebra isomorphism if and only if $\eta_{\mathfrak{B},R'}$ is an isomorphism, and this is the case if and only if $\Gamma_{\mathfrak{B}}(R') = H_{\mathfrak{B}}^1(R') = 0$.

12.3.4 Remark. Let $\mathfrak{C}$ be a subsystem of $\mathfrak{B}$, and suppose that both $H_{\mathfrak{B}}^0(R)$ and $H_{\mathfrak{B}}^1(R)$ are $\mathfrak{C}$-torsion. Then it follows from 12.3.3 that there is a unique R-algebra homomorphism $D_{\mathfrak{B}}(R) \longrightarrow D_{\mathfrak{C}}(R)$; also the facts that $\mathrm{Ker}\,\eta_{\mathfrak{C},R} \cong H_{\mathfrak{C}}^0(R)$ and $\mathrm{Coker}\,\eta_{\mathfrak{C},R} \cong H_{\mathfrak{C}}^1(R)$ are $\mathfrak{C}$-torsion, and therefore $\mathfrak{B}$-torsion, show that there is a unique R-algebra homomorphism $e' : D_{\mathfrak{C}}(R) \to D_{\mathfrak{B}}(R)$.

The uniqueness aspects of these statements therefore mean that e' is an R-algebra isomorphism.

Similarly, if $\mathfrak{d}$ is a member of $\mathfrak{B}$ and $H_{\mathfrak{B}}^0(R)$ and $H_{\mathfrak{B}}^1(R)$ are $\mathfrak{d}$-torsion, then there is a uniquely determined R-algebra isomorphism $D_{\mathfrak{d}}(R) \xrightarrow{\cong} D_{\mathfrak{B}}(R)$.

12.3.5 Remark. Suppose $D_{\mathfrak{B}}(R)$ is a finitely generated R-module. Then it follows from 2.2.6(i) that $H_{\mathfrak{B}}^1(R)$ is a finitely generated R-module; of course, $H_{\mathfrak{B}}^0(R)$ is finitely generated. Since $\mathfrak{B}$ is a system of ideals, there exists $\alpha \in \Lambda$ such that $\mathfrak{b}_\alpha$ annihilates both $H_{\mathfrak{B}}^0(R)$ and $H_{\mathfrak{B}}^1(R)$. Let $\mathfrak{C}$ be a subsystem of $\mathfrak{B}$ such that $\mathfrak{b}_\alpha$ belongs to the family $\mathfrak{C}$. Then it is immediate from 12.3.4 that there are uniquely determined R-algebra isomorphisms

$$D_{\mathfrak{b}_\alpha}(R) \xrightarrow{\cong} D_{\mathfrak{C}}(R) \xrightarrow{\cong} D_{\mathfrak{B}}(R).$$

12.3.6 Definition. A $\mathfrak{B}$-*closure of* R is a commutative R-algebra $\theta : R \longrightarrow A$ (with identity) such that

(i) the structural ring homomorphism θ makes A into a finitely generated R-module;

(ii) both $\operatorname{Ker} \theta$ and $\operatorname{Coker} \theta$ are $\mathfrak{B}$-torsion; and

(iii) whenever $e : R \longrightarrow R'$ is a commutative R-algebra such that both $\operatorname{Ker} e$ and $\operatorname{Coker} e$ are $\mathfrak{B}$-torsion, there is a unique R-algebra homomorphism $\psi' : R' \longrightarrow A$.

We are using the terminology '$\mathfrak{B}$-closure' in 12.3.6 because the concept has similarities with 'la Z-clôture' studied in Grothendieck [24, §5.9, §5.10].

12.3.7 Remarks. We can make the following comments about existence of a $\mathfrak{B}$-closure of R.

(i) It is immediate from the definition that a $\mathfrak{B}$-closure of R, if it exists, is uniquely determined up to R-algebra isomorphism.

(ii) Conditions (ii) and (iii) in Definition 12.3.6 require that a $\mathfrak{B}$-closure of R has to be a solution to a certain universal problem. As $\eta_{\mathfrak{B},R} : R \longrightarrow D_{\mathfrak{B}}(R)$ is a solution to that universal problem (by 12.3.3(iii)), it follows that there is a $\mathfrak{B}$-closure of R if and only if $D_{\mathfrak{B}}(R)$ is a finitely generated R-module, and then $\eta_{\mathfrak{B},R} : R \longrightarrow D_{\mathfrak{B}}(R)$ is the $\mathfrak{B}$-closure of R.

(iii) If there exists a $\mathfrak{B}$-closure of R, then, by part (ii), the R-module $D_{\mathfrak{B}}(R)$ is finitely generated, and so it is immediate from 12.3.5 that there is an ideal $\mathfrak{a}$ in the family $\mathfrak{B}$ for which $H_{\mathfrak{B}}^0(R)$ and $H_{\mathfrak{B}}^1(R)$ are $\mathfrak{a}$-torsion, and then, for any such $\mathfrak{a}$, and any subsystem $\mathfrak{C}$ of $\mathfrak{B}$ such that $\mathfrak{a}$ belongs to $\mathfrak{C}$, there are uniquely determined R-algebra isomorphisms

$$D_{\mathfrak{a}}(R) \xrightarrow{\cong} D_{\mathfrak{C}}(R) \xrightarrow{\cong} D_{\mathfrak{B}}(R).$$

12.3.8 Notation.   Throughout this section, $\mathfrak{H}$ will denote the system of ideals of R formed by the set of all ideals of R of height at least 2 (indexed by itself, partially ordered by reverse inclusion). (Recall that we interpret the height of the improper ideal R of R as ∞.)

We define the *non-S_2 locus of* R to be $\{\mathfrak{p} \in \operatorname{Spec}(R) : R_{\mathfrak{p}} \text{ is not } S_2\}$. Also throughout this section, $\mathfrak{S}$ will denote the system of ideals of R formed by the set of all ideals $\mathfrak{s}$ such that $\operatorname{Var}(\mathfrak{s})$ is contained in the non-S_2 locus of R. Note that an R-module M is $\mathfrak{S}$-torsion if and only if $(0 :_R m) \in \mathfrak{S}$ for all $m \in M$.

12.3.9 Definition.   We define an $\mathfrak{S}$-closure of R, in the sense of 12.3.6, to be an S_2-*ification of* R. It follows from 12.3.7(ii) that there is an S_2-ification of R if and only if $D_{\mathfrak{S}}(R)$ is a finitely generated R-module, and then $\eta_{\mathfrak{S},R} : R \longrightarrow D_{\mathfrak{S}}(R)$ provides the S_2-ification of R.

We shall reconcile this definition of S_2-ification with that made by Hochster and Huneke in [39, Discussion (2.3)] in the case where R is local, $\mathfrak{u}_R(0) = 0$ and ω_R exists.

12.3.10 Theorem.   *Suppose that* $(R, \mathfrak{m})$ *is local, that* $\mathfrak{u}_R(0) = 0$ *and that* ω_R *exists. Then*

(i) $\mathfrak{S}$ *is a subsystem of* $\mathfrak{H}$*, and the ring homomorphism*

$$h_R : R \longrightarrow \operatorname{Hom}_R(\omega_R, \omega_R)$$

of 12.2.4 has (kernel and) cokernel that are $\mathfrak{H}$*-torsion;*

(ii) *the ring* $\operatorname{Hom}_R(\omega_R, \omega_R)$ *is commutative;*

(iii) *there is a unique* R*-algebra isomorphism* $\psi' : \operatorname{Hom}_R(\omega_R, \omega_R) \overset{\cong}{\longrightarrow} D_{\mathfrak{H}}(R)$*;*

(iv) $D_{\mathfrak{H}}(R)$ *and* $H_{\mathfrak{H}}^1(R)$ *are finitely generated* R*-modules, and the ideal* $\mathfrak{a} := (0 :_R H_{\mathfrak{H}}^1(R))$ *belongs to the family* $\mathfrak{H}$*; also* $\operatorname{Var}(\mathfrak{a})$ *is the non-S_2 locus of* R*;*

(v) *there are unique* R*-algebra isomorphisms*

$$\operatorname{Hom}_R(\omega_R, \omega_R) \overset{\cong}{\longrightarrow} D_{\mathfrak{a}}(R) \overset{\cong}{\longrightarrow} D_{\mathfrak{S}}(R) \overset{\cong}{\longrightarrow} D_{\mathfrak{H}}(R),$$

and each of these R*-algebras provides the* S_2*-ification of* R*.*

Proof.   (i) Let $\mathfrak{a}$ be an ideal in $\mathfrak{S}$, and suppose that $\mathfrak{a} \subseteq \mathfrak{p}$, where $\mathfrak{p}$ is a prime ideal of height less than 2. Then $R_{\mathfrak{p}}$ is not S_2, and this is a contradiction because $\mathfrak{u}_R(0) = 0$. Now h_R is injective (since its kernel is $\mathfrak{u}_R(0)$), and $\operatorname{Coker} h_R$ is $\mathfrak{H}$-torsion by 12.2.7.

(ii) Note that $\operatorname{Im} h_R$ is contained in the centre of $H := \operatorname{Hom}_R(\omega_R, \omega_R)$; also $\Gamma_{\mathfrak{H}}(H) = 0$, since otherwise there would be an associated prime of the

R-module H having height at least 2, and this is not possible because H is a faithful R-module satisfying the condition S_2 (by 12.1.18). It now follows from 2.2.16 that H is a commutative ring.

(iii) It is now immediate from 2.2.17 that there is a unique R-algebra homomorphism $\psi' : \text{Hom}_R(\omega_R, \omega_R) =: H \longrightarrow D_{\mathfrak{H}}(R)$. Since H is (faithful and) S_2 by 12.1.18, we see that $H^0_{\mathfrak{a}}(H) = H^1_{\mathfrak{a}}(H) = 0$ for all ideals $\mathfrak{a}$ of R with ht $\mathfrak{a} \geq 2$. Hence $H^0_{\mathfrak{H}}(H) = H^1_{\mathfrak{H}}(H) = 0$ by 12.3.2. It therefore follows from 2.2.15 that ψ' is an isomorphism.

(iv) Since H is a finitely generated R-module, $D_{\mathfrak{H}}(R)$ must also be finitely generated, by part (iii). Therefore its epimorphic image $H^1_{\mathfrak{H}}(R)$ is finitely generated, and, since this module is $\mathfrak{H}$-torsion, its annihilator must be in $\mathfrak{H}$.

By 12.3.5, there is a uniquely determined R-algebra isomorphism

$$D_{\mathfrak{a}}(R) \overset{\cong}{\longrightarrow} D_{\mathfrak{H}}(R).$$

We show next that $\text{Var}(\mathfrak{a})$ is the non-S_2 locus of R. Let $\mathfrak{p} \in \text{Spec}(R)$. If $\mathfrak{p} \notin \text{Var}(\mathfrak{a})$, then there is an $R_{\mathfrak{p}}$-isomorphism $R_{\mathfrak{p}} \cong H_{\mathfrak{p}}$, so that $R_{\mathfrak{p}}$ is S_2 because H is. On the other hand, if $\mathfrak{p}$ is a minimal member of $\text{Var}(\mathfrak{a})$, then

$$0 \neq (H^1_{\mathfrak{a}}(R))_{\mathfrak{p}} \cong H^1_{\mathfrak{a}R_{\mathfrak{p}}}(R_{\mathfrak{p}})$$

and depth $R_{\mathfrak{p}} \leq 1$; since ht $\mathfrak{p} \geq 2$, this means that $R_{\mathfrak{p}}$ is not S_2.

(v) By part (iv), the ideal $\mathfrak{a}$ lies in the family $\mathfrak{S}$. Also, $\mathfrak{S}$ is a subsystem of $\mathfrak{H}$, by part (i). We can now use part (iii) and 12.3.5 to complete the proof. $\square$

12.3.11 Corollary. *Suppose $(R, \mathfrak{m})$ is local, ω_R exists and $\mathfrak{u}_R(0) = 0$. Let S denote the set of all non-zerodivisors of R, so that R can be considered as a subring of $S^{-1}R$, the total quotient ring of R. Then the S_2-ification of R (which exists by 12.3.10) is given by the R-algebra $A := \bigcup_{\mathfrak{b} \in \mathfrak{H}} (R :_{S^{-1}R} \mathfrak{b})$. Furthermore,*

(i) *A is finitely generated and S_2 as an R-module,*
(ii) *for all $a \in A$, we have $\text{ht}(R :_R a) \geq 2$, and*
(iii) *there are unique R-algebra isomorphisms*

$$\bigcup_{\mathfrak{b} \in \mathfrak{H}} (R :_{S^{-1}R} \mathfrak{b}) \overset{\cong}{\longrightarrow} \text{Hom}_R(\omega_R, \omega_R)$$
$$\overset{\cong}{\longrightarrow} D_{\mathfrak{a}}(R) \overset{\cong}{\longrightarrow} D_{\mathfrak{S}}(R) \overset{\cong}{\longrightarrow} D_{\mathfrak{H}}(R),$$

where $\mathfrak{a} := (0 :_R H^1_{\mathfrak{H}}(R))$.

Proof. Set $H := \text{Hom}_R(\omega_R, \omega_R)$. Since $\mathfrak{u}_R(0) = 0$, it follows from 12.1.10 that S consists entirely of non-zerodivisors on H, and that $S \cap \mathfrak{b} \neq \emptyset$ for all $\mathfrak{b} \in \mathfrak{H}$. We can now use 2.2.18 to see that there is a unique R-algebra isomorphism $A \xrightarrow{\cong} D_{\mathfrak{H}}(R)$. It therefore follows from 12.3.10 that $R \xrightarrow{\subseteq} A$ provides the S_2-ification of R. All the other claims are now immediate from 12.3.10. $\qquad\square$

12.3.12 Remark. Suppose that R is local and that $\mathfrak{u}_R(0) = 0$. Hochster and Huneke [39, Discussion (2.3)] defined an S_2-ification of R to be a subring A' of the total quotient ring of R such that $R \subseteq A'$, such that A', as R-module, is finitely generated and S_2, and such that, for all $a \in A'$, $\mathrm{ht}(R :_R a) \geq 2$. If ω_R exists, then the S_2-ification A of R found in 12.3.11 is an S_2-ification in the sense of Hochster and Huneke.

12.3.13 Remark. Suppose $(R, \mathfrak{m})$ is local and ω_R exists. By 12.1.13, the ideal $\mathfrak{u}_R(0)$ annihilates ω_R and ω_R is a canonical module for $R/\mathfrak{u}_R(0)$. It therefore follows from 12.3.11 that the endomorphism ring $\mathrm{Hom}_R(\omega_R, \omega_R)$ of ω_R is a commutative Noetherian semi-local ring.

12.3.14 Exercise (Goto [2, Example 3.3]). Let

$$R := K[[X, Y, Z, W]]/(X, Y) \cap (Z, W),$$

where K is a field and X, Y, Z, W are independent indeterminates. Show that the R-homomorphism $h_R : R \longrightarrow \mathrm{Hom}_R(\omega_R, \omega_R)$ of 12.2.4 is not an isomorphism, and that the (commutative) ring $\mathrm{Hom}_R(\omega_R, \omega_R)$ is not local.

12.3.15 Exercise. Suppose that $(R, \mathfrak{m})$ is local and ω_R exists. Assume that R is analytically irreducible; that is, assume that the completion $\widehat{R}$ is an integral domain. Show that $\mathrm{Hom}_R(\omega_R, \omega_R)$ is also an analytically irreducible local ring. (Here is a hint: you might find [50, Theorem 8.15] helpful.)

Note. In [39, Theorem (3.6)], Hochster and Huneke provide, in the case where $(R, \mathfrak{m})$ is local, complete and equidimensional, several conditions equivalent to the statement that the semi-local ring $\mathrm{Hom}_R(\omega_R, \omega_R)$ is local.

13

Foundations in the graded case

If our Noetherian ring R is ($\mathbb{Z}$-)graded, and our ideal $\mathfrak{a}$ is graded, then it is natural to ask whether the local cohomology modules $H_{\mathfrak{a}}^i(R)$ ($i \in \mathbb{N}_0$), and $H_{\mathfrak{a}}^i(M)$ ($i \in \mathbb{N}_0$) for a graded R-module M, also carry structures as graded R-modules. Some of the realizations of these local cohomology modules that we have obtained earlier in the book suggest that they should. For instance, if $a_1, \ldots, a_n$ (where $n > 0$) denote n *homogeneous* elements which generate $\mathfrak{a}$, then the Čech complex $C^\bullet(M)$ of M with respect to $a_1, \ldots, a_n$ is composed of graded R-modules and homogeneous homomorphisms, and so $H_{\mathfrak{a}}^i(M)$ (for $i \in \mathbb{N}_0$), which, by Theorem 5.1.20, is isomorphic to $H^i(C(M)^\bullet)$, inherits a grading. But is this grading independent of the choice of homogeneous generators for $\mathfrak{a}$?

Additional hopeful evidence is provided by the isomorphism

$$H_{\mathfrak{a}}^i(M) \cong \varinjlim_{n \in \mathbb{N}} \operatorname{Ext}_R^i(R/\mathfrak{a}^n, M)$$

of 1.3.8. For each $n \in \mathbb{N}$, since $R/\mathfrak{a}^n$ is a finitely generated graded R-module, $\operatorname{Ext}_R^i(R/\mathfrak{a}^n, M)$ is actually the graded R-module

$$^*\operatorname{Ext}_R^i(R/\mathfrak{a}^n, M)$$

(see [7, pp. 32–33]) with its grading forgotten, and, for $n, m \in \mathbb{N}$ with $n \geq m$, the natural homomorphism $h_m^n : R/\mathfrak{a}^n \to R/\mathfrak{a}^m$ is homogeneous, so that the induced homomorphism

$$^*\operatorname{Ext}_R^i(h_m^n, M) : {}^*\operatorname{Ext}_R^i(R/\mathfrak{a}^m, M) \longrightarrow {}^*\operatorname{Ext}_R^i(R/\mathfrak{a}^n, M)$$

is homogeneous; hence $\varinjlim_{n \in \mathbb{N}} \operatorname{Ext}_R^i(R/\mathfrak{a}^n, M)$ is graded, and $H_{\mathfrak{a}}^i(M)$ inherits a grading by virtue of the above isomorphism. But is this grading the same as that which comes from the approach using the Čech complex described in the preceding paragraph?

One could take another approach to local cohomology in this graded situation, an approach which, at first sight, seems substantially different from those described in the preceding two paragraphs. Again suppose that our Noetherian ring R is ($\mathbb{Z}$-)graded, and our ideal $\mathfrak{a}$ is graded. The category $*\mathcal{C}(R)$ of all graded R-modules and homogeneous R-homomorphisms is an Abelian category which has enough projective objects [7, p. 32] and enough injective objects [7, 3.6.2]: we can therefore carry out standard techniques of homological algebra in this category. In particular, the $\mathfrak{a}$-torsion functor $\Gamma_{\mathfrak{a}}$ can be viewed as a (left exact, additive) functor from $*\mathcal{C}(R)$ to itself, and so we can form its right derived functors $*H^i_{\mathfrak{a}}$ ($i \in \mathbb{N}_0$) on that category. This will produce, for a graded R-module M, graded local cohomology modules $*H^i_{\mathfrak{a}}(M)$ ($i \in \mathbb{N}_0$), which are constructed by the following procedure. Since $*\mathcal{C}(R)$ has enough injectives, we can construct an injective resolution of M in this category, that is, we can construct an exact sequence

$$0 \longrightarrow M \xrightarrow{\alpha} E^0 \xrightarrow{d^0} E^1 \longrightarrow \cdots \longrightarrow E^i \xrightarrow{d^i} E^{i+1} \longrightarrow \cdots$$

in $*\mathcal{C}(R)$ in which the E^i ($i \in \mathbb{N}_0$) are injective objects in that category, that is, they are *injective graded R-modules in the terminology of [7, §3.6]; then we apply the functor $\Gamma_{\mathfrak{a}}$ to the complex

$$0 \longrightarrow E^0 \xrightarrow{d^0} E^1 \longrightarrow \cdots \longrightarrow E^i \xrightarrow{d^i} E^{i+1} \longrightarrow \cdots ;$$

the i-th cohomology module of the resulting complex is $*H^i_{\mathfrak{a}}(M)$ (for each $i \in \mathbb{N}_0$). But if we forget the grading on $*H^i_{\mathfrak{a}}(M)$, is the resulting R-module isomorphic to $H^i_{\mathfrak{a}}(M)$, and, if so, is the grading on $H^i_{\mathfrak{a}}(M)$ induced by this isomorphism the same as that which comes from the approaches in the first two paragraphs of this chapter?

Our main purpose in this chapter is to reconcile these various approaches. We think it is desirable that it should be established that they all give (up to isomorphism in $*\mathcal{C}(R)$) the same object of $*\mathcal{C}(R)$. Thus we are going to show that the questions posed at the ends of the first three paragraphs of this chapter all have affirmative answers.

In fact, because many modern applications of local cohomology are to multigraded rings and modules, we are going to deal in this chapter with a $\mathbb{Z}^n$-graded commutative Noetherian ring and $\mathbb{Z}^n$-graded modules over it (note that n will always denote a positive integer in this context).

13.1 Basic multi-graded commutative algebra

13.1.1 Notation and Terminology. Throughout this chapter, G will denote a finitely generated, torsion-free Abelian group, written additively. (Thus G is either 0 or isomorphic to $\mathbb{Z}^n$ for some $n \in \mathbb{N}$.)

When we write that $R = \bigoplus_{g \in G} R_g$ is a G-graded ring, it is to be understood that the direct decomposition is as $\mathbb{Z}$-module, that $R_g R_{g'} \subseteq R_{g+g'}$ for all $g, g' \in G$, and that all other 'graded' objects related to R (such as R-modules, ideals of R) are graded by G. Thus, when R is G-graded as above and we write that M is a graded R-module, it is to be understood that M has a direct decomposition $M = \bigoplus_{g \in G} M_g$ as $\mathbb{Z}$-module and that $R_g M_{g'} \subseteq M_{g+g'}$ for all $g, g' \in G$. The elements of $\bigcup_{g \in G} M_g$ are called the *homogeneous elements* of M, and an element of $M_{g'} \setminus \{0\}$, where $g' \in G$, is said to have *degree* g'. We shall use 'deg' as an abbreviation for degree.

Suppose that $R = \bigoplus_{g \in G} R_g$ is G-graded. An R-homomorphism $f : M = \bigoplus_{g \in G} M_g \longrightarrow N = \bigoplus_{g \in G} N_g$ between graded R-modules will be said to be *homogeneous* precisely when $f(M_g) \subseteq N_g$ for all $g \in G$; we shall denote by $*\mathcal{C}(R)$ (or by $*\mathcal{C}^G(R)$ when it is desirable to specify the grading group G) the category of all graded R-modules and homogeneous R-homomorphisms between them.

For $g_0 \in G$, we shall denote the g_0-*th shift functor* by $(\bullet)(g_0) : *\mathcal{C}(R) \longrightarrow *\mathcal{C}(R)$: thus, for a graded R-module $M = \bigoplus_{g \in G} M_g$, we have $(M(g_0))_g = M_{g+g_0}$ for all $g \in G$; also, $f(g_0) \lceil (M(g_0))_g = f \lceil M_{g+g_0}$ for each morphism f in $*\mathcal{C}(R)$ and all $g \in G$.

If S is a multiplicatively closed subset of R consisting of non-zero homogeneous elements, then it is routine to check that the ring $S^{-1}R$ is also G-graded, with g-th component, for $g \in G$, equal to the set of all elements of $S^{-1}R$ that can be expressed in the form r/s with $s \in S$, r a homogeneous element of R and $(r = 0$ or$)$ $r \neq 0$ and $\deg r - \deg s = g$. Similarly, for a graded R-module M, the $S^{-1}R$-module $S^{-1}M$ is also graded.

13.1.2 Remark. A total order $\prec$ on G is said to be *compatible with addition* if, whenever $g, g', h \in G$ with $g \prec g'$, we have $g + h \prec g' + h$. Note that it is always possible to put a total order compatible with addition on G: this is trivial if $G = 0$, and otherwise $G \cong \mathbb{Z}^n$ for some $n \in \mathbb{N}$, and, for example, the lexicographical order on $\mathbb{Z}^n$ is compatible with addition. This lexicographical order $\prec$ is defined as follows: for $\mathbf{a} = (a_1, \ldots, a_n)$ and $\mathbf{b} = (b_1, \ldots, b_n) \in \mathbb{Z}^n$, we set $\mathbf{a} \prec \mathbf{b}$ if and only if $\mathbf{a} \neq \mathbf{b}$ and, if i is the least integer in $\{1, \ldots, n\}$ for which $a_i \neq b_i$, then $a_i < b_i$.

13.1.3 Elementary Reminders.   Assume that R is G-graded, and let M be a graded R-module.

(i) We have $1_R \in R_0$; furthermore, R_0 is a Noetherian subring of R and R_g is a finitely generated R_0-module, for each $g \in G$. (See [61, §2.11, Theorem 21, and §4.3, Theorem 13].)

(ii) We shall say that a submodule of M is *graded* precisely when it can be generated by homogeneous elements: see [61, §2.11, Proposition 28].

(iii) Recall from [61, §2.11, Proposition 29] that the sum and intersection of the members of an arbitrary family of graded submodules of M are again graded.

(iv) Let N be a graded submodule of M and assume that $\mathfrak{a}$ is graded. Then $(N :_R M)$, $\mathfrak{a}N$ and $(N :_M \mathfrak{a})$ are all graded: see [61, §2.11, Propositions 30 and 31].

(v) Assume that $\mathfrak{a}$ is graded. Then $\mathfrak{a}$ is prime if and only if it is proper and, whenever r and r' are homogeneous elements of $R \setminus \mathfrak{a}$, then $rr' \in R \setminus \mathfrak{a}$ too: see [61, §2.13, Lemma 13].

(vi) For an arbitrary ideal $\mathfrak{b}$ of R, we denote by $\mathfrak{b}^*$ the (necessarily graded) ideal generated by all homogeneous elements of $\mathfrak{b}$. Thus $\mathfrak{b}^*$ is the largest graded ideal of R contained in $\mathfrak{b}$. By [61, §2.13, Proposition 33], if $\mathfrak{b}$ is prime, then so too is $\mathfrak{b}^*$.

(vii) We denote by $*\operatorname{Spec}(R)$ (or $*\operatorname{Spec}^G(R)$) the set of all graded prime ideals of R.

(viii) The radical of a graded ideal of R is again graded: see [61, §2.13, Proposition 32].

13.1.4 Example.   Let R_0 be a commutative Noetherian ring, let $n \in \mathbb{N}$, and let $R := R_0[X_1, \ldots, X_n]$, the ring of polynomials over R_0.

(i) We shall often consider R as $\mathbb{Z}^n$-graded, with grading given by

$$R_{(i_1,\ldots,i_n)} = R_0 X_1^{i_1} \ldots X_n^{i_n} \quad \text{for all } (i_1, \ldots, i_n) \in \mathbb{Z}^n.$$

Thus $R_{(i_1,\ldots,i_n)} = 0$ unless $i_j \geq 0$ for all $j = 1, \ldots, n$. Observe that X_i is homogeneous of degree $(0, \ldots, 0, 1, 0, \ldots, 0)$ (where the '1' is in the i-th spot), for all $i = 1, \ldots, n$.

(ii) Consider the special case in which R_0 is a field. Then the graded ideals of R are just the ideals that can be generated by monomials $X_1^{i_1} \ldots X_n^{i_n}$, where $(i_1, \ldots, i_n) \in \mathbb{N}_0^{\ n}$. The graded prime ideals of R are just the ideals that can be generated by a subset of the set $\{X_1, \ldots, X_n\}$ of the variables. It follows that there are only finitely many graded prime ideals of R.

(iii) Now consider the case where $R_0 = S$ is a G-graded ring, with grading given by $S = \bigoplus_{g \in G} S_g$. Then $R = S[X_1, \ldots, X_n]$ is $(G \oplus \mathbb{Z}^n)$-graded, with $(g, (i_1, \ldots, i_n))$-th component equal to $S_g X_1^{i_1} \ldots X_n^{i_n}$ for all $g \in G$ and $(i_1, \ldots, i_n) \in \mathbb{Z}^n$.

13.1.5 Example. Let $n \in \mathbb{N}$. A *simplicial complex* on the vertex set $V :=$ $\{1, \ldots, n\}$ is a set Δ of subsets of V which is closed under passage to subsets, that is, whenever $F \in \Delta$ and $F' \subseteq F$, then $F' \in \Delta$ too.

Let R_0 be a commutative Noetherian ring and let $R := R_0[X_1, \ldots, X_n]$, the ring of polynomials over R_0 in indeterminates $X_1, \ldots, X_n$, considered to be $\mathbb{Z}^n$-graded as in 13.1.4. Let Δ be a simplicial complex on $\{1, \ldots, n\}$. Let $\mathfrak{a}_\Delta$ be the ideal of R generated by all square-free monomials $X_{i_1} \ldots X_{i_t}$ such that $\{i_1, \ldots, i_t\} \notin \Delta$. Observe that, if F' is a subset of V that does not belong to Δ, then no subset of V that contains F' can belong to Δ. Observe also that $\mathfrak{a}_\Delta$ is a graded ideal of R, as it is generated by homogeneous elements. Therefore the ring $R_0[\Delta] := R/\mathfrak{a}_\Delta$ inherits a $\mathbb{Z}^n$-grading from R. The ring $R_0[\Delta]$ is called the *Stanley–Reisner ring of* Δ *with respect to* R_0.

13.1.6 Lemma. *Assume that R is G-graded, let M be a graded R-module, and let $\mathfrak{p}$ be a prime ideal of R.*

(i) *If $\mathfrak{p} \in \operatorname{Supp} M$, then $\mathfrak{p}^* \in \operatorname{Supp} M$.*

(ii) *If $\mathfrak{p} \in \operatorname{Ass} M$, then $\mathfrak{p}$ is graded; in addition, $\mathfrak{p} = (0 :_R m)$ for some homogeneous element $m \in M$.*

(iii) *In particular, if $\mathfrak{a}$ is a graded ideal of R, then $\operatorname{ass} \mathfrak{a}$ consists of graded prime ideals.*

Proof. The special case of this lemma in which $G = \mathbb{Z}$ is proved in [7, Lemma 1.5.6(b)]. The same proof with minor modifications works in this G-graded case provided one puts a total order $\prec$ on G compatible with addition (see 13.1.2), and uses $\prec$ instead of the usual order $<$ on $\mathbb{Z}$.   □

13.1.7 Categorical Reminders. Assume that R is G-graded.

(i) The category $*\mathcal{C}(R)$ of all graded R-modules and homogeneous R-homomorphisms is Abelian.

(ii) If $f : M \longrightarrow N$ is a morphism in $*\mathcal{C}(R)$ (that is, if M and N are graded R-modules and f is a homogeneous R-homomorphism), then the ordinary kernel and image of f are graded submodules of M and N respectively, and act as $\operatorname{Ker} f$ and $\operatorname{Im} f$ in $*\mathcal{C}(R)$.

(iii) A sequence $M \longrightarrow N \longrightarrow P$ of objects and morphisms in $*\mathcal{C}(R)$ is exact in that category if and only if it is exact in $\mathcal{C}(R)$.

(iv) Projective (respectively injective) objects in the category $*\mathcal{C}(R)$ will be referred to as *projective (respectively *injective) graded R-modules. The category $*\mathcal{C}(R)$ 'has enough projectives', for if M is a graded R-module, then M is a homogeneous homomorphic image of a graded R-module $\bigoplus_{i \in I} R(g_i)$, where $(g_i)_{i \in I}$ is a family of elements of G, and it is easy to see that $\bigoplus_{i \in I} R(g_i)$ is *projective. A graded R-module of the form $\bigoplus_{i \in I} R(g_i)$ will be called *free*. That $*\mathcal{C}(R)$ 'has enough injectives' is proved in §13.2 below.

13.1.8 ♯Exercise and Definitions. Assume that R is G-graded, and let $M = \bigoplus_{g \in G} M_g$ and $N = \bigoplus_{g \in G} N_g$ be graded R-modules.

(i) Let $g_0 \in G$. We say that an R-homomorphism $f : M \longrightarrow N$ is *homogeneous of degree* g_0 precisely when $f(M_g) \subseteq N_{g+g_0}$ for all $g \in G$. We denote by $* \operatorname{Hom}_R(M, N)_{g_0}$ the set of all homogeneous R-homomorphisms from M to N of degree g_0. Show that

   (a) $* \operatorname{Hom}_R(M, N)_{g_0}$ is an R_0-submodule of $\operatorname{Hom}_R(M, N)$, and
   (b) the sum $\sum_{g' \in G} * \operatorname{Hom}_R(M, N)_{g'}$ is direct.

(ii) We set

$$* \operatorname{Hom}_R(M, N) := \sum_{g' \in G} * \operatorname{Hom}_R(M, N)_{g'} = \bigoplus_{g' \in G} * \operatorname{Hom}_R(M, N)_{g'}.$$

Show that this is an R-submodule of $\operatorname{Hom}_R(M, N)$, and that the above direct decomposition turns $* \operatorname{Hom}_R(M, N)$ into a graded R-module. Deduce that $* \operatorname{Hom}_R(\bullet , \bullet) : *\mathcal{C}(R) \times *\mathcal{C}(R) \longrightarrow *\mathcal{C}(R)$ is a left exact, additive functor.

(iii) Show that, if M is finitely generated, then $\operatorname{Hom}_R(M, N)$ is actually equal to $* \operatorname{Hom}_R(M, N)$ with its grading forgotten.

(iv) Let $i \in \mathbb{N}_0$. In order to avoid lengthy excursions into the homological algebra of the category $*\mathcal{C}(R)$, we shall define $* \operatorname{Ext}^i_R(\bullet , N)$ to be the i-th right derived functor in $*\mathcal{C}(R)$ of $* \operatorname{Hom}_R(\bullet , N)$.

Show that, if M is finitely generated, then $\operatorname{Ext}^i_R(M, N)$ is actually equal to $* \operatorname{Ext}^i_R(M, N)$ with its grading forgotten.

13.1.9 ♯Exercise. Suppose R is G-graded, and let $M = \bigoplus_{g \in G} M_g$ and $N = \bigoplus_{g \in G} N_g$ be graded R-modules. Let $(M \otimes_R N)_g$, for a $g \in G$, be the $\mathbb{Z}$-submodule of $M \otimes_R N$ generated by all elements $m_{g_1} \otimes n_{g_2}$, where $g_1, g_2 \in G$ are such that $g_1 + g_2 = g$, and $m_{g_1} \in M_{g_1}$, $n_{g_1} \in N_{g_1}$. It is clear that $M \otimes_R N = \sum_{g \in G} (M \otimes_R N)_g$; the aim of this exercise is to show that this sum is direct, and provides $M \otimes_R N$ with a structure as a graded R-module.

(i) Let F be a *free R-module; thus $F = \bigoplus_{i \in I} R(g_i)$, where $(g_i)_{i \in I}$ is a family of elements of G. Show that the sum $F \otimes_R N = \sum_{g \in G} (F \otimes_R N)_g$ is direct, and that the decomposition

$$F \otimes_R N = \bigoplus_{g \in G} (F \otimes_R N)_g$$

provides a grading for $F \otimes_R N$.

(ii) Consider an exact sequence $F_1 \longrightarrow F_0 \longrightarrow M \longrightarrow 0$ in the category $*\mathcal{C}(R)$, where F_1 and F_0 are *free R-modules. Use part (i) to show that the sum $M \otimes_R N = \sum_{g \in G} (M \otimes_R N)_g$ is direct; deduce that the decomposition $M \otimes_R N = \bigoplus_{g \in G} (M \otimes_R N)_g$ provides a grading for $M \otimes_R N$.

(iii) Deduce that $\otimes_R \lceil$ can be considered as a functor

$$\otimes_R \lceil \; : *\mathcal{C}(R) \times *\mathcal{C}(R) \longrightarrow *\mathcal{C}(R).$$

13.2 *Injective modules

One of the main aims of this section is to show that, when R is G-graded, the category $*\mathcal{C}(R)$ 'has enough injectives'. Our route to this will involve the concept of *injective envelope.

13.2.1 Definition. Assume that R is G-graded. Let M be a graded submodule of the graded R-module L.

(i) We say that L is a *essential extension* of M precisely when $B \cap M \neq 0$ for every non-zero graded submodule B of L. Such a *essential extension of M is said to be *proper* if and only if it is not equal to M.

(ii) We say that L is a *injective envelope* (or *injective hull*) of M precisely when L is a *injective R-module and also a *essential extension of M.

13.2.2 Lemma. *Assume that R is G-graded. Let M be a graded submodule of the graded R-module L such that L is a *essential extension of M. Then, with the gradings forgotten, L is an essential extension of M.*

Proof. Let $0 \neq x \in L$. Write x as a sum of homogeneous elements $x = x_{g_1} + \cdots + x_{g_r}$, where $g_1, \ldots, g_r$ are r different members of G, and $0 \neq x_{g_i}$ for all $i = 1, \ldots, r$. We show by induction on r that there exists a homogeneous element $a \in R$ such that $0 \neq ax \in M$; this will suffice to prove the lemma. This is clear when $r = 1$, since then x is itself homogeneous, and so $Rx \cap M \neq 0$ because L is a *essential extension of M.

Thus we suppose that $r > 1$ and make the obvious inductive assumption. There exists a homogeneous element $b \in R$ such that $0 \neq bx_{g_r} \in M$; let $x' := x - x_{g_r} = x_{g_1} + \cdots + x_{g_{r-1}}$. If $bx' = 0$, then $bx = bx_{g_r}$ is a non-zero element of M, as wanted; otherwise, $0 \neq bx'$, and bx' has fewer than r non-zero homogeneous components; therefore, by the inductive hypothesis, there is a homogeneous element $c \in R$ such that $0 \neq cbx' \in M$. Then, since b and c are both homogeneous, $0 \neq cbx = cbx' + cbx_{g_r} \in M$.   □

13.2.3 Proposition. *Assume that R is G-graded, and let M be a graded R-module. Then M is *injective if and only if M has no proper *essential extension.*

Proof. ($\Rightarrow$) Assume that M is *injective, and let $\iota : M \longrightarrow N$ be the inclusion homomorphism from M into a *essential extension N of M. (Thus N is graded and ι is homogeneous.) Since M is *injective, there exists a homogeneous R-homomorphism $\varphi : N \longrightarrow M$ such that $\varphi \circ \iota = \mathrm{Id}_M$, and $N = M \oplus \mathrm{Ker}\,\varphi$. Since N is a *essential extension of M, we must have $\mathrm{Ker}\,\varphi = 0$, so that $N = M$.

($\Leftarrow$) This can be proved just as in the ungraded case: see [7, Proposition 3.2.2].   □

In the next Theorem 13.2.4, we shall show that each graded R-module M (where R is G-graded) has a *injective envelope; in particular, this will show that M can be embedded, by means of a homogeneous R-homomorphism, into a *injective R-module, and so prove that the category $*\mathcal{C}(R)$ has enough injectives. The proof of 13.2.4 is modelled on Bruns' and Herzog's proof in [7, Theorem 3.6.2] of the particular case in which $G = \mathbb{Z}$.

13.2.4 Theorem. *Assume that R is G-graded, and let M be a graded R-module.*

  (i) *The graded R-module M has a *injective envelope, which, with its grading forgotten, is an R-submodule of $E(M)$, the ordinary injective envelope of M.*

  (ii) *Between any two *injective envelopes of M there is a homogeneous isomorphism which restricts to the identity map on M. We denote by $*E(M)$ or $*E_R(M)$ one choice of *injective envelope of M.*

  (iii) *Considered as (ungraded) R-module, $*E_R(M)$ is an essential extension of M.*

Proof. (i) Consider M as an ungraded R-module, and embed M into an ungraded injective R-module I; for example, we could take I to be $E(M)$. Let $\mathcal{S}$ denote the set of all R-submodules N of I that carry a grading with respect to which M is a graded submodule of N and N is a *essential extension of M. There is a partial order $\leq$ on $\mathcal{S}$ defined as follows: for $N_1, N_2 \in \mathcal{S}$, we declare that $N_1 \leq N_2$ if and only if N_1 is a graded submodule of N_2 (with the specified gradings). By Zorn's Lemma, $\mathcal{S}$ has a maximal member E; we note that E is a graded R-module, a *essential extension of M, and an R-submodule of I.

Suppose that E is not *injective; it then follows from 13.2.3 that E has a proper *essential extension $E \subset E'$. By 13.2.2, when the gradings are forgotten, E' is an essential extension of E. As I is injective, there exists an R-homomorphism (possibly not homogeneous) $\psi : E' \longrightarrow I$ that extends the inclusion homomorphism $E \overset{\subseteq}{\longrightarrow} I$. Since $\operatorname{Ker} \psi \cap E = 0$, and E' is an essential extension of E, we see that $\operatorname{Ker} \psi = 0$ and ψ is a monomorphism. We can now use the grading on E' and the monomorphism ψ to put a grading on $\operatorname{Im} \psi$ with respect to which E is a homogeneous submodule of $\operatorname{Im} \psi$. Since $E \subset E'$ was a proper *essential extension, it follows that $\operatorname{Im} \psi$ is a proper *essential extension of E, and so a proper *essential extension of M, and we have a contradiction to the maximality of E in $\mathcal{S}$. This contradiction shows that E is *injective.

(ii) Suppose that $M \subseteq E_1$ and $M \subseteq E_2$ are two *injective envelopes of M. Since E_2 is *injective, there is a homogeneous R-homomorphism $\beta : E_1 \longrightarrow E_2$ whose restriction to M is just the identity map on M. Since $\operatorname{Ker} \beta \cap M = 0$, we can deduce that $\operatorname{Ker} \beta = 0$ because $M \subseteq E_1$ is a *essential extension; therefore β is monomorphic, and $\operatorname{Im} \beta$ is a *injective graded submodule of E_2, and so has no proper *essential extension, by 13.2.3. But E_2 is a *essential extension of M, and therefore of $\operatorname{Im} \beta$; hence $\operatorname{Im} \beta = E_2$, and $\beta : E_1 \longrightarrow E_2$ is a homogeneous isomorphism.

(iii) This is immediate from 13.2.2, because $*E_R(M)$ is a *essential extension of M. $\qquad\square$

Now that we know that the category $*\mathcal{C}(R)$ has enough injectives, we can make progress with multi-graded local cohomology. We shall build on the above introduction of the concept of *injective envelope, but not until §14.2.

13.2.5 Lemma. *Assume that R is G-graded, and let I, M be graded R-modules with I *injective.*

(i) *We have* $* \operatorname{Ext}_R^i(M, I) = 0$ *for all $i \in \mathbb{N}$.*

(ii) *If M is finitely generated, then* $\operatorname{Ext}_R^i(M, I) = 0$ *for all $i \in \mathbb{N}$.*

Proof. (i) There is a *projective graded R-module F and a homogeneous R-epimorphism $\psi : F \longrightarrow M$; let $K := \operatorname{Ker} \psi$. Since F is *projective,

$^*\mathrm{Ext}^i_R(F, I) = 0$ for all $i \in \mathbb{N}$. The exact sequence

$$0 \longrightarrow K \longrightarrow F \overset{\psi}{\longrightarrow} M \longrightarrow 0$$

therefore induces an exact sequence

$$0 \longrightarrow {}^*\mathrm{Hom}_R(M, I) \longrightarrow {}^*\mathrm{Hom}_R(F, I) \longrightarrow {}^*\mathrm{Hom}_R(K, I)$$
$$\longrightarrow {}^*\mathrm{Ext}^1_R(M, I) \longrightarrow 0$$

(in the category $^*\mathcal{C}(R)$) and homogeneous isomorphisms

$$^*\mathrm{Ext}^i_R(K, I) \cong {}^*\mathrm{Ext}^{i+1}_R(M, I) \quad \text{for all } i \in \mathbb{N}.$$

The induced homomorphism $^*\mathrm{Hom}_R(F, I) \longrightarrow {}^*\mathrm{Hom}_R(K, I)$ is surjective because I is *injective. Hence $^*\mathrm{Ext}^1_R(M, I) = 0$. But M was an arbitrary graded R-module; therefore $^*\mathrm{Ext}^1_R(K, I) = 0$, and so $^*\mathrm{Ext}^2_R(M, I) = 0$. Use induction to complete the proof of (i).

(ii) This is now immediate, since if, for finitely generated M, we forget the grading on $^*\mathrm{Ext}^i_R(M, I)$, then, by 13.1.8(iv), we obtain the ordinary 'Ext' module $\mathrm{Ext}^i_R(M, I)$. $\qquad\square$

Lemma 13.2.5 already enables us to prove the following result, which will play a key rôle in this chapter.

13.2.6 Proposition. *Assume that R is G-graded. Let I be a *injective graded R-module. Let $(\Lambda, \leq)$ be a (non-empty) directed partially ordered set, and let $\mathfrak{B} = (\mathfrak{b}_\alpha)_{\alpha \in \Lambda}$ be a system of ideals of R over Λ (as in 2.1.10), with the property that all ideals in the system are graded. Then I is $\Gamma_\mathfrak{B}$-acyclic.*

In particular, if the ideal $\mathfrak{a}$ is graded, then I is $\Gamma_\mathfrak{a}$-acyclic.

Proof. For each $\alpha \in \Lambda$, the graded R-module $R/\mathfrak{b}_\alpha$ is finitely generated, and so $\mathrm{Ext}^i_R(R/\mathfrak{b}_\alpha, I) = 0$ for all $i \in \mathbb{N}$, by 13.2.5. Hence, by 1.3.7,

$$H^i_\mathfrak{B}(I) \cong \varinjlim_{\alpha \in \Lambda} \mathrm{Ext}^i_R(R/\mathfrak{b}_\alpha, I) = 0 \quad \text{for all } i \in \mathbb{N}. \qquad\square$$

The following exercise establishes a G-graded analogue of the well-known Baer Criterion (see [71, Theorem 3.20]) for a module to be injective.

13.2.7 ♯Exercise. Assume that R is G-graded, and let I be a graded R-module. Show that I is *injective if and only if, for each graded ideal $\mathfrak{b}$ of R, each $g_0 \in G$ and each homogeneous homomorphism $f : \mathfrak{b} \longrightarrow I$ of degree g_0, there exists a homogeneous homomorphism $f' : R \longrightarrow I$ of degree g_0 such that $f' \lceil_\mathfrak{b} = f$.

We made use of the (ungraded) Baer Criterion in our proof of Proposition 2.1.4, and the G-graded version in the above exercise can be used in a similar way to establish a G-graded version of that proposition. This is addressed in the next exercise.

13.2.8 ♯Exercise. Assume that R is G-graded and that the ideal $\mathfrak{a}$ is graded. Let I be a *injective graded R-module. Show that $\Gamma_{\mathfrak{a}}(I)$ is *injective.

Now let $(\Lambda, \leq)$ be a (non-empty) directed partially ordered set, and let $\mathfrak{B} = (\mathfrak{b}_{\alpha})_{\alpha \in \Lambda}$ be a system of graded ideals of R over Λ. Show that $\Gamma_{\mathfrak{B}}(I)$ is *injective.

13.2.9 ♯Exercise. Assume that R is G-graded; let I be a graded R-module. Show that the following conditions are equivalent:

(i) I is *injective;

(ii) $*\operatorname{Ext}^1_R(M, I) = 0$ for all graded R-modules M;

(iii) $\operatorname{Ext}^1_R(L, I) = 0$ for all finitely generated graded R-modules L.

13.3 The *restriction property

13.3.1 Hypotheses for the section. We shall assume, throughout §13.3, that $R = \bigoplus_{g \in G} R_g$ is G-graded. (Recall that we always assume that R is Noetherian.) Also throughout this section, we shall let G' be another finitely generated torsion-free Abelian group, and we shall let $R' = \bigoplus_{g' \in G'} R'_{g'}$ be a G'-graded commutative ring. It should be noted that we are not assuming that R' is Noetherian; we shall assume that the reader is familiar with the elementary properties of graded R'-modules expounded in [61, §2.11 and §2.13].

13.3.2 Definition. A sequence $(T^i)_{i \in \mathbb{N}_0}$ of covariant functors from $*\mathcal{C}^G(R)$ to $*\mathcal{C}^{G'}(R')$ is said to be a *negative connected sequence* (respectively, a *negative strongly connected sequence*) if the following conditions are satisfied.

(i) Whenever

$$0 \longrightarrow L \xrightarrow{f} M \xrightarrow{g} N \longrightarrow 0$$

is an exact sequence in $*\mathcal{C}^G(R)$, there are defined connecting homogeneous R'-homomorphisms

$$T^i(N) \longrightarrow T^{i+1}(L) \quad \text{for all } i \in \mathbb{N}_0$$

(in $^*\mathcal{C}^{G'}(R')$) such that the long sequence

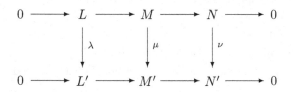

$$0 \longrightarrow T^0(L) \xrightarrow{T^0(f)} T^0(M) \xrightarrow{T^0(g)} T^0(N)$$
$$\longrightarrow T^1(L) \xrightarrow{T^1(f)} T^1(M) \xrightarrow{T^1(g)} T^1(N)$$
$$\longrightarrow \quad \cdots \qquad\qquad\qquad \cdots$$
$$\longrightarrow T^i(L) \xrightarrow{T^i(f)} T^i(M) \xrightarrow{T^i(g)} T^i(N)$$
$$\longrightarrow T^{i+1}(L) \longrightarrow \quad \cdots$$

is a complex (respectively, is exact).

(ii) Whenever

$$
\begin{array}{ccccccccc}
0 & \longrightarrow & L & \longrightarrow & M & \longrightarrow & N & \longrightarrow & 0 \\
& & \downarrow \lambda & & \downarrow \mu & & \downarrow \nu & & \\
0 & \longrightarrow & L' & \longrightarrow & M' & \longrightarrow & N' & \longrightarrow & 0
\end{array}
$$

is a commutative diagram of graded R-modules and homogeneous R-homomorphisms with exact rows, then there is induced, by λ, μ and ν, a chain map of the long complex of (i) for the top row into the corresponding long complex for the bottom row.

13.3.3 Examples. Let M be a graded R-module.

(i) Let $T : {}^*\mathcal{C}^G(R) \longrightarrow {}^*\mathcal{C}^{G'}(R')$ be an additive covariant functor. Since the category $^*\mathcal{C}^G(R)$ has enough injectives (by 13.2.4), we can carry out a standard procedure of homological algebra in that category and form the right derived functors $\mathcal{R}^i T$ ($i \in \mathbb{N}_0$) of T. In more detail, for the graded R-module M, we can construct an exact sequence

$$0 \longrightarrow M \xrightarrow{\alpha} E^0 \xrightarrow{d^0} E^1 \longrightarrow \cdots \longrightarrow E^i \xrightarrow{d^i} E^{i+1} \longrightarrow \cdots$$

in $^*\mathcal{C}^G(R)$ in which the E^i ($i \in \mathbb{N}_0$) are *injective graded R-modules; then we apply the functor T to the complex

$$0 \longrightarrow E^0 \xrightarrow{d^0} E^1 \longrightarrow \cdots \longrightarrow E^i \xrightarrow{d^i} E^{i+1} \longrightarrow \cdots ;$$

the i-th cohomology module of the resulting complex is the graded R'-module $\mathcal{R}^i T(M)$ (for each $i \in \mathbb{N}_0$).

It should be noted that $(\mathcal{R}^i T)_{i \in \mathbb{N}_0}$ is a negative strongly connected sequence of covariant functors from $^*\mathcal{C}^G(R)$ to $^*\mathcal{C}^{G'}(R')$; furthermore, if T is left exact, then $\mathcal{R}^0 T$ is naturally equivalent to T.

(ii) When $\mathfrak{a}$ is graded, the $\mathfrak{a}$-torsion functor $\Gamma_{\mathfrak{a}}$ can be viewed as a (left exact, additive) functor from $*\mathcal{C}^G(R)$ to itself, and so we can form the connected sequence $(*H_{\mathfrak{a}}^i)_{i \in \mathbb{N}_0}$ of its right derived functors on that category. For each $i \in \mathbb{N}_0$, this leads to a graded local cohomology module $*H_{\mathfrak{a}}^i(M)$: one of the aims of this chapter is to show that, if we forget the grading on $*H_{\mathfrak{a}}^i(M)$, then the resulting R-module is isomorphic to $H_{\mathfrak{a}}^i(M)$.

(iii) Another example is provided by a system of graded ideals of R. Let $(\Lambda, \leq)$ be a (non-empty) directed partially ordered set, and let $\mathfrak{B} = (\mathfrak{b}_\alpha)_{\alpha \in \Lambda}$ be a system of graded ideals of R over Λ. The $\mathfrak{B}$-torsion functor $\Gamma_{\mathfrak{B}}$ can be viewed as a (left exact, additive) functor from $*\mathcal{C}^G(R)$ to itself, and so we can form the connected sequence $(*H_{\mathfrak{B}}^i)_{i \in \mathbb{N}_0}$ of its right derived functors on that category. For each $i \in \mathbb{N}_0$, this leads to a graded generalized local cohomology module $*H_{\mathfrak{B}}^i(M)$: in this chapter, we shall show that, if we forget the grading on $*H_{\mathfrak{B}}^i(M)$, then the resulting R-module is isomorphic to $H_{\mathfrak{B}}^i(M)$.

13.3.4 Definition. Let $(T^i)_{i \in \mathbb{N}_0}$ and $(U^i)_{i \in \mathbb{N}_0}$ be two negative connected sequences of covariant functors from $*\mathcal{C}^G(R)$ to $*\mathcal{C}^{G'}(R')$. A *homomorphism* $\Psi : (T^i)_{i \in \mathbb{N}_0} \longrightarrow (U^i)_{i \in \mathbb{N}_0}$ *of connected sequences* is a family $(\psi^i)_{i \in \mathbb{N}_0}$ where, for each $i \in \mathbb{N}_0$, $\psi^i : T^i \longrightarrow U^i$ is a natural transformation of functors, and which is such that the following condition is satisfied: whenever $0 \longrightarrow L \longrightarrow M \longrightarrow N \longrightarrow 0$ is an exact sequence of graded R-modules and homogeneous R-homomorphisms, then, for each $i \in \mathbb{N}_0$, the diagram

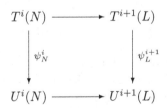

(in which the horizontal maps are the appropriate connecting homomorphisms arising from the connected sequences) commutes.

Such a homomorphism $\Psi = (\psi^i)_{i \in \mathbb{N}_0} : (T^i)_{i \in \mathbb{N}_0} \longrightarrow (U^i)_{i \in \mathbb{N}_0}$ of connected sequences is said to be an *isomorphism (of connected sequences)* precisely when $\psi^i : T^i \longrightarrow U^i$ is a natural equivalence of functors for each $i \in \mathbb{N}_0$.

13.3.5 ♯Exercise. Let $(T^i)_{i \in \mathbb{N}_0}$ and $(U^i)_{i \in \mathbb{N}_0}$ be negative connected sequences of covariant functors from $*\mathcal{C}^G(R)$ to $*\mathcal{C}^{G'}(R')$.

(i) Let $\psi^0 : T^0 \longrightarrow U^0$ be a natural transformation of functors. Assume that

(a) the sequence $(T^i)_{i \in \mathbb{N}_0}$ is strongly connected, and
(b) $T^i(I) = 0$ for all $i \in \mathbb{N}$ and *injective graded R-modules I.

Show that there are uniquely determined natural transformations ψ^i : $T^i \longrightarrow U^i$ ($i \in \mathbb{N}$) such that $(\psi^i)_{i \in \mathbb{N}_0} : (T^i)_{i \in \mathbb{N}_0} \longrightarrow (U^i)_{i \in \mathbb{N}_0}$ is a homomorphism of connected sequences.

(ii) Now let $\psi^0 : T^0 \longrightarrow U^0$ be a natural equivalence of functors. Assume that

(a) the sequence $(T^i)_{i \in \mathbb{N}_0}$ is strongly connected,
(b) the sequence $(U^i)_{i \in \mathbb{N}_0}$ is strongly connected, and
(c) for all $i \in \mathbb{N}$ and *injective graded R-modules I, we have

$$T^i(I) = U^i(I) = 0.$$

Show that there exist uniquely determined natural equivalences ψ^i : $T^i \longrightarrow U^i$ ($i \in \mathbb{N}$) such that $(\psi^i)_{i \in \mathbb{N}_0} : (T^i)_{i \in \mathbb{N}_0} \longrightarrow (U^i)_{i \in \mathbb{N}_0}$ is an isomorphism of connected sequences.

13.3.6 Definition. Let $T : \mathcal{C}(R) \longrightarrow \mathcal{C}(R')$ be a covariant functor. If

(i) whenever M is a graded R-module, the R'-module $T(M)$ is graded, and
(ii) the gradings in (i) are such that, whenever $f : M \longrightarrow N$ is a homogeneous homomorphism of graded R-modules, then $T(f) : T(M) \longrightarrow T(N)$ is homogeneous,

then we say that T *has the *restriction property* (with respect to the gradings specified in (i)).

Of course, the restriction of T to $*\mathcal{C}^G(R)$ is a functor $T\lceil: *\mathcal{C}^G(R) \longrightarrow \mathcal{C}(R')$. (We should, strictly speaking, denote $T\lceil$ by $T\lceil_{*\mathcal{C}^G(R)}$, but we shall use the shorter notation in the interests of simplicity.) Note that T has the *restriction property if and only if $T\lceil$ can be viewed as a functor from $*\mathcal{C}^G(R)$ to $*\mathcal{C}^{G'}(R')$.

13.3.7 Definition. Let $T, U : \mathcal{C}(R) \longrightarrow \mathcal{C}(R')$ be covariant functors. Let $\alpha : T \longrightarrow U$ be a natural transformation of functors (from $\mathcal{C}(R)$ to $\mathcal{C}(R')$). Of course, the restriction of α to $*\mathcal{C}^G(R)$ provides a natural transformation $\alpha\lceil: T\lceil \longrightarrow U\lceil$ of functors from $*\mathcal{C}^G(R)$ to $\mathcal{C}(R')$.

Now assume that both T and U have the *restriction property. We say that α *has the *restriction property* precisely when, for every graded R-module M, the map $\alpha_M : T(M) \longrightarrow U(M)$ is homogeneous.

Note that this is the case if and only if $\alpha\lceil$ can be viewed as a natural trans-formation $\alpha\lceil: T\lceil \longrightarrow U\lceil$ of functors from $*\mathcal{C}^G(R)$ to $*\mathcal{C}^{G'}(R')$.

13.3.8 Remark. Observe that, if, in the notation of 13.3.7, $\alpha : T \longrightarrow U$ is a natural equivalence of functors which has the *restriction property, then the inverse natural equivalence $\alpha^{-1} : U \longrightarrow T$ also has the *restriction property.

Note also that, if $V : \mathcal{C}(R) \longrightarrow \mathcal{C}(R')$ is a third covariant functor having the *restriction property, and both α and a natural transformation $\beta : U \longrightarrow V$ have the *restriction property, then the composition $\beta \circ \alpha : T \longrightarrow V$ again has the *restriction property.

13.3.9 Definition. Let $(T^i)_{i\in\mathbb{N}_0}$ be a negative strongly connected sequence of covariant functors from $\mathcal{C}(R)$ to $\mathcal{C}(R')$. We can regard $(T^i\lceil)_{i\in\mathbb{N}_0}$ as a negative strongly connected sequence of covariant functors from $*\mathcal{C}^G(R)$ to $\mathcal{C}(R')$.

We shall say that $(T^i)_{i\in\mathbb{N}_0}$ *has the *restriction property* precisely when

(i) T^i has the *restriction property for all $i \in \mathbb{N}_0$; and
(ii) whenever $0 \longrightarrow L \longrightarrow M \longrightarrow N \longrightarrow 0$ is an exact sequence in the category $*\mathcal{C}^G(R)$, the connecting homomorphisms $T^i(N) \longrightarrow T^{i+1}(L)$ ($i \in \mathbb{N}_0$) (which exist by virtue of the fact that $(T^i)_{i\in\mathbb{N}_0}$ is a connected sequence) are all homogeneous.

Note that this is the case if and only if $(T^i\lceil)_{i\in\mathbb{N}_0}$ can be viewed as a negative strongly connected sequence of covariant functors from $*\mathcal{C}^G(R)$ to $*\mathcal{C}^{G'}(R')$.

13.3.10 Definition. Let $(T^i)_{i\in\mathbb{N}_0}$ and $(U^i)_{i\in\mathbb{N}_0}$ be negative strongly connec-ted sequences of covariant functors from $\mathcal{C}(R)$ to $\mathcal{C}(R')$; assume that both these sequences have the *restriction property of 13.3.9. Let $\Psi := (\psi^i)_{i\in\mathbb{N}_0} : (T^i)_{i\in\mathbb{N}_0} \longrightarrow (U^i)_{i\in\mathbb{N}_0}$ be a homomorphism of connected sequences. We say that Ψ *has the *restriction property* if and only if, for all $i \in \mathbb{N}_0$, the natural transformation ψ^i has the *restriction property of 13.3.7.

13.3.11 Proposition. *Let V, M be graded R-modules with V finitely gen-erated. By 13.1.8(iv), for each $i \in \mathbb{N}_0$, $\mathrm{Ext}^i_R(V, M)$ is actually the graded R-module $* \mathrm{Ext}^i_R(V, M)$ with its grading forgotten; hence each $\mathrm{Ext}^i_R(V, M)$ has a natural structure as a graded R-module, and these structures are such that, if V' is a second finitely generated graded R-module and $h : V \longrightarrow V'$ is a homogeneous homomorphism, then $\mathrm{Ext}^i_R(h, M) : \mathrm{Ext}^i_R(V', M) \longrightarrow \mathrm{Ext}^i_R(V, M)$ is also homogeneous.*

*Moreover, with respect to these graded R-module structures, the negative strongly connected sequence $\left(\mathrm{Ext}^i_R(V, \bullet)\right)_{i\in\mathbb{N}_0}$ of covariant functors (from $\mathcal{C}(R)$ to itself) has the *restriction property (see 13.3.9).*

Proof. Only the claim in the second paragraph requires proof. Let $F_\bullet$ be a *free resolution of V in $*\mathcal{C}(R)$ by *free graded R-modules of finite rank.

An exact sequence $0 \longrightarrow L' \longrightarrow M' \longrightarrow N' \longrightarrow 0$ in $\mathcal{C}(R)$ induces a sequence

$$0 \longrightarrow \operatorname{Hom}_R(F_\bullet, L') \longrightarrow \operatorname{Hom}_R(F_\bullet, M') \longrightarrow \operatorname{Hom}_R(F_\bullet, N') \longrightarrow 0$$

of complexes of R-modules and chain maps of such complexes such that, for each $i \in \mathbb{N}_0$, the sequence

$$0 \longrightarrow \operatorname{Hom}_R(F_i, L') \longrightarrow \operatorname{Hom}_R(F_i, M') \longrightarrow \operatorname{Hom}_R(F_i, N') \longrightarrow 0$$

(where F_i denotes the i-th term of $F_\bullet$) is exact. It is straightforward to use the long exact sequences of cohomology modules induced by such sequences of complexes to turn $\left(H^i(\operatorname{Hom}_R(F_\bullet, \,\bullet\,)) \right)_{i \in \mathbb{N}_0}$ into a negative strongly connected sequence of covariant functors from $\mathcal{C}(R)$ to itself, and a standard application of Exercise 1.3.4(ii) then shows that this negative connected sequence is isomorphic to $\left(\operatorname{Ext}_R^i(V, \,\bullet\,) \right)_{i \in \mathbb{N}_0}$.

In particular, when we apply this to a morphism $f : M \longrightarrow M'$ in $*\mathcal{C}(R)$, and to an exact sequence $0 \longrightarrow L \longrightarrow M \longrightarrow N \longrightarrow 0$ of graded R-modules and homogeneous R-homomorphisms, there result, for each $i \in \mathbb{N}_0$, commutative diagrams

$$
\begin{array}{ccc}
H^i(\operatorname{Hom}_R(F_\bullet, M)) & \longrightarrow & H^i(\operatorname{Hom}_R(F_\bullet, M')) \\
\cong \downarrow & & \downarrow \cong \\
\operatorname{Ext}_R^i(V, M) & \xrightarrow{\operatorname{Ext}_R^i(V, f)} & \operatorname{Ext}_R^i(V, M')
\end{array}
$$

and

$$
\begin{array}{ccc}
H^i(\operatorname{Hom}_R(F_\bullet, N)) & \longrightarrow & H^{i+1}(\operatorname{Hom}_R(F_\bullet, L)) \\
\cong \downarrow & & \downarrow \cong \\
\operatorname{Ext}_R^i(V, N) & \longrightarrow & \operatorname{Ext}_R^{i+1}(V, L)
\end{array}
\quad .
$$

In these circumstances, $\operatorname{Hom}_R(F_\bullet, M)$, $\operatorname{Hom}_R(F_\bullet, M')$, $\operatorname{Hom}_R(F_\bullet, N)$ and $\operatorname{Hom}_R(F_\bullet, L)$ are complexes of graded R-modules and homogeneous R-homomorphisms, so that all their cohomology modules are graded, and it is easy to check that the top horizontal homomorphisms in the above two commutative diagrams are homogeneous. Furthermore, the graded module structures on $\operatorname{Ext}_R^i(V, M)$, $\operatorname{Ext}_R^i(V, M')$, $\operatorname{Ext}_R^i(V, N)$ and $\operatorname{Ext}_R^{i+1}(V, L)$ induced by the vertical isomorphisms in the diagrams are precisely the graded module structures referred to in the first paragraph of the statement of the proposition. All the claims follow easily from these observations. $\qquad\square$

13.3.12 Remark. Let $(\Omega, \leq)$ be a (non-empty) directed partially ordered set, and let $(W_\omega)_{\omega \in \Omega}$ be a direct system of graded R-modules and homogeneous R-homomorphisms over Ω, with constituent R-homomorphisms $h_\nu^\omega : W_\nu \to W_\omega$ (for each $(\omega, \nu) \in \Omega \times \Omega$ with $\omega \geq \nu$). We can forget the gradings and calculate the ordinary direct limit

$$W_\infty := \varinjlim_{\omega \in \Omega} W_\omega,$$

in $\mathcal{C}(R)$: let $h_\omega : W_\omega \longrightarrow W_\infty$ be the canonical map (for each $\omega \in \Omega$). It should be noted that W_∞ inherits a natural grading, for which all the h_ω ($\omega \in \Omega$) are homogeneous, and that W_∞ actually acts as the direct limit of the direct system $(W_\omega)_{\omega \in \Omega}$ in the category $^*\mathcal{C}(R)$.

13.3.13 Examples. The following examples will be important for us.

(i) Let $(\Lambda, \leq)$ be a (non-empty) directed partially ordered set, and let $\mathfrak{B} = (\mathfrak{b}_\alpha)_{\alpha \in \Lambda}$ be a system of graded ideals of R over Λ. For $\alpha, \beta \in \Lambda$ with $\alpha \geq \beta$, the natural homomorphism $h_\beta^\alpha : R/\mathfrak{b}_\alpha \longrightarrow R/\mathfrak{b}_\beta$ is homogeneous. It follows from 13.3.11 and 13.3.12 that the negative strongly connected sequence of covariant functors

$$\left(\varinjlim_{\alpha \in \Lambda} \operatorname{Ext}_R^i(R/\mathfrak{b}_\alpha, \, \bullet \,) \right)_{i \in \mathbb{N}_0}$$

from $\mathcal{C}(R)$ to itself has the *restriction property.

(ii) In particular, when the ideal $\mathfrak{a}$ is graded, the negative strongly connected sequence of covariant functors

$$\left(\varinjlim_{n \in \mathbb{N}} \operatorname{Ext}_R^i(R/\mathfrak{a}^n, \, \bullet \,) \right)_{i \in \mathbb{N}_0}$$

from $\mathcal{C}(R)$ to itself has the *restriction property.

(iii) Assume that $\mathfrak{a}$ is graded, and let $a_1, \ldots, a_n$ (where $n > 0$) denote n homogeneous elements which generate $\mathfrak{a}$. Let $C^\bullet$ denote the Čech complex of R with respect to $a_1, \ldots, a_n$, as in 5.1.5. Then it is straightforward to check that the negative strongly connected sequence of covariant functors $(H^i((\, \bullet \,) \otimes_R C^\bullet))_{i \in \mathbb{N}_0}$ from $\mathcal{C}(R)$ to itself has the *restriction property.

13.3.14 ♯Exercise. Let $(\Lambda, \leq)$ be a (non-empty) directed partially ordered set, and let $\mathfrak{B} = (\mathfrak{b}_\alpha)_{\alpha \in \Lambda}$ be a system of graded ideals of R over Λ. Show that the functor $D_\mathfrak{B}$ of 2.2.3 has the *restriction property, and that the natural transformation $\eta_\mathfrak{B}$ of 2.2.6(i) has the *restriction property.

Deduce that, when the ideal $\mathfrak{a}$ is graded, the functor $D_\mathfrak{a}$ of 2.2.1 has the *restriction property, and the natural transformation $\eta_\mathfrak{a}$ of 2.2.6(i) has the *restriction property.

13.3.15 Theorem. *Let* $(T^i)_{i \in \mathbb{N}_0}$ *be a negative strongly connected sequence of covariant additive functors from* $\mathcal{C}(R)$ *to* $\mathcal{C}(R')$. *Suppose that, for each graded* R-*module* M, *a grading is given on* $T^0(M)$, *and that, with respect to these gradings,* T^0 *has the *restriction property. Suppose further that* $T^i(I) = 0$ *for all* $i \in \mathbb{N}$ *whenever* I *is a *injective graded* R-*module.*

Then there is exactly one choice of gradings on the $T^i(M)$ *(i* $\in \mathbb{N}$, *M a graded* R-*module) with respect to which* $(T^i)_{i \in \mathbb{N}_0}$ *has the *restriction property.*

Furthermore, if we denote by *T^0 : *$\mathcal{C}^G(R) \longrightarrow$ *$\mathcal{C}^{G'}(R')$ *the (necessarily left exact) functor induced by* $T^0 \lceil$ *(see 13.3.6), then there is a unique isomorphism* $(\psi^i)_{i \in \mathbb{N}_0} : (T^i \lceil)_{i \in \mathbb{N}_0} \overset{\cong}{\longrightarrow} (\mathcal{R}^i(*T^0))_{i \in \mathbb{N}_0}$ *of negative connected sequences of covariant functors from* *$\mathcal{C}^G(R)$ *to* *$\mathcal{C}^{G'}(R')$ *for which* ψ^0 : $T^0 \lceil = \!*T^0 \longrightarrow \mathcal{R}^0(*T^0)$ *is the canonical natural equivalence.*

Proof. We first show, by induction on $t \in \mathbb{N}_0$, that, if $(T^i)_{i \in \mathbb{N}_0}$ has the *restriction property, then the gradings on the R'-modules $T^t(M)$ (M a graded R-module) are uniquely determined by the gradings on the R'-modules $T^0(L)$ (L a graded R-module). This statement is certainly true when $t = 0$, and so we suppose now that $t > 0$ and the statement is true for smaller values of t.

Let M be an arbitrary graded R-module. Since the category *$\mathcal{C}^G(R)$ has enough injective objects, there is an exact sequence

$$0 \longrightarrow M \longrightarrow I \overset{g}{\longrightarrow} N \longrightarrow 0$$

in *$\mathcal{C}^G(R)$ with I a *injective graded R-module. The hypotheses and the assumption that $(T^i)_{i \in \mathbb{N}_0}$ has the *restriction property therefore lead to an exact sequence

$$T^{t-1}(I) \xrightarrow{T^{t-1}(g)} T^{t-1}(N) \longrightarrow T^t(M) \longrightarrow 0$$

of graded R'-modules and homogeneous homomorphisms. Since, by the inductive hypothesis, the grading on $T^{t-1}(N)$ is uniquely determined by our assumptions, it follows that the grading on $T^t(M)$ is similarly uniquely determined.

Thus there is at most one choice of gradings on the $T^i(M)$ ($i \in \mathbb{N}$, M a graded R-module) with respect to which $(T^i)_{i \in \mathbb{N}_0}$ has the *restriction property. We still have to show that there is one such choice.

Since $(T^i)_{i \in \mathbb{N}_0}$ is a negative strongly connected sequence of covariant functors from $\mathcal{C}(R)$ to $\mathcal{C}(R')$, it is automatic that T^0 is left exact. Therefore $*T^0$: $*\mathcal{C}^G(R) \longrightarrow *\mathcal{C}^{G'}(R')$ is left exact (and additive), $(\mathcal{R}^i(*T^0))_{i \in \mathbb{N}_0}$ is a negative strongly connected sequence of functors from $*\mathcal{C}^G(R)$ to $*\mathcal{C}^{G'}(R')$, and there is a canonical natural equivalence $\psi^0 : T^0 \lceil = *T^0 \longrightarrow \mathcal{R}^0(*T^0)$. Note that ψ^0_M is homogeneous for each graded R-module M.

Now forget, just temporarily, the grading on R': then $(\mathcal{R}^i(*T^0))_{i \in \mathbb{N}_0}$ and $(T^i \lceil)_{i \in \mathbb{N}_0}$ are negative strongly connected sequences of covariant functors from $*\mathcal{C}^G(R)$ to $\mathcal{C}(R')$ and $\psi^0 : T^0 \lceil \longrightarrow \mathcal{R}^0(*T^0)$ is a natural equivalence (of functors from $*\mathcal{C}^G(R)$ to $\mathcal{C}(R')$). Moreover, $T^i(I) = \mathcal{R}^i(*T^0)(I) = 0$ for all $i \in \mathbb{N}$ and all *injective graded R-modules I.

At this point, regard, again temporarily, R' as trivially graded, that is, graded by 0: we can use 13.3.5(ii) to see that there exist uniquely determined natural equivalences $\psi^i : T^i \lceil \longrightarrow \mathcal{R}^i(*T^0)$ ($i \in \mathbb{N}$) such that $(\psi^i)_{i \in \mathbb{N}_0}$: $(T^i \lceil)_{i \in \mathbb{N}_0} \longrightarrow (\mathcal{R}^i(*T^0))_{i \in \mathbb{N}_0}$ is an isomorphism of connected sequences (of functors from $*\mathcal{C}^G(R)$ to $\mathcal{C}(R') = *\mathcal{C}^0(R')$).

Now remember the original G'-gradings on R' and its graded modules: for each graded R-module M and each $i \in \mathbb{N}$, the R'-module $\mathcal{R}^i(*T^0)(M)$ is graded. For such i and M, we can define a grading on $T^i(M)$ in such a way that the R'-isomorphism

$$\psi^i_M : T^i(M) \xrightarrow{\cong} \mathcal{R}^i(*T^0)(M)$$

is homogeneous. With these gradings in place, all the remaining claims of the theorem follow routinely. $\qquad\qquad\square$

13.3.16 Corollary. *Let T be a left exact additive covariant functor from $\mathcal{C}(R)$ to $\mathcal{C}(R')$. Suppose that, for each graded R-module M, a grading is given on $T(M)$, and that, with respect to these gradings, T has the *restriction property. Suppose further that each *injective graded R-module I is T-acyclic, that is, $\mathcal{R}^i T(I) = 0$ for all $i \in \mathbb{N}$.*

*Then there is exactly one choice of gradings on the $\mathcal{R}^i T(M)$ ($i \in \mathbb{N}$, M a graded R-module) with respect to which $(\mathcal{R}^i T)_{i \in \mathbb{N}_0}$ has the *restriction property.*

Furthermore, if we denote by $*T : *\mathcal{C}^G(R) \longrightarrow *\mathcal{C}^{G'}(R')$ the functor induced by $T \lceil$, then there is a unique isomorphism

$$(\psi^i)_{i \in \mathbb{N}_0} : ((\mathcal{R}^i T) \lceil)_{i \in \mathbb{N}_0} \xrightarrow{\cong} (\mathcal{R}^i(*T))_{i \in \mathbb{N}_0}$$

of negative connected sequences of functors from $*\mathcal{C}^G(R)$ *to* $*\mathcal{C}^{G'}(R')$ *for which* $\psi^0 : T\lceil = *T \longrightarrow *T$ *is the identity.* □

13.3.17 Theorem. *Let*

$$(T^i)_{i \in \mathbb{N}_0} \quad and \quad (U^i)_{i \in \mathbb{N}_0}$$

be negative strongly connected sequences of covariant additive functors from $\mathcal{C}(R)$ *to* $\mathcal{C}(R')$ *which have the* *restriction property of 13.3.9, and suppose that* $T^i(I) = 0$ *for all* $i \in \mathbb{N}$ *whenever* I *is a* *injective graded R-module.*

Let $\Psi := (\psi^i)_{i \in \mathbb{N}_0} : (T^i)_{i \in \mathbb{N}_0} \longrightarrow (U^i)_{i \in \mathbb{N}_0}$ *be a homomorphism of connected sequences. Suppose that* ψ^0 *has the* *restriction property of 13.3.7. Then it is automatic that* ψ^i *has the* *restriction property for all* $i \in \mathbb{N}_0$, *so that* Ψ *has the* *restriction property of 13.3.10.*

Furthermore, $(\psi^i\lceil)_{i \in \mathbb{N}_0} : (T^i\lceil)_{i \in \mathbb{N}_0} \longrightarrow (U^i\lceil)_{i \in \mathbb{N}_0}$ *is the unique extension of the natural transformation* $\psi^0\lceil : T^0\lceil \longrightarrow U^0\lceil$ *(of functors from* $*\mathcal{C}^G(R)$ *to* $*\mathcal{C}^{G'}(R')$*) to a homomorphism of connected sequences of functors from* $*\mathcal{C}^G(R)$ *to* $*\mathcal{C}^{G'}(R')$.

Proof. We prove by induction on t that ψ^t has the *restriction property for all $t \in \mathbb{N}_0$. We know that ψ^0 has the *restriction property, and so we suppose now that $t > 0$ and that ψ^i has the *restriction property for all $i < t$.

Let M be an arbitrary graded R-module. Since the category $*\mathcal{C}^G(R)$ has enough injective objects, there is an exact sequence

$$0 \longrightarrow M \longrightarrow I \overset{g}{\longrightarrow} N \longrightarrow 0$$

in $*\mathcal{C}^G(R)$ with I a *injective graded R-module. The hypotheses therefore lead to a commutative diagram

$$
\begin{array}{ccccccc}
T^{t-1}(I) & \overset{T^{t-1}(g)}{\longrightarrow} & T^{t-1}(N) & \longrightarrow & T^t(M) & \longrightarrow & 0 \\
\downarrow{\scriptstyle \psi_I^{t-1}} & & \downarrow{\scriptstyle \psi_N^{t-1}} & & \downarrow{\scriptstyle \psi_M^t} & & \\
U^{t-1}(I) & \overset{U^{t-1}(g)}{\longrightarrow} & U^{t-1}(N) & \longrightarrow & U^t(M) & \longrightarrow & U^t(I)
\end{array}
$$

of graded R'-modules and homogeneous homomorphisms with exact rows. It follows from the inductive hypothesis, and the fact that $(T^i)_{i \in \mathbb{N}_0}$ and $(U^i)_{i \in \mathbb{N}_0}$ have the *restriction property, that all the homomorphisms in the above diagram, with the possible exception of ψ_M^t, are homogeneous. It is immediate from the fact that the top row is exact that ψ_M^t must be homogeneous too.

The claim in the final paragraph is immediate from 13.3.5, 13.3.7 and 13.3.9. □

13.4 The reconciliation

13.4.1 Hypotheses for the section. We shall assume, throughout §13.4, that $R = \bigoplus_{g \in G} R_g$ is G-graded, and that the ideal $\mathfrak{a}$ is graded.

Also, we shall assume that $(\Lambda, \leq)$ is a (non-empty) directed partially ordered set, and that $\mathfrak{B} = (\mathfrak{b}_\alpha)_{\alpha \in \Lambda}$ is a system of graded ideals of R over Λ.

13.4.2 Theorem. *There is a unique choice of gradings on the $H_{\mathfrak{B}}^i(M)$ ($i \in \mathbb{N}$, M a graded R-module) with respect to which $(H_{\mathfrak{B}}^i)_{i \in \mathbb{N}_0}$ has the *restriction property; furthermore, when these gradings are imposed, there is a unique isomorphism*

$$(\widetilde{\phi}^i)_{i \in \mathbb{N}_0} : (H_{\mathfrak{B}}^i \lceil)_{i \in \mathbb{N}_0} \xrightarrow{\cong} (*H_{\mathfrak{B}}^i)_{i \in \mathbb{N}_0}$$

of connected sequences of covariant functors from *$\mathcal{C}(R)$ to itself for which $\widetilde{\phi}^0$ is the identity.

Proof. By 13.2.6, each *injective graded R-module I is $\Gamma_{\mathfrak{B}}$-acyclic; also, it is clear that $\Gamma_{\mathfrak{B}} : \mathcal{C}(R) \longrightarrow \mathcal{C}(R)$ has the *restriction property. The result therefore follows from 13.3.16. □

We record separately a very important special case of 13.4.2.

13.4.3 Corollary. *There is a unique choice of gradings on the $H_{\mathfrak{a}}^i(M)$ ($i \in \mathbb{N}$, M a graded R-module) with respect to which $(H_{\mathfrak{a}}^i)_{i \in \mathbb{N}_0}$ has the *restriction property; furthermore, when these gradings are imposed, there is a unique isomorphism*

$$(\widetilde{\psi}^i)_{i \in \mathbb{N}_0} : (H_{\mathfrak{a}}^i \lceil)_{i \in \mathbb{N}_0} \xrightarrow{\cong} (*H_{\mathfrak{a}}^i)_{i \in \mathbb{N}_0}$$

of connected sequences of covariant functors from *$\mathcal{C}(R)$ to itself for which $\widetilde{\psi}^0$ is the identity. □

13.4.4 Remark. For a graded R-module M, we do, of course, grade the $H_{\mathfrak{B}}^i(M)$ ($i \in \mathbb{N}$) and the $H_{\mathfrak{a}}^i(M)$ ($i \in \mathbb{N}$) using the unique gradings for which the conclusions of 13.4.2 and 13.4.3 are satisfied.

Thus, during the rest of the book, whenever our current assumptions about gradings (that R is G-graded, and $\mathfrak{a}$ and all the ideals in $\mathfrak{B}$ are graded) are in force, we shall, without further ado, utilize the gradings on the $H_{\mathfrak{a}}^i(M)$ and $H_{\mathfrak{B}}^i(M)$ whenever M is a graded R-module; for each $g \in G$, we shall denote the g-th component of the graded R-module $H_{\mathfrak{a}}^i(M)$ (respectively $H_{\mathfrak{B}}^i(M)$) by $H_{\mathfrak{a}}^i(M)_g$ (respectively $H_{\mathfrak{B}}^i(M)_g$). (In many of the examples that we shall consider, G will actually be the additive group $\mathbb{Z}$ of integers.) We shall also make much use of the fact that, whenever $0 \longrightarrow L \longrightarrow M \longrightarrow N \longrightarrow 0$ is an exact sequence of graded R-modules and homogeneous homomorphisms, not only are all the terms in the induced long exact sequence of local cohomology modules with respect to $\mathfrak{a}$ (respectively generalized local cohomology modules with respect to $\mathfrak{B}$) graded, but all the homomorphisms in this sequence, including the connecting homomorphisms, are homogeneous. These are really crucial facts about graded local cohomology.

In the introduction to this chapter, we posed three questions about various possible approaches to the construction of gradings on local cohomology modules. The next theorem will enable us to answer those questions.

13.4.5 Theorem. *Let $(T^i)_{i \in \mathbb{N}_0}$ be a negative strongly connected sequence of covariant additive functors from $\mathcal{C}(R)$ to itself. Suppose that*

$$\Omega = (\omega^i)_{i \in \mathbb{N}_0} : (T^i)_{i \in \mathbb{N}_0} \xrightarrow{\cong} (H^i_{\mathfrak{B}})_{i \in \mathbb{N}_0}$$

is an isomorphism of connected sequences (from $\mathcal{C}(R)$ to itself). Suppose that, for each graded R-module M, a grading is given on $T^0(M)$, and that, with respect to these gradings, T^0 has the *restriction property. Suppose also that ω^0 has the *restriction property.

Then there is exactly one choice of gradings on the $T^i(M)$ ($i \in \mathbb{N}$, M a graded R-module) with respect to which $(T^i)_{i \in \mathbb{N}_0}$ has the *restriction property; with these gradings (and those of 13.4.4) imposed, Ω has the *restriction property of 13.3.10.

Proof. By 13.2.6, we have $T^i(I) = H^i_{\mathfrak{B}}(I) = 0$ for all $i \in \mathbb{N}$ whenever I is a *injective graded R-module. Therefore, by 13.3.15, there is exactly one choice of gradings on the $T^i(M)$ ($i \in \mathbb{N}$, M a graded R-module) with respect to which $(T^i)_{i \in \mathbb{N}_0}$ has the *restriction property.

By 13.4.4, the connected sequence $(H^i_{\mathfrak{B}})_{i \in \mathbb{N}_0}$ has the *restriction property. The final claim therefore follows from 13.3.17. □

13.4.6 Remarks. Here, we finally answer the three questions posed in the introduction to this chapter.

(i) We apply Theorem 13.4.5 to the isomorphism

$$\Phi_{\mathfrak{B}} = \left(\phi^i_{\mathfrak{B}}\right)_{i \in \mathbb{N}_0} : \left(\varinjlim_{\alpha \in \Lambda} \operatorname{Ext}^i_R(R/\mathfrak{b}_\alpha, \, \bullet)\right)_{i \in \mathbb{N}_0} \xrightarrow{\cong} \left(H^i_{\mathfrak{B}}\right)_{i \in \mathbb{N}_0}$$

of negative strongly connected sequences of functors (from $\mathcal{C}(R)$ to itself) of 2.2.2. It is clear from the definition of $\phi^0_{\mathfrak{B}}$ in 1.2.11(ii) that it has the *restriction property. Hence Theorem 13.4.5 can be applied: there is exactly one choice of gradings (and that was described in 13.3.13(i)) on the $\varinjlim_{\alpha \in \Lambda} \operatorname{Ext}_R^i(R/\mathfrak{b}_\alpha, M)$ ($i \in \mathbb{N}$, M a graded R-module) with respect to which

$$\left(\varinjlim_{\alpha \in \Lambda} \operatorname{Ext}_R^i(R/\mathfrak{b}_\alpha, \bullet) \right)_{i \in \mathbb{N}_0}$$

has the *restriction property; with respect to these gradings, $\Phi_{\mathfrak{B}}$ has the *restriction property.

(ii) The following special case of part (i) is worthy of separate mention. We can apply Theorem 13.4.5 to the isomorphism

$$\Phi_{\mathfrak{a}} = \left(\phi_{\mathfrak{a}}^i \right)_{i \in \mathbb{N}_0} : \left(\varinjlim_{n \in \mathbb{N}} \operatorname{Ext}_R^i(R/\mathfrak{a}^n, \bullet) \right)_{i \in \mathbb{N}_0} \xrightarrow{\cong} \left(H_{\mathfrak{a}}^i \right)_{i \in \mathbb{N}_0}$$

of negative strongly connected sequences of functors (from $\mathcal{C}(R)$ to itself) of 1.3.8. It is clear from the definition of $\phi_{\mathfrak{a}}^0$ in 1.2.11(iii) that it has the *restriction property. Hence Theorem 13.4.5 can be applied: there is exactly one choice of gradings (and that was described in 13.3.13(ii)) on the $\varinjlim_{n \in \mathbb{N}} \operatorname{Ext}_R^i(R/\mathfrak{a}^n, M)$ ($i \in \mathbb{N}$, M a graded R-module) with respect to which

$$\left(\varinjlim_{n \in \mathbb{N}} \operatorname{Ext}_R^i(R/\mathfrak{a}^n, \bullet) \right)_{i \in \mathbb{N}_0}$$

has the *restriction property; with these gradings, $\Phi_{\mathfrak{a}}$ has the *restriction property.

(iii) Let $a_1, \ldots, a_n$ (where $n > 0$) denote n homogeneous elements which generate $\mathfrak{a}$. Let $C^\bullet$ denote the Čech complex of R with respect to the elements $a_1, \ldots, a_n$, as in 5.1.5. It is easy to check that Theorem 13.4.5 can be applied to the isomorphism

$$((\gamma^i)^{-1})_{i \in \mathbb{N}_0} : (H^i((\bullet) \otimes_R C^\bullet))_{i \in \mathbb{N}_0} \xrightarrow{\cong} (H_{\mathfrak{a}}^i)_{i \in \mathbb{N}_0}$$

of negative strongly connected sequences of functors (from $\mathcal{C}(R)$ to itself) of 5.1.20. We conclude that there is exactly one choice of gradings (and that was described in 13.3.13(iii)) on the $H^i(M \otimes_R C^\bullet)$ ($i \in \mathbb{N}$, M a graded R-module) with respect to which $(H^i((\bullet) \otimes_R C^\bullet))_{i \in \mathbb{N}_0}$ has the *restriction property; with these gradings, $((\gamma^i)^{-1})_{i \in \mathbb{N}_0}$ has the *restriction property.

(iv) In short, if we use the isomorphism of connected sequences

$$\Phi_{\mathfrak{a}} = \left(\phi_{\mathfrak{a}}^i \right)_{i \in \mathbb{N}_0} : \left(\varinjlim_{n \in \mathbb{N}} \operatorname{Ext}_R^i(R/\mathfrak{a}^n, \bullet) \right)_{i \in \mathbb{N}_0} \xrightarrow{\cong} \left(H_{\mathfrak{a}}^i \right)_{i \in \mathbb{N}_0}$$

of 1.3.8 to define gradings on the $H^i_{\mathfrak{a}}(M)$ ($i \in \mathbb{N}$) (for a graded R-module M), or if we use the isomorphism of connected sequences

$$((\gamma^i)^{-1})_{i \in \mathbb{N}_0} : (H^i((\bullet) \otimes_R C^\bullet))_{i \in \mathbb{N}_0} \xrightarrow{\cong} (H^i_{\mathfrak{a}})_{i \in \mathbb{N}_0}$$

of 5.1.20, with *any* choice of homogeneous generators for $\mathfrak{a}$, to define gradings on the $H^i_{\mathfrak{a}}(M)$ ($i \in \mathbb{N}$), the resulting gradings are always the same, and are precisely those with respect to which $(H^i_{\mathfrak{a}})_{i \in \mathbb{N}_0}$ has the *restriction property: see 13.4.3. (Note also that it would make no difference if we used $C(M)^\bullet$, the Čech complex of M with respect to $a_1, \ldots, a_n$, instead of $M \otimes_R C^\bullet$: the constituent isomorphisms in the isomorphism of complexes of 5.1.11 are all homogeneous.) Furthermore, as, when these gradings are imposed, there is a unique isomorphism $(\widetilde{\psi}^i)_{i \in \mathbb{N}_0} : (H^i_{\mathfrak{a}}\lceil)_{i \in \mathbb{N}_0} \xrightarrow{\cong} (*H^i_{\mathfrak{a}})_{i \in \mathbb{N}_0}$ of connected sequences of covariant functors from $*\mathcal{C}(R)$ to itself for which $\widetilde{\psi}^0$ is the identity (see again 13.4.3), all the questions posed in the introduction to this chapter have been answered affirmatively!

13.5 Some examples and applications

In our discussion of examples, we shall sometimes wish to change gradings and use a 'less fine' or 'coarser' grading than one that occurs naturally. In this connection, the notation and terminology introduced in the following definition will be helpful.

13.5.1 Definition. Let $\phi : G \longrightarrow H$ be a homomorphism of finitely generated torsion-free Abelian groups, and suppose that $R = \bigoplus_{g \in G} R_g$ is G-graded.

For each $h \in H$, let $R^\phi_h := \bigoplus_{g \in \phi^{-1}(\{h\})} R_g$. Then

$$R^\phi := \bigoplus_{h \in H} R^\phi_h = \bigoplus_{h \in H} \left(\bigoplus_{g \in \phi^{-1}(\{h\})} R_g \right)$$

provides an H-grading on R, and we denote R by R^ϕ when considering it as an H-graded ring in this manner.

Furthermore, for each G-graded R-module $M = \bigoplus_{g \in G} M_g$, let $M^\phi_h := \bigoplus_{g \in \phi^{-1}(\{h\})} M_g$ and $M^\phi := \bigoplus_{h \in H} M^\phi_h$; then M^ϕ is an H-graded R^ϕ-module. Also, if $f : M \longrightarrow N$ is a G-homogeneous homomorphism of G-graded R-modules, then the same map f becomes an H-homogeneous homomorphism of H-graded R^ϕ-modules $f^\phi : M^\phi \longrightarrow N^\phi$.

In this way, $(\bullet)^\phi$ becomes an exact additive covariant functor from $*\mathcal{C}^G(R)$ to $*\mathcal{C}^H(R)$. We refer to it as the ϕ-*coarsening functor*, and we shall consider the H-gradings constructed in this way as a *coarsening* of the given G-gradings.

The following lemma will help us to exploit coarsenings of graded local cohomology modules.

13.5.2 Lemma. *Let the situation be as in* 13.5.1, *so that R is G-graded and* $\phi : G \longrightarrow H$ *is a homomorphism of finitely generated torsion-free Abelian groups. Assume that $\mathfrak{a}$ is graded. Then*

$$\left(H^i_{\mathfrak{a}}(\bullet)^\phi\right)_{i \in \mathbb{N}_0} \quad and \quad \left(H^i_{\mathfrak{a}^\phi}((\bullet)^\phi)\right)_{i \in \mathbb{N}_0}$$

are isomorphic connected sequences of functors from $*\mathcal{C}^G(R)$ to $*\mathcal{C}^H(R)$.

Consequently, for each G-graded R-module M and each $i \in \mathbb{N}_0$, there is an H-homogeneous isomorphism $H^i_{\mathfrak{a}}(M)^\phi \cong H^i_{\mathfrak{a}^\phi}(M^\phi)$.

Proof. Because $(\bullet)^\phi$ is an exact additive covariant functor from $*\mathcal{C}^G(R)$ to $*\mathcal{C}^H(R)$, one easily sees that $\left(H^i_{\mathfrak{a}}(\bullet)^\phi\right)_{i \in \mathbb{N}_0}$ and $\left(H^i_{\mathfrak{a}^\phi}((\bullet)^\phi)\right)_{i \in \mathbb{N}_0}$ are both negative strongly connected sequences of covariant functors from $*\mathcal{C}^G(R)$ to $*\mathcal{C}^H(R)$. Now $\Gamma_{\mathfrak{a}}(\bullet)^\phi$ and $\Gamma_{\mathfrak{a}^\phi}((\bullet)^\phi)$ are the same functor. Also, whenever I is a *injective G-graded R-module, we can see that $H^i_{\mathfrak{a}}(I)^\phi = 0 = H^i_{\mathfrak{a}^\phi}(I^\phi)$ for all $i > 0$: the first of these claims is immediate, whereas the second follows from 13.4.3 because, without the H-grading, $H^i_{\mathfrak{a}^\phi}(I^\phi) \cong H^i_{\mathfrak{a}}(I)$, and I is $\Gamma_{\mathfrak{a}}$-acyclic (by 13.2.6). The result therefore follows from 13.3.5. $\square$

An illustration of the idea of coarsening occurs with our first example in this section; in this example, we show that the Čech complex approach can quickly lead to important information about graded local cohomology over a polynomial ring.

13.5.3 Example. Let $S = \bigoplus_{g \in G} S_g$ be a G-graded commutative Noetherian ring, let $n \in \mathbb{N}$, and let $R := S[X_1, \ldots, X_n]$, the ring of polynomials over S, graded by $G \oplus \mathbb{Z}^n$ as in 13.1.4(iii). The reader should note the special case in which $S = S_0$ is trivially graded, in which case the grading on R is (essentially) as described in 13.1.4(i), with S_0 playing the rôle of R_0.

We propose to use the Čech complex of R with respect to the homogeneous elements $X_1, \ldots, X_n$,

$$C^\bullet : 0 \longrightarrow C^0 \xrightarrow{d^0} C^1 \longrightarrow \cdots \longrightarrow C^i \xrightarrow{d^i} C^{i+1} \longrightarrow \cdots \xrightarrow{d^{n-1}} C^n \longrightarrow 0,$$

in conjunction with 13.4.6 above, to calculate the graded R-module

$$H^n_{(X_1,\ldots,X_n)}(S[X_1, \ldots, X_n]) = H^n_{(X_1,\ldots,X_n)}(R).$$

There is a homogeneous R-isomorphism between this local cohomology module and

$$R_{X_1 \ldots X_n} / \left(\sum_{t=1}^{n} d^{n-1} \left(R_{X_1 \ldots X_{t-1} X_{t+1} \ldots X_n} \right) \right).$$

Now $R_{X_1 \ldots X_n}$ is a free S-module with $\left(X_1^{i_1} \ldots X_n^{i_n} \right)_{(i_1, \ldots, i_n) \in \mathbb{Z}^n}$ as a base; for each $t = 1, \ldots, n$, its R-submodule $d^{n-1} \left(R_{X_1 \ldots X_{t-1} X_{t+1} \ldots X_n} \right)$ is again free as S-module, with base $\left(X_1^{i_1} \ldots X_n^{i_n} \right)_{(i_1, \ldots, i_n) \in \mathbb{Z}^n, \, i_t \geq 0}$. Use $-\mathbb{N}$ to denote the set $\{ n \in \mathbb{Z} : n < 0 \}$. It follows that the graded R-module

$$R_{X_1 \ldots X_n} / \left(\sum_{t=1}^{n} d^{n-1} \left(R_{X_1 \ldots X_{t-1} X_{t+1} \ldots X_n} \right) \right)$$

can be considered as a free S-module with base

$$\left(X_1^{i_1} \ldots X_n^{i_n} \right)_{(i_1, \ldots, i_n) \in (-\mathbb{N})^n}$$

and R-module structure such that, for $(i_1, \ldots, i_n) \in (-\mathbb{N})^n$ and $t \in \mathbb{N}$ with $1 \leq t \leq n$,

$$X_t(X_1^{i_1} \ldots X_n^{i_n}) = \begin{cases} X_1^{i_1} \ldots X_{t-1}^{i_{t-1}} X_t^{i_t+1} X_{t+1}^{i_{t+1}} \ldots X_n^{i_n} & \text{if } i_t < -1, \\ 0 & \text{if } i_t = -1. \end{cases}$$

The $(G \oplus \mathbb{Z}^n)$-grading is such that, for $g \in G$ and $s_g \in S_g \setminus \{0\}$,

$$\deg(s_g X_1^{i_1} \ldots X_n^{i_n}) = (g, (i_1, \ldots, i_n)) \quad \text{for all } (i_1, \ldots, i_n) \in (-\mathbb{N})^n.$$

We refer to this graded R-module as *the module of inverse polynomials in $X_1, \ldots, X_n$ over S*, and denote it by $S[X_1^-, \ldots, X_n^-]$. To summarize, the Čech complex approach to the calculation of graded local cohomology modules quickly yields a $(G \oplus \mathbb{Z}^n)$-homogeneous $S[X_1, \ldots, X_n]$-isomorphism

$$H^n_{(X_1, \ldots, X_n)}(S[X_1, \ldots, X_n]) \cong S[X_1^-, \ldots, X_n^-].$$

We could also regard $R = S[X_1, \ldots, X_n]$ as $(G \oplus \mathbb{Z})$-graded, with

$$R_{(g,m)} = \bigoplus_{\substack{(i_1, \ldots, i_n) \in \mathbb{N}_0^n \\ i_1 + \cdots + i_n = m}} R_{(g,(i_1, \ldots, i_n))} \quad \text{for all } (g, m) \in G \oplus \mathbb{Z}.$$

However, with this grading, our polynomial ring is just the result R^ϕ of applying the ϕ-coarsening functor of 13.5.1 to R, where $\phi : G \oplus \mathbb{Z}^n \longrightarrow G \oplus \mathbb{Z}$ is the Abelian group homomorphism for which

$$\phi((g, (i_1, \ldots, i_n))) = (g, i_1 + \cdots + i_n) \quad \text{for all } (g, (i_1, \ldots, i_n)) \in G \oplus \mathbb{Z}^n.$$

Thus, in view of 13.5.2, there are $(G \oplus \mathbb{Z})$-homogeneous R-isomorphisms

$$H^n_{(X_1,\ldots,X_n)^\phi}(S[X_1,\ldots,X_n]^\phi) \cong \left(H^n_{(X_1,\ldots,X_n)}(S[X_1,\ldots,X_n]) \right)^\phi$$

$$\cong \left(S[X_1^-,\ldots,X_n^-] \right)^\phi,$$

where, in the right-hand module, for $(i_1,\ldots,i_n) \in (-\mathbb{N})^n$ and $s_g \in S_g \setminus \{0\}$ (where $g \in G$), the element $s_g X_1^{i_1} \ldots X_n^{i_n}$ has degree $(g, i_1 + \cdots + i_n)$.

The reader should note the special case of the above in which S is trivially graded, so that $S = S_0 = R_0$: when R is considered to be $\mathbb{Z}^n$-graded as described in 13.1.4(i), the isomorphism

$$H^n_{(X_1,\ldots,X_n)}(R_0[X_1,\ldots,X_n]) \cong R_0[X_1^-,\ldots,X_n^-]$$

is $\mathbb{Z}^n$-homogeneous and, in the module of inverse polynomials,

$$\deg(X_1^{i_1}\ldots X_n^{i_n}) = (i_1,\ldots,i_n) \quad \text{for all } (i_1,\ldots,i_n) \in (-\mathbb{N})^n;$$

when R is considered to be $\mathbb{Z}$-graded, with $\deg X_i = 1$ for all $i = 1,\ldots,n$, then the isomorphism

$$H^n_{(X_1,\ldots,X_n)}(R_0[X_1,\ldots,X_n]) \cong R_0[X_1^-,\ldots,X_n^-]$$

is $\mathbb{Z}$-homogeneous and, in the module of inverse polynomials, $X_1^{i_1}\ldots X_n^{i_n}$ has degree $i_1 + \cdots + i_n$ for all $(i_1,\ldots,i_n) \in (-\mathbb{N})^n$.

13.5.4 ♯Exercise. Assume that R is G-graded and that the ideal $\mathfrak{a}$ is graded. Let $(\Lambda, \leq)$ be a (non-empty) directed partially ordered set, and $\mathfrak{B} = (\mathfrak{b}_\alpha)_{\alpha \in \Lambda}$ be a system of graded ideals of R over Λ. By 13.3.14, the functor $D_\mathfrak{B}$ of 2.2.3 has the *restriction property, and the natural transformation $\eta_\mathfrak{B}$ of 2.2.6(i) has the *restriction property. In particular, $D_\mathfrak{a}$ has the *restriction property and $\eta_\mathfrak{a}$ has the *restriction property.

(i) Show that the natural transformation $\zeta^0_\mathfrak{B}$ of 2.2.6 also has the *restriction property, so that, for each graded R-module M, all the homomorphisms in the exact sequence

$$0 \longrightarrow \Gamma_\mathfrak{B}(M) \xrightarrow{\xi_{\mathfrak{B}M}} M \xrightarrow{\eta_{\mathfrak{B}M}} D_\mathfrak{B}(M) \xrightarrow{\zeta^0_{\mathfrak{B}M}} H^1_\mathfrak{B}(M) \longrightarrow 0$$

are homogeneous. Note that, as a special case, this exercise shows that $\zeta^0_\mathfrak{a}$ (see 2.2.6(i)) has the *restriction property.

(ii) Let $e : M \longrightarrow M'$ be a homogeneous homomorphism of graded R-modules such that $\operatorname{Ker} e$ and $\operatorname{Coker} e$ are both $\mathfrak{B}$-torsion. Show that the unique R-homomorphism $\psi' : M' \to D_{\mathfrak{B}}(M)$ for which the diagram

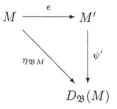

commutes (see 2.2.15) is homogeneous.

The special case of this result in which $\mathfrak{B}$ is taken to be the system formed by the powers of $\mathfrak{a}$ should be noted.

13.5.5 ♯Exercise. Assume that R is G-graded and let a be a homogeneous element of R. Show that the natural equivalence of functors

$$\omega' : D_{Ra} = \varinjlim_{n \in \mathbb{N}} \operatorname{Hom}_R(Ra^n, \bullet) \longrightarrow (\bullet)_a$$

of 2.2.19 has the *restriction property, and deduce that, for a graded R-module M, the isomorphism

$$H^1_{Ra}(M) \cong M_a/(M/\Gamma_{Ra}(M))$$

of 2.2.21(i) is homogeneous.

13.5.6 Exercise. Assume that R is G-graded and that the ideal $\mathfrak{a}$ is graded. Let $a_1, \ldots, a_n$ (where $n > 0$) denote n homogeneous elements which generate $\mathfrak{a}$. Let $u \in \mathbb{N}$ and let $K(a^u)^\bullet$ denote the Koszul complex of R with respect to $a_1^u, \ldots, a_n^u$, as in 5.2.1.

(i) Show that $K(a^u)^\bullet$ has the structure of a complex of graded R-modules and homogeneous R-homomorphisms in which $e_1 \wedge \ldots \wedge e_n$ has degree 0, the element $1 \in K(a^u)_0$ has degree $-u \sum_{i=1}^{n} \deg a_i$, and, for $k \in \{1, \ldots, n\}$ with $k < n$ and $i \in \mathcal{I}(k, n)$ (the notation is as in 5.1.4), the degree of $e_{i(1)} \wedge \ldots \wedge e_{i(k)}$ is $-u \sum_{h=1}^{n-k} \deg a_{j(h)}$, where $j \in \mathcal{I}(n-k, n)$ is the n-complement of i (see 5.1.4).

(ii) Let

$$(\delta^i)_{i \in \mathbb{N}_0} : \left(\varinjlim_{u \in \mathbb{N}} H_{n-i}(K(a^u, \bullet)^\bullet) \right)_{i \in \mathbb{N}_0} \xrightarrow{\cong} \left(H^i_{\mathfrak{a}} \right)_{i \in \mathbb{N}_0}$$

be the isomorphism of connected sequences of functors (from $\mathcal{C}(R)$ to itself) of Theorem 5.2.9. Show that $\left(\varinjlim_{u \in \mathbb{N}} H_{n-i}(K(a^u, \bullet)^\bullet) \right)_{i \in \mathbb{N}_0}$ has the *restriction property, and that $(\delta^i)_{i \in \mathbb{N}_0}$ has the *restriction property, so that, for each graded R-module M, the isomorphism

$$\delta^i_M : \varinjlim_{u \in \mathbb{N}} H_{n-i}(K(\mathfrak{a}^u, M)^\bullet) \xrightarrow{\cong} H^i_\mathfrak{a}(M)$$

is homogeneous.

13.5.7 ♯Exercise. Assume that R is G-graded and that the ideal $\mathfrak{a}$ is graded. It follows from 13.3.14 that $D_\mathfrak{a}$ has the *restriction property.

(i) Use 13.3.11 and 13.3.12 to show that the negative strongly connected sequence of covariant functors

$$\left(\varinjlim_{n \in \mathbb{N}} \operatorname{Ext}^i_R(\mathfrak{a}^n, \bullet) \right)_{i \in \mathbb{N}_0}$$

from $\mathcal{C}(R)$ to itself has the *restriction property.

(ii) Recall from 2.2.4 that there is a unique isomorphism of connected sequences (of functors from $\mathcal{C}(R)$ to itself)

$$\Psi_\mathfrak{a} = \left(\psi^i_\mathfrak{a} \right)_{i \in \mathbb{N}_0} : \left(\mathcal{R}^i D_\mathfrak{a} \right)_{i \in \mathbb{N}_0} \xrightarrow{\cong} \left(\varinjlim_{n \in \mathbb{N}} \operatorname{Ext}^i_R(\mathfrak{a}^n, \bullet) \right)_{i \in \mathbb{N}_0}$$

which extends the identity natural equivalence from $D_\mathfrak{a}$ to itself.

Let M be a graded R-module. We define structures as graded R-modules on the $\mathcal{R}^i D_\mathfrak{a}(M)$ ($i \in \mathbb{N}$) so that the R-isomorphism $\psi^i_{\mathfrak{a}\, M}$ is homogeneous for all $i \in \mathbb{N}$. A consequence of this definition is that the connected sequence $(\mathcal{R}^i D_\mathfrak{a})_{i \in \mathbb{N}_0}$ from $\mathcal{C}(R)$ to itself has the *restriction property.

Denote by $^*D_\mathfrak{a} : {}^*\mathcal{C}(R) \longrightarrow {}^*\mathcal{C}(R)$ the functor induced by $D_\mathfrak{a} \lceil$. Show that the three strongly connected sequences of covariant functors (from $^*\mathcal{C}(R)$ to itself)

$$\left(\mathcal{R}^i (^*D_\mathfrak{a}) \right)_{i \in \mathbb{N}_0}, \quad \left((\mathcal{R}^i D_\mathfrak{a}) \lceil \right)_{i \in \mathbb{N}_0} \quad \text{and} \quad \left(\varinjlim_{n \in \mathbb{N}} \operatorname{Ext}^i_R(\mathfrak{a}^n, \bullet) \lceil \right)_{i \in \mathbb{N}_0}$$

are isomorphic.

(iii) Let $i \in \mathbb{N}$. Show that the natural equivalence $\gamma^i : \mathcal{R}^i D_\mathfrak{a} \xrightarrow{\cong} H^{i+1}_\mathfrak{a}$ of 2.2.6(ii) has the *restriction property, so that, for each graded R-module M, there is a homogeneous isomorphism $\mathcal{R}^i D_\mathfrak{a}(M) \cong H^{i+1}_\mathfrak{a}(M)$ of graded R-modules.

13.5.8 Exercise. Generalize 13.5.7 to systems of graded ideals.

In detail, assume that R is G-graded, let $(\Lambda, \leq)$ be a (non-empty) directed partially ordered set, and let $\mathfrak{B} = (\mathfrak{b}_\alpha)_{\alpha \in \Lambda}$ be a system of graded ideals of R over Λ. By 13.3.14, the functor $D_{\mathfrak{B}}$ of 2.2.3 has the *restriction property.

(i) Recall from 2.2.4 that

$$\left(\varinjlim_{\alpha \in \Lambda} \operatorname{Ext}_R^i(\mathfrak{b}_\alpha, \bullet) \right)_{i \in \mathbb{N}_0}$$

is a negative strongly connected sequence of covariant functors from $\mathcal{C}(R)$ to itself; show that it has the *restriction property.

(ii) Recall from 2.2.4 that there is a unique isomorphism of connected sequences (of functors from $\mathcal{C}(R)$ to itself)

$$\Psi_{\mathfrak{B}} = \left(\psi_{\mathfrak{B}}^i \right)_{i \in \mathbb{N}_0} : \left(\mathcal{R}^i D_{\mathfrak{B}} \right)_{i \in \mathbb{N}_0} \xrightarrow{\cong} \left(\varinjlim_{\alpha \in \Lambda} \operatorname{Ext}_R^i(\mathfrak{b}_\alpha, \bullet) \right)_{i \in \mathbb{N}_0}$$

which extends the identity natural equivalence from $D_{\mathfrak{B}}$ to itself. Deduce that the first connected sequence here has the *restriction property.

Denote by $^*D_{\mathfrak{B}} : {}^*\mathcal{C}(R) \longrightarrow {}^*\mathcal{C}(R)$ the functor induced by $D_{\mathfrak{B}} \lceil$. Show that the three strongly connected sequences of covariant functors (from $^*\mathcal{C}(R)$ to itself)

$$\left(\mathcal{R}^i({}^*D_{\mathfrak{B}}) \right)_{i \in \mathbb{N}_0}, \quad \left((\mathcal{R}^i D_{\mathfrak{B}}) \lceil \right)_{i \in \mathbb{N}_0} \quad \text{and} \quad \left(\varinjlim_{\alpha \in \Lambda} \operatorname{Ext}_R^i(\mathfrak{b}_\alpha, \bullet) \lceil \right)_{i \in \mathbb{N}_0}$$

are isomorphic.

(iii) Let $i \in \mathbb{N}$. Show that there is a natural equivalence

$$\gamma_{\mathfrak{B}}^i : \mathcal{R}^i D_{\mathfrak{B}} \xrightarrow{\cong} H_{\mathfrak{B}}^{i+1}$$

that has the *restriction property.

Up to this point, this chapter has been fairly technical. In contrast, we end the chapter with some concrete illustrations of some of the ideas developed so far: the next exercise introduces the important concept of Veronesean subring, and this idea is involved in Example 13.5.12 and Exercise 13.5.13.

13.5.9 ♮Exercise: Veronesean subrings and functors. Assume that $R = \bigoplus_{n \in \mathbb{Z}} R_n$ is $\mathbb{Z}$-graded and that the ideal $\mathfrak{a}$ is graded. Let $r \in \mathbb{N}$ and $s \in \mathbb{Z}$ be fixed. Define $R^{(r)} := \bigoplus_{n \in \mathbb{Z}} R_{rn}$: then $R^{(r)}$ is a subring of R, and is itself a $\mathbb{Z}$-graded (commutative Noetherian) ring with grading given by $(R^{(r)})_n = R_{rn}$

for all $n \in \mathbb{Z}$. We refer to $R^{(r)}$, with this grading, as the *r-th Veronesean subring of R*.

Let $M = \bigoplus_{n \in \mathbb{Z}} M_n$ and $L = \bigoplus_{n \in \mathbb{Z}} L_n$ be general graded R-modules, and let $f : M \longrightarrow L$ be a homogeneous homomorphism, with n-th component $f_n : M_n \longrightarrow L_n$ for all $n \in \mathbb{Z}$. We define $M^{(r,s)} := \bigoplus_{n \in \mathbb{Z}} M_{rn+s}$, an $R^{(r)}$-submodule of $M \lceil_{R^{(r)}}$; in fact, $M^{(r,s)}$ is a graded $R^{(r)}$-module with grading given by $(M^{(r,s)})_n = M_{rn+s}$ for all $n \in \mathbb{Z}$. We refer to $M^{(r,s)}$ as the (r,s)-*th Veronesean submodule of* $M \lceil_{R^{(r)}}$. Note that $M^{(r,s)} = (M(s))^{(r,0)}$. Also, we denote by $f^{(r,s)} : M^{(r,s)} \longrightarrow L^{(r,s)}$ the homogeneous homomorphism of graded $R^{(r)}$-modules for which $(f^{(r,s)})_n = f_{rn+s} : M_{rn+s} \longrightarrow L_{rn+s}$ for all $n \in \mathbb{Z}$. With these assignments, $(\bullet)^{(r,s)} : {}^*\mathcal{C}(R) \longrightarrow {}^*\mathcal{C}(R^{(r)})$ becomes an exact additive covariant functor, which we refer to as the (r,s)-*th Veronesean functor*.

Note that, since $\mathfrak{a}$ is graded, $\mathfrak{a}^{(r)} := \mathfrak{a}^{(r,0)}$ is a graded ideal of $R^{(r)}$.

(i) Show that there is an isomorphism of $R^{(r)}$-modules

$$\bigoplus_{i=0}^{r-1} M^{(r,s+i)} \xrightarrow{\cong} M \lceil_{R^{(r)}}.$$

(ii) Show that $\sqrt{\mathfrak{a}^{(r)}R} = \sqrt{\mathfrak{a}}$.

(iii) Show that $\Gamma_{\mathfrak{a}^{(r)}}(M^{(r,s)}) = (\Gamma_{\mathfrak{a}}(M))^{(r,s)}$.

(iv) Let $\mathfrak{b}$ be a graded ideal of $R^{(r)}$, and let I be a *injective graded R-module. Let $j \in \mathbb{N}$. Show that there is an isomorphism of $R^{(r)}$-modules

$$\bigoplus_{i=0}^{r-1} H_{\mathfrak{b}}^j(I^{(r,s+i)}) \xrightarrow{\cong} H_{\mathfrak{b}R}^j(I),$$

and deduce that the $R^{(r)}$-module $I^{(r,s)}$ is $\Gamma_{\mathfrak{b}}$-acyclic.

(v) Show that there is a unique isomorphism

$$\Phi = \left(\phi^i\right)_{i \in \mathbb{N}_0} : \left(H_{\mathfrak{a}^{(r)}}^i((\bullet)^{(r,s)})\right)_{i \in \mathbb{N}_0} \longrightarrow \left((H_{\mathfrak{a}}^i(\bullet))^{(r,s)}\right)_{i \in \mathbb{N}_0}$$

of negative connected sequences of covariant functors from ${}^*\mathcal{C}(R)$ to ${}^*\mathcal{C}(R^{(r)})$ for which ϕ^0 is the identity natural equivalence.

(vi) Use 2.2.15 to show that there is a natural equivalence of functors

$$(D_{\mathfrak{a}}(\bullet))^{(r,s)} \xrightarrow{\cong} D_{\mathfrak{a}^{(r)}}((\bullet)^{(r,s)})$$

from ${}^*\mathcal{C}(R)$ to ${}^*\mathcal{C}(R^{(r)})$.

It is natural to ask whether one can generalize the concept of Veronesean subring to multi-graded situations. This is the subject of the next exercise.

13.5.10 Exercise. Let $n \in \mathbb{N}$ and assume that $G = \mathbb{Z}^n$ and R is G-graded; let $\mathfrak{a}$ be a graded ideal of R. Let G' be a subgroup of finite index t in G, so that rank $G' = n$. Let $g_0 := 0_G, g_1, \ldots, g_{t-1}$ be representatives of the distinct cosets of G' in G. Let $M = \bigoplus_{g \in G} M_g$ be a $(G\text{-})$graded R-module.

  (i) Define $R^{G'} := \bigoplus_{g' \in G'} R_{g'}$, and show that this is a Noetherian subring of R; the decomposition of its definition provides $R^{G'}$ with a grading by G'. (Here is a hint to help you show that $R^{G'}$ is Noetherian: note that $R^{G'}$ is a direct summand of R as $R^{G'}$-module; consider an ascending chain of ideals of $R^{G'}$, extend the ideals to R and then contract back to $R^{G'}$.)

  (ii) Let $k \in \{0, 1, \ldots, t-1\}$. Set $M^{G', g_k} := \bigoplus_{g' \in G'} M_{g'+g_k}$, and show that this is an $R^{G'}$-submodule of M which is G'-graded (by the decomposition given in its definition). Show further that there is an isomorphism of $R^{G'}$-modules $\bigoplus_{j=0}^{t-1} M^{G', g_j} \xrightarrow{\cong} M \upharpoonright_{R^{G'}}$.

  (iii) Let $\mathfrak{a}^{G'} := \mathfrak{a}^{G', 0_G}$. Show that $\sqrt{\mathfrak{a}^{G'} R} = \sqrt{\mathfrak{a}}$.

  (iv) Let $k \in \{0, 1, \ldots, t-1\}$. Generalize arguments from 13.5.9 to produce an isomorphism

  $$\Theta = \left(\theta^i\right)_{i \in \mathbb{N}_0} : \left(H^i_{\mathfrak{a}^{G'}}((\,\bullet\,)^{G', g_k})\right)_{j \in \mathbb{N}_0} \longrightarrow \left((H^i_{\mathfrak{a}}(\,\bullet\,))^{G', g_k}\right)_{i \in \mathbb{N}_0}$$

  of negative connected sequences of covariant functors from $*\mathcal{C}^G(R)$ to $*\mathcal{C}^{G'}(R^{G'})$ for which θ^0 is the identity natural equivalence.

  (v) Again for $k \in \{0, 1, \ldots, t-1\}$, generalize arguments from 13.5.9 to produce a natural equivalence of functors

  $$(D_{\mathfrak{a}}(\,\bullet\,))^{G', g_k} \xrightarrow{\cong} D_{\mathfrak{a}^{G'}}((\,\bullet\,)^{G', g_k})$$

  from $*\mathcal{C}^G(R)$ to $*\mathcal{C}^{G'}(R^{G'})$.

13.5.11 Definition. When $R = \bigoplus_{n \in \mathbb{Z}} R_n$ is $\mathbb{Z}$-graded, we shall say that R is *positively graded* precisely when $R_n = 0$ for all $n < 0$.

More generally, when $R = \bigoplus_{g \in \mathbb{Z}^n} R_g$ is $\mathbb{Z}^n$-graded for some positive integer n, we shall say that R is *positively graded* if and only if $R_g = 0$ for all $g = (g_1, \ldots, g_n) \in \mathbb{Z}^n \setminus \mathbb{N}_0^n$.

13.5.12 Example. Let K be a field, and consider the ring $K[X, Y]$ of polynomials over K in two indeterminates X and Y to be $\mathbb{Z}$-graded so that K is the component of degree 0 and $\deg X = \deg Y = 1$. (See 13.5.3.) Let $d \in \mathbb{N}$ with $d \geq 3$, and let $A_{(d)}$ be the subring of $K[X, Y]$ given by

$$A_{(d)} := K[X^d, X^{d-1}Y, XY^{d-1}, Y^d].$$

This is a subring of the d-th Veronesean subring $K[X,Y]^{(d)}$ of $K[X,Y]$, described in 13.5.9; in fact, $A_{(d)}$ inherits a $\mathbb{Z}$-grading from $K[X,Y]^{(d)}$.

In this example, it will be convenient, when considering a positively graded commutative Noetherian ring $R' = \bigoplus_{n \in \mathbb{N}_0} R'_n$, to denote the graded ideal $\bigoplus_{n \in \mathbb{N}} R'_n$ by R'_+.

Note that $A_{(d)+}$ is the unique graded maximal ideal of $A_{(d)}$. We shall now illustrate some of the ideas of this chapter by showing that the ideal transform $D_{A_{(d)+}}(A_{(d)})$ can be naturally identified with $K[X,Y]^{(d)}$, and then exploiting this fact to obtain information about $H^1_{A_{(d)+}}(A_{(d)})$.

Let $\phi : A_{(d)} \longrightarrow K[X,Y]^{(d)}$ denote the inclusion homomorphism. If $m := X^i Y^j$ with $i,j \in \mathbb{N}_0$, $i+j \equiv 0 \pmod{d}$ and $i+j > 0$, then we can write $m = X^{du} Y^{dv} X^r Y^{d-r}$ for some $u,v,r \in \mathbb{N}_0$ with $0 \le r \le d$. If $r = 0, 1, d-1$ or d, then $m \in A_{(d)}$. Now suppose that $2 \le r \le d-2$. Then

$$Y^{d(r-1)} m = X^{du} Y^{dv} (XY^{d-1})^r \in A_{(d)},$$
$$X^{d(d-r-1)} m = X^{du} Y^{dv} (X^{d-1}Y)^{d-r} \in A_{(d)},$$
$$(X^{d-1}Y)^r m = X^{du} Y^{dv} X^{dr} Y^d \in A_{(d)},$$
$$(XY^{d-1})^{d-r} m = X^{du} Y^{dv} X^d Y^{d(d-r)} \in A_{(d)}.$$

Hence $\operatorname{Ker}\phi$ and $\operatorname{Coker}\phi$ are both $A_{(d)+}$-torsion. Note that

$$\sqrt{A_{(d)+} K[X,Y]^{(d)}} = \sqrt{(X^d, X^{d-1}Y, XY^{d-1}, Y^d) K[X,Y]^{(d)}}$$
$$= (K[X,Y]^{(d)})_+ = (K[X,Y]_+)^{(d)}.$$

Therefore, by 13.5.9 and the Independence Theorem 4.2.1, we have, for each $i \in \mathbb{N}_0$,

$$H^i_{A_{(d)+}}(K[X,Y]^{(d)}) \cong H^i_{A_{(d)+} K[X,Y]^{(d)}}(K[X,Y]^{(d)})$$
$$= H^i_{(K[X,Y]_+)^{(d)}}(K[X,Y]^{(d)})$$
$$\cong (H^i_{(X,Y)}(K[X,Y]))^{(d,0)},$$

and this is zero for $i = 0, 1$.

We can now use 2.2.15 to see that there is a unique $A_{(d)}$-isomorphism $\phi' : K[X,Y]^{(d)} \longrightarrow D_{A_{(d)+}}(A_{(d)})$ such that the diagram

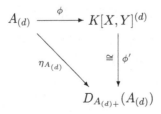

commutes, and we can use 13.5.4(ii) to see that ϕ' is homogeneous. So, by 13.5.4(i), there is a homogeneous isomorphism $\operatorname{Coker} \phi \cong H^1_{A_{(d)+}}(A_{(d)})$ of $A_{(d)}$-modules. For each $n \in \mathbb{N}_0$, let A_n (respectively B_n) denote the n-th component of the graded $A_{(d)}$-module $A_{(d)}$ (respectively $K[X, Y]^{(d)}$). We now compare A_n and B_n.

First, when $1 \leq n < d - 2$ the monomial $X^{d-2}Y^{nd-d+2}$ cannot be expressed as a product of n factors taken from $\{X^d, X^{d-1}Y, XY^{d-1}, Y^d\}$, and so $A_n \subset B_n$.

Next, we consider the case where $n \geq d - 2$. Of course, $X^{dn} \in A_n$; we therefore consider a monomial of the form $X^{rd+s}Y^{(n-r)d-s}$, where $0 \leq r \leq n-1$ and $0 \leq s \leq d-1$. We claim that either (a) $r+s \leq n$ or (b) $r+s \geq d-1$: if this were not the case, then we should have

$$n \leq r + s - 1 \quad \text{and} \quad r + s \leq d - 2,$$

which would imply that $n \leq d - 3$, a contradiction. In case (a), we have

$$X^{rd+s}Y^{(n-r)d-s} = (X^d)^r(XY^{d-1})^s(Y^d)^{n-r-s} \in A_n,$$

while in case (b) we have

$$X^{rd+s}Y^{(n-r)d-s} = (X^d)^{r+s+1-d}(X^{d-1}Y)^{d-s}(Y^d)^{n-r-1} \in A_n.$$

Thus $A_n = B_n$ in this case.

It follows that the n-th component of $H^1_{A_{(d)+}}(A_{(d)})$ is non-zero if and only if $1 \leq n < d - 2$.

13.5.13 ♯Exercise. Let K be a field and let $R := K[X_1, \ldots, X_n]$, the ring of polynomials over K in n indeterminates (where $n \in \mathbb{N}$), $\mathbb{Z}$-graded so that X_i has degree 1, for all $i = 1, \ldots, n$. Let $r \in \mathbb{N}$, and consider the r-th Veronesean subring $R^{(r)}$ of R, as in 13.5.9. Thus $R^{(r)}$ is the K-subspace of R generated by

$$\{X_1^{v_1} \ldots X_n^{v_n} : v_1, \ldots, v_n \in \mathbb{N}_0, \ v_1 + \cdots + v_n \equiv 0 \ (\operatorname{mod} r)\}.$$

Let $R_+^{(r)}$ denote the unique graded maximal ideal of $R^{(r)}$. Show that

$$H^i_{R_+^{(r)}}(R^{(r)}) = 0 \quad \text{for all } i \in \mathbb{N}_0 \setminus \{n\},$$

and deduce from [7, 1.5.8 and 1.5.9] that $R^{(r)}$ is Cohen–Macaulay.

14

Graded versions of basic theorems

We have now laid the foundations of multi-graded local cohomology theory in Chapter 13, where the gradings on R and R-modules M are by a finitely generated, torsion-free Abelian group G. Indeed, in the case where R is G-graded and the ideal $\mathfrak{a}$ is G-graded, we now know that, for a G-graded R-module M, there is a natural way in which to define G-gradings on the local cohomology modules $H_{\mathfrak{a}}^i(M)$ ($i \in \mathbb{N}_0$); furthermore, whenever $f : M \longrightarrow N$ is a morphism in $*\mathcal{C}^G(R)$, then $H_{\mathfrak{a}}^i(f)$ is a homogeneous homomorphism for all $i \in \mathbb{N}_0$; also, whenever $0 \longrightarrow L \longrightarrow M \longrightarrow N \longrightarrow 0$ is an exact sequence in the 'G-graded' category $*\mathcal{C}^G(R)$, then all the homomorphisms, including the connecting homomorphisms, in the induced long exact sequence of local cohomology modules (with respect to $\mathfrak{a}$) are G-homogeneous.

This chapter is concerned with refinements available in this multi-graded case of such fundamental results as the Independence Theorem 4.2.1, the Flat Base Change Theorem 4.3.2, Faltings' Annihilator Theorem 9.5.1, Grothendieck's Finiteness Theorem 9.5.2 and the Local Duality Theorem 11.2.6, and of the theory of canonical modules developed in Chapter 12. However, although it is true that part of this chapter is a retracing of steps through earlier chapters, revisiting many of the highlights in order to 'add graded frills', we have felt it necessary to include quite a bit of the underlying algebra of multi-graded commutative Noetherian rings.

For example, there is a multi-graded analogue of Matlis's decomposition theory for injective modules over a commutative Noetherian ring, and this multi-graded analogue has both strong similarities to, and fascinating links with, the ungraded theory. We present the multi-graded version in §14.2, and make much use of it in our treatment of graded local duality in §14.4 and *canonical modules in §14.5.

In §14.3, we present some results due to S. Goto and K.-i. Watanabe [22, §1.2] that also concern the algebra of a commutative Noetherian ring R that is graded by a finitely generated torsion-free Abelian group. Let M be a finitely generated graded R-module, and let $\mathfrak{p} \in \operatorname{Supp} M$. We showed in 13.1.6(i) that $\mathfrak{p}^* \in \operatorname{Supp} M$. Our principal aims in §14.3 are to present Goto's and Watanabe's results that

$$\dim M_{\mathfrak{p}} = \dim M_{\mathfrak{p}^*} + \operatorname{ht} \mathfrak{p}/\mathfrak{p}^* \quad \text{and} \quad \operatorname{depth} M_{\mathfrak{p}} = \operatorname{depth} M_{\mathfrak{p}^*} + \operatorname{ht} \mathfrak{p}/\mathfrak{p}^*,$$

and to apply them to prove graded versions of Faltings' Annihilator Theorem 9.5.1 and Grothendieck's Finiteness Theorem 9.5.2. However, we also use the Goto–Watanabe results later in the chapter.

Limitations on space mean that almost all of our treatment in §14.5 of *canonical modules is concerned with the case where R is a Cohen–Macaulay multigraded ring with a unique maximal graded proper ideal $\mathfrak{m}$. We define a *canonical module for R to be a finitely generated graded R-module C for which there is a homogeneous isomorphism $* \operatorname{Hom}_R(C, *E_R(R/\mathfrak{m})) \cong H^n_{\mathfrak{m}}(R)$. This definition is the obvious graded analogue of our definition of canonical module in the ungraded local case. However, in the special case in which R is $\mathbb{Z}$-graded, Bruns and Herzog give an alternative definition in [7, Definition 3.6.8] which is not obviously equivalent to ours. The work involved in our reconciliation of these two approaches (see 14.5.12) has contributed significantly to the length of this chapter.

Readers who are only interested in the case of $\mathbb{Z}$-graded rings rather than the full multi-graded case should be able to pass quickly over some parts of this chapter. For example, the special cases of the above-mentioned results of Goto and Watanabe in which the grading group is $\mathbb{Z}$ are essentially covered by [7, Theorems 1.5.8 and 1.5.9].

14.1 Fundamental theorems

14.1.1 Notation and Terminology. Throughout this chapter, G will denote a finitely generated, torsion-free Abelian group, written additively, and we shall assume that our commutative Noetherian ring R is G-graded, with grading given by $R = \oplus_{g \in G} R_g$. Occasionally, we shall consider particular cases in which R is $\mathbb{Z}$-graded, that is, in which G is taken to be $\mathbb{Z}$.

We shall employ the notation, conventions and terminology concerning G-graded rings and modules described in 13.1.1. In addition, when the ideal $\mathfrak{a}$ is graded, and M is a graded R-module, we use $H^i_{\mathfrak{a}}(M)_g$ to denote the g-th component of the graded R-module $H^i_{\mathfrak{a}}(M)$ (for $i \in \mathbb{N}_0$ and $g \in G$).

A maximal member of the set of proper graded ideals of R is referred to as a *maximal graded ideal of R. We shall say that R is *local precisely when it has exactly one *maximal graded ideal. The statement '$(R, \mathfrak{m})$ is *local' is to be interpreted as meaning that $\mathfrak{m}$ is the unique *maximal graded ideal of the G-graded ring R. For an ideal $\mathfrak{b}$ of R, we shall denote by $\mathfrak{b}^*$ the graded ideal generated by all homogeneous elements of $\mathfrak{b}$.

Let S be a multiplicatively closed subset of R consisting of non-zero homogeneous elements. We pointed out in 13.1.1 that the ring $S^{-1}R$ is also G-graded, and that, if M is a graded R-module, then the $S^{-1}R$-module $S^{-1}M$ is also graded.

If $\mathfrak{p}$ is a prime ideal of R and we take S to be the set of all homogeneous elements of R that lie outside $\mathfrak{p}$, then the resulting G-graded ring $S^{-1}R$ (respectively module $S^{-1}M$) is called the *homogeneous localization* of R (respectively M) *at* $\mathfrak{p}$, and is denoted by $R_{(\mathfrak{p})}$ (respectively $M_{(\mathfrak{p})}$). This concept should not be confused with a different concept for which Hartshorne, in [30, p. 18], uses the notation '$_{(\mathfrak{p})}$'. Note that $R_{(\mathfrak{p})}$ is a *local ring, with $\mathfrak{p}R_{(\mathfrak{p})}$ as its unique *maximal graded ideal.

Some of our work will be particularly concerned with the case where $G = \mathbb{Z}$. Suppose that R is $\mathbb{Z}$-graded (that is, that $G = \mathbb{Z}$), and let $M = \bigoplus_{n \in \mathbb{Z}} M_n$ be a graded R-module. We define the *end of* M by

$$\operatorname{end}(M) := \sup \{n \in \mathbb{Z} : M_n \neq 0\}$$

if this supremum exists, and ∞ otherwise. (We adopt the convention that the supremum of the empty set of integers is to be taken as $-\infty$, and we interpret $-\infty + t$ as $-\infty$ for all $t \in \mathbb{Z}$.) With analogous conventions, we similarly define the *beginning of* M, denoted by $\operatorname{beg}(M)$, to be $\inf \{n \in \mathbb{Z} : M_n \neq 0\}$ if this infimum exists, and $-\infty$ otherwise.

14.1.2 Remark. Suppose that M is a graded R-module.

(i) Recall from 13.1.6(i) that, if $\mathfrak{p} \in \operatorname{Supp} M$, then the graded prime $\mathfrak{p}^*$ also belongs to $\operatorname{Supp} M$. Hence if M has no graded prime ideal in its support, then $M = 0$.

(ii) Consequently, if $(R, \mathfrak{m})$ is *local (see 14.1.1) and $M_{\mathfrak{m}} = 0$, then $M = 0$.

14.1.3 Definition and Remark. Let $R' = \bigoplus_{g \in G} R'_g$ be a commutative G-graded ring, and let $f : R \longrightarrow R'$ be a ring homomorphism.

We say that f is *homogeneous* precisely when $f(R_g) \subseteq R'_g$ for all $g \in G$. Assume that this is the case. Let $M' = \bigoplus_{g \in G} M'_g$ be a graded R'-module. Then the same direct sum decomposition provides the R-module $M' \lceil_R$ with a structure as a graded R-module. In fact, in the terminology of 13.3.6, the functor $\lceil_R : \mathcal{C}^G(R') \longrightarrow \mathcal{C}^G(R)$ has the *restriction property.

Whenever we regard, in such circumstances, a graded R'-module M' as a graded R-module, it is to be understood that the same grading is used for the two structures. Thus we can write $(M' \lceil_R)_g = (M'_g) \lceil_{R_0}$ for all $g \in G$.

14.1.4 Proposition. *Assume that $\mathfrak{a}$ is graded, and let $R' = \bigoplus_{g \in G} R'_g$ be a second commutative Noetherian G-graded ring. Let $f : R \longrightarrow R'$ be a ring homomorphism which is homogeneous (see 14.1.3).*

*The natural equivalence of functors $\varepsilon : D_{\mathfrak{a}R'}(\bullet) \lceil_R \longrightarrow D_{\mathfrak{a}}(\bullet \lceil_R)$ of 2.2.24 (from $\mathcal{C}(R')$ to $\mathcal{C}(R)$) has the *restriction property.*

Proof. Since $\mathfrak{a}R'$ is a graded ideal of R', it follows from 13.3.14 and 14.1.3 that both $D_{\mathfrak{a}R'}(\bullet) \lceil_R$ and $D_{\mathfrak{a}}(\bullet \lceil_R)$ have the *restriction property.

Let M' be a graded R'-module. The homomorphism

$$\eta_{\mathfrak{a}R',M'} \lceil_R : M' \lceil_R \longrightarrow D_{\mathfrak{a}R'}(M') \lceil_R$$

is homogeneous by 13.3.14, and has kernel and cokernel which are $\mathfrak{a}$-torsion by 2.2.6(i)(c). The result therefore follows from 13.5.4(ii). $\square$

14.1.5 ♯Exercise: Graded Mayer–Vietoris sequence. Assume that $\mathfrak{a}$ is graded; let $\mathfrak{b}$ be a second graded ideal of R. Show that, for a graded R-module M, all the homomorphisms in the Mayer–Vietoris sequence (see 3.2.3)

$$0 \longrightarrow H^0_{\mathfrak{a}+\mathfrak{b}}(M) \longrightarrow H^0_{\mathfrak{a}}(M) \oplus H^0_{\mathfrak{b}}(M) \longrightarrow H^0_{\mathfrak{a}\cap\mathfrak{b}}(M)$$

$$\longrightarrow H^1_{\mathfrak{a}+\mathfrak{b}}(M) \longrightarrow H^1_{\mathfrak{a}}(M) \oplus H^1_{\mathfrak{b}}(M) \longrightarrow H^1_{\mathfrak{a}\cap\mathfrak{b}}(M)$$

$$\longrightarrow \quad \cdots \qquad\qquad\qquad\qquad\qquad \cdots$$

$$\longrightarrow H^i_{\mathfrak{a}+\mathfrak{b}}(M) \longrightarrow H^i_{\mathfrak{a}}(M) \oplus H^i_{\mathfrak{b}}(M) \longrightarrow H^i_{\mathfrak{a}\cap\mathfrak{b}}(M)$$

$$\longrightarrow H^{i+1}_{\mathfrak{a}+\mathfrak{b}}(M) \longrightarrow \qquad \cdots$$

are homogeneous.

14.1.6 ♯Exercise. Assume that $\mathfrak{a}$ is graded; let $i \in \mathbb{N}_0$.

(i) Let $(\Lambda, \leq)$ be a (non-empty) directed partially ordered set; let $(W_\alpha)_{\alpha \in \Lambda}$ be a direct system of graded R-modules and homogeneous R-homomorphisms over Λ, with constituent R-homomorphisms $h^\alpha_\beta : W_\beta \to W_\alpha$ (for each $(\alpha, \beta) \in \Lambda \times \Lambda$ with $\alpha \geq \beta$). (See 13.3.12.)

Show that the isomorphism

$$\varinjlim_{\alpha \in \Lambda} H^i_{\mathfrak{a}}(W_\alpha) \overset{\cong}{\longrightarrow} H^i_{\mathfrak{a}}\left(\varinjlim_{\alpha \in \Lambda} W_\alpha\right)$$

given by Theorem 3.4.10 (see also 3.4.1) is homogeneous.

(ii) Let $(L_\theta)_{\theta \in \Omega}$ be a non-empty family of graded R-modules. Show that the isomorphism

$$H_\mathfrak{a}^i \left(\bigoplus_{\theta \in \Omega} L_\theta \right) \xrightarrow{\cong} \bigoplus_{\theta \in \Omega} H_\mathfrak{a}^i(L_\theta)$$

given by (Theorem 3.4.10 and) Exercise 3.4.5 is homogeneous.

Next we use Theorem 13.3.15 to establish quickly G-graded versions of the Independence Theorem 4.2.1 and the Flat Base Change Theorem 4.3.2.

14.1.7 Graded Independence Theorem. *Assume that $\mathfrak{a}$ is graded, and let $R' = \bigoplus_{g \in G} R'_g$ be a second commutative Noetherian G-graded ring. Let $f : R \longrightarrow R'$ be a ring homomorphism which is homogeneous (see 14.1.3).*

(i) *Both the negative (strongly) connected sequences of covariant functors*

$$(H_{\mathfrak{a}R'}^i(\bullet) \lceil_R)_{i \in \mathbb{N}_0} \quad and \quad (H_\mathfrak{a}^i(\bullet \lceil_R))_{i \in \mathbb{N}_0}$$

from $\mathcal{C}(R')$ to $\mathcal{C}(R)$ have the *restriction property.

(ii) *The isomorphism of connected sequences*

$$\Lambda = (\lambda^i)_{i \in \mathbb{N}_0} : (H_{\mathfrak{a}R'}^i(\bullet) \lceil_R)_{i \in \mathbb{N}_0} \xrightarrow{\cong} (H_\mathfrak{a}^i(\bullet \lceil_R))_{i \in \mathbb{N}_0}$$

of 4.2.1 has the *restriction property. Consequently, for each $i \in \mathbb{N}_0$ and each graded R'-module M', there is a homogeneous R-isomorphism

$$\lambda_{M'}^i : H_{\mathfrak{a}R'}^i(M') \xrightarrow{\cong} H_\mathfrak{a}^i(M').$$

Proof. (i) Since f is homogeneous, the restriction functor $\lceil_R : \mathcal{C}(R') \longrightarrow \mathcal{C}(R)$ has the *restriction property of 13.3.6: see 14.1.3. Since, by 13.4.3 and 13.4.4, the connected sequence $(H_\mathfrak{a}^i)_{i \in \mathbb{N}_0}$ (from $\mathcal{C}(R)$ to $\mathcal{C}(R)$) has the *restriction property, it is immediate that $(H_\mathfrak{a}^i(\bullet \lceil_R))_{i \in \mathbb{N}_0}$ has the *restriction property. Since the extension $\mathfrak{a}R'$ of $\mathfrak{a}$ to R' under f is a graded ideal, it is just as easy to see that $(H_{\mathfrak{a}R'}^i(\bullet) \lceil_R)_{i \in \mathbb{N}_0}$ has the *restriction property.

(ii) Since λ^0 is the identity natural equivalence from $\Gamma_{\mathfrak{a}R'}(\bullet) \lceil_R = \Gamma_\mathfrak{a}(\bullet \lceil_R)$ to itself, λ^0 has the *restriction property.

For each $i \in \mathbb{N}_0$ and each graded R'-module M', we can use the grading on $H_\mathfrak{a}^i(M' \lceil_R)$ of (i) to define a grading on $H_{\mathfrak{a}R'}^i(M') \lceil_R$ in such a way that the isomorphism $\lambda_{M'}^i : H_{\mathfrak{a}R'}^i(M') \lceil_R \xrightarrow{\cong} H_\mathfrak{a}^i(M' \lceil_R)$ is homogeneous: we recover the grading of part (i) on the $\Gamma_{\mathfrak{a}R'}(M') \lceil_R$. With respect to *these* gradings, $(H_{\mathfrak{a}R'}^i(\bullet) \lceil_R)_{i \in \mathbb{N}_0}$ also has the *restriction property. Since $H_{\mathfrak{a}R'}^i(I') \lceil_R = 0$ for all $i \in \mathbb{N}$ whenever I' is a *injective graded R'-module (by 13.2.6), it follows from 13.3.15 that these gradings coincide with the natural ones used in part (i). All the remaining claims follow from this. $\square$

Below, we shall use a similar argument to establish a graded version of the Flat Base Change Theorem 4.3.2. However, the reader might find the following preparatory remark helpful.

14.1.8 Remark. Let $M = \bigoplus_{g \in G} M_g$ be a graded R-module. Let $R' = \bigoplus_{g \in G} R'_g$ be a second commutative Noetherian G-graded ring, and let $f : R \longrightarrow R'$ be a ring homomorphism which is homogeneous (see 14.1.3). Then $R' \lceil_R$ is a graded R-module (as is explained in 14.1.3), and therefore $M \otimes_R R'$ has a structure as a graded R-module. In fact, the direct decomposition

$$M \otimes_R R' = \bigoplus_{g \in G} (M \otimes_R R')_g$$

described in 13.1.9 actually provides $M \otimes_R R'$ with a structure as a graded R'-module, and $(\bullet) \otimes_R R' : \mathcal{C}(R) \longrightarrow \mathcal{C}(R')$ has the *restriction property.

14.1.9 Graded Flat Base Change Theorem. *Assume that $\mathfrak{a}$ is graded, and let $R' = \bigoplus_{g \in G} R'_g$ be a second commutative Noetherian G-graded ring.*

Let $f : R \longrightarrow R'$ be a ring homomorphism which is homogeneous (see 14.1.3) and flat.

(i) *Both the negative (strongly) connected sequences of covariant functors*

$$(H^i_{\mathfrak{a}}(\bullet) \otimes_R R')_{i \in \mathbb{N}_0} \quad and \quad (H^i_{\mathfrak{a}R'}((\bullet) \otimes_R R'))_{i \in \mathbb{N}_0}$$

*from $\mathcal{C}(R)$ to $\mathcal{C}(R')$ have the *restriction property.*

(ii) *The isomorphism of connected sequences*

$$(\rho^i)_{i \in \mathbb{N}_0} : (H^i_{\mathfrak{a}}(\bullet) \otimes_R R')_{i \in \mathbb{N}_0} \stackrel{\cong}{\longrightarrow} (H^i_{\mathfrak{a}R'}((\bullet) \otimes_R R'))_{i \in \mathbb{N}_0}$$

*of 4.3.2 has the *restriction property, so that, for each $i \in \mathbb{N}_0$ and each graded R-module M, there is a* homogeneous *R'-isomorphism*

$$\rho^i_M : H^i_{\mathfrak{a}}(M) \otimes_R R' \stackrel{\cong}{\longrightarrow} H^i_{\mathfrak{a}R'}(M \otimes_R R').$$

Proof. (i) Since f is homogeneous, the functor $(\bullet) \otimes_R R' : \mathcal{C}(R) \longrightarrow \mathcal{C}(R')$ has the *restriction property: see 14.1.8. Since the extension $\mathfrak{a}R'$ of $\mathfrak{a}$ to R' under f is a graded ideal, it follows from 13.4.3 and 13.4.4 that the connected sequence $(H^i_{\mathfrak{a}R'})_{i \in \mathbb{N}_0}$ (from $\mathcal{C}(R')$ to $\mathcal{C}(R')$) has the *restriction property. Hence $(H^i_{\mathfrak{a}R'}((\bullet) \otimes_R R'))_{i \in \mathbb{N}_0}$ has the *restriction property. It is just as easy to see that $(H^i_{\mathfrak{a}}(\bullet) \otimes_R R')_{i \in \mathbb{N}_0}$ has the *restriction property.

(ii) Note that, by 4.3.1, the natural equivalence ρ^0 has the *restriction property.

For each $i \in \mathbb{N}_0$ and each graded R-module M, we can use the grading on $H^i_{\mathfrak{a}R'}(M \otimes_R R')$ of (i) to define a grading on $H^i_{\mathfrak{a}}(M) \otimes_R R'$ in such a way that the isomorphism $\rho_M^i \; : \; H_\mathfrak{a}^i(M) \otimes_R R' \xrightarrow{\cong} H_{\mathfrak{a}R'}^i(M \otimes_R R')$ is homogeneous: we recover the grading of (i) on the $\Gamma_\mathfrak{a}(M) \otimes_R R'$. With respect to *these* gradings, $(H_\mathfrak{a}^i(\,\bullet\,) \otimes_R R')_{i \in \mathbb{N}_0}$ also has the *restriction property. Since $H_\mathfrak{a}^i(I) \otimes_R R' = 0$ for all $i \in \mathbb{N}$ whenever I is a *injective graded R-module (by 13.2.6), it follows from 13.3.15 that these gradings coincide with the natural ones used in part (i). All the remaining claims follow from this. $\qquad\square$

Next, we explore the behaviour of graded local cohomology with respect to a shift functor.

14.1.10 Remarks. Let $g_0 \in G$ and $j \in \mathbb{N}_0$. Let $r_{g_0} \in R_{g_0}$ be a homogeneous element of degree g_0. Let L, M be graded R-modules.

(i) The graded R-modules

$$* \operatorname{Hom}_R(L, M(g_0)) \quad \text{and} \quad (* \operatorname{Hom}_R(L, M))(g_0)$$

are equal. Since $* \operatorname{Ext}_R^j(\,\bullet\,, N)$ is the j-th right derived functor in $*\mathcal{C}(R)$ of $* \operatorname{Hom}_R(\,\bullet\,, N)$, we see that the graded R-modules $* \operatorname{Ext}_R^j(L, M(g_0))$ and $(* \operatorname{Ext}_R^j(L, M))(g_0)$ are again equal (and not just isomorphic in $*\mathcal{C}(R)$). Hence the graded R-modules

$$\varinjlim_{n \in \mathbb{N}} \operatorname{Ext}_R^j(R/\mathfrak{a}^n, M(g_0)) \quad \text{and} \quad \left(\varinjlim_{n \in \mathbb{N}} \operatorname{Ext}_R^j(R/\mathfrak{a}^n, M) \right)(g_0)$$

are equal.

(ii) If we forget the gradings on M and $M(g_0)$, we obtain the same ungraded R-module M. By 13.4.6(ii), the grading on $H_\mathfrak{a}^j(M)$ can be defined from the grading of 13.3.13(ii) on $\varinjlim_{n \in \mathbb{N}} \operatorname{Ext}_R^j(R/\mathfrak{a}^n, M)$ simply by requiring that the isomorphism $\phi_{\mathfrak{a}\,M}^j$ be homogeneous. Similarly, we can obtain the grading on $H_\mathfrak{a}^j(M(g_0))$ by using $\phi_{\mathfrak{a}\,M}^j$ to 'lift across' the grading of 13.3.13(ii) on $\varinjlim_{n \in \mathbb{N}} \operatorname{Ext}_R^j(R/\mathfrak{a}^n, M(g_0))$. It therefore follows from part (i) that the graded modules $H_\mathfrak{a}^j(M(g_0))$ and $(H_\mathfrak{a}^j(M))(g_0)$ are equal.

(iii) The fact that $\mathcal{R}^j D_\mathfrak{a}(M(g_0))$ and $(\mathcal{R}^j D_\mathfrak{a}(M))(g_0)$ are the same graded R-module can be deduced in a similar way from 13.5.7(ii).

14.1.11 ♯Exercise. Obtain a 'graded' version of Proposition 8.1.2.

In detail, assume that $\mathfrak{a}$ is graded; let b be a homogeneous element of R. Let $f : M \longrightarrow N$ be a homogeneous homomorphism of graded R-modules.

(i) Show that there is a long exact sequence of graded R-modules and homogeneous R-homomorphisms

$$0 \longrightarrow H^0_{\mathfrak{a}+Rb}(M) \longrightarrow H^0_{\mathfrak{a}}(M) \longrightarrow H^0_{\mathfrak{a}}(M_b)$$
$$\longrightarrow H^1_{\mathfrak{a}+Rb}(M) \longrightarrow H^1_{\mathfrak{a}}(M) \longrightarrow H^1_{\mathfrak{a}}(M_b)$$
$$\longrightarrow \quad \cdots \qquad\qquad\qquad \cdots$$
$$\longrightarrow H^i_{\mathfrak{a}+Rb}(M) \longrightarrow H^i_{\mathfrak{a}}(M) \longrightarrow H^i_{\mathfrak{a}}(M_b)$$
$$\longrightarrow H^{i+1}_{\mathfrak{a}+Rb}(M) \longrightarrow \quad \cdots$$

such that the diagram

$$
\begin{array}{ccccccc}
H^i_{\mathfrak{a}+Rb}(M) & \longrightarrow & H^i_{\mathfrak{a}}(M) & \longrightarrow & H^i_{\mathfrak{a}}(M_b) & \longrightarrow & H^{i+1}_{\mathfrak{a}+Rb}(M) \\
\downarrow{\scriptstyle H^i_{\mathfrak{a}+Rb}(f)} & & \downarrow{\scriptstyle H^i_{\mathfrak{a}}(f)} & {\scriptstyle H^i_{\mathfrak{a}}(f_b)}\downarrow & & \downarrow{\scriptstyle H^{i+1}_{\mathfrak{a}+Rb}(f)} \\
H^i_{\mathfrak{a}+Rb}(N) & \longrightarrow & H^i_{\mathfrak{a}}(N) & \longrightarrow & H^i_{\mathfrak{a}}(N_b) & \longrightarrow & H^{i+1}_{\mathfrak{a}+Rb}(N)
\end{array}
$$

commutes for all $i \in \mathbb{N}_0$.

(ii) Let $i \in \mathbb{N}_0$. Show that there is a commutative diagram

$$
\begin{array}{ccccccccc}
0 & \longrightarrow & H^1_{Rb}(H^i_{\mathfrak{a}}(M)) & \longrightarrow & H^{i+1}_{\mathfrak{a}+Rb}(M) & \longrightarrow & \Gamma_{Rb}(H^{i+1}_{\mathfrak{a}}(M)) & \longrightarrow & 0 \\
& & \downarrow{\scriptstyle H^1_{Rb}(H^i_{\mathfrak{a}}(f))} & & \downarrow{\scriptstyle H^{i+1}_{\mathfrak{a}+Rb}(f)} & & \downarrow{\scriptstyle \Gamma_{Rb}(H^{i+1}_{\mathfrak{a}}(f))} & & \\
0 & \longrightarrow & H^1_{Rb}(H^i_{\mathfrak{a}}(N)) & \longrightarrow & H^{i+1}_{\mathfrak{a}+Rb}(N) & \longrightarrow & \Gamma_{Rb}(H^{i+1}_{\mathfrak{a}}(N)) & \longrightarrow & 0
\end{array}
$$

(in the category $*\mathcal{C}(R)$) with exact rows. The top row is referred to as the *comparison exact sequence for M*.

14.1.12 Lemma. *Assume that $G = \mathbb{Z}^n$ and $R = \bigoplus_{g \in \mathbb{N}_0^n} R_g$ is positively graded, and that the ideal $\mathfrak{a}$ is generated by homogeneous elements of degree 0. Let $\mathfrak{a}_0 = \mathfrak{a} \cap R_0$. Let $h \in G$. We denote by $(\bullet)_h : *\mathcal{C}(R) \longrightarrow \mathcal{C}(R_0)$ the functor which assigns, to each graded R-module, and to each homogeneous homomorphism of graded R-modules, the h-th component.*

There is an isomorphism $(H^i_{\mathfrak{a}}(\bullet)_h)_{i \in \mathbb{N}_0} \overset{\cong}{\longrightarrow} (H^i_{\mathfrak{a}_0}((\bullet)_h))_{i \in \mathbb{N}_0}$ of negative strongly connected sequences of covariant functors from $*\mathcal{C}(R)$ to $\mathcal{C}(R_0)$.

Proof. Observe that $\Gamma_{\mathfrak{a}}(\bullet)_h$ and $\Gamma_{\mathfrak{a}_0}((\bullet)_h)$ are the same functor. Let $I = \bigoplus_{g \in \mathbb{Z}^n} I_g$ be a *injective graded R-module. Impose the trivial $\mathbb{Z}^n$-grading on the commutative Noetherian ring R_0. Since $\mathfrak{a} = \mathfrak{a}_0 R$, it follows from the Graded Independence Theorem 14.1.7 and Exercise 3.4.5 that, for all $i \in \mathbb{N}$,

$$0 = H_{\mathfrak{a}}^i(I) = H_{\mathfrak{a}_0 R}^i(I) \cong H_{\mathfrak{a}_0}^i(I \lceil_{R_0}) = H_{\mathfrak{a}_0}^i\left(\bigoplus_{g \in G} I_g\right) \cong \bigoplus_{g \in G} H_{\mathfrak{a}_0}^i(I_g).$$

Hence I_h is $\Gamma_{\mathfrak{a}_0}$-acyclic. The result is now an easy consequence of 13.3.5. $\square$

14.1.13 Example. Assume that that $G = \mathbb{Z}^n$ and $R = \bigoplus_{g \in \mathbb{N}_0^n} R_g$ is positively graded. Let $h \in \mathbb{Z}^n$ and let L be an R_0-module. We can define a graded R-module ${}^h L$ such that, for all $g \in G$,

$$({}^h L)_g = \begin{cases} L & \text{if } g = h, \\ 0 & \text{if } g \neq h. \end{cases}$$

(These conditions necessitate that $r_g m = 0$ for all $m \in {}^h L$ and all $r_g \in R_g$ whenever $g \in G \setminus \{0\}$.)

Assume that the ideal $\mathfrak{a}$ is graded, and let $\mathfrak{a}_0 = \mathfrak{a} \cap R_0$. Let $i \in \mathbb{N}_0$. We show how to use Lemma 14.1.12 to calculate $H_{\mathfrak{a}}^i({}^h L)$. Let $R_+ = \bigoplus_{g \in \mathbb{N}_0^n \setminus \{0\}} R_g$; observe that R_+ annihilates ${}^h L$ and that $\mathfrak{a} + R_+ = \mathfrak{a}_0 R + R_+$. The argument of Example 4.2.2 can be modified to our '$\mathbb{Z}^n$-graded' situation to produce homogeneous R-isomorphisms $H_{\mathfrak{a}}^i({}^h L) \cong H_{\mathfrak{a}+R_+}^i({}^h L) \cong H_{\mathfrak{a}_0 R}^i({}^h L)$. We can now use 14.1.12 to deduce that, for $g \in \mathbb{Z}^n$,

$$H_{\mathfrak{a}}^i({}^h L)_g \cong H_{\mathfrak{a}_0 R}^i({}^h L)_g \cong H_{\mathfrak{a}_0}^i(({}^h L)_g) = \begin{cases} H_{\mathfrak{a}_0}^i(L) & \text{if } g = h, \\ 0 & \text{if } g \neq h. \end{cases}$$

Hence there is a homogeneous R-isomorphism ${}^h H_{\mathfrak{a}_0}^i(L) \cong H_{\mathfrak{a}}^i({}^h L)$.

We end this section with some results that concern the special case of $\mathbb{Z}$-graded rings, that is, the special case where $G = \mathbb{Z}$. In this case, much basic theory has been developed by Bruns and Herzog in [7, §1.5 and §3.6]. We now prepare for a $\mathbb{Z}$-graded analogue of (a special case of) the local Lichtenbaum–Hartshorne Vanishing Theorem 8.2.1. Before we come to the theorem itself, however, we provide two preparatory exercises.

14.1.14 ♯Exercise. Assume that $G = \mathbb{Z}$ and that $(R, \mathfrak{m})$ is *local; assume further that $\mathfrak{m}$ is actually a maximal ideal of R. Let M be a graded R-module. Use [7, 1.5.6 and 1.5.8] to show that $\dim M = \dim_{R_{\mathfrak{m}}} M_{\mathfrak{m}}$.

14.1.15 ♯Exercise. Assume that $G = \mathbb{Z}$, that $R = \bigoplus_{j \in \mathbb{N}_0} R_j$ is positively graded, and that the subring R_0 is a local ring having maximal ideal $\mathfrak{m}_0$.

(i) Show that R is *local with unique *maximal ideal

$$\mathfrak{m} := \mathfrak{m}_0 \oplus R_1 \oplus R_2 \oplus \cdots \oplus R_n \oplus \cdots .$$

(ii) Show that, for all $j, i \in \mathbb{N}$, the j-th component of $\mathfrak{m}^{j+i}$ is contained in $\mathfrak{m}_0^i R_j$.

(iii) Assume that $R \neq R_0$. By [7, 1.5.4], there exist non-zero homogeneous elements $y_1, \ldots, y_t$ of R of positive degrees such that

$$R = R_0[y_1, \ldots, y_t].$$

Let $d := \max\{\deg y_i : 1 \leq i \leq t\}$. Show that $R_{(i-1)d+j} \subseteq \mathfrak{m}^i$ for all $i, j \in \mathbb{N}$.

(iv) Show that the multiplication in R induces a natural ring structure on the Abelian group $\prod_{j \in \mathbb{N}_0} R_j$. (The following hint can ease the checking of the ring axioms. For $a := (a_0, a_1, \ldots, a_j, \ldots) \in \prod_{j \in \mathbb{N}_0} R_j$ and $h \in \mathbb{N}_0$, define

$$a_{\leq h} := (a_0, a_1, \ldots, a_h, 0, 0, \ldots);$$

show that $(a_{\leq h} b_{\leq h})_{\leq h} = (ab)_{\leq h}$ for $a, b \in \prod_{j \in \mathbb{N}_0} R_j$.)

(v) Assume in addition that R_0 is complete. Prove that the inclusion map $R \longrightarrow \prod_{j \in \mathbb{N}_0} R_j$ provides the $\mathfrak{m}$-adic completion of R.

14.1.16 Graded Lichtenbaum–Hartshorne Vanishing Theorem. *Assume that $G = \mathbb{Z}$, and that $R = \bigoplus_{j \in \mathbb{N}_0} R_j$ is positively graded and an integral domain; assume also that the subring R_0 is a complete local ring having maximal ideal $\mathfrak{m}_0$.*

Assume that the ideal $\mathfrak{a}$ is graded and proper, and that $\dim R/\mathfrak{a} > 0$. Set $d := \dim R$. Then $H_{\mathfrak{a}}^d(R) = 0$.

Proof. Let $\mathfrak{m} := \mathfrak{m}_0 \oplus R_1 \oplus R_2 \oplus \cdots \oplus R_j \oplus \cdots$. By 14.1.15, our graded ring R is *local with unique *maximal ideal $\mathfrak{m}$. Let $\widehat{R_{\mathfrak{m}}}$ denote the completion of the local ring $R_{\mathfrak{m}}$. In view of the natural ring isomorphisms

$$R/\mathfrak{m}^j \xrightarrow{\cong} R_{\mathfrak{m}}/(\mathfrak{m}R_{\mathfrak{m}})^j \quad \text{for } j \in \mathbb{N},$$

$\widehat{R_{\mathfrak{m}}}$ is isomorphic to the $\mathfrak{m}$-adic completion of R, which is an integral domain by 14.1.15.

Note that $\dim R_{\mathfrak{m}} = d$, by 14.1.14, since $\mathfrak{m}$ is actually a maximal ideal of R. Since $\mathfrak{a}$ is graded, all its minimal primes are graded, and so are contained in $\mathfrak{m}$. Since $\dim R/\mathfrak{a} > 0$, it therefore follows that $\mathfrak{m}$ is not a minimal prime of $\mathfrak{a}$. Hence $\dim\left(\widehat{R_{\mathfrak{m}}}/\mathfrak{a}\widehat{R_{\mathfrak{m}}}\right) = \dim\left(R_{\mathfrak{m}}/\mathfrak{a}R_{\mathfrak{m}}\right) > 0$. We can therefore use the local Lichtenbaum–Hartshorne Vanishing Theorem 8.2.1 to deduce that $H_{\mathfrak{a}R_{\mathfrak{m}}}^d(R_{\mathfrak{m}}) = 0$. Thus $(H_{\mathfrak{a}}^d(R))_{\mathfrak{m}} = 0$, by 4.3.3. Therefore $H_{\mathfrak{a}}^d(R) = 0$, by 14.1.2(ii). $\square$

Two other results to which we would like to add 'graded frills' are Faltings' Annihilator Theorem 9.5.1 and Grothendieck's Finiteness Theorem 9.5.2. We shall have to defer graded versions of these results until the end of §14.3, by which point we shall have presented more results about the behaviour of G-graded modules.

14.2 *Indecomposable *injective modules

One of our major aims in this chapter is to provide versions of local duality which apply in 'graded' situations, including versions which involve 'graded canonical modules'. To prepare for this, we shall develop the 'graded' analogue of the decomposition theory for injective modules over a commutative Noetherian ring.

Recall that we are assuming that R is G-graded throughout this chapter. In 13.2.4, we showed that each graded R-module M has a *injective envelope, and between any two *injective envelopes of M there is a homogeneous isomorphism which restricts to the identity map on M; we agreed to denote by $*E(M)$ or $*E_R(M)$ one choice of *injective envelope of M; and we proved that $*E(M)$, with its grading forgotten, is an essential extension of M.

14.2.1 ♯Exercise. Consider a non-empty family $(M_\lambda)_{\lambda \in \Lambda}$ of graded R-modules.

(i) Show that $\bigoplus_{\lambda \in \Lambda} M_\lambda$ is *injective if and only if M_λ is *injective for all $\lambda \in \Lambda$. (Here is a hint: you might find the 'graded Baer criterion' 13.2.7 useful.)

(ii) Show that the obvious map $\bigoplus_{\lambda \in \Lambda} M_\lambda \longrightarrow \bigoplus_{\lambda \in \Lambda} *E(M_\lambda)$ provides the *injective envelope of $\bigoplus_{\lambda \in \Lambda} M_\lambda$.

14.2.2 ♯Exercise. Let I be a graded submodule of the graded R-module M, and suppose that I is *injective. Show that I is a direct summand of M with graded complement.

14.2.3 Definition. A graded R-module is said to be *indecomposable* precisely when it is non-zero and cannot be written as the direct sum of two proper graded submodules.

14.2.4 Proposition. *Let* $\mathfrak{p} \in {}^*\operatorname{Spec}(R)$.

(i) *The *injective graded R-module* $*E(R/\mathfrak{p})(g)$ *is *indecomposable for each* $g \in G$.

(ii) *A non-zero *injective graded R-module I has a *indecomposable *injective graded submodule which must, by 14.2.2, be a direct summand (with graded complement). In fact, for each $\mathfrak{q} \in \operatorname{Ass} I$, there exists $g_0 \in G$ and a homogeneous element m of I of degree g_0 for which $(0 :_R m) = \mathfrak{q}$, and then I has a graded submodule that is homogeneously isomorphic to *$E(R/\mathfrak{q})(-g_0)$.*

(iii) *Each *indecomposable *injective graded R-module is isomorphic (in the category *$\mathcal{C}(R)$) to *$E(R/\mathfrak{q})(-g_0)$ for some $\mathfrak{q} \in$ * $\operatorname{Spec}(R)$ and $g_0 \in G$.*

(iv) *Let r be a homogeneous element of degree g in $R \setminus \mathfrak{p}$. Then multiplication by r provides a homogeneous automorphism of degree g of *$E(R/\mathfrak{p})$. Also, each element of *$E(R/\mathfrak{p})$ is annihilated by some power of $\mathfrak{p}$.*

(v) *Let $\mathfrak{q} \in$ * $\operatorname{Spec}(R)$. If *$E(R/\mathfrak{p}) \cong$ *$E(R/\mathfrak{q})(-g)$ (in *$\mathcal{C}(R)$) for some $g \in G$, then $\mathfrak{p} = \mathfrak{q}$.*

Proof. (i) Let m be a homogeneous generator of the graded submodule $R/\mathfrak{p}$ of *$E(R/\mathfrak{p})$. Suppose that L, N are non-zero graded submodules of *$E(R/\mathfrak{p})$ such that *$E(R/\mathfrak{p}) = L \oplus N$. Then $L \cap Rm \neq 0$ and $N \cap Rm \neq 0$; thus there exist homogeneous elements $a, b \in R$ such that $0 \neq am \in L$ and $0 \neq bm \in N$. Since $(0 :_R m) = \mathfrak{p}$, prime, we must have $a, b \in R \setminus \mathfrak{p}$, so that $ab \notin \mathfrak{p}$. Therefore $0 \neq abm \in L \cap N$, and this is a contradiction. Therefore *$E(R/\mathfrak{p})$ is *indecomposable. It follows easily that *$E(R/\mathfrak{p})(g)$ is *indecomposable for each $g \in G$.

(ii) Since $I \neq 0$, it must have an associated prime, $\mathfrak{q}$ say, which must be graded, by 13.1.6(ii); moreover, $\mathfrak{q} = (0 :_R m)$ for some homogeneous element m of I. Let $\deg m = g_0$. Thus there is a homogeneous isomorphism $\varphi :$ $(R/\mathfrak{q})(-g_0) \overset{\cong}{\longrightarrow} Rm$ such that $\varphi(1 + \mathfrak{q}) = m$.

Since I is *injective, we can extend φ to a homogeneous homomorphism $\psi :$ *$E((R/\mathfrak{q})(-g_0)) =$ *$E(R/\mathfrak{q})(-g_0) \longrightarrow I$; as

$$\operatorname{Ker} \psi \cap (R/\mathfrak{q})(g_0) = 0,$$

it follows that ψ is monomorphic; therefore $J := \operatorname{Im} \psi$ is a *injective graded submodule of I and there is a homogeneous isomorphism

$$J \cong \ ^*E(R/\mathfrak{q})(-g_0).$$

It follows from (i) that J is *indecomposable.

Since J is *injective, there is a homogeneous R-homomorphism $\xi : I \longrightarrow J$ that extends the identity map Id_J on J; then $\operatorname{Ker} \xi$ is a graded submodule of I and $J \oplus \operatorname{Ker} \xi = I$. Thus J is a direct summand of I with graded complement.

(iii) Apply (ii) to a *indecomposable *injective graded R-module I, and the desired conclusion is immediate.

(iv) Multiplication by r provides a homogeneous R-homomorphism μ_r : $*E(R/\mathfrak{p}) \longrightarrow *E(R/\mathfrak{p})(g)$; since $\operatorname{Ker} \mu_r \cap (R/\mathfrak{p}) = 0$, the map μ_r is monomorphic. Therefore $\operatorname{Im} \mu_r$ is a *injective graded submodule of the *indecomposable *injective graded R-module $*E(R/\mathfrak{p})(g)$, and it follows from 14.2.2 that μ_r must be surjective.

For the second claim, it is enough for us to show that an arbitrary non-zero homogeneous element $m \in *E(R/\mathfrak{p})$ is annihilated by some power of $\mathfrak{p}$. Let $\mathfrak{q} \in \operatorname{Ass} Rm$, and recall from 13.1.6(ii) that $\mathfrak{q}$ is graded. Since $Rm \cap (R/\mathfrak{p}) \neq 0$, we see that $\mathfrak{q} \in \operatorname{Ass}(R/\mathfrak{p}) = \{\mathfrak{p}\}$, so that $\mathfrak{q} = \mathfrak{p}$. Therefore $(0 :_R Rm)$ is $\mathfrak{p}$-primary.

(v) Let $h(R)$ denote the set of non-zero homogeneous elements of R, and let $g' \in G$. By (iv), the set of homogeneous elements r of R for which multiplication by r provides an automorphism (of some degree) of $*E(R/\mathfrak{p})(-g')$ is precisely $h(R) \setminus \mathfrak{p}$. The desired conclusion is now immediate. □

The next lemma can be proved by making straightforward modifications to the proof of the corresponding 'ungraded' result in 10.1.12.

14.2.5 Lemma. *Let S be a multiplicatively closed subset of homogeneous elements of R, and let M be a $(G$-)graded $S^{-1}R$-module. Then M is *injective over R if and only if it is *injective over $S^{-1}R$.*

14.2.6 Lemma. *Let S be a multiplicatively closed subset of homogeneous elements of R, and let $\mathfrak{p} \in \operatorname{Spec}(R)$ be such that $\mathfrak{p} \cap S = \emptyset$. By 14.2.4(iv), the *indecomposable *injective graded R-module $*E_R(R/\mathfrak{p})$ has a natural structure as a G-graded $S^{-1}R$-module.*

In the category $*\mathcal{C}(S^{-1}R)$, we have

$$*E_R(R/\mathfrak{p}) \cong *E_{S^{-1}R}(S^{-1}R/S^{-1}\mathfrak{p}).$$

Furthermore, $*E_{S^{-1}R}(S^{-1}R/S^{-1}\mathfrak{p})$, when considered as a G-graded R-module by means of the natural homomorphism $R \longrightarrow S^{-1}R$, is homogeneously isomorphic to $*E_R(R/\mathfrak{p})$.

Proof. By Lemma 14.2.5, the graded $S^{-1}R$-module $*E_R(R/\mathfrak{p})$ is *injective over $S^{-1}R$. Since a graded $S^{-1}R$-submodule of $*E_R(R/\mathfrak{p})$ is automatically a graded R-submodule, it is immediate from 14.2.4(i) that $*E_R(R/\mathfrak{p})$ is *indecomposable as $S^{-1}R$-module.

There is a homogeneous generator m of degree 0 of the graded R-submodule $R/\mathfrak{p}$ of $*E_R(R/\mathfrak{p})$. When the latter module is considered as a graded $S^{-1}R$-module, $(0 :_{S^{-1}R} m) = S^{-1}\mathfrak{p}$, and m still has degree 0. It therefore follows from 14.2.4(ii) that there is a homogeneous $S^{-1}R$-isomorphism $*E_R(R/\mathfrak{p}) \cong *E_{S^{-1}R}(S^{-1}R/S^{-1}\mathfrak{p})$.

The final claim is now immediate. $\qquad\square$

14.2.7 ♯Exercise. Suppose that our G-graded ring R is *simple*, that is, R has exactly two graded ideals, namely 0 and R. Use the 'graded Baer criterion' 13.2.7 to show that every graded R-module is *injective.

14.2.8 ♯Exercise. Assume that $\mathfrak{a}$ is graded, and that $\mathfrak{p} \in *\operatorname{Spec}(R)$ is such that $\mathfrak{a} \subseteq \mathfrak{p}$. Show that the graded $R/\mathfrak{a}$-module $(0 :_{*E_R(R/\mathfrak{p})} \mathfrak{a})$ is homogeneously isomorphic to $*E_{R/\mathfrak{a}}((R/\mathfrak{a})/(\mathfrak{p}/\mathfrak{a}))$.

14.2.9 ♯Exercise. Assume that $(R, \mathfrak{m})$ is *local. Show that there are homogeneous isomorphisms

$$*\operatorname{Hom}_R(R/\mathfrak{m}, *E(R/\mathfrak{m})) \cong (0 :_{*E(R/\mathfrak{m})} \mathfrak{m}) \cong *E_{R/\mathfrak{m}}(R/\mathfrak{m}) = R/\mathfrak{m}.$$

14.2.10 ♯Exercise. Show that each *injective graded R-module I is a direct sum of *indecomposable *injective graded submodules. (Here is a hint: adapt the argument in the proof of 10.1.8 to our G-graded situation, that is, apply Zorn's Lemma to the set of all sets of *indecomposable *injective graded submodules of I whose sum is direct.)

14.2.11 Definition and ♯Exercise. Let M be a graded R-module. A *minimal *injective resolution of* M is a *injective resolution

$$I^\bullet : 0 \longrightarrow I^0 \xrightarrow{d^0} I^1 \longrightarrow \cdots \longrightarrow I^i \xrightarrow{d^i} I^{i+1} \longrightarrow \cdots$$

of M (in the category $*\mathcal{C}(R)$) such that I^i is a *essential extension of $\operatorname{Ker} d^i$ for every $i \in \mathbb{N}_0$.

Show that M has a minimal *injective resolution, and that the i-th term in such a resolution is uniquely determined, up to isomorphism in $*\mathcal{C}(R)$, by M. We denote this i-th term by $*E^i(M)$, or by $*E^i_R(M)$.

With the notation of 14.2.11, and for $i \in \mathbb{N}_0$, it follows from 14.2.4(iii) and 14.2.10 that there is a family $(\mathfrak{p}_\alpha)_{\alpha \in \Lambda}$ of graded prime ideals of R and a family $(g_\alpha)_{\alpha \in \Lambda}$ of elements of G for which there is a homogeneous isomorphism

$$*E^i(M) \xrightarrow{\cong} \bigoplus_{\alpha \in \Lambda} *E(R/\mathfrak{p}_\alpha)(-g_\alpha).$$

Of course, we would like, as in the analogous ungraded situation (see 11.1.4) to be able to show that, for $\mathfrak{p} \in *\operatorname{Spec}(R)$, the cardinality of the set

$$\{\alpha \in \Lambda : \mathfrak{p}_\alpha = \mathfrak{p}\}$$

depends only on $*E^i(M)$ and $\mathfrak{p}$ (and therefore only on i, M and $\mathfrak{p}$) and not on the particular decomposition of $*E^i(M)$ (as a direct sum of *indecomposable *injective submodules) chosen. This is indeed the case, and, interestingly, the cardinality in question turns out to be equal to the ordinary Bass number $\mu^i(\mathfrak{p}, M)$ of 11.1.4. We are, in the spirit of this book, going to establish this; however, instead of trying to imitate the standard 'ungraded' argument, we shall employ Theorem 13.2.4(iii).

14.2.12 Proposition. *Let M be a graded R-module. By 14.2.10 and 14.2.4, there is a family $(\mathfrak{p}_\alpha)_{\alpha \in \Lambda}$ of graded prime ideals of R and a family $(g_\alpha)_{\alpha \in \Lambda}$ of elements of G for which there is a homogeneous isomorphism $*E(M) \xrightarrow{\cong} \bigoplus_{\alpha \in \Lambda} *E(R/\mathfrak{p}_\alpha)(-g_\alpha)$.*

*Let $\mathfrak{p} \in *\mathrm{Spec}(R)$. Then the cardinality of the set $\{\alpha \in \Lambda : \mathfrak{p}_\alpha = \mathfrak{p}\}$ is equal to the ordinary Bass number $\mu^0(\mathfrak{p}, M)$ (see 11.1.4), and so depends only on M and $\mathfrak{p}$ and not on the particular decomposition of $*E(M)$ (as a direct sum of *indecomposable *injective submodules) chosen.*

Proof. If $(N_\alpha)_{\alpha \in \Lambda}$ is a family of R-modules, then, by 14.2.1(ii) (applied in the case where R is considered to be trivially G-graded, that is, $R_0 = R$ and $R_g = 0$ for all $g \in G \setminus \{0\}$),

$$E\left(\bigoplus_{\alpha \in \Lambda} N_\alpha\right) \cong \bigoplus_{\alpha \in \Lambda} E(N_\alpha).$$

Now consider the given G-grading on R. For a graded R-module U and $g \in G$, it follows from 13.2.4(iii) that $E(*E(U)(g)) \cong E(U)$. Hence, on use of this and the preceding paragraph, we see that

$$E(M) \cong E(*E(M)) \cong E\left(\bigoplus_{\alpha \in \Lambda} *E(R/\mathfrak{p}_\alpha)(g_\alpha)\right)$$
$$\cong \bigoplus_{\alpha \in \Lambda} E(*E(R/\mathfrak{p}_\alpha)(g_\alpha)) \cong \bigoplus_{\alpha \in \Lambda} E(R/\mathfrak{p}_\alpha),$$

from which the claim is clear. $\square$

We want to extend the result of 14.2.12 to the 'higher' terms in minimal *injective resolutions. The following exercise concerns an important point in the theory of Bass numbers for ungraded situations.

14.2.13 ♯Exercise. Let R' be a commutative Noetherian ring, let M be an R'-module, and let

$$0 \longrightarrow E^0(M) \xrightarrow{d^0} E^1(M) \longrightarrow \cdots \longrightarrow E^i(M) \xrightarrow{d^i} E^{i+1}(M) \longrightarrow \cdots$$

be the minimal injective resolution of M, so that there is an augmentation R'-homomorphism $\alpha : M \to E^0(M)$ such that the sequence

$$0 \longrightarrow M \xrightarrow{\alpha} E^0(M) \longrightarrow \cdots \longrightarrow E^i(M) \xrightarrow{d^i} E^{i+1}(M) \longrightarrow \cdots$$

is exact.

Deduce from 11.1.7 that, for each $\mathfrak{p} \in \operatorname{Spec}(R')$, the induced homomorphism $\operatorname{Hom}_{R'_\mathfrak{p}}(R'_\mathfrak{p}/\mathfrak{p}R'_\mathfrak{p}, \alpha_\mathfrak{p})$ is an isomorphism.

It will also be convenient if we record now some further consequences of Lemma 13.2.4(iii).

14.2.14 Remarks. Let M be a graded R-module. Let

$$0 \longrightarrow {}^*E^0(M) \xrightarrow{e^0} {}^*E^1(M) \longrightarrow \cdots \longrightarrow {}^*E^i(M) \xrightarrow{e^i} {}^*E^{i+1}(M) \longrightarrow \cdots$$

be the minimal *injective resolution of M, with associated (necessarily homogeneous) augmentation homomorphism $\beta : M \longrightarrow {}^*E^0(M)$. Also, let $\alpha : M \longrightarrow E^0(M)$ provide the injective envelope of M.

(i) Since $E^0(M)$ is injective, there is a commutative diagram

$$
\begin{array}{ccc}
M & \xrightarrow{\ \beta\ } & {}^*E^0(M) \\
\| & & \Big\downarrow{\phi^0} \\
M & \xrightarrow{\ \alpha\ } & E^0(M)
\end{array}
$$

of R-modules and R-homomorphisms. Now ${}^*E^0(M)$ is an essential extension of $\operatorname{Im}\beta$, by 13.2.4(iii), and so, since $\operatorname{Ker}\phi^0 \cap \operatorname{Im}\beta = 0$, it follows that ϕ^0 is actually a monomorphism.

(ii) Let $\mathfrak{p} \in \operatorname{Spec}(R)$ and set $k(\mathfrak{p}) := R_\mathfrak{p}/\mathfrak{p}R_\mathfrak{p}$. Now

$$\operatorname{Hom}_{R_\mathfrak{p}}(k(\mathfrak{p}), \phi^0_\mathfrak{p}) \circ \operatorname{Hom}_{R_\mathfrak{p}}(k(\mathfrak{p}), \beta_\mathfrak{p}) = \operatorname{Hom}_{R_\mathfrak{p}}(k(\mathfrak{p}), \alpha_\mathfrak{p}),$$

which is an isomorphism by 14.2.13; however, the left-exactness of the 'Hom' functor ensures that $\operatorname{Hom}_{R_\mathfrak{p}}(k(\mathfrak{p}), \phi^0_\mathfrak{p})$ and $\operatorname{Hom}_{R_\mathfrak{p}}(k(\mathfrak{p}), \beta_\mathfrak{p})$ are both monomorphisms, and so it follows that they are both isomorphisms.

Our need for a proof of the following lemma is explained by our 'first variable only' approach to the functors ${}^*\operatorname{Ext}^i_R(\,\bullet\,, N)$ $(i \in \mathbb{N}_0)$: see 13.1.8.

14.2.15 Lemma. *Let L, M be graded R-modules with M finitely generated. Let*

$$I^\bullet : 0 \longrightarrow I^0 \xrightarrow{f^0} I^1 \longrightarrow \cdots \longrightarrow I^i \xrightarrow{f^i} I^{i+1} \longrightarrow \cdots$$

*be a *injective resolution of L (in the category ${}^*\mathcal{C}(R)$). Then, for each $i \in \mathbb{N}_0$, there is a homogeneous isomorphism*

$${}^*\operatorname{Ext}^i_R(M, L) \cong H^i({}^*\operatorname{Hom}_R(M, I^\bullet)).$$

Proof. For each $i \in \mathbb{N}_0$, set $K^i := \operatorname{Ker} f^i$. Note that there is a homogeneous isomorphism $K^0 \cong L$. Suppose that $j \in \mathbb{N}$ with $j > 1$. For all $i = 1, \ldots, j-1$, there is an exact sequence

$$0 \longrightarrow K^{i-1} \longrightarrow I^{i-1} \longrightarrow K^i \longrightarrow 0$$

of graded R-modules and homogeneous homomorphisms. Since M is a finitely generated graded R-module, it follows from (13.2.5 and) 13.3.11 that there are homogeneous isomorphisms

$$* \operatorname{Ext}_R^j(M, L) \cong * \operatorname{Ext}_R^j(M, K^0) \cong * \operatorname{Ext}_R^{j-1}(M, K^1) \cong \cdots$$
$$\cong * \operatorname{Ext}_R^2(M, K^{j-2}) \cong * \operatorname{Ext}_R^1(M, K^{j-1}).$$

This means that it is now enough for us to prove the claim in the statement of the lemma in the special cases in which $i = 0$ and $i = 1$. However, the claims in these two cases follow easily from the exact sequences

$$0 \longrightarrow * \operatorname{Hom}_R(M, L) \longrightarrow * \operatorname{Hom}_R(M, I^0) \longrightarrow * \operatorname{Hom}_R(M, K^1)$$
$$\longrightarrow * \operatorname{Ext}_R^1(M, L) \longrightarrow 0$$

and $0 \longrightarrow * \operatorname{Hom}_R(M, K^1) \longrightarrow * \operatorname{Hom}_R(M, I^1) \longrightarrow * \operatorname{Hom}_R(M, K^2)$ of graded R-modules and homogeneous homomorphisms which can also be obtained from 13.2.5 and 13.3.11.   □

14.2.16 Theorem. *Let M be a graded R-module. Let $j \in \mathbb{N}_0$, and denote the j-th term in the minimal *injective resolution of M by $*E^j(M)$, as in 14.2.11. By 14.2.10 and 14.2.4(iii), there is a family $(\mathfrak{p}_\alpha)_{\alpha \in \Lambda}$ of graded prime ideals of R and a family $(g_\alpha)_{\alpha \in \Lambda}$ of elements of G for which there is a homogeneous isomorphism*

$$*E^j(M) \xrightarrow{\cong} \bigoplus_{\alpha \in \Lambda} *E(R/\mathfrak{p}_\alpha)(-g_\alpha).$$

Let $\mathfrak{p} \in * \operatorname{Spec}(R)$. Then the cardinality of the set $\{\alpha \in \Lambda : \mathfrak{p}_\alpha = \mathfrak{p}\}$ is equal to the ordinary Bass number $\mu^j(\mathfrak{p}, M)$.

Proof. The claim was proved in 14.2.12 in the special case in which $j = 0$. We consider next the case in which $j = 1$.

Use the notation introduced in 14.2.14, and set $K^i = \operatorname{Ker} e^i$ for each $i \in \mathbb{N}_0$. Also, let

$$0 \longrightarrow E^0(M) \xrightarrow{d^0} E^1(M) \longrightarrow \cdots \longrightarrow E^i(M) \xrightarrow{d^i} E^{i+1}(M) \longrightarrow \cdots$$

be the (ordinary, ungraded) minimal injective resolution of M (with associated augmentation homomorphism $\alpha : M \longrightarrow E^0(M)$). For each $i \in \mathbb{N}_0$, set $C^i := \operatorname{Ker} d^i$. Consider the commutative diagram

$$
\begin{array}{ccccccccc}
0 & \longrightarrow & M & \longrightarrow & {*E^0(M)} & \longrightarrow & K^1 & \longrightarrow & 0 \\
& & \| & & \downarrow{\phi^0} & & \downarrow{\overline{\phi^0}} & & \\
0 & \longrightarrow & M & \longrightarrow & E^0(M) & \longrightarrow & C^1 & \longrightarrow & 0
\end{array}
$$

with exact rows, in which $\overline{\phi^0}$ is the homomorphism induced by ϕ^0. Localize this diagram at $\mathfrak{p}$, and then apply the functor $\operatorname{Hom}_{R_{\mathfrak{p}}}(k(\mathfrak{p}), \bullet)$, noting that $\operatorname{Ext}_R^1(R/\mathfrak{p}, *E^0(M)) = 0$ by 13.2.5, and that $\operatorname{Hom}_{R_{\mathfrak{p}}}(k(\mathfrak{p}), \phi^0_{\mathfrak{p}})$ is an isomorphism by 14.2.14(ii). The Five Lemma shows that

$$\operatorname{Hom}_{R_{\mathfrak{p}}}(k(\mathfrak{p}), \overline{\phi^0}_{\mathfrak{p}}) : \operatorname{Hom}_{R_{\mathfrak{p}}}(k(\mathfrak{p}), (K^1)_{\mathfrak{p}}) \xrightarrow{\cong} \operatorname{Hom}_{R_{\mathfrak{p}}}(k(\mathfrak{p}), (C^1)_{\mathfrak{p}})$$

is an isomorphism. Now $*E^1(M) = *E(K^1)$, and so, by 14.2.12, the cardinality of the set $\{\alpha \in \Lambda : \mathfrak{p}_\alpha = \mathfrak{p}\}$ is equal to $\mu^0(\mathfrak{p}, K^1)$; we have just shown that this is $\mu^0(\mathfrak{p}, C^1)$, which is equal to $\mu^1(\mathfrak{p}, M)$ by the theory of Bass numbers in the ungraded case.

Now suppose that $j > 1$. By 14.2.15, there is a homogeneous isomorphism $*\operatorname{Ext}_R^1(R/\mathfrak{p}, K^{j-1}) \cong *\operatorname{Ext}_R^j(R/\mathfrak{p}, M)$, so that $\mu^1(\mathfrak{p}, K^{j-1}) = \mu^j(\mathfrak{p}, M)$. However,

$$0 \longrightarrow K^{j-1} \xrightarrow{\subseteq} *E^{j-1}(M) \xrightarrow{d^{j-1}} *E^j(M)$$

is the start of the minimal *injective resolution of K^{j-1}, and so it follows from what we have already proved in the case in which $j = 1$ that the cardinality of the set $\{\alpha \in \Lambda : \mathfrak{p}_\alpha = \mathfrak{p}\}$ is equal to $\mu^1(\mathfrak{p}, K^{j-1})$, which we now know to be equal to $\mu^j(\mathfrak{p}, M)$. $\qquad\square$

14.3 A graded version of the Annihilator Theorem

The purpose of this section is to present graded versions of Faltings' Annihilator Theorem 9.5.1 and Grothendieck's Finiteness Theorem 9.5.2. To prepare for this, we are going to present some results due to S. Goto and K.-i. Watanabe [22]. Recall that we are assuming that R is G-graded throughout this chapter. The key points are that, if M is a non-zero finitely generated graded R-module, then, for each $\mathfrak{p} \in \operatorname{Supp} M$, we have $\mathfrak{p}^* \in \operatorname{Supp} M$ and

$$\dim M_{\mathfrak{p}} = \dim M_{\mathfrak{p}^*} + \operatorname{ht} \mathfrak{p}/\mathfrak{p}^* \quad \text{and} \quad \operatorname{depth} M_{\mathfrak{p}} = \operatorname{depth} M_{\mathfrak{p}^*} + \operatorname{ht} \mathfrak{p}/\mathfrak{p}^*.$$

We begin with a description of the structure of *simple G-graded commutative Noetherian rings.

14.3.1 Example. Let K be a field and let G' be a subgroup of G. Then the group ring $K[G']$ is a G-graded commutative ring with

$$K[G']_g = \begin{cases} Kg & \text{if } g \in G', \\ 0 & \text{if } g \notin G'. \end{cases}$$

Let $e_1, \ldots, e_h$ be a free base for G' as Abelian group, and, for each $i = 1, \ldots, h$, use T_i to denote the element $e_i = 1e_i$ in $K[G']$. Then $T_1, \ldots, T_h$ are algebraically independent over K, and

$$K[G'] \cong K[T_1, \ldots, T_h, T_1^{-1}, \ldots, T_h^{-1}]$$

under an isomorphism which maps e_i to T_i (for $i \in \{1, \ldots, h\}$). Thus $K[G']$ is a (Noetherian) regular unique factorization domain. Note also that every non-zero homogeneous element of $K[G']$ is a unit, and so $K[G']$ is a *simple G-graded commutative Noetherian ring. Our first aim in this section is to establish the converse statement.

14.3.2 Theorem. (See S. Goto and K.-i. Watanabe [22, Theorem 1.1.4].) *(Recall that R is G-graded.) The following statements are equivalent.*

(i) *There is a field K and a subgroup G' of G such that R is homogeneously isomorphic to $K[G']$, where the latter is G-graded in the manner described in* 14.3.1.

(ii) *The G-graded ring R is *simple.*

(iii) *Every graded R-module is *injective and *free and R is non-trivial.*

Proof. (i) $\Rightarrow$ (ii) This was proved in 14.3.1.

(ii) $\Rightarrow$ (iii) Assume that R is *simple. This means that the only graded ideals of R are 0 and R itself. Let M be a graded R-module. It is immediate from the 'graded Baer criterion' 13.2.7 that M is *injective. Therefore, by 14.2.10 and 14.2.4(iii), there is a family $(\mathfrak{p}_\alpha)_{\alpha \in \Lambda}$ of graded prime ideals of R and a family $(g_\alpha)_{\alpha \in \Lambda}$ of elements of G for which there is a homogeneous isomorphism $M \xrightarrow{\cong} \bigoplus_{\alpha \in \Lambda} {}^*E(R/\mathfrak{p}_\alpha)(-g_\alpha)$. However, the only proper graded ideal of R is 0, and so, since every graded R-module is *injective, we have ${}^*E(R/\mathfrak{p}_\alpha) = R$ for all $\alpha \in \Lambda$. Therefore M is *free.

(iii) $\Rightarrow$ (i) Assume that every graded R-module is *injective and *free. Let $\mathfrak{a}$ be a proper graded ideal of R. By assumption, the graded R-module $R/\mathfrak{a}$ is free, and so has zero annihilator. Therefore $\mathfrak{a} = 0$. Thus R is *simple.

Let $g \in G$. Each non-zero homogeneous element $u_g \in R_g$ must be invertible in R, with $u_g^{-1} \in R_{-g}$. In particular, each non-zero element of R_0 is invertible in R_0, so that R_0 is a field, K say, and, for each $g' \in G$ for which $R_{g'} \neq 0$, the R_0-module $R_{g'}$, that is, the K-vector space $R_{g'}$, is generated by any one of its non-zero elements, for if $u, v \in R_{g'} \setminus \{0\}$, then $vu^{-1} \in R_0 = K$, so that $v \in Ku$. Also in particular, the product of two non-zero homogeneous elements of R is again non-zero. Let $G' := \{g \in G : R_g \neq 0\}$, a subgroup of G. Let $e_1, \ldots, e_h$ be a free base for G' as Abelian group, and, for each $i = 1, \ldots, h$, let T_i be a non-zero element of R_{e_i}.

Consider $K[G']$ as G-graded in the manner described in 14.3.1. There is a homogeneous ring homomorphism $\phi : K[G'] \longrightarrow R$ which acts as the identity on $K = R_0$ and maps e_i to T_i (for each $i = 1, \ldots, h$). The choice of G' ensures that ϕ is surjective, and since Ker ϕ is a proper graded ideal of the *simple G-graded ring $K[G']$, we must have that ϕ is an isomorphism. □

14.3.3 Remark. It follows from 14.3.1 and 14.3.2 that, if R is *simple, then R is a regular unique factorization domain with dim $R \leq \operatorname{rank} G$.

14.3.4 Corollary. *Each *maximal ideal of R is *prime.* ·

Proof. Let $\mathfrak{m}$ be a *maximal ideal of R. Then the G-graded ring $R/\mathfrak{m}$ has exactly two graded ideals, and so is *simple. Therefore $R/\mathfrak{m}$ is an integral domain, by 14.3.3. □

14.3.5 Proposition. *Assume that the G-graded ring R is an integral domain, and let $\mathfrak{p} \in \operatorname{Spec}(R)$. Then* ht $\mathfrak{p} = $ ht $\mathfrak{p}^* + $ ht $\mathfrak{p}/\mathfrak{p}^*$.

Proof. Recall that $\mathfrak{p}^* \in \operatorname{Spec}(R)$. It is clear that ht $\mathfrak{p} \geq $ ht $\mathfrak{p}^* + $ ht $\mathfrak{p}/\mathfrak{p}^*$; we shall prove the opposite inequality by induction on $h := $ ht $\mathfrak{p}^*$. When $h = 0$, we have $\mathfrak{p}^* = 0$ (because R is an integral domain), so that ht $\mathfrak{p} = $ ht $\mathfrak{p}/\mathfrak{p}^*$ and the desired inequality is clear.

So suppose, inductively, that $h > 0$ and that the result has been proved in situations where ht $\mathfrak{p}^* < h$. Denote ht $\mathfrak{p}$ by n. Let $0 \neq r$ be a homogeneous element of R contained in $\mathfrak{p}^*$. Then dim $R_\mathfrak{p}/rR_\mathfrak{p} = n - 1$ because R is an integral domain, and so there exists a chain of prime ideals $\mathfrak{p}_0 \subset \mathfrak{p}_1 \subset \cdots \subset \mathfrak{p}_{n-1} = \mathfrak{p}$ of R with $r \in \mathfrak{p}_0$. Now $\mathfrak{p}_0$, being a minimal prime ideal of the graded ideal Rr, must be graded, and so $\mathfrak{p}_0 \subseteq \mathfrak{p}^*$. In fact, $(\mathfrak{p}/\mathfrak{p}_0)^* = \mathfrak{p}^*/\mathfrak{p}_0$. Note that ht $\mathfrak{p}/\mathfrak{p}_0 = n - 1$ and ht$(\mathfrak{p}/\mathfrak{p}_0)^* \leq h - 1$.

Therefore, if we apply the inductive hypothesis to the prime ideal $\mathfrak{p}/\mathfrak{p}_0$ in the G-graded integral domain $R/\mathfrak{p}_0$, we see that

$$\text{ht } \mathfrak{p}/\mathfrak{p}_0 = \text{ht}(\mathfrak{p}/\mathfrak{p}_0)^* + \text{ht}(\mathfrak{p}/\mathfrak{p}_0)/(\mathfrak{p}/\mathfrak{p}_0)^* = \text{ht}(\mathfrak{p}/\mathfrak{p}_0)^* + \text{ht } \mathfrak{p}/\mathfrak{p}^*,$$

so that $n - 1 \leq h - 1 + \operatorname{ht} \mathfrak{p}/\mathfrak{p}^*$. Therefore $\operatorname{ht} \mathfrak{p} \leq \operatorname{ht} \mathfrak{p}^* + \operatorname{ht} \mathfrak{p}/\mathfrak{p}^*$, and the inductive step is complete. $\square$

The next theorem, again due to Goto and Watanabe, is a strengthening of 14.3.5.

14.3.6 Theorem. (See S. Goto and K.-i. Watanabe [22, Proposition 1.2.2].) *(The reader is reminded that R is G-graded.) Let $\mathfrak{p} \in \operatorname{Supp} M$ where M is a graded R-module. Recall from 13.1.6(i) that $\mathfrak{p}^* \in \operatorname{Supp} M$. We have*

$$\dim M_{\mathfrak{p}} = \dim M_{\mathfrak{p}^*} + \operatorname{ht} \mathfrak{p}/\mathfrak{p}^*.$$

Proof. For $\mathfrak{q} \in \operatorname{Supp} M$, we shall denote $\dim M_{\mathfrak{q}}$ by $\operatorname{ht}_M \mathfrak{q}$. It is clear that $\operatorname{ht}_M \mathfrak{p} \geq \operatorname{ht}_M \mathfrak{p}^* + \operatorname{ht} \mathfrak{p}/\mathfrak{p}^*$; we shall prove the opposite inequality by induction on $h := \operatorname{ht}_M \mathfrak{p}^*$.

First suppose that $h = 0$. After homogeneous localization at $\mathfrak{p}$, we can, and do, assume that $(R, \mathfrak{p}^*)$ is *local. Since $h = 0$, it follows that $\mathfrak{p}^*$ is the one and only graded prime ideal in $\operatorname{Supp} M$; therefore $\operatorname{Ass} M = \{\mathfrak{p}^*\}$. Hence each $\mathfrak{q} \in \operatorname{Supp} M$ such that $\mathfrak{q} \subseteq \mathfrak{p}$ must have $\mathfrak{q}^* = \mathfrak{p}^*$, and it is clear from this that $\operatorname{ht}_M \mathfrak{p} = \operatorname{ht} \mathfrak{p}/\mathfrak{p}^*$.

Now suppose, inductively, that $h > 0$ and the desired result has been proved for smaller values of h. We can again assume that $(R, \mathfrak{p}^*)$ is *local. Denote $\operatorname{ht}_M \mathfrak{p}$ by n and $\operatorname{ht} \mathfrak{p}/\mathfrak{p}^*$ by d. Since $n \geq h + d$, we have $n > d$.

There exists a chain of prime ideals $\mathfrak{p}_0 \subset \mathfrak{p}_1 \subset \cdots \subset \mathfrak{p}_n = \mathfrak{p}$ in $\operatorname{Supp} M$. Set $\mathfrak{q} := \mathfrak{p}_{n-d}$; then $\operatorname{ht}_M \mathfrak{q} = n - d$ and $\operatorname{ht} \mathfrak{p}/\mathfrak{q} = d$. If $\mathfrak{q} \subseteq \mathfrak{p}^*$, then $n - d = \operatorname{ht}_M \mathfrak{q} \leq \operatorname{ht}_M \mathfrak{p}^* = h$ and $n \leq d + h$, as required. So we suppose that $\mathfrak{q} \not\subseteq \mathfrak{p}^*$, so that $\mathfrak{q}$ is not graded because $(R, \mathfrak{p}^*)$ is *local. We do have $\mathfrak{q}^* \subseteq \mathfrak{p}^*$ and $\mathfrak{q}^* \in \operatorname{Supp} M$. Let $t := \operatorname{ht} \mathfrak{q}/\mathfrak{q}^*$ and $u := \operatorname{ht} \mathfrak{p}^*/\mathfrak{q}^*$. Note that $t > 0$ because $\mathfrak{q}$ is not graded.

By 14.3.5 applied to the G-graded integral domain $R/\mathfrak{q}^*$, we have

$$\operatorname{ht} \mathfrak{p}/\mathfrak{q}^* = \operatorname{ht}(\mathfrak{p}/\mathfrak{q}^*)^* + \operatorname{ht}(\mathfrak{p}/\mathfrak{q}^*)/(\mathfrak{p}/\mathfrak{q}^*)^* = \operatorname{ht} \mathfrak{p}^*/\mathfrak{q}^* + \operatorname{ht} \mathfrak{p}/\mathfrak{p}^* = u + d.$$

But $\operatorname{ht} \mathfrak{p}/\mathfrak{q}^* \geq \operatorname{ht} \mathfrak{p}/\mathfrak{q} + \operatorname{ht} \mathfrak{q}/\mathfrak{q}^* = d + t$. Therefore $u \geq t > 0$. This means that $\mathfrak{q}^*$ is strictly contained in $\mathfrak{p}^*$, so that $\operatorname{ht}_M \mathfrak{q}^* < h$. The inductive hypothesis therefore yields that $n - d = \operatorname{ht}_M \mathfrak{q} \leq \operatorname{ht}_M \mathfrak{q}^* + \operatorname{ht} \mathfrak{q}/\mathfrak{q}^* = \operatorname{ht}_M \mathfrak{q}^* + t$. Now $\operatorname{ht}_M \mathfrak{q}^* + u = \operatorname{ht}_M \mathfrak{q}^* + \operatorname{ht} \mathfrak{p}^*/\mathfrak{q}^* \leq \operatorname{ht}_M \mathfrak{p}^* = h$. Therefore

$$n - d \leq \operatorname{ht}_M \mathfrak{q}^* + t \leq h - u + t,$$

so that

$$n \leq d + h + (t - u) \leq d + h$$

since $u \geq t$. This completes the inductive step. $\square$

In addition to Theorem 14.3.6, we would like to have available the companion result, also due to Goto and Watanabe, that, if M is a finitely generated graded R-module and $\mathfrak{p} \in \operatorname{Supp} M$, then depth $M_{\mathfrak{p}} = $ depth $M_{\mathfrak{p}^*} + \operatorname{ht} \mathfrak{p}/\mathfrak{p}^*$. We shall obtain this as a corollary of the following result about Bass numbers.

14.3.7 Theorem.   (See S. Goto and K.-i. Watanabe [22, Theorem 1.2.3].) *(Recall that R is G-graded.) Let M be a graded R-module. Let $\mathfrak{p} \in \operatorname{Spec}(R)$ and let $d := \operatorname{ht} \mathfrak{p}/\mathfrak{p}^*$. Then*

$$\mu^i(\mathfrak{p}, M) = \begin{cases} 0 & \text{if } 0 \leq i < d, \\ \mu^{i-d}(\mathfrak{p}^*, M) & \text{if } i \geq d. \end{cases}$$

Proof.   By homogeneous localization at $\mathfrak{p}$, and use of the 'invariance' of Bass numbers under fraction formation (see [50, p.150]), we reduce to the case where $(R, \mathfrak{p}^*)$ is *local. Then $R/\mathfrak{p}^*$ is a *simple G-graded ring, and so, by 14.3.2, every graded $R/\mathfrak{p}^*$-module is *free. Also, d is the dimension of the local ring $R_{\mathfrak{p}}/\mathfrak{p}^* R_{\mathfrak{p}}$, and this is regular, by 14.3.3. Let $u_1, \ldots, u_d \in \mathfrak{p}$ be such that their natural images in $R_{\mathfrak{p}}/\mathfrak{p}^* R_{\mathfrak{p}}$, which we denote by $\alpha_1, \ldots, \alpha_d$ respectively, generate the maximal ideal of that regular local ring. Thus

$$\mathfrak{p} R_{\mathfrak{p}} = \mathfrak{p}^* R_{\mathfrak{p}} + (\alpha_1, \ldots, \alpha_d) R_{\mathfrak{p}}.$$

For each $j = 0, \ldots, d$, set $\mathfrak{Q}_j := \mathfrak{p}^* R_{\mathfrak{p}} + (\alpha_1, \ldots, \alpha_j) R_{\mathfrak{p}}$, and note that this is a prime ideal of $R_{\mathfrak{p}}$. Therefore, if $j < d$ there is an exact sequence of (ungraded) $R_{\mathfrak{p}}$-modules and $R_{\mathfrak{p}}$-homomorphisms

$$0 \longrightarrow R_{\mathfrak{p}}/\mathfrak{Q}_j \xrightarrow{\alpha_{j+1}} R_{\mathfrak{p}}/\mathfrak{Q}_j \longrightarrow R_{\mathfrak{p}}/\mathfrak{Q}_{j+1} \longrightarrow 0,$$

and this induces a long exact sequence

$$0 \longrightarrow \operatorname{Hom}_{R_{\mathfrak{p}}}(R_{\mathfrak{p}}/\mathfrak{Q}_{j+1}, M_{\mathfrak{p}}) \longrightarrow \operatorname{Hom}_{R_{\mathfrak{p}}}(R_{\mathfrak{p}}/\mathfrak{Q}_j, M_{\mathfrak{p}})$$

$$\xrightarrow{\alpha_{j+1}} \operatorname{Hom}_{R_{\mathfrak{p}}}(R_{\mathfrak{p}}/\mathfrak{Q}_j, M_{\mathfrak{p}}) \longrightarrow \operatorname{Ext}^1_{R_{\mathfrak{p}}}(R_{\mathfrak{p}}/\mathfrak{Q}_{j+1}, M_{\mathfrak{p}})$$

$$\longrightarrow \cdots$$

$$\longrightarrow \operatorname{Ext}^i_{R_{\mathfrak{p}}}(R_{\mathfrak{p}}/\mathfrak{Q}_{j+1}, M_{\mathfrak{p}}) \longrightarrow \operatorname{Ext}^i_{R_{\mathfrak{p}}}(R_{\mathfrak{p}}/\mathfrak{Q}_j, M_{\mathfrak{p}})$$

$$\xrightarrow{\alpha_{j+1}} \operatorname{Ext}^i_{R_{\mathfrak{p}}}(R_{\mathfrak{p}}/\mathfrak{Q}_j, M_{\mathfrak{p}}) \longrightarrow \operatorname{Ext}^{i+1}_{R_{\mathfrak{p}}}(R_{\mathfrak{p}}/\mathfrak{Q}_{j+1}, M_{\mathfrak{p}})$$

$$\longrightarrow \cdots.$$

Note that $\mathfrak{Q}_0 = \mathfrak{p}^* R_{\mathfrak{p}}$. Let $i \in \mathbb{N}_0$. Now $\operatorname{Ext}^i_R(R/\mathfrak{p}^*, M)$ is the R-module underlying the graded R-module $^* \operatorname{Ext}^i_R(R/\mathfrak{p}^*, M)$. The latter has a natural structure as graded $R/\mathfrak{p}^*$-module, and we observed above that every graded

$R/\mathfrak{p}^*$-module is *free. Consequently, $\mathrm{Ext}^i_R(R/\mathfrak{p}^*, M)$ is a free $R/\mathfrak{p}^*$-module; therefore α_1 is a non-zerodivisor on $\mathrm{Ext}^i_{R_\mathfrak{p}}(R_\mathfrak{p}/\mathfrak{Q}_0, M_\mathfrak{p})$ and, in view of the above long exact sequence (in the case where $j = 0$), we have

$$\mathrm{Ext}^i_{R_\mathfrak{p}}(R_\mathfrak{p}/\mathfrak{Q}_1, M_\mathfrak{p}) \cong \mathrm{Ext}^{i-1}_{R_\mathfrak{p}}(R_\mathfrak{p}/\mathfrak{Q}_0, M_\mathfrak{p})/\alpha_1 \, \mathrm{Ext}^{i-1}_{R_\mathfrak{p}}(R_\mathfrak{p}/\mathfrak{Q}_0, M_\mathfrak{p})$$

(as $R_\mathfrak{p}$-modules) for all $i \in \mathbb{Z}$. (When i is negative, the statement is obviously true.) Our immediate aim is to show that

$$\mathrm{Ext}^i_{R_\mathfrak{p}}(R_\mathfrak{p}/\mathfrak{Q}_j, M_\mathfrak{p})$$
$$\cong \mathrm{Ext}^{i-j}_{R_\mathfrak{p}}(R_\mathfrak{p}/\mathfrak{Q}_0, M_\mathfrak{p})/(\alpha_1, \ldots, \alpha_j) \, \mathrm{Ext}^{i-j}_{R_\mathfrak{p}}(R_\mathfrak{p}/\mathfrak{Q}_0, M_\mathfrak{p})$$

for all $i \in \mathbb{Z}$ for all $j = 2, \ldots, d$.

So suppose, inductively, that $j \in \{1, \ldots, d-1\}$ and we have shown that

$$\mathrm{Ext}^i_{R_\mathfrak{p}}(R_\mathfrak{p}/\mathfrak{Q}_j, M_\mathfrak{p})$$
$$\cong \mathrm{Ext}^{i-j}_{R_\mathfrak{p}}(R_\mathfrak{p}/\mathfrak{Q}_0, M_\mathfrak{p})/(\alpha_1, \ldots, \alpha_j) \, \mathrm{Ext}^{i-j}_{R_\mathfrak{p}}(R_\mathfrak{p}/\mathfrak{Q}_0, M_\mathfrak{p})$$

(as $R_\mathfrak{p}$-modules) for all $i \in \mathbb{Z}$. Since $\mathrm{Ext}^k_{R_\mathfrak{p}}(R_\mathfrak{p}/\mathfrak{Q}_0, M_\mathfrak{p})$ is a free $R_\mathfrak{p}/\mathfrak{Q}_0$-module for all $k \in \mathbb{Z}$, it follows that $\mathrm{Ext}^i_{R_\mathfrak{p}}(R_\mathfrak{p}/\mathfrak{Q}_j, M_\mathfrak{p})$ is a free $R_\mathfrak{p}/\mathfrak{Q}_j$-module for all $i \in \mathbb{Z}$. For each $i \in \mathbb{Z}$ it therefore follows that α_{j+1} is a non-zerodivisor on $\mathrm{Ext}^{i-1}_{R_\mathfrak{p}}(R_\mathfrak{p}/\mathfrak{Q}_j, M_\mathfrak{p})$, so that, in view of the above long exact sequence,

$$\mathrm{Ext}^i_{R_\mathfrak{p}}(R_\mathfrak{p}/\mathfrak{Q}_{j+1}, M_\mathfrak{p}) \cong \mathrm{Ext}^{i-1}_{R_\mathfrak{p}}(R_\mathfrak{p}/\mathfrak{Q}_j, M_\mathfrak{p})/\alpha_{j+1} \, \mathrm{Ext}^{i-1}_{R_\mathfrak{p}}(R_\mathfrak{p}/\mathfrak{Q}_j, M_\mathfrak{p})$$

(as $R_\mathfrak{p}$-modules). We can now use the inductive hypothesis to deduce that

$$\mathrm{Ext}^i_{R_\mathfrak{p}}(R_\mathfrak{p}/\mathfrak{Q}_{j+1}, M_\mathfrak{p})$$
$$\cong \mathrm{Ext}^{i-j-1}_{R_\mathfrak{p}}(R_\mathfrak{p}/\mathfrak{Q}_0, M_\mathfrak{p})/(\alpha_1, \ldots, \alpha_{j+1}) \, \mathrm{Ext}^{i-j-1}_{R_\mathfrak{p}}(R_\mathfrak{p}/\mathfrak{Q}_0, M_\mathfrak{p})$$

(as $R_\mathfrak{p}$-modules) for all $i \in \mathbb{Z}$. This completes the inductive step.

Since $\mathfrak{Q}_d = \mathfrak{p}R_\mathfrak{p}$, it therefore follows, by induction, that

$$\mathrm{Ext}^i_{R_\mathfrak{p}}(R_\mathfrak{p}/\mathfrak{p}R_\mathfrak{p}, M_\mathfrak{p})$$
$$\cong \mathrm{Ext}^{i-d}_{R_\mathfrak{p}}(R_\mathfrak{p}/\mathfrak{Q}_0, M_\mathfrak{p})/(\alpha_1, \ldots, \alpha_d) \, \mathrm{Ext}^{i-d}_{R_\mathfrak{p}}(R_\mathfrak{p}/\mathfrak{Q}_0, M_\mathfrak{p})$$
$$= \mathrm{Ext}^{i-d}_{R_\mathfrak{p}}(R_\mathfrak{p}/\mathfrak{p}^*R_\mathfrak{p}, M_\mathfrak{p})/(\alpha_1, \ldots, \alpha_d) \, \mathrm{Ext}^{i-d}_{R_\mathfrak{p}}(R_\mathfrak{p}/\mathfrak{p}^*R_\mathfrak{p}, M_\mathfrak{p})$$

(as $R_\mathfrak{p}$-modules) for all $i \in \mathbb{Z}$.

For each prime ideal $\mathfrak{q}$ of R, denote the field $R_\mathfrak{q}/\mathfrak{q}R_\mathfrak{q}$ by $k(\mathfrak{q})$. Let $i \in \mathbb{Z}$. Recall from [50, Theorem 18.7] that

$$\mu^i(\mathfrak{q}, M) = \dim_{k(\mathfrak{q})} \mathrm{Ext}^i_{R_\mathfrak{q}}(k(\mathfrak{q}), M_\mathfrak{q}) = \dim_{k(\mathfrak{q})} \left(\mathrm{Ext}^i_R(R/\mathfrak{q}, M) \right)_\mathfrak{q}.$$

(We interpret $\mu^j(\mathfrak{q}, M)$ for negative j as zero.) Now $\mathrm{Ext}^{i-d}_R(R/\mathfrak{p}^*, M)$ is a free $R/\mathfrak{p}*$-module, and it follows from that that $\operatorname{Ext}_{R_{\mathfrak{p}}}^{i-d}(R_{\mathfrak{p}}/\mathfrak{p}*R_{\mathfrak{p}}, M_{\mathfrak{p}})$ is a free $R_{\mathfrak{p}}/\mathfrak{p}*R_{\mathfrak{p}}$-module of the same rank. At this point, recall also that $\mathfrak{p}R_{\mathfrak{p}} = \mathfrak{p}*R_{\mathfrak{p}} + (\alpha_1, \ldots, \alpha_d)R_{\mathfrak{p}}$. It therefore follows from the above inductive argument that

$$
\begin{aligned}
\mu^{i-d}(\mathfrak{p}*, M) &= \dim_{k(\mathfrak{p}*)} \left(\operatorname{Ext}_R^{i-d}(R/\mathfrak{p}*, M) \right)_{\mathfrak{p}*} \\
&= \operatorname{rank}_{R/\mathfrak{p}*} \operatorname{Ext}_R^{i-d}(R/\mathfrak{p}*, M) \\
&= \operatorname{rank}_{R_{\mathfrak{p}}/\mathfrak{p}*R_{\mathfrak{p}}} \operatorname{Ext}_{R_{\mathfrak{p}}}^{i-d}(R_{\mathfrak{p}}/\mathfrak{p}*R_{\mathfrak{p}}, M_{\mathfrak{p}}) \\
&= \operatorname{rank}_{k(\mathfrak{p})} \operatorname{Ext}_{R_{\mathfrak{p}}}^{i}(R_{\mathfrak{p}}/\mathfrak{p}R_{\mathfrak{p}}, M_{\mathfrak{p}}) \\
&= \dim_{k(\mathfrak{p})} \operatorname{Ext}_{R_{\mathfrak{p}}}^{i}(k(\mathfrak{p}), M_{\mathfrak{p}}) = \mu^i(\mathfrak{p}, M).
\end{aligned}
$$

In particular, $\mu^i(\mathfrak{p}, M) = 0$ for all $i < d$. $\qquad\square$

When discussing depths of localizations of a finitely generated R-module, we shall follow the notation and conventions of 9.2.1.

14.3.8 Corollary. *Let M be a finitely generated graded R-module. Let $\mathfrak{p} \in \operatorname{Spec}(R)$. Then*

$$\operatorname{depth} M_{\mathfrak{p}} = \operatorname{depth} M_{\mathfrak{p}*} + \operatorname{ht} \mathfrak{p}/\mathfrak{p}*.$$

Proof. Recall from from [50, Theorems 16.7 and 18.7] that, for $\mathfrak{q} \in \operatorname{Spec}(R)$, $\operatorname{depth} M_{\mathfrak{q}}$ is the least integer i such that $\mu^i(\mathfrak{q}, M) \neq 0$ (if any such integers exist, and ∞ otherwise). The claim is therefore immediate from 14.3.7. $\qquad\square$

14.3.9 ‡Exercise. (Recall that R is G-graded.)

(i) Show that the following statements are equivalent:

    (a) R is Cohen–Macaulay;

    (b) $R_{(\mathfrak{p})}$ is Cohen–Macaulay for all $\mathfrak{p} \in {}^*\operatorname{Spec}(R)$;

    (c) $R_{\mathfrak{p}}$ is Cohen–Macaulay for all $\mathfrak{p} \in {}^*\operatorname{Spec}(R)$.

(ii) Show that the following statements are equivalent:

    (a) R is Gorenstein;

    (b) $R_{(\mathfrak{p})}$ is Gorenstein for all $\mathfrak{p} \in {}^*\operatorname{Spec}(R)$;

    (c) $R_{\mathfrak{p}}$ is Gorenstein for all $\mathfrak{p} \in {}^*\operatorname{Spec}(R)$.

We are now in a position to present G-graded versions of Faltings' Annihilator Theorem 9.5.1 and Grothendieck's Finiteness Theorem 9.5.2. Loosely, these graded versions say that, in the G-graded case, both theorems can be reformulated so that only graded prime ideals need be considered. As the Graded Finiteness Theorem has important geometric significance in connection with the cohomology of projective schemes, we provide a proof.

14.3.10 Graded Annihilator and Finiteness Theorems. *(Recall that R is G-graded.) Assume that R is a homomorphic image of a regular (commutative Noetherian) ring. Assume that the ideal $\mathfrak{a}$ is graded; let $\mathfrak{b}$ be a second graded ideal of R. Let M be a finitely generated graded R-module. Then*

$$f_\mathfrak{a}^\mathfrak{b}(M) = \inf\{\mathrm{adj}_\mathfrak{a} \operatorname{depth} M_\mathfrak{p} : \mathfrak{p} \in {}^*\operatorname{Spec}(R) \setminus \operatorname{Var}(\mathfrak{b})\}$$
$$= \inf\{\operatorname{depth} M_\mathfrak{p} + \mathrm{ht}(\mathfrak{a} + \mathfrak{p})/\mathfrak{p} : \mathfrak{p} \in {}^*\operatorname{Spec}(R) \setminus \operatorname{Var}(\mathfrak{b})\}.$$

In particular,

$$f_\mathfrak{a}(M) = \inf\{\operatorname{depth} M_\mathfrak{p} + \mathrm{ht}(\mathfrak{a} + \mathfrak{p})/\mathfrak{p} : \mathfrak{p} \in {}^*\operatorname{Spec}(R) \setminus \operatorname{Var}(\mathfrak{a})\}.$$

Proof. By Faltings' Annihilator Theorem 9.5.1,

$$f_\mathfrak{a}^\mathfrak{b}(M) = \inf\{\operatorname{depth} M_\mathfrak{p} + \mathrm{ht}(\mathfrak{a} + \mathfrak{p})/\mathfrak{p} : \mathfrak{p} \in \operatorname{Spec}(R) \setminus \operatorname{Var}(\mathfrak{b})\}.$$

It is therefore sufficient for us to show that, for each non-graded prime ideal $\mathfrak{p} \in \operatorname{Spec}(R) \setminus \operatorname{Var}(\mathfrak{b})$ for which $\operatorname{depth} M_\mathfrak{p}$ is finite, the graded ideal $\mathfrak{p}^*$ is such that $\mathfrak{p}^* \in {}^*\operatorname{Spec}(R) \setminus \operatorname{Var}(\mathfrak{b})$ and

$$\operatorname{depth} M_{\mathfrak{p}^*} + \mathrm{ht}(\mathfrak{a} + \mathfrak{p}^*)/\mathfrak{p}^* \leq \operatorname{depth} M_\mathfrak{p} + \mathrm{ht}(\mathfrak{a} + \mathfrak{p})/\mathfrak{p}.$$

This we do.

Of course $\mathfrak{p}^*$ is a graded prime ideal of R (see 13.1.3(vi)), and it is clear that $\mathfrak{p}^* \not\supseteq \mathfrak{b}$. Let $d := \mathrm{ht}\, \mathfrak{p}/\mathfrak{p}^*$; by 14.3.8, $\operatorname{depth} M_\mathfrak{p} = \operatorname{depth} M_{\mathfrak{p}^*} + d$.

Let $t := \mathrm{ht}(\mathfrak{a}+\mathfrak{p})/\mathfrak{p}$ and let $\mathfrak{q}$ be a minimal prime of $\mathfrak{a}+\mathfrak{p}$ such that $\mathrm{ht}\, \mathfrak{q}/\mathfrak{p} = \mathrm{ht}(\mathfrak{a} + \mathfrak{p})/\mathfrak{p}$. Since R is a homomorphic image of a regular ring, it is catenary, and therefore $\mathrm{ht}\, \mathfrak{q}/\mathfrak{p}^* = \mathrm{ht}\, \mathfrak{q}/\mathfrak{p} + \mathrm{ht}\, \mathfrak{p}/\mathfrak{p}^* = t + d$. Since $\mathfrak{a} + \mathfrak{p}^* \subseteq \mathfrak{a} + \mathfrak{p} \subseteq \mathfrak{q}$, it follows that $\mathrm{ht}(\mathfrak{a} + \mathfrak{p}^*)/\mathfrak{p}^* \leq \mathrm{ht}\, \mathfrak{q}/\mathfrak{p}^* = t + d$. Therefore

$$\operatorname{depth} M_{\mathfrak{p}^*} + \mathrm{ht}(\mathfrak{a} + \mathfrak{p}^*)/\mathfrak{p}^* \leq \operatorname{depth} M_\mathfrak{p} - d + t + d$$
$$= \operatorname{depth} M_\mathfrak{p} + \mathrm{ht}(\mathfrak{a} + \mathfrak{p})/\mathfrak{p},$$

and this completes the proof. $\square$

14.4 Graded local duality

The purpose of this section is to develop a G-graded analogue of the Local Duality Theorem 11.2.6. This will concern the situation where $(R, \mathfrak{m})$ is a *local G-graded ring that can be expressed as a homomorphic image of a Gorenstein *local G-graded (commutative Noetherian) ring by means of a homogeneous homomorphism. Many of the G-graded rings that occur in applications are finitely generated algebras over fields, and so we are only imposing a mild restriction.

Recall that we are assuming that R is G-graded throughout this chapter.

14.4.1 Graded Local Duality Theorem. *Assume that* $(R, \mathfrak{m})$ *is* *local with* ht $\mathfrak{m} = n$. *Assume also that there is a Gorenstein* *local G-graded commutative Noetherian ring* $(R', \mathfrak{m}')$ *and a surjective homogeneous ring homomorphism* $f : R' \longrightarrow R$. *Let* ht $\mathfrak{m}' = n'$. *Let* *D *denote the functor* *$\operatorname{Hom}_R(\bullet , {}^*E(R/\mathfrak{m}))$ *from* *$\mathcal{C}(R)$ *to itself.*

Let M *be a graded R-module, let* N' *be a graded R'-module, and let* $j \in \mathbb{N}_0$. *Now* M *can be regarded as a graded R'-module by means of* f; *then,* *$\operatorname{Ext}^j_{R'}(M, N')$ *has a natural structure as graded R-module.*

There exists $g \in G$ *such that* $H^{n'}_{\mathfrak{m}'}(R') \cong {}^*E_{R'}(R'/\mathfrak{m}')(-g)$ *in* *$\mathcal{C}(R')$. *For any such* g, *there is a homomorphism*

$$\Psi := (\psi^i)_{i \in \mathbb{N}_0} : \left(H^i_{\mathfrak{m}} \right)_{i \in \mathbb{N}_0} \longrightarrow \left({}^*D({}^*\operatorname{Ext}^{n'-i}_{R'}(\bullet , R'(g))) \right)_{i \in \mathbb{N}_0}$$

of negative connected sequences of covariant functors from *$\mathcal{C}(R)$ *to* *$\mathcal{C}(R)$ *which is such that* ψ^i_M *is a (necessarily homogeneous) isomorphism for all* $i \in \mathbb{N}_0$ *whenever* M *is a finitely generated graded R-module.*

Proof. We shall first deal with the special case where $R = R'$ (and f is the identity map), so that $(R, \mathfrak{m})$ itself is a Gorenstein *local G-graded ring. Since $\dim R_{\mathfrak{m}} = n$, it follows from Grothendieck's Vanishing Theorem 6.1.2 that $(H^j_{\mathfrak{m}}(N))_{\mathfrak{m}} \cong H^j_{\mathfrak{m} R_{\mathfrak{m}}}(N_{\mathfrak{m}}) = 0$ for all $j > n$ and all R-modules N. Therefore, by 14.1.2(ii), we have $H^j_{\mathfrak{m}}(M) = 0$ for all $j > n$ and all graded R-modules M. Consequently, $H^n_{\mathfrak{m}}$ is a right exact functor from *$\mathcal{C}(R)$ to itself, and $\left(H^{n-i}_{\mathfrak{m}} \right)_{i \in \mathbb{N}_0}$ is a positive strongly connected sequence of covariant functors from *$\mathcal{C}(R)$ to *$\mathcal{C}(R)$.

Let $g \in G$. Note that the long complex resulting from application of the positive connected sequence $\left({}^*D({}^*\operatorname{Ext}^i_R(\bullet , R(g))) \right)_{i \in \mathbb{N}_0}$ to a short exact sequence of graded R-modules and homogeneous homomorphisms

$$0 \longrightarrow L \longrightarrow M \longrightarrow N \longrightarrow 0$$

is exact. Thus $\left({}^*D({}^*\operatorname{Ext}^i_R(\bullet , R(g))) \right)_{i \in \mathbb{N}_0}$ is a positive strongly connected sequence of covariant functors from *$\mathcal{C}(R)$ to itself.

Our next aim is to find a natural equivalence between the functors $H^n_{\mathfrak{m}}$ and *$D({}^*\operatorname{Hom}_R(\bullet , R(g)))$ (from *$\mathcal{C}(R)$ to itself) for a suitable $g \in G$. Since R is Gorenstein, it follows from 14.2.16 that there is a family $(g_{\mathfrak{p}})_{\mathfrak{p} \in {}^*\operatorname{Spec}(R)}$ of elements of G such that there is a homogeneous isomorphism

$$
{}^*E^i_R(R) \cong \bigoplus_{\substack{\mathfrak{p} \in {}^*\operatorname{Spec}(R) \\ \operatorname{ht} \mathfrak{p} = i}} {}^*E_R(R/\mathfrak{p})(-g_{\mathfrak{p}}) \quad \text{for each } i \in \mathbb{N}_0.
$$

We can calculate the graded module $H^n_{\mathfrak{m}}(R)$ by applying $\Gamma_{\mathfrak{m}}$ to the minimal *injective resolution of R and then taking cohomology; in view of 14.2.4(iv), the result is that $H^n_{\mathfrak{m}}(R) \cong {}^*E_R(R/\mathfrak{m})(-g_{\mathfrak{m}})$ in $^*\mathcal{C}(R)$.

We can now modify the ideas of 11.2.4 to establish the existence of a natural transformation

$$\phi : H^n_{\mathfrak{m}} \longrightarrow {}^*\mathrm{Hom}_R({}^*\mathrm{Hom}_R(\,\bullet\,, R), H^n_{\mathfrak{m}}(R))$$

of functors from $^*\mathcal{C}(R)$ to itself, for which, for each graded R-module M, we have $(\phi_M(y))(f) = H^n_{\mathfrak{m}}(f)(y)$ for all $y \in H^n_{\mathfrak{m}}(M)$ and $f \in {}^*\mathrm{Hom}_R(M, R)$. One easily checks that ϕ_R is an isomorphism.

Choose $g \in G$ such that $H^n_{\mathfrak{m}}(R) \cong {}^*E_R(R/\mathfrak{m})(-g)$ in $^*\mathcal{C}(R)$. (For example, we could take $g := g_{\mathfrak{m}}$.) We obtain from ϕ a natural transformation

$$\psi_0 : H^n_{\mathfrak{m}} \longrightarrow {}^*\mathrm{Hom}_R({}^*\mathrm{Hom}_R(\,\bullet\,, R), {}^*E_R(R/\mathfrak{m})(-g))$$
$$= {}^*\mathrm{Hom}_R({}^*\mathrm{Hom}_R(\,\bullet\,, R(g)), {}^*E_R(R/\mathfrak{m}))$$

of functors from $^*\mathcal{C}(R)$ to itself, for which $\psi_{0\,R}$ is an isomorphism. Note that, since R is Gorenstein, $(H^{n-i}_{\mathfrak{m}}(R))_{\mathfrak{m}} \cong H^{n-i}_{\mathfrak{m}R_{\mathfrak{m}}}(R_{\mathfrak{m}}) = 0$ for all $i > 0$, so that $H^{n-i}_{\mathfrak{m}}(R) = 0$ for all $i > 0$ by 14.1.2(ii). Also $^*\mathrm{Ext}^i_{R'}(P, R'(g)) = 0$ for all $i > 0$ for all *projective graded R-modules P. It follows from the graded version of the analogue of 1.3.4 for positive connected sequences that ψ_0 can be incorporated into a (uniquely determined) homomorphism

$$\Phi := (\phi_i)_{i \in \mathbb{N}_0} : \left(H^{n-i}_{\mathfrak{m}}\right)_{i \in \mathbb{N}_0} \longrightarrow \left({}^*D({}^*\mathrm{Ext}^i_R(\,\bullet\,, R(g)))\right)_{i \in \mathbb{N}_0}$$

of positive connected sequences of covariant functors from $^*\mathcal{C}(R)$ to $^*\mathcal{C}(R)$. Furthermore, it is easy to prove by induction that, for each $i \in \mathbb{N}$, the homomorphism $\phi_{i\,M}$ is an isomorphism whenever M is a finitely generated graded R-module: use the fact that such an M can be included in an exact sequence $0 \longrightarrow K \longrightarrow F \longrightarrow M \longrightarrow 0$ in $^*\mathcal{C}(R)$ in which F is a finitely generated *free R-module. Note that we can interpret

$$(\phi_{n-i})_{i \in \mathbb{N}_0} : \left(H^i_{\mathfrak{m}}\right)_{i \in \mathbb{N}_0} \longrightarrow \left({}^*D({}^*\mathrm{Ext}^{n-i}_R(\,\bullet\,, R(g)))\right)_{i \in \mathbb{N}_0}$$

as a homomorphism of negative connected sequences.

We have thus established the claims of the theorem in the case where $R = R'$. We now deal with the general case, where there is a Gorenstein *local G-graded commutative Noetherian ring $(R', \mathfrak{m}')$ and a surjective homogeneous ring homomorphism $f : R' \longrightarrow R$. Let $^*D' := {}^*\mathrm{Hom}_{R'}(\,\bullet\,, {}^*E_{R'}(R'/\mathfrak{m}'))$. Our work so far in this proof establishes the existence of a $g \in G$ such that $H^{n'}_{\mathfrak{m}'}(R') \cong {}^*E_{R'}(R'/\mathfrak{m}')(-g)$ in $^*\mathcal{C}(R')$ and, for such a g, the existence of a homomorphism

$$\Psi := \left(\psi^i\right)_{i\in\mathbb{N}_0} : \left(H^i_{\mathfrak{m}'}\right)_{i\in\mathbb{N}_0} \longrightarrow \left(*D'(*\mathrm{Ext}^{n'-i}_{R'}(\bullet, R'(g)))\right)_{i\in\mathbb{N}_0}$$

of negative connected sequences of covariant functors from $*\mathcal{C}(R')$ to $*\mathcal{C}(R')$ such that $\phi_{i\,M}$ is a (necessarily homogeneous) isomorphism whenever M is a finitely generated graded R'-module. To complete the proof, we use the argument in the proof of the Local Duality Theorem 11.2.6, modified for the G-graded context. Important ingredients include the Graded Independence Theorem 14.1.7, and the fact that the graded submodule $\left(0 :_{*E_{R'}(R'/\mathfrak{m}')} \mathrm{Ker}\, f\right)$ of $*E_{R'}(R'/\mathfrak{m}')$ is, when viewed as a graded R-module, homogeneously isomorphic to $*E_R(R/\mathfrak{m})$. $\qquad\qquad\square$

Note. It should be noted that the element g in the statement of the Graded Local Duality Theorem 14.4.1 need not be uniquely determined. The interested reader is referred to [6, Lemma 1.5].

In the situation of 14.4.1, the functor $*D = *\mathrm{Hom}_R(\bullet, *E(R/\mathfrak{m}))$ would seem to be an obvious graded analogue of the functor of 10.2.1 used to construct Matlis duals over a local ring. However, there is another approach to $*D$ which is particularly useful when the 0-th component R_0 of the *local ring R is a field, and this approach is the subject of the next exercise. The exercise is based on [7, Proposition 3.6.16], but our approach is a little different from that of Bruns and Herzog; also, we are working in the G-graded context, whereas Bruns' and Herzog's treatment is for $\mathbb{Z}$-graded rings.

14.4.2 ♯Exercise: Graded Matlis Duality. Suppose $(R, \mathfrak{m})$ is *local. Note that R_0 is local with maximal ideal $\mathfrak{m}_0 := \mathfrak{m} \cap R_0$; let $E_0 := E_{R_0}(R_0/\mathfrak{m}_0)$.

(i) Let $M = \bigoplus_{g\in G} M_g$ be a graded R-module. Now R_0 can be considered as a G-graded ring with trivial grading, and any R_0-module can be considered as a graded R_0-module concentrated in degree 0. Also, the grading on our graded R-module M provides a structure as graded R_0-module on M.

    Show that $M^\vee := *\mathrm{Hom}_{R_0}(M, E_0)$ has a natural structure as a graded R-module with g-th component $(M^\vee)_g = \mathrm{Hom}_{R_0}(M_{-g}, E_0)$ for all $g \in G$. (Remember that $*\mathrm{Hom}_{R_0}(M, E_0)$ is a submodule of $\mathrm{Hom}_{R_0}(M, E_0)$, and the latter has a natural structure as an R-module.)

    Deduce that $(\bullet)^\vee$ is an exact, additive functor from $*\mathcal{C}(R)$ to itself.

(ii) Show that the functors $(\bullet)^\vee$ and $*\mathrm{Hom}_R(\bullet, R^\vee)$, that is, the functors $*\mathrm{Hom}_{R_0}(\bullet, E_0)$ and $*\mathrm{Hom}_R(\bullet, *\mathrm{Hom}_{R_0}(R, E_0))$, from $*\mathcal{C}(R)$ to itself, are naturally equivalent. (Here is a hint: if we forget the gradings, there is a natural equivalence

$$\kappa : \operatorname{Hom}_{R_0}(\bullet, E_0) \longrightarrow \operatorname{Hom}_R(\bullet, \operatorname{Hom}_{R_0}(R, E_0));$$

for a graded R-module M, consider an appropriate restriction of κ_M.)

(iii) Let $\phi_0 : R \longrightarrow E_0$ be the R_0-homomorphism whose restriction to R_0 is the composition of the natural epimorphism $R_0 \to R_0/\mathfrak{m}_0$ and the inclusion $R_0/\mathfrak{m}_0 \to E_0 = E_{R_0}(R_0/\mathfrak{m}_0)$, and whose restriction to R_g for all $g \in G \setminus \{0\}$ is zero. Show that $R^\vee$ is a *essential extension of its graded submodule $R\phi_0$.

(iv) Show that $R^\vee \cong {}^*E_R(R/\mathfrak{m})$ (in the category ${}^*\mathcal{C}(R)$), and deduce that the functors $(\bullet)^\vee$ and ${}^*D := {}^*\operatorname{Hom}_R(\bullet, {}^*E(R/\mathfrak{m}))$ (from ${}^*\mathcal{C}(R)$ to itself) are naturally equivalent.

(v) Show that, if the local ring R_0 is complete, then, whenever M is a finitely generated graded R-module, there is a homogeneous R-isomorphism $M \cong (M^\vee)^\vee =: M^{\vee\vee}$.

Thus, in the situation of Exercise 14.4.2, it is reasonable for us to regard the functor $(\bullet)^\vee = {}^*\operatorname{Hom}_{R_0}(\bullet, E_{R_0}(R_0/\mathfrak{m}_0))$ as the 'graded Matlis duality functor'. In the particular case in which R_0 is a field K (and this is the situation in many practical applications of graded ring theory), $E_{R_0}(R_0/\mathfrak{m}_0) = K$, and, for a graded R-module $M = \bigoplus_{g \in G} M_g$, the grading of the 'graded Matlis dual' is given by the attractively simple formula

$$M^\vee = {}^*\operatorname{Hom}_K(M, K) = \bigoplus_{g \in G} \operatorname{Hom}_K(M_{-g}, K).$$

14.5 *Canonical modules

Recall that we are assuming that R is G-graded throughout this chapter. Also, recall from 12.1.2 that a canonical module for a local ring $(R', \mathfrak{m}')$ is a finitely generated R'-module K such that $\operatorname{Hom}_{R'}(K, E_{R'}(R'/\mathfrak{m}')) \cong H_{\mathfrak{m}'}^{\dim R'}(R')$. The obvious graded analogue is given in the following definition.

14.5.1 Definition. Suppose that $(R, \mathfrak{m})$ is *local; let ht $\mathfrak{m} = n$. A *canonical module for R is a finitely generated graded R-module C for which there is a homogeneous isomorphism

$${}^*\operatorname{Hom}_R(C, {}^*E_R(R/\mathfrak{m})) \cong H_{\mathfrak{m}}^n(R).$$

14.5.2 Example. Assume that $(R, \mathfrak{m})$ is *local with ht $\mathfrak{m} = n$, and that there is a Gorenstein *local G-graded commutative Noetherian ring $(R', \mathfrak{m}')$ with ht $\mathfrak{m}' = n'$ and a surjective homogeneous ring homomorphism $f : R' \longrightarrow R$.

It follows from the Graded Local Duality Theorem 14.4.1 that, for some $g \in G$, there is a homogeneous isomorphism

$$* \mathrm{Hom}(* \mathrm{Ext}_{R'}^{n'-n}(R, R'(g)), *E_R(R/\mathfrak{m})) \cong H_{\mathfrak{m}}^n(R).$$

Therefore $* \mathrm{Ext}_{R'}^{n'-n}(R, R'(g))$ is a *canonical module for R.

14.5.3 Proposition. *Assume that* $(R, \mathfrak{m})$ *is *local with a *canonical module* C. *Then* $C_{\mathfrak{m}}$ *is a canonical module for* $R_{\mathfrak{m}}$.

Proof. By 13.2.4(iii), the R-module $*E_R(R/\mathfrak{m})$ (with its grading forgotten) is an essential extension of $R/\mathfrak{m}$, and so there is a monomorphism α : $*E_R(R/\mathfrak{m}) \longrightarrow E_R(R/\mathfrak{m})$. Denote $*E_R(R/\mathfrak{m})$ by $*E$ and $E_R(R/\mathfrak{m})$ by E. Let $K := \mathrm{Coker}\,\alpha$, and let $k(\mathfrak{m})$ denote the residue field of the local ring $R_{\mathfrak{m}}$. When the grading on $*E$ is forgotten, we have

$$\mathrm{Ext}_{R_{\mathfrak{m}}}^1(k(\mathfrak{m}), (*E)_{\mathfrak{m}}) \cong (\mathrm{Ext}_R^1(R/\mathfrak{m}, *E))_{\mathfrak{m}} = 0,$$

by 13.2.5. The exact sequence $0 \longrightarrow *E \xrightarrow{\alpha} E \longrightarrow K \longrightarrow 0$ therefore induces an exact sequence

$$0 \longrightarrow \mathrm{Hom}_{R_{\mathfrak{m}}}(k(\mathfrak{m}), (*E)_{\mathfrak{m}}) \longrightarrow \mathrm{Hom}_{R_{\mathfrak{m}}}(k(\mathfrak{m}), E_{\mathfrak{m}})$$
$$\longrightarrow \mathrm{Hom}_{R_{\mathfrak{m}}}(k(\mathfrak{m}), K_{\mathfrak{m}}) \longrightarrow 0.$$

Since $\mathrm{Hom}_{R_{\mathfrak{m}}}(k(\mathfrak{m}), E_{\mathfrak{m}})$ is a 1-dimensional vector space over $k(\mathfrak{m})$ and

$$\mathrm{Hom}_{R_{\mathfrak{m}}}(k(\mathfrak{m}), (*E)_{\mathfrak{m}}) \neq 0,$$

we must have $\mathrm{Hom}_{R_{\mathfrak{m}}}(k(\mathfrak{m}), K_{\mathfrak{m}}) = 0$; therefore $K_{\mathfrak{m}} = 0$, since $K_{\mathfrak{m}}$ is an Artinian $R_{\mathfrak{m}}$-module.

Therefore $\alpha_{\mathfrak{m}}$ is an isomorphism.

Since C is a *canonical module for R, there is a homogeneous R-isomorphism $* \mathrm{Hom}_R(C, *E_R(R/\mathfrak{m})) \cong H_{\mathfrak{m}}^n(R)$, where $n = \mathrm{ht}\,\mathfrak{m}$. Now forget the gradings, localize at $\mathfrak{m}$, and use the isomorphism $\alpha_{\mathfrak{m}}$: there result isomorphisms of $R_{\mathfrak{m}}$-modules

$$\mathrm{Hom}_{R_{\mathfrak{m}}}(C_{\mathfrak{m}}, E_{R_{\mathfrak{m}}}(R_{\mathfrak{m}}/\mathfrak{m}R_{\mathfrak{m}})) \cong \mathrm{Hom}_{R_{\mathfrak{m}}}(C_{\mathfrak{m}}, (E_R(R/\mathfrak{m}))_{\mathfrak{m}})$$
$$\cong \mathrm{Hom}_{R_{\mathfrak{m}}}(C_{\mathfrak{m}}, (*E_R(R/\mathfrak{m}))_{\mathfrak{m}})$$
$$\cong (\mathrm{Hom}_R(C, *E_R(R/\mathfrak{m})))_{\mathfrak{m}}$$
$$\cong (H_{\mathfrak{m}}^n(R))_{\mathfrak{m}} \cong H_{\mathfrak{m}R_{\mathfrak{m}}}^n(R_{\mathfrak{m}}).$$

Therefore $C_{\mathfrak{m}}$ is a canonical module for $R_{\mathfrak{m}}$. $\square$

Limitations on space mean that we are not able, in this book, to develop the theory of *canonical modules in generality similar to that of Chapter 12. As a fairly short treatment is possible in the Cohen–Macaulay case, we content ourselves with that.

14.5.4 Theorem. *Assume that* $(R, \mathfrak{m})$ *is* *local and Cohen–Macaulay and that* C *is a finitely generated graded R-module such that* $\mu^i(\mathfrak{m}, C) = \delta_{i,\mathrm{ht}\,\mathfrak{m}}$ *for all* $i \in \mathbb{N}_0$. *Then* $C_\mathfrak{p}$ *is a canonical module for* $R_\mathfrak{p}$ *for all (graded or ungraded)* $\mathfrak{p} \in \mathrm{Spec}(R)$; *that is, C is a canonical module for R in the sense of* 12.1.28. *Consequently, for all* $\mathfrak{p} \in \mathrm{Spec}(R)$, *we have* $\mu^i(\mathfrak{p}, C) = \delta_{i,\mathrm{ht}\,\mathfrak{p}}$ *for all* $i \in \mathbb{N}_0$.

Proof. It follows from 12.1.27 that $C_\mathfrak{m}$ is a canonical module for $R_\mathfrak{m}$. Let $\mathfrak{p} \in \mathrm{Spec}(R)$, so that $\mathfrak{p}^* \in {}^*\mathrm{Spec}(R)$ and $\mathfrak{p}^* \subseteq \mathfrak{m}$. It also follows from 12.1.27 that $C_{\mathfrak{p}^*}$ is a canonical module for $R_{\mathfrak{p}^*}$, so that $\mu^i(\mathfrak{p}^*, C) = \delta_{i,\mathrm{ht}\,\mathfrak{p}^*}$ for all $i \in \mathbb{N}_0$. It now follows from the Goto–Watanabe Theorems 14.3.6 and 14.3.7 that $\mu^i(\mathfrak{p}, C) = \delta_{i,\mathrm{ht}\,\mathfrak{p}}$ for all $i \in \mathbb{N}_0$. Therefore $C_\mathfrak{p}$ is a canonical module for $R_\mathfrak{p}$, by 12.1.27 again. $\qquad\square$

14.5.5 Corollary. *Assume that* $(R, \mathfrak{m})$ *is* *local and Cohen–Macaulay and that C is a* *canonical module for R. Then C is a canonical module for R in the sense of* 12.1.28.

Proof. It follows from 14.5.3 that $C_\mathfrak{m}$ is a canonical module for $R_\mathfrak{m}$. Therefore, by 12.1.26, we have $\mu^i(\mathfrak{m}, C) = \mu^i(\mathfrak{m} R_\mathfrak{m}, C_\mathfrak{m}) = \delta_{i,\mathrm{ht}\,\mathfrak{m}}$ for all $i \in \mathbb{N}_0$. Now 14.5.4 shows that C is a canonical module for R. $\qquad\square$

14.5.6 Lemma. *Assume that R is Cohen–Macaulay and has a canonical module C. We saw in* 12.1.30 *that the trivial extension $R' = R \propto C$ of R by C is a Gorenstein ring, so that R is a homomorphic image of a Gorenstein ring.*

If $C = \oplus_{g \in G} C_g$ is graded, then R' is G-graded in such a way that the canonical homomorphism $\phi : R' \longrightarrow R$ is homogeneous. If, in addition, R is *local, then so too is R'.*

Proof. The decomposition $R' = \oplus_{g \in G}(R_g \oplus C_g)$ provides a grading on R', and ϕ is homogeneous with respect to this grading and the original grading on R. Furthermore, if $\mathfrak{m}$ is the unique *maximal graded ideal of R, then a routine check shows that $\mathfrak{m} \oplus C$ is the unique *maximal graded ideal of R'. $\qquad\square$

The following exercise will be very helpful in our development of the theory of *canonical modules.

14.5.7 ‡Exercise. Let $\alpha : M \longrightarrow N$ be a homogeneous homomorphism of graded R-modules.

(i) Show that, if $\alpha_\mathfrak{p} : M_\mathfrak{p} \longrightarrow N_\mathfrak{p}$ is an isomorphism for all $\mathfrak{p} \in {}^*\mathrm{Spec}(R)$, then α is an isomorphism.

(ii) Show that, if $(R, \mathfrak{m})$ is *local and $\alpha_{\mathfrak{m}} : M_{\mathfrak{m}} \longrightarrow N_{\mathfrak{m}}$ is an isomorphism, then α is an isomorphism.

When $(R, \mathfrak{m})$ is *local and Cohen–Macaulay, a *canonical module C for R is a canonical module for R (by 14.5.5), and so satisfies $\mu^i(\mathfrak{p}, C) = \delta_{i, \text{ht } \mathfrak{p}}$ for all $i \in \mathbb{N}_0$ and $\mathfrak{p} \in \text{Spec}(R)$. We are now going to study a general finitely generated graded R-module with the property that, when its grading is forgotten, it is a canonical module for R.

14.5.8 Lemma. *Assume* $(R, \mathfrak{m})$ *is *local and Cohen–Macaulay; set* $n :=$ $\text{ht } \mathfrak{m}$. *Suppose that* C *is a finitely generated graded R-module which is a canonical module for R in the sense of* 12.1.28, *so that, for all* $\mathfrak{p} \in {}^* \text{Spec}(R)$, *we have* $\mu^i(\mathfrak{p}, C) = \delta_{i, \text{ht } \mathfrak{p}}$.

(i) *There is a family* $(g_{\mathfrak{p}})_{\mathfrak{p} \in {}^* \text{Spec}(R)}$ *of elements of G for which there exist homogeneous isomorphisms*

$$*E^i(C) \cong \bigoplus_{\substack{\mathfrak{p} \in {}^* \text{Spec}(R) \\ \text{ht } \mathfrak{p} = i}} *E(R/\mathfrak{p})(-g_{\mathfrak{p}}) \quad \text{for all } i \in \mathbb{N}_0.$$

(ii) *For any family* $(g_{\mathfrak{p}})_{\mathfrak{p} \in {}^* \text{Spec}(R)}$ *of elements of G as in part* (i), *there are homogeneous isomorphisms*

 (a) $* \text{Ext}_R^i(R/\mathfrak{m}, C) \cong 0$ *for* $i \neq n$,
 (b) $* \text{Ext}_R^n(R/\mathfrak{m}, C) \cong (R/\mathfrak{m})(-g_{\mathfrak{m}})$, *and*
 (c) $H_{\mathfrak{m}}^n(C) \cong *E(R/\mathfrak{m})(-g_{\mathfrak{m}})$.

Proof. (i) This is immediate from Theorem 14.2.16.

(ii) Let $E^{\bullet}$ denote the minimal *injective resolution of C.

For each $\mathfrak{p} \in * \text{Spec}(R) \setminus \{\mathfrak{m}\}$, there exists a homogeneous element $r_{\mathfrak{p}} \in \mathfrak{m} \setminus \mathfrak{p}$, and so it follows from 14.2.4(iv) that $* \text{Hom}_R(R/\mathfrak{m}, *E(R/\mathfrak{p})) = 0$ and $\Gamma_{\mathfrak{m}}(*E(R/\mathfrak{p})) = 0$. Hence the complex $* \text{Hom}_R(R/\mathfrak{m}, E^{\bullet})$ has all terms other than its n-th equal to 0, while its n-th term is isomorphic (in $*\mathcal{C}(R)$) to $* \text{Hom}_R(R/\mathfrak{m}, *E(R/\mathfrak{m})(-g_{\mathfrak{m}}))$. Parts (a) and (b) therefore follow from 14.2.15 and 14.2.9, while part (c) is a consequence of the fact (see 13.4.3) that we can calculate the graded R-module $H_{\mathfrak{m}}^n(C)$ by application of the functor $\Gamma_{\mathfrak{m}}$ to $E^{\bullet}$. $\qquad \square$

Recall that, in 12.1.6, we proved that any two canonical modules for a local ring are isomorphic. We are now going to address the analogous issue in the Cohen–Macaulay G-graded *local case.

14.5.9 Theorem. *Suppose* $(R, \mathfrak{m})$ *is Cohen–Macaulay and *local,* *and let C, C' be graded R-modules that are canonical modules for R in the sense of* 12.1.28. *Let* $n := \operatorname{ht} \mathfrak{m}$. *Then*

(i) *the map* $\beta : R \longrightarrow {}^*\operatorname{Hom}_R(C, C)$ *defined by* $\beta(r) = r \operatorname{Id}_C$ *for all* $r \in R$ *is a homogeneous isomorphism; and*

(ii) *there exist* $g \in G$ *and a homogeneous isomorphism* $\phi : C \xrightarrow{\cong} C'(g)$.

Proof. (i) For each $\mathfrak{p} \in \operatorname{Spec}(R)$, the $R_{\mathfrak{p}}$-module $C_{\mathfrak{p}}$ is a canonical module for $R_{\mathfrak{p}}$. We can now deduce from 12.2.6 that $\beta_{\mathfrak{p}}$ is an isomorphism for all $\mathfrak{p} \in \operatorname{Spec}(R)$. Hence β is an isomorphism, and it is clearly homogeneous.

(ii) Let H denote the graded R-module ${}^*\operatorname{Hom}_R(C, C')$. It follows from part (i) above and 12.1.6 that $H_{\mathfrak{p}}$ is a free $R_{\mathfrak{p}}$-module of rank 1, for all (graded or ungraded) $\mathfrak{p} \in \operatorname{Spec}(R)$.

Let $\{\phi_1, \ldots, \phi_t\}$ be a generating set, consisting of t homogeneous elements, for H, that is minimal in the sense that no proper subset of it also generates H. Let $\deg \phi_i = g_i$ for $i = 1, \ldots, t$.

There is a graded *free R-module F, of the form $\bigoplus_{i=1}^{t} R(-g_i)$, and a homogeneous R-epimorphism $\psi : F \longrightarrow H$ which maps the generator, e_i say, of degree g_i in $R(-g_i)$ to ϕ_i.

Let $K := \operatorname{Ker} \psi$, a graded submodule of F. Our immediate aim is to show that $K \subseteq \mathfrak{m}F$. Suppose that this is not the case; then there exists a homogeneous element $k \in K \setminus \mathfrak{m}F$. Since k is homogeneous, we can write $k = \sum_{i=1}^{t} r_i e_i$, where, for $i \in \{1, \ldots, t\}$, the element $r_i \in R$ is homogeneous of degree $\deg k - g_i$. Then there exists $j \in \{1, \ldots, t\}$ such that $r_j \notin \mathfrak{m}$, so that, since r_j is homogeneous, it must be a unit of R. Application of ψ therefore yields that $0 = \psi(k) = \sum_{i=1}^{t} r_i \phi_i$, so that ϕ_j is an R-linear combination of the other ϕ_i and we have a contradiction to the minimality.

Therefore $K \subseteq \mathfrak{m}F$, and it follows from this that the maps $F/\mathfrak{m}F \longrightarrow H/\mathfrak{m}H$ and $F_{\mathfrak{m}}/\mathfrak{m}R_{\mathfrak{m}}F_{\mathfrak{m}} \longrightarrow H_{\mathfrak{m}}/\mathfrak{m}R_{\mathfrak{m}}H_{\mathfrak{m}}$ induced by ψ are both isomorphisms. Therefore $F_{\mathfrak{m}}/\mathfrak{m}R_{\mathfrak{m}}F_{\mathfrak{m}}$ and $H_{\mathfrak{m}}/\mathfrak{m}R_{\mathfrak{m}}H_{\mathfrak{m}}$ have equal dimensions as vector spaces over $R_{\mathfrak{m}}/\mathfrak{m}R_{\mathfrak{m}}$.

Now $\operatorname{Ext}^1_R(H, K)_{\mathfrak{p}} \cong \operatorname{Ext}^1_{R_{\mathfrak{p}}}(H_{\mathfrak{p}}, K_{\mathfrak{p}}) = 0$ for all $\mathfrak{p} \in \operatorname{Spec}(R)$, and so ${}^*\operatorname{Ext}^1_R(H, K) = 0$. Therefore the exact sequence

$$0 \longrightarrow K \xrightarrow{\subseteq} F \xrightarrow{\psi} H \longrightarrow 0$$

splits in ${}^*\mathcal{C}(R)$. It follows that there is an $R_{\mathfrak{m}}$-isomorphism

$$F_{\mathfrak{m}}/\mathfrak{m}R_{\mathfrak{m}}F_{\mathfrak{m}} \cong K_{\mathfrak{m}}/\mathfrak{m}R_{\mathfrak{m}}K_{\mathfrak{m}} \oplus H_{\mathfrak{m}}/\mathfrak{m}R_{\mathfrak{m}}H_{\mathfrak{m}},$$

and therefore our calculations above with vector space dimensions show that

$K_{\mathfrak{m}}/\mathfrak{m}R_{\mathfrak{m}}K_{\mathfrak{m}} = 0$. Therefore $K_{\mathfrak{m}} = 0$ by Nakayama's Lemma, so that $K = 0$ by 14.1.2(ii).

Hence H is *free. Its rank must be 1, so that there exists a homogeneous element $\phi \in H$, of degree g say, which forms a base for H. Thus $\phi : C \longrightarrow C'(g)$ is a homogeneous R-homomorphism. For each $\mathfrak{p} \in \mathrm{Spec}(R)$, the $R_{\mathfrak{p}}$-homomorphism $\phi_{\mathfrak{p}} : C_{\mathfrak{p}} \longrightarrow C'_{\mathfrak{p}}$ generates $\mathrm{Hom}_{R_{\mathfrak{p}}}(C_{\mathfrak{p}}, C'_{\mathfrak{p}})$. By 12.1.6, there is an $R_{\mathfrak{p}}$-isomorphism $\lambda : C_{\mathfrak{p}} \longrightarrow C'_{\mathfrak{p}}$ in $\mathrm{Hom}_{R_{\mathfrak{p}}}(C_{\mathfrak{p}}, C'_{\mathfrak{p}})$; therefore, there exist $r \in R$ and $s \in R \setminus \mathfrak{p}$ such that $\lambda = (r/s)\phi_{\mathfrak{p}}$. It follows that $\phi_{\mathfrak{p}}$ is surjective, so that $\lambda^{-1} \circ \phi_{\mathfrak{p}} : C_{\mathfrak{p}} \longrightarrow C_{\mathfrak{p}}$ is a surjective endomorphism, and therefore an isomorphism. Consequently, $\phi_{\mathfrak{p}}$ is an isomorphism. As this is true for each $\mathfrak{p} \in \mathrm{Spec}(R)$, it follows that $\phi : C \longrightarrow C'(g)$ is a homogeneous isomorphism.   □

14.5.10 Theorem. *Assume that $(R, \mathfrak{m})$ is Cohen–Macaulay and *local with $\mathrm{ht}\,\mathfrak{m} = n$, and admits a *canonical module C. Set $*E := *E(R/\mathfrak{m})$, and let $*D := *\mathrm{Hom}_R(\bullet, *E)$.*

There is a natural transformation

$$\phi_0 : H_{\mathfrak{m}}^n \longrightarrow *D(*\mathrm{Hom}_R(\bullet, C))$$

of functors from $*\mathcal{C}(R)$ to itself which is such that $\phi_{0\,M}$ is an isomorphism whenever M is a finitely generated graded R-module.

There is a unique extension of ϕ_0 to a homomorphism

$$\Phi := (\phi_i)_{i \in \mathbb{N}_0} : \left(H_{\mathfrak{m}}^{n-i}\right)_{i \in \mathbb{N}_0} \longrightarrow \left(*D(*\mathrm{Ext}_R^i(\bullet, C))\right)_{i \in \mathbb{N}_0}$$

of (positive strongly) connected sequences of covariant functors from $*\mathcal{C}(R)$ to $*\mathcal{C}(R)$. Furthermore, $\phi_{i\,M}$ is an isomorphism for all $i \in \mathbb{N}_0$ whenever M is a finitely generated graded R-module.

In particular, for each finitely generated graded R-module M, there are homogeneous isomorphisms

$$H_{\mathfrak{m}}^{n-i}(M) \cong *D(*\mathrm{Ext}_R^i(M, C)) \quad \text{for all } i \in \mathbb{Z}.$$

Proof.   As in the proof of the Graded Local Duality Theorem 14.4.1, we can show that $H_{\mathfrak{m}}^j(M) = 0$ for all $j > n$ and all graded R-modules M, that $H_{\mathfrak{m}}^n$ is a right exact functor from $*\mathcal{C}(R)$ to itself, and that $\left(H_{\mathfrak{m}}^{n-i}\right)_{i \in \mathbb{N}_0}$ is a positive strongly connected sequence of covariant functors from $*\mathcal{C}(R)$ to $*\mathcal{C}(R)$.

Next, the ideas of 6.1.9 and 6.1.10 can be modified to show that the functors $H_{\mathfrak{m}}^n$ and $(\bullet) \otimes_R H_{\mathfrak{m}}^n(R)$, from $*\mathcal{C}(R)$ to itself, are naturally equivalent. Since C is a *canonical module for R, there is a homogeneous isomorphism $*\mathrm{Hom}_R(C, *E) \cong H_{\mathfrak{m}}^n(R)$; therefore, the functors

$$H_{\mathfrak{m}}^n \quad \text{and} \quad (\bullet) \otimes_R *\mathrm{Hom}_R(C, *E),$$

from $*\mathcal{C}(R)$ to itself, are naturally equivalent.

Next, recall the natural transformation of functors

$$\xi_{\bullet,\bullet,\bullet} : (\bullet) \otimes_R \operatorname{Hom}_R(\bullet,\bullet) \longrightarrow \operatorname{Hom}_R(\operatorname{Hom}_R(\bullet,\bullet),\bullet)$$

(from $\mathcal{C}(R) \times \mathcal{C}(R) \times \mathcal{C}(R)$ to $\mathcal{C}(R)$) of 10.2.16: it is such that, for R-modules M, I and J, we have $(\xi_{M,I,J}(m \otimes f))(g) = f(g(m))$ for $m \in M$, $f \in \operatorname{Hom}_R(I,J)$ and $g \in \operatorname{Hom}_R(M,I)$. Take C for I and $*E$ for J; then, when M is graded, $\xi_{M,C,*E}$ maps $M \otimes_R *\operatorname{Hom}_R(C,*E)$ into

$$*\operatorname{Hom}_R(*\operatorname{Hom}_R(M,C),*E);$$

one can easily check that degrees are preserved. Set $\psi_{0\,M} := \xi_{M,C,*E}$ for each graded R-module M. Then

$$\psi_0 : (\bullet) \otimes_R *\operatorname{Hom}_R(C,*E) \longrightarrow *\operatorname{Hom}_R(*\operatorname{Hom}_R(\bullet,C),*E)$$

is a natural transformation of functors from $*\mathcal{C}(R)$ to itself. Moreover, one can modify the argument in the proof of 10.2.16 to show that $\psi_{0\,M}$ is an isomorphism whenever M is a finitely generated graded R-module: use of 13.2.5(i) enables one to see that, whenever $A \longrightarrow B \longrightarrow L$ is an exact sequence in the category $*\mathcal{C}(R)$, then the induced sequence

$$*\operatorname{Hom}_R(L,*E) \longrightarrow *\operatorname{Hom}_R(B,*E) \longrightarrow *\operatorname{Hom}_R(A,*E)$$

is again exact.

We can compose ψ_0 with a natural equivalence from the second paragraph of this proof to obtain a natural transformation

$$\phi_0 : H_{\mathfrak{m}}^n \longrightarrow *D(*\operatorname{Hom}_R(\bullet,C))$$

of functors (from $*\mathcal{C}(R)$ to itself) with the property that $\phi_{0\,M}$ is an isomorphism whenever M is a finitely generated graded R-module.

We now reason as in the proof of 14.4.1. Since R is Cohen–Macaulay, $(H_{\mathfrak{m}}^{n-i}(R))_{\mathfrak{m}} \cong H_{\mathfrak{m}R_{\mathfrak{m}}}^{n-i}(R_{\mathfrak{m}}) = 0$ for all $i > 0$, so that $H_{\mathfrak{m}}^{n-i}(R) = 0$ for all $i > 0$ by 14.1.2(ii). Also $*\operatorname{Ext}_R^i(P,C) = 0$ for all $i > 0$ for all *projective graded R-modules P. It follows from the graded version of the analogue of 1.3.4 for positive connected sequences that ϕ_0 can be incorporated into a (uniquely determined) homomorphism

$$\Phi := (\phi_i)_{i\in\mathbb{N}_0} : \left(H_{\mathfrak{m}}^{n-i}\right)_{i\in\mathbb{N}_0} \longrightarrow \left(*D(*\operatorname{Ext}_R^i(\bullet,C))\right)_{i\in\mathbb{N}_0}$$

of positive connected sequences of covariant functors from $*\mathcal{C}(R)$ to $*\mathcal{C}(R)$. Furthermore, it is easy to prove by induction that, for each $i \in \mathbb{N}$, the homomorphism $\phi_{i\,M}$ is an isomorphism whenever M is a finitely generated graded R-module: use the fact that such an M can be included in an exact sequence

$0 \longrightarrow K \longrightarrow F \longrightarrow M \longrightarrow 0$ in $*\mathcal{C}(R)$ in which F is a finitely generated *free R-module.

In order to complete the proof of the final claim, one should note that, in view of 14.1.2(ii), the graded module $* \operatorname{Ext}_R^j(M, C) = 0$ for all $j > n$ for each finitely generated graded R-module M, because $(* \operatorname{Ext}_R^j(M, C))_\mathfrak{m} = 0$ since $\operatorname{inj} \dim_{R_\mathfrak{m}} C_\mathfrak{m} = n$ by 12.1.21. $\qquad\square$

14.5.11 Corollary. *Assume that* $(R, \mathfrak{m})$ *is Cohen–Macaulay and* *local with* $\operatorname{ht} \mathfrak{m} = n$, *and that it admits a* *canonical module* C. *Then there is a homogeneous isomorphism* $H_\mathfrak{m}^n(C) \overset{\cong}{\longrightarrow} *E(R/\mathfrak{m})$.

Proof. By 14.5.10, there is a homogeneous isomorphism

$$H_\mathfrak{m}^n(C) \cong * \operatorname{Hom}_R(* \operatorname{Hom}_R(C, C), *E(R/\mathfrak{m})).$$

Since C is a canonical module for R (by 14.5.5), it follows from 14.5.9(i) that there is a homogeneous isomorphism $R \cong * \operatorname{Hom}_R(C, C)$, and the desired result follows. $\qquad\square$

In the case where $(R, \mathfrak{m})$ is a Cohen–Macaulay *local $\mathbb{Z}$-graded ring, Bruns and Herzog in [7, 3.6.8] gave a definition of *canonical module for R different from ours. We are now in a position to reconcile these two approaches.

14.5.12 Corollary. *Assume the* G-graded ring $(R, \mathfrak{m})$ *is* *local and Cohen–Macaulay; set* $\operatorname{ht} \mathfrak{m} = n$. *Let* C *be a finitely generated graded* R-module. *Then* C *is a* *canonical module for* R *if and only if there are homogeneous isomorphisms*

$$* \operatorname{Ext}_R^i(R/\mathfrak{m}, C) \cong \begin{cases} 0 & \text{for } i \neq n, \\ R/\mathfrak{m} & \text{for } i = n. \end{cases}$$

Proof. ($\Rightarrow$) When C is a *canonical module for R, it follows from 14.5.10 that there are homogeneous isomorphisms

$$H_\mathfrak{m}^{n-i}(R/\mathfrak{m}) \cong * \operatorname{Hom}_R(* \operatorname{Ext}_R^i(R/\mathfrak{m}, C), *E(R/\mathfrak{m})) \quad \text{for all } i \in \mathbb{Z}.$$

Since, for $i \neq n$, we have $(H_\mathfrak{m}^{n-i}(R/\mathfrak{m}))_\mathfrak{m} = 0$, it follows from 14.1.2(ii) that $H_\mathfrak{m}^{n-i}(R/\mathfrak{m}) = 0$, so that $* \operatorname{Ext}_R^i(R/\mathfrak{m}, C) = 0$. Also, there is a homogeneous isomorphism

$$* \operatorname{Hom}_R(* \operatorname{Ext}_R^n(R/\mathfrak{m}, C), *E(R/\mathfrak{m})) \cong R/\mathfrak{m}.$$

Since the G-graded ring $R/\mathfrak{m}$ is *simple, every graded $R/\mathfrak{m}$-module is *free and *injective, by 14.3.2. Now $(0 :_{*E(R/\mathfrak{m})} \mathfrak{m}) = R/\mathfrak{m}$ by 14.2.9, and so the graded $R/\mathfrak{m}$-module $* \operatorname{Ext}_R^n(R/\mathfrak{m}, C)$ must be *free of rank 1. Thus

$$* \operatorname{Ext}_R^n(R/\mathfrak{m}, C) \cong (R/\mathfrak{m})(g)$$

in $*\mathcal{C}(R)$ for some $g \in G$, so that

$$R/\mathfrak{m} \cong {}^* \operatorname{Hom}_R({}^* \operatorname{Ext}_R^n(R/\mathfrak{m}, C), {}^*E(R/\mathfrak{m}))$$
$$\cong {}^* \operatorname{Hom}_R((R/\mathfrak{m})(g), {}^*E(R/\mathfrak{m})) \cong (R/\mathfrak{m})(-g)$$

in $*\mathcal{C}(R)$. Thus there is a homogeneous isomorphism $(R/\mathfrak{m})(-g) \cong R/\mathfrak{m}$; application of the shift functor $(\,\bullet\,)(g)$ then shows that $R/\mathfrak{m} \cong (R/\mathfrak{m})(g)$, and this part of the proof is complete.

($\Leftarrow$) Suppose that there are homogeneous isomorphisms

$$^* \operatorname{Ext}_R^i(R/\mathfrak{m}, C) \cong \begin{cases} 0 & \text{for } i \neq n, \\ R/\mathfrak{m} & \text{for } i = n. \end{cases}$$

It follows from this that $\mu^i(\mathfrak{m}, C) = \delta_{i,\operatorname{ht}\mathfrak{m}}$ for all $i \in \mathbb{N}_0$, so that C is a canonical module for R by 14.5.4. Therefore, by 14.5.6, there is a Gorenstein *local G-graded ring R' and a homogeneous surjective ring homomorphism $R' \longrightarrow R$. It now follows from 14.5.2 that there is a *canonical module C' for R. Since a *canonical module for R is automatically a canonical module for R (by 14.5.5), we can now use 14.5.9(ii) to see that there exists $g \in G$ for which there is a homogeneous isomorphism $C' \cong C(g)$.

It follows from the '($\Rightarrow$)' part of this proof that there is a homogeneous isomorphism $*\operatorname{Ext}_R^n(R/\mathfrak{m}, C') \cong R/\mathfrak{m}$. We can now make use of 13.3.11 to see that there are homogeneous isomorphisms

$$R/\mathfrak{m} \cong {}^* \operatorname{Ext}_R^n(R/\mathfrak{m}, C') \cong {}^* \operatorname{Ext}_R^n(R/\mathfrak{m}, C(g))$$
$$= {}^* \operatorname{Ext}_R^n(R/\mathfrak{m}, C)(g) \cong (R/\mathfrak{m})(g).$$

The homogeneous isomorphism $R/\mathfrak{m} \cong R/\mathfrak{m}(g)$ leads to a homogeneous isomorphism $*E(R/\mathfrak{m}) \cong {}^*E(R/\mathfrak{m})(g)$, and application of the shift functor $(\,\bullet\,)(-g)$ yields a homogeneous isomorphism $*E(R/\mathfrak{m})(-g) \cong {}^*E(R/\mathfrak{m})$. By 14.5.10, there are homogeneous isomorphisms

$$H_\mathfrak{m}^n(R) \cong {}^* \operatorname{Hom}_R({}^* \operatorname{Hom}_R(R, C'), {}^*E(R/\mathfrak{m})) \cong {}^* \operatorname{Hom}_R(C', {}^*E(R/\mathfrak{m})),$$

and use of our homogeneous isomorphisms obtained above then yield further homogeneous isomorphisms

$$H_\mathfrak{m}^n(R) \cong {}^* \operatorname{Hom}_R(C(g), {}^*E(R/\mathfrak{m})) = {}^* \operatorname{Hom}_R(C, {}^*E(R/\mathfrak{m})(-g))$$
$$\cong {}^* \operatorname{Hom}_R(C, {}^*E(R/\mathfrak{m})).$$

(We have used 13.3.11 again.) Therefore C is a *canonical module for R. $\square$

14.5.13 ♯Exercise. Assume that $(R, \mathfrak{m})$ is *local.

(i) Show that $G_R := \{g \in G : R/\mathfrak{m} \cong (R/\mathfrak{m})(g) \text{ in } {}^*\mathcal{C}(R)\}$ is a subgroup of G.

(ii) Suppose that R is Cohen–Macaulay and that C is a *canonical module for R. Show that

$$G_R = \{g \in G : C(g) \text{ is a *canonical module for } R\}.$$

(iii) In the case where $G = \mathbb{Z}^n$ for some $n \in \mathbb{N}$ and R is positively graded, show that $G_R = 0$.

(iv) Let H be an arbitrary subgroup of $\mathbb{Z}$. Give an example of a (commutative Noetherian) $\mathbb{Z}$-graded *local ring S such that $G_S = H$.

14.5.14 Theorem. *Suppose that $G = \mathbb{Z}^n$, and that $(R, \mathfrak{m})$ is *local, Cohen–Macaulay, and positively ($\mathbb{Z}^n$-)graded, and has a *canonical module. Then any two *canonical modules for R are isomorphic in the graded category $^*\mathcal{C}(R)$.*

Proof. Let C, C' be *canonical modules for R. It follows from 14.5.5 that C and C' are canonical modules for R in the sense of 12.1.28. Therefore, by 14.5.9(ii), there exist $g \in G$ and a homogeneous isomorphism $\phi : C \xrightarrow{\cong} C'(g)$. Hence $C'(g)$ is a *canonical module for R, so that, since R is *positively* ($\mathbb{Z}^n$-)graded, we must have $g = 0$ by 14.5.13. □

14.5.15 Lemma. *Assume that $(R, \mathfrak{m})$ is *local and Cohen–Macaulay with $\operatorname{ht} \mathfrak{m} = n$, and let C be a finitely generated graded R-module such that $\mu^i(\mathfrak{m}, C) = \delta_{i,n}$ for all $i \in \mathbb{N}_0$, that is (by 12.1.25), such that $C_\mathfrak{m}$ is canonical for the Cohen–Macaulay local ring $R_\mathfrak{m}$.*

*Then there exists $g \in G$ such that $^*E^n(C) \cong {}^*E(R/\mathfrak{m})(-g)$ (in $^*\mathcal{C}(R)$); furthermore, for any such g, the shifted graded module $C(g)$ is a *canonical module for R.*

Proof. It follows from 14.5.4 that C is a canonical module for R in the sense of 12.1.28. It follows from Lemma 14.5.8(i) that there exists $g \in G$ such that $^*E^n(C) \cong {}^*E(R/\mathfrak{m})(-g)$ (in $^*\mathcal{C}(R)$); furthermore, for any such g, parts (ii)(a),(b) of the same lemma show that there are homogeneous isomorphisms

$$^*\operatorname{Ext}_R^i(R/\mathfrak{m}, C) \cong \begin{cases} 0 & \text{for } i \neq n, \\ (R/\mathfrak{m})(-g) & \text{for } i = n. \end{cases}$$

The result therefore follows from 14.5.12 and 14.1.10(i). □

14.5.16 Corollary. *Suppose $(R, \mathfrak{m})$ is *local and Gorenstein. Then there exists $g \in G$ such that $R(g)$ is a *canonical module for R.*

Proof. This is immediate from 14.5.15 because $R_\mathfrak{m}$ is a canonical module for $R_\mathfrak{m}$. □

In the case where $G = \mathbb{Z}^n$ and $(R, \mathfrak{m})$ is *local, Gorenstein and positively $(\mathbb{Z}^n\text{-})$graded, it follows from 14.5.16 and 14.5.13 that there is a unique $g \in \mathbb{Z}^n$ such that $R(g)$ is a *canonical module for R. Note that, since $\mathfrak{m} = (\mathfrak{m} \cap R_0) \bigoplus \left(\bigoplus_{g \in \mathbb{N}_0{}^n \setminus \{0\}} R_g \right)$, the *simple R-module $R/\mathfrak{m}$ is concentrated in degree 0. By 14.5.11, there is a homogeneous isomorphism $H_\mathfrak{m}^n(R(g)) \cong {}^*E(R/\mathfrak{m})$, so that $H_\mathfrak{m}^n(R) \cong {}^*E(R/\mathfrak{m})(-g)$ in *$\mathcal{C}(R)$. This means that g can be identified as the degree in which the *simple submodule of $H_\mathfrak{m}^n(R)$ is concentrated. This g is an important invariant that we shall discuss further (in special cases) below; for the present we content ourselves with a calculation of the invariant in the case of a polynomial ring over a field.

14.5.17 Example. Let K be a field and let $R := K[X_1, \ldots, X_n]$, the ring of polynomials over K in n indeterminates (where $n \in \mathbb{N}$). Consider R to be $\mathbb{Z}^n$-graded as in 13.1.4(ii) (so we are taking G to be $\mathbb{Z}^n$ here).

Note that R is Gorenstein and *local, and is positively graded by $\mathbb{Z}^n$, with unique *maximal ideal $\mathfrak{m} := (X_1, \ldots, X_n)$. As was explained just above, there is a unique $g \in \mathbb{Z}^n$ such that $R(g)$ is a *canonical module for R. We now calculate g.

We use 14.5.11, 14.1.10(ii) and 13.5.3 to see that there are homogeneous isomorphisms

$$ {}^*E(R/\mathfrak{m}) \cong H_\mathfrak{m}^n(R(g)) = (H_\mathfrak{m}^n(R))(g) \cong K[X_1^-, \ldots, X_n^-](g). $$

Now the graded submodule $R/\mathfrak{m}$ of *$E(R/\mathfrak{m})$ is generated by a homogeneous element of degree 0 which has annihilator $\mathfrak{m}$; furthermore, the only homogeneous elements of $K[X_1^-, \ldots, X_n^-]$ which have annihilator $\mathfrak{m}$ are the elements $\alpha X_1^{-1} \ldots X_n^{-1}$ where $\alpha \in K \setminus \{0\}$. It follows that $g = (-1, \ldots, -1)$.

As in 13.5.3, we can also regard $K[X_1, \ldots, X_n]$ as $\mathbb{Z}$-graded, where $\deg X_i = 1$ for all $i = 1, \ldots, n$; then our polynomial ring is just the result R^ϕ of applying the ϕ-coarsening functor of 13.5.1 to R, where $\phi : \mathbb{Z}^n \longrightarrow \mathbb{Z}$ is the Abelian group homomorphism for which

$$ \phi((i_1, \ldots, i_n)) = i_1 + \cdots + i_n \quad \text{for all } (i_1, \ldots, i_n) \in \mathbb{Z}^n. $$

Since $H_{\mathfrak{m}^\phi}^n(R^\phi) \cong H_\mathfrak{m}^n(R)^\phi$ in *$\mathcal{C}^\mathbb{Z}(R)$ by 13.5.2, it follows that, when we regard R as $\mathbb{Z}$-graded in this way, the unique integer a for which $R(a)$ is a *canonical module for R is the degree in which the $\mathbb{Z}$-graded *simple submodule of $H_\mathfrak{m}^n(R)^\phi$ is concentrated, namely $-n$.

14.5.18 Example. Here we review graded local duality for a ring of polynomials over a field K in n indeterminates $X_1, \ldots, X_n$. Let $R = \bigoplus_{g \in \mathbb{Z}^n} R_g = K[X_1, \ldots, X_n]$, considered to be $\mathbb{Z}^n$-graded as in 13.1.4, and let $\mathfrak{m}$ denote the unique *maximal ideal $(X_1, \ldots, X_n)$ of R. We apply 14.4.1 to R: we can take $R' = R$ and $f : R' \longrightarrow R$ to be the identity ring homomorphism; also, $R_0 = K$ and $\mathfrak{m}_0 := \mathfrak{m} \cap R_0 = 0$, so that $E_0 := E_{R_0}(R_0/\mathfrak{m}_0)$ is just K. We saw in 14.5.17 that $-\mathbf{1} = (-1, \ldots, -1)$ is the unique $g \in \mathbb{Z}^n$ such that $R(g)$ is a *canonical module for R, that is, the unique $g \in \mathbb{Z}^n$ such that $H_\mathfrak{m}^n(R) \cong {}^*E_R(R/\mathfrak{m})(-g)$ in *$\mathcal{C}(R)$.

It therefore follows from 14.4.1 and 14.4.2 that graded local duality for this R takes the following form: there is a homomorphism

$$\Psi := (\psi^i)_{i \in \mathbb{N}_0} : \left(H_\mathfrak{m}^i\right)_{i \in \mathbb{N}_0} \longrightarrow \left({}^*\mathrm{Hom}_K({}^*\mathrm{Ext}_R^{n-i}(\,\bullet\,, R(-\mathbf{1})), K)\right)_{i \in \mathbb{N}_0}$$

of (negative strongly) connected sequences of covariant functors from *$\mathcal{C}(R)$ to itself which is such that ψ_M^i is a (homogeneous) isomorphism for all $i \in \mathbb{N}_0$ whenever M is a finitely generated graded R-module. In particular, for such i and M, there are homogeneous isomorphisms

$$H_\mathfrak{m}^i(M) \cong {}^*\mathrm{Hom}_K({}^*\mathrm{Ext}_R^{n-i}(M, R(-\mathbf{1})), K)$$

and (since $R_0 = K$ is a complete local ring)

$$ {}^*\mathrm{Hom}_K(H_\mathfrak{m}^i(M), K) \cong {}^*\mathrm{Ext}_R^{n-i}(M, R(-\mathbf{1})).$$

Similarly, when we regard $R = K[X_1, \ldots, X_n]$ as $\mathbb{Z}$-graded, where $\deg X_i = 1$ for all $i = 1, \ldots, n$ (so that $\mathfrak{m} = (X_1, \ldots, X_n)$ is again the unique *maximal ideal of R), graded local duality yields a homomorphism

$$\Phi := (\phi^i)_{i \in \mathbb{N}_0} : \left(H_\mathfrak{m}^i\right)_{i \in \mathbb{N}_0} \longrightarrow \left({}^*\mathrm{Hom}_K({}^*\mathrm{Ext}_R^{n-i}(\,\bullet\,, R(-n)), K)\right)_{i \in \mathbb{N}_0}$$

of (negative strongly) connected sequences of covariant functors from *$\mathcal{C}^{\mathbb{Z}}(R)$ to itself which is such that ϕ_M^i is a (homogeneous) isomorphism for all $i \in \mathbb{N}_0$ whenever M is a finitely generated ($\mathbb{Z}$-)graded R-module.

14.5.19 Example. Let K and $R = K[X_1, \ldots, X_n]$, considered to be $\mathbb{Z}^n$-graded, be as in 14.5.17. We now describe the structure of the *indecomposable *injective R-modules. Recall that each of these is homogeneously isomorphic to a shift of *$E_R(R/\mathfrak{p})$ for some graded prime ideal $\mathfrak{p}$ of R, and that we calculated *$E_R(R/\mathfrak{m})$ (where $\mathfrak{m} = (X_1, \ldots, X_n)$) in 14.5.17.

It follows immediately from 14.2.6 and 14.2.7 that there are homogeneous R-isomorphisms

$$ {}^*E_R(R/0) \cong {}^*E_{R_{((0))}}(R_{((0))}/0) \cong K[X_1, \ldots, X_n, X_1^{-1}, \ldots, X_n^{-1}].$$

Now let $t \in \{1, \ldots, n-1\}$ and $\mathfrak{p} := (X_{t+1}, \ldots, X_n)$. We now describe *$E_R(R/\mathfrak{p})$, and note that similar calculations (and shifts) will then provide a complete description of all *indecomposable *injective R-modules.

Note that $\operatorname{ht} \mathfrak{p} = n - t$. Since R is Gorenstein, $\mu^i(\mathfrak{q}R_{(\mathfrak{p})}, R_{(\mathfrak{p})}) = \delta_{i,\operatorname{ht} \mathfrak{q}}$ (Kronecker delta) for all $i \in \mathbb{N}_0$ and $\mathfrak{q} \in {}^*\operatorname{Spec}(R)$ with $\mathfrak{q} \subseteq \mathfrak{p}$. Thus, if we use the minimal *injective resolution to calculate $H^{n-t}_{\mathfrak{p}R_{(\mathfrak{p})}}(R_{(\mathfrak{p})})$, the properties described in 14.2.4(iv) yield a homogeneous $R_{(\mathfrak{p})}$-isomorphism $H^{n-t}_{\mathfrak{p}R_{(\mathfrak{p})}}(R_{(\mathfrak{p})}) \cong$ ${}^*E_{R_{(\mathfrak{p})}}(R_{(\mathfrak{p})}/\mathfrak{p}R_{(\mathfrak{p})})(-g)$ for some $g \in \mathbb{Z}^n$. By 14.1.9, there is a homogeneous $R_{(\mathfrak{p})}$-isomorphism $H^{n-t}_{\mathfrak{p}R_{(\mathfrak{p})}}(R_{(\mathfrak{p})}) \cong (H^{n-t}_{\mathfrak{p}}(R))_{(\mathfrak{p})}$. Also, when viewed as an R-module, ${}^*E_{R_{(\mathfrak{p})}}(R_{(\mathfrak{p})}/\mathfrak{p}R_{(\mathfrak{p})})(-g)$ is homogeneously isomorphic to ${}^*E_R(R/\mathfrak{p})(-g)$. There is therefore a homogeneous R-isomorphism

$$(H^{n-t}_{\mathfrak{p}}(R))_{(\mathfrak{p})} \cong {}^*E_R(R/\mathfrak{p})(-g),$$

and so our strategy is to calculate the $\mathbb{Z}^n$-graded R-module $H^{n-t}_{\mathfrak{p}}(R)$ and then homogenously localize it at $\mathfrak{p}$.

We consider R as $K[X_1, \ldots, X_t][X_{t+1}, \ldots, X_n]$ and use the calculations in 13.5.3, with S taken to be the $\mathbb{Z}^t$-graded ring $K[X_1, \ldots, X_t]$. There results a $\mathbb{Z}^n$-homogeneous R-isomorphism

$$H^{n-t}_{\mathfrak{p}}(K[X_1, \ldots, X_n]) \cong K[X_1, \ldots, X_t][X^-_{t+1}, \ldots, X^-_n],$$

where $\deg(X^{i_1}_1 \ldots X^{i_t}_t X^{j_{t+1}}_{t+1} \ldots X^{j_n}_n) = (i, j)$ for all $i = (i_1, \ldots, i_t) \in \mathbb{N}_0{}^t$ and $j = (j_{t+1}, \ldots, j_n) \in (-\mathbb{N})^{n-t}$. Thus, after homogenous localization at $\mathfrak{p}$, we see that a K-basis for ${}^*E_R(R/\mathfrak{p})(-g)$ is

$$\left(X^{i_1}_1 \ldots X^{i_t}_t X^{j_{t+1}}_{t+1} \ldots X^{j_n}_n \right)_{(i_1, \ldots, i_t) \in \mathbb{Z}^t, (j_{t+1}, \ldots, j_n) \in (-\mathbb{N})^{n-t}},$$

where $\deg(X^{i_1}_1 \ldots X^{i_t}_t X^{j_{t+1}}_{t+1} \ldots X^{j_n}_n) = (i, j)$ for all $i = (i_1, \ldots, i_t) \in \mathbb{Z}^t$ and $j = (j_{t+1}, \ldots, j_n) \in (-\mathbb{N})^{n-t}$, and that the $\mathbb{Z}^n$-graded R-module structure is such that

$$X_k(X^{i_1}_1 \ldots X^{i_t}_t X^{j_{t+1}}_{t+1} \ldots X^{j_n}_n)$$
$$= \begin{cases} X^{i_1}_1 \ldots X^{i_k+1}_k \ldots X^{i_t}_t X^{j_{t+1}}_{t+1} \ldots X^{j_n}_n & \text{if } 1 \leq k \leq t, \\ X^{i_1}_1 \ldots X^{i_t}_t X^{j_{t+1}}_{t+1} \ldots X^{j_k+1}_k \ldots X^{j_n}_n & \text{if } t+1 \leq k \leq n, j_k < -1, \\ 0 & \text{if } t+1 \leq k \leq n, j_k = -1. \end{cases}$$

Furthermore, such an $X^{i_1}_1 \ldots X^{i_t}_t X^{j_{t+1}}_{t+1} \ldots X^{j_n}_n$ is annihilated by $\mathfrak{p}$ if and only if $j = (-1, \ldots, -1) \in \mathbb{Z}^{n-t}$, that is, if and only if it has the form

$$X^{i_1}_1 \ldots X^{i_t}_t X^{-1}_{t+1} \ldots X^{-1}_n.$$

Note that multiplication by X_1 provides a homogeneous automorphism of degree $(1, 0, \ldots, 0)$ of ${}^*E_R(R/\mathfrak{p})$; similar comments apply to $X_2, \ldots, X_t$. It follows that there is a $\mathbb{Z}^n$-homogeneous isomorphism

$${}^*E_R(R/\mathfrak{p}) \cong K[X_1, \ldots, X_t][X^-_{t+1}, \ldots, X^-_n]((k, (-1, \ldots, -1)))$$

for any $k \in \mathbb{Z}^t$. A more combinatorial approach to this *indecomposable *injective module is provided by Miller and Sturmfels in [53, Chapter 11].

We can use 14.2.8 and 14.5.19 to find the *indecomposable *injective modules over a homomorphic image of the polynomial ring of 14.5.19. In particular, this applies to the Stanley–Reisner rings (with respect to the field K) of simplicial complexes on $\{1, \ldots, n\}$ that were introduced in 13.1.5.

14.5.20 Exercise. Let Δ be the simplicial complex on $\{1, 2, 3, 4\}$ consisting of all the subsets of $\{1, 2, 3\}$ and all the subsets of $\{1, 4\}$. Let K be a field and work in the polynomial ring $K[X_1, X_2, X_3, X_4]$.

  (i) Show that $\mathfrak{a}_\Delta = (X_2 X_4, X_3 X_4)$.
 (ii) Identify $*E_{K[\Delta]}\left(K[\Delta]/\left((X_2, X_3, X_4)/\mathfrak{a}_\Delta\right)\right)$ with a shift of a submodule of $K[X_1][X_2^-, X_3^-, X_4^-]$ via 14.2.8 and 14.5.19. Show that

$$\mathcal{B} := \{X_1^{i_1} X_2^{i_2} X_3^{i_3} X_4^{i_4} \,:\, i_1 \in \mathbb{Z}, i_2, i_3, i_4 \in -\mathbb{N}$$
$$\text{and } i_2 = i_3 = -1 \text{ or } i_4 = -1\}$$

is a K-basis for $*E_{K[\Delta]}\left(K[\Delta]/\left((X_2, X_3, X_4)/\mathfrak{a}_\Delta\right)\right)$. What can you say about the degrees of the elements of $\mathcal{B}$?

14.5.21 Definition and Remarks. Suppose $G = \mathbb{Z}$ and $(R, \mathfrak{m})$ is Cohen–Macaulay and *local, and is a positively ($\mathbb{Z}$-)graded ring; assume also that R has a *canonical module. By 14.5.14, this is uniquely determined up to homogeneous isomorphism: we denote by ω_R one choice of *canonical module for R. The *a-invariant of* R is defined to be

$$a(R) := -\operatorname{beg}(\omega_R) = -\min\{n \in \mathbb{Z} : (\omega_R)_n \neq 0\}.$$

(See 14.1.1 for the definition of the beginning of a $\mathbb{Z}$-graded R-module.)

  (i) Note that, when R is Gorenstein, $a(R)$ is the unique integer a for which $R(a)$ is a *canonical module for R. Thus the a-invariant of a polynomial ring over a field, $\mathbb{Z}$-graded so that each variable has degree 1, was calculated in 14.5.17.
 (ii) In the general case, we can use graded local duality, as described in 14.5.10, to see that (with the notation $*D$ of that result) there are homogeneous isomorphisms

$$H_{\mathfrak{m}}^n(R) \cong *D(*\operatorname{Hom}_R(R, \omega_R)) \cong *D(\omega_R),$$

and so it follows from Graded Matlis Duality 14.4.2 that

$$a(R) = -\operatorname{beg}(\omega_R) = \operatorname{end}(*D(\omega_R)) = \operatorname{end}(H_{\mathfrak{m}}^n(R)).$$

The following proposition gives some hints about ways in which graded local duality and graded Matlis duality can be used in tandem.

14.5.22 Proposition. *Suppose that $(R, \mathfrak{m})$ is Cohen–Macaulay and *local, and has a *canonical module. Set $n := \operatorname{ht} \mathfrak{m}$.*

Then R is Gorenstein if and only if there is a homogeneous isomorphism $(0 :_{H_{\mathfrak{m}}^n(R)} \mathfrak{m}) \cong (R/\mathfrak{m})(g)$ for some $g \in G$.

Proof. ($\Rightarrow$) There exists $g \in G$ such that $R(g)$ is a *canonical module for R, by 14.5.16. By 14.5.5 and 14.5.8, there exist $h \in G$ and a homogeneous isomorphism $H_{\mathfrak{m}}^n(R(g)) \cong {}^*E(R/\mathfrak{m})(-h)$. It follows that there are homogeneous isomorphisms $H_{\mathfrak{m}}^n(R) \cong {}^*E(R/\mathfrak{m})(-h-g)$ and

$$(0 :_{H_{\mathfrak{m}}^n(R)} \mathfrak{m}) \cong (0 :_{{}^*E(R/\mathfrak{m})(-h-g)} \mathfrak{m}) = (R/\mathfrak{m})(-h-g).$$

(We have used 14.2.9 here.)

($\Leftarrow$) Let C be a *canonical module for R, and let *D denote the functor $^*\operatorname{Hom}_R(\bullet, {}^*E(R/\mathfrak{m}))$ from $^*\mathcal{C}(R)$ to itself. By definition, there is a homogeneous isomorphism $^*D(C) \cong H_{\mathfrak{m}}^n(R)$. There are therefore homogeneous isomorphisms

$$
\begin{aligned}
(0 :_{H_{\mathfrak{m}}^n(R)} \mathfrak{m}) &\cong {}^*\operatorname{Hom}_R(R/\mathfrak{m}, H_{\mathfrak{m}}^n(R)) \\
&\cong {}^*\operatorname{Hom}_R(R/\mathfrak{m}, {}^*\operatorname{Hom}_R(C, {}^*E(R/\mathfrak{m}))) \\
&\cong {}^*\operatorname{Hom}_R(R/\mathfrak{m} \otimes_R C, {}^*E(R/\mathfrak{m})) \\
&\cong {}^*\operatorname{Hom}_R(C/\mathfrak{m}C, {}^*E(R/\mathfrak{m})).
\end{aligned}
$$

Application of *D and use of 14.2.9 therefore yield further homogeneous isomorphisms

$$^*D(^*D(C/\mathfrak{m}C)) \cong {}^*D((0 :_{H_{\mathfrak{m}}^n(R)} \mathfrak{m})) \cong {}^*D((R/\mathfrak{m})(g)) \cong (R/\mathfrak{m})(-g).$$

However, the canonical R-homomorphism $C/\mathfrak{m}C \longrightarrow {}^*D(^*D(C/\mathfrak{m}C))$ is homogeneous and monomorphic. Since $C_{\mathfrak{m}}/\mathfrak{m}R_{\mathfrak{m}}C_{\mathfrak{m}} \neq 0$ (because $C_{\mathfrak{m}}$ is canonical for $R_{\mathfrak{m}}$ by 14.5.3), it follows that $C_{\mathfrak{m}}$ is a cyclic $R_{\mathfrak{m}}$-module, so that $R_{\mathfrak{m}}$ is Gorenstein. Therefore $R_{\mathfrak{p}}$ is Gorenstein for all $\mathfrak{p} \in {}^*\operatorname{Spec}(R)$, so that R is Gorenstein by 14.3.9(ii). $\square$

In §12.3, we discussed the concept of S_2-ification, and we showed that an S_2-ification exists in a local ring R that has a faithful canonical module. We are now going to guide the reader to some results about S_2-ifications in graded situations.

14.5.23 Notation. Suppose that $(R, \mathfrak{m})$ is G-graded and *local, and that C

is a *canonical module for R. Recall that there is a minimal primary decomposition $0 = \bigcap_{i=1}^{n} \mathfrak{q}_i$ for the zero ideal in which each term is graded. Suppose that the $\mathfrak{q}_i$ are indexed so that $\operatorname{ht} \mathfrak{m}/\mathfrak{q}_i = \operatorname{ht} \mathfrak{m}$ for all $i = 1, \ldots, t$ and $\operatorname{ht} \mathfrak{m}/\mathfrak{q}_i < \operatorname{ht} \mathfrak{m}$ for all $i = t + 1, \ldots, n$. (Of course, t could be n.) Set $\mathfrak{u}_R(0) = \bigcap_{i=1}^{t} \mathfrak{q}_i$. Note that $\mathfrak{u}_R(0)$ is graded, so that $\mathfrak{u}_R(0) = 0$ if and only if $\mathfrak{u}_R(0)R_\mathfrak{m} = 0$, and that $\mathfrak{u}_R(0)R_\mathfrak{m} = \mathfrak{u}_{R_\mathfrak{m}}(0)$ in the notation of 12.1.12.

Note also that, by 14.5.3, the localization $C_\mathfrak{m}$ is a canonical module for $R_\mathfrak{m}$. We have $(0 :_{R_\mathfrak{m}} C_\mathfrak{m}) = \mathfrak{u}_{R_\mathfrak{m}}(0)$, by 12.1.15. Therefore the graded ideal $(0 :_R C)$ is equal to $\mathfrak{u}_R(0)$.

As in 12.3.8, we shall denote by $\mathfrak{S}$ the system of all ideals $\mathfrak{s}$ of R such that $\operatorname{Var}(\mathfrak{s})$ is contained in the non-S_2 locus of R. We shall guide the reader, in Exercise 14.5.24 below, to the result that, when the *canonical R-module C is faithful, there is an S_2-ification $\eta_R^{\mathfrak{S}} : R \longrightarrow D_{\mathfrak{S}}(R)$ in the sense of 12.3.9, and this R-algebra is G-graded with homogeneous structural homomorphism.

14.5.24 ♯Exercise. Let the situation and notation be as in 14.5.23, and denote by *$\mathfrak{H}$ the system of all graded ideals of R of height at least 2. Assume that $\mathfrak{u}_R(0) = 0$, so that $\mathfrak{u}_{R_\mathfrak{m}}(0) = 0$. Let $h_R : R \longrightarrow {}^*\operatorname{Hom}_R(C, C) =: H$ denote the natural homogeneous R-homomorphism for which $h_R(r) = r \operatorname{Id}_C$ for all $r \in R$.

(i) Show that there is an isomorphism of $R_\mathfrak{m}$-algebras

$$H_\mathfrak{m} \xrightarrow{\cong} \operatorname{Hom}_{R_\mathfrak{m}}(C_\mathfrak{m}, C_\mathfrak{m}),$$

and recall that $\operatorname{Hom}_{R_\mathfrak{m}}(C_\mathfrak{m}, C_\mathfrak{m})$ is the endomorphism ring of the canonical $R_\mathfrak{m}$-module $C_\mathfrak{m}$.

(ii) Show that the natural R-homomorphism $H \longrightarrow H_\mathfrak{m}$ is injective. (Here is a hint: note that an associated prime of the kernel of the specified homomorphism would have to be an associated prime of H.)

(iii) Show that $\operatorname{Ker} h_R = 0$; use 12.2.7 to show that $\operatorname{Coker} h_R$ is *$\mathfrak{H}$-torsion.

(iv) Use the Goto–Watanabe results 14.3.6 and 14.3.8, part (i) and 12.1.18(i) to show that the R-module H is S_2 (see 12.1.16).

(v) Show that $\Gamma_{*\mathfrak{H}}(H) = 0$ and use part (ii) together with 12.3.10(ii) to show that H is a commutative G-graded Noetherian ring.

(vi) Use 13.5.4(ii), 2.2.15, 2.2.17 and 12.3.2 to show that there is a unique homogeneous R-algebra isomorphism

$$\psi' : H = {}^*\operatorname{Hom}_R(C, C) \xrightarrow{\cong} D_{*\mathfrak{H}}(R).$$

(vii) Use part (vi) to show that $H^1_{*\mathfrak{H}}(R)$ is a finitely generated R-module; let $\mathfrak{a}$ be its annihilator, and note that $\mathfrak{a}$ is graded. Show that

(a) $\operatorname{ht} \mathfrak{a} \geq 2$;

(b) there is a uniquely determined homogeneous R-algebra isomorphism $D_{\mathfrak{a}}(R) \xrightarrow{\cong} D_{*\mathfrak{H}}(R)$; and

(c) $\operatorname{Var}(\mathfrak{a})$ is equal to the non-S_2 locus of R, so that every $\mathfrak{S}$-torsion R-module is $\mathfrak{a}$-torsion.

(viii) Deduce that R has an S_2-ification $\eta_R^{\mathfrak{S}} : R \longrightarrow D_{\mathfrak{S}}(R)$, that $D_{\mathfrak{S}}(R)$ is a G-graded commutative Noetherian ring, that $\eta_R^{\mathfrak{S}}$ is homogeneous, and that there are unique homogeneous R-algebra isomorphisms

$$* \operatorname{Hom}_R(C, C) \xrightarrow{\cong} D_{\mathfrak{a}}(R) \xrightarrow{\cong} D_{*\mathfrak{H}}(R) \xrightarrow{\cong} D_{\mathfrak{S}}(R).$$

We shall consider some examples of S_2-ifications in graded situations in §15.2.

The result established in the following exercise is due to J. Herzog and E. Kunz [34].

14.5.25 Exercise. Let $a_1, \ldots, a_h \in \mathbb{N} \setminus \{1\}$ satisfy $\operatorname{GCD}(a_1, \ldots, a_h) = 1$. Let $S := a_1 \mathbb{N}_0 + \cdots + a_h \mathbb{N}_0$, the additive subsemigroup of $\mathbb{N}_0$ generated by $a_1, \ldots, a_h$. Let K be a field and let R be the subring of the polynomial ring $K[X]$ given by $R := K[X^{a_1}, \ldots, X^{a_h}]$. Of course, R is a 1-dimensional Cohen–Macaulay ring; furthermore, it is positively $\mathbb{Z}$-graded, by virtue of the grading inherited from the usual $\mathbb{Z}$-grading on $K[X]$ in which $\deg X = 1$, and, with this grading, R is *local. We denote by R_+ the unique graded maximal ideal of R.

(i) Show that there exists $c \in \mathbb{N}$ such that $n \in S$ for all $n \in \mathbb{N}$ with $n \geq c$. Thus $L := \mathbb{N}_0 \setminus S$ is a non-empty finite set, which we refer to as *the set of non-members of S*, and so has a greatest member, e say. The semigroup S is said to be *symmetric* precisely when, for all integers n with $0 \leq n \leq e$, we have $n \in S$ if and only if $e - n \notin S$.

(ii) Show that R has a *canonical module.

(iii) Show that $K[X]_X = K[X, X^{-1}]$ can be naturally identified with the ideal transform $D_{R_+}(R)$ in the sense that there is a unique homogeneous R-isomorphism $\phi' : K[X, X^{-1}] \longrightarrow D_{R_+}(R)$ such that the diagram

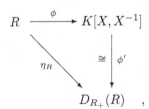

in which ϕ is the inclusion map, commutes.

(iv) Use 13.5.4(i) to show that, for $n \in \mathbb{Z}$,

$$\dim_K(H^1_{R_+}(R)_n) = \begin{cases} 1 & \text{if } n < 0 \text{ or } n \in L, \\ 0 & \text{if } n \in S, \end{cases}$$

and deduce that $\text{end}(H^1_{R_+}(R)) = e$.

(v) Show that, if R is Gorenstein, then S is symmetric.

(vi) Now suppose that S is symmetric. Show that, if $n \in L \cup \{-i : i \in \mathbb{N}\}$, then $s := e - n \in S$ and $X^s H^1_{R_+}(R)_n \neq 0$. Deduce that R is Gorenstein.

14.5.26 ♯Exercise. Let K be a field and let $R := K[X_1, \ldots, X_n]$, the ring of polynomials over K in n indeterminates (where $n \in \mathbb{N}$), graded by $\mathbb{Z}$ so that $R_0 = K$ and $\deg X_i = 1$ for all $i = 1, \ldots, n$. Let $r \in \mathbb{N}$, and consider the r-th Veronesean subring $R^{(r)}$ of R, as in 13.5.9. In Exercise 13.5.13, the reader was asked to show that $R^{(r)}$ is Cohen–Macaulay. Prove that $R^{(r)}$ is Gorenstein if and only if $n \equiv 0 \pmod{r}$.

14.5.27 Exercise. Assume that $G = \mathbb{Z}$ and that $(R, \mathfrak{m})$ is Cohen–Macaulay and *local, and positively ($\mathbb{Z}$-)graded; assume also that R has a *canonical module. Let $b \in R$ be a homogeneous element of positive degree d which is a non-zerodivisor on R. Show that R/bR has a *canonical module and that $a(R/bR) = a(R) + d$.

15

Links with projective varieties

One of the reasons for the interest in graded local cohomology is provided by the numerous applications to projective algebraic geometry. This short chapter is intended to provide a little geometric insight, with the aim of motivating the work on Castelnuovo regularity in Chapter 16.

In 2.3.2, we saw that the ideal transform has a geometric meaning in certain cases: if V is an affine variety over K, an algebraically closed field, $\mathfrak{b}$ is a non-zero ideal of $\mathcal{O}(V)$, and U denotes the open subset of V determined by $\mathfrak{b}$, then the ideal transform $D_{\mathfrak{b}}(\mathcal{O}(V))$ is isomorphic, as an $\mathcal{O}(V)$-algebra, to the ring of regular functions on U. One of our first aims for this chapter is the establishment of a graded analogue of this result. This graded analogue applies to irreducible affine algebraic cones.

Throughout this chapter, all graded rings and modules are to be understood to be $\mathbb{Z}$-graded.

15.1 Affine algebraic cones

15.1.1 Notation and Terminology. We shall employ the notation and terminology concerning graded rings and modules described in 13.1.1, 13.1.3 and 14.1.1, but, in accordance with our convention for the whole of this chapter, restricted to the special case in which $G = \mathbb{Z}$. In addition, when R is graded and the ideal $\mathfrak{a}$ is graded, and M is a graded R-module, so that (see 13.4.3 and 13.4.4) the $H_{\mathfrak{a}}^i(M)$ ($i \in \mathbb{N}_0$) are all graded R-modules, we use $H_{\mathfrak{a}}^i(M)_n$ to denote the n-th component of the graded module $H_{\mathfrak{a}}^i(M)$ (for $i \in \mathbb{N}_0$ and $n \in \mathbb{Z}$); also $D_{\mathfrak{a}}(M)$ is graded (by 13.3.14), and we use $D_{\mathfrak{a}}(M)_n$ to denote its n-th component.

Now assume that $R = \bigoplus_{n \in \mathbb{N}_0} R_n$ is positively graded. We set

$$R_+ := \bigoplus_{n \in \mathbb{N}} R_n = \bigoplus_{n > 0} R_n = 0 \oplus R_1 \oplus R_2 \oplus \ldots \oplus R_n \oplus \ldots,$$

the *irrelevant ideal* of R. Of course, R_+ is graded, and so, for a graded R-module M, we can define the R_0-modules $H^i_{R_+}(M)_n$ ($i \in \mathbb{N}_0$, $n \in \mathbb{Z}$).

15.1.2 Geometric Notation and Reminders. Let K be an algebraically closed field, and let $r \in \mathbb{N}$. We shall use the notation of 2.3.1, and we shall denote the origin $(0, \ldots, 0)$ of $\mathbb{A}^r(K)$ simply by 0. For a subset C of a quasi-affine variety W over K, we shall extend the notation of 6.4.1 to denote the *vanishing ideal*

$$\{f \in \mathcal{O}(W) : f(q) = 0 \text{ for all } q \in C\}$$

of C by $I_W(C)$.

(i) By a *cone* (with vertex 0) in $\mathbb{A}^r(K)$ we mean a set $C \subseteq \mathbb{A}^r(K)$ such that $0 \in C$ and, whenever $q \in C$, then $\lambda q \in C$ for all $\lambda \in K$. Such a cone C is called an *affine algebraic cone in $\mathbb{A}^r(K)$* if and only if it is also an affine algebraic set; also C is said to be *non-degenerate* precisely when $C \neq \{0\}$.

   Since a graded ideal can be generated by homogeneous elements, it is clear that, if $\mathfrak{b}$ is a proper graded ideal of $K[X_1, \ldots, X_r]$, then $V_{\mathbb{A}^r(K)}(\mathfrak{b})$ is an affine algebraic cone in $\mathbb{A}^r(K)$. It is easy to see that, conversely, if C is an affine algebraic cone in $\mathbb{A}^r(K)$, then $I_{\mathbb{A}^r(K)}(C)$ is a proper graded ideal. Thus the affine algebraic cones in $\mathbb{A}^r(K)$ are precisely the algebraic sets in $\mathbb{A}^r(K)$ which have proper graded vanishing ideals.

(ii) Let C be an irreducible affine algebraic cone in $\mathbb{A}^r(K)$. Since $I_{\mathbb{A}^r(K)}(C)$ is a graded prime ideal of $K[X_1, \ldots, X_r]$, the ring $\mathcal{O}(C)$ of regular functions on C inherits a grading from $K[X_1, \ldots, X_r]$, in such a way that the restriction homomorphism $K[X_1, \ldots, X_r] \longrightarrow \mathcal{O}(C)$ is homogeneous (see 14.1.3). Of course, for each $n \in \mathbb{N}_0$, we denote the n-th component of $\mathcal{O}(C)$ by $\mathcal{O}(C)_n$. Note that $\mathcal{O}(C)_0$ can be identified with K. It is also worth noting that $\mathcal{O}(C)$ is *local with unique *maximal graded ideal $\mathcal{O}(C)_+$; furthermore, $\mathcal{O}(C)_+$ is actually a maximal ideal in this case.

(iii) With the notation of part (ii), let $\mathfrak{b}$ be a non-zero graded ideal of $\mathcal{O}(C)$, let $V_C(\mathfrak{b})$ denote the closed subset of C determined by $\mathfrak{b}$, and let U be the open subset $C \setminus V_C(\mathfrak{b})$ of C. In fact, if $\mathfrak{c}$ denotes the inverse image of $\mathfrak{b}$ in $K[X_1, \ldots, X_r]$ under the restriction homomorphism, then $V_C(\mathfrak{b}) = V_{\mathbb{A}^r(K)}(\mathfrak{c})$, and U is obtained from C by removal of another affine algebraic cone (or the empty set).

Let $n \in \mathbb{Z}$. A regular function $f \in \mathcal{O}(U)$ is said to be *homogeneous of degree* n precisely when, for each $p \in U$, there exists an open set $W \subseteq U$ with $p \in W$, an integer $d \in \mathbb{N}_0$, and $g \in \mathcal{O}(C)_d$, $h \in \mathcal{O}(C)_{n+d}$ such that, for each $q \in W$, we have $g(q) \neq 0$ and $f(q) = h(q)/g(q)$. The set of all regular functions on U which are homogeneous of degree n is denoted by $\mathcal{O}(U)_n$. The fact that this definition is not ambiguous in the case when $U = C$ is one consequence of the next proposition.

15.1.3 Proposition. *Let the situation be as in 15.1.2(iii). Thus K is an algebraically closed field, $r \in \mathbb{N}$, C is an irreducible affine algebraic cone in $\mathbb{A}^r(K)$, $\mathfrak{b}$ is a non-zero graded ideal of $\mathcal{O}(C)$, and $U = C \setminus V_C(\mathfrak{b})$.*

The subsets $\mathcal{O}(U)_n$ $(n \in \mathbb{Z})$ defined in 15.1.2(iii) provide $\mathcal{O}(U)$ with a structure as a graded ring with respect to which the homomorphisms $\lceil_U$ and $\nu_{C,\mathfrak{b}}$ in the commutative diagram

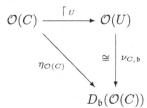

of 2.3.2 are homogeneous. (It should be noted that, since $\mathfrak{b}$ is graded, it follows from 13.3.14 that $D_{\mathfrak{b}}(\mathcal{O}(C))$ is a graded $\mathcal{O}(C)$-module and that $\eta_{\mathcal{O}(C)}$ is homogeneous.)

Proof. Let us abbreviate $\nu_{C,\mathfrak{b}}$ by ν and $\eta_{\mathcal{O}(C)}$ by η. Since ν is a ring isomorphism (by 2.3.2), it is enough for us to show that $\mathcal{O}(U)_n = \nu^{-1}(D_{\mathfrak{b}}(\mathcal{O}(C))_n)$ for each $n \in \mathbb{Z}$, and this is what we shall do.

Let $y \in D_{\mathfrak{b}}(\mathcal{O}(C))_n$. Since $\operatorname{Coker}\eta$, being isomorphic to $H^1_{\mathfrak{b}}(\mathcal{O}(C))$, is $\mathfrak{b}$-torsion, there exists $t \in \mathbb{N}$ such that $\mathfrak{b}^t y \subseteq \operatorname{Im}\eta$. Let $p \in U$. Since $U = C \setminus V_C(\mathfrak{b})$ and $\mathfrak{b}$ is graded, there exists a homogeneous element $g \in \mathfrak{b}^t$, of degree $d \in \mathbb{N}_0$ say, such that $g(p) \neq 0$. Then $W := U \setminus V_C(g\mathcal{O}(C))$ is an open subset of U which contains p.

As $gy \in \operatorname{Im}\eta$, there is an element $h \in \mathcal{O}(C)$ with $gy = \eta(h)$. As $gy \in D_{\mathfrak{b}}(\mathcal{O}(C))_{n+d}$ and as η is homogeneous and injective, we must have $h \in \mathcal{O}(C)_{n+d}$. Now

$$g \lceil_U \nu^{-1}(y) = \nu^{-1}(\eta(g))\nu^{-1}(y) = \nu^{-1}(\eta(g)y) = \nu^{-1}(gy)$$
$$= \nu^{-1}(\eta(h)) = h \lceil_U .$$

Therefore, for each $q \in W$, we have $g(q) \neq 0$ and $\nu^{-1}(y)(q) = h(q)/g(q)$. Hence $\nu^{-1}(y) \in \mathcal{O}(U)_n$. We have proved that $\mathcal{O}(U)_n \supseteq \nu^{-1}(D_{\mathfrak{b}}(\mathcal{O}(C))_n)$.

Now let $f \in \mathcal{O}(U)_n$. Choose $p \in U$; then there exists an open set $W \subseteq U$ with $p \in W$, an integer $d \in \mathbb{N}_0$, and $g \in \mathcal{O}(C)_d$, $h \in \mathcal{O}(C)_{n+d}$ such that, for each $q \in W$, we have $g(q) \neq 0$ and $f(q) = h(q)/g(q)$. This shows, in particular, that $g\lceil_W.f\lceil_W = h\lceil_W$. As W is a non-empty open subset of the irreducible topological space U, it follows that $g\lceil_U.f = h\lceil_U$. Now apply ν: we obtain

$$\eta(g)\nu(f) = \nu(g\lceil_U)\nu(f) = \nu(g\lceil_U.f) = \nu(h\lceil_U) = \eta(h).$$

As η is homogeneous, $\eta(g) \in D_\mathfrak{b}(\mathcal{O}(C))_d$ and $\eta(h) \in D_\mathfrak{b}(\mathcal{O}(C))_{n+d}$. As η is injective and $g \neq 0$, we have $\eta(g) \neq 0$; also, as $D_\mathfrak{b}(\mathcal{O}(C))$ is a domain, it follows that $\nu(f)$ must be homogeneous of degree n. We have therefore shown that $\mathcal{O}(U)_n \subseteq \nu^{-1}(D_\mathfrak{b}(\mathcal{O}(C))_n)$, and so the proof is complete. $\square$

Theorem 2.3.2, and its graded refinement 15.1.3, provide a link between local cohomology and algebraic varieties. Towards the end of the book, we shall encounter more general versions of these two results, because they are related to the Deligne Isomorphism Theorem 20.1.14 (see 20.1.17) and its graded version 20.2.7.

We explore now the special case of Proposition 15.1.3 in which $\dim C > 0$ and the graded ideal $\mathfrak{b}$ is the irrelevant ideal $\mathcal{O}(C)_+$ (see 15.1.1) of $\mathcal{O}(C)$: then the open set $U = C \setminus \{0\}$ is just the *punctured cone* $\overset{\circ}{C}$ of the given irreducible affine algebraic cone C.

15.1.4 Corollary. *Consider the special case of 15.1.3 where the irreducible affine algebraic cone C has $\dim C > 1$ and $\mathfrak{b} = \mathcal{O}(C)_+$. Then*

$$U = C \setminus V_C(\mathcal{O}(C)_+) = C \setminus \{0\} = \overset{\circ}{C},$$

the punctured cone of C; also

(i) *the restriction ring homomorphism $\lceil_{\overset{\circ}{C}} : \mathcal{O}(C) \longrightarrow \mathcal{O}(\overset{\circ}{C})$ makes $\mathcal{O}(\overset{\circ}{C})$ into a finitely generated graded $\mathcal{O}(C)$-module;*

(ii) *$\mathrm{end}(H^1_{\mathcal{O}(C)_+}(\mathcal{O}(C))) < \infty$ (see 14.1.1 for the definition of the end of a graded module);*

(iii) *$H^1_{\mathcal{O}(C)_+}(\mathcal{O}(C))_n$ is a finite-dimensional vector space over K, for all $n \in \mathbb{Z}$; and*

(iv) *$H^1_{\mathcal{O}(C)_+}(\mathcal{O}(C))_n = 0$ for all $n \leq 0$.*

Proof. Set $R := \mathcal{O}(C)$. Now (R, R_+) is *local, by 15.1.2(ii), and a domain. Hence, by 14.1.14, we have $\mathrm{ht}\, R_+ = \dim R = \dim C > 1$. Therefore, for all $\mathfrak{p} \in {}^*\mathrm{Spec}(R) \setminus \mathrm{Var}(R_+)$, we have $\mathfrak{p} \subset R_+$ and

$$\mathrm{depth}\, R_\mathfrak{p} + \mathrm{ht}\, R_+/\mathfrak{p} \geq 2 > 1.$$

It thus follows from the Graded Finiteness Theorem 14.3.10 that $H^1_{R_+}(R)$ is finitely generated. By 2.2.6(i)(c), 13.5.4 and 15.1.3, there is an exact sequence $R \xrightarrow{\lceil_{\overset{\circ}{C}}} \mathcal{O}(\overset{\circ}{C}) \longrightarrow H^1_{R_+}(R) \longrightarrow 0$ of graded R-modules and homogeneous R-homomorphisms. It is now immediate that $\lceil_{\overset{\circ}{C}}$ makes $\mathcal{O}(\overset{\circ}{C})$ into a finitely generated graded R-module.

Note also that, since $H^1_{R_+}(R)$ is finitely generated and R_+-torsion, we must have $H^1_{R_+}(R)_n = 0$ for all $n \gg 0$ (that is, for all n greater than some fixed integer n_0), and so $\operatorname{end}(H^1_{R_+}(R)) < \infty$. We have therefore completed the proofs of parts (i) and (ii), while part (iii) is now immediate from the facts that $R_0 = K$ and, for each $n \in \mathbb{Z}$, the component $H^1_{R_+}(R)_n$ is a finitely generated R_0-module.

It remains for us to prove part (iv). Use $\lceil_{\overset{\circ}{C}}$ to identify R as a subring of $\mathcal{O}(\overset{\circ}{C})$. It is enough for us to show that $K = R_0 = \mathcal{O}(C)_0 = \mathcal{O}(\overset{\circ}{C})_0$ and that $\mathcal{O}(\overset{\circ}{C})_n = 0$ for all $n < 0$, and this we do.

First, by part (i), the integral domain $\mathcal{O}(\overset{\circ}{C})_0$ is an integral extension of the algebraically closed field $K = R_0$, and so $R_0 = \mathcal{O}(\overset{\circ}{C})_0$. Second, since $\mathcal{O}(\overset{\circ}{C})$ is a finitely generated graded R-module and R is positively graded, $\mathcal{O}(\overset{\circ}{C})_n = 0$ for all $n \ll 0$. Hence, for each $n < 0$ and each $\gamma \in \mathcal{O}(\overset{\circ}{C})_n$, there exists $t \in \mathbb{N}$ such that $\gamma^t = 0$, so that $\gamma = 0$ since $\mathcal{O}(\overset{\circ}{C})$ is a domain. This completes the proof. $\qquad\square$

15.1.5 Exercise. Calculate $\mathcal{O}(C)$, $\mathcal{O}(\overset{\circ}{C})$ and $H^1_{\mathcal{O}(C)_+}(\mathcal{O}(C))$ for an irreducible affine algebraic cone C of dimension 1 in $\mathbb{A}^r(K)$, where K is an algebraically closed field and $r \in \mathbb{N}$. (Here is a hint. Let $c := (c_1, \ldots, c_r) \in C \setminus \{0\}$. Without loss of generality, one can assume that $c_1 \neq 0$: with this assumption, show that

$$I_{\mathbb{A}^r(K)}(C) = (c_2 X_1 - c_1 X_2, c_3 X_1 - c_1 X_3, \ldots, c_r X_1 - c_1 X_r).)$$

15.1.6 Exercise. Assume that $R = \bigoplus_{n \in \mathbb{N}_0} R_n$ is positively graded and an integral domain, and that the ideal $\mathfrak{a}$ is graded and non-zero.

(i) Show that, if $\mathfrak{a} \cap R_0 \neq 0$, then $D_{\mathfrak{a}}(R)_n = H^1_{\mathfrak{a}}(R)_n = 0$ for all $n < 0$.

(ii) Assume that the subring R_0 is a homomorphic image of a regular ring, and that $\operatorname{ht} \mathfrak{a} > 1$. Show that $\eta_{\mathfrak{a}}$ makes $D_{\mathfrak{a}}(R)$ into a finitely generated R-module, and deduce that $D_{\mathfrak{a}}(R)_n$ and $H^1_{\mathfrak{a}}(R)_n$ are finitely generated R_0-modules for all $n \in \mathbb{Z}$. Show further that $D_{\mathfrak{a}}(R)_n = H^1_{\mathfrak{a}}(R)_n = 0$ for all $n < 0$.

(iii) Assume that R_0 is an algebraically closed field K, and that $\operatorname{ht} \mathfrak{a} > 1$.

Using $\eta_\mathfrak{a}$ to identify R as a subring of $D_\mathfrak{a}(R)$, show that $D_\mathfrak{a}(R)_0 = K$ and $H^1_\mathfrak{a}(R)_0 = 0$.

15.1.7 Exercise. Let the situation and notation be as in 15.1.3, and assume that ht $\mathfrak{b} > 1$. Use 15.1.6(ii),(iii) to show that $\mathcal{O}(U)$ is a finitely generated $\mathcal{O}(C)$-module for which $\mathcal{O}(U)_0 = K$.

15.2 Projective varieties

In the situation of 15.1.3, we have seen (see 2.3.3) that we can regard non-zero elements of the local cohomology module $H^1_\mathfrak{b}(\mathcal{O}(C))$ as obstructions to the extension of regular functions on U to regular functions on C; we have also obtained, in 15.1.4, in the case when U is the punctured cone $\overset{\circ}{C}$, information about some of the components $H^1_{\mathcal{O}(C)_+}(\mathcal{O}(C))_n = H^1_\mathfrak{b}(\mathcal{O}(C))_n$. In the light of the connections between irreducible affine algebraic cones and projective varieties (reviewed in 15.2.1 below), it would be reasonable for one to suspect that there are links between graded local cohomology and projective varieties. Such suspicions would be well founded, and we plan to expose some of the links in this and subsequent chapters.

15.2.1 Reminders and Notation. Here we specify the notation and terminology that we shall use for discussion of projective varieties. Let K be an algebraically closed field, and let $r \in \mathbb{N}$. We shall find it convenient to vary slightly the notation of 2.3.1 and regard the polynomial ring $K[X_0, \ldots, X_r]$ in $r + 1$ indeterminates $X_0, X_1, \ldots, X_r$ as the coordinate ring $\mathcal{O}(\mathbb{A}^{r+1}(K))$ of affine $(r+1)$-space $\mathbb{A}^{r+1}(K)$ over K. As in 15.1.2, we shall denote the origin $(0, \ldots, 0)$ of $\mathbb{A}^{r+1}(K)$ simply by 0.

(i) For $c := (c_0, \ldots, c_r) \in \mathbb{A}^{r+1}(K) \setminus \{0\}$, we use $(c_0 : \cdots : c_r)$ to denote the line $\{\lambda c : \lambda \in K\}$ in $\mathbb{A}^{r+1}(K)$ through c and 0. We shall use $\mathbb{P}^r(K)$ to denote projective r-space over K, that is, the set

$$\left\{ (c_0 : \cdots : c_r) : (c_0, \ldots, c_r) \in \mathbb{A}^{r+1}(K) \setminus \{0\} \right\}$$

of all lines through the origin in $\mathbb{A}^{r+1}(K)$, endowed with the Zariski topology. We remind the reader that the closed sets in this topology are precisely the projective algebraic sets, that is, the sets of the form

$$V_{\mathbb{P}^r(K)}(\mathfrak{a}) := \left\{ (c_0 : \cdots : c_r) \in \mathbb{P}^r(K) : f(c_0, \ldots, c_r) = 0 \text{ for all } f \in \mathfrak{a} \right\},$$

where $\mathfrak{a}$ is a proper graded ideal of $K[X_0, \ldots, X_r]$. (When $\mathfrak{a} = (X_0, \ldots, X_r)$, the corresponding projective algebraic set is the empty one.) If $f_1, \ldots, f_t$ are non-constant homogeneous polynomials in $K[X_0, \ldots, X_r]$, then the projective algebraic set $V_{\mathbb{P}^r(K)}((f_1, \ldots, f_t))$ is denoted by $V_{\mathbb{P}^r(K)}(f_1, \ldots, f_t)$. Of course, every projective algebraic set in $\mathbb{P}^r(K)$ can be represented in this form.

We shall use $\mathbb{P}^r$ to denote complex projective r-space $\mathbb{P}^r(\mathbb{C})$. All unexplained mentions of topological notions, including 'open' and 'closed' subsets, in connection with projective spaces will refer to the Zariski topology.

(ii) By the statement '$V \subseteq \mathbb{P}^r(K)$ is a *projective variety*' we shall mean that V is an irreducible closed subset of $\mathbb{P}^r(K)$ (with the induced topology), and by the statement '$W \subseteq \mathbb{P}^r(K)$ is a *quasi-projective variety*' we shall mean that W is a non-empty open subset of a projective variety $V \subseteq \mathbb{P}^r(K)$ (again with the induced topology). (It is wise for us to include the '$\subseteq \mathbb{P}^r(K)$' in the notation because (and some of our examples below will remind the reader of this) isomorphic projective varieties, embedded in projective spaces in different ways, can have non-isomorphic homogeneous coordinate rings! (See parts (vi) and (viii) below.) This problem does not occur with affine varieties.)

By a *variety* we shall mean an affine, quasi-affine, projective, or quasi-projective variety.

(iii) Let V be the closed subset of $\mathbb{P}^r(K)$ given by $V = V_{\mathbb{P}^r(K)}(\mathfrak{a})$, where $\mathfrak{a}$ is a proper graded ideal of $K[X_0, \ldots, X_r]$. The *affine cone* $\mathrm{Cone}(V) \subseteq \mathbb{A}^{r+1}(K)$ *over V in $\mathbb{A}^{r+1}(K)$* is defined by

$$\mathrm{Cone}(V) = \left\{ (c_0, \ldots, c_r) \in \mathbb{A}^{r+1}(K) \setminus \{0\} : (c_0 : \cdots : c_r) \in V \right\} \cup \{0\}$$
$$= V_{\mathbb{A}^{r+1}(K)}(\mathfrak{a}),$$

an affine algebraic cone in $\mathbb{A}^{r+1}(K)$.

On the other hand, for the affine algebraic cone C in $\mathbb{A}^{r+1}(K)$ defined by $C = V_{\mathbb{A}^{r+1}(K)}(\mathfrak{a}')$, where $\mathfrak{a}'$ is a proper graded ideal of $K[X_0, \ldots, X_r]$, we define the *projectivization* C^+ *of C* to be the closed subset

$$C^+ = \left\{ (c_0 : \cdots : c_r) \in \mathbb{P}^r(K) : (c_0, \ldots, c_r) \in C \setminus \{0\} \right\} = V_{\mathbb{P}^r(K)}(\mathfrak{a}')$$

of $\mathbb{P}^r(K)$. Thus $V \mapsto \mathrm{Cone}(V)$ provides a bijective map from the set $T := \{V \subseteq \mathbb{P}^r(K) : V \text{ is closed}\}$ of all closed subsets of $\mathbb{P}^r(K)$ to the set

$$\left\{ C \subseteq \mathbb{A}^{r+1}(K) : C \text{ is an affine algebraic cone} \right\}$$

of all affine algebraic cones in $\mathbb{A}^{r+1}(K)$; the inverse map is given by $C \mapsto C^+$. Note that these two maps both preserve inclusion relations.

(iv) It follows from part (iii) and standard facts from affine algebraic geometry that there is a bijective correspondence between the set T of all closed subsets of $\mathbb{P}^r(K)$ and the set of all radical proper graded ideals of $K[X_0, \ldots, X_r]$

under which a closed subset $V \in T$ corresponds to

$$
\begin{aligned}
I_{\mathbb{A}^{r+1}(K)}(\mathrm{Cone}(V)) &= \{f \in K[X_0, \ldots, X_r] : f(c_0, \ldots, c_r) = 0 \text{ for all} \\
&\qquad (c_0, \ldots, c_r) \in \mathrm{Cone}(V)\} \\
&= \{f \in K[X_0, \ldots, X_r]_+ : f(c_0, \ldots, c_r) = 0 \text{ for all} \\
&\qquad (c_0, \ldots, c_r) \in \mathbb{A}^{r+1}(K) \setminus \{0\} \\
&\qquad \text{with } (c_0 : \cdots : c_r) \in V\}.
\end{aligned}
$$

This is also denoted by $I_{\mathbb{P}^r(K)}(V)$, and is called the *vanishing ideal of V*.

(v) Note that, if $V_1, \ldots, V_t$ are closed subsets in $\mathbb{P}^r(K)$, then

$$
\mathrm{Cone}\left(\bigcup_{i=1}^{t} V_i\right) = \bigcup_{i=1}^{t} \mathrm{Cone}(V_i) \quad \text{and} \quad \mathrm{Cone}\left(\bigcap_{i=1}^{t} V_i\right) = \bigcap_{i=1}^{t} \mathrm{Cone}(V_i).
$$

Also, the minimal prime ideals of a proper graded ideal of $K[X_0, \ldots, X_r]$ are again graded (by 13.1.6(ii)), and so it follows that a closed subset V of $\mathbb{P}^r(K)$ is irreducible if and only if $\mathrm{Cone}(V)$ is irreducible and non-degenerate, and that this is the case if and only if $I_{\mathbb{A}^{r+1}(K)}(\mathrm{Cone}(V))$ is a graded prime ideal of $K[X_0, \ldots, X_r]$ properly contained in $(X_0, \ldots, X_r)$.

Thus, in the bijective correspondence of part (iv), the projective varieties in $\mathbb{P}^r(K)$ correspond to the graded prime ideals of $K[X_0, \ldots, X_r]$ properly contained in $(X_0, \ldots, X_r)$.

(vi) Let $V \subseteq \mathbb{P}^r(K)$ be a projective variety. By parts (iv) and (v), the ideal $\mathfrak{p} := I_{\mathbb{A}^{r+1}(K)}(\mathrm{Cone}(V))$ is a graded prime ideal of $K[X_0, \ldots, X_r]$ with $\mathfrak{p} \subset (X_0, \ldots, X_r)$. We refer to the positively graded *local ring

$$
\mathcal{O}(\mathrm{Cone}(V)) = K[X_0, \ldots, X_r]/\mathfrak{p} = K[X_0, \ldots, X_r]/I_{\mathbb{A}^{r+1}(K)}(\mathrm{Cone}(V))
$$

as the *homogeneous coordinate ring of $V \subseteq \mathbb{P}^r(K)$*.

(vii) Remember that the *dimension* $\dim V$ of a variety V is defined as the maximum length l of a strictly descending chain $C_0 \supset C_1 \supset \cdots \supset C_l$ of closed irreducible subsets of V, and that such a V is called a *curve* if and only if $\dim V = 1$ and a *surface* if and only if $\dim V = 2$. Using the correspondence described in part (iii) and the observation made in part (v), we obtain, for a projective variety $V' \subseteq \mathbb{P}^r(K)$, that

$$
\dim V' = \dim(\mathrm{Cone}(V')) - 1 = \dim(\mathcal{O}(\mathrm{Cone}(V'))) - 1.
$$

(viii) We mentioned in (ii) above that isomorphic projective varieties, embedded in projective spaces in different ways, can have non-isomorphic homogeneous coordinate rings. Properties of the homogeneous coordinate ring $\mathcal{O}(\mathrm{Cone}(V))$ of a projective variety $V \subseteq \mathbb{P}^r(K)$ are referred to as *arithmetic properties of V*. For example, we say that $V \subseteq \mathbb{P}^r(K)$ is *arithmetically*

Cohen–Macaulay (respectively arithmetically Gorenstein, . . .) if and only if $\mathcal{O}(\mathrm{Cone}(V))$ is a Cohen–Macaulay (respectively Gorenstein, . . .) ring. We also define the *arithmetic depth* arithdepth V *of* V to be the grade of the unique *maximal graded ideal of the homogeneous coordinate ring of V. Thus arithdepth $V = \mathrm{grade}\, \mathcal{O}(\mathrm{Cone}(V))_+$.

15.2.2 Reminder and ♯Exercise: Veronesean tranformations. Let K be an algebraically closed field, and let $r, d \in \mathbb{N}$. Let $V \subseteq \mathbb{P}^r(K)$ be a projective variety, and let $\mathfrak{p}$ denote (the graded prime) kernel $I_{\mathbb{P}^r(K)}(V)$ of the restriction homomorphism $K[X_0, \ldots, X_r] \longrightarrow \mathcal{O}(\mathrm{Cone}(V))$. Set

$$\mathcal{T} = \mathcal{T}_r^{(d)} = \left\{ (\nu_0, \ldots, \nu_r) \in \mathbb{N}_0^{r+1} : \nu_0 + \cdots + \nu_r = d \right\},$$

a set with cardinality $\binom{r+d}{r}$. We shall use a family of (algebraically independent) indeterminates $(Y_\nu)_{\nu \in \mathcal{T}}$ indexed by $\mathcal{T}$. Let

$$\phi : K[Y_\nu : \nu \in \mathcal{T}] \longrightarrow K[X_0, \ldots, X_r]$$

be the K-algebra homomorphism for which $\phi(Y_{(\nu_0,\ldots,\nu_r)}) = X_0^{\nu_0} \ldots X_r^{\nu_r}$ for all $(\nu_0, \ldots, \nu_r) \in \mathcal{T}$. Let θ be the composition

$$K[Y_\nu : \nu \in \mathcal{T}] \xrightarrow{\phi} K[X_0, \ldots, X_r] \xrightarrow{\lceil \mathrm{Cone}(V)} \mathcal{O}(\mathrm{Cone}(V)),$$

and let $\mathfrak{q} := \mathrm{Ker}\,\theta$, a graded prime ideal of $K[Y_\nu : \nu \in \mathcal{T}]$.

(i) Show that, with the notation of 13.5.9, $\mathrm{Im}\,\phi = K[X_0, \ldots, X_r]^{(d)}$, the d-th Veronesean subring of $K[X_0, \ldots, X_r]$, that $\phi(\mathfrak{q}) = \mathfrak{p}^{(d)}$, and that $\mathrm{Im}\,\theta = \mathcal{O}(\mathrm{Cone}(V))^{(d)}$.

Note that, when Veronesean subrings are given the grading described at the beginning of 13.5.9, both $\phi : K[Y_\nu : \nu \in \mathcal{T}] \longrightarrow K[X_0, \ldots, X_r]^{(d)}$ and

$$\theta : K[Y_\nu : \nu \in \mathcal{T}] \longrightarrow \mathcal{O}(\mathrm{Cone}(V))^{(d)}$$

are homogeneous ring homomorphisms.

If we consider $K[Y_\nu : \nu \in \mathcal{T}]$ as $\mathcal{O}\left(\mathbb{A}^{\binom{r+d}{r}}(K)\right)$, then the graded prime ideal $\mathfrak{q}$ of this polynomial ring defines a projective variety

$$V^{(d)} := V_{\mathbb{P}^{\binom{r+d}{r}-1}(K)}(\mathfrak{q}) \subseteq \mathbb{P}^{\binom{r+d}{r}-1}(K),$$

called the *d-th Veronesean of V*. Note that there is a homogeneous isomorphism of graded K-algebras $\mathcal{O}(\mathrm{Cone}(V^{(d)})) \xrightarrow{\cong} \mathcal{O}(\mathrm{Cone}(V))^{(d)}$.

The main aim of this exercise is to show that the varieties V and $V^{(d)}$ are isomorphic. Some additional notation will be helpful. Given an element $\beta = (\beta_\nu)_{\nu \in \mathcal{T}} \in \mathbb{A}^{\binom{r+d}{r}}(K) \setminus \{0\}$, we use $\overline{\beta} = \overline{(\beta_\nu)_{\nu \in \mathcal{T}}}$ to denote the line

$\{\lambda\beta : \lambda \in K\}$, considered as an element of $\mathbb{P}^{\binom{r+d}{r}-1}(K)$. Also, if $y = (y_0,\ldots,y_r)$ is an $(r+1)$-tuple of elements of some commutative K-algebra and $\nu = (\nu_0,\ldots,\nu_r) \in \mathbb{N}_0^{r+1}$, then we shall use y^ν to denote $y_0^{\nu_0}\ldots y_r^{\nu_r}$. For each $i = 0,\ldots,r$, let $e_i = (0,\ldots,0,1,0,\ldots,0)$ be the element of $\mathbb{N}_0^{r+1}$ whose only non-zero component is a '1' in the i-th position. In order to simplify notation, we shall also employ the $\mathbb{Z}$-module structure on $\mathbb{Z}^{r+1}$. For example, with this notation, we can write that $de_i \in \mathcal{T}$ for all $i = 0,\ldots,r$.

(ii) Let $s \in \mathbb{N}$ and let $\omega^{(1)},\ldots,\omega^{(s)},\mu^{(1)},\ldots,\mu^{(s)} \in \mathcal{T}$ be such that

$$\sum_{i=1}^s \omega^{(i)} = \sum_{i=1}^s \mu^{(i)}.$$

Show that $Y_{\omega^{(1)}}\ldots Y_{\omega^{(s)}} - Y_{\mu^{(1)}}\ldots Y_{\mu^{(s)}} \in \mathfrak{q}$.

(iii) Let $\nu = (\nu_0,\ldots,\nu_r) \in \mathcal{T}$. Deduce from part (ii) that, for all $i,j,k \in \{0,\ldots,r\}$, we have

    (a) $Y_{e_k+(d-1)e_i}Y_{e_i+(d-1)e_j} - Y_{de_i}Y_{e_k+(d-1)e_j} \in \mathfrak{q}$;

    (b) $\prod_{\alpha=0}^r (Y_{e_\alpha+(d-1)e_i})^{\nu_\alpha} - Y_\nu (Y_{de_i})^{d-1} \in \mathfrak{q}$; and

    (c) $Y_\nu^d - \prod_{\alpha=0}^r (Y_{de_\alpha})^{\nu_\alpha} \in \mathfrak{q}$.

(iv) Show that there is a morphism of varieties $\vartheta^{(d)} : V \longrightarrow V^{(d)}$ which is such that

$$\vartheta^{(d)}((\alpha_0 : \cdots : \alpha_r)) = \overline{(\alpha^\nu)_{\nu \in \mathcal{T}}} \quad \text{for all } \alpha = (\alpha_0 : \cdots : \alpha_r) \in V.$$

(v) For $i = 0,\ldots,r$, let $U_i := V^{(d)} \setminus V_{\mathbb{P}^{\binom{r+d}{r}-1}(K)}(Y_{de_i})$. Use part (iii)(c) to show that $U_0,\ldots,U_r$ form an open covering of $V^{(d)}$.

(vi) Show that there is a morphism of varieties $\omega : V^{(d)} \longrightarrow \mathbb{P}^r(K)$ which is such that, for each $i = 0,\ldots,r$ and for all $\overline{(\beta_\nu)_{\nu \in \mathcal{T}}} \in U_i$,

$$\omega(\overline{(\beta_\nu)_{\nu \in \mathcal{T}}}) = (\beta_{e_0+(d-1)e_i} : \cdots : \beta_{e_j+(d-1)e_i} : \cdots : \beta_{e_r+(d-1)e_i}).$$

(vii) Show that $\mathrm{Im}\,\omega \subseteq V$. (Here are some hints. Suppose $i \in \{0,\ldots,r\}$ and $\overline{(\beta_\nu)_{\nu \in \mathcal{T}}} \in U_i$. Let $\alpha := (\beta_{e_0+(d-1)e_i},\ldots,\beta_{e_r+(d-1)e_i})$. It is enough to show that, for each homogeneous $g \in \mathfrak{p}$, we have $g^d(\alpha) = 0$ (as this would imply that $g(\alpha) = 0$). Now $g^d \in \mathfrak{p}^{(d)}$; by part (i), we have $\mathfrak{p}^{(d)} = \phi(\mathfrak{q})$, and so there exists a homogeneous $f \in \mathfrak{q}$ such that $\phi(f) = g^d$. Now use part (iii)(b).)

(viii) Show that $\omega : V^{(d)} \longrightarrow V$ and $\vartheta^{(d)} : V \longrightarrow V^{(d)}$ are inverse isomorphisms of varieties. The isomorphism $\vartheta^{(d)}$ is called the *d-th Veronesean transformation of V* or the *d-th Veronesean map on V*.

15.2.3 Example: rational normal curves. Consider the particular case of Exercise 15.2.2 in which $r = 1$, $V = \mathbb{P}^1(K)$, but d is still arbitrary. The d-th

Veronesean $\mathbb{P}^1(K)^{(d)}$ of $\mathbb{P}^1(K)$ has the property that there is a homogeneous isomorphism

$$\mathcal{O}(\mathbb{P}^1(K)^{(d)}) \cong \mathcal{O}(\mathbb{P}^1(K))^{(d)} = K[X_0, X_1]^{(d)}$$
$$= K[X_0^d, X_0^{d-1}X_1, \ldots, X_0X_1^{d-1}, X_1^d] \subseteq K[X_0, X_1].$$

The Veronesean transformation $\vartheta^{(d)} : \mathbb{P}^1(K) \longrightarrow \mathbb{P}^1(K)^{(d)}$ is an isomorphism of varieties, given by $\vartheta^{(d)}((\sigma : \tau)) = (\sigma^d : \sigma^{d-1}\tau : \cdots : \sigma\tau^{d-1} : \tau^d)$ for all $(\sigma : \tau) \in \mathbb{P}^1(K)$. Thus $\mathbb{P}^1(K)^{(d)} \subseteq \mathbb{P}^d(K)$ is a curve which is isomorphic to the projective line $\mathbb{P}^1(K)$; it is called the *rational normal curve in projective d-space*, and we shall denote it by $N_{(d)} \subseteq \mathbb{P}^d(K)$. The curve $N_{(3)} \subseteq \mathbb{P}^3(K)$ is also called the *twisted cubic*.

The phenomenon mentioned in 15.2.1(ii),(viii) is illustrated by these rational normal curves: it follows from 14.5.26 that, provided $d \geq 3$, the rational normal curve $N_{(d)} \subseteq \mathbb{P}^d(K)$ is not arithmetically Gorenstein, whereas $\mathbb{P}^1(K)$, to which it is isomorphic, is.

15.2.4 Proposition. *Let K be an algebraically closed field, and let $r, d \in \mathbb{N}$. Let $V \subseteq \mathbb{P}^r(K)$ be a projective variety of positive dimension, and consider the d-th Veronesean $V^{(d)} \subseteq \mathbb{P}^{\binom{r+d}{r}-1}(K)$ of V. Set*

$$e := \mathrm{end}(H^1_{\mathcal{O}(\mathrm{Cone}(V))_+}(\mathcal{O}(\mathrm{Cone}(V)))).$$

Then

$$\mathrm{arithdepth}\, V^{(d)} \begin{cases} > 1 & \text{if } d > e, \\ = 1 & \text{if } d = e. \end{cases}$$

In particular, if the projective variety $V \subseteq \mathbb{P}^r(K)$ is a curve, then $V^{(d)}$ is arithmetically Cohen–Macaulay if $d > e$, and is not arithmetically Cohen–Macaulay if $d = e$.

Proof. Let $R := \mathcal{O}(\mathrm{Cone}(V))$. By 15.1.4(ii),(iv), we have $e < \infty$ and $H^1_{R_+}(R)_n = 0$ for all $n \leq 0$. Hence the $(d, 0)$-th Veronesean submodule $H^1_{R_+}(R)^{(d,0)}$ of $H^1_{R_+}(R)\lceil_{R^{(d)}}$ (see 13.5.9) is zero if $d > e$ and is non-zero if $d = e$. By 13.5.9(v), there is a homogeneous $R^{(d)}$-isomorphism

$$H^1_{R_+}(R)^{(d,0)} \xrightarrow{\cong} H^1_{(R_+)^{(d)}}(R^{(d)});$$

furthermore, $(R_+)^{(d)} = (R^{(d)})_+$. Note also that $H^0_{(R^{(d)})_+}(R^{(d)}) = 0$ since $R^{(d)}$ is a domain and $R_+^{(d)} \neq 0$. Since there is a homogeneous ring isomorphism $\mathcal{O}(\mathrm{Cone}(V^{(d)})) \xrightarrow{\cong} R^{(d)}$ by 15.2.2, it follows that arithdepth $V^{(d)} > 1$ if $d > e$ and arithdepth $V^{(d)} = 1$ if $d = e$.

If V is a curve, then $\dim R = 2$, so that, since R is an integral extension domain of $R^{(d)}$, it follows that the *local graded ring $\mathcal{O}(\mathrm{Cone}(V^{(d)}))$ is Cohen–Macaulay if and only if arithdepth $V^{(d)} > 1$. □

15.2.5 ♯Exercise. Let K be an algebraically closed field, let $d \in \mathbb{N}$ with $d \geq 3$, and let $A_{(d)}$ be the subring of the ring $K[X, Y]$ of polynomials over K in two indeterminates X and Y described in 13.5.12 and given by $A_{(d)} := K[X^d, X^{d-1}Y, XY^{d-1}, Y^d]$. Recall from 13.5.12 that $A_{(d)}$ inherits a grading from $K[X, Y]^{(d)}$, the d-th Veronesean subring of $K[X, Y]$.

Let Y_0, Y_1, Y_2, Y_3 be independent indeterminates over K and consider the polynomial ring $K[Y_0, Y_1, Y_2, Y_3]$ as the coordinate ring $\mathcal{O}(\mathbb{A}^4(K))$. Let

$$\psi : K[Y_0, Y_1, Y_2, Y_3] \longrightarrow A_{(d)}$$

be the (surjective, homogeneous) K-algebra homomorphism for which

$$\psi(Y_0) = X^d, \quad \psi(Y_1) = X^{d-1}Y, \quad \psi(Y_2) = XY^{d-1} \quad \text{and} \quad \psi(Y_3) = Y^d.$$

Let $\mathfrak{p} := \mathrm{Ker}\,\psi$; this is a graded prime ideal of $K[Y_0, Y_1, Y_2, Y_3]$ and so defines a projective variety $\Sigma_{(d)} := V_{\mathbb{P}^3(K)}(\mathfrak{p}) \subseteq \mathbb{P}^3(K)$.

(i) Show that there is a morphism $\rho : \mathbb{P}^1(K) \longrightarrow \Sigma_{(d)}$ of varieties for which $\rho((\sigma : \tau)) = (\sigma^d : \sigma^{d-1}\tau : \sigma\tau^{d-1} : \tau^d)$ for all $(\sigma : \tau) \in \mathbb{P}^1(K)$.

(ii) Show that $Y_2^d - Y_3^{d-1}Y_0 \in \mathfrak{p}$.

(iii) Set $U := \Sigma_{(d)} \setminus V_{\mathbb{P}^3(K)}((Y_0))$ and $U' := \Sigma_{(d)} \setminus V_{\mathbb{P}^3(K)}((Y_3))$. Show that U and U' form an open covering of $\Sigma_{(d)}$.

(iv) Show that there is a morphism $\mu : \Sigma_{(d)} \longrightarrow \mathbb{P}^1(K)$ of varieties for which

$$\mu((\alpha : \beta : \gamma : \delta)) = \begin{cases} (\alpha : \beta) & \text{if } (\alpha : \beta : \gamma : \delta) \in U, \\ (\gamma : \delta) & \text{if } (\alpha : \beta : \gamma : \delta) \in U'. \end{cases}$$

(v) Show that ρ and μ are inverse isomorphisms of varieties, so that $\Sigma_{(d)}$ is again a curve isomorphic to $\mathbb{P}^1(K)$.

(vi) Let $t \in \mathbb{N}$. Use 13.5.12 to show that, when $d > 3$, the curve $\Sigma_{(d)} \subseteq \mathbb{P}^3(K)$ is not arithmetically Cohen–Macaulay, but that the t-th Veronesean $(\Sigma_{(d)})^{(t)} \subseteq \mathbb{P}^{\binom{t+3}{3}-1}(K)$ is arithmetically Cohen–Macaulay provided that $t \geq d-2$. Is $(\Sigma_{(d)})^{(d-3)} \subseteq \mathbb{P}^{\binom{d}{3}-1}(K)$ arithmetically Cohen–Macaulay?

15.2.6 Remarks. Let the situation and notation be as in Exercise 15.2.5, where the curve $\Sigma_{(d)} \subseteq \mathbb{P}^3(K)$ was constructed, for each integer $d \geq 3$. As a variety, $\Sigma_{(d)}$ is isomorphic to the projective line $\mathbb{P}^1(K)$. Note that $\Sigma_{(3)} \subseteq$

$\mathbb{P}^3(K)$ is just the twisted cubic $N_{(3)} \subseteq \mathbb{P}^3(K)$ of 15.2.3. The curve $\Sigma_{(4)} \subseteq$ $\mathbb{P}^3(K)$ is called the *twisted quartic* or *Macaulay's curve*.

Exercise 15.2.5(vi) shows that the arithmetic structure of $\Sigma_{(d)} \subseteq \mathbb{P}^3(K)$, which reflects the way in which the curve is embedded in $\mathbb{P}^3(K)$, is more complicated for $d > 3$ then it is when $d = 3$. This is another illustration of the phenomenon mentioned in 15.2.1(ii),(viii).

15.2.7 Exercise. Let K be a field, and let U, W, S, T be indeterminates over K. In the 5-th Veronesean subring $K[U, W, S, T]^{(5)}$ of the polynomial ring $K[U, W, S, T]$, set $f_1 := U^4 T$, $f_2 := W S^4$ and $f_3 := U^4 S - W T^4$. Consider the four K-subalgebras A, B, C and R of $K[U, W, S, T]^{(5)}$ given by

$$A := K[U^4 S, U^4 T, W S^4, W S^3 T, W S^2 T^2, W S T^3, W T^4],$$

$$B := K[U^4 S, U^4 T, W S^4, W S^3 T, W S T^3, W T^4] \subseteq A,$$

$$C := K[U^4 S, U^4 T, W S^4, W S^2 T^2, W S T^3, W T^4] \subseteq A,$$

$$R := K[f_1, f_2, f_3] = K[U^4 T, W S^4, U^4 S - W T^4] \subseteq B \cap C \subseteq A.$$

Observe that each of these subrings inherits a grading from $K[U, W, S, T]^{(5)}$ that turns it into a homogeneous positively graded *local integral domain. It follows from 14.5.2 and 14.5.24 that each of them has a *canonical module and an S_2-ification.

(i) Show (by induction on n) that, for all $n \in \mathbb{N}_0$, the set of monomials

$$\{U^{4i} W^{n-i} S^k T^{4n-3i-k} : 0 \le i \le n, \ 0 \le k \le 4n - 3i\}$$

is a K-basis of A_n and conclude that $\dim_K A_n = (n + 1)(5n + 2)/2$.

(ii) Show that

$$A_2 = (f_1 K + f_2 K + f_3 K) A_1$$

and deduce that

$$A_n = (f_1 K + f_2 K + f_3 K) A_{n-1} \quad \text{for all } n \ge 2.$$

Deduce that the R-module A is generated by the five homogeneous elements

$$g_1 := 1, \ g_2 := U^4 S, \ g_3 := W S^3 T, \ g_4 := W S^2 T^2, \ g_5 := W S T^3.$$

(iii) Show that $\dim A = 3 = \dim R$, that f_1, f_2, f_3 are algebraically independent over K, that f_1, f_2, f_3 is an R-sequence, and that $\dim_K R_n = (n + 1)(n + 2)/2$ for all $n \in \mathbb{N}_0$.

(iv) For each $n \in \mathbb{N}_0$ let

$$\mathcal{M}_n := \{f_1^{\nu_1} f_2^{\nu_2} f_3^{\nu_3} : \nu_1, \nu_2, \nu_3 \in \mathbb{N}_0, \ \nu_1 + \nu_2 + \nu_3 = n\},$$

the set of all monomials of degree n in f_1, f_2, f_3, and let $\mathcal{M}_n g_i$, for $n \in \mathbb{N}_0$ and $i \in \{1, \ldots, 5\}$, denote $\{m_n g_i : m_n \in \mathcal{M}_n\}$. Use part (ii) to show that, for each $n \in \mathbb{N}$, the members of the set

$$\mathcal{M}_n g_1 \cup \mathcal{M}_{n-1} g_2 \cup \mathcal{M}_{n-1} g_3 \cup \mathcal{M}_{n-1} g_4 \cup \mathcal{M}_{n-1} g_5$$

span the K-space A_n, and use the formula in part (i) for $\dim_K A_n$ to deduce that the above-displayed set is actually a K-basis for A_n. Deduce that the five homogeneous elements g_1, g_2, g_3, g_4, g_5 form a base for the R-module A, so that A is a *free graded R-module of rank 5.

(v) Deduce that f_1, f_2, f_3 is an A-sequence and conclude that the ring A is Cohen–Macaulay.

(vi) Show that $A = B + WS^2T^2B$ and $B_+A \subseteq B$. Deduce that there is a homogeneous B-isomorphism $A/B \cong {}^1K$, where ${}^1K$ is as defined in 14.1.13.

(vii) Show that that there is a homogeneous B-isomorphism $H^1_{B_+}(B) \cong {}^1K$; deduce that grade $B_+ = 1$ and that the non-S_2 locus of B is $\{B_+\} = \mathrm{Var}(B_+)$.

(viii) Use 2.2.15(iii), 2.2.17 and 13.5.4(ii) to show that there is a homogeneous isomorphism of B-algebras $A \cong D_{B_+}(B)$, and that both these B-algebras are isomorphic to the S_2-ification of B.

Let $\mathfrak{p}$ be the ideal $(U^4S, U^4T, WS^2T^2, WST^3, WT^4)$ of

$$C = K[U^4S, U^4T, WS^4, WS^2T^2, WST^3, WT^4].$$

(ix) Show that $A = C + WS^3TC$, and that $\mathfrak{p}A \subseteq C$.

(x) Show that, for all $n \in \mathbb{N}$, we have $(WS^4)^n WS^3T \notin C$.

(xi) Show that $C/\mathfrak{p} = K[WS^4 + \mathfrak{p}]$, and that $WS^4 + \mathfrak{p}$ is transcendental over K (in the usual sense that the only polynomial $f \in K[X]$ for which $f(WS^4 + \mathfrak{p}) = 0$ is the zero polynomial). Deduce that $\mathfrak{p} \in {}^*\mathrm{Spec}(C)$, that $\dim C/\mathfrak{p} = 1$ and that $\mathrm{ht}\,\mathfrak{p} = 2$.

(xii) Show that the C-module A/C is generated by $WS^3T + C$, and that the annihilator of this element is $\mathfrak{p}$. Deduce that there is a homogeneous C-isomorphism $H^1_{\mathfrak{p}}(C) \cong (C/\mathfrak{p})(-1)$. Deduce that grade $\mathfrak{p} = 1$ and that the non-S_2 locus of C is $\mathrm{Var}(\mathfrak{p})$.

(xiii) Show that there is a homogeneous isomorphism of C-algebras $A \cong D_{\mathfrak{p}}(C)$, and that both these C-algebras are isomorphic to the S_2-ification of C.

15.2.8 Exercise. Let K be an algebraically closed field, and consider the homogeneous, $\mathbb{N}_0$-graded K-algebras A, B and C defined in 15.2.7. Let $V_A \subset \mathbb{P}^6(K)$ be the irreducible projective variety whose homogeneous coordinate ring satisfies $\mathcal{O}(\mathrm{Cone}(V_A)) = A$; also let $V_B, V_C \subset \mathbb{P}^5(K)$ be the irreducible projective varieties such that $\mathcal{O}(\mathrm{Cone}(V_B)) = B$ and $\mathcal{O}(\mathrm{Cone}(V_C)) = C$ respectively.

(i) Show that

$$V_A = \{(\alpha s : \alpha t : \beta s^4 : \beta s^3 t : \beta s^2 t^2 : \beta s t^3 : \beta t^4) \in \mathbb{P}^6(K) :$$
$$(\alpha, \beta), (s, t) \in K^2 \setminus \{(0,0)\}\},$$

so that V_A is the disjoint union of the lines in $\mathbb{P}^6(K)$ joining the points $(s : t : 0 : 0 : 0 : 0 : 0)$ and $(0 : 0 : s^4 : s^3 t : s^2 t^2 : s t^3 : t^4)$, where $(s : t)$ runs through the projective line $\mathbb{P}^1(K)$. Readers with some background in classical algebraic geometry might recognize V_A as the standard rational normal surface scroll $S(1,4) \subset \mathbb{P}^6(K)$ (see [27, p. 94]).

(ii) Find descriptions for V_B and V_C similar to that for V_A in part (a).

(iii) Let $e_4 = (0 : 0 : 0 : 0 : 1 : 0 : 0) \in \mathbb{P}^6(K)$. Consider the projection map (see [30, p. 22]) $\pi_4 : \mathbb{P}^6(K) \setminus \{e_4\} \longrightarrow \mathbb{P}^5(K)$ for which

$$\pi_4((x_0 : x_1 : x_2 : x_3 : x_4 : x_5 : x_6)) = (x_0 : x_1 : x_2 : x_3 : x_5 : x_6)$$

for all $(x_0 : x_1 : x_2 : x_3 : x_4 : x_5 : x_6) \in \mathbb{P}^6(K) \setminus \{e_4\}$. Show that $\pi_4(V_A) = V_B$.

(iv) Let $e_3 = (0 : 0 : 0 : 1 : 0 : 0 : 0) \in \mathbb{P}^6(K)$. Consider the projection map $\pi_3 : \mathbb{P}^6(K) \setminus \{e_3\} \longrightarrow \mathbb{P}^5(K)$ for which

$$\pi_3((x_0 : x_1 : x_2 : x_3 : x_4 : x_5 : x_6)) = (x_0 : x_1 : x_2 : x_4 : x_5 : x_6)$$

for all $(x_0 : x_1 : x_2 : x_3 : x_4 : x_5 : x_6) \in \mathbb{P}^6(K) \setminus \{e_3\}$. Show that $\pi_3(V_A) = V_C$.

Although the two varieties V_B and V_C are both obtained by projecting V_A from a point, the two projections turn out to be quite different, as can be seen from the properties of their homogeneous coordinate rings developed in 15.2.7.

In the next Chapter 16, we shall discuss the situation where $R = \bigoplus_{n \in \mathbb{N}_0} R_n$ is positively graded; we shall study in considerable depth the components $H^i_{R_+}(M)_n$ ($n \in \mathbb{Z}$) of the i-th ($i \in \mathbb{N}_0$) local cohomology module, with respect to the irrelevant ideal R_+ of R, of a finitely generated graded R-module M. There is very strong motivation from projective algebraic geometry for such study, and we hope that our discussions of projective varieties in this chapter have given the reader some hints of this motivation.

16

Castelnuovo regularity

In Chapter 15, we have seen that, when (K is an algebraically closed field, $r \in \mathbb{N}$ and) R is the homogeneous coordinate ring of a projective variety $V \subseteq \mathbb{P}^r(K)$ of positive dimension, the end of the (necessarily graded) first local cohomology module $H^1_{R_+}(R)$ is of interest: see 15.2.4. This is one motivation for our work in this chapter, where we shall study, in the case when $R = \bigoplus_{n \in \mathbb{N}_0} R_n$ is positively graded, the ends of the local cohomology modules $H^i_{R_+}(M)$ for a finitely generated graded R-module M. Perhaps the most important invariant related to these ends is the so-called (Castelnuovo–Mumford) regularity of M. This invariant is of great significance in algebraic geometry, and, as we shall see in 16.3.7 and 16.3.8, it provides links between local cohomology theory and the syzygies of finitely generated graded modules over a polynomial ring over a field.

Throughout this chapter, all graded rings and modules are to be understood to be $\mathbb{Z}$-graded.

16.1 Finitely generated components

Our first goal in this chapter is to establish the basic facts that, in the notation of the above introduction, for each $i \in \mathbb{N}_0$, the R_0-module $H^i_{R_+}(M)_n$ is finitely generated for all $n \in \mathbb{Z}$, and is zero for all sufficiently large n. These facts, which generalize 15.1.4(iii),(ii), are the basis for much of the work in this and the next chapter.

16.1.1 Notation and Terminology. We shall employ the notation and terminology concerning graded rings and modules described in 13.1.1, 13.1.3, 14.1.1 and 15.1.1, but, in accordance with our convention for the whole of this chapter, restricted to the special case in which $G = \mathbb{Z}$. In particular, all polynomial rings $R_0[X_1, \ldots, X_n]$ (in n indeterminates $X_1, \ldots, X_n$ over a commutative Noetherian ring R_0) considered in this chapter will be positively $\mathbb{Z}$-graded so that R_0 is the component of degree 0 and $\deg X_i = 1$ for all $i = 1, \ldots, n$.

Recall [7, p. 29] that, when $R = \bigoplus_{n \in \mathbb{N}_0} R_n$ is positively graded, we say that R is *homogeneous* if and only if R is generated as an R_0-algebra by its forms of degree 1, that is, if and only if $R = R_0[R_1]$.

Also, when R is merely positively graded, $L = \bigoplus_{n \in \mathbb{Z}} L_n$ is a graded R-module and $t \in \mathbb{Z}$ is fixed, we define

$$L_{\geq t} := \bigoplus_{\substack{n \in \mathbb{Z} \\ n \geq t}} L_n,$$

a graded submodule of L. For example, in 16.1.6 below, we shall be concerned with the graded R-modules $D_{R_+}(M)_{\geq t}$ (for a graded R-module M).

We shall need to use the following graded version of the Prime Avoidance Theorem; for a proof, we refer the reader to [7].

16.1.2 Homogeneous Prime Avoidance Lemma. (See [7, 1.5.10].) *Let R be graded and assume the ideal $\mathfrak{a}$ is graded and generated by elements of positive degree. Let $\mathfrak{p}_1, \ldots, \mathfrak{p}_n \in \operatorname{Spec}(R)$ be such that, for all $i = 1, \ldots, n$, we have $\mathfrak{a} \not\subseteq \mathfrak{p}_i$. Then there exists a homogeneous element in $\mathfrak{a} \setminus (\mathfrak{p}_1 \cup \cdots \cup \mathfrak{p}_n)$.* □

We can improve on the result of 16.1.2 when R_0 is local with infinite residue field and $\mathfrak{a}$ is generated by elements of degree 1.

16.1.3 Lemma. (See [7, 1.5.12].) *Assume that $R = \bigoplus_{n \in \mathbb{Z}} R_n$ is graded, that R_0 is local with infinite residue field, and that the ideal $\mathfrak{a}$ is graded and generated by elements of degree 1. Let $\mathfrak{b}_1, \ldots, \mathfrak{b}_n$ be ideals of R such that, for all $i = 1, \ldots, n$, we have $\mathfrak{a} \not\subseteq \mathfrak{b}_i$. Then there exists a homogeneous element of degree 1 in $\mathfrak{a} \setminus (\mathfrak{b}_1 \cup \cdots \cup \mathfrak{b}_n)$.*

Proof. Set $\mathfrak{a}_1 := \mathfrak{a} \cap R_1$, and let $\mathfrak{m}_0$ denote the maximal ideal of R_0. The hypotheses ensure that, for all $i = 1, \ldots, n$,

$$(\mathfrak{b}_i \cap \mathfrak{a}_1 + \mathfrak{m}_0 \mathfrak{a}_1)/\mathfrak{m}_0 \mathfrak{a}_1 \neq \mathfrak{a}_1/\mathfrak{m}_0 \mathfrak{a}_1.$$

Since $R_0/\mathfrak{m}_0$ is infinite, there exists $r_1 \in \mathfrak{a}_1 \setminus \bigcup_{i=1}^n (\mathfrak{b}_i \cap \mathfrak{a}_1 + \mathfrak{m}_0 \mathfrak{a}_1)$. □

16.1.4 Lemma. *Assume that $R = \bigoplus_{n \in \mathbb{Z}} R_n$ is graded and that the ideal $\mathfrak{a}$ is graded and generated by elements of positive degree. Let M be a finitely generated R-module such that $\mathfrak{a}M \neq M$ and $\operatorname{grade}_M \mathfrak{a} > 0$.*

(i) *There is a homogeneous element in $\mathfrak{a}$ which is a non-zerodivisor on M.*

(ii) *If, in addition, R_0 is local with infinite residue field and $\mathfrak{a}$ is generated by elements of degree 1, then there exists a homogeneous element of degree 1 in $\mathfrak{a}$ which is a non-zerodivisor on M.*

Proof. Since $\mathfrak{a} \not\subseteq \bigcup_{\mathfrak{p} \in \mathrm{Ass}\, M} \mathfrak{p}$, part (i) is immediate from 16.1.2, while part (ii) follows from 16.1.3. □

We are now ready to prove the basic finiteness and vanishing theorem which was mentioned in the first paragraph of this section.

16.1.5 Theorem. *Assume that $R = \bigoplus_{n \in \mathbb{N}_0} R_n$ is positively graded; let M be a finitely generated graded R-module.*

(i) *For all $i \in \mathbb{N}_0$ and all $n \in \mathbb{Z}$, the R_0-module $H^i_{R_+}(M)_n$ is finitely generated.*

(ii) *There exists $r \in \mathbb{Z}$ such that $H^i_{R_+}(M)_n = 0$ for all $i \in \mathbb{N}_0$ and all $n \geq r$.*

Proof. Let $i \in \mathbb{N}_0$. We prove, by induction on i, that $H^i_{R_+}(M)_n$ is a finitely generated R_0-module for all $n \in \mathbb{Z}$, and is zero for all sufficiently large values of n. This will prove not only part (i) but also part (ii), because, in view of 3.3.3, there can only be finitely many integers i for which $H^i_{R_+}(M) \neq 0$, since $H^i_{R_+}(M) = 0$ for all $i > \mathrm{ara}(R_+)$.

We consider first the case where $i = 0$. Since $H^0_{R_+}(M)$ is a submodule of M, it is finitely generated as an R-module, and so there exists $u \in \mathbb{N}$ such that $(R_+)^u H^0_{R_+}(M) = 0$. Now $(R_+)^i H^0_{R_+}(M)/(R_+)^{i+1} H^0_{R_+}(M)$ is a Noetherian R/R_+-module, and so is a Noetherian R_0-module (for each $i = 0, \ldots, u - 1$). Therefore $H^0_{R_+}(M)$ is a Noetherian R_0-module. Hence $H^0_{R_+}(M)_n$ is a finitely generated R_0-module for all $n \in \mathbb{Z}$, and only finitely many of the $H^0_{R_+}(M)_n$ can be non-zero.

Now suppose that $i > 0$ and our desired result has been proved for smaller values of i (and for all choices of the finitely generated graded R-module M). Since $\Gamma_{R_+}(M)$ is a graded submodule of M, it follows from 2.1.7(iii) that, for all $i \in \mathbb{N}$, there is a (homogeneous) isomorphism

$$H^i_{R_+}(M) \cong H^i_{R_+}(M/\Gamma_{R_+}(M)).$$

Thus, for the purpose of this inductive step, we can replace M by $M/\Gamma_{R_+}(M)$ and so (in view of 2.1.2 and 2.1.1) assume that R_+ contains a non-zerodivisor on M. This we do.

We can assume that $M \neq 0$; we therefore assume that $M \neq R_+ M$. Then, by Lemma 16.1.4, there exists a homogeneous element $r \in R_+$ which is a non-zerodivisor on M. Let t denote the degree of r.

The exact sequence $0 \longrightarrow M \stackrel{r}{\longrightarrow} M(t) \longrightarrow (M/rM)(t) \longrightarrow 0$ of graded R-modules and homogeneous homomorphisms induces a long exact sequence (in $*\mathcal{C}(R)$) of local cohomology modules, from which we deduce (with the aid of 14.1.10(ii)) an exact sequence

$$H_{R_+}^{i-1}(M/rM)_{n+t} \longrightarrow H_{R_+}^i(M)_n \stackrel{r}{\longrightarrow} H_{R_+}^i(M)_{n+t}$$

of R_0-modules for all $n \in \mathbb{Z}$.

By the inductive hypothesis, there exists $s \in \mathbb{Z}$ such that

$$H_{R_+}^{i-1}(M/rM)_j = 0 \quad \text{for all } j \geq s.$$

Hence, for all $n \geq s - t$, the sequence $0 \longrightarrow H_{R_+}^i(M)_n \stackrel{r}{\longrightarrow} H_{R_+}^i(M)_{n+t}$ is exact; therefore, since $H_{R_+}^i(M)$ is R_+-torsion and $r \in R_+$, we have

$$H_{R_+}^i(M)_n = 0 \quad \text{for all } n \geq s - t.$$

The inductive hypothesis also yields that $H_{R_+}^{i-1}(M/rM)_j$ is a finitely generated R_0-module for all $j \in \mathbb{Z}$. Fix $n \in \mathbb{Z}$ and let $k \in \mathbb{N}_0$ be such that $n + kt \geq s - t$, so that $H_{R_+}^i(M)_{n+kt} = 0$ by the last paragraph. Now, for each $j = 0, \ldots, k - 1$, there is an exact sequence

$$H_{R_+}^{i-1}(M/rM)_{n+(j+1)t} \longrightarrow H_{R_+}^i(M)_{n+jt} \stackrel{r}{\longrightarrow} H_{R_+}^i(M)_{n+(j+1)t}$$

of R_0-modules, and $H_{R_+}^{i-1}(M/rM)_{n+(j+1)t}$ is finitely generated over R_0. We can therefore deduce, successively, that $H_{R_+}^i(M)_{n+jt}$ is finitely generated over R_0 for $j = k - 1, k - 2, \ldots, 1, 0$; we thus conclude that $H_{R_+}^i(M)_n$ is finitely generated over R_0, and the inductive step is complete. $\square$

16.1.6 Corollary. *Assume that* $R = \bigoplus_{n \in \mathbb{N}_0} R_n$ *is positively graded; let* $M = \bigoplus_{n \in \mathbb{Z}} M_n$ *be a finitely generated graded R-module. Then (with the notation of 16.1.1)*

(i) $D_{R_+}(M)_{\geq t}$ *is a finitely generated R-module for all* $t \in \mathbb{Z}$;

(ii) $D_{R_+}(M)_n$ *is a finitely generated R_0-module for all* $n \in \mathbb{Z}$; *and*

(iii) *for all sufficiently large n, the restriction to M_n of the map* $\eta_M : M \longrightarrow D_{R_+}(M)$ *of 2.2.6(i)(c) provides an R_0-isomorphism*

$$(\eta_M)_n : M_n \longrightarrow D_{R_+}(M)_n.$$

Proof. These claims are all immediate consequences of Theorem 16.1.5 and the exact sequence

$$0 \longrightarrow \Gamma_{R_+}(M) \stackrel{\xi_M}{\longrightarrow} M \stackrel{\eta_M}{\longrightarrow} D_{R_+}(M) \stackrel{\zeta_M}{\longrightarrow} H_{R_+}^1(M) \longrightarrow 0$$

of graded R-modules and homogeneous homomorphisms: see 13.5.4(i). $\square$

Theorem 16.1.5 shows that, when R is positively graded, the theory of local cohomology with respect to the irrelevant ideal of R is particularly satisfactory. It is natural to ask whether local cohomology with respect to other homogeneous ideals exhibits similar properties. The next three exercises concern examples where the answer is clearly negative: the conclusions of these exercises should be compared with those of Theorem 16.1.5.

16.1.7 Exercise. Let $(R_0, \pi R_0)$ be a discrete valuation ring, and let R denote the polynomial ring $R_0[X]$. Let $\mathfrak{m} := \pi R + XR$, the unique maximal graded ideal of R. Let M be a non-zero, torsion-free, finitely generated, graded R-module.

(i) Use the Graded Finiteness Theorem 14.3.10 to show that $H^1_{\mathfrak{m}}(M)$ is finitely generated.

(ii) Show that $H^2_{\mathfrak{m}}(M/XM) = 0$.

(iii) Use 13.5.5 to see that there exists $n_0 \in \mathbb{Z}$ for which $H^1_{\mathfrak{m}}(M/XM)_{n_0}$ is not finitely generated as R_0-module.

(iv) Use the exact sequence $0 \to M \xrightarrow{X} M(1) \to (M/XM)(1) \to 0$ to show that there exists $n_1 \in \mathbb{Z}$ such that $H^2_{\mathfrak{m}}(M)_n$ is not finitely generated for all $n < n_1$ and there exists $n_2 \in \mathbb{Z}$ such that $H^2_{\mathfrak{m}}(M)_n = 0$ for all $n > n_2$.

16.1.8 Exercise. Let R_0 be a commutative Noetherian ring, and let R denote the polynomial ring $R_0[X, Y]$. Use Exercise 13.5.5 to show that, for all $n \in \mathbb{Z}$, the R_0-module $H^1_{XR}(R)_n$ is free but not finitely generated.

Compare this with the conclusions of Theorem 16.1.5.

16.1.9 Exercise. Let $(R_0, \pi R_0)$ be a discrete valuation ring as in Exercise 16.1.7, and let R denote the graded ring $R_0[X, Y]$, as in Exercise 16.1.8. Let $\mathfrak{a} := \pi R + XR$. Use 16.1.8 and 14.1.11 to show that, for all $n \in \mathbb{Z}$, the R_0-module $H^2_{\mathfrak{a}}(R)_n$ is not finitely generated.

16.1.10 ♯Exercise. Assume that $R = \bigoplus_{n \in \mathbb{N}_0} R_n$ is positively graded and such that $(R_0, \mathfrak{m}_0)$ is local. Set $\mathfrak{m} = \mathfrak{m}_0 R + R_+$, the unique graded maximal ideal of R. Let M be a finitely generated graded R-module such that $d := \dim(M/\Gamma_{R_+}(M)) > 0$.

(i) Show that there is a homogeneous element a of positive degree t which is a non-zerodivisor on $\overline{M} := M/\Gamma_{R_+}(M)$.

(ii) Use Theorem 6.1.4 to show that $H^d_{\mathfrak{m}}(\overline{M}) \neq 0$, and part (i) to deduce that there are infinitely many negative $n \in \mathbb{Z}$ such that $H^d_{\mathfrak{m}}(\overline{M})_n \neq 0$.

(iii) Use 14.1.7 and 14.1.12 to show that, for each $i \in \mathbb{N}_0$, there are only finitely many integers $n \in \mathbb{Z}$ for which $H^i_{\mathfrak{m}}(\Gamma_{R_+}(M))_n \neq 0$.

(iv) Prove that $H^d_{\mathfrak{m}}(M)_n \neq 0$ for infinitely many negative integers n.

16.2 The basics of Castelnuovo regularity

Assume that R is positively graded and homogeneous, and let M be a finitely generated graded R-module. It is clear from 3.3.3 and 16.1.5 that there is an integer r such that, for all $s \in \mathbb{Z}$ with $s > r$, we have $H^i_{R_+}(M)_{s-i} = 0$ for all $i \in \mathbb{Z}$ (or, equivalently, for all $i \in \mathbb{N}_0$). The Castelnuovo regularity of M is the infimum of the set of integers r with this property (interpreted as $-\infty$ if this infimum does not exist). This is a very important invariant of M, and a topic which has featured in much recent research. In this section, we develop the basic theory of this and some related, slightly more complicated, invariants.

16.2.1 Definitions. Assume that R is positively graded and homogeneous (see 16.1.1). Let M be a finitely generated graded R-module, and let $r \in \mathbb{Z}$ and $l \in \mathbb{N}_0$.

(i) We follow A. Ooishi in [65, Definition 1] and say that M is *r-regular in the sense of Castelnuovo–Mumford* if and only if $H^i_{R_+}(M)_{s-i} = 0$ for all $i, s \in \mathbb{Z}$ with $s > r$. In practice, the phrase 'in the sense of Castelnuovo–Mumford' is usually omitted.

As pointed out just before this definition, there does exist an $r \in \mathbb{Z}$ such that M is r-regular. Note that M is r-regular if and only if $H^i_{R_+}(M)_{s-i} = 0$ for all $i, s \in \mathbb{Z}$ with $s > r$ and $i \geq 0$. This observation leads to the following more general definition.

(ii) We say that M is *r-regular at and above level l* if and only if

$$H^i_{R_+}(M)_{s-i} = 0 \quad \text{for all } i, s \in \mathbb{Z} \text{ with } s > r \text{ and } i \geq l.$$

Thus M is r-regular if and only if it is r-regular at and above level 0.

Motivation for Definition 16.2.1 comes from Theorem 16.2.5 below, which is a version of a proposition of D. Mumford [54, p. 99], adapted to the context of local cohomology. (See also Ooishi [65, Theorem 2].) However, before presenting Theorem 16.2.5, we provide some technical comments which will help us to make appropriate reductions in the proof of the theorem and other results.

16.2.2 Remarks. Assume that $R = \bigoplus_{n \in \mathbb{Z}} R_n$ is graded, and that the ideal $\mathfrak{a}$ is graded. Let R'_0 be a commutative Noetherian ring, and let $f_0 : R_0 \longrightarrow R'_0$

be a flat ring homomorphism. Set $R' := R \otimes_{R_0} R'_0$, and let $f : R \longrightarrow R'$ be
the natural ring homomorphism.

(i) Since R is a finitely generated R_0-algebra by [7, Theorem 1.5.5], it
     follows that R' is Noetherian. Furthermore, R' has the structure of a
     graded ring given by $R' = \bigoplus_{n \in \mathbb{Z}} R'_n$, where R'_n is the natural image of
     $R_n \otimes_{R_0} R'_0$ in R' for each $n \in \mathbb{N}$. Thus the ring homomorphism f is
     homogeneous (in the sense of 14.1.3), and flat.

(ii) It therefore follows from the Graded Flat Base Change Theorem 14.1.9
     that the sequences $(H^i_{\mathfrak{a}}(\bullet) \otimes_R R')_{i \in \mathbb{N}_0}$ and $(H^i_{\mathfrak{a}R'}((\bullet) \otimes_R R'))_{i \in \mathbb{N}_0}$
     (restricted to the 'graded' category $*\mathcal{C}(R)$) are isomorphic negative
     connected sequences of covariant functors from $*\mathcal{C}(R)$ to $*\mathcal{C}(R')$.

(iii) Let $L = \bigoplus_{n \in \mathbb{Z}} L_n$ be a graded R-module. Then $L \otimes_R R'$ is a graded
      R'-module: if we use the natural isomorphisms

$$L \otimes_R R' = L \otimes_R (R \otimes_{R_0} R'_0) \xrightarrow{\cong} (L \otimes_R R) \otimes_{R_0} R'_0 \xrightarrow{\cong} L \otimes_{R_0} R'_0$$

      to identify $L \otimes_R R'$ with $L \otimes_{R_0} R'_0 =: L'$, then the grading on L' is
      given by $L' = \bigoplus_{n \in \mathbb{Z}} L'_n$, where L'_n is the natural image of $L_n \otimes_{R_0} R'_0$
      in L'.

(iv) It follows from part (ii), and part (iii) applied to the graded R-module
     $H^i_{\mathfrak{a}}(M)$ (where M is a graded R-module), that, for all $i \in \mathbb{N}_0$ and all
     $n \in \mathbb{Z}$, there is a natural equivalence of functors (from $*\mathcal{C}(R)$ to $\mathcal{C}(R'_0)$)

$$H^i_{\mathfrak{a}}(\bullet)_n \otimes_{R_0} R'_0 \longrightarrow H^i_{\mathfrak{a}R'}((\bullet) \otimes_R R')_n.$$

(v) We can deduce from 4.3.5, 13.3.14 and the formula in 2.2.15(ii) that
     there is a natural equivalence of functors

$$\varepsilon : D_{\mathfrak{a}}(\bullet) \otimes_R R' \longrightarrow D_{\mathfrak{a}R'}((\bullet) \otimes_R R')$$

     from the 'graded' category $*\mathcal{C}(R)$ to $*\mathcal{C}(R')$.

(vi) If we now use part (v), and part (iii) applied to the graded R-module
     $D_{\mathfrak{a}}(M)$ (where M is a graded R-module), we obtain, for each $n \in \mathbb{Z}$, a
     natural equivalence of functors (from $*\mathcal{C}(R)$ to $\mathcal{C}(R'_0)$)

$$D_{\mathfrak{a}}(\bullet)_n \otimes_{R_0} R'_0 \longrightarrow D_{\mathfrak{a}R'}((\bullet) \otimes_R R')_n.$$

We give two examples where the ideas of 16.2.2 can be used.

16.2.3 Example. Assume that $R = \bigoplus_{n \in \mathbb{N}_0} R_n$ is positively graded and homogeneous (see 16.1.1), let M be a graded R-module, and let $\mathfrak{p}_0 \in \operatorname{Spec}(R_0)$.

We can apply the techniques (and notation) of 16.2.2 with the choices $\mathfrak{a} = R_+$ and $R'_0 = (R_0)_{\mathfrak{p}_0}$, a flat R_0-algebra. The ring $R' = R \otimes_{R_0} (R_0)_{\mathfrak{p}_0}$ is
again positively graded and homogeneous, and $M' = M \otimes_{R_0} (R_0)_{\mathfrak{p}_0}$ is a

graded R'-module. Also, $R_+R' = R'_+$, the irrelevant ideal of R'. It follows from 16.2.2(iv) that, for each $i \in \mathbb{N}_0$ and each $n \in \mathbb{Z}$, there is an isomorphism of $(R_0)_{\mathfrak{p}_0}$-modules $\left(H^i_{R_+}(M)_n\right)_{\mathfrak{p}_0} \cong H^i_{R'_+}(M')_n$. Finally, note that the 0-th component of R' is a local ring isomorphic to $(R_0)_{\mathfrak{p}_0}$.

16.2.4 Example. Assume that $R = \bigoplus_{n \in \mathbb{N}_0} R_n$ is positively graded and homogeneous, and such that $(R_0, \mathfrak{m}_0)$ is local; let M be a graded R-module.

We can apply the techniques (and notation) of 16.2.2 with the choices $\mathfrak{a} = R_+$ and $R'_0 = R_0[X]_{\mathfrak{m}_0 R_0[X]}$, the localization of the polynomial ring $R_0[X]$ at the prime ideal $\mathfrak{m}_0 R_0[X]$. This R'_0 is a flat R_0-algebra, and

$$R' = R \otimes_{R_0} R_0[X]_{\mathfrak{m}_0 R_0[X]}$$

is positively graded and homogeneous, $M' = M \otimes_{R_0} R_0[X]_{\mathfrak{m}_0 R_0[X]}$ is a graded R'-module, and $R_+R' = R'_+$, the irrelevant ideal of R'.

It follows from 16.2.2(iv) that, for each $i \in \mathbb{N}_0$ and each $n \in \mathbb{Z}$, there is an isomorphism of $R_0[X]_{\mathfrak{m}_0 R_0[X]}$-modules

$$H^i_{R_+}(M)_n \otimes_{R_0} R_0[X]_{\mathfrak{m}_0 R_0[X]} \cong H^i_{R'_+}(M')_n;$$

thus, since $R_0[X]_{\mathfrak{m}_0 R_0[X]}$ is a faithfully flat R_0-algebra, $H^i_{R_+}(M)_n = 0$ if and only if $H^i_{R'_+}(M')_n = 0$.

Finally, note that the 0-th component of R' is a local ring isomorphic to $R_0[X]_{\mathfrak{m}_0 R_0[X]}$, having infinite residue field.

16.2.5 Theorem. *Assume that $R = \bigoplus_{n \in \mathbb{N}_0} R_n$ is positively graded and homogeneous; let M be a finitely generated graded R-module. Let $r \in \mathbb{Z}$ and $l \in \mathbb{N}$: the reader should note that we are assuming that l is* positive.

Assume that $H^i_{R_+}(M)_{r+1-i} = 0$ for all $i \geq l$. Then M is r-regular at and above level l (see 16.2.1(ii)).

Proof. It suffices to show that, for each $\mathfrak{p}_0 \in \operatorname{Spec}(R_0)$, the $(R_0)_{\mathfrak{p}_0}$-module $(H^i_{R_+}(M)_{s-i})_{\mathfrak{p}_0}$ vanishes for all $i \geq l$ and all $s > r$.

We can apply Examples 16.2.3 and 16.2.4 in turn to see that it is enough for us to establish the claim in the statement of the theorem under the additional assumption that $(R_0, \mathfrak{m}_0)$ is a local ring with infinite residue field, and we shall make this assumption in what follows.

It follows from [7, 1.5.4] that $\dim M$ is finite, and we are going to argue by induction on $\dim M$. When $\dim M = -1$ there is nothing to prove, and when $\dim M = 0$ the result is an easy consequence of Grothendieck's Vanishing Theorem 6.1.2.

Now suppose that $\dim M > 0$ and our desired result has been proved for all finitely generated graded R-modules of smaller dimension. Since $\Gamma_{R_+}(M)$

is a graded submodule of M, it follows from 2.1.7(iii) that, for each $i \in \mathbb{N}$, there is a homogeneous isomorphism $H^i_{R_+}(M) \cong H^i_{R_+}(M/\Gamma_{R_+}(M))$. In particular, since l is positive, $H^i_{R_+}(M/\Gamma_{R_+}(M))_{r+1-i} = 0$ for all $i \geq l$. Now we must have $\dim(M/\Gamma_{R_+}(M)) \leq \dim M$: if this inequality is strict, we can use the inductive assumption to achieve our aim; otherwise, for the purpose of this inductive step, we can replace M by $M/\Gamma_{R_+}(M)$ and so (in view of 2.1.2 and 2.1.1) assume that M is R_+-torsion-free and that R_+ contains a non-zerodivisor on M. This we do.

Now $M \neq R_+M$ (since otherwise $M = 0$), and so, by Lemma 16.1.4(ii), there exists a homogeneous element $a \in R_1$ (and so of degree 1) which is a non-zerodivisor on M.

The exact sequence $0 \longrightarrow M \xrightarrow{a} M(1) \longrightarrow (M/aM)(1) \longrightarrow 0$ of graded R-modules and homogeneous homomorphisms induces a long exact sequence (in $^*\mathcal{C}(R)$) of local cohomology modules, from which we deduce (with the aid of 14.1.10(ii)) an exact sequence

$$H^i_{R_+}(M)_n \longrightarrow H^i_{R_+}(M/aM)_n \longrightarrow H^{i+1}_{R_+}(M)_{n-1} \xrightarrow{a} H^{i+1}_{R_+}(M)_n$$

of R_0-modules, for all $i \in \mathbb{N}$ and all $n \in \mathbb{Z}$. Use of this with $n = r+1-i$ shows that $H^i_{R_+}(M/aM)_{r+1-i} = 0$ for all $i \geq l$. Now $\dim(M/aM) < \dim M$ (since a lies outside every minimal prime ideal of $\operatorname{Supp} M$), and so it follows from the inductive hypothesis that M/aM is r-regular at and above level l.

Fix an integer $i \geq l$. For each $n \in \mathbb{Z}$, there is an exact sequence

$$H^i_{R_+}(M)_{n-1-i} \xrightarrow{a} H^i_{R_+}(M)_{n-i} \longrightarrow H^i_{R_+}(M/aM)_{n-i}$$

of R_0-modules. We know that $H^i_{R_+}(M)_{r+1-i} = H^i_{R_+}(M/aM)_{r+2-i} = 0$, and so it follows from the above exact sequence (with $n = r + 2$) that $H^i_{R_+}(M)_{r+2-i} = 0$. We can now repeat this argument, using induction, to deduce that $H^i_{R_+}(M)_{r+j-i} = 0$ for all $j \in \mathbb{N}$. It follows that M is r-regular at and above level l. $\qquad\square$

In the statement (and proof) of Theorem 16.2.5, we stressed that the integer l is assumed to be *positive*. In fact, if l in that theorem is replaced by 0, then the resulting statement is no longer always true: this is illustrated by the following exercise.

16.2.6 Exercise. Let the situation and notation be as in Theorem 16.2.5. Find a counterexample to the statement 'if $H^i_{R_+}(M)_{r+1-i} = 0$ for all $i \geq 0$, then M is r-regular at and above level 0 (that is, M is r-regular)' by considering $R := R_0[X_1, \ldots, X_n]$, the ring of polynomials in n (≥ 1) indeterminates over a commutative Noetherian ring R_0, and $M := {}^tR_0 \oplus R$ for an appropriate value of the integer t, where ${}^tR_0$ is as defined in 14.1.13.

The result of Theorem 16.2.5 can be rephrased thus: for a fixed positive integer l, if $H^i_{R_+}(M)_{r+1-i} = 0$ for all $i \geq l$, then $H^i_{R_+}(M)_{s-i} = 0$ for all $i \geq l$ and $s > r$. Some readers might find the interest in 'reverse diagonal vanishing' surprising, and they might find the following diagram helpful. The diagram concerns the special case of Theorem 16.2.5 in which R_0 is Artinian, so that, for all $i, n \in \mathbb{Z}$, the R_0-module $H^i_{R_+}(M)_n$ has finite length $h^i_M(n)$. In the diagram, $h^i_M(n)$ is plotted at the position (n, i) in the Oni co-ordinate plane. Theorem 16.2.5 shows that, if in the diagram

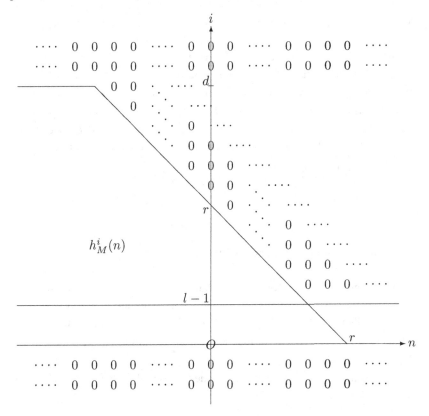

(in which d denotes dim M) there is a line of zeros on the line $i + n = r + 1$ above the line $i = l - 1$, then there must be a similar line of zeros on the line $i + n = s$ above the line $i = l - 1$ for every integer $s > r$.

Some readers may wonder whether 'vertical vanishing' works just as well in this context as 'reverse diagonal vanishing'. It doesn't: the following example provides a counterexample to the statement 'if $H^i_{R_+}(M)_{r+1} = 0$ for all $i \in \mathbb{N}$

(where R is positively graded and homogeneous and M is a finitely generated graded R-module), then $H^i_{R_+}(M)_s = 0$ for all $i \in \mathbb{N}$ and $s > r$'.

16.2.7 Example. We consider again the example studied in 13.5.12. Thus K is a field, $d \in \mathbb{N}$ with $d \geq 3$, and $A_{(d)}$ is the subring of the d-th Veronesean subring $K[X,Y]^{(d)}$ of the ring $K[X,Y]$ of polynomials over K in two indeterminates X and Y given by $A_{(d)} := K[X^d, X^{d-1}Y, XY^{d-1}, Y^d]$, with grading inherited from $K[X,Y]^{(d)}$.

Since $A_{(d)}$ is a 2-dimensional domain, we have $H^i_{A_{(d)+}}(A_{(d)}) = 0$ if $i = 0$ or $i > 2$. By 13.5.12, for $n \in \mathbb{Z}$, we have $H^1_{A_{(d)+}}(A_{(d)})_n \neq 0$ if and only if $1 \leq n \leq d - 3$.

Let $\phi : A_{(d)} \longrightarrow K[X,Y]^{(d)}$ denote the inclusion homomorphism. We showed in 13.5.12 that there is a homogeneous $A_{(d)}$-isomorphism Coker $\phi \cong H^1_{A_{(d)+}}(A_{(d)})$. It therefore follows from 2.1.7(i) that there is a homogeneous $A_{(d)}$-isomorphism $H^2_{A_{(d)+}}(A_{(d)}) \cong H^2_{A_{(d)+}}(K[X,Y]^{(d)})$. Further, arguments we provided in 13.5.12, together with 13.5.9 and the Graded Independence Theorem 14.1.7, show that there are homogeneous $A_{(d)}$-isomorphisms

$$H^2_{A_{(d)+}}(K[X,Y]^{(d)}) \cong H^2_{A_{(d)+}K[X,Y]^{(d)}}(K[X,Y]^{(d)})$$
$$= H^2_{(K[X,Y]_+)^{(d)}}(K[X,Y]^{(d)})$$
$$\cong (H^2_{(X,Y)}(K[X,Y]))^{(d,0)}.$$

It now follows from 13.5.3 and 13.5.9 that $H^2_{A_{(d)+}}(A_{(d)})_n \neq 0$ if and only if $n < 0$.

We can now conclude, for $r, s \in \mathbb{Z}$, that $H^i_{A_{(d)+}}(A_{(d)})_{s-i} = 0$ for all $i \in \mathbb{Z}$ if and only if $s > d - 2$. Thus $A_{(d)}$ is r-regular if and only if $r \geq d - 2$.

Notice also that $H^i_{A_{(d)+}}(A_{(d)})_0 = 0$ for all $i \in \mathbb{N}$, whereas, when $d > 3$, we have $H^1_{A_{(d)+}}(A_{(d)})_1 \neq 0$: thus this example shows that the statement 'if $H^i_{R_+}(M)_{r+1} = 0$ for all $i \in \mathbb{N}$ (where R is positively graded and homogeneous and M is a finitely generated graded R-module), then $H^i_{R_+}(M)_s = 0$ for all $i \in \mathbb{N}$ and $s > r$' is false.

16.2.8 Remarks. Assume that R is positively graded and homogeneous; let M be a finitely generated graded R-module. Let $r \in \mathbb{Z}$ and let l be a *positive* integer.

(i) It follows from Theorem 16.2.5 that M is r-regular at and above level l if and only if $H^i_{R_+}(M)_{r+1-i} = 0$ for all $i \geq l$.

(ii) However, M is r-regular at and above level 0 (that is, M is r-regular) if and only if $H^i_{R_+}(M)_{r+1-i} = 0$ for all $i \in \mathbb{Z}$ and $H^0_{R_+}(M)_s = 0$ for all

$s > r$. This is precisely the condition used to define the condition 'M is r-regular' by D. Eisenbud and S. Goto [11, p. 95].

We are now ready to define Castelnuovo regularity, and also some refinements which have been studied by several authors, including U. Nagel [57], Nagel and P. Schenzel [58, Definition 6.1] and L. T. Hoa and C. Miyazaki [36, §2].

16.2.9 Definition. Assume that $R = \bigoplus_{n \in \mathbb{N}_0} R_n$ is positively graded and homogeneous, and let M be a finitely generated graded R-module. Let $l \in \mathbb{N}_0$. The end of a graded R-module was defined in 14.1.1.

We define the *(Castelnuovo–Mumford) regularity* $\mathrm{reg}(M)$ *of* M by

$$\mathrm{reg}(M) := \sup \left\{ \mathrm{end}(H^i_{R_+}(M)) + i : i \in \mathbb{N}_0 \right\}$$

$$= \sup \left\{ \mathrm{end}(H^i_{R_+}(M)) + i : 0 \le i \le \dim M \right\}.$$

Also, the *(Castelnuovo–Mumford) regularity* $\mathrm{reg}^l(M)$ *of* M *at and above level* l is defined by $\mathrm{reg}^l(M) := \sup \left\{ \mathrm{end}(H^i_{R_+}(M)) + i : i \ge l \right\}$. Thus the regularity $\mathrm{reg}^0(M)$ of M at and above level 0 is the regularity $\mathrm{reg}(M)$ of M. Furthermore, $\mathrm{reg}^l(M) = \inf \{ r \in \mathbb{Z} : M$ is r-regular at and above level $l \}$.

Since there are only finitely many integers i for which $H^i_{R_+}(M) \ne 0$, it follows from 16.1.5(ii) that $\mathrm{reg}(M)$ is either an integer or $-\infty$. Note also that $\mathrm{reg}^l(M) \le \mathrm{reg}^{l-1}(M) \le \cdots \le \mathrm{reg}^0(M) = \mathrm{reg}(M)$.

Observe also that, by 14.1.10(ii), $\mathrm{reg}^l(M(t)) = \mathrm{reg}^l(M) - t$ for all $t \in \mathbb{Z}$.

16.2.10 ♯Exercise. Let the situation be as in 16.2.9.

(i) Show that $\mathrm{reg}^l(M) = -\infty$ if and only if $H^i_{R_+}(M) = 0$ for all $i \ge l$.

(ii) Let N be an R_+-torsion graded submodule of M. Show that $\mathrm{reg}^l(M) = \mathrm{reg}^l(M/N)$ for all $l > 0$.

(iii) Let $\mathfrak{p}$ be a minimal prime of $R_+ + (0 :_R M)$ and assume that $l \le \mathrm{ht}(\mathfrak{p}/(0 :_R M))$. Show that $\mathrm{reg}^l(M) > -\infty$.

(iv) Assume that R_0 is Artinian and local, and that $M \ne 0$. Show that $\mathrm{reg}^l(M) = -\infty$ if and only if $l > \mathrm{ht}((R_+ + (0 :_R M))/(0 :_R M))$.

16.2.11 ♯Exercise. Let R_0 be a commutative Noetherian ring, let $n \in \mathbb{N}$, and let $R := R_0[X_1, \ldots, X_n]$, the ring of polynomials over R_0. Use Example 13.5.3 to show that $\mathrm{reg}(R) = 0$.

16.2.12 Proposition. *Assume that $R = \bigoplus_{n \in \mathbb{N}_0} R_n$ is positively graded and homogeneous, and let $M = \bigoplus_{n \in \mathbb{Z}} M_n$ be a finitely generated graded R-module. Then $\mathrm{reg}^1(M) = -\infty$ if and only if $M_n = 0$ for all $n \gg 0$.*

Proof. ($\Leftarrow$) Suppose $n_0 \in \mathbb{N}$ is such that $M_n = 0$ for all $n \in \mathbb{Z}$ with $|n| \geq n_0$. Then $(R_+)^{2n_0} M = 0$, so that M is R_+-torsion. Therefore, by 2.1.7(i), we have $H^i_{R_+}(M) = 0$ for all $i > 0$, and so $\mathrm{reg}^1(M) = -\infty$.

($\Rightarrow$) Assume that $\mathrm{reg}^1(M) = -\infty$. This means that $H^i_{R_+}(M) = 0$ for all $i > 0$. It therefore follows from 2.1.7(iii) and 2.1.2 that $H^i_{R_+}(M/\Gamma_{R_+}(M)) = 0$ for all $i \in \mathbb{N}_0$. Hence $M/\Gamma_{R_+}(M) = R_+(M/\Gamma_{R_+}(M))$, by 6.2.7 and 6.2.4. It follows from this that, since $M/\Gamma_{R_+}(M)$ is finitely generated and graded, we must have $M/\Gamma_{R_+}(M) = 0$. Hence $M = \Gamma_{R_+}(M)$, and the desired result follows from this and the fact that M is finitely generated. $\square$

16.2.13 ♯Exercise. Let the situation be as in Proposition 16.2.12. Show that $\mathrm{reg}(M) = -\infty$ if and only if $M = 0$.

16.2.14 Exercise. Assume that $R = \bigoplus_{n \in \mathbb{N}_0} R_n$ is positively graded and homogeneous. Let $M = \bigoplus_{n \in \mathbb{Z}} M_n$ be a finitely generated graded R-module, let $l, l' \in \mathbb{N}_0$ with $l \geq l'$, and let $r \in \mathbb{Z}$.

It is clear that, if M is r-regular at and above level l', then M is r-regular at and above level l. Give an example to show that the converse statement is not always true.

16.2.15 ♯Exercise. Assume that $R = \bigoplus_{n \in \mathbb{N}_0} R_n$ is positively graded and homogeneous, let $l \in \mathbb{N}_0$, and let $0 \longrightarrow L \longrightarrow M \longrightarrow N \longrightarrow 0$ be an exact sequence of finitely generated graded R-modules and homogeneous homomorphisms. Show that

  (i) $\mathrm{reg}(L) \leq \max\{\mathrm{reg}(M), \mathrm{reg}(N) + 1\}$,
  (ii) $\mathrm{reg}^{l+1}(L) \leq \max\{\mathrm{reg}^{l+1}(M), \mathrm{reg}^l(N) + 1\}$,
  (iii) $\mathrm{reg}^l(M) \leq \max\{\mathrm{reg}^l(L), \mathrm{reg}^l(N)\}$, and
  (iv) $\mathrm{reg}^l(N) \leq \max\{\mathrm{reg}^{l+1}(L) - 1, \mathrm{reg}^l(M)\}$.

16.3 Degrees of generators

The following theorem is one of the main reasons for our introduction of the concept of regularity. Among its consequences are the links, hinted at in the introduction to this chapter, between local cohomology theory and the syzygies of finitely generated graded modules over a polynomial ring over a field.

16.3.1 Theorem. *Assume that $R = \bigoplus_{n \in \mathbb{N}_0} R_n$ is positively graded and homogeneous, and let $M = \bigoplus_{n \in \mathbb{Z}} M_n$ be a non-zero finitely generated graded R-module. Then M can be generated by homogeneous elements of degrees not exceeding $\mathrm{reg}(M)$.*

Proof. Our strategy in this proof has some similarities to that for our proof of Theorem 16.2.5 above, inasmuch as we apply 16.2.3 and 16.2.4 to reduce to the case in which R_0 is local with infinite residue field.

Let N be the graded submodule of M generated by $\bigoplus_{n \leq \mathrm{reg}(M)} M_n$: we must show that $M = N$.

It suffices to show that $M_{\mathfrak{p}_0} = N_{\mathfrak{p}_0}$ for each $\mathfrak{p}_0 \in \mathrm{Spec}(R_0)$. We use 16.2.3: consider the positively graded, homogeneous ring $R' = R \otimes_{R_0} (R_0)_{\mathfrak{p}_0}$ and the graded R'-module $M' = M \otimes_{R_0} (R_0)_{\mathfrak{p}_0}$, which is finitely generated. By 16.2.3, $\mathrm{reg}(M') \leq \mathrm{reg}(M)$; also, under the canonical isomorphism between M' and $M_{\mathfrak{p}_0}$, the R'-submodule of M' generated by all homogeneous elements of degrees not exceeding $\mathrm{reg}(M)$ is mapped onto $N_{\mathfrak{p}_0}$. It is therefore enough for us to establish the claim in the statement of the theorem under the additional assumption that R_0 is local. We can then use 16.2.4 in a similar way to see that it is enough for us to establish the claim under the additional assumption that $(R_0, \mathfrak{m}_0)$ is a local ring with infinite residue field, and we shall make this assumption in what follows.

Now $d := \dim M$ is finite, and $\mathrm{reg}(M) \neq -\infty$ since $M \neq 0$ (by 16.2.13). We argue by induction on d. Consider first the case where $d = 0$. Then $\mathrm{Ass}\, M = \{\mathfrak{m}_0 \oplus R_+\}$, so that there exists $t \in \mathbb{N}$ such that $(R_+)^t M = 0$. In this case, $M = \Gamma_{R_+}(M)$ and so $H^i_{R_+}(M) = 0$ for all $i \in \mathbb{N}$, by 2.1.7(ii). Thus $\mathrm{reg}(M) = \mathrm{end}(M)$, and it is obvious that M can be generated by homogeneous elements of degrees not exceeding $\mathrm{end}(M)$.

Now suppose that $\dim M > 0$ and our desired result has been proved for all non-zero, finitely generated graded R-modules of smaller dimensions. Of course, $\Gamma_{R_+}(M)$, if non-zero, can be generated by homogeneous elements of degrees not exceeding $\mathrm{end}(\Gamma_{R_+}(M))$; since $\mathrm{end}(\Gamma_{R_+}(M)) \leq \mathrm{reg}(M)$, it follows from the exact sequence

$$0 \longrightarrow \Gamma_{R_+}(M) \longrightarrow M \longrightarrow M/\Gamma_{R_+}(M) \longrightarrow 0$$

that it is enough for us to show that $M/\Gamma_{R_+}(M)$ can be generated by homogeneous elements of degrees not exceeding $\mathrm{reg}(M)$. Note that

$$\mathrm{reg}(M/\Gamma_{R_+}(M)) = \mathrm{reg}^1(M) \leq \mathrm{reg}(M).$$

Now $\dim(M/\Gamma_{R_+}(M)) \leq \dim M$: if this inequality is strict, we can use the inductive assumption to achieve our aim; otherwise, for the purpose of this inductive step, we can replace M by $M/\Gamma_{R_+}(M)$ and so (in view of 2.1.2 and 2.1.1) assume that M is R_+-torsion-free and that R_+ contains a non-zerodivisor on M. This we do.

Note that $M \neq R_+ M$. Then, by Lemma 16.1.4(ii), there exists a homogeneous element $a \in R_1$ (and so of degree 1) which is a non-zerodivisor on M.

Apply 16.2.15(iv) to the exact sequence

$$0 \longrightarrow M(-1) \overset{a}{\longrightarrow} M \longrightarrow M/aM \longrightarrow 0$$

of graded R-modules and homogeneous homomorphisms to see that

$$\operatorname{reg}(M/aM) \leq \max\left\{\operatorname{reg}^1(M(-1)) - 1, \operatorname{reg}(M)\right\} = \operatorname{reg}(M).$$

Now $\dim(M/aM) < \dim M$, and so it follows from the inductive hypothesis that M/aM can be generated by homogeneous elements of degrees not exceeding $\operatorname{reg}(M/aM)$. Hence $M = N + aM$. Now R is *local with unique *maximal graded ideal $\mathfrak{m} := \mathfrak{m}_0 \oplus R_+$; it follows from Nakayama's Lemma that $M_\mathfrak{m} = N_\mathfrak{m}$, so that $M = N$ by 14.1.2(ii) applied to M/N. □

16.3.2 ♯Exercise. Let $R := K[X_1, \ldots, X_n]$, the ring of polynomials in n (≥ 1) indeterminates over a field K. Let $\mathfrak{a}$ be a proper graded ideal of R such that $H^0_{R_+}(R/\mathfrak{a}) = H^n_{R_+}(R/\mathfrak{a}) = 0$. Use 16.2.11, 16.2.15 and 16.3.1 to show that $0 \leq \operatorname{reg}^1(R/\mathfrak{a}) = \operatorname{reg}^2(\mathfrak{a}) - 1$.

In the following definition and subsequent exercises, we consider a geometric significance of the Castelnuovo regularity.

16.3.3 Definition and Remark. Let K be an algebraically closed field, and let $r \in \mathbb{N}$. Let $V \subset \mathbb{P}^r(K)$ be a projective variety. Let $R := K[X_0, \ldots, X_r]$, and consider R as the coordinate ring $\mathcal{O}(\mathbb{A}^{r+1}(K))$. Then the graded prime ideal $I_{\mathbb{P}^r(K)}(V) := I_{\mathbb{A}^{r+1}(K)}(\operatorname{Cone}(V))$ of R is non-zero: we define the *Castelnuovo–Mumford regularity* $\operatorname{reg}(V)$ *of* V by $\operatorname{reg}(V) := \operatorname{reg}(I_{\mathbb{P}^r(K)}(V))$.

Note that $H^0_{R_+}(R) = H^1_{R_+}(R) = 0$, and the homogeneous coordinate ring $\mathcal{O}(\operatorname{Cone}(V))$ of V is an integral domain. It follows from these observations and 16.3.2 that

$$\operatorname{reg}(V) = \operatorname{reg}(I_{\mathbb{P}^r(K)}(V)) = \operatorname{reg}^1(I_{\mathbb{P}^r(K)}(V)) = \operatorname{reg}^2(I_{\mathbb{P}^r(K)}(V))$$
$$= \operatorname{reg}^1(\mathcal{O}(\operatorname{Cone}(V))) + 1 = \operatorname{reg}(\mathcal{O}(\operatorname{Cone}(V))) + 1.$$

Note that, by 16.3.1, the vanishing ideal $I_{\mathbb{P}^r(K)}(V)$ can be generated by homogeneous polynomials of degrees not exceeding $\operatorname{reg}(V)$.

16.3.4 Exercise. Let K be an algebraically closed field, and let $r \in \mathbb{N}$. Let $V \subset \mathbb{P}^r(K)$ be a projective variety such that $\dim V > 0$. Let $n \in \mathbb{N}$ be such that $\operatorname{reg}(V) \leq n + 1$.

(i) Show that the n-th Veronesean $V^{(n)} \subseteq \mathbb{P}^{\binom{r+n}{r}-1}(K)$ of V has

$$\operatorname{arithdepth} V^{(n)} > 1.$$

(ii) Show that $\operatorname{reg}(V^{(n)}) \leq \dim V + 2$.

16.3.5 Exercise. Let K be an algebraically closed field, and let $d \in \mathbb{N}$ with $d > 1$. Here we study again the rational normal curve in projective d-space $N_{(d)} \subseteq \mathbb{P}^d(K)$ of 15.2.3.

Consider the Veronesean subring $K[X_0, X_1]^{(d)}$ of the ring $K[X_0, X_1]$ of polynomials over K in two indeterminates X_0, X_1, and the polynomial ring $K[Y_0, \ldots, Y_d]$ in $d + 1$ indeterminates $Y_0, \ldots, Y_d$ as $\mathcal{O}(\mathbb{A}^{d+1}(K))$, the co-ordinate ring of affine $(d + 1)$-space over K. Let $\phi : K[Y_0, \ldots, Y_d] \longrightarrow K[X_0, X_1]^{(d)}$ be the K-algebra homomorphism for which

$$\phi(Y_0) = X_0^d, \ \phi(Y_1) = X_0^{d-1}X_1, \ \ldots, \ \phi(Y_{d-1}) = X_0 X_1^{d-1}, \ \phi(Y_d) = X_1^d.$$

Then $\operatorname{Ker} \phi = I_{\mathbb{P}^d(K)}(N_{(d)})$.

(i) Calculate $\operatorname{reg}(N_{(d)})$, and deduce that $I_{\mathbb{P}^d(K)}(N_{(d)})$ can be generated by quadratics.

(ii) Calculate the dimension of $I_{\mathbb{P}^d(K)}(N_{(d)})_2$, the component of degree 2 of $I_{\mathbb{P}^d(K)}(N_{(d)})$, as a vector space over K.

(iii) Deduce that $I_{\mathbb{P}^d(K)}(N_{(d)})$ can be generated by the 2×2 minors of the matrix

$$\begin{bmatrix} Y_0 & Y_1 & Y_2 & \ldots & Y_{d-2} & Y_{d-1} \\ Y_1 & Y_2 & Y_3 & \ldots & Y_{d-1} & Y_d \end{bmatrix}.$$

16.3.6 Exercise. Use the notation of 15.2.5, but in the special case in which $d = 4$. Thus K is an algebraically closed field, and $A_{(4)}$ is the subring

$$K[X^4, X^3Y, XY^3, Y^4]$$

of the ring $K[X, Y]$ of polynomials over K in two indeterminates X and Y. Also, $\psi : K[Y_0, Y_1, Y_2, Y_3] \longrightarrow A_{(4)}$ is the K-algebra homomorphism for which $\psi(Y_0) = X^4$, $\psi(Y_1) = X^3Y$, $\psi(Y_2) = XY^3$ and $\psi(Y_3) = Y^4$, and $\operatorname{Ker} \psi$ is the vanishing ideal of the projective variety $\Sigma_{(4)} \subseteq \mathbb{P}^3(K)$.

(i) Calculate $\operatorname{reg}(\Sigma_{(4)})$, and deduce that $I_{\mathbb{P}^3(K)}(\Sigma_{(4)})$ can be generated by homogeneous polynomials of degrees not exceeding 3.

(ii) For $i = 2, 3$, calculate the dimension of $I_{\mathbb{P}^3(K)}(\Sigma_{(4)})_i$, the component of degree i of $I_{\mathbb{P}^3(K)}(\Sigma_{(4)})$, as a vector space over K. Find a quadratic and three cubics which generate $I_{\mathbb{P}^3(K)}(\Sigma_{(4)})$.

The next result uses Theorem 16.3.1 to establish connections between the regularity of a finitely generated graded module M over a polynomial ring over a field and the syzygies of M: we hinted at these connections in the intro-duction to this chapter. The result is presented by Eisenbud and Goto in [11, p. 89], although they state that it is not new and that it has origins in ideas of

Castelnuovo and Mumford. We refer the reader to [7, pp. 36–37] for fundamental facts concerning the minimal graded free resolution of M.

16.3.7 Theorem: syzygetic characterization of regularity. *Let K be a field and let $R = \bigoplus_{j \in \mathbb{N}_0} R_j := K[X_1, \ldots, X_n]$, the ring of polynomials over K in n indeterminates (where $n \in \mathbb{N}$). Let M be a non-zero finitely generated graded R-module having $\operatorname{proj\,dim} M = p$, and let*

$$0 \longrightarrow F_p \xrightarrow{f_p} F_{p-1} \longrightarrow \ldots \longrightarrow F_1 \xrightarrow{f_1} F_0 \longrightarrow 0,$$

be the minimal graded free resolution of M. Thus there exist $b_0, \ldots, b_p \in \mathbb{N}$ and, for each $j = 0, \ldots, p$, integers $a_i^{(j)}$ $(i = 1, \ldots, b_j)$ such that $a_1^{(j)} \geq \cdots \geq a_{b_j}^{(j)}$ and $F_j = \bigoplus_{i=1}^{b_j} R(a_i^{(j)})$. Then

$$\operatorname{reg}(M) = \max\left\{ -a_{b_j}^{(j)} - j : j = 0, \ldots, p \right\}.$$

Proof. The f_i $(i = 1, \ldots, p)$ are homogeneous, and there is a homogeneous R-epimorphism $f_0 : F_0 \longrightarrow M$ such that $\operatorname{Ker} f_0 = \operatorname{Im} f_1$.

In view of 16.2.11 and 16.2.9, we have

$$\operatorname{reg}(F_0) = \operatorname{reg}\left(\bigoplus_{i=1}^{b_0} R(a_i^{(0)}) \right) = \max\left\{ \operatorname{reg}\left(R(a_i^{(0)}) \right) : i = 0, \ldots, b_0 \right\}$$

$$= \max\left\{ -a_i^{(0)} : i = 0, \ldots, b_0 \right\} = -a_{b_0}^{(0)}.$$

We argue by induction on p. When $p = 0$, we have a homogeneous isomorphism $M \cong F_0$, and the claim follows from the above equations. So suppose, inductively, that $p > 0$ and that the result has been proved for smaller values of p. Set $L := \operatorname{Ker} f_0 = \operatorname{Im} f_1$.

Now F_0, and therefore M, can be generated by homogeneous elements of degrees not exceeding $\operatorname{reg}(F_0) = -a_{b_0}^{(0)}$, by 16.3.1. By the minimality, M cannot be generated by homogeneous elements all of whose degrees are less than $-a_{b_0}^{(0)}$; therefore $\operatorname{reg}(M) \geq -a_{b_0}^{(0)} = \operatorname{reg}(F_0)$ by 16.3.1.

There is an exact sequence $0 \longrightarrow L \xrightarrow{\subseteq} F_0 \longrightarrow M \longrightarrow 0$ in the category $^*\mathcal{C}(R)$. Application of 16.2.15(i) to this yields that

$$\operatorname{reg}(L) \leq \max\{\operatorname{reg}(F_0), \operatorname{reg}(M) + 1\} = \operatorname{reg}(M) + 1,$$

so that $\operatorname{reg}(L) - 1 \leq \operatorname{reg}(M)$; on the other hand, application of 16.2.15(iv) to the same sequence yields that

$$\operatorname{reg}(M) \leq \max\{\operatorname{reg}^1(L) - 1, \operatorname{reg}(F_0)\} \leq \max\{\operatorname{reg}(L) - 1, -a_{b_0}^{(0)}\},$$

and so we see that $\operatorname{reg}(M) = \max\{\operatorname{reg}(L) - 1, -a_{b_0}^{(0)}\}$.

The exact sequence $0 \longrightarrow F_p \xrightarrow{f_p} F_{p-1} \longrightarrow \ldots \longrightarrow F_1 \xrightarrow{f_1} L \longrightarrow 0$

yields the minimal graded free resolution of L, so that $\operatorname{proj\,dim} L = p - 1$. Therefore, by the inductive hypothesis,

$$\operatorname{reg}(L) = \max\left\{-a_{b_j}^{(j)} - (j - 1) : j = 1, \dots, p\right\}$$
$$= \max\left\{-a_{b_j}^{(j)} - j : j = 1, \dots, p\right\} + 1.$$

Therefore

$$\operatorname{reg}(M) = \max\left\{\operatorname{reg}(L) - 1, -a_{b_0}^{(0)}\right\}$$
$$= \max\left\{\max\left\{-a_{b_j}^{(j)} - j : j = 1, \dots, p\right\}, -a_{b_0}^{(0)}\right\},$$

and the inductive step can be completed. $\qquad\square$

16.3.8 Example. Consider the situation of Theorem 16.3.7 above in the special case in which $M = R/\mathfrak{b}$, where $\mathfrak{b}$ is a proper graded ideal of R. Since $R/\mathfrak{b}$ can be generated by one homogeneous element of degree 0, and since $\operatorname{Ker} f_j \subseteq \mathfrak{m}F_j$ for all $j = 0, \dots, p$, it follows that, in this case, $b_0 = 1$, $a_1^{(0)} = 0$, and $-a_1^{(j+1)} \geq -a_1^{(j)} + 1$ for all $j = 0, \dots, p - 1$. Hence $-a_1^{(j)} \geq j$ for all $j = 0, \dots, p$.

On the other hand, by 16.3.7, we have $-a_{b_j}^{(j)} \leq \operatorname{reg}(M) + j$ for all $j = 0, \dots, p$. Thus, for a given integer j such that $0 \leq j \leq p$, the $-a_i^{(j)}$ $(1 \leq i \leq b_j)$ are constrained to lie in the interval $\{r \in \mathbb{N}_0 : j \leq r \leq \operatorname{reg}(M) + j\}$ of length $\operatorname{reg}(M)$.

16.3.9 ♯Exercise. Assume that $R = \bigoplus_{n \in \mathbb{N}_0} R_n$ is positively graded and homogeneous, and let M be a non-zero finitely generated graded R-module. Let $a_1, \dots, a_h$ be homogeneous elements of R with $\deg a_i = n_i > 0$ for $i = 1, \dots, h$, and suppose that $a_1, \dots, a_h$ is an M-sequence. Use Exercise 16.2.15 to show that

$$\operatorname{reg}^1\left(M/\textstyle\sum_{i=1}^h a_i M\right) \leq \operatorname{reg}^1(M) + \textstyle\sum_{i=1}^h n_i - h \leq \operatorname{reg}\left(M/\textstyle\sum_{i=1}^h a_i M\right),$$

and that both inequalities in this display are equalities if $\operatorname{grade}_M(R_+) > h$.

17

Hilbert polynomials

Suppose that $R = \bigoplus_{n \in \mathbb{N}_0} R_n$ is positively $\mathbb{Z}$-graded and homogeneous and such that R_0 is an Artinian ring, and let $M = \bigoplus_{n \in \mathbb{Z}} M_n$ be a non-zero finitely generated graded R-module of dimension d. By a classical theorem of Hilbert (see [7, Theorem 4.1.3]), there is a polynomial $P_M \in \mathbb{Q}[X]$ of degree $d - 1$, and an $n_0 \in \mathbb{Z}$, such that $P_M(n) = \ell_{R_0}(M_n)$ for all $n \geq n_0$. The polynomial P_M is necessarily uniquely determined, and known as the *Hilbert polynomial of M*. The classical 'postulation' problem asked for an explanation of the difference $P_M(n) - \ell_{R_0}(M_n)$ for $n < n_0$. Serre solved this in [77] using sheaf cohomology. In this chapter we present, via graded local cohomology, the ideas behind Serre's approach; the reader will find a translation into the language of sheaf cohomology in 20.4.16.

We study, in §17.1 below, the so-called characteristic function $\chi_M : \mathbb{Z} \longrightarrow \mathbb{Z}$ of M and show that this is completely represented by a polynomial. In more detail, it turns out that, for each $i \in \mathbb{N}_0$ and $n \in \mathbb{Z}$, the n-th component $\mathcal{R}^i D_{R_+}(M)_n$ of $\mathcal{R}^i D_{R_+}(M)$ has finite length as an R_0-module, and we denote this length by $d_M^i(n)$; it makes sense for us to define the *characteristic function* $\chi_M : \mathbb{Z} \longrightarrow \mathbb{Z}$ *of M* by setting

$$\chi_M(n) = \sum_{i \in \mathbb{N}_0} (-1)^i d_M^i(n) \quad \text{for all } n \in \mathbb{Z}.$$

We shall see in 17.1.7 that there is a polynomial $q \in \mathbb{Q}[X]$ of degree $d - 1$ such that $q(n) = \chi_M(n)$ for *all* $n \in \mathbb{Z}$. The fact that $H_{R_+}^i(M)_n = 0$ for all $i \geq 0$ and $n \gg 0$ (see 16.1.5) means that we can conclude that $\chi_M(n) = \ell_{R_0}(M_n)$ for all $n \gg 0$; this not only provides a proof of Hilbert's Theorem mentioned above (q turns out to be the Hilbert polynomial P_M), but also yields precise information about the difference $P_M(n) - \ell_{R_0}(M_n)$ for small n.

Also in §17.1, we use Hilbert's Theorem and graded local duality to establish the existence of 'cohomological Hilbert polynomials': for each $i \in \mathbb{N}_0$, there is a polynomial $p_M^i \in \mathbb{Q}[X]$ of degree less than i such that

$$\ell_{R_0}(H_{R_+}^i(M)_n) = p_M^i(n) \quad \text{for all } n \ll 0.$$

In §17.2, we study the invariant $\operatorname{reg}^2(M)$ (where M is as above). We show that both $D_{R_+}(M)$ and $H_{R_+}^1(M)$ can be generated by homogeneous elements of degrees not exceeding $\operatorname{reg}^2(M)$ (provided $\operatorname{reg}^2(M)$ is finite). This will enable us to prove (in 20.4.13) an important result on coherent sheaves due to Serre [80]. This is a small hint about the significance of the invariant reg^2. Furthermore, in the case where there is a homogeneous element a of R of degree 1 that is a non-zerodivisor on M, we show that, as n increases beyond $\operatorname{reg}^2(M/aM) - 1$, the integers $\ell_{R_0}(H_{R_+}^2(M)_n)$ decrease *strictly* to 0 and then remain at 0. This is a generalization of a result, actually formulated in terms of sheaf cohomology, of Mumford [54, p. 99].

To appreciate fully the significance of the invariant reg^2, the reader will need to become aware of the links between local cohomology and sheaf cohomology, which will be treated in Chapter 20. In this chapter, we content ourselves with the hints given above, and some elementary comments at the beginning of §17.2 about the significance of reg^2 in the framework of defining equations of projective varieties.

In §17.3, we generalize an important result of D. Mumford [54, p. 101], actually formulated in terms of sheaf cohomology, which shows that, for a non-zero graded ideal $\mathfrak{b}$ of the polynomial ring $K[X_1, \ldots, X_d]$ over a field K, the regularity $\operatorname{reg}^2(\mathfrak{b})$ of $\mathfrak{b}$ at and above level 2 is bounded in terms of the Hilbert polynomial $P_{\mathfrak{b}}$ of $\mathfrak{b}$. We generalize this in 17.3.6 and prove that, for a polynomial ring $R = R_0[X_1, \ldots, X_d]$ over an Artinian local ring R_0, for integers $a_1, \ldots, a_r$ and a non-zero graded submodule M of the finitely generated *free graded R-module $\bigoplus_{i=1}^r R(a_i)$, the regularity $\operatorname{reg}^2(M)$ is bounded in terms of P_M. The reader will find a sheaf-theoretic formulation of Mumford's Regularity Bound in 20.4.18.

In §17.4, we apply the main result of §17.3 to give upper bounds on the invariants reg^1 and reg^0 in certain circumstances. We prove that the regularity $\operatorname{reg}^1(M)$ at and above level 1 is bounded in terms of P_M and the number and degrees of homogeneous generators of M. We also prove in the same section that the regularity $\operatorname{reg}^0(M) = \operatorname{reg}(M)$ is bounded in terms of P_M, the degrees of homogeneous generators of M and the so-called postulation number $\operatorname{pstln}(M)$ of M, defined as the greatest integer n for which $\ell_{R_0}(M_n) \neq P_M(n)$ if any such integers exist, and $-\infty$ otherwise.

Once again, throughout this chapter, all graded rings and modules are to be understood to be $\mathbb{Z}$-graded, and all polynomial rings $R_0[X_1, \ldots, X_d]$ (over a commutative Noetherian ring R_0) are to be understood to be (positively)

$\mathbb{Z}$-graded so that R_0 is the component of degree 0 and $\deg X_i = 1$ for all $i = 1, \ldots, d$.

17.1 The characteristic function

17.1.1 Reminders, Notation and Terminology. We shall use notation and terminology employed by Bruns and Herzog in [7, §4.1]. Consider a function $f : \mathbb{Z} \longrightarrow \mathbb{Z}$.

(i) We say that f is *of polynomial type (of degree d)* if and only if there exists a polynomial $p \in \mathbb{Q}[X]$ (of degree d) such that $f(n) = p(n)$ for all $n \gg 0$ (that is, for all n greater than some fixed constant integer n_0). (When this is the case, the polynomial p will be uniquely determined, of course, since two rational polynomials which take the same values at infinitely many integers must coincide. We adopt the convention that the zero polynomial has degree -1.)

(ii) We define $\Delta f : \mathbb{Z} \longrightarrow \mathbb{Z}$ by $\Delta f(n) = f(n+1) - f(n)$ for all $n \in \mathbb{Z}$. Thus Δ is a mapping from the set of functions from $\mathbb{Z}$ to $\mathbb{Z}$ to itself. For $n \in \mathbb{N}$, we use Δ^n to denote the result of repeating Δ n times; we also write $\Delta^0 f = f$.

(iii) Let $d \in \mathbb{N}_0$. Recall from [7, Lemma 4.1.2] that f is of polynomial type of degree d if and only if there exists $c \in \mathbb{Z}$ with $c \neq 0$ such that $\Delta^d f(n) = c$ for all $n \gg 0$.

(iv) Recall that, for $i \in \mathbb{N}$,

$$\binom{X+i}{i} := \frac{(X+i)(X+i-1)\ldots(X+1)}{1.2.3.\,\ldots\,.i} \in \mathbb{Q}[X],$$

and that $\binom{X+0}{0}$ denotes the constant polynomial $1 \in \mathbb{Q}[X]$.

(v) Let $d \in \mathbb{N}$. Let $P \in \mathbb{Q}[X]$ be a non-zero polynomial of degree $d - 1$. Recall from [7, Lemma 4.1.4] that $P(n) \in \mathbb{Z}$ for all $n \in \mathbb{Z}$ if and only if there exist integers $e_0, \ldots, e_{d-1}$ such that

$$P(X) = \sum_{i=0}^{d-1} (-1)^i e_i \binom{X+d-i-1}{d-i-1}.$$

When this is the case, we say that P is a *numerical polynomial*; also, the integers $e_0, \ldots, e_{d-1}$ are uniquely determined by P, and we denote them by $e_0(P), \ldots, e_{d-1}(P)$.

Also, given $c := (c_0, \ldots, c_{d-1}) \in \mathbb{Z}^d$, we shall denote by $p_c =$

$p_{(c_0,\ldots,c_{d-1})}$ the polynomial in $\mathbb{Q}[X]$ given by

$$p_c(X) = \sum_{i=0}^{d-1} (-1)^i c_i \binom{X + d - i - 1}{d - i - 1}.$$

Note that, when $d > 1$, we have

$$p_c(X) - p_c(X - 1)$$

$$= \sum_{i=0}^{d-1} (-1)^i c_i \left(\binom{X + d - i - 1}{d - i - 1} - \binom{(X-1) + d - i - 1}{d - i - 1} \right)$$

$$= \sum_{i=0}^{d-2} (-1)^i c_i \binom{X + d - i - 2}{d - i - 2} = p_{(c_0,\ldots,c_{d-2})}(X).$$

(vi) We shall say that f is *of reverse polynomial type (of degree d)* if and only if there exists a polynomial $p \in \mathbb{Q}[X]$ (of degree d) such that $f(n) = p(n)$ for all $n \ll 0$ (that is, for all n less than some fixed constant integer n_0'). Thus f is of reverse polynomial type of degree d if and only if the function $g : \mathbb{Z} \longrightarrow \mathbb{Z}$ defined by $g(n) = f(-n)$ for all $n \in \mathbb{Z}$ is of polynomial type of degree d in the sense of (i).

17.1.2 ♯Exercise. Let $0 \neq P \in \mathbb{Q}[X]$ have $\deg P = d - 1$. Assume that P takes integer values at d consecutive integers. Show that $P(n) \in \mathbb{Z}$ for *all* $n \in \mathbb{Z}$.

17.1.3 ♯Exercise. Suppose that the function $f : \mathbb{Z} \longrightarrow \mathbb{Z}$ is such that there exists $Q \in \mathbb{Q}[X]$ for which $\Delta f(n) = Q(n)$ for all $n \in \mathbb{Z}$. By using, for various values of the integer n_0, the fact that

$$f(n) = f(n_0) + \sum_{i=n_0}^{n-1} Q(i) \quad \text{for all } n \geq n_0,$$

show that there exists $P \in \mathbb{Q}[X]$ such that $f(n) = P(n)$ for all $n \in \mathbb{Z}$.

17.1.4 Further Notation and Terminology. Assume that $R = \bigoplus_{n \in \mathbb{N}_0} R_n$ is positively graded and homogeneous and such that R_0 is an Artinian ring, and let $M = \bigoplus_{n \in \mathbb{Z}} M_n$ be a finitely generated graded R-module. It follows from 16.1.5(i) that the R_0-module $H^i_{R_+}(M)_n$ has finite length, for all $i \in \mathbb{N}_0$ and all $n \in \mathbb{Z}$: we denote this length by $h^i_M(n)$. Similarly, it follows from 16.1.6(ii) that the R_0-module $D_{R_+}(M)_n$ has finite length, for all $n \in \mathbb{Z}$. Let $i \in \mathbb{N}$. For $n \in \mathbb{Z}$, we denote the n-th component of $\mathcal{R}^i D_{R_+}(M)$ by $\mathcal{R}^i D_{R_+}(M)_n$; by 13.5.7(iii), this R_0-module is isomorphic to $H^{i+1}_{R_+}(M)_n$, and so has finite length.

We denote $\ell_{R_0}(\mathcal{R}^i D_{R_+}(M)_n)$ by $d^i_M(n)$, for all $i \in \mathbb{N}_0$ and all $n \in \mathbb{Z}$. Note that $d^i_M(n) = h^{i+1}_M(n)$ for $i > 0$, and that this is zero for $i \geq \max\{1, \dim M\}$.

It therefore makes sense for us to define the *characteristic function* χ_M : $\mathbb{Z} \longrightarrow \mathbb{Z}$ of M by setting $\chi_M(n) = \sum_{i \in \mathbb{N}_0} (-1)^i d_M^i(n)$ for all $n \in \mathbb{Z}$. Note that, for all $n \in \mathbb{Z}$, we have $d_M^0(n) = \ell_{R_0}(M_n) + h_M^1(n) - h_M^0(n)$ by 13.5.4, so that, for any integer $d \geq \dim M$,

$$\chi_M(n) = \sum_{i=0}^{d} (-1)^i d_M^i(n) = d_M^0(n) - \sum_{i=2}^{d} (-1)^i h_M^i(n)$$

$$= \ell_{R_0}(M_n) - \sum_{i=0}^{d} (-1)^i h_M^i(n).$$

Thus it follows from 16.1.5(ii) that $\chi_M(n) = \ell_{R_0}(M_n) \geq 0$ for all $n \gg 0$. Note also that, by 14.1.10(ii), we have $\chi_{M(t)}(n) = \chi_M(t+n)$ for all $t, n \in \mathbb{Z}$.

17.1.5 ♯Exercise. Assume that $R = \bigoplus_{n \in \mathbb{N}_0} R_n$ is positively graded and homogeneous and such that R_0 is an Artinian ring. Let

$$0 \longrightarrow L \longrightarrow M \longrightarrow N \longrightarrow 0$$

be an exact sequence of finitely generated graded R-modules and homogeneous homomorphisms. Show that, for each $n \in \mathbb{Z}$, we have

$$\chi_M(n) = \chi_L(n) + \chi_N(n).$$

17.1.6 Remark. Assume that $R = \bigoplus_{n \in \mathbb{N}_0} R_n$ is positively graded and homogeneous, and such that R_0 is Artinian. Let $\mathfrak{m}_0^{(1)}, \ldots, \mathfrak{m}_0^{(t)}$ be the maximal ideals of R_0. Let M be a finitely generated graded R-module. For each $j = 1, \ldots, t$, set $R'^{(j)} := R \otimes_{R_0} (R_0)_{\mathfrak{m}_0^{(j)}}$ and $M'^{(j)} := M \otimes_{R_0} (R_0)_{\mathfrak{m}_0^{(j)}}$.

(i) For each R_0-module N, the natural R_0-homomorphism $\omega_N : N \longrightarrow \bigoplus_{j=1}^{t} N_{\mathfrak{m}_0^{(j)}}$ (for which $\omega_N(y) = (y/1, \ldots, y/1)$ for all $y \in N$) is an isomorphism; therefore, when N is finitely generated, its length $\ell_{R_0}(N)$ satisfies $\ell_{R_0}(N) = \sum_{j=1}^{t} \ell_{(R_0)_{\mathfrak{m}_0^{(j)}}} (N_{\mathfrak{m}_0^{(j)}})$.

(ii) Since $\mathfrak{m}_0^{(1)} + R_+, \ldots, \mathfrak{m}_0^{(t)} + R_+$ are the only *maximal graded ideals of R and, for $j = 1, \ldots, t$, the ring $R'^{(j)}$ is isomorphic, as a graded ring, to the homogeneous localization (see 14.1.1) $R_{(\mathfrak{m}_0^{(j)} + R_+)}$, it follows that

$$\dim M = \max \left\{ \dim_{R'^{(j)}} M'^{(j)} : j = 1, \ldots, t \right\}.$$

(iii) Fix an integer j between 1 and t. Note that, by 16.2.3, the ring $R'^{(j)}$ is positively graded and homogeneous, and has 0-th component isomorphic to the Artinian local ring $(R_0)_{\mathfrak{m}_0^{(j)}}$; moreover, $R_+ R'^{(j)} = R_+'^{(j)}$ and there is an $(R_0)_{\mathfrak{m}_0^{(j)}}$-isomorphism

$$\left(H_{R_+}^i (M)_n \right)_{\mathfrak{m}_0^{(j)}} \cong H_{R_+'^{(j)}}^i (M'^{(j)})_n \quad \text{for each } i \in \mathbb{N}_0 \text{ and each } n \in \mathbb{Z}.$$

Also, by 16.2.2(vi), there is an $(R_0)_{\mathfrak{m}_0^{(j)}}$-isomorphism

$$\left(D_{R_+}(M)_n \right)_{\mathfrak{m}_0^{(j)}} \cong D_{R_+'^{(j)}}(M'^{(j)})_n \quad \text{for each } n \in \mathbb{Z}.$$

(iv) It follows from parts (i) and (iii) that

$$\operatorname{reg}^l(M) = \max \left\{ \operatorname{reg}^l(M'^{(j)}) : j = 1, \ldots, t \right\} \quad \text{for all } l \in \mathbb{N}_0,$$

that $h_M^i(n) = \sum_{j=1}^{t} h_{M'^{(j)}}^i(n)$ for all $i \in \mathbb{N}_0$ and $n \in \mathbb{Z}$, and that (with the notation of 17.1.4)

$$d_M^0(n) = \sum_{j=1}^{t} d_{M'^{(j)}}^0(n) \quad \text{and} \quad \chi_M(n) = \sum_{j=1}^{t} \chi_{M'^{(j)}}(n) \text{ for all } n \in \mathbb{Z}.$$

17.1.7 Theorem. *Assume that $R = \bigoplus_{n \in \mathbb{N}_0} R_n$ is positively graded and homogeneous, and such that R_0 is Artinian; let M be a non-zero finitely generated graded R-module. Then there is a (necessarily uniquely determined) polynomial $P_M \in \mathbb{Q}[X]$ of degree $\dim M - 1$ such that $P_M(n) = \chi_M(n)$ for all $n \in \mathbb{Z}$.*

Proof. It follows from 17.1.6 (and the fact that $\chi_M(n) \geq 0$ for all $n \gg 0$) that it is sufficient for us to prove this result under the additional assumption that the Artinian ring R_0 is local. We make this assumption in what follows, and we let $\mathfrak{m}_0$ be the maximal ideal of R_0.

We now use 16.2.4 and 16.2.2(vi) to reduce to the case where the residue field of R_0 is infinite. We therefore assume that $R_0/\mathfrak{m}_0$ is infinite for the remainder of this proof.

We now use induction on $\dim M$. Consider first the case when $\dim M = 0$. Then $\operatorname{Ass} M = \{\mathfrak{m}_0 \oplus R_+\}$, so that there exists $t \in \mathbb{N}$ such that $(R_+)^t M = 0$. In this case, $M = \Gamma_{R_+}(M)$ and so $H_{R_+}^i(M) = 0$ for all $i \in \mathbb{N}$, by 2.1.7(ii), and $D_{R_+}(M) = 0$ by 2.2.10(i). Thus $\chi_M(n) = 0$ for all $n \in \mathbb{Z}$, and our claim is proved in this case.

Now suppose that $\dim M > 0$ and our desired result has been proved for all non-zero, finitely generated graded R-modules of smaller dimension. Since $\Gamma_{R_+}(M)$ is a graded submodule of M, it follows from 2.1.7(iii) that there is a (homogeneous) isomorphism

$$H_{R_+}^i(M) \cong H_{R_+}^i(M/\Gamma_{R_+}(M)) \quad \text{for each } i \in \mathbb{N},$$

and from 13.3.14 and 2.2.10(ii) that there is a (homogeneous) isomorphism $D_{R_+}(M) \cong D_{R_+}(M/\Gamma_{R_+}(M))$. Hence $\chi_M = \chi_{M/\Gamma_{R_+}(M)}$. Since $\Gamma_{R_+}(M)$ is annihilated by some power of R_+, it follows that

$$\mathrm{Supp}(\Gamma_{R_+}(M)) \subseteq \{\mathfrak{m}_0 \oplus R_+\},$$

so that $\dim M = \dim(M/\Gamma_{R_+}(M))$. Hence, for the purpose of this inductive step, we can replace M by $M/\Gamma_{R_+}(M)$ and so (in view of 2.1.2, 2.1.1 and 16.1.4(ii)) assume that M is R_+-torsion-free and that R_1 contains a nonzerodivisor a on M. This we do.

Application of 17.1.5 to the exact sequence

$$0 \longrightarrow M(-1) \xrightarrow{a} M \longrightarrow M/aM \longrightarrow 0$$

of graded R-modules and homogeneous homomorphisms, together with the observation that $\chi_{M(-1)}(n) = \chi_M(n-1)$ for all $n \in \mathbb{Z}$, yields that

$$\chi_M(n) - \chi_M(n-1) = \chi_{M/aM}(n) \quad \text{for all } n \in \mathbb{Z}.$$

Now $\dim M/aM = \dim M - 1$ (by 14.1.14), and so it follows from the inductive hypothesis that there is a polynomial $P_{M/aM} \in \mathbb{Q}[X]$ of degree $\dim M - 2$ such that $P_{M/aM}(n) = \chi_{M/aM}(n)$ for all $n \in \mathbb{Z}$. Hence, by 17.1.3, there exists a polynomial $P_M \in \mathbb{Q}[X]$ such that $P_M(n) = \chi_M(n)$ for all $n \in \mathbb{Z}$. Furthermore, if $P_{M/aM} \neq 0$, then P_M must have degree $\dim M - 1$. However, if $P_{M/aM} = 0$, then the above argument shows only that $\deg P_M \leq 0$, and we must show that $\chi_M(n) \neq 0$ for some $n \in \mathbb{Z}$ in order to complete the inductive step.

The assumption that $P_{M/aM} = 0$, in conjunction with the inductive hypothesis, implies that $\dim M = 1$. Hence $H^i_{R_+}(M) = 0$ for all $i > 1$, and so $\chi_M(n) = d^0_M(n)$ for all $n \in \mathbb{Z}$. However, the fact that $\Gamma_{R_+}(M) = 0$ ensures that the map $\eta_M : M \longrightarrow D_{R_+}(M)$ of 2.2.6(i) is a monomorphism, so that $D_{R_+}(M) \neq 0$ and $d^0_M(n) \neq 0$ for some $n \in \mathbb{Z}$. $\qquad\square$

17.1.8 Remark. Assume that $R = \bigoplus_{n \in \mathbb{N}_0} R_n$ is positively graded and homogeneous, and such that R_0 is Artinian; let $M = \bigoplus_{n \in \mathbb{Z}} M_n$ be a non-zero finitely generated graded R-module of dimension d.

It was pointed out in 17.1.4 that $\chi_M(n) = \ell_{R_0}(M_n)$ for all $n \gg 0$. Thus Hilbert's Theorem (see [7, Theorem 4.1.3]), that the function

$$\ell_{R_0}(M_{(\,\bullet\,)}) : \mathbb{Z} \longrightarrow \mathbb{Z}$$

is of polynomial type of degree $d - 1$, is a corollary of Theorem 17.1.7, and the polynomial P_M of that theorem is just the *Hilbert polynomial of M* of [7, 4.1.5].

With the notation of 17.1.1(v), for each $j = 0, \ldots, d - 1$, we set $e_j(M) :=$ $e_j(P_M)$ and refer to this as the *j-th Hilbert coefficient of M*. Note that, in the terminology of [7, 4.1.5], the integer $e_0(M)$ is the *multiplicity of M* if $d > 0$. To sum up, we can write

$$P_M(X) = \sum_{i=0}^{d-1} (-1)^i e_i(M) \binom{X + d - i - 1}{d - i - 1}$$

and

$$\chi_M(n) = \sum_{i=0}^{d-1} (-1)^i e_i(M) \binom{n + d - i - 1}{d - i - 1} \quad \text{for all } n \in \mathbb{Z}.$$

17.1.9 ♯Exercise. Assume that $R = \bigoplus_{n \in \mathbb{N}_0} R_n$ is positively graded and homogeneous, and such that R_0 is Artinian; let M be a non-zero finitely generated graded R-module of dimension d. Suppose that $a \in R_1$ is a nonzerodivisor on M. Show that

(i) $P_{M/aM}(X) = P_M(X) - P_M(X - 1)$; and
(ii) if $d > 1$, then, with the notation of 17.1.1(v),

$$P_{M/aM}(X) = p_{(e_0(M), \ldots, e_{d-2}(M))}(X),$$

and $e_i(M/aM) = e_i(M)$ for all $i = 0, \ldots, d - 2$.

17.1.10 Definition and ♯Exercise. Assume that $R = \bigoplus_{n \in \mathbb{N}_0} R_n$ is positively graded and homogeneous, and such that R_0 is Artinian; let M be a finitely generated graded R-module. The *postulation number* $\mathrm{pstln}(M)$ *of M* is defined by $\mathrm{pstln}(M) = \sup\{n \in \mathbb{Z} : \ell_{R_0}(M_n) \neq P_M(n)\}$. Thus $\mathrm{pstln}(M)$ is an integer or $-\infty$. Show that

(i) $\mathrm{pstln}(M) \leq \mathrm{reg}(M) \leq \max\{\mathrm{reg}^1(M), \mathrm{pstln}(M)\}$, and
(ii) if $\dim M \leq 0$, then $\mathrm{pstln}(M) = \mathrm{reg}(M) = \mathrm{end}(M)$.

17.1.11 Theorem and Definition. *Assume $R = \bigoplus_{n \in \mathbb{N}_0} R_n$ is positively graded and homogeneous, and such that R_0 is Artinian; let M be a finitely generated graded R-module. Let $i \in \mathbb{N}_0$. Then the function $h_M^i : \mathbb{Z} \longrightarrow \mathbb{N}_0$ is of reverse polynomial type of degree less than i (in the sense of 17.1.1(vi)); in other words, there is a polynomial $p_M^i \in \mathbb{Q}[X]$ of degree less than i such that*

$$\ell_{R_0}(H_{R_+}^i(M)_n) = p_M^i(n) \quad \text{for all } n \ll 0.$$

The function h_M^i is called the i-th cohomological Hilbert function *of M, while the (uniquely determined) polynomial p_M^i is called the i-th* cohomological Hilbert polynomial *of M.*

Proof.   By 17.1.6, it is sufficient for us to prove the result under the additional hypothesis that R_0 is local. We therefore assume for the remainder of this proof that R_0 is an Artinian local ring, with maximal ideal $\mathfrak{m}_0$, say. With this assumption, R is *local, and $\sqrt{R_+} = \mathfrak{m}_0 + R_+ =: \mathfrak{m}$, the unique *maximal ideal of R.

Since $(R_0, \mathfrak{m}_0)$ is a complete local ring, we can use Cohen's Structure Theorem for such rings (see [50, Theorem 29.4(ii)], for example) in conjunction with standard facts about the structure of positively graded homogeneous commutative Noetherian rings (see [7, Proposition 1.5.4]) to see that there is a Gorenstein graded *local commutative Noetherian ring R' and a surjective homogeneous ring homomorphism $f : R' \longrightarrow R$. Let $d' := \dim R'$.

We are going to use the Graded Local Duality Theorem 14.4.1. Let $^*D$ denote the functor $^*\operatorname{Hom}_R(\,\bullet\,, {}^*E(R/\mathfrak{m}))$ from $^*\mathcal{C}(R)$ to itself. By 14.4.1, there exists $a' \in \mathbb{Z}$ such that there is a natural transformation

$$\psi^i : H^i_{R_+} \longrightarrow {}^*D({}^*\operatorname{Ext}^{d'-i}_{R'}(\,\bullet\,, R'(a')))$$

of covariant functors from $^*\mathcal{C}(R)$ to $^*\mathcal{C}(R)$ which is such that $\psi^i_{M'}$ is a (necessarily homogeneous) isomorphism for all $i \in \mathbb{N}_0$ whenever M' is a finitely generated graded R-module.

Let $N = {}^*\operatorname{Ext}^{d'-i}_{R'}(M, R'(a'))$, a finitely generated graded R-module. We now use Graded Matlis Duality 14.4.2. We use the notation $(\,\bullet\,)^\vee$ of 14.4.2 to denote the functor $^*\operatorname{Hom}_{R_0}(\,\bullet\,, E_{R_0}(R_0/\mathfrak{m}_0))$ from $^*\mathcal{C}(R)$ to itself. It follows from 14.4.2 that, for each $n \in \mathbb{Z}$ (and with an obvious notation), there are R_0-isomorphisms

$$H^i_{R_+}(M)_n \cong ({}^*D(N))_n \cong (N^\vee)_n \cong ({}^*\operatorname{Hom}_{R_0}(N, E_{R_0}(R_0/\mathfrak{m}_0)))_n$$
$$= \operatorname{Hom}_{R_0}(N_{-n}, E_{R_0}(R_0/\mathfrak{m}_0)),$$

so that $h^i_M(n) = \ell_{R_0}(\operatorname{Hom}_{R_0}(N_{-n}, E_{R_0}(R_0/\mathfrak{m}_0))) = \ell_{R_0}(N_{-n})$ by 10.2.13. Thus, in order to prove that h^i_M is of reverse polynomial type of degree less than i, it is now sufficient for us to prove that the function $\ell_{R_0}(N_{(\bullet)}) : \mathbb{Z} \longrightarrow \mathbb{N}_0$ is of polynomial type of degree less than i; hence, by Hilbert's Theorem (see 17.1.8 above and [7, Theorem 4.1.3]), it is enough for us to prove that $\dim N \leq i$.

To do this, let $\mathfrak{p}' \in \operatorname{Supp}_{R'} N$. Then $\operatorname{Ext}^{d'-i}_{R'_{\mathfrak{p}'}}(M_{\mathfrak{p}'}, R'_{\mathfrak{p}'}) \neq 0$, so that, since $\operatorname{inj\,dim}_{R'_{\mathfrak{p}'}} R'_{\mathfrak{p}'} = \operatorname{ht} \mathfrak{p}'$, we must have $\operatorname{ht} \mathfrak{p}' \geq d' - i$. Therefore $\dim R'/\mathfrak{p}' \leq i$, so that $\dim_R N \leq i$. $\qquad\square$

17.1.12 Exercise.   Let the notation be as in 15.2.7 and 15.2.8.

(i) Compute the Hilbert polynomial, all Hilbert coefficients, the postulation number, all cohomological Hilbert functions and the Castelnuovo–Mumford regularity for each of the three rings A, B and C.

(ii) Conclude that the vanishing ideal $I_{\mathbb{P}^6(K)}(V_A) \subset K[X_0, \ldots, X_6]$ of V_A can be generated by homogeneous polynomials of degree 2, and that the vanishing ideals $I_{\mathbb{P}^5(K)}(V_B), I_{\mathbb{P}^5(K)}(V_C) \subset K[X_0, \ldots, X_5]$ of V_B and V_C can be generated by homogeneous polynomials of degrees 2 and 3. (You might find 16.3.2 and 16.3.3 helpful.)

17.1.13 ♯Exercise. Let the situation be as in 17.1.11, and let M be a non-zero finitely generated graded R-module of dimension d. Show that the d-th cohomological Hilbert polynomial p_M^d of M has degree exactly $d - 1$.

17.1.14 Exercise. Assume that $R = \bigoplus_{n \in \mathbb{N}_0} R_n$ is positively graded and homogeneous, and such that R_0 is Artinian; let M be a non-zero finitely generated graded R-module of positive dimension. This exercise involves the finiteness dimension $f_{R_+}(M)$ of M relative to R_+ of 9.1.3, and, for $r \in \mathbb{N}$ and $s \in \mathbb{Z}$, the notation $(\,\bullet\,)^{(r,s)} : {}^*\mathcal{C}(R) \longrightarrow {}^*\mathcal{C}(R^{(r)})$ of 13.5.9 for the (r,s)-th Veronesean functor.

(i) Show that $f_{R_+}(M)$ is finite.
(ii) Show that $f_{R_+}(M) = \max \{ \mathrm{grade}_{M^{(r,s)}}((R^{(r)})_+) : r \in \mathbb{N}, \ s \in \mathbb{Z} \}$.

17.2 The significance of reg^2

In this section, we shall show that, if R is positively graded and homogeneous, M is a finitely generated graded R-module, and $t \in \mathbb{Z}$ is such that $t \geq \mathrm{reg}^2(M)$, then $D_{R_+}(M)$ and $H^1_{R_+}(M)$ can be generated by homogeneous elements of degrees not exceeding t. We shall also start to prepare the ground for a result (to be presented in the next section) that bounds $\mathrm{reg}^2(M)$ in certain circumstances.

As promised in the introduction to this chapter, we now give another hint about the significance of the invariant reg^2. Let K be an algebraically closed field, let $r \in \mathbb{N}$ and let $V \subset \mathbb{P}^r(K)$ be a projective variety. Let $\mathfrak{p}$ denote the vanishing ideal $I_{\mathbb{P}^r(K)}(V)$ of V, so that $\mathfrak{p}$ is the (non-zero) graded prime ideal $I_{\mathbb{A}^{r+1}(K)}(\mathrm{Cone}(V))$ of $K[X_0, \ldots, X_r] = \mathcal{O}(\mathbb{A}^{r+1}(K))$. In 16.3.3, we introduced the Castelnuovo–Mumford regularity reg(V) of V, and pointed out that $\mathrm{reg}(V) = \mathrm{reg}(\mathfrak{p}) = \mathrm{reg}^1(\mathfrak{p}) = \mathrm{reg}^2(\mathfrak{p})$ and that $\mathfrak{p}$ can be generated by homogeneous polynomials of degrees not exceeding reg(V). Since $\mathfrak{p}$ can be used to define V, in the sense that $V = V_{\mathbb{P}^r(K)}(\mathfrak{p})$, the regularity $\mathrm{reg}^2(\mathfrak{p})$ of $\mathfrak{p}$ at and above level 2 provides an upper bound on the degrees of homogeneous polynomials needed to define V. This is another reason why one could be interested in upper bounds on $\mathrm{reg}^2(\mathfrak{p})$.

17.2.1 Proposition. *Assume that $R = \bigoplus_{n \in \mathbb{N}_0} R_n$ is positively graded and homogeneous, and let $M = \bigoplus_{n \in \mathbb{Z}} M_n$ be a finitely generated graded R-module.*

If $\mathrm{reg}^2(M) \neq -\infty$, then each of $D_{R_+}(M)$ and $H^1_{R_+}(M)$ can be generated by homogeneous elements of degrees not exceeding $\mathrm{reg}^2(M)$.

If $\mathrm{reg}^2(M) = -\infty$, then, for each $t \in \mathbb{Z}$, both $D_{R_+}(M)$ and $H^1_{R_+}(M)$ can be generated by homogeneous elements of degrees not exceeding t.

Proof. Since $\mathrm{reg}^2(M) = \mathrm{reg}^2(M/\Gamma_{R_+}(M))$ (by 16.2.10(ii)) and there are homogeneous isomorphisms

$$D_{R_+}(M) \xrightarrow{\cong} D_{R_+}(M/\Gamma_{R_+}(M)) \text{ and } H^1_{R_+}(M) \xrightarrow{\cong} H^1_{R_+}(M/\Gamma_{R_+}(M))$$

(by 13.3.14, 2.2.10(ii) and 2.1.7(iii)), we can replace M by $M/\Gamma_{R_+}(M)$ and so (in view of 2.1.2) assume that M is R_+-torsion-free. We shall make this simplification.

Since M is finitely generated, there exists $h \in \mathbb{Z}$ such that $M_n = 0$ for all $n < h$. If $\mathrm{reg}^2(M) \neq -\infty$, choose $t \in \mathbb{Z}$ such that $t \leq \min\{h, \mathrm{reg}^2(M)\}$; if $\mathrm{reg}^2(M) = -\infty$, let t be any integer such that $t \leq h$. With the notation of 16.1.1, let $N := D_{R_+}(M)_{\geq t}$. Note that N is a finitely generated R-module, by 16.1.6(i). Also, by choice of t, the homogeneous monomorphism $\eta_M : M \longrightarrow D_{R_+}(M)$ satisfies $\mathrm{Im}\,\eta_M \subseteq N$; thus $N/\mathrm{Im}\,\eta_M \subseteq D_{R_+}(M)/\mathrm{Im}\,\eta_M \cong H^1_{R_+}(M)$, and so $N/\mathrm{Im}\,\eta_M$ is finitely generated and R_+-torsion. It therefore follows from 2.1.7(i) that there is a (homogeneous) isomorphism $H^i_{R_+}(M) \xrightarrow{\cong} H^i_{R_+}(N)$ for each $i > 1$, so that $\mathrm{reg}^2(M) = \mathrm{reg}^2(N)$.

Let $C = \bigoplus_{n \in \mathbb{Z}} C_n$ be the graded R-module $D_{R_+}(M)/N$. Since

$$H^i_{R_+}(D_{R_+}(M)) = 0 \quad \text{for } i = 0, 1$$

by 2.2.10(iv), it follows from the exact sequence

$$0 \longrightarrow N \longrightarrow D_{R_+}(M) \longrightarrow C \longrightarrow 0$$

(in the category $^*\mathcal{C}(R)$) that $H^0_{R_+}(N) = 0$ and there is a homogeneous isomorphism $\Gamma_{R_+}(C) \xrightarrow{\cong} H^1_{R_+}(N)$. Now $\Gamma_{R_+}(C)$ is a graded submodule of C, and $C_n = 0$ for all $n \geq t$. Therefore $\mathrm{end}(H^1_{R_+}(N)) + 1 \leq t - 1 + 1 = t$. In the case where $\mathrm{reg}^2(M) \neq -\infty$, it follows that

$$\mathrm{reg}(N) = \mathrm{reg}^2(N) = \mathrm{reg}^2(M) \geq t,$$

whereas, in the case where $\mathrm{reg}^2(M) = -\infty$, we have $\mathrm{reg}(N) \leq t$.

It follows from 16.3.1 that $N = D_{R_+}(M)_{\geq t}$, if non-zero, can be generated by homogeneous elements of degrees not exceeding $\max\{t, \mathrm{reg}^2(M)\}$, so that, since $H^1_{R_+}(M)$ is a homomorphic image of $D_{R_+}(M)$ by a homogeneous homomorphism, all the claims are proved. $\qquad\square$

17.2.2 ♯Exercise. Let the notation and hypotheses be as in 17.2.1. Let $t \in \mathbb{Z}$ with $t \geq \mathrm{reg}^2(M)$. Let $\mathfrak{p} \in \mathrm{Proj}(R) := {}^*\mathrm{Spec}(R) \setminus \mathrm{Var}(R_+)$. Recall that $\bullet_{(\mathfrak{p})}$ denotes the homogeneous localization functor with respect to $\mathfrak{p}$.

Show that the natural homomorphism $\eta := \eta_M : M \longrightarrow D_{R_+}(M)$ induces a homogeneous isomorphism $\eta_{(\mathfrak{p})} : M_{(\mathfrak{p})} \overset{\cong}{\longrightarrow} D_{R_+}(M)_{(\mathfrak{p})}$ of graded $R_{(\mathfrak{p})}$-modules.

Let $\phi_{(\mathfrak{p})} : D_{R_+}(M) \longrightarrow D_{R_+}(M)_{(\mathfrak{p})}$ be the natural homogeneous homomorphism of graded R-modules, and consider the composition $\beta_{(\mathfrak{p})} = \eta_{(\mathfrak{p})}^{-1} \circ \phi_{(\mathfrak{p})} : D_{R_+}(M) \longrightarrow M_{(\mathfrak{p})}$. As usual, the t-th component of $\beta_{(\mathfrak{p})}$ will be denoted by $(\beta_{(\mathfrak{p})})_t$. Let $\mathcal{S}$ be a generating set for the R_0-module $D_{R_+}(M)_t$. Use 17.2.1 and the fact that R is homogeneous to show that

$$M_{(\mathfrak{p})} = \sum_{m \in \mathcal{S}} R_{(\mathfrak{p})} \beta_{(\mathfrak{p})}(m)$$

and $(M(t)_{(\mathfrak{p})})_0 = (M_{(\mathfrak{p})})_t = \sum_{m \in \mathcal{S}} (R_{(\mathfrak{p})})_0 (\beta_{(\mathfrak{p})})_t(m)$.

Note. Exercise 17.2.2 will be used in an application to sheaf cohomology in Theorem 20.4.13; the aim of that theorem is to establish a (generalization of a) result of Serre that certain sheaves are generated by their global sections.

17.2.3 Proposition. *Assume that $R = \bigoplus_{n \in \mathbb{N}_0} R_n$ is positively graded and homogeneous, and let M be a finitely generated graded R-module. Suppose that $a \in R_1$ is a non-zerodivisor on M. Then*

(i) *for all integers $m \geq \mathrm{reg}^2(M/aM) - 1$, the multiplication map*

$$H^2_{R_+}(M)_{m-1} \overset{a}{\longrightarrow} H^2_{R_+}(M)_m$$

is surjective; and

(ii) *for all integers $m \geq \mathrm{reg}^2(M/aM)$ such that $H^2_{R_+}(M)_{m-1} \neq 0$, the multiplication map $H^2_{R_+}(M)_{m-1} \overset{a}{\longrightarrow} H^2_{R_+}(M)_m$ is not injective.*

Proof. (i) By 14.1.10(ii), the exact sequence

$$0 \longrightarrow M \overset{a}{\longrightarrow} M(1) \longrightarrow (M/aM)(1) \longrightarrow 0$$

of graded R-modules and homogeneous homomorphisms induces an exact sequence $H^2_{R_+}(M)_{m-1} \overset{a}{\longrightarrow} H^2_{R_+}(M)_m \longrightarrow H^2_{R_+}(M/aM)_m$ of R_0-modules, for all $m \in \mathbb{Z}$, and $H^2_{R_+}(M/aM)_m = 0$ when $m + 1 \geq \mathrm{reg}^2(M/aM)$.

(ii) For each $m \in \mathbb{Z}$, let

$$\beta_m : R_1 \otimes_{R_0} H^1_{R_+}(M/aM)_m \longrightarrow H^1_{R_+}(M/aM)_{m+1}$$

be the R_0-homomorphism for which $\beta_m(r_1 \otimes z) = r_1 z$ for all $r_1 \in R_1$ and $z \in H^1_{R_+}(M/aM)_m$, and let $\gamma_m : R_1 \otimes_{R_0} H^1_{R_+}(M)_m \longrightarrow H^1_{R_+}(M)_{m+1}$ be defined similarly.

Let $m_0 \in \mathbb{Z}$ be such that $H^1_{R_+}(M/aM)$ can be generated by homogeneous elements of degrees not exceeding m_0: by Proposition 17.2.1, we can take $m_0 = \mathrm{reg}^2(M/aM)$ if $\mathrm{reg}^2(M/aM) \neq -\infty$, and we can take m_0 to be an arbitrary integer if $\mathrm{reg}^2(M/aM) = -\infty$. Since R is homogeneous, it follows that, for all $m \geq m_0$, each element $y \in H^1_{R_+}(M/aM)_{m+1}$ can be expressed in the form $\sum_{i=1}^t a_i z_i$ for suitable $a_1, \ldots, a_t \in R_1$ and $z_1, \ldots, z_t \in H^1_{R_+}(M/aM)_m$. In other words, β_m is surjective for all $m \geq m_0$.

Let $\pi : M \longrightarrow M/aM$ be the canonical epimorphism, and, for each $m \in \mathbb{Z}$, let $\alpha_m : H^1_{R_+}(M)_m \longrightarrow H^1_{R_+}(M/aM)_m$ be the m-th component of the homogeneous homomorphism $H^1_{R_+}(\pi)$. Observe that the diagram

$$
\begin{array}{ccc}
R_1 \otimes_{R_0} H^1_{R_+}(M)_m & \xrightarrow{\;\gamma_m\;} & H^1_{R_+}(M)_{m+1} \\
\big\downarrow {\scriptstyle \mathrm{Id}_{R_1} \otimes \alpha_m} & & \big\downarrow {\scriptstyle \alpha_{m+1}} \\
R_1 \otimes_{R_0} H^1_{R_+}(M/aM)_m & \xrightarrow{\;\beta_m\;} & H^1_{R_+}(M/aM)_{m+1}
\end{array}
$$

commutes. It follows that, if, for some $m \geq m_0$, we know that α_m is surjective, then α_{m+1} is surjective too.

Consider an $m \geq m_0$ and suppose that $H^2_{R_+}(M)_{m-1} \xrightarrow{a} H^2_{R_+}(M)_m$ is injective. By 14.1.10(ii), the exact sequence

$$0 \longrightarrow M(-1) \xrightarrow{a} M \longrightarrow M/aM \longrightarrow 0$$

induces an exact sequence

$$H^1_{R_+}(M)_n \xrightarrow{\alpha_n} H^1_{R_+}(M/aM)_n \longrightarrow H^2_{R_+}(M)_{n-1} \xrightarrow{a} H^2_{R_+}(M)_n$$

of R_0-modules for all $n \in \mathbb{Z}$. This exact sequence (in the case where $n = m$) shows that α_m is surjective, and so it follows, as explained in the immediately preceding paragraph, that α_{m+1} is surjective and α_n is surjective for all $n \geq m$. It follows from the above exact sequence that $H^2_{R_+}(M)_{n-1} \xrightarrow{a} H^2_{R_+}(M)_n$ is injective for all $n \geq m$. It therefore follows from 16.1.5(ii) that $H^2_{R_+}(M)_{m-1} = 0$. $\qquad \square$

We point out that, in the situation and with the notation of 17.2.3, and with the additional assumption that R_0 is Artinian, the proposition shows that, as n increases beyond $\text{reg}^2(M/aM) - 1$, the integers $h_M^2(n)$ decrease *strictly* to 0 and then remain at 0.

The following corollary is a consequence of Propositions 17.2.1 and 17.2.3, and it provides the key for some crucial arguments later in this chapter.

17.2.4 Corollary. *Assume that* $R = \bigoplus_{n \in \mathbb{N}_0} R_n$ *is positively graded and homogeneous, and that* R_0 *is Artinian; let* M *be a finitely generated graded* R-*module. Suppose that* $a \in R_1$ *is a non-zerodivisor on* M.

Let $n_0, q \in \mathbb{Z}$ *with* $q \geq \max\{n_0, \text{reg}^2(M/aM) - 1\}$; *for each* $n \geq n_0$, *set*

$$s(n) := h_M^2(n_0) + \sum_{m=n_0+1}^{n} h_{M/aM}^2(m).$$

Then

(i) $h_M^2(n) \leq s(n)$ *for all* $n \geq n_0$;

(ii) $s(n) = s(q)$ *for all* $n \geq q$, *and* $s(q) = s(q-1)$ *if* $q > n_0$; *and*

(iii) $h_M^2(n) \leq \max\{0, s(q) + q - n\}$ *for all* $n > q$.

Proof. (i) By 14.1.10(ii), the exact sequence

$$0 \longrightarrow M \overset{a}{\longrightarrow} M(1) \longrightarrow (M/aM)(1) \longrightarrow 0$$

of graded R-modules and homogeneous homomorphisms induces, for each $m \in \mathbb{Z}$, an exact sequence

$$H_{R_+}^2(M)_{m-1} \overset{a}{\longrightarrow} H_{R_+}^2(M)_m \longrightarrow H_{R_+}^2(M/aM)_m$$

of R_0-modules, from which we deduce that

$$h_M^2(m) = \ell_{R_0}(H_{R_+}^2(M)_m) \leq \ell_{R_0}(H_{R_+}^2(M)_{m-1}) + \ell_{R_0}(H_{R_+}^2(M/aM)_m)$$
$$= h_M^2(m-1) + h_{M/aM}^2(m),$$

and the claim follows easily from this.

(ii) For $m > q \geq \text{reg}^2(M/aM) - 1$, we have

$$q \geq \text{end}(H_{R_+}^2(M/aM)) + 1,$$

so that $H_{R_+}^2(M/aM)_m = H_{R_+}^2(M/aM)_q = 0$. Hence

$$h_{M/aM}^2(m) = h_{M/aM}^2(q) = 0.$$

Therefore $s(n) = s(q)$ for all $n \geq q$, and $s(q) = s(q-1)$ if $q > n_0$.

(iii) For $m \geq q + 1 \geq \text{reg}^2(M/aM)$, we have, by Proposition 17.2.3, that the multiplication map $H_{R_+}^2(M)_{m-1} \overset{a}{\longrightarrow} H_{R_+}^2(M)_m$ is surjective, and is not injective unless $H^2_{R_+}(M)_{m-1} = 0$; therefore, if $h^2_M(m-1) \neq 0$, then $h^2_M(m) \leq h^2_M(m-1) - 1$. Thus $h^2_M(m) \leq \max\left\{0, h^2_M(m-1) - 1\right\}$ for all $m \geq q + 1$. Repeated use of this then shows that

$$h^2_M(n) \leq \max\left\{0, h^2_M(q) - (n - q)\right\} \quad \text{for all } n \geq q + 1,$$

and the claim follows from part (i). $\square$

17.3 Bounds on reg^2 in terms of Hilbert coefficients

17.3.1 Definition. Assume that $R = \bigoplus_{n \in \mathbb{N}_0} R_n$ is graded. Let $\mathcal{D}$ be a sub-class of the class of all graded R-modules. By a *numerical invariant for R-modules in $\mathcal{D}$* (or *on $\mathcal{D}$*) we mean an assignment μ which, to each R-module M in $\mathcal{D}$, assigns $\mu(M) \in \mathbb{Z} \cup \{-\infty\}$, and which is such that $\mu(M) = \mu(N)$ whenever M and N are modules in $\mathcal{D}$ for which there is a homogeneous iso-morphism $M \cong N$. We say that such a numerical invariant μ is *finite* if and only if μ takes only finite values.

Let $\mu_1, \ldots, \mu_s, \rho$ be numerical invariants for R-modules in $\mathcal{D}$, such that $\mu_1, \ldots, \mu_s$ are finite. We say that $\mu_1, \ldots, \mu_s$ form a *bounding system for ρ* (for R-modules in $\mathcal{D}$) (or *on $\mathcal{D}$*) if and only if there is a function $B : \mathbb{Z}^s \longrightarrow \mathbb{Z}$ such that $\rho(M) \leq B(\mu_1(M), \ldots, \mu_s(M))$ for each R-module M in $\mathcal{D}$. Furthermore, we say that $\mu_1, \ldots, \mu_s$ form a *minimal* bounding system for ρ (for R-modules in $\mathcal{D}$) precisely when they form a bounding system for ρ on $\mathcal{D}$ but no $s - 1$ of $\mu_1, \ldots, \mu_s$ form a bounding system for ρ on $\mathcal{D}$.

The main result of this section is a generalization of a ring-theoretic formu-lation of a classical result of D. Mumford on sheaves of ideals on projective spaces. A consequence is that, over the polynomial ring $S = K[X_0, \ldots, X_r]$ where K is an algebraically closed field, $e_0(\bullet), \ldots, e_r(\bullet)$ form a bounding system for reg^2 for non-zero graded ideals of S.

The following exercise shows that, in general, for an Artinian local ring R_0, the Hilbert coefficients do not form a bounding system for reg^2 on the class of *all* finitely generated graded $R_0[X_1, \ldots, X_n]$-modules.

17.3.2 Exercise. Suppose that $R = R_0[X_1, X_2]$ is a polynomial ring in 2 indeterminates X_1, X_2 over an Artinian ring R_0.

For each $t \in \mathbb{N}_0$, set $M^{(t)} := R(t) \oplus R(-t)$, and calculate $e_0(M^{(t)})$, $e_1(M^{(t)})$, $\mathrm{reg}^2(M^{(t)})$. Conclude that the numerical invariants $e_0(\bullet), e_1(\bullet)$ do not form a bounding system for reg^2 on the class of *all* finitely generated graded R-modules.

17.3.3 Notation and Remark. For $(n,i) \in \mathbb{Z} \times \mathbb{N}_0$, we set

$$\binom{n}{i}^+ := \begin{cases} \binom{n}{i} & \text{if } n \geq i, \\ 0 & \text{if } n < i. \end{cases}$$

This notation is helpful in the following situation. Suppose that

$$R = R_0[X_1, \ldots, X_d]$$

is a polynomial ring in d (> 0) indeterminates $X_1, \ldots, X_d$ over an Artinian ring R_0, and let $a \in \mathbb{Z}$. Then, for all $n \in \mathbb{Z}$, the n-th component $R(a)_n$ of the graded R-module $R(a)$ has length as an R_0-module given by

$$\ell_{R_0}\left(R(a)_n\right) = \ell_{R_0}(R_0) \binom{n+a+d-1}{d-1}^+.$$

17.3.4 Lemma. *Suppose that* $R = R_0[X_1, \ldots, X_d]$ *is a polynomial ring in* $d > 1$ *indeterminates* $X_1, \ldots, X_d$ *over an Artinian ring* R_0. *Let* $r \in \mathbb{N}$ *and let* $a_1, \ldots, a_r \in \mathbb{Z}$. *Let* M *be a graded submodule of* $\bigoplus_{i=1}^r R(a_i)$, *and let* f *be an integer such that* $f \geq \mathrm{reg}^2(M/X_dM)$. *Then*

$$\mathrm{reg}^2(M) \leq \ell_{R_0}(R_0) \sum_{i=1}^r \binom{f + a_i + d - 3}{d-1}^+ - \chi_M(f-2) + f.$$

Proof. Note that X_d is a non-zerodivisor on M, because M is a submodule of $G := \bigoplus_{i=1}^r R(a_i)$. Let $i, n \in \mathbb{Z}$ be such that $i > 2$ and $n > f - i$. Since $f \geq \mathrm{reg}^2(M/X_dM) \geq \mathrm{end}\, H_{R_+}^{i-1}(M/X_dM) + i - 1$, it follows that $H_{R_+}^{i-1}(M/X_dM)_{n+1} = 0$. Therefore, the exact sequence

$$0 \longrightarrow M(-1) \overset{X_d}{\longrightarrow} M \longrightarrow M/X_dM \longrightarrow 0$$

of graded R-modules and homogeneous homomorphisms induces an exact sequence

$$0 \longrightarrow H_{R_+}^i(M)_n \overset{X_d}{\longrightarrow} H_{R_+}^i(M)_{n+1}$$

of R_0-modules. Hence $H_{R_+}^i(M)_n = 0$ for all $i > 2$ and $n > f - i$. Thus, by Definition 17.1.4,

$$\chi_M(f-2) = d_M^0(f-2) - h_M^2(f-2),$$

so that $h_M^2(f-2) = d_M^0(f-2) - \chi_M(f-2)$.

We are now going to use Corollary 17.2.4 with $n_0 = f - 2$ and $q = f - 1$. Observe that, with the notation of that result, $h_{M/X_dM}^2(m) = 0$ for all $m > n_0 = f - 2$, so that $s(n) = h_M^2(f-2)$ for all $n \geq n_0$. It follows from

17.2.4 that $h_M^2(n) \leq \max\{0, h_M^2(f-2) + f - 1 - n\}$ for all $n \geq f - 1$. Hence $h_M^2(n) = 0$ for all $n > h_M^2(f-2) + f - 2$. As we know already that $h_M^i(n) = 0$ for all $i > 2$ and $n > f - i$, it follows that

$$\operatorname{reg}^2(M) \leq h_M^2(f-2) + f = d_M^0(f-2) - \chi_M(f-2) + f.$$

As $\operatorname{grade}_G R_+ = d > 1$, it follows from 6.2.7, 2.2.6(i)(c) and 13.3.14 that $\eta_{R_+} : G \longrightarrow D_{R_+}(G)$ is a homogeneous isomorphism. It therefore follows from the fact that the functor D_{R_+} is left exact that there is a homogeneous monomorphism $D_{R_+}(M) \longrightarrow G$; we thus deduce, on use of 17.3.3, that (with an obvious notation)

$$d_M^0(f-2) \leq \ell_{R_0}(G_{f-2}) = \ell_{R_0}(R_0) \sum_{i=1}^{r} \binom{f - 2 + a_i + d - 1}{d - 1}^+ .$$

The claim now follows. □

Our next theorem is the main result of this chapter. We need one preliminary lemma.

17.3.5 Lemma. *Suppose that $R = R_0[X_1, \ldots, X_d]$ is a polynomial ring in d (> 0) indeterminates $X_1, \ldots, X_d$ over an Artinian local ring $(R_0, \mathfrak{m}_0)$.*

Let M be a non-zero finitely generated graded R-module, and suppose that there exists $a \in R_1$ which is a non-zerodivisor on M. Then there exist $X_1', \ldots, X_d' \in R_1$ such that X_d' is a non-zerodivisor on M, the family $(X_i')_{i=1}^d$ is algebraically independent over R_0, and $R = R_0[X_1', \ldots, X_d']$.

Proof. We have $a = b_1 X_1 + \cdots + b_d X_d$ for some $b_1, \ldots, b_d \in R_0$. Now $\mathfrak{m}_0$ is nilpotent, and so there exists $j \in \{1, \ldots, d\}$ such that $b_j \notin \mathfrak{m}_0$. Hence b_j is a unit of R_0, and so, after multiplication by b_j^{-1}, we can, and do, assume that $b_j = 1$. Let $c := a - X_j \in R_1$.

There is a homogeneous R_0-algebra homomorphism $\phi : R \longrightarrow R$ for which $\phi(X_j) = X_j + c = a$ and $\phi(X_i) = X_i$ for all $i = 1, \ldots, d$ with $i \neq j$. Similarly, there is a homogeneous R_0-algebra homomorphism $\psi : R \longrightarrow R$ for which $\psi(X_j) = X_j - c$ and $\psi(X_i) = X_i$ for all $i = 1, \ldots, d$ with $i \neq j$. Since $\phi(c) = \psi(c) = c$, it follows that ϕ and ψ are inverse isomorphisms. If we now take $X_j' = a = X_j + c$ and $X_i' = X_i$ for all $i = 1, \ldots, d$ with $i \neq j$, then $(X_i')_{i=1}^d$ is algebraically independent over R_0 and $R = R_0[X_1', \ldots, X_d']$; since X_j' is a non-zerodivisor on M, we can now reorder the X_i' (if necessary) to complete the proof. □

17.3.6 Theorem. *Suppose $R = R_0[X_1, \ldots, X_d]$ is a polynomial ring in d (> 0) indeterminates $X_1, \ldots, X_d$ over an Artinian local ring $(R_0, \mathfrak{m}_0)$.*

Let G be a non-zero finitely generated *free graded R-module. Then

$$e_0(\bullet), \ldots, e_{d-1}(\bullet)$$

form a bounding system for reg^2 *on the class of all non-zero graded submodules of G.*

Proof. Write $G = \bigoplus_{i=1}^{r} R(a_i)$, where $a := (a_1, \ldots, a_r) \in \mathbb{Z}^r$, and set $s := \min\{-a_i : 1 \leq i \leq r\}$. We start by defining, for each integer $h \geq 2$, a numerical function $F_a^{(h)} : \mathbb{Z}^h \longrightarrow \mathbb{Z}$. The definition will be made by induction on h.

First, for $(e_0, e_1) \in \mathbb{Z}^2$, define (with the notation of 17.1.1(v))

$$F_a^{(2)}(e_0, e_1) = s + 1 - p_{(e_0, e_1)}(s - 1).$$

Now suppose that $h > 2$ and that the function $F_a^{(h-1)}$ has already been defined. Then, for $e := (e_0, \ldots, e_{h-1}) \in \mathbb{Z}^h$ and with $e' := (e_0, \ldots, e_{h-2})$, we write $f = F_a^{(h-1)}(e')$ and set (again with the notation of 17.1.1(v))

$$F_a^{(h)}(e) := \ell_{R_0}(R_0) \sum_{i=1}^{r} \binom{f + a_i + h - 3}{h - 1}^{+} - p_e(f - 2) + f.$$

Consider first the case where $d = 1$, and let M be a non-zero graded submodule of G. Then $\dim M = 1$, so that $H_{R_+}^i(M) = 0$ for all $i > 1$ and reg$^2(M) = -\infty$. We therefore assume henceforth in this proof that $d > 1$.

As in the proof of Theorem 17.1.7, we can use the ideas of Example 16.2.4 to show that it is enough for us to prove the theorem under the additional assumption that $R_0/\mathfrak{m}_0$ is infinite, and we shall make this assumption in what follows. We shall prove the result by induction on d. We shall make frequent use of the fact that, as each associated prime $\mathfrak{p}$ of a free R-module has $\dim R/\mathfrak{p} = d$, every non-zero submodule of a free R-module has dimension d.

Consider now the case where $d = 2$, and let M be a non-zero graded submodule of G. Then $\dim M/X_2 M < 2$, and so reg$^2(M/X_2 M) = -\infty$. Use Lemma 17.3.4 with the choice $f = s+1$: since $f + a_i + d - 3 = s + a_i + d - 2 \leq d - 2 < d - 1$ for each $i = 1, \ldots, r$, we obtain that

$$\text{reg}^2(M) \leq \ell_{R_0}(R_0) \sum_{i=1}^{r} \binom{s + a_i + d - 2}{d - 1}^{+} - \chi_M(s - 1) + s + 1$$

$$= s + 1 - P_M(s - 1) = s + 1 - p_{(e_0(M), e_1(M))}(s - 1)$$

$$= F_a^{(2)}(e_0(M), e_1(M)).$$

Therefore $e_0(\bullet), e_1(\bullet)$ form a bounding system for reg^2 on the class of all non-zero graded submodules of G.

Now suppose $d > 2$ and we have proved, for $\overline{R} := R_0[X_1, \ldots, X_{d-1}]$ and for every non-zero graded submodule N of $\bigoplus_{i=1}^r \overline{R}(a_i)$, that $\operatorname{reg}^2(N) \leq F_a^{(d-1)}(e_0(N), \ldots, e_{d-2}(N))$.

Now let M be a non-zero graded submodule of G. As $\operatorname{grade}_G R_+ = d > 1$, it follows from 6.2.7, 2.2.6(i)(c) and 13.3.14 that $\eta_{R_+} : G \longrightarrow D_{R_+}(G)$ is a homogeneous isomorphism. It therefore follows from the fact that the functor D_{R_+} is left exact that there is a homogeneous monomorphism $\varepsilon : D_{R_+}(M) \longrightarrow G$: let $L := \varepsilon(D_{R_+}(M))$, a d-dimensional graded submodule of G. Since there is a homogeneous isomorphism $L \xrightarrow{\cong} D_{R_+}(M)$, it follows from 13.3.14 and 2.2.10(iii) that there is a homogeneous isomorphism $D_{R_+}(L) \xrightarrow{\cong} D_{R_+}(M)$, and from 13.4.3, 13.4.4 and 2.2.10(v) that there are homogeneous isomorphisms $H_{R_+}^i(L) \xrightarrow{\cong} H_{R_+}^i(M)$ for all $i > 1$. Hence $\chi_L = \chi_M$, so that $P_L = P_M$ and $e_i(L) = e_i(M)$ for all $i = 0, \ldots, d-1$; also $\operatorname{reg}^2(L) = \operatorname{reg}^2(M)$. Now $H_{R_+}^1(L) \cong H_{R_+}^1(D_{R_+}(M)) = 0$ by 2.2.10(iv), so that, since $H_{R_+}^0(G) = 0$, we have $H_{R_+}^0(G/L) = 0$. It is therefore sufficient for us to complete this inductive step under the additional assumption that $H_{R_+}^0(G/M) = 0$, and so we shall make this assumption in the remainder of this proof. In view of 2.1.1 and 16.1.4(ii), this means that R_1 contains a non-zerodivisor on G/M, and it follows from Lemma 17.3.5 that we can assume without loss of generality that X_d is a non-zerodivisor on G/M. This we do.

Set $\overline{R} := R_0[X_1, \ldots, X_{d-1}]$, $\overline{G} := G/X_d G$, and $\overline{M} := M/X_d M$. We can view $\overline{G}$ and $\overline{M}$ as $\overline{R}$-modules by means of the inclusion homomorphism $\overline{R} \longrightarrow R$, and the natural map $\overline{M} \longrightarrow \overline{G}$ is an $\overline{R}$-monomorphism because X_d is a non-zerodivisor on G/M. Note also that there is a homogeneous $\overline{R}$-isomorphism $\overline{G} \xrightarrow{\cong} \bigoplus_{i=1}^r \overline{R}(a_i)$. It therefore follows from the inductive hypothesis that

$$\operatorname{reg}^2(\overline{M}) \leq F_a^{(d-1)}(e_0(\overline{M} \lceil \overline{R}), \ldots, e_{d-2}(\overline{M} \lceil \overline{R})).$$

A straightforward adaptation of 4.2.2 to the graded case, with use of the Graded Independence Theorem 14.1.7 instead of the Independence Theorem 4.2.1, will show that there are homogeneous $\overline{R}$-isomorphisms

$$H_{\overline{R}_+}^i(\overline{M} \lceil \overline{R}) \xrightarrow{\cong} H_{R_+}^i(\overline{M}) \lceil \overline{R} \quad \text{for all } i \in \mathbb{N}_0,$$

and a similar argument based on 14.1.4 will show that there is a homogeneous $\overline{R}$-isomorphism

$$D_{\overline{R}_+}(\overline{M} \lceil \overline{R}) \xrightarrow{\cong} D_{R_+}(\overline{M}) \lceil \overline{R}.$$

It follows that $\operatorname{reg}^2(\overline{M} \lceil \overline{R}) = \operatorname{reg}^2(M/X_d M)$ and, in view of 17.1.9, that

$$e_i(\overline{M} \lceil \overline{R}) = e_i(M/X_d M) = e_i(M) \quad \text{for all } i = 0, \ldots, d-2.$$

Set $e := (e_0(M), \ldots, e_{d-1}(M)) \in \mathbb{Z}^d$ and $e' := (c_0(M), \ldots, e_{d-2}(M))$. Thus $\operatorname{reg}^2(M/X_dM) \leq F_a^{(d-1)}(e')$. Recall that $\chi_M(n) = P_M(n) = p_e(n)$ for all $n \in \mathbb{Z}$. We can now deduce from Lemma 17.3.4 with the choice $f = F_a^{(d-1)}(e')$ that

$$
\begin{aligned}
\operatorname{reg}^2(M) &\leq \ell_{R_0}(R_0) \sum_{i=1}^{r} \binom{f + a_i + d - 3}{d - 1}^+ - \chi_M(f - 2) + f \\
&= \ell_{R_0}(R_0) \sum_{i=1}^{r} \binom{f + a_i + d - 3}{d - 1}^+ - p_e(f - 2) + f \\
&= F_a^{(d)}(e_0(M), \ldots, e_{d-1}(M)).
\end{aligned}
$$

This completes the inductive step. $\qquad\square$

Theorem 17.3.6 is based on work of D. Mumford: see [54, p. 101]. In our notation (of 17.3.6), Mumford's work is concerned with the case where $G = R$; thus Mumford showed that $e_0(\,\bullet\,), \ldots, e_{d-1}(\,\bullet\,)$ form a bounding system for reg^2 on the class of all non-zero graded ideals of R. In fact, Mumford showed that there exists a rational polynomial q in d indeterminates such that, for every non-zero graded ideal $\mathfrak{b}$ of R, $\operatorname{reg}^2(\mathfrak{b})$ is bounded above by $q(e_0(\mathfrak{b}), \ldots, e_{d-1}(\mathfrak{b}))$.

17.4 Bounds on reg^1 and reg^0

In this section we show that the bounding result for reg^2 obtained in the last section leads to bounding results for reg^1 and reg^0 in which the Hilbert coefficients play an important rôle.

17.4.1 Theorem. *Assume that $R = \bigoplus_{n \in \mathbb{N}_0} R_n$ is positively graded and homogeneous, and such that R_0 is Artinian and local, with maximal ideal $\mathfrak{m}_0$; assume also that R is generated as R_0-algebra by d homogeneous elements of degree 1, where $d > 1$. Let $r, t \in \mathbb{N}$ and $a := (a_1, \ldots, a_r) \in \mathbb{Z}^r$ be fixed. Let $\mathcal{D}$ denote the class of all graded R-modules M of dimension t which can be written in the form $M = \sum_{j=1}^{r} Rm_j$ with $m_j \in M_{-a_j}$ for all $j \in \{1, \ldots, r\}$. Then $e_0(\,\bullet\,), \ldots, e_{t-1}(\,\bullet\,)$ form a bounding system for reg^1 on $\mathcal{D}$.*

Proof. We begin by defining a function that we shall use to bound reg^1. Let $\mathcal{E}$ denote the class of non-zero graded submodules of $G = \bigoplus_{i=1}^{r} R(a_i)$. By 17.3.6, there is a function $B : \mathbb{Z}^d \longrightarrow \mathbb{Z}$ such that

$$
\operatorname{reg}^2(L) \leq B(e_0(L), \ldots, e_{d-1}(L))
$$

for each R-module L in $\mathcal{E}$. Let $c := (c_0, \ldots, c_{t-1}) \in \mathbb{Z}^t$ and let p_c be the polynomial of degree $t - 1$ in $\mathbb{Q}[X]$ given by

$$p_c(X) = \sum_{i=0}^{t-1} (-1)^i c_i \binom{X + t - i - 1}{t - i - 1}.$$

(See 17.1.1(v).) Also, define $p_c^{(a)} \in \mathbb{Q}[X]$ by

$$p_c^{(a)}(X) := \sum_{j=1}^{r} \ell_{R_0}(R_0) \binom{X + a_j + d - 1}{d - 1} - p_c(X).$$

With the notation of 17.1.1(v), if $\deg(p_c^{(a)}) = d - 1$, set

$$C_a^{(d)}(c) := \max\{B(e_0(p_c^{(a)}), \ldots, e_{d-1}(p_c^{(a)})) - 1, -a_1, \ldots, -a_r\},$$

but let $C_a^{(d)}(c) := \max\{-a_1, \ldots, -a_r\}$ if $\deg(p_c^{(a)}) < d - 1$.

The hypotheses ensure that there is a surjective homogeneous homomorphism $\phi : S := R_0[X_1, \ldots, X_d] \longrightarrow R$ of graded rings. On use of the Graded Independence Theorem 14.1.7, we see that we can replace R by S; thus we assume that $R = R_0[X_1, \ldots, X_d]$. Let M be a module in $\mathcal{D}$; then there is an exact sequence $0 \longrightarrow N \longrightarrow G \longrightarrow M \longrightarrow 0$ in the category $^*\mathcal{C}(R)$. Write $e := (e_0(M), \ldots, e_{t-1}(M))$. The Hilbert polynomials P_N and P_M satisfy

$$P_N(X) = P_G(X) - P_M(X) = \sum_{j=1}^{r} \ell_{R_0}(R_0) \binom{X + a_j + d - 1}{d - 1} - p_e(X)$$

$$= p_e^{(a)}(X).$$

If $N = 0$, then $\deg(p_e^{(a)}) = -1 < d - 1$, and

$$\operatorname{reg}^1(M) = \operatorname{reg}^1(\bigoplus_{j=1}^{r} R(a_j)) \leq \operatorname{reg}(\bigoplus_{j=1}^{r} R(a_j))$$

$$= \max\{-a_1, \ldots, -a_r\} = C_a^{(d)}(e).$$

Now consider the case where $N \neq 0$. Then $\dim N = d$, and so $\deg(p_e^{(a)}) = d - 1$. Notice that N lies in the class $\mathcal{E}$. Therefore, in view of 17.3.6 and 16.2.15(iv), we have

$$\operatorname{reg}^1(M) \leq \max\{\operatorname{reg}^2(N) - 1, \operatorname{reg}(\bigoplus_{j=1}^{r} R(a_j))\}$$

$$\leq \max\{B(e_0(N), \ldots, e_{d-1}(N)) - 1, \operatorname{reg}(\bigoplus_{j=1}^{r} R(a_j))\}$$

$$= \max\{B(e_0(p_e^{(a)}), \ldots, e_{d-1}(p_e^{(a)})) - 1, -a_1, \ldots, -a_r\}$$

$$= C_a^{(d)}(e).$$

Observe that

$$e_i(p_e^{(a)})$$

$$= \begin{cases} e_i\left(\bigoplus_{j=1}^r R(a_j)\right) & \text{if } 0 \le i \le d-t-1, \\ e_i\left(\bigoplus_{j=1}^r R(a_j)\right) - (-1)^{d-t}e_{i-(d-t)}(M) & \text{if } d-t \le i \le d-1, \end{cases}$$

and that each such $e_i\left(\bigoplus_{j=1}^r R(a_j)\right)$ depends only on $\ell_{R_0}(R_0), a_1, \dots, a_r$ and d. Thus $e_0(\,\bullet\,), \dots, e_{t-1}(\,\bullet\,)$ form a bounding system for reg^1 on $\mathcal{D}$.   □

17.4.2 Corollary. *Let* $R = \bigoplus_{n \in \mathbb{N}_0} R_n$ *be as in 17.4.1. Let* $W = \bigoplus_{n \in \mathbb{Z}} W_n$ *be a finitely generated graded* R-*module and let* $0 \ne P \in \mathbb{Q}[x]$. *Then there is an integer* G *such that, for each homogeneous* R-*homomorphism* $f : W \longrightarrow M$ *of finitely generated graded* R-*modules that is surjective in all large degrees and is such that* $P_M = P$, *we have* $\mathrm{reg}^1(M) \le G$.

Proof Choose $r \in \mathbb{N}$ and $a := (a_1, \dots, a_r) \in \mathbb{Z}^r$ such that $W = \sum_{j=1}^r Rw_j$ with $w_j \in W_{-a_j}$ for all $j \in \{1, \dots, r\}$. Let $t := \deg P + 1$, and define $\mathcal{D}$ as in 17.4.1, that is, as the class of all graded R-modules N of dimension t which can be written in the form $N = \sum_{j=1}^r Ry_j$ with $y_j \in N_{-a_j}$ for all $j \in \{1, \dots, r\}$. By 17.4.1, there is a function $C : \mathbb{Z}^t \longrightarrow \mathbb{Z}$ such that $\mathrm{reg}^1(N) \le C(e_0(N), \dots, e_{t-1}(N))$ for each R-module N in $\mathcal{D}$.

Let $f : W \longrightarrow M$ be a homomorphism as described in the statement, and set $\overline{W} := W/\mathrm{Ker}\, f$. Then $\overline{W} = \sum_{j=1}^r R\overline{w}_j$ with $\overline{w}_j \in \overline{W}_{-a_j}$ for all $j \in \{1, \dots, r\}$. Moreover, f induces an exact sequence

$$0 \longrightarrow \overline{W} \longrightarrow M \longrightarrow C \longrightarrow 0$$

(in the category $^*\mathcal{C}(R)$), where $C_n = 0$ for all except finitely many $n \in \mathbb{Z}$. In particular, this means that $P_{\overline{W}} = P_M = P$ and $\mathrm{reg}^1(C) = -\infty$. Therefore, by 16.2.15(iii) and 17.4.1, we have

$$\mathrm{reg}^1(M) \le \mathrm{reg}^1(\overline{W}) \le C(e_0(P), \dots, e_{\deg P}(P)) =: G. \qquad □$$

Note, in particular, that Corollary 17.4.2 tells us (with the notation of that corollary) that, for all graded homomorphic images M, with specified Hilbert polynomial P, of a fixed finitely generated graded R-module W, the invariant reg^1 is bounded in terms of P.

Next, we bound the regularity in terms of the Hilbert coefficients, the postulation number and the degrees of generators.

17.4.3 Corollary. *Let* $R = \bigoplus_{n \in \mathbb{N}_0} R_n$ *be as in 17.4.1. Let* $g, \pi, t \in \mathbb{Z}$ *with* $t > 0$.

Let $\mathcal{G}$ denote the class of all finitely generated graded R-modules of dimension t that can be generated by homogeneous elements of degrees not exceeding g and whose postulation number does not exceed π. Then the invariants $e_0(\bullet),\ldots,e_{t-1}(\bullet)$ form a bounding system for reg on $\mathcal{G}$.

Proof Let $\bar{g} := \max\{g, \pi + 1\}$, and, for a graded module $M \in \mathcal{G}$, consider the graded submodule $\overline{M} := \bigoplus_{n \geq \bar{g}} M_n$ of M. Observe that $P_{\overline{M}} = P_M = p_e$, where $e := (e_0(M),\ldots,e_{t-1}(M))$. Observe also that $\ell_{R_0}(\overline{M}_{\bar{g}}) = P_{\overline{M}}(\bar{g}) = p_e(\bar{g})$, and that $\overline{M}$ is generated by $p_e(\bar{g}) > 0$ homogeneous elements of degree $\bar{g}$. This time write $a := (-\bar{g},\ldots,-\bar{g}) \in \mathbb{Z}^{p_e(\bar{g})}$, and define $\mathcal{D}'$ to be the class of all graded R-modules N of dimension t that can be generated by $p_e(\bar{g})$ homogeneous elements all of degree $\bar{g}$. By 17.4.1, there is a function C' : $\mathbb{Z}^t \longrightarrow \mathbb{Z}$ such that $\text{reg}^1(M) \leq C'(e_0(M),\ldots,e_{t-1}(M))$ for each R-module M in $\mathcal{D}'$.

Note that there is an exact sequence $0 \longrightarrow \overline{M} \longrightarrow M \longrightarrow C \longrightarrow 0$ (in the category $^*\mathcal{C}(R)$) in which C is finitely generated and R_+-torsion. Therefore, by 16.2.15(iii), $\text{reg}^1(M) \leq \text{reg}^1(\overline{M})$. Since $\overline{M} \in \mathcal{D}'$, we have $\text{reg}^1(\overline{M}) \leq C'(e_0(M),\ldots,e_{t-1}(M))$. However, it follows from 17.1.10(i) that $\text{reg}(M) \leq \max\{\text{reg}^1(M), \text{pstln}(M)\}$. Putting these inequalities together, we have

$$\text{reg}(M) \leq \max\{C'(e_0(M),\ldots,e_{t-1}(M)), \pi\},$$

and this is enough to complete the proof. $\qquad\square$

17.4.4 Exercise. Let $R = K[X_1, X_2]$ be a polynomial ring in two indeterminates X_1, X_2 over a field K. By the *generating degree* $\text{gendeg}(M)$ of a non-zero finitely generated graded R-module M we mean the smallest integer g such that M is generated by homogeneous elements of degrees not exceeding g; thus

$$\text{gendeg}(M) = \inf\left\{g \in \mathbb{Z} : M = R\textstyle\sum_{n \leq g} M_n\right\}.$$

For each $h \in \mathbb{Z}$, let ^hK be the graded R-module (see 14.1.13) such that, for all $n \in \mathbb{Z}$,

$$(^hK)_n = \begin{cases} K & \text{if } n = h, \\ 0 & \text{if } n \neq h. \end{cases}$$

(i) For each $t \in \mathbb{N}$, let $M^{(t)} := X_1^t(R/X_2R) \oplus {}^0K \oplus {}^1K \oplus \cdots \oplus {}^{t-1}K$, and determine $\text{end}(\Gamma_{R_+}(M^{(t)}))$, $\text{end}(H^1_{R_+}(M^{(t)}))$, $P_{M^{(t)}}$ and $\text{pstln}(M^{(t)})$. Conclude that the multiplicity $e_0(\bullet)$ and the postulation number $\text{pstln}(\bullet)$ do not form a bounding system of invariants for either reg^1 or reg on the class of all non-zero finitely generated graded R-modules.

(ii) Use the graded R-modules $N^{(t)} := R \oplus (R/R_+^t)$ $(t \in \mathbb{N})$ to show that $e_0(\,\bullet\,)$, $e_1(\,\bullet\,)$ and gendeg$(\,\bullet\,)$ do not form a bounding system of invariants for reg on the class $\mathcal{D}$ of all finitely generated graded R-modules of dimension 2.

(iii) For each $t \in \mathbb{N}$, let $L^{(t)} := \bigoplus_{h=0}^{t-1} ({}^hK)^{h+1} \bigoplus R_{\geq t}$, where $R_{\geq t}$ is as defined in 16.1.1 and $({}^hK)^{h+1}$ denotes the direct sum of $h+1$ copies of ${}^hK$. Use the $L^{(t)}$ $(t \in \mathbb{N})$ to show that $e_0(\,\bullet\,)$, $e_1(\,\bullet\,)$ and pstln$(\,\bullet\,)$ do not form a bounding system of invariants for reg^1 on the class $\mathcal{D}$ of part (ii).

17.4.5 Definition and Exercise. Assume that $R = \bigoplus_{n \in \mathbb{N}_0} R_n$ is positively graded and that the ideal $\mathfrak{a}$ is graded. The *saturation of* $\mathfrak{a}$ is defined as $\mathfrak{a}^{\mathrm{sat}} := \bigcup_{n \in \mathbb{N}} (\mathfrak{a} :_R (R_+)^n)$. We say that $\mathfrak{a}$ is *saturated* precisely when $\mathfrak{a} = \mathfrak{a}^{\mathrm{sat}}$. Show that

(i) $\mathfrak{a}^{\mathrm{sat}}$ is a graded ideal of R that contains $\mathfrak{a}$;

(ii) $\mathfrak{a}^{\mathrm{sat}}/\mathfrak{a} = \Gamma_{R_+}(R/\mathfrak{a})$;

(iii) $\mathfrak{a}^{\mathrm{sat}}$ is the largest graded ideal of R which coincides with $\mathfrak{a}$ in all large degrees;

(iv) $(\mathfrak{a}^{\mathrm{sat}})^{\mathrm{sat}} = \mathfrak{a}^{\mathrm{sat}}$; and

(v) if $\Gamma_{R_+}(R) = H_{R_+}^1(R) = 0$, then there is a homogeneous isomorphism $\mathfrak{a}^{\mathrm{sat}}/\mathfrak{a} \cong H_{R_+}^1(\mathfrak{a})$.

17.4.6 Exercise. Let $R = R_0[X_1, \ldots, X_d]$ be the polynomial ring over the commutative Noetherian ring R_0 in $d > 1$ indeterminates. Assume that $\mathfrak{a}$ is graded.

Show that $\mathfrak{a}$ is saturated if and only if $H_{R_+}^1(\mathfrak{a}) = 0$, and that $\mathrm{reg}(\mathfrak{a}) = \mathrm{reg}^2(\mathfrak{a})$ if $\mathfrak{a}$ is saturated.

Now assume in addition that the base ring R_0 is local and Artinian. For $P \in \mathbb{Q}[T]$, let $\mathcal{A}_P := \{\mathfrak{b}$ is a graded ideal of $R : P_{\mathfrak{b}} = P\}$. Show that the maximal members of $\mathcal{A}_P$ are precisely the saturated ideals in $\mathcal{A}_P$. Show also that the Hilbert coefficients $e_0(\,\bullet\,), \ldots, e_{d-1}(\,\bullet\,)$ form a bounding system for reg on the class of all non-zero saturated graded ideals of R.

18

Applications to reductions of ideals

Graded local cohomology theory has played a substantial rôle in the study of Rees rings and associated graded rings of proper ideals in local rings. We do not have enough space in this book to include all we would like about the applications of local cohomology to this area, and so we have decided to select a small portion of the theory which gives some idea of the flavour. The part we have chosen to present in this chapter concerns links between the theory of reductions of ideals in local rings and the concept of Castelnuovo regularity, discussed in Chapter 16. The highlight will be a theorem of L. T. Hoa which asserts that, if $\mathfrak{b}$ is a proper ideal in a local ring having infinite residue field, then there exist $t_0 \in \mathbb{N}$ and $c \in \mathbb{N}_0$ such that, for all $t > t_0$ and every minimal reduction $\mathfrak{a}$ of $\mathfrak{b}^t$, the reduction number $r_{\mathfrak{a}}(\mathfrak{b}^t)$ of $\mathfrak{b}^t$ with respect to $\mathfrak{a}$ is equal to c. This statement of Hoa's Theorem is satisfyingly simple, and makes no mention of local cohomology, and yet Hoa's proof, which we present towards the end of this chapter, makes significant use of graded local cohomology.

Throughout this chapter, all graded rings and modules are to be understood to be $\mathbb{Z}$-graded, and all polynomial rings $R[X_1, \ldots, X_t]$ (and $R[T]$) over R are to be understood to be (positively) $\mathbb{Z}$-graded so that each indeterminate has degree 1 and $\deg a = 0$ for all $a \in R \setminus \{0\}$.

18.1 Reductions and integral closures

Reductions of ideals of local rings were first considered by D. G. Northcott and D. Rees in [63]. Nowadays, the concept is recognized as being of major importance in commutative algebra. The original paper [63] of Northcott and Rees was written under unnecessarily restrictive hypotheses, and so we begin this chapter with a rapid development of the links between the two concepts of reduction and integral closure of ideals over a general commutative Noetherian ring. Throughout this chapter, $\mathfrak{b}$ will denote a second ideal of R.

18.1.1 Definitions. (See D. G. Northcott and D. Rees [63].)

(i) We say that $\mathfrak{a}$ is a *reduction of* $\mathfrak{b}$ precisely when $\mathfrak{a} \subseteq \mathfrak{b}$ and there exists $s \in \mathbb{N}_0$ such that $\mathfrak{a}\mathfrak{b}^s = \mathfrak{b}^{s+1}$; then the least such s is denoted by $r_\mathfrak{a}(\mathfrak{b})$ and called the *reduction number* of $\mathfrak{b}$ with respect to $\mathfrak{a}$. Note that, if $\mathfrak{a}$ is a reduction of $\mathfrak{b}$, then $\mathfrak{a}^m\mathfrak{b}^j = \mathfrak{b}^{m+j}$ for all $m \in \mathbb{N}$ and $j \geq r_\mathfrak{a}(\mathfrak{b})$.

(ii) We say that $\mathfrak{a}$ is a *minimal reduction of* $\mathfrak{b}$ if and only if $\mathfrak{a}$ is a reduction of $\mathfrak{b}$ and there is no reduction $\mathfrak{c}$ of $\mathfrak{b}$ with $\mathfrak{c} \subset \mathfrak{a}$.

(iii) We say that $r \in R$ is *integrally dependent on* $\mathfrak{a}$ if and only if there exist $n \in \mathbb{N}$ and $c_1, \ldots, c_n \in R$ with $c_i \in \mathfrak{a}^i$ for $i = 1, \ldots, n$ such that

$$r^n + c_1 r^{n-1} + \cdots + c_{n-1}r + c_n = 0.$$

18.1.2 ♯Exercise. Assume that $\mathfrak{a}$ is a reduction of $\mathfrak{b}$.

(i) Show that $\sqrt{\mathfrak{a}} = \sqrt{\mathfrak{b}}$.
(ii) Let $\mathfrak{c}$ be a third ideal of R such that $\mathfrak{b}$ is a reduction of $\mathfrak{c}$. Show that $\mathfrak{a}$ is a reduction of $\mathfrak{c}$.

18.1.3 ♯Exercise. Show that, if $\mathfrak{a}$ is a reduction of $\mathfrak{b}$ and also a reduction of another ideal $\mathfrak{b}'$ of R, then $\mathfrak{a}$ is a reduction of $\mathfrak{b} + \mathfrak{b}'$.

18.1.4 Notation and ♯Exercise. Let $\{a_1, \ldots, a_h\}$ be a generating set for $\mathfrak{a}$, and let T be an indeterminate. We use $R[\mathfrak{a}T, T^{-1}]$ to denote the subring

$$R[a_1 T, \ldots, a_h T, T^{-1}]$$

of $R[T, T^{-1}] = R[T]_T$, and refer to this as the *extended Rees ring of* $\mathfrak{a}$. (Note that $R[a_1 T, \ldots, a_h T, T^{-1}]$ is independent of the choice of finite generating set $\{a_1, \ldots, a_h\}$ for $\mathfrak{a}$.) Note that $R[\mathfrak{a}T, T^{-1}]$ inherits a $\mathbb{Z}$-grading from $R[T, T^{-1}] = R[T]_T$.

Let $r \in R$. Show that r is integrally dependent on $\mathfrak{a}$ if and only if the element rT of $R[T, T^{-1}]$ is integral over $R[\mathfrak{a}T, T^{-1}]$.

18.1.5 Proposition. *Assume that $\mathfrak{a} \subseteq \mathfrak{b}$ and let $r \in R$.*

(i) *The element r is integrally dependent on $\mathfrak{a}$ if and only if $\mathfrak{a}$ is a reduction of $\mathfrak{a} + Rr$.*
(ii) *The ideal $\mathfrak{a}$ is a reduction of $\mathfrak{b}$ if and only if each element of $\mathfrak{b}$ is integrally dependent on $\mathfrak{a}$.*

Proof. First we suppose that $\mathfrak{a}$ is a reduction of $\mathfrak{b}$, so that there exists $s \in \mathbb{N}$ such that $\mathfrak{a}^m \mathfrak{b}^s = \mathfrak{b}^{m+s}$ for all $m \in \mathbb{N}$. Hence, provided we interpret $\mathfrak{a}^i$ as R when the integer i is negative, $\mathfrak{b}^n \subseteq \mathfrak{a}^{n-s}$ for all $n \in \mathbb{N}_0$. Therefore, within the ring $R[T, T^{-1}]$, we have $R[\mathfrak{b}T, T^{-1}] \subseteq T^s R[\mathfrak{a}T, T^{-1}]$, a finitely generated module over the Noetherian ring $R[\mathfrak{a}T, T^{-1}]$. It follows that $R[\mathfrak{b}T, T^{-1}]$ is integral over its subring $R[\mathfrak{a}T, T^{-1}]$, and so, by 18.1.4, each element of $\mathfrak{b}$ is integrally dependent on $\mathfrak{a}$.

(i) It follows from the above paragraph that, if $\mathfrak{a}$ is a reduction of $\mathfrak{a} + Rr$, then r is integrally dependent on $\mathfrak{a}$. Conversely, if r is integrally dependent on $\mathfrak{a}$, then there exist $n \in \mathbb{N}$ and $c_1, \ldots, c_n \in R$ with $c_i \in \mathfrak{a}^i$ for $i = 1, \ldots, n$ such that $r^n + c_1 r^{n-1} + \cdots + c_{n-1} r + c_n = 0$. Then

$$(\mathfrak{a} + Rr)^n = \mathfrak{a}(\mathfrak{a} + Rr)^{n-1} + Rr^n = \mathfrak{a}(\mathfrak{a} + Rr)^{n-1}$$

and $\mathfrak{a}$ is a reduction of $\mathfrak{a} + Rr$.

(ii) Suppose each element of $\mathfrak{b}$ is integrally dependent on $\mathfrak{a}$. Let $\{b_1, \ldots, b_t\}$ be a generating set for $\mathfrak{b}$. By part (i), for each $i = 1, \ldots, t$, the ideal $\mathfrak{a}$ is a reduction of $\mathfrak{a} + Rb_i$. Hence $\mathfrak{a}$ is a reduction of $\mathfrak{a} + \sum_{i=1}^t Rb_i = \mathfrak{b}$, by 18.1.3. The converse statement has already been proved in the first paragraph of this proof. □

18.1.6 Corollary and Definition. *By 18.1.3, the set $\mathcal{I}$ of all ideals of R that have $\mathfrak{a}$ as a reduction has a unique maximal member, $\bar{\mathfrak{a}}$ say: $\bar{\mathfrak{a}}$ is the union of the members of $\mathcal{I}$. By 18.1.5, this ideal $\bar{\mathfrak{a}}$ is precisely the set of all elements of R which are integrally dependent on $\mathfrak{a}$ (and so the latter set is an ideal of R): we refer to $\bar{\mathfrak{a}}$ as the* integral closure *of $\mathfrak{a}$.* □

18.1.7 Exercise. Show that

(i) the integral closure $\bar{\mathfrak{a}}$ is not a reduction of any ideal of R which properly contains it, and

(ii) $\bar{\mathfrak{a}} = \bar{\mathfrak{b}}$ if and only if $\mathfrak{a}$ and $\mathfrak{b}$ are both reductions of $\mathfrak{a} + \mathfrak{b}$.

We now reproduce the classical argument of Northcott and Rees which shows that, when R is local, every reduction of $\mathfrak{b}$ contains a minimal reduction of $\mathfrak{b}$.

18.1.8 Lemma. *Assume that $(R, \mathfrak{m})$ is local and that $\mathfrak{a} \subseteq \mathfrak{b} + \mathfrak{a}\mathfrak{m}$. Then $\mathfrak{a} \subseteq \mathfrak{b}$.*

Proof. Since $\mathfrak{b} + \mathfrak{a} = \mathfrak{b} + \mathfrak{a}\mathfrak{m}$, we have $(\mathfrak{b} + \mathfrak{a})/\mathfrak{b} = \mathfrak{m}((\mathfrak{b} + \mathfrak{a})/\mathfrak{b})$, and so the result follows from Nakayama's Lemma. □

18.1.9 Lemma (Northcott and Rees [63, §2, Lemma 2]). *Assume that* $(R, \mathfrak{m})$ *is local and that* $\mathfrak{a} \subseteq \mathfrak{b}$. *Then* $\mathfrak{a}$ *is a reduction of* $\mathfrak{b}$ *if and only if* $\mathfrak{a} + \mathfrak{b}\mathfrak{m}$ *is a reduction of* $\mathfrak{b}$.

Proof. It is clear that, if $\mathfrak{a}$ is a reduction of $\mathfrak{b}$, then $\mathfrak{a} + \mathfrak{b}\mathfrak{m}$ is a reduction of $\mathfrak{b}$. Now assume that $\mathfrak{a} + \mathfrak{b}\mathfrak{m}$ is a reduction of $\mathfrak{b}$. Then there exists $n \in \mathbb{N}$ such that $(\mathfrak{a} + \mathfrak{b}\mathfrak{m})\mathfrak{b}^n = \mathfrak{b}^{n+1}$, that is, $\mathfrak{b}^{n+1} = \mathfrak{a}\mathfrak{b}^n + \mathfrak{b}^{n+1}\mathfrak{m}$. It therefore follows from Lemma 18.1.8 that $\mathfrak{b}^{n+1} = \mathfrak{a}\mathfrak{b}^n$. □

18.1.10 Lemma. *Assume that* $(R, \mathfrak{m})$ *is local and that* $\mathfrak{a}$ *is a reduction of* $\mathfrak{b}$. *Let* $a_1, \ldots, a_t \in \mathfrak{a}$ *be such that their natural images in* $(\mathfrak{a} + \mathfrak{b}\mathfrak{m})/\mathfrak{b}\mathfrak{m}$ *form a basis for this* $R/\mathfrak{m}$-*space, and set* $\mathfrak{a}' = \sum_{i=1}^{t} Ra_i$. *Then*

(i) $\mathfrak{a}'$ *is a reduction of* $\mathfrak{b}$ *contained in* $\mathfrak{a}$ *and* $\mathfrak{a}' \cap \mathfrak{b}\mathfrak{m} = \mathfrak{a}'\mathfrak{m}$; *and*

(ii) *if* $\mathfrak{a}$ *is a minimal reduction of* $\mathfrak{b}$, *then* $\mathfrak{a} = \mathfrak{a}'$ *and* $\mathfrak{a} \cap \mathfrak{b}\mathfrak{m} = \mathfrak{a}\mathfrak{m}$.

Proof. (i) Note that $\mathfrak{a}' \subseteq \mathfrak{a} \subseteq \mathfrak{b}$ and $\mathfrak{a}' + \mathfrak{b}\mathfrak{m} = \mathfrak{a} + \mathfrak{b}\mathfrak{m}$. Since $\mathfrak{a}$ is a reduction of $\mathfrak{b}$, it follows from Lemma 18.1.9 that $\mathfrak{a} + \mathfrak{b}\mathfrak{m} = \mathfrak{a}' + \mathfrak{b}\mathfrak{m}$ is a reduction of $\mathfrak{b}$, so that $\mathfrak{a}'$ is a reduction of $\mathfrak{b}$ by the same lemma. Also, if $r_1, \ldots, r_t \in R$ are such that $\sum_{i=1}^{t} r_i a_i \in \mathfrak{b}\mathfrak{m}$, then, by choice of $a_1, \ldots, a_t$, we must have $r_1, \ldots, r_t \in \mathfrak{m}$.

(ii) Now suppose that $\mathfrak{a}$ is a minimal reduction of $\mathfrak{b}$. Then $\mathfrak{a}' = \mathfrak{a}$ by part (i), so that $\mathfrak{a} \cap \mathfrak{b}\mathfrak{m} = \mathfrak{a}\mathfrak{m}$, again by part (i). □

18.1.11 ♯Exercise. Assume $(R, \mathfrak{m})$ is local and that $\mathfrak{a}$ is a minimal reduction of $\mathfrak{b}$. Let $\mathfrak{d}$ be an ideal of R such that $\mathfrak{a} \subseteq \mathfrak{d} \subseteq \mathfrak{b}$. Show that every minimal generating set for $\mathfrak{a}$ can be extended to a minimal generating set for $\mathfrak{d}$.

18.1.12 Theorem (Northcott and Rees [63, §2, Theorem 1]). *Assume that* $(R, \mathfrak{m})$ *is local and that* $\mathfrak{a}$ *is a reduction of* $\mathfrak{b}$. *Then* $\mathfrak{a}$ *contains a minimal reduction of* $\mathfrak{b}$.

Proof. Let Σ be the set of all ideals of R of the form $\mathfrak{d} + \mathfrak{b}\mathfrak{m}$, where $\mathfrak{d}$ is a reduction of $\mathfrak{b}$ and is contained in $\mathfrak{a}$. Note that $\mathfrak{a} + \mathfrak{b}\mathfrak{m} \in \Sigma$. Since $\mathfrak{b}/\mathfrak{b}\mathfrak{m}$ is a finite-dimensional vector space over $R/\mathfrak{m}$, the set Σ has a minimal member, and so there exists a reduction $\mathfrak{c}'$ of $\mathfrak{b}$ contained in $\mathfrak{a}$ such that $\mathfrak{c}' + \mathfrak{b}\mathfrak{m}$ is a minimal member of Σ. Let $c_1, \ldots, c_t \in \mathfrak{c}'$ be such that their natural images in $(\mathfrak{c}' + \mathfrak{b}\mathfrak{m})/\mathfrak{b}\mathfrak{m}$ form a basis for this $R/\mathfrak{m}$-space, and set $\mathfrak{c} = \sum_{i=1}^{t} Rc_i$. Now $\mathfrak{c}$ is a reduction of $\mathfrak{b}$ contained in $\mathfrak{c}'$, and $\mathfrak{c} \cap \mathfrak{b}\mathfrak{m} = \mathfrak{c}\mathfrak{m}$ by 18.1.10(i); we shall prove that $\mathfrak{c}$ is a minimal reduction of $\mathfrak{b}$.

Suppose that $\mathfrak{c}_0$ is a reduction of $\mathfrak{b}$ with $\mathfrak{c}_0 \subseteq \mathfrak{c}$. Then $\mathfrak{c}_0 + \mathfrak{b}\mathfrak{m} \in \Sigma$ and $\mathfrak{c}_0 + \mathfrak{b}\mathfrak{m} \subseteq \mathfrak{c}' + \mathfrak{b}\mathfrak{m}$. Hence, by the choice of $\mathfrak{c}'$, we must have

$$\mathfrak{c}_0 + \mathfrak{b}\mathfrak{m} = \mathfrak{c}' + \mathfrak{b}\mathfrak{m} = \mathfrak{c} + \mathfrak{b}\mathfrak{m}.$$

It follows that $\mathfrak{c} \subseteq \mathfrak{c}_0 + \mathfrak{b}\mathfrak{m}$; therefore $\mathfrak{c} \subseteq \mathfrak{c}_0 + (\mathfrak{b}\mathfrak{m} \cap \mathfrak{c}) \subseteq \mathfrak{c}_0 + \mathfrak{c}\mathfrak{m}$ by the immediately preceding paragraph in this proof. It now follows from 18.1.8 that $\mathfrak{c} \subseteq \mathfrak{c}_0$, so that $\mathfrak{c} = \mathfrak{c}_0$. $\square$

There is an important connection between minimal reductions of ideals in local rings and analytical independence.

18.1.13 Definition. Assume that $(R, \mathfrak{m})$ is local. Then $v_1, \ldots, v_t \in \mathfrak{b}$ are said to be *analytically independent in* $\mathfrak{b}$ if and only if, whenever $h \in \mathbb{N}$ and $f \in R[X_1, \ldots, X_t]$ (the ring of polynomials over R in t indeterminates) is a homogeneous polynomial of degree h such that $f(v_1, \ldots, v_t) \in \mathfrak{b}^h\mathfrak{m}$, then all the coefficients of f lie in $\mathfrak{m}$.

Note that, if $v_1, \ldots, v_t \in \mathfrak{b}$ are analytically independent in $\mathfrak{b}$, and $\mathfrak{c} := \sum_{i=1}^t Rv_i$, then $\mathfrak{c}^h \cap \mathfrak{b}^h\mathfrak{m} = \mathfrak{c}^h\mathfrak{m}$ for all $h \in \mathbb{N}$.

18.1.14 Lemma (Northcott and Rees [63, §4, Lemma 1]). *Assume $(R, \mathfrak{m})$ is local and $\mathfrak{b}$ is proper, and let $v_1, \ldots, v_t \in \mathfrak{b}$ be analytically independent in $\mathfrak{b}$. Then $v_1, \ldots, v_t$ form a minimal generating set for $\sum_{i=1}^t Rv_i$, and, furthermore, if $\{w_1, \ldots, w_t\}$ is a minimal generating set for $\sum_{i=1}^t Rv_i$, then $w_1, \ldots, w_t$ are analytically independent in $\mathfrak{b}$.*

Proof. Set $\mathfrak{c} := \sum_{i=1}^t Rv_i$. If $r_1, \ldots, r_t \in R$ are such that $\sum_{i=1}^t r_i v_i \in \mathfrak{c}\mathfrak{m}$, then, since $\mathfrak{c} \subseteq \mathfrak{b}$, it follows that $r_1, \ldots, r_t \in \mathfrak{m}$. Hence $v_1, \ldots, v_t$ form a minimal generating set for $\mathfrak{c}$.

Next, let $h \in \mathbb{N}$ and $f \in R[X_1, \ldots, X_t]$ be a homogeneous polynomial of degree h such that $f(w_1, \ldots, w_t) \in \mathfrak{b}^h\mathfrak{m}$. There exist $r_{ij} \in R$ $(1 \leq i, j \leq t)$ such that $w_i = \sum_{j=1}^t r_{ij}v_j$ for $i = 1, \ldots, t$, and so

$$f\left(\sum_{j=1}^t r_{1j}v_j, \ldots, \sum_{j=1}^t r_{tj}v_j\right) \in \mathfrak{b}^h\mathfrak{m}.$$

Since $v_1, \ldots, v_t \in \mathfrak{b}$ are analytically independent in $\mathfrak{b}$, it follows that all the coefficients of the homogeneous polynomial

$$f\left(\sum_{j=1}^t r_{1j}X_j, \ldots, \sum_{j=1}^t r_{tj}X_j\right)$$

lie in $\mathfrak{m}$.

Denote the natural image in $k := R/\mathfrak{m}$ (respectively $k[X_1, \ldots, X_t]$) of $r \in R$ (respectively $q \in R[X_1, \ldots, X_t]$) by $\overline{r}$ (respectively $\overline{q}$). Now $[\overline{r_{ij}}]$ is an invertible $t \times t$ matrix over k, and so there exists a $t \times t$ matrix $[s_{ij}]$ over R such that $[\overline{r_{ij}}][\overline{s_{ij}}] = I_t$, the $t \times t$ identity matrix. Since

$$\overline{f}\left(\sum_{j=1}^t \overline{r_{1j}}X_j, \ldots, \sum_{j=1}^t \overline{r_{tj}}X_j\right) = 0$$

in $k[X_1, \ldots, X_t]$, it follows that

$$\overline{f}\left(\sum_{j=1}^t \overline{r_{1j}}\left(\sum_{l=1}^t \overline{s_{jl}}X_l\right), \ldots, \sum_{j=1}^t \overline{r_{tj}}\left(\sum_{l=1}^t \overline{s_{jl}}X_l\right)\right) = 0,$$

and so $\overline{f}(X_1, \ldots, X_t) = 0$. Thus all the coefficients of f lie in $\mathfrak{m}$. □

18.1.15 Proposition (Northcott and Rees [63, §4, Lemma 2]). *Assume that* $(R, \mathfrak{m})$ *is local and that* $k := R/\mathfrak{m}$ *is infinite. Suppose that* $\mathfrak{b}$ *is proper and* $\mathfrak{a}$ *is a minimal reduction of* $\mathfrak{b}$, *and let* $v_1, \ldots, v_t$ *form a minimal generating set for* $\mathfrak{a}$. *Then* $v_1, \ldots, v_t$ *are analytically independent in* $\mathfrak{b}$.

Proof. Let $h \in \mathbb{N}$ and $f \in R[X_1, \ldots, X_t]$ be a homogeneous polynomial of degree h such that $f(v_1, \ldots, v_t) \in \mathfrak{b}^h\mathfrak{m}$. Suppose that the coefficient u of X_1^h is a unit of R. Then $v_1^h \in \mathfrak{b}^h\mathfrak{m} + v_2\mathfrak{a}^{h-1} + \cdots + v_t\mathfrak{a}^{h-1}$, and so

$$\mathfrak{a}^h = Rv_1^h + (Rv_2 + \cdots + Rv_t)\mathfrak{a}^{h-1} \subseteq \mathfrak{b}^h\mathfrak{m} + (Rv_2 + \cdots + Rv_t)\mathfrak{a}^{h-1}.$$

Set $\mathfrak{a}' := Rv_2 + \cdots + Rv_t$. Now there exists $n \in \mathbb{N}$ such that $\mathfrak{a}\mathfrak{b}^n = \mathfrak{b}^{n+1}$. Hence $\mathfrak{b}^{h+n} = \mathfrak{a}^h\mathfrak{b}^n \subseteq \mathfrak{b}^{h+n}\mathfrak{m} + \mathfrak{a}'\mathfrak{a}^{h-1}\mathfrak{b}^n = \mathfrak{b}^{h+n}\mathfrak{m} + \mathfrak{a}'\mathfrak{b}^{h+n-1}$. Hence $\mathfrak{b}^{h+n} \subseteq \mathfrak{a}'\mathfrak{b}^{h+n-1}$ by 18.1.8, so that $\mathfrak{b}^{h+n} = \mathfrak{a}'\mathfrak{b}^{h+n-1}$ and $\mathfrak{a}'$ is a reduction of $\mathfrak{b}$. This contradicts the fact that $\mathfrak{a}$ is a minimal reduction of $\mathfrak{b}$, since $\mathfrak{a}' \subset \mathfrak{a}$. Hence $u \in \mathfrak{m}$.

Next let $[r_{ij}]$ be a $t \times t$ matrix over R such that all the entries in the first column, that is $r_{11}, r_{21}, \ldots, r_{t1}$, are units of R and $\det[r_{ij}]$ is also a unit. Then there is a minimal generating set $\{w_1, \ldots, w_t\}$ for $\mathfrak{a}$ such that $v_i = \sum_{j=1}^t r_{ij}w_j$ for all $i = 1, \ldots, t$. Since

$$f\left(\sum_{j=1}^t r_{1j}w_j, \ldots, \sum_{j=1}^t r_{tj}w_j\right) \in \mathfrak{b}^h\mathfrak{m},$$

it follows from the first paragraph of this proof that $f(r_{11}, \ldots, r_{t1})$, which is the coefficient of X_1^h in the form $f\left(\sum_{j=1}^t r_{1j}X_j, \ldots, \sum_{j=1}^t r_{tj}X_j\right)$, lies in $\mathfrak{m}$. Thus, if we denote the natural image in $k = R/\mathfrak{m}$ of $r \in R$ by $\overline{r}$, and the natural image in $k[X_1, \ldots, X_t]$ of $q \in R[X_1, \ldots, X_t]$ by $\overline{q}$, then

$$\overline{f}(\overline{r_{11}}, \ldots, \overline{r_{t1}}) = 0.$$

Therefore $\overline{f}(\alpha_1, \ldots, \alpha_t) = 0$ for all choices of $(\alpha_1, \ldots, \alpha_t) \in (k \setminus \{0\})^t$; hence, since k is infinite, $\overline{f} = 0$. □

18.2 The analytic spread

We now consider the graded rings which will provide the framework for connections between reductions of ideals and graded local cohomology.

18.2.1 Notation and Definition. Suppose that $\mathfrak{b}$ is proper and $\mathfrak{a}$ is a reduction of $\mathfrak{b}$.

(i) By 18.1.6 and 18.1.4, the extended Rees ring $R[\mathfrak{b}T, T^{-1}]$ is integral over its subring $R[\mathfrak{a}T, T^{-1}]$. The associated graded ring $\bigoplus_{i\in\mathbb{N}_0} \mathfrak{b}^i/\mathfrak{b}^{i+1}$ will be denoted by $\mathcal{G}(\mathfrak{b}) = \bigoplus_{i\in\mathbb{N}_0} \mathcal{G}(\mathfrak{b})_i$. Since there is a homogeneous isomorphism of graded rings $R[\mathfrak{b}T, T^{-1}]/T^{-1}R[\mathfrak{b}T, T^{-1}] \xrightarrow{\cong} \mathcal{G}(\mathfrak{b})$ (and a similar one for $\mathcal{G}(\mathfrak{a})$), it follows that $\mathcal{G}(\mathfrak{b})$ is integral over the natural image of $\mathcal{G}(\mathfrak{a})$.

(ii) Now suppose, in addition, that $(R, \mathfrak{m})$ is local. The extension $\mathfrak{m}\mathcal{G}(\mathfrak{b})$ of $\mathfrak{m}$ to $\mathcal{G}(\mathfrak{b})$ under the natural ring homomorphism is a graded ideal: in fact, $\mathfrak{m}\mathcal{G}(\mathfrak{b}) = \bigoplus_{i\in\mathbb{N}_0} \mathfrak{m}\mathfrak{b}^i/\mathfrak{b}^{i+1}$. The *analytic spread* $\mathrm{spr}(\mathfrak{b})$ of $\mathfrak{b}$ is defined by

$$\mathrm{spr}(\mathfrak{b}) := \dim\left(\mathcal{G}(\mathfrak{b})/\mathfrak{m}\mathcal{G}(\mathfrak{b})\right).$$

Note that $\mathcal{G}(\mathfrak{b})/\mathfrak{m}\mathcal{G}(\mathfrak{b})$ is (homogeneously) isomorphic to the graded ring $\bigoplus_{i\in\mathbb{N}_0} \mathfrak{b}^i/\mathfrak{m}\mathfrak{b}^i$, in which $(r_i + \mathfrak{m}\mathfrak{b}^i)(r'_j + \mathfrak{m}\mathfrak{b}^j) = r_i r'_j + \mathfrak{m}\mathfrak{b}^{i+j}$ for $(i, j \in \mathbb{N}_0$ and) $r_i \in \mathfrak{b}^i$, $r'_j \in \mathfrak{b}^j$. Some authors refer to the graded ring $\mathcal{G}(\mathfrak{b})/\mathfrak{m}\mathcal{G}(\mathfrak{b})$ as the *fibre cone of* $\mathfrak{b}$.

(iii) Note that, since $\mathfrak{a}$ is a reduction of $\mathfrak{b}$, it follows that $\mathcal{G}(\mathfrak{b})/\mathfrak{m}\mathcal{G}(\mathfrak{b})$ is integral over the natural image of $\mathcal{G}(\mathfrak{a})/\mathfrak{m}\mathcal{G}(\mathfrak{a})$. Observe also that, if $\mathfrak{a}$ can be generated by elements which are analytically independent in $\mathfrak{b}$, then $\mathfrak{a}^i \cap \mathfrak{b}^i\mathfrak{m} = \mathfrak{a}^i\mathfrak{m}$ for all $i \in \mathbb{N}$, so that the natural homogeneous ring homomorphism $\mathcal{G}(\mathfrak{a})/\mathfrak{m}\mathcal{G}(\mathfrak{a}) \longrightarrow \mathcal{G}(\mathfrak{b})/\mathfrak{m}\mathcal{G}(\mathfrak{b})$ is injective.

18.2.2 Lemma. *Suppose* $(R, \mathfrak{m})$ *is local and* $\mathfrak{b}$ *is proper. Then* $\mathrm{spr}(\mathfrak{b}) = \mathrm{spr}(\mathfrak{b}^t)$ *for all* $t \in \mathbb{N}$.

Proof. By Definition 18.2.1(ii),

$$\mathrm{spr}(\mathfrak{b}) := \dim\left(\mathcal{G}(\mathfrak{b})/\mathfrak{m}\mathcal{G}(\mathfrak{b})\right) \quad \text{and} \quad \mathrm{spr}(\mathfrak{b}^t) := \dim\left(\mathcal{G}(\mathfrak{b}^t)/\mathfrak{m}\mathcal{G}(\mathfrak{b}^t)\right).$$

Now there are homogeneous ring isomorphisms

$$\mathcal{G}(\mathfrak{b})/\mathfrak{m}\mathcal{G}(\mathfrak{b}) \xrightarrow{\cong} \bigoplus_{i\in\mathbb{N}_0} \mathfrak{b}^i/\mathfrak{m}\mathfrak{b}^i \quad \text{and} \quad \mathcal{G}(\mathfrak{b}^t)/\mathfrak{m}\mathcal{G}(\mathfrak{b}^t) \xrightarrow{\cong} \bigoplus_{i\in\mathbb{N}_0} \mathfrak{b}^{ti}/\mathfrak{m}\mathfrak{b}^{ti}.$$

It is clear from these that $\mathcal{G}(\mathfrak{b}^t)/\mathfrak{m}\mathcal{G}(\mathfrak{b}^t)$ is homogeneously isomorphic to the t-th Veronesean subring (see 13.5.9) of $\mathcal{G}(\mathfrak{b})/\mathfrak{m}\mathcal{G}(\mathfrak{b})$. The claim now follows from the fact that a commutative Noetherian graded ring has the same dimension as its t-th Veronesean subring because the former is integral over the latter. $\square$

The observations in 18.2.1(iii) have some very important consequences.

18.2.3 Remark. Suppose that $(R, \mathfrak{m})$ is local, that $\mathfrak{a} \subseteq \mathfrak{b} \subset R$ and that $\mathfrak{a}$ is generated by $v_1, \ldots, v_t$ which are analytically independent in $\mathfrak{b}$. Set $k :=$ $R/\mathfrak{m}$. Then $v_1, \ldots, v_t$ are, of course, analytically independent in $\mathfrak{a}$, and it is immediate from Definition 18.1.13 that there is a homogeneous isomorphism ϕ : $k[X_1, \ldots, X_t] \xrightarrow{\cong} \bigoplus_{i \in \mathbb{N}_0} \mathfrak{a}^i/\mathfrak{m}\mathfrak{a}^i$ of graded k-algebras (where $X_1, \ldots, X_t$ are independent indeterminates) such that $\phi(X_i) = v_i + \mathfrak{m}\mathfrak{a}$ for all $i = 1, \ldots, t$. Since $\mathcal{G}(\mathfrak{a})/\mathfrak{m}\mathcal{G}(\mathfrak{a}) \cong \bigoplus_{i \in \mathbb{N}_0} \mathfrak{a}^i/\mathfrak{m}\mathfrak{a}^i$, we have $\dim(\mathcal{G}(\mathfrak{a})/\mathfrak{m}\mathcal{G}(\mathfrak{a})) = t$.

18.2.4 Theorem (Northcott and Rees [63, §4, Theorems 1 and 2]). *Assume that $(R, \mathfrak{m})$ is local and that $k := R/\mathfrak{m}$ is infinite. Suppose that $\mathfrak{b}$ is proper and that $\mathfrak{a}$ is a reduction of $\mathfrak{b}$, and let $t := \dim_k(\mathfrak{a}/\mathfrak{m}\mathfrak{a})$, the number of elements in each minimal generating set for $\mathfrak{a}$. Let $\{v_1, \ldots, v_t\}$ be a minimal generating set for $\mathfrak{a}$. Then*

(i) $\mathrm{spr}(\mathfrak{b}) \leq t$;

(ii) $\mathfrak{a}$ *is a minimal reduction of $\mathfrak{b}$ if and only if $\mathrm{spr}(\mathfrak{b}) = t$;*

(iii) *hence, at least $\mathrm{spr}(\mathfrak{b})$ elements are needed to generate $\mathfrak{a}$, and, if $\mathfrak{a}$ can be generated by $\mathrm{spr}(\mathfrak{b})$ elements, then it is a minimal reduction of $\mathfrak{b}$;*

(iv) $\mathfrak{a}$ *is a minimal reduction of $\mathfrak{b}$ if and only if $v_1, \ldots, v_t$ are analytically independent in $\mathfrak{b}$.*

Proof. Note that $\mathfrak{a}$ contains a minimal reduction $\mathfrak{c}$ of $\mathfrak{b}$, by Theorem 18.1.12; let $w_1, \ldots, w_s$ form a minimal generating set for $\mathfrak{c}$. Then $w_1, \ldots, w_s$ are analytically independent in $\mathfrak{b}$ by 18.1.15. By 18.2.1(iii), we can view the graded ring $\mathcal{G}(\mathfrak{b})/\mathfrak{m}\mathcal{G}(\mathfrak{b})$ as an integral extension ring of $\mathcal{G}(\mathfrak{c})/\mathfrak{m}\mathcal{G}(\mathfrak{c})$, and so

$$\dim(\mathcal{G}(\mathfrak{c})/\mathfrak{m}\mathcal{G}(\mathfrak{c})) = \dim(\mathcal{G}(\mathfrak{b})/\mathfrak{m}\mathcal{G}(\mathfrak{b})) = \mathrm{spr}(\mathfrak{b}).$$

However, it is immediate from 18.2.3 that $\dim(\mathcal{G}(\mathfrak{c})/\mathfrak{m}\mathcal{G}(\mathfrak{c})) = s$, and so $s = \mathrm{spr}(\mathfrak{b})$.

(i) By 18.1.11, every minimal generating set for $\mathfrak{c}$ can be extended to a minimal generating set for $\mathfrak{a}$. Hence $\mathrm{spr}(\mathfrak{b}) = s \leq t$.

(ii) If $\mathfrak{a}$ is a minimal reduction of $\mathfrak{b}$, then $\mathfrak{a} = \mathfrak{c}$ and $t = s = \mathrm{spr}(\mathfrak{b})$.

Now suppose that $t = \mathrm{spr}(\mathfrak{b})$. By 18.1.11, every minimal generating set for $\mathfrak{c}$ can be extended to a minimal generating set for $\mathfrak{a}$, and so $\mathrm{spr}(\mathfrak{b}) = s \leq t = \mathrm{spr}(\mathfrak{b})$. Therefore $s = t$ and $w_1, \ldots, w_s$ actually generate $\mathfrak{a}$. It follows that $\mathfrak{a} = \mathfrak{c}$, and $\mathfrak{a}$ is a minimal reduction of $\mathfrak{b}$.

(iii) This is a restatement of most of parts (i) and (ii).

(iv) If $\mathfrak{a}$ is a minimal reduction of $\mathfrak{b}$, then $v_1, \ldots, v_t$ are analytically independent in $\mathfrak{b}$ by 18.1.15. Conversely, if $v_1, \ldots, v_t$ are analytically independent in $\mathfrak{b}$, then $t = \dim(\mathcal{G}(\mathfrak{a})/\mathfrak{m}\mathcal{G}(\mathfrak{a}))$ by 18.2.3, and $\mathcal{G}(\mathfrak{b})/\mathfrak{m}\mathcal{G}(\mathfrak{b})$ can be viewed as an integral extension ring of $\mathcal{G}(\mathfrak{a})/\mathfrak{m}\mathcal{G}(\mathfrak{a})$, by 18.2.1(iii); therefore

$t = \dim(\mathcal{G}(\mathfrak{b})/\mathfrak{m}\mathcal{G}(\mathfrak{b})) = \mathrm{spr}(\mathfrak{b})$, so that we can deduce from part (ii) that $\mathfrak{a}$ is a minimal reduction of $\mathfrak{b}$. $\square$

The next Lemma 18.2.5 is included to help the reader solve Exercise 18.2.6; that exercise presents an important result of Northcott and Rees.

18.2.5 Lemma.   *Suppose that* $(R, \mathfrak{m})$ *is local and that* $v_1, \ldots, v_t$ *are analytically independent in* $\mathfrak{b}$. *Then* $t \leq \mathrm{spr}(\mathfrak{b})$.

Proof.   Set $k := R/\mathfrak{m}$ and $\mathfrak{c} := \sum_{i=1}^{t} Rv_i$. Since $\mathfrak{c}^i \cap \mathfrak{b}^i\mathfrak{m} = \mathfrak{c}^i\mathfrak{m}$ for all $i \in \mathbb{N}$, the natural homogeneous ring homomorphism $\mathcal{G}(\mathfrak{c})/\mathfrak{m}\mathcal{G}(\mathfrak{c}) \longrightarrow \mathcal{G}(\mathfrak{b})/\mathfrak{m}\mathcal{G}(\mathfrak{b})$ is injective. Therefore, by 18.2.3, and with an obvious notation,

$$\dim_k \left(\mathcal{G}(\mathfrak{b})/\mathfrak{m}\mathcal{G}(\mathfrak{b})\right)_i \geq \dim_k \left(\mathcal{G}(\mathfrak{c})/\mathfrak{m}\mathcal{G}(\mathfrak{c})\right)_i = \binom{i+t-1}{t-1} \quad \text{for all } i \in \mathbb{N}_0.$$

Hence, with the notation of 17.1.8, $\deg P_{\mathcal{G}(\mathfrak{b})/\mathfrak{m}\mathcal{G}(\mathfrak{b})} \geq t - 1$, and therefore $t \leq \mathrm{spr}(\mathfrak{b})$ by 17.1.7. $\square$

18.2.6 Exercise   (Northcott and Rees [63, §4, Theorem 3]). Assume $(R, \mathfrak{m})$ is local and that $k := R/\mathfrak{m}$ is infinite. Suppose that $\mathfrak{b}$ is proper. Show that $\mathrm{spr}(\mathfrak{b})$ is the maximum number of elements of $\mathfrak{b}$ which are analytically independent in $\mathfrak{b}$.

18.2.7 Exercise.   Assume $(R, \mathfrak{m})$ is local and let $v_1, \ldots, v_t \in R$. Recall from [50, pp. 106–107] that $v_1, \ldots, v_t$ are *analytically independent* if and only if, whenever $f \in R[X_1, \ldots, X_t]$ is a homogeneous polynomial such that

$$f(v_1, \ldots, v_t) = 0,$$

then all the coefficients of f lie in $\mathfrak{m}$.

Show that $v_1, \ldots, v_t$ are analytically independent if and only if they are analytically independent in the ideal $\sum_{i=1}^{t} Rv_i$ which they generate.

18.2.8 Exercise   (Northcott and Rees [63, §4, Lemma 4, Theorems 4, 5, and §6, Theorem 1]). Assume that $(R, \mathfrak{m})$ is local and that $k := R/\mathfrak{m}$ is infinite. Suppose that $\mathfrak{b}$ is proper. We say that $\mathfrak{b}$ is *basic* precisely when it has no reduction other than itself. Let $v_1, \ldots, v_t$ form a minimal generating set for $\mathfrak{b}$.

(i) Show that $\mathrm{ht}\, \mathfrak{b} \leq \mathrm{spr}(\mathfrak{b}) \leq \dim_k(\mathfrak{b}/\mathfrak{m}\mathfrak{b}) = t$.

(ii) Show that the following statements are equivalent:

    (a) $\mathfrak{b}$ is basic;

    (b) $v_1, \ldots, v_t$ are analytically independent (see Exercise 18.2.7);

    (c) $\mathrm{spr}(\mathfrak{b}) = \dim_k(\mathfrak{b}/\mathfrak{m}\mathfrak{b})$.

(iii) Show that, if $\mathfrak{b}$ can be generated by t elements and has height t, then $\mathfrak{b}$ is basic and the members of each minimal generating set for $\mathfrak{b}$ are analytically independent.

(iv) Deduce that the members of a system of parameters for R are analytically independent.

(v) Let $\mathfrak{q}$ be an $\mathfrak{m}$-primary ideal of R. Show that $\mathrm{spr}(\mathfrak{q}) = \dim R$. (Here is a hint: $\ell_R(\mathfrak{q}^n/\mathfrak{q}^{n+1}) = \ell_R(R/\mathfrak{q}^{n+1}) - \ell_R(R/\mathfrak{q}^n)$ for all $n \in \mathbb{N}$, where 'ℓ' denotes length.) Deduce that each minimal reduction of $\mathfrak{q}$ is an ideal generated by a system of parameters of R.

18.3 Links with Castelnuovo regularity

Local cohomology has still not made an appearance in this chapter! We are soon going to make, in the case where $(R, \mathfrak{m})$ is local, some calculations with local cohomology over the associated graded ring $\mathcal{G}(\mathfrak{b}) = \bigoplus_{n \in \mathbb{N}_0} \mathcal{G}(\mathfrak{b})_n$ of a proper ideal $\mathfrak{b}$ of R. We note that $\mathcal{G}(\mathfrak{b})$ is a positively graded, homogeneous commutative Noetherian ring: the theory in the next part of this chapter concerns such rings and will eventually be applied to $\mathcal{G}(\mathfrak{b})$.

18.3.1 Definition. Assume that $R = \bigoplus_{n \in \mathbb{N}_0} R_n$ is positively graded and homogeneous. We say that an ideal $\mathfrak{A}$ of R is a **reduction of R_+* if and only if $\mathfrak{A}$ is graded, can be generated by homogeneous elements of R of degree 1, and is a reduction of R_+.

18.3.2 Lemma. *Assume that $R = \bigoplus_{n \in \mathbb{N}_0} R_n$ is positively graded and homogeneous, and let $\mathfrak{A} = \bigoplus_{n \in \mathbb{N}} \mathfrak{A}_n$ be a graded ideal of R generated by homogeneous elements of degree 1. Then $\mathfrak{A}$ is a *reduction of R_+ if and only if there exists $m \in \mathbb{N}_0$ such that $R_{m+1} = \mathfrak{A}_{m+1}$; when this is the case, $R_{i+1} = \mathfrak{A}_{i+1}$ for all $i \geq m$, and the least $i \in \mathbb{N}_0$ such that $R_{i+1} = \mathfrak{A}_{i+1}$ is equal to the reduction number $r_{\mathfrak{A}}(R_+)$ (see 18.1.1), so that $r_{\mathfrak{A}}(R_+) = \mathrm{end}(R/\mathfrak{A})$.*

Proof. Suppose that $R = R_0[x_1, \ldots, x_t]$, where $x_1, \ldots, x_t \in R_1$, and that $\mathfrak{A}$ is generated by $a_1, \ldots, a_h \in \mathfrak{A}_1$. We use the notation $R_{\geq i}$, for $i \in \mathbb{N}$, of 16.1.1.

Let $i \in \mathbb{N}_0$. Then a typical element of R_i is a sum of finitely many elements of the form $r_0 x_1^{j_1} \ldots x_t^{j_t}$, where $r_0 \in R_0$ and $j_1, \ldots, j_t \in \mathbb{N}_0$ are such that $\sum_{k=1}^{t} j_k = i$, and a typical element of $\mathfrak{A}_{i+1}$ is a sum of finitely many elements of the form $a_l r_i$, where $l \in \{1, \ldots, h\}$ and $r_i \in R_i$. It follows that, if $R_{i+1} = \mathfrak{A}_{i+1}$, then $R_{i+2} = \mathfrak{A}_{i+2}$, that $(R_+)^i = R_{\geq i}$, and that $\mathfrak{A}(R_+)^i = \mathfrak{A}_{\geq i+1}$. All the claims now follow easily. $\square$

18.3.3 Remark. Assume that $R = \bigoplus_{n \in \mathbb{N}_0} R_n$ is positively graded and that $(R_0, \mathfrak{m}_0)$ is local; set $k := R_0/\mathfrak{m}_0$. Let $\mathfrak{A} = \bigoplus_{n \in \mathbb{N}} \mathfrak{A}_n$ be a graded ideal of R generated by homogeneous elements of degree 1. Then the minimum number of homogeneous elements (of degree 1) needed to generate $\mathfrak{A}$ is

$$\dim_k(\mathfrak{A}_1/\mathfrak{m}_0\mathfrak{A}_1).$$

18.3.4 Lemma. *Suppose that* $(R, \mathfrak{m})$ *is local, and that* $\mathfrak{a} \subseteq \mathfrak{b} \subset R$. *Let* $\mathfrak{G}(\mathfrak{a})$ *denote the ideal of the associated graded ring* $\mathcal{G}(\mathfrak{b}) = \bigoplus_{i \in \mathbb{N}_0} \mathcal{G}(\mathfrak{b})_i$ *of* $\mathfrak{b}$ *generated by* $\{a + \mathfrak{b}^2 : a \in \mathfrak{a}\} \subseteq \mathcal{G}(\mathfrak{b})_1$.

Then $\mathfrak{a}$ *is a reduction of* $\mathfrak{b}$ *if and only if* $\mathfrak{G}(\mathfrak{a})$ *is a *reduction of* $\mathcal{G}(\mathfrak{b})_+$. *Furthermore, when this is the case,* $r_\mathfrak{a}(\mathfrak{b}) = r_{\mathfrak{G}(\mathfrak{a})}(\mathcal{G}(\mathfrak{b})_+)$.

Proof. Let $i \in \mathbb{N}_0$. Then the $(i + 1)$-th component $\mathfrak{G}(\mathfrak{a})_{i+1}$ of $\mathfrak{G}(\mathfrak{a})$ is given by $\mathfrak{G}(\mathfrak{a})_{i+1} = (\mathfrak{a}\mathfrak{b}^i + \mathfrak{b}^{i+2})/\mathfrak{b}^{i+2}$. By 18.1.8 and the fact that $\mathfrak{b}$ is proper, $\mathfrak{a}\mathfrak{b}^i = \mathfrak{b}^{i+1}$ if and only if $\mathfrak{a}\mathfrak{b}^i + \mathfrak{b}^{i+2} = \mathfrak{b}^{i+1}$, that is, if and only if $\mathfrak{G}(\mathfrak{a})_{i+1} = \mathcal{G}(\mathfrak{b})_{i+1}$. The claims therefore follow from 18.3.2. □

18.3.5 Lemma (L. T. Hoa [35, Lemma 2.3]). *Assume that* $(R, \mathfrak{m})$ *is local and that* $k := R/\mathfrak{m}$ *is infinite. Suppose that* $\mathfrak{b}$ *is proper. Then* $\mathrm{spr}(\mathfrak{b}) = \mathrm{ara}(\mathcal{G}(\mathfrak{b})_+)$, *and their common value is the greatest integer* i *such that*

$$H^i_{\mathcal{G}(\mathfrak{b})_+}(\mathcal{G}(\mathfrak{b})) \neq 0.$$

Proof. By 18.1.12 and 18.2.4, there exists a minimal reduction $\mathfrak{a}$ of $\mathfrak{b}$, and this can be generated by $s := \mathrm{spr}(\mathfrak{b})$ elements. Therefore, by 18.3.4, and with the notation of that Lemma, $\mathfrak{G}(\mathfrak{a})$ is a *reduction of $\mathcal{G}(\mathfrak{b})_+$. Hence, by 18.1.2(i), we have $\sqrt{\mathfrak{G}(\mathfrak{a})} = \sqrt{\mathcal{G}(\mathfrak{b})_+}$, and so $\mathrm{ara}(\mathcal{G}(\mathfrak{b})_+) \leq s$. Let M be an arbitrary $\mathcal{G}(\mathfrak{b})$-module. By 3.3.3, we have $H^i_{\mathcal{G}(\mathfrak{b})_+}(M) = 0$ for every $i > \mathrm{ara}(\mathcal{G}(\mathfrak{b})_+)$, and so, in particular, for every $i > s$.

Consequently, the epimorphism $\mathcal{G}(\mathfrak{b}) \longrightarrow \mathcal{G}(\mathfrak{b})/\mathfrak{m}\mathcal{G}(\mathfrak{b})$ induces an epimorphism $H^s_{\mathcal{G}(\mathfrak{b})_+}(\mathcal{G}(\mathfrak{b})) \longrightarrow H^s_{\mathcal{G}(\mathfrak{b})_+}(\mathcal{G}(\mathfrak{b})/\mathfrak{m}\mathcal{G}(\mathfrak{b}))$, and so it is enough for us to show that $H^s_{\mathcal{G}(\mathfrak{b})_+}(\mathcal{G}(\mathfrak{b})/\mathfrak{m}\mathcal{G}(\mathfrak{b})) \neq 0$ in order to complete the proof. This we do.

The extension of $\mathcal{G}(\mathfrak{b})_+$ to the *local graded ring $\mathcal{G}(\mathfrak{b})/\mathfrak{m}\mathcal{G}(\mathfrak{b})$ is the unique *maximal ideal, and is, in fact, also maximal; it is therefore of height s, by 14.1.14. It therefore follows from Theorem 6.1.4 and the Graded Independence Theorem 14.1.7 that $H^s_{\mathcal{G}(\mathfrak{b})_+}(\mathcal{G}(\mathfrak{b})/\mathfrak{m}\mathcal{G}(\mathfrak{b})) \neq 0$. □

18.3.6 ♯Exercise. Let the situation be as in Lemma 18.3.5, and let $\mathfrak{a}$ be a minimal reduction of $\mathfrak{b}$ (such an $\mathfrak{a}$ certainly exists, by 18.1.12). Show that the graded ideal $\mathfrak{G}(\mathfrak{a})$ of $\mathcal{G}(\mathfrak{b})$ defined in 18.3.4 can be generated by $\mathrm{spr}(\mathfrak{b})$ homogeneous elements of degree 1, and not by fewer.

In order to make further progress, we wish to use the fact that, when $R = \bigoplus_{n \in \mathbb{N}_0} R_n$ is positively graded, homogeneous, and such that $(R_0, \mathfrak{m}_0)$ is local with infinite residue field, and the non-zero graded ideal $\mathfrak{A} = \bigoplus_{n \in \mathbb{N}} \mathfrak{A}_n$ of R is a *reduction of R_+, it is possible to generate $\mathfrak{A}$ by the members of an R_+-filter-regular sequence of length $\dim_{R_0/\mathfrak{m}_0}(\mathfrak{A}_1/\mathfrak{m}_0\mathfrak{A}_1)$. We therefore explain the concept of R_+-filter-regular sequence.

18.3.7 Definition. Assume that R is positively graded, and let M be a non-zero finitely generated graded R-module. Let $f_1, \ldots, f_h$ be a sequence of homogeneous elements of R. We say that $f_1, \ldots, f_h$ is an R_+-*filter-regular sequence with respect to* M if and only if, for all $i = 1, \ldots, h$, we have $f_i \notin \bigcup_{\mathfrak{P} \in \mathrm{Ass}(M/(f_1, \ldots, f_{i-1})M) \setminus \mathrm{Var}(R_+)} \mathfrak{P}$.

We collect some elementary properties of R_+-filter-regular sequences together in the next exercise.

18.3.8 ♯Exercise. Assume that R is positively graded, and let M be a non-zero finitely generated graded R-module and $f_1, \ldots, f_h$ be a sequence of homogeneous elements of R. Show that the following statements are equivalent:

(i) $f_1, \ldots, f_h$ is an R_+-filter-regular sequence with respect to M;
(ii) for all $\mathfrak{P} \in \mathrm{Spec}(R) \setminus \mathrm{Var}(R_+)$, the sequence $f_1/1, \ldots, f_h/1$ of natural images in $R_{\mathfrak{P}}$ is a poor $M_{\mathfrak{P}}$-sequence;
(iii) $((f_1, \ldots, f_{i-1})M :_M f_i)/(f_1, \ldots, f_{i-1})M$ is R_+-torsion for all $i = 1, \ldots, h$;
(iv) $\mathrm{end}\left(((f_1, \ldots, f_{i-1})M :_M f_i)/(f_1, \ldots, f_{i-1})M\right) < \infty$ for $i = 1, \ldots, h$ (the end of a graded R-module was defined in 14.1.1);
(v) f_i is a non-zerodivisor on $\overline{M/(f_1, \ldots, f_{i-1})M}$ for each $i = 1, \ldots, h$, where $\overline{L}$, for an R-module L, denotes $L/\Gamma_{R_+}(L)$.

18.3.9 Exercise. Let the situation be as in 18.3.8, and assume that $f_1, \ldots, f_h$ is an R_+-filter-regular sequence with respect to M. Let $\deg f_i = n_i$ for $i = 1, \ldots, h$. Show that

$$\mathrm{reg}^1\left(M/\textstyle\sum_{i=1}^h f_i M\right) \leq \mathrm{reg}^1(M) + \textstyle\sum_{i=1}^h n_i - h.$$

We now turn to the construction of R_+-filter-regular sequences.

18.3.10 Proposition. *Assume that $R = \bigoplus_{n \in \mathbb{N}_0} R_n$ is positively graded, that $(R_0, \mathfrak{m}_0)$ is local with infinite residue field $k := R_0/\mathfrak{m}_0$, and that the non-zero graded ideal $\mathfrak{A} = \bigoplus_{n \in \mathbb{N}} \mathfrak{A}_n$ of R is a *reduction of R_+. Let $t := \dim_k(\mathfrak{A}_1/\mathfrak{m}_0\mathfrak{A}_1)$. Then there exist $f_1, \ldots, f_t \in \mathfrak{A}_1$ such that $\mathfrak{A} = \sum_{i=1}^t R f_i$ and $f_1, \ldots, f_t$ is an R_+-filter-regular sequence with respect to R.*

Proof. Suppose, inductively, that $i \in \mathbb{N}_0$ with $i < t$ and that an R_+-filter-regular sequence (with respect to R) $f_1, \ldots, f_i$ of elements of $\mathfrak{A}_1$ has been constructed such that the natural images of $f_1, \ldots, f_i$ in $\mathfrak{A}_1/\mathfrak{m}_0\mathfrak{A}_1$ are linearly independent in this k-space. This is certainly the case when $i = 0$. Set $\mathfrak{f} := \sum_{j=1}^{i} Rf_j$. Note that $\mathfrak{m}_0\mathfrak{A} + \mathfrak{f} \neq \mathfrak{A}$ and that $\sqrt{\mathfrak{A}} = \sqrt{R_+}$ by 18.1.2(i). We can use 16.1.3 to see that there exists

$$f_{i+1} \in \mathfrak{A}_1 \setminus \left((\mathfrak{m}_0\mathfrak{A} + \mathfrak{f}) \cup \left(\bigcup_{\mathfrak{P} \in \mathrm{Ass}(R/\mathfrak{f}) \setminus \mathrm{Var}(R_+)} \mathfrak{P} \right) \right).$$

Then the extended sequence $f_1, \ldots, f_i, f_{i+1}$ is R_+-filter-regular (with respect to R), and the choice of f_{i+1} ensures that the natural images of $f_1, \ldots, f_i, f_{i+1}$ in $\mathfrak{A}_1/\mathfrak{m}_0\mathfrak{A}_1$ are linearly independent. This completes the inductive step.

In this way, we can construct an R_+-filter-regular sequence (with respect to R) $f_1, \ldots, f_t$ of elements of $\mathfrak{A}_1$ whose natural images in $\mathfrak{A}_1/\mathfrak{m}_0\mathfrak{A}_1$ form a basis for this k-space. Since $\mathfrak{A}$ can be generated by elements of degree 1, it follows that $f_1, \ldots, f_t$ generate $\mathfrak{A}$. $\qquad\square$

We now show that R_+-filter-regular sequences lend themselves to satisfactory calculations with various Castelnuovo regularities. We remind the reader that, when R is positively graded and homogeneous, the regularity $\mathrm{reg}^l(M)$ of M at and above level l (for $l \in \mathbb{N}_0$) of a finitely generated graded R-module M was defined in 16.2.9.

18.3.11 Proposition. *Assume that $R = \bigoplus_{n \in \mathbb{N}_0} R_n$ is positively graded and homogeneous, let M be a non-zero finitely generated graded R-module, and let $f_1, \ldots, f_h \in R_1$ be an R_+-filter-regular sequence with respect to M. Then $\mathrm{reg}^l(M) \leq \mathrm{reg}^{l-h}(M/(f_1, \ldots, f_h)M)$ for all $l \geq h$ and*

$$\mathrm{reg}^h(M) \leq \mathrm{reg}(M/(f_1, \ldots, f_h)M) \leq \mathrm{reg}(M).$$

Proof. We prove this by induction on h, there being nothing to prove when $h = 0$. Suppose, inductively, that $h > 0$, and that both statements in the claim have been proved for smaller values of h. Set $\mathfrak{f} := \sum_{i=1}^{h-1} Rf_i$. By our inductive hypothesis, $\mathrm{reg}^l(M) \leq \mathrm{reg}^{l-h+1}(M/\mathfrak{f}M)$ for all $l \geq h-1$ and $\mathrm{reg}^{h-1}(M) \leq \mathrm{reg}(M/\mathfrak{f}M) \leq \mathrm{reg}(M)$.

Since $(\mathfrak{f}M :_M f_h)/\mathfrak{f}M$ is R_+-torsion, we have $H^j_{R_+}((\mathfrak{f}M :_M f_h)/\mathfrak{f}M) = 0$ for all $j \in \mathbb{N}$, by 2.1.7(i); therefore, the canonical epimorphism $\pi : M/\mathfrak{f}M \longrightarrow M/(\mathfrak{f}M :_M f_h)$ induces homogeneous isomorphisms

$$H^j_{R_+}(\pi) : H^j_{R_+}(M/\mathfrak{f}M) \overset{\cong}{\longrightarrow} H^j_{R_+}(M/(\mathfrak{f}M :_M f_h)) \quad \text{for all } j \in \mathbb{N}.$$

Hence $\mathrm{reg}^l(M/\mathfrak{f}M) = \mathrm{reg}^l(M/(\mathfrak{f}M :_M f_h))$ for all $l \in \mathbb{N}$.

Multiplication by f_h leads to an exact sequence

$$0 \longrightarrow (M/(\mathfrak{f}M :_M f_h))(-1) \longrightarrow M/\mathfrak{f}M \longrightarrow M/(\mathfrak{f} + Rf_h)M \longrightarrow 0$$

of graded R-modules and homogeneous homomorphisms. By 16.2.15(iv) and the inductive hypothesis,

$$\begin{aligned}
\operatorname{reg}(M/(\mathfrak{f} + Rf_h)M) \\
\leq \max\left\{\operatorname{reg}^1((M/(\mathfrak{f}M :_M f_h))(-1)) - 1, \operatorname{reg}(M/\mathfrak{f}M)\right\} \\
= \max\left\{\operatorname{reg}^1(M/\mathfrak{f}M), \operatorname{reg}(M/\mathfrak{f}M)\right\} = \operatorname{reg}(M/\mathfrak{f}M) \\
\leq \operatorname{reg}(M).
\end{aligned}$$

We can also apply 16.2.15(ii) to deduce that, for $l \geq h$,

$$\begin{aligned}
\operatorname{reg}^{l-h+1}(M/\mathfrak{f}M) \\
= \operatorname{reg}^{l-h+1}((M/(\mathfrak{f}M :_M f_h))(-1)) - 1 \\
\leq \max\left\{\operatorname{reg}^{l-h+1}(M/\mathfrak{f}M) - 1, \operatorname{reg}^{l-h}(M/(\mathfrak{f} + Rf_h)M)\right\},
\end{aligned}$$

so that $\operatorname{reg}^{l-h+1}(M/\mathfrak{f}M) \leq \operatorname{reg}^{l-h}(M/(\mathfrak{f} + Rf_h)M)$. We can therefore use the inductive hypothesis to see that

$$\operatorname{reg}^l(M) \leq \operatorname{reg}^{l-h+1}(M/\mathfrak{f}M) \leq \operatorname{reg}^{l-h}(M/(\mathfrak{f} + Rf_h)M) \quad \text{for all } l \geq h.$$

In particular, $\operatorname{reg}^h(M) \leq \operatorname{reg}(M/(\mathfrak{f} + Rf_h)M)$, and so this completes the inductive step. $\qquad\Box$

18.3.12 Theorem. (See N. V. Trung [86, Proposition 3.2].) *Assume* $(R, \mathfrak{m})$ *is local and that* $k := R/\mathfrak{m}$ *is infinite. Suppose that* $\mathfrak{b}$ *is proper, and set* $s := \operatorname{spr}(\mathfrak{b})$. *Let* $\mathfrak{a}$ *be a minimal reduction of* $\mathfrak{b}$. *Then*

$$\operatorname{reg}^s(\mathcal{G}(\mathfrak{b})) \leq r_\mathfrak{a}(\mathfrak{b}) \leq \operatorname{reg}(\mathcal{G}(\mathfrak{b})).$$

Proof. By 18.3.4, and with the notation of that lemma, $\mathfrak{G}(\mathfrak{a})$ is a *reduction of $\mathcal{G}(\mathfrak{b})_+$, so that $\mathcal{G}(\mathfrak{b})/\mathfrak{G}(\mathfrak{a})$ is $\mathcal{G}(\mathfrak{b})_+$-torsion and (by 18.3.2)

$$r_\mathfrak{a}(\mathfrak{b}) = r_{\mathfrak{G}(\mathfrak{a})}(\mathcal{G}(\mathfrak{b})_+) = \operatorname{end}(\mathcal{G}(\mathfrak{b})/\mathfrak{G}(\mathfrak{a})).$$

Note that, by 2.1.7(i), we have $H^i_{\mathcal{G}(\mathfrak{b})_+}(\mathcal{G}(\mathfrak{b})/\mathfrak{G}(\mathfrak{a})) = 0$ for all $i \in \mathbb{N}$, and so

$$r_\mathfrak{a}(\mathfrak{b}) = \operatorname{end}(\mathcal{G}(\mathfrak{b})/\mathfrak{G}(\mathfrak{a})) = \operatorname{reg}(\mathcal{G}(\mathfrak{b})/\mathfrak{G}(\mathfrak{a})).$$

Next, we observe from 18.3.6 that $\mathfrak{G}(\mathfrak{a})$ can be generated by $s := \operatorname{spr}(\mathfrak{b})$ homogeneous elements of degree 1, and not by fewer. Therefore, by 18.3.3 and 18.3.10, there exists in $\mathfrak{G}(\mathfrak{a})$ a $\mathcal{G}(\mathfrak{b})_+$-filter-regular sequence (with respect to $\mathcal{G}(\mathfrak{b})$) $f_1, \ldots, f_s$ of elements of degree 1 which generate this ideal. Consequently, by Proposition 18.3.11, we have

$$\text{reg}^s(\mathcal{G}(\mathfrak{b})) \leq \text{reg}(\mathcal{G}(\mathfrak{b})/\mathfrak{G}(\mathfrak{a})) \leq \text{reg}(\mathcal{G}(\mathfrak{b}))$$

and this completes the proof because we have already shown that $r_\mathfrak{a}(\mathfrak{b}) = \text{reg}(\mathcal{G}(\mathfrak{b})/\mathfrak{G}(\mathfrak{a}))$. □

We are now going to use Trung's Theorem 18.3.12 to derive a theorem due to L. T. Hoa, which shows that, in the situation of Trung's Theorem, for $t \in \mathbb{N}$ sufficiently large, the reduction number of $\mathfrak{b}^t$ with respect to a minimal reduction of $\mathfrak{b}^t$ is independent of the choice of minimal reduction and independent of t. We need one preparatory result, also due to Hoa.

18.3.13 Proposition (L. T. Hoa [35, Lemma 2.4]). *Assume that $(R, \mathfrak{m})$ is local and that $k := R/\mathfrak{m}$ is infinite. Suppose that $\mathfrak{b}$ is proper, and assume that $s := \text{spr}(\mathfrak{b}) \geq 1$. For $q \in \mathbb{Q}$, we use $\lfloor q \rfloor$ to denote $\max\{i \in \mathbb{Z} : i \leq q\}$; we interpret $\lfloor q \rfloor$ as $-\infty$ when $q = -\infty$. Let $t \in \mathbb{N}$. Then*

(i) $\text{end}\left(H^j_{\mathcal{G}(\mathfrak{b}^t)_+}(\mathcal{G}(\mathfrak{b}^t)) \right) \leq \left\lfloor \text{end}\left(H^j_{\mathcal{G}(\mathfrak{b})_+}(\mathcal{G}(\mathfrak{b})) \right) \Big/ t \right\rfloor$ *for all $j \in \mathbb{N}_0$;*

(ii) $\text{end}\left(H^s_{\mathcal{G}(\mathfrak{b}^t)_+}(\mathcal{G}(\mathfrak{b}^t)) \right) = \left\lfloor \text{end}\left(H^s_{\mathcal{G}(\mathfrak{b})_+}(\mathcal{G}(\mathfrak{b})) \right) \Big/ t \right\rfloor$.

Proof. By 18.2.2 and 18.3.5, we have

$$\text{end}\left(H^j_{\mathcal{G}(\mathfrak{b}^t)_+}(\mathcal{G}(\mathfrak{b}^t)) \right) = \text{end}\left(H^j_{\mathcal{G}(\mathfrak{b})_+}(\mathcal{G}(\mathfrak{b})) \right) = -\infty \quad \text{for all } j > s.$$

The inequality in part (i) is therefore certainly true when $j > s$.

Now consider the case where $0 \leq j \leq s$, and let $i \in \mathbb{N}_0$ with $i \leq t$. The extension $\mathfrak{b}^i \mathcal{G}(\mathfrak{b}^t)$ of $\mathfrak{b}^i$ to $\mathcal{G}(\mathfrak{b}^t)$ under the composition $R \to R/\mathfrak{b}^t \to \mathcal{G}(\mathfrak{b}^t)$ of canonical ring homomorphisms is a graded ideal of $\mathcal{G}(\mathfrak{b}^t)$ with grading given by

$$\mathfrak{b}^i \mathcal{G}(\mathfrak{b}^t) = \bigoplus_{n \in \mathbb{N}_0} \mathfrak{b}^{tn+i}/\mathfrak{b}^{t(n+1)}.$$

Note that $\mathfrak{b}^t \mathcal{G}(\mathfrak{b}^t) = 0$. Our strategy is to consider the chain of submodules

$$\mathcal{G}(\mathfrak{b}^t) = \mathfrak{b}^0 \mathcal{G}(\mathfrak{b}^t) \supseteq \cdots \supseteq \mathfrak{b}^r \mathcal{G}(\mathfrak{b}^t) \supseteq \mathfrak{b}^{r+1} \mathcal{G}(\mathfrak{b}^t) \supseteq \cdots \supseteq \mathfrak{b}^t \mathcal{G}(\mathfrak{b}^t) = 0,$$

and to obtain information about the 'subquotients' in the chain by use of Veronesean functors (see 13.5.9).

There is a natural homogeneous surjective ring homomorphism $\theta : \mathcal{G}(\mathfrak{b}^t) \to (\mathcal{G}(\mathfrak{b}))^{(t)}$, and so any graded $\mathcal{G}(\mathfrak{b})^{(t)}$-module can be regarded as a graded $\mathcal{G}(\mathfrak{b}^t)$-module by means of θ. This applies, in particular, to

$$(\mathcal{G}(\mathfrak{b}))^{(t,i)} = \bigoplus_{n \in \mathbb{N}_0} \mathfrak{b}^{tn+i}/\mathfrak{b}^{tn+i+1}.$$

It follows that, for $i < t$, there is an exact sequence

$$0 \longrightarrow \mathfrak{b}^{i+1}\mathcal{G}(\mathfrak{b}^t) \longrightarrow \mathfrak{b}^i\mathcal{G}(\mathfrak{b}^t) \longrightarrow (\mathcal{G}(\mathfrak{b}))^{(t,i)} \longrightarrow 0$$

of graded $\mathcal{G}(\mathfrak{b}^t)$-modules and homogeneous homomorphisms.

Let $r \in \mathbb{Z}$ with $r > \left\lfloor \mathrm{end}\left(H^j_{\mathcal{G}(\mathfrak{b})_+}(\mathcal{G}(\mathfrak{b}))\right)\Big/t \right\rfloor$. By 13.5.9(v) and the Graded Independence Theorem 14.1.7, there are homogeneous $\mathcal{G}(\mathfrak{b}^t)$-isomorphisms

$$H^j_{\mathcal{G}(\mathfrak{b}^t)_+}((\mathcal{G}(\mathfrak{b}))^{(t,i)}) \xrightarrow{\cong} H^j_{\mathcal{G}(\mathfrak{b})^{(t)}_+}((\mathcal{G}(\mathfrak{b}))^{(t,i)}) \xrightarrow{\cong} \left(H^j_{\mathcal{G}(\mathfrak{b})_+}(\mathcal{G}(\mathfrak{b}))\right)^{(t,i)}.$$

Hence

$$H^j_{\mathcal{G}(\mathfrak{b}^t)_+}((\mathcal{G}(\mathfrak{b}))^{(t,i)})_r \cong H^j_{\mathcal{G}(\mathfrak{b})_+}(\mathcal{G}(\mathfrak{b}))_{rt+i} = 0,$$

since $rt + i > \mathrm{end}(H^j_{\mathcal{G}(\mathfrak{b})_+}(\mathcal{G}(\mathfrak{b})))$. Therefore the short exact sequence displayed in the last paragraph induces an $R/\mathfrak{b}^t$-epimorphism

$$H^j_{\mathcal{G}(\mathfrak{b}^t)_+}(\mathfrak{b}^{i+1}\mathcal{G}(\mathfrak{b}^t))_r \longrightarrow H^j_{\mathcal{G}(\mathfrak{b}^t)_+}(\mathfrak{b}^i\mathcal{G}(\mathfrak{b}^t))_r.$$

Since $\mathfrak{b}^t\mathcal{G}(\mathfrak{b}^t) = 0$, we can therefore deduce by descending induction that $H^j_{\mathcal{G}(\mathfrak{b}^t)_+}(\mathfrak{b}^i\mathcal{G}(\mathfrak{b}^t))_r = 0$ for $i = t-1, \ldots, 1, 0$. Hence $H^j_{\mathcal{G}(\mathfrak{b}^t)_+}(\mathcal{G}(\mathfrak{b}^t))_r = 0$. Hence $\mathrm{end}\left(H^j_{\mathcal{G}(\mathfrak{b}^t)_+}(\mathcal{G}(\mathfrak{b}^t))\right) \leq \left\lfloor \mathrm{end}\left(H^j_{\mathcal{G}(\mathfrak{b})_+}(\mathcal{G}(\mathfrak{b}))\right)\Big/t \right\rfloor$.

Now consider the case where $j = s$. Let $u := \left\lfloor \mathrm{end}\left(H^s_{\mathcal{G}(\mathfrak{b})_+}(\mathcal{G}(\mathfrak{b}))\right)\Big/t \right\rfloor$. Thus $\mathrm{end}(H^s_{\mathcal{G}(\mathfrak{b})_+}(\mathcal{G}(\mathfrak{b}))) = ut + i$ for some integer i with $0 \leq i \leq t-1$. Recall from 18.3.5 that s is the greatest integer i such that $H^i_{\mathcal{G}(\mathfrak{b})_+}(\mathcal{G}(\mathfrak{b})) \neq 0$. It therefore follows from 16.2.5 that $H^s_{\mathcal{G}(\mathfrak{b})_+}(\mathcal{G}(\mathfrak{b}))_n \neq 0$ for all $n \leq ut + i$, and so, in particular, for $n = ut$. Recall also, from 18.2.2 and 18.3.5, that $s = \mathrm{ara}(\mathcal{G}(\mathfrak{b}^t)_+)$. It therefore follows from the exact sequence

$$0 \longrightarrow \mathfrak{b}\mathcal{G}(\mathfrak{b}^t) \longrightarrow \mathcal{G}(\mathfrak{b}^t) \longrightarrow (\mathcal{G}(\mathfrak{b}))^{(t,0)} \longrightarrow 0$$

of graded $\mathcal{G}(\mathfrak{b}^t)$-modules and homogeneous homomorphisms that there is an exact sequence of $R/\mathfrak{b}^t$-modules

$$H^s_{\mathcal{G}(\mathfrak{b}^t)_+}(\mathcal{G}(\mathfrak{b}^t))_u \longrightarrow H^s_{\mathcal{G}(\mathfrak{b})_+}(\mathcal{G}(\mathfrak{b}))_{ut} \longrightarrow 0.$$

Hence $H^s_{\mathcal{G}(\mathfrak{b}^t)_+}(\mathcal{G}(\mathfrak{b}^t))_u \neq 0$, and $\mathrm{end}\left(H^s_{\mathcal{G}(\mathfrak{b}^t)_+}(\mathcal{G}(\mathfrak{b}^t))\right) \geq u$. This, in conjunction with the result of part (i), completes the proof. $\square$

We are now ready to present Hoa's proof of his theorem, mentioned in the introduction to this chapter, about the asymptotic behaviour, with respect to reduction numbers, of powers of a proper ideal of a local ring having infinite residue field. We remark again that there is no mention of local cohomology in the statement of the theorem, but the powerful tool of graded local cohomology plays a major rôle in the proof.

18.3.14 Theorem (L. T. Hoa [35, Theorem 2.1]). *Assume that $(R, \mathfrak{m})$ is local and that $R/\mathfrak{m}$ is infinite. Suppose that $\mathfrak{b}$ is proper. Then there exist $t_0 \in \mathbb{N}$ and $c \in \mathbb{N}_0$ such that, for all $t > t_0$ and every minimal reduction $\mathfrak{a}$ of $\mathfrak{b}^t$, we have $r_{\mathfrak{a}}(\mathfrak{b}^t) = c$.*

Proof. Let $s := \mathrm{spr}(\mathfrak{b})$. If $s = 0$, then $\mathfrak{b}$ is nilpotent, and the result is obvious because the only minimal reduction of the zero ideal of R is the zero ideal itself. We therefore assume that $s \geq 1$ for the remainder of this proof. Set $t_0 := \max\{|\,\mathrm{end}(H^i_{\mathcal{G}(\mathfrak{b})_+}(\mathcal{G}(\mathfrak{b})))|\, : \, i \in \mathbb{N}_0 \text{ and } H^i_{\mathcal{G}(\mathfrak{b})_+}(\mathcal{G}(\mathfrak{b})) \neq 0\}$. Note that $t_0 \in \mathbb{N}_0$, by 18.3.5.

Suppose that $t \in \mathbb{N}$ with $t > t_0$. By Proposition 18.3.13, we have

$$\mathrm{end}\left(H^i_{\mathcal{G}(\mathfrak{b}^t)_+}(\mathcal{G}(\mathfrak{b}^t))\right) \leq 0 \quad \text{for } i = 0, \ldots, s-1$$

and

$$\mathrm{end}\left(H^s_{\mathcal{G}(\mathfrak{b}^t)_+}(\mathcal{G}(\mathfrak{b}^t))\right) = \begin{cases} 0 & \text{if } \mathrm{end}\left(H^s_{\mathcal{G}(\mathfrak{b})_+}(\mathcal{G}(\mathfrak{b}))\right) \geq 0, \\ -1 & \text{if } \mathrm{end}\left(H^s_{\mathcal{G}(\mathfrak{b})_+}(\mathcal{G}(\mathfrak{b}))\right) < 0. \end{cases}$$

Of course, $H^i_{\mathcal{G}(\mathfrak{b}^t)_+}(\mathcal{G}(\mathfrak{b}^t)) = 0$ for all $i > s$, by 18.2.2 and 18.3.5. Hence

$$\mathrm{reg}^s(\mathcal{G}(\mathfrak{b}^t)) = \mathrm{reg}(\mathcal{G}(\mathfrak{b}^t)) = \begin{cases} s & \text{if } \mathrm{end}\left(H^s_{\mathcal{G}(\mathfrak{b})_+}(\mathcal{G}(\mathfrak{b}))\right) \geq 0, \\ s-1 & \text{if } \mathrm{end}\left(H^s_{\mathcal{G}(\mathfrak{b})_+}(\mathcal{G}(\mathfrak{b}))\right) < 0. \end{cases}$$

The result now follows from Trung's Theorem 18.3.12.   □

18.3.15 Exercise. Assume that $(R, \mathfrak{m})$ is local and that $R/\mathfrak{m}$ is infinite; let $\dim R = d$. Suppose that $\mathfrak{q}$ is an $\mathfrak{m}$-primary ideal of R such that $\mathcal{G}(\mathfrak{q})$ is Cohen–Macaulay. Show that each minimal reduction $\mathfrak{q}'$ of $\mathfrak{q}$ has $r_{\mathfrak{q}'}(\mathfrak{q}) = \mathrm{end}\left(H^d_{\mathcal{G}(\mathfrak{q})_+}(\mathcal{G}(\mathfrak{q}))\right) + d$.

As was mentioned in the introduction to this chapter, we have only been able to present a small portion of the body of work linking graded local cohomology and reductions of ideals. An interested reader might like to consult, in addition to papers already cited in this chapter, [48] by T. Marley, [87] by N. V. Trung, [31] by M. Herrmann, E. Hyry and T. Korb, [9] by C. D'Cruz, V. Kodiyalam and J. K. Verma, [12] by J. Elias, and some of the papers cited by these authors.

19

Connectivity in algebraic varieties

The study of the topological connectivity of algebraic sets is a fundamental subject in algebraic geometry. Local cohomology is a powerful tool in this field. In this chapter we shall use this tool to prove some results on connectivity which are of basic significance. Our main result will be the Connectedness Bound for Complete Local Rings, a refinement of Grothendieck's Connectedness Theorem. We shall apply this result to projective varieties in order to obtain a refined version of the Bertini–Gothendieck Connectivity Theorem. Another central result of this chapter will be the Intersection Inequality for Connectedness Dimensions of Affine Algebraic Cones. As an application it will furnish a refined version of the Connectedness Theorem for Projective Varieties due to W. Barth, to W. Fulton and J. Hansen, and to G. Faltings. The final goal of the chapter will be a ring-theoretic version of Zariski's Main Theorem on the Connectivity of Fibres of Blowing-up.

The crucial appearances of local cohomology in this chapter are just in two proofs, but the resulting far-reaching consequences in algebraic geometry illustrate again the power of local cohomology as a tool in the subject. We shall use little more from local cohomology than the Mayer–Vietoris sequence 3.2.3 and its graded version 14.1.5, the Lichtenbaum–Hartshorne Vanishing Theorem 8.2.1 and the graded version 14.1.16, and the vanishing result of 3.3.3. See the proofs of Proposition 19.2.8 and Lemma 19.7.2. The use of these techniques in this context originally goes back to Hartshorne [29] and has been pushed further by J. Rung (see [5]).

Throughout this chapter, all graded rings and modules are to be understood to be $\mathbb{Z}$-graded, and all polynomial rings $K[X_1, \ldots, X_d]$ (over a field K) are to be understood to be (positively) $\mathbb{Z}$-graded so that K is the component of degree 0 and $\deg X_i = 1$ for all $i = 1, \ldots, d$.

19.1 The connectedness dimension

To begin, we have to introduce a measure for the connectivity of an algebraic set, or, more generally, of a Noetherian topological space. We start with some reminders of topological concepts which are fundamental to our work in this chapter.

19.1.1 Reminders.   Here we are concerned with a general topological space.

(i) Recall that a non-empty topological space is said to be *disconnected* precisely when it can be expressed as the disjoint union of two proper open (or closed) subsets. Otherwise the space is said to be *connected*. We adopt the convention whereby the empty set is considered to be disconnected.

(ii) Recall also that a topological space T is said to be *quasi-compact* precisely when every open covering of T has a finite subcovering, that is, if and only if, whenever $(U_\alpha)_{\alpha \in \Lambda}$ is a family of open subsets of T such that $T = \bigcup_{\alpha \in \Lambda} U_\alpha$, then there is a finite subset Φ of Λ such that $T = \bigcup_{\alpha \in \Phi} U_\alpha$.

(iii) Recall that a non-empty topological space T is said to be irreducible precisely when T is not the union of two proper closed subsets, that is, if and only if every pair of non-empty open subsets of T has non-empty intersection.

19.1.2 ‡Exercise.   Let T be a non-empty topological space.

(i) Show that the following statements are equivalent:

  (a) T is irreducible;
  (b) every non-empty open subset of T is dense in T;
  (c) every non-empty open subset of T is connected.

(ii) Let S be an irreducible subset of T (that is, a subset of T which is an irreducible space in the topology induced from T), and let $(C_i)_{1 \le i \le n}$ be a finite covering of S by closed subsets of T (so that $C_1, \ldots, C_n$ are closed subsets of T such that $S \subseteq \bigcup_{i=1}^n C_i$). Show that $S \subseteq C_i$ for some i with $1 \le i \le n$.

(iii) Show that, if T is irreducible, then every non-empty open subset of T is irreducible.

(iv) Let $(U_i)_{1 \le i \le n}$ be a finite open covering of T, with $U_i \ne \emptyset$ for all $i = 1, \ldots, n$. Prove that T is irreducible if and only if U_i is irreducible for all $i = 1, \ldots, n$ and $U_i \cap U_j \ne \emptyset$ for all $i, j = 1, \ldots, n$.

(v) Let S be a subset of T. Show that S is irreducible if and only if its closure $\overline{S}$ is irreducible.

(vi) Let T' be a second topological space and let $f : T \to T'$ be a continuous map. Show that, if T is irreducible, then so too is $f(T)$.

19.1.3 ♯Exercise and Definition. Use Zorn's Lemma to show that a non-empty topological space T has maximal irreducible subsets.

The maximal irreducible subsets of T are called its *irreducible components*.

Show that the irreducible components of T are closed and that they cover T; show also that every irreducible subset of T is contained in an irreducible component of T.

19.1.4 ♯Exercise. Let K be an algebraically closed field, and let $r \in \mathbb{N}$.

(i) Show that the irreducible components of an affine algebraic cone in $\mathbb{A}^r(K)$ (see 15.1.2(i)) are again affine algebraic cones.

(ii) Let W be a non-empty closed subset of $\mathbb{P}^r(K)$; suppose that the distinct irreducible components of W are $W_1, \ldots, W_n$. Show that $\mathrm{Cone}(W)$ (see 15.2.1(iii)) has $\mathrm{Cone}(W_1), \ldots, \mathrm{Cone}(W_n)$ as its (distinct) irreducible components.

19.1.5 Definition. Let T be a topological space. We say that T is a *Noetherian* topological space precisely when it satisfies the following equivalent conditions.

(i) Whenever $(C_i)_{i \in \mathbb{N}}$ is a family of closed subsets of T such that

$$C_1 \supseteq C_2 \supseteq \cdots \supseteq C_i \supseteq C_{i+1} \supseteq \cdots ,$$

then there exists $k \in \mathbb{N}$ such that $C_k = C_{k+i}$ for all $i \in \mathbb{N}$.

(ii) Every non-empty set of closed subsets of T contains a minimal element with respect to inclusion.

19.1.6 ♯Exercise. Let T be a topological space.

(i) Show that T is Noetherian if and only if every open subset of T is quasi-compact.

(ii) Show that the spectrum of a commutative Noetherian ring, furnished with its Zariski topology, is a Noetherian topological space.

(iii) Show that a quasi-affine variety over an algebraically closed field (see 2.3.1) is a Noetherian topological space.

19.1.7 Lemma. *Let T be a non-empty Noetherian topological space. Then*

T has only finitely many irreducible components. Also, if $T_1, \ldots, T_n$ are the distinct irreducible components of T, then

$$T_j \not\subseteq \bigcup_{\substack{i=1 \\ i \neq j}}^{n} T_i \quad \text{for all } j = 1, \ldots, n.$$

Proof. Suppose that T has infinitely many irreducible components. Let $\mathcal{S}$ be the set of non-empty closed subsets of T which have infinitely many irreducible components. Since T is Noetherian, $\mathcal{S}$ has a minimal member: let C be one such. Then C itself cannot be irreducible, so that C can be written as $C = C_1 \cup C_2$ for some proper closed subsets C_1 and C_2 of C. Note that C_1 and C_2 are closed in T, and so, by the minimality of C, each of C_1 and C_2 has only finitely many irreducible components. But, by 19.1.2(ii), each irreducible subset of C must be contained in C_1 or C_2, and so each irreducible component of C must be an irreducible component of C_1 or C_2 (by 19.1.3). Hence there can only be finitely many irreducible components of C, a contradiction.

The final claim follows easily from another use of 19.1.2(ii) since, by 19.1.3, each T_i $(1 \leq i \leq n)$ is closed in T. □

19.1.8 Definition. Let T be a non-empty Noetherian topological space. The *dimension of T*, denoted by $\dim T$, is defined as the supremum of the lengths n of all strictly descending chains $Z_0 \supset Z_1 \supset \cdots \supset Z_n$ of closed irreducible subsets of T if this supremum exists, and ∞ otherwise. Thus, by 19.1.3, $\dim T$ is a non-negative integer or ∞.

The dimension of the empty space is defined to be -1.

Note that, for our Noetherian ring R, we have $\dim(\operatorname{Spec}(R)) = \dim R$; thus, in view of [56, Appendix, Example 1], the dimension of a Noetherian topological space can be ∞.

19.1.9 Definition. Let T be a Noetherian topological space. The *connectedness dimension $c(T)$* of T is defined to be the minimum of the dimensions of those closed subsets Z of T for which $T \setminus Z$ is disconnected. (Observe that $T \setminus T$ is certainly disconnected!) Thus

$$c(T) := \min \{\dim Z : Z \subseteq T, \ Z \text{ is closed and } T \setminus Z \text{ is disconnected}\}.$$

For our Noetherian ring R, we write $c(R) := c(\operatorname{Spec}(R))$.

19.1.10 Examples. Let T be a Noetherian topological space.

(i) Note that T is disconnected, that is, $T \setminus \emptyset$ is disconnected, if and only if $c(T) = -1$. We can thus conclude that $c(T)$ is negative if and only if T is disconnected.

(ii) It follows from part (i) that, if $(R, \mathfrak{m})$ is local, then $c(R) \geq 0$.

(iii) Suppose that T is irreducible. Then if Z is a proper closed subset of T, it is impossible for $T \setminus Z$ to be disconnected (by 19.1.2(i)); on the other hand, $T \setminus T$ is disconnected. Thus $c(T) = \dim T$ in this case.

(iv) It follows from part (iii) that, if R is an integral domain, then $c(R) = \dim R$. Thus, in view of [56, Appendix, Example 1], the connectedness dimension of a Noetherian topological space can be ∞.

19.1.11 Example. Assume that $(R, \mathfrak{m})$ is local, and has exactly two minimal prime ideals, $\mathfrak{p}$ and $\mathfrak{q}$. Then $c(R) = 0$ if and only if $\dim R/(\mathfrak{p} + \mathfrak{q}) = 0$.

Proof. Note that the hypotheses ensure that $\dim R > 0$.

($\Leftarrow$) We have $\mathrm{Spec}(R) = \mathrm{Var}(\mathfrak{p}) \cup \mathrm{Var}(\mathfrak{q})$. Since $\mathfrak{m}$ is the only prime ideal of R which contains both $\mathfrak{p}$ and $\mathfrak{q}$, it follows that

$$\mathrm{Var}(\mathfrak{p}) \cap (\mathrm{Spec}(R) \setminus \{\mathfrak{m}\}) \quad \text{and} \quad \mathrm{Var}(\mathfrak{q}) \cap (\mathrm{Spec}(R) \setminus \{\mathfrak{m}\})$$

are two non-empty disjoint closed subsets of $\mathrm{Spec}(R) \setminus \{\mathfrak{m}\}$ which cover this space. Therefore, bearing in mind 19.1.10(ii), we see that $c(R) = 0$.

($\Rightarrow$) Assume that $c(R) = 0$. Thus there is a closed subset Z of $\mathrm{Spec}(R)$ for which $\dim Z = 0$ and $\mathrm{Spec}(R) \setminus Z$ is disconnected. We must have $Z = \{\mathfrak{m}\}$. Set $T := \mathrm{Spec}(R) \setminus \{\mathfrak{m}\}$. Thus there exist ideals $\mathfrak{a}, \mathfrak{b}$ of R such that $T \cap \mathrm{Var}(\mathfrak{a})$ and $T \cap \mathrm{Var}(\mathfrak{b})$ are non-empty disjoint subsets of T which cover T.

Then $\mathfrak{p} \supseteq \mathfrak{a}$ or $\mathfrak{p} \supseteq \mathfrak{b}$; also $\mathfrak{q} \supseteq \mathfrak{a}$ or $\mathfrak{q} \supseteq \mathfrak{b}$. For the sake of argument, let us assume that $\mathfrak{p} \supseteq \mathfrak{a}$.

Then $\mathfrak{q} \not\supseteq \mathfrak{a}$, since otherwise $\mathrm{Spec}(R) = \mathrm{Var}(\mathfrak{a})$ and $T \cap \mathrm{Var}(\mathfrak{b}) = \emptyset$. Therefore $\mathfrak{q} \supseteq \mathfrak{b}$, and it follows that $(T \cap \mathrm{Var}(\mathfrak{p})) \cap (T \cap \mathrm{Var}(\mathfrak{q})) = \emptyset$, so that $\dim R/(\mathfrak{p} + \mathfrak{q}) = 0$. $\square$

19.1.12 Exercise. Let f, g be non-constant and irreducible polynomials in $\mathbb{C}[X, Y]$ which are not associates of each other. Let $T := V_{\mathbb{A}^2}(fg)$. Show that

$$c(T) = \begin{cases} -1 & \text{if } V_{\mathbb{A}^2}(f) \cap V_{\mathbb{A}^2}(g) = \emptyset, \\ 0 & \text{otherwise.} \end{cases}$$

19.1.13 Notation and ♯Exercise. Assume that $(R, \mathfrak{m})$ is local. The topological space $\mathrm{Spec}(R) \setminus \{\mathfrak{m}\}$, with the topology induced from the Zariski topology on $\mathrm{Spec}(R)$, is called the *punctured spectrum* of R, and denoted by $\mathrm{Sp\overset{\circ}{e}c}(R)$. Show that $c(\mathrm{Sp\overset{\circ}{e}c}(R)) = c(R) - 1$.

19.1.14 Notation. For $r \in \mathbb{N}$, denote by $\mathcal{S}(r)$ the set of all ordered pairs (A, B) of non-empty subsets of $\{1, \ldots, r\}$ for which $A \cup B = \{1, \ldots, r\}$.

19.1.15 Lemma. *Let T be a non-empty Noetherian topological space with (distinct) irreducible components $T_1, \ldots, T_r$. With the notation of* 19.1.14, *we have* $c(T) = \min \left\{ \dim \left(\left(\bigcup_{i \in A} T_i \right) \cap \left(\bigcup_{j \in B} T_j \right) \right) : (A, B) \in \mathcal{S}(r) \right\}$.

Proof. We write $c = c(T)$ and m for the minimum that occurs on the right-hand side of the equation in the above statement.

We first show that $c \geq m$. To achieve this, let Z be a closed subset of T with $\dim Z = c$ and such that $T \setminus Z$ is disconnected. If $T_i \subseteq Z$ for some $i \in \{1, \ldots, r\}$, then $c = \dim Z \geq \dim T_i \geq m$. Thus we can, and do, assume that $T_i \cap (T \setminus Z) \neq \emptyset$ for all $i \in \{1, \ldots, r\}$. As $T \setminus Z$ is disconnected we can write $T \setminus Z = U_1 \cup U_2$, where U_1 and U_2 are non-empty open sets in T such that $U_1 \cap U_2 = \emptyset$. Set

$$A := \{i \in \{1, \ldots, r\} : T_i \cap U_1 \neq \emptyset\},$$

$$B := \{j \in \{1, \ldots, r\} : T_j \cap U_2 \neq \emptyset\}.$$

Then, the pair $(A, B) \in \mathcal{S}(r)$. Moreover, A and B are disjoint, as otherwise for any index $i \in A \cap B$ the irreducible space T_i would contain the two non-empty and disjoint open subsets $T_i \cap U_1$ and $T_i \cap U_2$. Thus $\left(\bigcup_{i \in A} T_i \right) \cap \left(\bigcup_{j \in B} T_j \right)$ has no point in common with $U_1 \cup U_2$, so that it is contained in Z and has dimension not exceeding c. This proves that $c \geq m$.

To prove the inequality $m \geq c$, let $(A, B) \in \mathcal{S}(r)$ be a pair such that $Z := \left(\bigcup_{i \in A} T_i \right) \cap \left(\bigcup_{j \in B} T_j \right)$ is of dimension m. If $r = 1$, we have $m = \dim T \geq c$, as required. Assume therefore that $r > 1$. Then, by the minimality in the definition of m, we can, and do, assume that A and B are disjoint. Then the open sets $U_1 = T \setminus \bigcup_{i \in A} T_i$ and $U_2 = T \setminus \bigcup_{j \in B} T_j$ are non-empty. Note that $U_1 \cap U_2 = \emptyset$, since $\left(\bigcup_{i \in A} T_i \right) \cup \left(\bigcup_{j \in B} T_j \right) = T$. Hence $T \setminus Z = U_1 \cup U_2$ is disconnected. Consequently $c \leq \dim Z = m$. □

19.1.16 ♯Exercise. Let V be an affine variety over the algebraically closed field K. Let $\mathfrak{b}$ be an ideal of $\mathcal{O}(V)$, and let $V(\mathfrak{b})$ denote the closed subset of V determined by $\mathfrak{b}$. Show that $c(V(\mathfrak{b})) = c(\mathcal{O}(V)/\mathfrak{b})$.

19.2 Complete local rings and connectivity

We now introduce another invariant of Noetherian spaces.

19.2.1 Definition. Let T be a non-empty Noetherian topological space. The *subdimension* $\mathrm{sdim}\, T$ of T is defined as the minimum of the dimensions of the irreducible components of T. For our Noetherian ring R, we write sdim R instead of sdim(Spec R).

Notice the following easy fact.

19.2.2 Lemma. *Let T be a non-empty Noetherian topological space of finite dimension. Then $c(T) \leq$ sdim T. Moreover, equality holds here if and only if T is irreducible.*

Proof. Let $T_1, \ldots, T_r$ be the irreducible components of T. Then

$$\text{sdim } T = \min \{\dim T_i : i = 1, \ldots, r\};$$

also, we have, by 19.1.7, for each $j = 1, \ldots, r$, that

$$T_j \cap \left(\bigcup_{\substack{i=1 \\ i \neq j}}^{r} T_i \right) \subset T_j,$$

so that, since T has finite dimension,

$$\dim \left(T_j \cap \left(\bigcup_{\substack{i=1 \\ i \neq j}}^{r} T_i \right) \right) < \dim T_j.$$

The claims now follow easily from Lemma 19.1.15. □

19.2.3 Exercise. Calculate $c(T)$ and sdim T for the following choices of the Noetherian topological space T:

(i) $T := V_{\mathbb{A}^3}(X_1^2 - X_1, X_2^2 - X_2)$;
(ii) $T := V_{\mathbb{A}^4}(X_1X_3, X_1X_4, X_2X_3, X_2X_4)$;
(iii) $T := V_{\mathbb{A}^3}((X_1 - 1)(X_1^2 + X_2^2 + X_3^2 - 1))$.

19.2.4 ♯Exercise. Assume that $(R, \mathfrak{m})$ is local and that $\dim R > 0$. Show that the subdimension of the punctured spectrum $\text{Sp\breve{e}c}(R)$ of R (see 19.1.13) is given by $\text{sdim}(\text{Sp\breve{e}c}(R)) = \text{sdim } R - 1$.

19.2.5 Remark. We shall frequently consider connectedness dimensions and subdimensions of spectra of Noetherian rings. Therefore it will be helpful to translate the previous lemmas into ring-theoretic terms. So, let $\mathfrak{p}_1, \ldots, \mathfrak{p}_r$ be the distinct minimal prime ideals of R. Then, again using the notation of 19.1.14, it follows from 19.1.15 that

$$c(R) = \min \left\{ \dim \left(R \big/ \left(\left(\bigcap_{i \in A} \mathfrak{p}_i \right) + \left(\bigcap_{j \in B} \mathfrak{p}_j \right) \right) \right) : (A, B) \in \mathcal{S}(r) \right\}.$$

Also, 19.2.2 shows that $\operatorname{sdim} R = \min\{\dim R/\mathfrak{p}_i : i = 1, \ldots, r\} \geq c(R)$, with equality if and only if $r = 1$.

19.2.6 Exercise.   Suppose that $(R, \mathfrak{m})$ is local, catenary and S_2, and that R has more than one minimal prime ideal; set $d := \dim R$. Use Proposition 12.2.8 to show that $c(R) = d - 1$.

Our applications of local cohomology to connectedness dimensions will involve use of the concepts of arithmetic rank and cohomological dimension of an ideal, introduced in 3.3.2 and 3.3.4. Recall that $\operatorname{cohd}(\mathfrak{a}) \leq \operatorname{ara}(\mathfrak{a})$. The reader should be aware of the elementary properties of arithmetic rank described in the following exercise.

19.2.7 ♯Exercise.   Assume that the ideal $\mathfrak{a}$ is proper.

(i) Show that $\operatorname{ht} \mathfrak{a} \leq \operatorname{ara}(\mathfrak{a})$.
(ii) Let R' be a second commutative Noetherian ring and let $f : R \longrightarrow R'$ be a ring homomorphism. Prove that $\operatorname{ara}(\mathfrak{a}R') \leq \operatorname{ara}(\mathfrak{a})$.

The next result will play a crucial rôle in our approach to connectivity. Its proof uses most of the main ingredients from local cohomology theory that we shall need in this chapter. The result relates, in certain circumstances, the cohomological dimension $\operatorname{cohd}(\mathfrak{a} \cap \mathfrak{b})$ of the intersection of two ideals $\mathfrak{a}$ and $\mathfrak{b}$ in a complete local domain R with the dimensions of R and $R/(\mathfrak{a} + \mathfrak{b})$.

19.2.8 Proposition.   *Assume that $(R, \mathfrak{m})$ is a complete local domain. Let $\mathfrak{b}$ be a second ideal of R, and assume that $\mathfrak{a}$ and $\mathfrak{b}$ are both proper and that* $\min\{\dim R/\mathfrak{a}, \dim R/\mathfrak{b}\} > \dim R/(\mathfrak{a} + \mathfrak{b})$. *Then*

$$\operatorname{cohd}(\mathfrak{a} \cap \mathfrak{b}) \geq \dim R - \dim R/(\mathfrak{a} + \mathfrak{b}) - 1,$$

so that $\operatorname{ara}(\mathfrak{a} \cap \mathfrak{b}) \geq \dim R - \dim R/(\mathfrak{a} + \mathfrak{b}) - 1$.

Proof.   Set $d := \dim R$ and $\delta := \dim R/(\mathfrak{a} + \mathfrak{b})$. We proceed by induction on δ. First, let $\delta = 0$. Then, we have to show that $\operatorname{cohd}(\mathfrak{a} \cap \mathfrak{b}) \geq d - 1$. By the Mayer–Vietoris sequence 3.2.3, there is an exact sequence

$$H_{\mathfrak{a} \cap \mathfrak{b}}^{d-1}(R) \longrightarrow H_{\mathfrak{a}+\mathfrak{b}}^{d}(R) \longrightarrow H_{\mathfrak{a}}^{d}(R) \oplus H_{\mathfrak{b}}^{d}(R).$$

It follows from the local Lichtenbaum–Hartshorne Vanishing Theorem 8.2.1 that $H_{\mathfrak{a}}^{d}(R) = H_{\mathfrak{b}}^{d}(R) = 0$. As $\delta = 0$, the ideal $\mathfrak{a} + \mathfrak{b}$ is $\mathfrak{m}$-primary, and so $H_{\mathfrak{a}+\mathfrak{b}}^{d}(R) \neq 0$ by 1.2.3 and 6.1.4. Altogether we obtain that $H_{\mathfrak{a} \cap \mathfrak{b}}^{d-1}(R) \neq 0$, and so $\operatorname{cohd}(\mathfrak{a} \cap \mathfrak{b}) \geq d - 1$. This proves the claim when $\delta = 0$.

So, let $\delta > 0$ and make the obvious inductive assumption. Set $\operatorname{cohd}(\mathfrak{a} \cap \mathfrak{b}) =: r$. As $\mathfrak{a} + \mathfrak{b}$ is not $\mathfrak{m}$-primary we can find an element $y \in \mathfrak{m}$ which lies outside all the minimal prime ideals of $\mathfrak{a}$, $\mathfrak{b}$ and $\mathfrak{a} + \mathfrak{b}$. We write $\mathfrak{a}' = \mathfrak{a} + Ry$ and $\mathfrak{b}' = \mathfrak{b} + Ry$ and note that $\dim R/(\mathfrak{a}' + \mathfrak{b}') = \delta - 1$,

$$\dim R/\mathfrak{a}' = \dim R/\mathfrak{a} - 1 > \delta - 1 \quad \text{and} \quad \dim R/\mathfrak{b}' = \dim R/\mathfrak{b} - 1 > \delta - 1.$$

As

$$\sqrt{\mathfrak{a}' \cap \mathfrak{b}'} = \sqrt{(\mathfrak{a} + Ry) \cap (\mathfrak{b} + Ry)} = \sqrt{(\mathfrak{a} \cap \mathfrak{b}) + Ry},$$

we have $\operatorname{cohd}(\mathfrak{a}' \cap \mathfrak{b}') \leq \operatorname{cohd}(\mathfrak{a} \cap \mathfrak{b}) + 1 = r + 1$ by 8.1.3. Therefore, by the inductive hypothesis, we have $r + 1 \geq d - (\delta - 1) - 1$; hence $r \geq d - \delta - 1$. This completes the inductive step, and the proof. $\qquad\square$

As an application of this we can now prove the following.

19.2.9 Lemma. *Assume that $(R, \mathfrak{m})$ is a complete local ring. Let $\mathfrak{b}$ be a second ideal of R, and assume that $\mathfrak{a}$ and $\mathfrak{b}$ are both proper and that*

$$\min\{\dim R/\mathfrak{a}, \dim R/\mathfrak{b}\} > \dim R/(\mathfrak{a} + \mathfrak{b}).$$

Then

$$\dim R/(\mathfrak{a} + \mathfrak{b}) \geq \min\{c(R), \operatorname{sdim} R - 1\} - \operatorname{cohd}(\mathfrak{a} \cap \mathfrak{b}),$$

so that $\dim R/(\mathfrak{a} + \mathfrak{b}) \geq \min\{c(R), \operatorname{sdim} R - 1\} - \operatorname{ara}(\mathfrak{a} \cap \mathfrak{b})$.

Proof. Set $\delta := \dim R/(\mathfrak{a} + \mathfrak{b})$. Let $\mathfrak{p}_1, \ldots, \mathfrak{p}_n$ be the distinct minimal prime ideals of R.

First, we treat the case where, for all $i \in \{1, \ldots, n\}$, either

$$\dim R/(\mathfrak{a} + \mathfrak{p}_i) \leq \delta \quad \text{or} \quad \dim R/(\mathfrak{b} + \mathfrak{p}_i) \leq \delta.$$

After an appropriate reordering of the $\mathfrak{p}_i$, there will be an $s \in \mathbb{N}_0$ such that $s \leq n$ and $\dim R/(\mathfrak{a} + \mathfrak{p}_i) \leq \delta$ for $1 \leq i \leq s$ and $\dim R/(\mathfrak{b} + \mathfrak{p}_j) \leq \delta$ for $s + 1 \leq j \leq n$. As $\max\{\dim R/(\mathfrak{a} + \mathfrak{p}_k) : 1 \leq k \leq n\} = \dim R/\mathfrak{a} > \delta$, we see that $s < n$. As $\max\{\dim R/(\mathfrak{b} + \mathfrak{p}_k) : 1 \leq k \leq n\} = \dim R/\mathfrak{b} > \delta$, we see that $1 \leq s$.

Now, let $\mathfrak{p}$ be a minimal prime ideal of the ideal

$$\mathfrak{c} := (\mathfrak{p}_1 \cap \cdots \cap \mathfrak{p}_s) + (\mathfrak{p}_{s+1} \cap \cdots \cap \mathfrak{p}_n)$$

such that $\dim R/\mathfrak{p} = \dim R/\mathfrak{c}$. By 19.2.5, we have $\dim R/\mathfrak{p} \geq c(R)$. Moreover we can choose indices i and j with $1 \leq i \leq s < j \leq n$ and such that $\mathfrak{p}_i, \mathfrak{p}_j \subseteq \mathfrak{p}$. It follows that

$$\delta \geq \dim R/(\mathfrak{a} + \mathfrak{p}_i) \geq \dim R/(\mathfrak{a} + \mathfrak{p})$$

and $\delta \geq \dim R/(\mathfrak{b} + \mathfrak{p}_j) \geq \dim R/(\mathfrak{b} + \mathfrak{p})$; hence

$$\delta \geq \dim R/((\mathfrak{a} + \mathfrak{p}) \cap (\mathfrak{b} + \mathfrak{p})) = \dim R/((\mathfrak{a} \cap \mathfrak{b}) + \mathfrak{p}).$$

As $R/\mathfrak{p}$ is catenary (see [50, Theorem 29.4(ii)]), we can write

$$\dim R/((\mathfrak{a} \cap \mathfrak{b}) + \mathfrak{p}) = \dim R/\mathfrak{p} - \mathrm{ht}((\mathfrak{a} \cap \mathfrak{b}) + \mathfrak{p})/\mathfrak{p}.$$

As $\dim R/\mathfrak{p} \geq c(R)$ and

$$\mathrm{ht}((\mathfrak{a} \cap \mathfrak{b}) + \mathfrak{p})/\mathfrak{p} \leq \mathrm{cohd}(((\mathfrak{a} \cap \mathfrak{b}) + \mathfrak{p})/\mathfrak{p}) \leq \mathrm{cohd}(\mathfrak{a} \cap \mathfrak{b})$$

(by 6.1.6 and 4.2.3), we thus obtain that $\delta \geq c(R) - \mathrm{cohd}(\mathfrak{a} \cap \mathfrak{b})$.

Therefore, it remains for us to treat the case in which there exists $i \in \{1, \ldots, n\}$ such that

$$\dim R/(\mathfrak{a} + \mathfrak{p}_i) > \delta \quad \text{and} \quad \dim R/(\mathfrak{b} + \mathfrak{p}_i) > \delta.$$

As $\dim R/(\mathfrak{a} + \mathfrak{b} + \mathfrak{p}_i) =: \delta' \leq \delta$, it follows from 19.2.8 that

$$\delta \geq \delta' \geq \dim R/\mathfrak{p}_i - \mathrm{cohd}\left(((\mathfrak{a} + \mathfrak{p}_i) \cap (\mathfrak{b} + \mathfrak{p}_i))/\mathfrak{p}_i\right) - 1.$$

Observing that $\dim R/\mathfrak{p}_i \geq \mathrm{sdim}\, R$ and (by 4.2.3)

$$\mathrm{cohd}\left(((\mathfrak{a} + \mathfrak{p}_i) \cap (\mathfrak{b} + \mathfrak{p}_i))/\mathfrak{p}_i\right) = \mathrm{cohd}\left(((\mathfrak{a} \cap \mathfrak{b}) + \mathfrak{p}_i)/\mathfrak{p}_i\right) \leq \mathrm{cohd}(\mathfrak{a} \cap \mathfrak{b}),$$

we thus deduce that $\delta \geq \mathrm{sdim}\, R - 1 - \mathrm{cohd}(\mathfrak{a} \cap \mathfrak{b})$. $\quad\square$

Note. In the First Edition of this book, only the statements involving arithmetic rank appeared in the results corresponding to 19.2.8 and 19.2.9. We are very grateful to M. Varbaro for pointing out to us that those statements can be strengthened by replacement of 'arithmetic rank' by 'cohomological dimension'. We have left to the interested reader the formulation of similar strengthenings of some subsequent results in this chapter, such as 19.2.10, 19.2.11 and 19.2.12.

We are now in a position to prove the first main result of this chapter, namely the Connectedness Bound for Complete Local Rings.

19.2.10 Connectedness Bound for Complete Local Rings. *Suppose that* $(R, \mathfrak{m})$ *is a complete local ring, and let* $\mathfrak{a}$ *be proper. Then*

$$c(R/\mathfrak{a}) \geq \min\left\{c(R), \mathrm{sdim}\, R - 1\right\} - \mathrm{ara}(\mathfrak{a}).$$

Proof. Without loss of generality we can, and do, assume that $\mathfrak{a} = \sqrt{\mathfrak{a}}$. Let $\mathfrak{p}_1, \ldots, \mathfrak{p}_n$ be the distinct minimal prime ideals of $\mathfrak{a}$, and set $c := c(R/\mathfrak{a})$.

If $n = 1$, we have $\mathfrak{a} = \mathfrak{p}_1$ and $c = \dim R/\mathfrak{p}_1$ by 19.2.5. Choose a minimal prime ideal $\mathfrak{p}$ of R with $\mathfrak{p} \subseteq \mathfrak{p}_1$, and observe that $\operatorname{ht}\mathfrak{p}_1/\mathfrak{p} \leq \operatorname{cohd}(\mathfrak{a})$ by 4.2.3 and 6.1.6. Since R is catenary, we deduce that

$$c = \dim R/\mathfrak{p} - \operatorname{ht}\mathfrak{p}_1/\mathfrak{p} \geq \operatorname{sdim} R - \operatorname{cohd}(\mathfrak{a}),$$

from which our claim follows (since $\operatorname{cohd}(\mathfrak{a}) \leq \operatorname{ara}(\mathfrak{a})$).

Consider now the case where $n > 1$. By 19.2.5, there exist two non-empty subsets A, B of $\{1, \ldots, n\}$ for which $A \cup B = \{1, \ldots, n\}$ and

$$c = \dim\left(R \Big/ \left(\left(\bigcap_{i \in A}\mathfrak{p}_i \right) + \left(\bigcap_{j \in B}\mathfrak{p}_j \right) \right) \right);$$

moreover, we can, and do, assume that A and B are disjoint. Set $\mathfrak{r} := \bigcap_{i \in A}\mathfrak{p}_i$ and $\mathfrak{s} := \bigcap_{j \in B}\mathfrak{p}_j$; then $\dim R/\mathfrak{r} > c$ and $\dim R/\mathfrak{s} > c$ (by the final comment of 19.2.5), and $\mathfrak{r} \cap \mathfrak{s} = \mathfrak{a}$. We can now use 19.2.9 to complete the proof. $\square$

19.2.11 Corollary. *Let $(R, \mathfrak{m})$ and $\mathfrak{a}$ be as in 19.2.10. Then*

$$c(R/\mathfrak{a}) \geq c(R) - \operatorname{ara}(\mathfrak{a}) - 1.$$

If R has more than one minimal prime ideal, then the inequality is strict.

Proof. By 19.2.5, we have $c(R) \leq \operatorname{sdim} R$, with strict inequality if R has more than one minimal prime ideal. The claim therefore follows from 19.2.10. $\square$

As another application of 19.2.10 we now prove Grothendieck's Connectedness Theorem.

19.2.12 Grothendieck's Connectedness Theorem. (See [26, Exposé XIII, Théorème 2.1].) *Assume that $(R, \mathfrak{m})$ is a complete local ring, and let $\mathfrak{a}$ be proper. Let $k \in \mathbb{N}_0$ be such that $c(R) \geq k$ and $\operatorname{sdim} R \geq k+1$. Then $c(R/\mathfrak{a}) \geq k - \operatorname{ara}(\mathfrak{a})$.*

Proof. By 19.2.10, we have

$$c(R/\mathfrak{a}) \geq \min\{c(R), \operatorname{sdim} R - 1\} - \operatorname{ara}(\mathfrak{a}) \geq k - \operatorname{ara}(\mathfrak{a}). \quad \square$$

19.2.13 Remark. Let the situation be as in 19.2.12. In Grothendieck's original version of that result, connectivity and subdimension are considered on the punctured spectrum $\operatorname{Sp\overset{8}{e}c}(R)$ of R (see 19.1.13). However, since

$$c(\operatorname{Sp\overset{8}{e}c}(R)) = c(R) - 1 \quad \text{and} \quad \operatorname{sdim}(\operatorname{Sp\overset{8}{e}c}(R)) = \operatorname{sdim} R - 1$$

(by 19.1.13 and 19.2.4), Grothendieck's version can be recovered from ours.

19.3 Some local dimensions

Up to now, we have obtained a certain understanding of the connectivity in the spectrum of a complete local Noetherian ring. In order to apply this knowledge to the non-complete case, we prove the following lemma. The reader is warned that our proof of part (iv) of the lemma uses L. J. Ratliff's Theorem [67] (see also [50, Theorem 31.7]) that a local ring $(R, \mathfrak{m})$ is universally catenary only if, for every $\mathfrak{p} \in \operatorname{Spec}(R)$ and for every minimal prime $\mathfrak{P}$ of the ideal $\mathfrak{p}\widehat{R}$ of $\widehat{R}$, we have $\dim \widehat{R}/\mathfrak{P} = \dim R/\mathfrak{p}$.

19.3.1 Lemma. *Assume that* $(R, \mathfrak{m})$ *is local. The following hold:*

(i) $c(R) \geq c(\widehat{R});$

(ii) *if* $\mathfrak{p}\widehat{R} \in \operatorname{Spec}(\widehat{R})$ *for all minimal prime ideals* $\mathfrak{p}$ *of R, then equality holds in* (i);

(iii) $\operatorname{sdim} R \geq \operatorname{sdim} \widehat{R};$ *and*

(iv) *if R is universally catenary, then equality holds in* (iii).

Proof. Let $\mathcal{P}$ be the set of minimal prime ideals of R and $\mathcal{Q}$ be the set of minimal prime ideals of $\widehat{R}$. Note that, by [50, Theorem 7.3(i)] for example, each $\mathfrak{p} \in \mathcal{P}$ is the contraction to R of some member of $\mathcal{Q}$. Also, $\mathfrak{Q} \cap R \in \mathcal{P}$ for all $\mathfrak{Q} \in \mathcal{Q}$ (by [50, Theorem 15.1(ii)], for example).

(i) By 19.2.5, we can find two non-empty subsets $\mathcal{P}_1$ and $\mathcal{P}_2$ of $\mathcal{P}$ such that $\mathcal{P}_1 \cup \mathcal{P}_2 = \mathcal{P}$ and, if $\mathfrak{a}_1 := \bigcap_{\mathfrak{p}\in\mathcal{P}_1}\mathfrak{p}$ and $\mathfrak{a}_2 := \bigcap_{\mathfrak{p}\in\mathcal{P}_2}\mathfrak{p}$, then

$$c := c(R) = \dim R/(\mathfrak{a}_1 + \mathfrak{a}_2).$$

Let $\mathcal{Q}_i := \{\mathfrak{Q} \in \mathcal{Q} : \mathfrak{Q} \cap R \in \mathcal{P}_i\}$ for $i = 1, 2$; then $\mathcal{Q}_1$ and $\mathcal{Q}_2$ are non-empty and such that $\mathcal{Q}_1 \cup \mathcal{Q}_2 = \mathcal{Q}$. Let $\mathfrak{B}_i = \bigcap_{\mathfrak{Q}\in\mathcal{Q}_i}\mathfrak{Q}$ $(i = 1, 2)$. Then, by 19.2.5, we have $c(\widehat{R}) \leq \dim \widehat{R}/(\mathfrak{B}_1 + \mathfrak{B}_2)$. Since $\mathfrak{a}_i\widehat{R} \subseteq \mathfrak{B}_i$ for $i = 1, 2$, it follows that $c(\widehat{R}) \leq \dim \widehat{R}/(\mathfrak{B}_1 + \mathfrak{B}_2) \leq \dim \widehat{R}/(\mathfrak{a}_1\widehat{R} + \mathfrak{a}_2\widehat{R}) = c$.

(ii) Suppose that $\mathfrak{p}\widehat{R} \in \operatorname{Spec}(\widehat{R})$ for all $\mathfrak{p} \in \mathcal{P}$. By 19.2.5, we can find two non-empty subsets $\mathcal{Q}_3$ and $\mathcal{Q}_4$ of $\mathcal{Q}$ such that $\mathcal{Q}_3 \cup \mathcal{Q}_4 = \mathcal{Q}$ and, if $\mathfrak{B}_i = \bigcap_{\mathfrak{Q}\in\mathcal{Q}_i}\mathfrak{Q}$ for $i = 3, 4$, then $\widehat{c} := c(\widehat{R}) = \dim \widehat{R}/(\mathfrak{B}_3 + \mathfrak{B}_4)$.

Set $\mathcal{P}_i := \{\mathfrak{Q} \cap R : \mathfrak{Q} \in \mathcal{Q}_i\}$ and $\mathfrak{a}_i = \bigcap_{\mathfrak{p}\in\mathcal{P}_i}\mathfrak{p}$ for $i = 3, 4$. Since each $\mathfrak{p} \in \mathcal{P}$ is the contraction to R of some minimal prime ideal of $\widehat{R}$, we have $\mathcal{P}_3 \cup \mathcal{P}_4 = \mathcal{P}$. In particular, we have $(\mathfrak{Q} \cap R)\widehat{R} = \mathfrak{Q}$ for all $\mathfrak{Q} \in \mathcal{Q}$, and so we can use [50, Theorem 7.4(ii)] to see that $\mathfrak{a}_i\widehat{R} = \mathfrak{B}_i$ for $i = 3, 4$. Hence, by 19.2.5,

$$\widehat{c} = \dim \widehat{R}/(\mathfrak{B}_3 + \mathfrak{B}_4) = \dim \widehat{R}/(\mathfrak{a}_3\widehat{R} + \mathfrak{a}_4\widehat{R}) = \dim R/(\mathfrak{a}_3 + \mathfrak{a}_4) \geq c(R).$$

The claim follows from this and part (i).

(iii) Let $\mathfrak{p} \in \mathcal{P}$ be such that $\dim R/\mathfrak{p} = \operatorname{sdim} R$. Now there exists $\mathfrak{Q} \in \mathcal{Q}$ such that $\mathfrak{Q} \cap R = \mathfrak{p}$. Then

$$\operatorname{sdim} \widehat{R} \leq \dim \widehat{R}/\mathfrak{Q} \leq \dim \widehat{R}/\mathfrak{p}\widehat{R} = \dim R/\mathfrak{p} = \operatorname{sdim} R.$$

(iv) Assume now that R is universally catenary. Let $\mathfrak{Q} \in \mathcal{Q}$ be such that $\operatorname{sdim} \widehat{R} = \dim \widehat{R}/\mathfrak{Q}$. Now $\mathfrak{p} := \mathfrak{Q} \cap R \in \mathcal{P}$; by Ratliff's Theorem [50, Theorem 31.7], we have $\dim \widehat{R}/\mathfrak{Q} = \dim R/\mathfrak{p}$. Hence

$$\operatorname{sdim} \widehat{R} = \dim \widehat{R}/\mathfrak{Q} = \dim R/\mathfrak{p} \geq \operatorname{sdim} R.$$

The claim follows from this and part (iii). $\qquad\square$

19.3.2 Definitions. Let T be a Noetherian topological space, and let $p \in T$. The *local dimension of T at p*, denoted by $\dim_p T$, is defined as the supremum of the lengths of all strictly descending chains $Z_0 \supset Z_1 \supset \cdots \supset Z_n$ of closed irreducible subsets of T which all contain p if this supremum exists, and ∞ otherwise. Thus, by 19.1.3, $\dim_p T$ is a non-negative integer or ∞. Note that, for $\mathfrak{p} \in \operatorname{Spec}(R)$, we have $\dim_{\mathfrak{p}} \operatorname{Spec}(R) = \dim R_{\mathfrak{p}}$.

Let V be an affine variety over the algebraically closed field K. Since, in an integral domain R' which is a finitely generated K-algebra, every maximal ideal has height equal to $\dim R'$ (see, for example, [81, 14.33]), it follows that $\dim_q V = \dim \mathcal{O}_{V,q} = \dim \mathcal{O}(V) = \dim V$ for all $q \in V$.

If $T_1, \ldots, T_r$ are the irreducible components of T which contain p, we write $T^p := T_1 \cup \cdots \cup T_r$, and call this subspace T^p of T the *p-component of T*. The *local connectedness dimension of T at p*, denoted by $c_p(T)$, is defined to be the minimum of the local dimensions at p of those closed subsets Z of the p-component T^p of T which contain p and for which $T^p \setminus Z$ is disconnected. Thus

$$c_p(T) := \min \left\{ \dim_p Z : Z \subseteq T^p,\ p \in Z,\ Z \text{ is closed} \right.$$
$$\left. \text{and } T^p \setminus Z \text{ is disconnected} \right\}.$$

The *local subdimension* $\operatorname{sdim}_p T$ *of T at p* is defined as the minimum of the local dimensions at p of the irreducible components of T which contain p. (Of course, p does belong to at least one irreducible component of T, by 19.1.3.)

19.3.3 ♯Exercise. Let T be a Noetherian topological space and let $p \in T$.

(i) Show that $\dim_p Z = \dim_p(T^p \cap Z)$ and $\operatorname{sdim}_p Z = \operatorname{sdim}_p(T^p \cap Z)$ for each closed subset Z of T for which $p \in Z$.

(ii) Let $T_1, \ldots, T_r$ be the (distinct) irreducible components of T that contain p. Show that, with the notation of 19.1.14,

$$c_p(T) = \min \left\{ \dim_p\left(\left(\bigcup_{i \in A} T_i \right) \cap \left(\bigcup_{j \in B} T_j \right) \right) : (A, B) \in \mathcal{S}(r) \right\}.$$

(Use the ideas of the proof of Theorem 19.1.15.)

The next exercise provides justification for the appearance of the word 'local' in the definitions in 19.3.2.

19.3.4 ♯Exercise. Let T be a Noetherian topological space and let $p \in T$; let U be an open subset of T which contains p. Show that

(i) $U^p = T^p \cap U$;
(ii) $\dim_p T = \dim_p U$;
(iii) $\mathrm{sdim}_p T = \mathrm{sdim}_p U$; and
(iv) $c_p(T) = c_p(U)$.

(Use Exercise 19.1.2(i),(iii),(v) to establish the existence of a bijection between the set of irreducible closed subsets of T containing p and the set of irreducible closed subsets of U containing p; then use Exercise 19.3.3(ii).)

19.3.5 Exercise. Let $p \in V := V_{\mathbb{A}^3}(X_1 X_2 X_3)$. Calculate $c_p(V)$

(i) when $p = (0, 0, 0)$;
(ii) when exactly two of the co-ordinates of p are 0; and
(iii) in all other cases.

19.3.6 ♯Exercise. Let $\mathfrak{p} \in \mathrm{Spec}(R)$. Show that $c_{\mathfrak{p}}(\mathrm{Spec}(R)) = c(R_{\mathfrak{p}})$ and $\mathrm{sdim}_{\mathfrak{p}}(\mathrm{Spec}(R)) = \mathrm{sdim}(R_{\mathfrak{p}})$.

We wish to study the connectivity of varieties over an algebraically closed field. We remind the reader about some elementary facts concerning such varieties.

19.3.7 Reminders. Let K be an algebraically closed field, and let $r \in \mathbb{N}$. Let $V \subseteq \mathbb{P}^r(K)$ be a quasi-projective variety.

(i) Regard the polynomial ring $K[X_0, X_1, \ldots, X_r]$ as the coordinate ring $\mathcal{O}(\mathbb{A}^{r+1}(K))$, and let $i \in \{0, \ldots, r\}$. Let $U_i \mathbb{P}^r(K)$ denote the open subset of $\mathbb{P}^r(K)$ given by

$$U_i \mathbb{P}^r(K) = \mathbb{P}^r(K) \setminus V_{\mathbb{P}^r(K)}(X_i)$$
$$= \{(c_0 : \cdots : c_i : \cdots : c_r) \in \mathbb{P}^r(K) : c_i \neq 0\}.$$

There is an isomorphism of varieties $\sigma_i : \mathbb{A}^r(K) \xrightarrow{\cong} U_i \mathbb{P}^r(K)$ given by $\sigma_i((a_1, \ldots, a_r)) = (a_1 : \cdots : a_i : 1 : a_{i+1} : \cdots : a_r)$ for all $(a_1, \ldots, a_r) \in \mathbb{A}^r(K)$. Note that $\mathbb{P}^r(K) = \bigcup_{j=0}^r U_j \mathbb{P}^r(K)$; it follows that the quasi-projective variety $V \subseteq \mathbb{P}^r(K)$ has a finite covering by open sets each of which is *quasi-affine* (in the sense that it is isomorphic to a quasi-affine variety over K). As any quasi-affine variety has a finite covering by open sets which are affine, we thus see that quasi-projective varieties also have finite coverings by affine open sets.

The above isomorphisms show that affine varieties, and also quasi-affine varieties, are quasi-projective: the reader should remember that, for us, the word 'variety', as introduced in 15.2.1(ii), is synonymous with 'quasi-projective variety' (and does *not* mean the same as '(abstract) variety' in the sense of Hartshorne [30, p. 105]).

(ii) Let $p \in V$. By part (i), there is an open subset U of V such that $p \in U$ and U is an affine variety. We can, and do, identify the local ring $\mathcal{O}_{V,p}$ of p on V with $\mathcal{O}_{U,p} = \mathcal{O}(U)_{I_U(p)}$ (see 6.4.1). Let C be a closed subset of V such that $p \in C$. The vanishing ideal $I_U(C \cap U)$ was defined in 15.1.2. Recall that the *local vanishing ideal* $I_{V,p}(C)$ *of* C *at* p is the (radical) ideal of $\mathcal{O}_{V,p}$ consisting of all germs of regular functions $f \in \mathcal{O}(W)$ defined on some open neighbourhood W of p in V and such that $f(W \cap C) = 0$; hence

$$I_{V,p}(C) = I_{U,p}(C \cap U) = I_U(C \cap U)\mathcal{O}(U)_{I_U(p)} = I_U(C \cap U)\mathcal{O}_{U,p}.$$

19.3.8 ♯Exercise. Let K be an algebraically closed field.

(i) Let V be an affine variety over K. Let $\mathfrak{b}$ be a proper ideal of $\mathcal{O}(V)$, and let $V(\mathfrak{b})$ denote the closed subset of V determined by $\mathfrak{b}$; let $p \in V(\mathfrak{b})$. Show that $c_p(V(\mathfrak{b})) = c(\mathcal{O}_{V,p}/\mathfrak{b}\mathcal{O}_{V,p})$ and

$$\mathrm{sdim}_p V(\mathfrak{b}) = \mathrm{sdim}(\mathcal{O}_{V,p}/\mathfrak{b}\mathcal{O}_{V,p})$$
$$= \min\{\dim \mathcal{O}(V)/\mathfrak{q} : \mathfrak{q} \text{ is a minimal prime of } \mathfrak{b}$$
$$\text{such that } p \in V(\mathfrak{q})\}.$$

(ii) Let V' be a variety over K. Let W be a non-empty closed subset of V', and let $p \in W$. Show that, with the notation of 19.3.7(ii), $\mathrm{sdim}_p W = \mathrm{sdim}(\mathcal{O}_{V',p}/I_{V',p}(W))$ and $c_p(W) = c(\mathcal{O}_{V',p}/I_{V',p}(W))$. (Use part (i) and 19.3.4.)

19.3.9 Definition. Let V be a variety over the algebraically closed field K. Let W be a non-empty closed subset of V, and let $p \in W$. We define the *formal connectedness dimension of* W *at* p, denoted $\widehat{c}_p(W)$, by

$$\widehat{c}_p(W) = c\left(\widehat{\mathcal{O}_{V,p}/I_{V,p}(W)\mathcal{O}_{V,p}}\right) = c\left((\mathcal{O}_{V,p}/I_{V,p}(W))^{\widehat{}}\right).$$

Here again, $I_{V,p}(W)$ is the local vanishing ideal of W at p: see 19.3.7(ii).

19.3.10 Remark. Let V, W and p be as in 19.3.9. Then it follows from 19.3.1(i) and 19.3.8(ii) that $c_p(W) \geq \widehat{c}_p(W)$. Moreover, if V is affine and $\mathfrak{b}$ is any ideal of $\mathcal{O}(V)$ for which $W = V(\mathfrak{b})$, then $\widehat{c}_p(W) = c\left((\mathcal{O}_{V,p}/\mathfrak{b}\mathcal{O}_{V,p})^{\widehat{\ }}\right)$.

19.3.11 Exercise. Let

$$V_1 := V_{\mathbb{A}^2}(X^3 - Y^2) \quad \text{and} \quad V_2 := V_{\mathbb{A}^2}(X^2 + X^3 - Y^2);$$

also let

$$V := V_{\mathbb{A}^4}(X_1 X_4 - X_2 X_3, \ X_1^2 X_3 + X_1 X_2 - X_2^2, \ X_3^3 + X_3 X_4 - X_4^2),$$

as in Example 2.3.7.

   (i) Calculate $c_{(0,0)}(V_i)$ and $\widehat{c}_{(0,0)}(V_i)$ for $i = 1, 2$.
  (ii) Show that $\widehat{c}_{(0,0,0,0)}(V) = 0$. (You might find 8.2.13(iv) and 19.1.11 helpful.)

19.3.12 Definition. Let V be a variety over the algebraically closed field K. Let W and Z be closed subsets of V with $Z \subseteq W$, and let $p \in Z$. Note that $I_{V,p}(W) \subseteq I_{V,p}(Z)$.

We define the *(local) arithmetic rank of Z at p with respect to W*, denoted $\mathrm{ara}_{W,p}(Z)$, by

$$\mathrm{ara}_{W,p}(Z) = \mathrm{ara}\left(I_{V,p}(Z)/I_{V,p}(W)\right).$$

(Here, $I_{V,p}(Z)/I_{V,p}(W)$ is considered as an ideal of $\mathcal{O}_{V,p}/I_{V,p}(W)$.)

19.3.13 Definition. Let V be a variety over the algebraically closed field K. Let W be a non-empty closed subset of V, and let $p \in W$. We extend the terminology of 8.2.15 and say that W is *analytically reducible at p* precisely when $\widehat{\mathcal{O}_{V,p}}/I_{V,p}(W)\widehat{\mathcal{O}_{V,p}}$ has more than one minimal prime ideal. Otherwise, W is said to be *analytically irreducible at p*. Note that $\widehat{\mathcal{O}_{V,p}}/I_{V,p}(W)\widehat{\mathcal{O}_{V,p}} \cong (\mathcal{O}_{V,p}/I_{V,p}(W))^{\widehat{\ }}$, and by [50, Theorem 32.2(i) and p. 259, Remark 1], W is analytically irreducible at p if and only if $(\mathcal{O}_{V,p}/I_{V,p}(W))^{\widehat{\ }}$ is a domain.

19.3.14 Remark. Note that, in the situation of 19.3.9, 19.3.12 and 19.3.13, the invariants $\widehat{c}_p(W)$ and $\mathrm{ara}_{W,p}(Z)$, and the notion (for W) of analytical reducibility at p, do not depend on the ambient variety V, and, indeed, remain unchanged if V, W and Z are replaced, respectively, by U, $W \cap U$ and $Z \cap U$, where U is any open subset of V containing p.

We can now deduce the following from 19.2.10 and 19.2.11.

19.3.15 Proposition. *Let V be a variety over the algebraically closed field K. Let W and Z be closed subsets of V with $Z \subseteq W$, and let $p \in Z$. Then*

(i) $\widehat{c}_p(Z) \geq \min \{\widehat{c}_p(W), \operatorname{sdim}_p W - 1\} - \operatorname{ara}_{W,p}(Z)$;

(ii) $\widehat{c}_p(Z) \geq \widehat{c}_p(W) - \operatorname{ara}_{W,p}(Z) - 1$; *and*

(iii) *if W is analytically reducible at p, then the inequality in part* (ii) *is strict.*

Proof. Let $R := \mathcal{O}_{V,p}/I_{V,p}(W)$ and $\mathfrak{a} := I_{V,p}(Z)/I_{V,p}(W)$. Then

$$\widehat{c}_p(W) = c(\widehat{R}) \quad \text{and} \quad \widehat{c}_p(Z) = c\left(\widehat{\mathcal{O}_{V,p}}\Big/ I_{V,p}(Z)\widehat{\mathcal{O}_{V,p}}\right) = c(\widehat{R}/\mathfrak{a}\widehat{R}).$$

Next, $\operatorname{ara}_{W,p}(Z) = \operatorname{ara}(\mathfrak{a})$ (see 19.3.12), and so it follows from 19.2.7(ii) that $\operatorname{ara}_{W,p}(Z) \geq \operatorname{ara}(\mathfrak{a}\widehat{R})$. Also, $\operatorname{sdim}_p W = \operatorname{sdim} R = \operatorname{sdim} \widehat{R}$, by 19.3.8 and 19.3.1(iv).

We can now apply the Connectedness Bound for Complete Local Rings 19.2.10 to prove part (i), and 19.2.11 to prove parts (ii) and (iii). $\qquad\square$

19.3.16 Exercise. Let V be the affine variety in $\mathbb{A}^4$ of Example 2.3.7 given by

$$V := V_{\mathbb{A}^4}(X_1X_4 - X_2X_3, \; X_1^2X_3 + X_1X_2 - X_2^2, \; X_3^3 + X_3X_4 - X_4^2).$$

Use 19.3.11(ii) and 19.3.15 to show that $\operatorname{ara}_{\mathbb{A}^4,(0,0,0,0)}(V) \geq 3$.

It seems natural to ask whether, if the formal connectedness dimensions in the inequality of 19.3.15(ii) are replaced by the corresponding local connectedness dimensions, the resulting statement is still true. We shall now provide an example which shows that this is not always the case.

19.3.17 Example. Let R be the subring of $R' := \mathbb{C}[X_1, X_2, X_3] = \mathcal{O}(\mathbb{A}^3)$ given by $R := \mathbb{C}[X_1, X_2, X_1X_3, X_2X_3, X_3^2 - 1, X_3(X_3^2 - 1)]$.

Let $Y_1, \ldots, Y_6$ be independent indeterminates over $\mathbb{C}$, and let

$$f : \mathcal{O}(\mathbb{A}^6) = \mathbb{C}[Y_1, Y_2, Y_3, Y_4, Y_5, Y_6] \longrightarrow R$$

be the $\mathbb{C}$-algebra homomorphism such that $f(Y_1) = X_1$, $f(Y_2) = X_2$, $f(Y_3) = X_1X_3$, $f(Y_4) = X_2X_3$, $f(Y_5) = X_3^2 - 1$ and $f(Y_6) = X_3(X_3^2 - 1)$. Then $\mathfrak{p} := \operatorname{Ker} f$ is a prime ideal of $\mathcal{O}(\mathbb{A}^6)$ (since R is an integral domain); let $V := V_{\mathbb{A}^6}(\mathfrak{p})$ denote the affine variety determined by $\mathfrak{p}$, so that there is a natural isomorphism of $\mathbb{C}$-algebras $\mathcal{O}(V) \cong R$.

The inclusion mapping $R \to R' = \mathcal{O}(\mathbb{A}^3)$, which makes R' integral over its subring R, therefore gives rise to a finite morphism of varieties $\alpha : \mathbb{A}^3 \to V$ such that

$$\alpha((c_1, c_2, c_3)) = (c_1, c_2, c_1c_3, c_2c_3, c_3^2 - 1, c_3(c_3^2 - 1))$$

for all $(c_1, c_2, c_3) \in \mathbb{A}^3$. Let $p = (0, 0, 1), q = (0, 0, -1) \in \mathbb{A}^3$, and let

0 denote $(0,0,0,0,0,0) \in \mathbb{A}^6$. Now it is straightforward to check that $\alpha\lceil$: $\mathbb{A}^3 \setminus \{p,q\} \longrightarrow V \setminus \{0\}$ is an isomorphism of (quasi-affine) varieties, with inverse $\beta : V \setminus \{0\} \longrightarrow \mathbb{A}^3 \setminus \{p,q\}$ given by

$$\beta((d_1, d_2, d_3, d_4, d_5, d_6)) = \begin{cases} (d_1, d_2, d_3/d_1) & \text{if } d_1 \neq 0, \\ (d_1, d_2, d_4/d_2) & \text{if } d_2 \neq 0, \\ (d_1, d_2, d_6/d_5) & \text{if } d_5 \neq 0 \end{cases}$$

(for all $(d_1, d_2, d_3, d_4, d_5, d_6) \in V \setminus \{0\}$).

Let $W := V \cap V_{\mathbb{A}^6}(Y_5)$, a closed subset of V such that $0 \in W$. Note that $\operatorname{ara}_{V,0}(W) = \operatorname{ara}(I_V(W)\mathcal{O}_{V,0}) = 1$.

Let $E_1 := V_{\mathbb{A}^3}(X_3 - 1)$ and $E_2 := V_{\mathbb{A}^3}(X_3 + 1)$. Since α is finite, it is a closed map, and so $\alpha(E_1)$ and $\alpha(E_2)$ are closed subsets of W and we have $\dim(\alpha(E_i)) = \dim E_i = 2$ for $i = 1, 2$.

Since $\alpha^{-1}(W) = V_{\mathbb{A}^3}(X_3^2 - 1) = E_1 \cup E_2$, it follows that W can be expressed as $W = \alpha(E_1) \cup \alpha(E_2)$, where $\alpha(E_1)$ and $\alpha(E_2)$ are closed irreducible subsets of dimension 2. Therefore $\alpha(E_1)$ and $\alpha(E_2)$ must be the irreducible components of W. Hence, by 19.3.3(ii), we have

$$c_0(W) = \dim_0(\alpha(E_1) \cap \alpha(E_2)) = \dim_0\{0\} = 0.$$

On the other hand, since V is irreducible, $c_0(V) = \dim_0 V = 3$. We therefore have the strict inequality

$$c_0(W) = 0 < 1 = 3 - 1 - 1 = c_0(V) - \operatorname{ara}_{V,0}(W) - 1.$$

Thus, if the formal connectedness dimensions in the inequality of 19.3.15(ii) are replaced by the corresponding local connectedness dimensions, the resulting statement is not always true.

19.4 Connectivity of affine algebraic cones

Connectedness dimensions of affine algebraic cones behave particularly satisfactorily, and the next two lemmas provide the key to this good behaviour. We shall use the notation of 15.1.2 for affine algebraic cones.

19.4.1 Lemma. *Let K be an algebraically closed field, let $r \in \mathbb{N}$, and let $C \subseteq \mathbb{A}^r(K)$ be an affine algebraic cone. Then C is irreducible (that is, C is a variety) if and only if C is analytically irreducible at 0.*

Proof. $(\Rightarrow)$ Assume that C is irreducible. Then, by 15.1.2(ii), the ring $\mathcal{O}(C)$

is a positively graded *local homogeneous domain with $\mathcal{O}(C)_0 = K$; furthermore, $\mathcal{O}(C)_+$ is the unique *maximal graded ideal of $\mathcal{O}(C)$, and is actually maximal and equal to $I_C(0)$. Hence $\mathcal{O}_{C,0} \cong \mathcal{O}(C)_{\mathcal{O}(C)_+}$. Therefore, by 14.1.15(iv),(v), the completion $\widehat{\mathcal{O}_{C,0}}$ is a domain, and so C is analytically irreducible at 0.

($\Leftarrow$) Assume that C is reducible, so that C has more than one irreducible component. By 19.1.4(i), these irreducible components all contain 0. This means that the local vanishing ideal $I_{\mathbb{A}^r(K),0}(C)$ of C at 0 has more than one minimal prime. Hence $\mathcal{O}_{C,0}$ is not a domain, and neither is its completion. $\quad\square$

19.4.2 Lemma. *Let V be an affine variety over the algebraically closed field K, let $W \subseteq V$ be a closed subset, let $p \in W$, and assume that all the irreducible components of W which contain p are analytically irreducible at p. Then $\widehat{c}_p(W) = c_p(W)$.*

Proof. Let $\mathfrak{b} := I_V(W)$ and $R := \mathcal{O}_{V,p}/\mathfrak{b}\mathcal{O}_{V,p}$. By 19.3.8(ii) and 19.3.10, it is enough for us to show that $c(R) = c(\widehat{R})$; therefore, by 19.3.1(ii), it is enough for us to show that, for each minimal prime ideal $\mathfrak{p}$ of R, we have $\mathfrak{p}\widehat{R} \in \mathrm{Spec}(\widehat{R})$, and this is what we shall do. Now $\mathfrak{p} = \mathfrak{q}\mathcal{O}_{V,p}/\mathfrak{b}\mathcal{O}_{V,p}$ for some minimal prime ideal $\mathfrak{q}$ of $\mathfrak{b}$ such that $\mathfrak{q} \subseteq I_V(p)$. But then $Z := V(\mathfrak{q})$, the closed subset of V determined by $\mathfrak{q}$, is an irreducible component of W which contains p. By hypothesis, Z is analytically irreducible at p, and so, by 19.3.13, the ring $(\mathcal{O}_{V,p}/I_{V,p}(Z))\widehat{} = (\mathcal{O}_{V,p}/\mathfrak{q}\mathcal{O}_{V,p})\widehat{}$ is a domain. Hence $(R/\mathfrak{p})\widehat{}$ is a domain, so that $\mathfrak{p}\widehat{R} \in \mathrm{Spec}(\widehat{R})$. $\quad\square$

The next lemma establishes a very useful fact about the connectedness dimensions of an affine algebraic cone C: the formal connectedness dimension of C at the origin and the local connectedness dimension of C at the origin are equal, and they are both equal to the connectedness dimension $c(C)$.

19.4.3 Proposition. *Let K be an algebraically closed field, let $r \in \mathbb{N}$, and let $C \subseteq \mathbb{A}^r(K)$ be an affine algebraic cone. As in 15.1.2, we use 0 to denote the origin of $\mathbb{A}^r(K)$. Then*

(i) $\dim C = \dim_0 C$,
(ii) $\mathrm{sdim}\, C = \mathrm{sdim}_0 C$, *and*
(iii) $c(C) = c_0(C) = \widehat{c}_0(C)$.

Proof. By 19.1.4(i), all the irreducible components of C are themselves affine algebraic cones in $\mathbb{A}^r(K)$, and so contain 0. Now the dimension of an irreducible affine algebraic cone C' satisfies $\dim C' = \dim_0 C'$ (by 19.3.2). The claims in parts (i) and (ii) now follow immediately, while the equality $c(C) = c_0(C)$ follows from these considerations, 19.1.15 and 19.3.3(ii).

Finally, Lemma 19.4.1 shows that all the irreducible components of C are analytically irreducible at 0, and so Lemma 19.4.2 shows that $c_0(C) = \widehat{c}_0(C)$. $\square$

We can now deduce the following corollary from 19.3.15.

19.4.4 Corollary. *Let K be an algebraically closed field, let $r \in \mathbb{N}$, let $D, E \subseteq \mathbb{A}^r(K)$ be affine algebraic cones such that $E \subseteq D$. Then*

(i) $c(E) \geq \min\{c(D), \operatorname{sdim} D - 1\} - \operatorname{ara}_{D,0}(E)$;

(ii) $c(E) \geq c(D) - \operatorname{ara}_{D,0}(E) - 1$; *and*

(iii) *if D is reducible, then the inequality in part* (ii) *is strict.*

Proof. In view of Proposition 19.4.3, statements (i) and (ii) follow immediately from the corresponding statements of 19.3.15 (used with $V = \mathbb{A}^r(K)$, $W = D$ and $Z = E$). Statement (iii) follows from 19.3.15(iii) and Lemma 19.4.1. $\square$

19.5 Connectivity of projective varieties

In view of the close relationship between affine algebraic cones and projective algebraic sets (see 15.2.1(iii)), we can exploit 19.4.4 to study the connectivity of closed sets in projective varieties. We intend to do this, but first we need a few preliminaries.

19.5.1 Lemma. *Let K be an algebraically closed field, let $r \in \mathbb{N}$, let $W \subseteq \mathbb{P}^r(K)$ be a non-empty closed subset of $\mathbb{P}^r(K)$, and consider the affine cone $\operatorname{Cone}(W) \subseteq \mathbb{A}^{r+1}(K)$ over W, as in 15.2.1(iii). Then*

(i) $\dim W = \dim(\operatorname{Cone}(W)) - 1 = \dim_0(\operatorname{Cone}(W)) - 1$;

(ii) $\operatorname{sdim} W = \operatorname{sdim}(\operatorname{Cone}(W)) - 1 = \operatorname{sdim}_0(\operatorname{Cone}(W)) - 1$; *and*

(iii) $c(W) = c(\operatorname{Cone}(W)) - 1 = c_0(\operatorname{Cone}(W)) - 1 = \widehat{c}_0(\operatorname{Cone}(W)) - 1$.

Proof. Let the distinct irreducible components of W be $W_1, \ldots, W_n$. By 19.1.4(ii), $\operatorname{Cone}(W_1), \ldots, \operatorname{Cone}(W_n)$ are the irreducible components (again distinct) of $\operatorname{Cone}(W)$. Furthermore, it follows from 15.2.1(vii) that

$$\dim W_i = \dim(\operatorname{Cone}(W_i)) - 1 \quad \text{for } i = 1, \ldots, n.$$

In view of 19.4.3, the claims in statements (i) and (ii) are now immediate.

(iii) With the notation of 19.1.14, let $(A, B) \in \mathcal{S}(r)$. By 15.2.1(v), we have

$$\operatorname{Cone}\left(\left(\bigcup_{i \in A} W_i\right) \cap \left(\bigcup_{j \in B} W_j\right)\right)$$
$$= \left(\bigcup_{i \in A} \operatorname{Cone}(W_i)\right) \cap \left(\bigcup_{j \in B} \operatorname{Cone}(W_j)\right).$$

Since $\mathrm{Cone}(W_i)$ $(i = 1, \ldots, n)$ are the irreducible components of $\mathrm{Cone}(W)$, it now follows from part (i) and 19.1.15 that $c(W) = c(\mathrm{Cone}(W)) - 1$. Finally, we can use 19.4.3 to complete the proof. $\qquad\square$

19.5.2 Definition. Let K be an algebraically closed field and let $r \in \mathbb{N}$. Let W and Z be non-empty closed subsets of $\mathbb{P}^r(K)$ with $Z \subseteq W$. Note that $I_{\mathbb{A}^{r+1}(K)}(\mathrm{Cone}(W)) \subseteq I_{\mathbb{A}^{r+1}(K)}(\mathrm{Cone}(Z))$, and that

$$I_{\mathbb{A}^{r+1}(K)}(\mathrm{Cone}(Z))/I_{\mathbb{A}^{r+1}(K)}(\mathrm{Cone}(W))$$

is an ideal of $\mathcal{O}(\mathbb{A}^{r+1}(K))/I_{\mathbb{A}^{r+1}(K)}(\mathrm{Cone}(W))$.

We define the *arithmetic rank of Z with respect to W*, denoted $\mathrm{ara}_W(Z)$, by

$$\mathrm{ara}_W(Z) = \mathrm{ara}\left(I_{\mathbb{A}^{r+1}(K)}(\mathrm{Cone}(Z))/I_{\mathbb{A}^{r+1}(K)}(\mathrm{Cone}(W))\right).$$

It should be noted that, by 19.2.7(ii) and 19.3.7(ii), we have

$$\mathrm{ara}_W(Z) \geq \mathrm{ara}_{\mathrm{Cone}(W),0}(\mathrm{Cone}(Z)).$$

We are now able to state and prove a form of the Bertini–Grothendieck Connectivity Theorem.

19.5.3 The Bertini–Grothendieck Connectivity Theorem. (See [26, Exposé XIII, Corollaire 2.3].) *Let K be an algebraically closed field and let $r \in \mathbb{N}$. Let W and Z be non-empty closed subsets of $\mathbb{P}^r(K)$ with $Z \subseteq W$. Then*

(i) $c(Z) \geq \min\{c(W), \mathrm{sdim}\, W - 1\} - \mathrm{ara}_W(Z)$;
(ii) $c(Z) \geq c(W) - \mathrm{ara}_W(Z) - 1$; *and*
(iii) *if W is reducible, then the inequality in part* (ii) *is strict.*

Proof. By Lemma 19.5.1, we have $\mathrm{sdim}\, W = \mathrm{sdim}(\mathrm{Cone}(W)) - 1$,

$$c(W) = c(\mathrm{Cone}(W)) - 1 \quad \text{and} \quad c(Z) = c(\mathrm{Cone}(Z)) - 1.$$

Moreover, $\mathrm{ara}_W(Z) \geq \mathrm{ara}_{\mathrm{Cone}(W),0}(\mathrm{Cone}(Z))$ by 19.5.2, while 15.2.1(v) shows that $\mathrm{Cone}(W)$ is reducible if W is reducible. Therefore, all three statements follow from the corresponding statements of Corollary 19.4.4. $\qquad\square$

Let K be an algebraically closed field and let $r \in \mathbb{N}$. Recall that a *hypersurface* in $\mathbb{P}^r(K)$ is a closed set $V_{\mathbb{P}^r(K)}(f)$ defined by a single homogeneous polynomial $f \in K[X_0, X_1, \ldots, X_r]$ of positive degree. We can now deduce the following corollary from the Bertini–Grothendieck Connectivity Theorem 19.5.3.

19.5.4 Corollary. *Let K be an algebraically closed field, let $r \in \mathbb{N}$, let W be a non-empty closed subset of $\mathbb{P}^r(K)$, and let $H_1, \ldots, H_t \subseteq \mathbb{P}^r(K)$ be hypersurfaces. Then*

(i) $c(W \cap H_1 \cap \cdots \cap H_t) \geq \min\{c(W), \operatorname{sdim} W - 1\} - t;$
(ii) $c(W \cap H_1 \cap \cdots \cap H_t) \geq c(W) - t - 1;$ *and*
(iii) *if W is reducible, then the inequality in part* (ii) *is strict.*

Proof. For $i = 1, \ldots, t$, there is a homogeneous polynomial (of positive degree) $f_i \in K[X_0, X_1, \ldots, X_r]$ such that $H_i = V_{\mathbb{P}^r(K)}(f_i)$. Set

$$Z := W \cap H_1 \cap \cdots \cap H_t = V_{\mathbb{P}^r(K)}\left(I_{\mathbb{P}^r(K)}(W) + (f_1, \ldots, f_t)\right),$$

so that $I_{\mathbb{P}^r(K)}(Z) = \sqrt{I_{\mathbb{P}^r(K)}(W) + (f_1, \ldots, f_t)}$. This equation shows that $\operatorname{ara}_W(Z) = \operatorname{ara}\left(I_{\mathbb{P}^r(K)}(Z)/I_{\mathbb{P}^r(K)}(W)\right) \leq t$. The claims now follow from application of 19.5.3. $\qquad\square$

19.5.5 Exercise. Let K be an algebraically closed field and let $r \in \mathbb{N}$.

(i) Prove the 'classical' form of Bertini's Connectivity Theorem, that, if $V \subseteq \mathbb{P}^r(K)$ is a projective variety such that $\dim V > 1$ and $H \subseteq \mathbb{P}^r(K)$ is a hypersurface, then $V \cap H$ is connected.

(ii) Provide an example which shows that if the irreducibility of V is dropped from the statement in part (i) above, then the resulting statement is no longer always true.

(iii) Provide an example of an affine variety $V \subseteq \mathbb{A}^r(K)$ with $\dim V > 1$ such that $V \cap H$ is disconnected for a hyperplane $H \subseteq \mathbb{A}^r(K)$.

19.6 Connectivity of intersections

Our next aim is the study of the connectivity of the intersection of two affine algebraic cones. For this, we recall, in 19.6.1 and 19.6.2 below, some elementary facts about products of affine algebraic sets.

19.6.1 Reminder and Remark. Let K be an algebraically closed field and let $r, s \in \mathbb{N}$. We consider polynomial rings $K[X_1, \ldots, X_r] = \mathcal{O}(\mathbb{A}^r(K))$, $K[Y_1, \ldots, Y_s] = \mathcal{O}(\mathbb{A}^s(K))$ and

$$K[X_1, \ldots, X_r; Y_1, \ldots, Y_s] = \mathcal{O}(\mathbb{A}^{r+s}(K)).$$

Let $\mathfrak{a}$ be an ideal of $K[X_1, \ldots, X_r]$ and $\mathfrak{b}$ be an ideal of $K[Y_1, \ldots, Y_s]$, and set $V := V_{\mathbb{A}^r(K)}(\mathfrak{a})$ and $W := V_{\mathbb{A}^s(K)}(\mathfrak{b})$. Recall that the *product of V and*

W is just the Cartesian product $V \times W \subseteq \mathbb{A}^{r+s}(K)$; it is an affine algebraic set because

$$V \times W$$
$$= V_{\mathbb{A}^{r+s}(K)}(\mathfrak{a}K[X_1, \ldots, X_r; Y_1, \ldots, Y_s] + \mathfrak{b}K[X_1, \ldots, X_r; Y_1, \ldots, Y_s]).$$

(i) Recall from Hartshorne [30, Chapter I, Exercise 3.15] that, when $V \subseteq \mathbb{A}^r(K)$ and $W \subseteq \mathbb{A}^s(K)$ are irreducible, then $V \times W \subseteq \mathbb{A}^{r+s}(K)$ is again irreducible and, moreover, $\dim(V \times W) = \dim V + \dim W$.

(ii) Now suppose $r = s$. Let $\Delta^{(r)}$ be the *diagonal*

$$\{(c, c) : c \in \mathbb{A}^r(K)\} \subseteq \mathbb{A}^{2r}(K).$$

Note that $\Delta^{(r)} = V_{\mathbb{A}^{2r}(K)}(X_1 - Y_1, \ldots, X_r - Y_r)$, so that $\Delta^{(r)}$ is irreducible, and also that there is the *diagonal isomorphism* of varieties $\delta^{(r)} : \mathbb{A}^r(K) \longrightarrow \Delta^{(r)}$ for which $\delta^{(r)}(c) = (c, c)$ for all $c \in \mathbb{A}^r(K)$. If $V, W \subseteq \mathbb{A}^r(K)$ are closed subsets of $\mathbb{A}^r(K)$, then

$$\delta^{(r)}(V \cap W) = (V \times W) \cap \Delta^{(r)},$$

and $\delta^{(r)}$ gives rise to a homeomorphism

$$\delta^{(r)} \restriction : V \cap W \xrightarrow{\approx} (V \times W) \cap \Delta^{(r)}.$$

19.6.2 ♯Exercise. Let K be an algebraically closed field, let $r, s \in \mathbb{N}$, and let $V \subseteq \mathbb{A}^r(K)$ and $W \subseteq \mathbb{A}^s(K)$ be non-empty closed sets. Consider their product $V \times W \subseteq \mathbb{A}^{r+s}(K)$, as in 19.6.1.

(i) Show that, if V and W are affine algebraic cones, then $V \times W$ is again an affine algebraic cone.

(ii) Let $V_1, \ldots, V_p$ (respectively $W_1, \ldots, W_q$) be the distinct irreducible components of V (respectively W). Show that the products

$$V_i \times W_j \quad (i = 1, \ldots, p, \ j = 1, \ldots, q)$$

are the (distinct) irreducible components of $V \times W$.

(iii) Show that $\dim(V \times W) = \dim V + \dim W$ and $\operatorname{sdim}(V \times W) = \operatorname{sdim} V + \operatorname{sdim} W$.

19.6.3 Lemma. *Let K be an algebraically closed field, let $r, s \in \mathbb{N}$, and let $V \subseteq \mathbb{A}^r(K)$ and $W \subseteq \mathbb{A}^s(K)$ be non-empty closed sets. Then*

(i) $c(V \times W) \geq \min\{\operatorname{sdim} V + c(W), \operatorname{sdim} W + c(V)\}$;

(ii) $c(V \times W) \geq c(V) + c(W)$; *and*

(iii) *if V and W are both reducible, then the inequality in part* (ii) *is strict.*

Proof. Let $V_1, \ldots, V_p$ be the distinct irreducible components of V and let $W_1, \ldots, W_q$ be the distinct irreducible components of W. By 19.6.2(ii), the products $V_i \times W_j$ ($i = 1, \ldots, p$, $j = 1, \ldots, q$) are the (distinct) irreducible components of $V \times W$. By 19.1.15, there are two non-empty subsets $A, B \subseteq \{1, \ldots, p\} \times \{1, \ldots, q\}$ such that $A \cup B = \{1, \ldots, p\} \times \{1, \ldots, q\}$ and

$$c(V \times W) = \dim \left(\left(\bigcup_{(i,j) \in A} V_i \times W_j \right) \cap \left(\bigcup_{(k,l) \in B} V_k \times W_l \right) \right).$$

The argument now splits into two cases. Suppose first that there exists $i_0 \in \{1, \ldots, p\}$ such that there are indices $j, l \in \{1, \ldots, q\}$ for which $(i_0, j) \in A$ and $(i_0, l) \in B$. Then

$$\overline{A} := \{ j \in \mathbb{N} : (i_0, j) \in A \} \quad \text{and} \quad \overline{B} := \{ l \in \mathbb{N} : (i_0, l) \in B \}$$

are non-empty sets such that $\overline{A} \cup \overline{B} = \{1, \ldots, q\}$. Set

$$Z := \left(\bigcup_{j \in \overline{A}} W_j \right) \cap \left(\bigcup_{l \in \overline{B}} W_l \right).$$

By 19.1.15, $\dim Z \geq c(W)$. Let Y be an irreducible component of Z such that $\dim Y = \dim Z$. Then there exist $j \in \overline{A}$ and $l \in \overline{B}$ such that $Y \subseteq W_j$ and $Y \subseteq W_l$. Therefore $V_{i_0} \times Y \subseteq (V_{i_0} \times W_j) \cap (V_{i_0} \times W_l)$, and so (see 19.6.2(iii))

$$c(V \times W) \geq \dim ((V_{i_0} \times W_j) \cap (V_{i_0} \times W_l))$$
$$\geq \dim(V_{i_0} \times Y) = \dim V_{i_0} + \dim Y \geq \operatorname{sdim} V + c(W).$$

We now deal with the remaining case, when there is no index i_0 with the properties described above. Choose $i \in \{1, \ldots, p\}$ such that $(i, j') \in A$ for some $j' \in \{1, \ldots, q\}$. Then, for all $j \in \{1, \ldots, q\}$, we must have $(i, j) \notin B$, so that $(i, j) \in A$. But there is also a pair $(k, j_0) \in B$; thus $(i, j_0) \in A$ and $(k, j_0) \in B$, and we can use the argument of the previous paragraph, with the rôles of V and W interchanged, to deduce that $c(V \times W) \geq \operatorname{sdim} W + c(V)$. This proves statement (i).

Parts (ii) and (iii) are now immediate from part (i) and 19.2.2. $\square$

19.6.4 Exercise. Let $V' := V_{\mathbb{A}^2}(XY)$ and $W' := V_{\mathbb{A}^2}(X)$ (with the notation of 2.3.1).

(i) Calculate $\operatorname{sdim}(V' \times V')$ and $c(V' \times V')$.

(ii) Calculate $\operatorname{sdim}(V' \times W')$ and $c(V' \times W')$.

(iii) Show that it is possible for the inequality in 19.6.3(ii) to be an equality when just one of V and W is reducible.

19.6.5 Proposition: the Intersection Inequality for the Connectedness Dimensions of Affine Algebraic Cones. *Let K be an algebraically closed field, let $r \in \mathbb{N}$, and let $C, D \subseteq \mathbb{A}^r(K)$ be affine algebraic cones.*

(i) *We have*

$$c(C \cap D)$$
$$\geq \min \{\operatorname{sdim} C + \operatorname{sdim} D - 1, \operatorname{sdim} C + c(D), \operatorname{sdim} D + c(C)\} - r.$$

(ii) *Furthermore,*

$$c(C \cap D) \geq \min \{\operatorname{sdim} C + c(D), \operatorname{sdim} D + c(C)\} - r - 1,$$

with strict inequality if C or D is reducible.

(iii) *Consequently, $c(C \cap D) \geq c(C) + c(D) - r - 1 + \varepsilon$, where $\varepsilon = 0, 1$ or 2 according as none, one or both of C and D are reducible.*

Proof. By 19.6.2(i), the product $C \times D \subseteq \mathbb{A}^{2r}(K)$ is an affine algebraic cone, and by 19.6.2(iii) we have $\operatorname{sdim}(C \times D) = \operatorname{sdim} C + \operatorname{sdim} D$. Also, by 19.6.3, we have $c(C \times D) \geq \min \{\operatorname{sdim} C + c(D), \operatorname{sdim} D + c(C)\}$.

As in 19.6.1, write $K[X_1, \ldots, X_r; Y_1, \ldots, Y_r] = \mathcal{O}(\mathbb{A}^{2r}(K))$, and note that the diagonal $\Delta^{(r)} \subseteq \mathbb{A}^{2r}(K)$ is an affine algebraic cone in $\mathbb{A}^{2r}(K)$. Since $I_{\mathbb{A}^{2r}(K)}(\Delta^{(r)}) = (X_1 - Y_1, \ldots, X_r - Y_r)$, we have

$$I_{\mathbb{A}^{2r}(K)}((C \times D) \cap \Delta^{(r)}) = \sqrt{I_{\mathbb{A}^{2r}(K)}(C \times D) + (X_1 - Y_1, \ldots, X_r - Y_r)}.$$

It therefore follows from 19.2.7(ii) that

$$\operatorname{ara}_{C \times D, 0}((C \times D) \cap \Delta^{(r)})$$
$$\leq \operatorname{ara} \left(I_{\mathbb{A}^{2r}(K)}((C \times D) \cap \Delta^{(r)}) / I_{\mathbb{A}^{2r}(K)}(C \times D) \right)$$
$$\leq r.$$

If we now apply 19.4.4(i) to the two affine algebraic cones $(C \times D) \cap \Delta^{(r)} \subseteq C \times D$ in $\mathbb{A}^{2r}(K)$, we obtain

$$c((C \times D) \cap \Delta^{(r)})$$
$$\geq \min \{\operatorname{sdim} C + \operatorname{sdim} D - 1, \operatorname{sdim} C + c(D), \operatorname{sdim} D + c(C)\} - r.$$

In view of the homeomorphism provided by 19.6.1(ii), we have

$$c((C \times D) \cap \Delta^{(r)}) = c(C \cap D),$$

and so the claim in (i) is proved.

If C or D is reducible, then so too is $C \times D$ (by 19.6.2(ii)); we can therefore deduce part (ii) from 19.4.4(ii),(iii).

In view of 19.2.2, part (iii) follows from part (ii) if at least one of C, D is irreducible, and so we deal now with the case where both C and D are reducible. Then, by 19.2.2, we have $\operatorname{sdim} C \geq c(C) + 1$ and also $\operatorname{sdim} D \geq c(D) + 1$, so that

$$\operatorname{sdim} C + c(D) \geq c(C) + c(D) + 1 \quad \text{and} \quad \operatorname{sdim} D + c(C) \geq c(C) + c(D) + 1.$$

We use these inequalities in conjunction with part (ii) to see that

$$c(C \cap D) \geq \min\{\operatorname{sdim} C + c(D), \operatorname{sdim} D + c(C)\} - r - 1 + 1$$
$$\geq c(C) + c(D) + 1 - r - 1 + 1 = c(C) + c(D) - r - 1 + 2. \,\square$$

We now apply Proposition 19.6.5 to affine cones over projective algebraic sets to deduce the following theorem.

19.6.6 Theorem: the Intersection Inequality for the Connectedness Dimensions of Projective Algebraic Sets. *Let K be an algebraically closed field, let $r \in \mathbb{N}$, and let $V, W \subseteq \mathbb{P}^r(K)$ be non-empty closed sets.*

(i) *We have*

$$c(V \cap W) + r$$
$$\geq \min\{\operatorname{sdim} V + \operatorname{sdim} W - 1, \operatorname{sdim} V + c(W), \operatorname{sdim} W + c(V)\}.$$

(ii) *Furthermore,*

$$c(V \cap W) + r \geq \min\{\operatorname{sdim} V + c(W), \operatorname{sdim} W + c(V)\} - 1,$$

with strict inequality if V or W is reducible.

(iii) *Consequently, $c(V \cap W) + r \geq c(V) + c(W) - 1 + \varepsilon$, where $\varepsilon = 0, 1$ or 2 according as none, one or both of V and W are reducible.*

Proof. Let $C := \operatorname{Cone}(V) \subseteq \mathbb{A}^{r+1}(K)$ and $D := \operatorname{Cone}(W) \subseteq \mathbb{A}^{r+1}(K)$ be the affine cones in $\mathbb{A}^{r+1}(K)$ over V and W respectively. Observe that $C \cap D = \operatorname{Cone}(V) \cap \operatorname{Cone}(W) = \operatorname{Cone}(V \cap W)$. By Lemma 19.5.1, we have $\operatorname{sdim} C = \operatorname{sdim} V + 1$, $\operatorname{sdim} D = \operatorname{sdim} W + 1$,

$$c(C) = c(V) + 1, \quad c(D) = c(W) + 1, \quad c(C \cap D) = c(V \cap W) + 1.$$

Moreover, it follows from 15.2.1(v) that C is reducible if and only if V is, and D is reducible if and only if W is. All three parts of the theorem now follow from the corresponding parts of Proposition 19.6.5 applied to the affine algebraic cones $C, D \subseteq \mathbb{A}^{r+1}(K)$. $\square$

Part of Theorem 19.6.6(iii) amounts to the following formulation of the Connectivity Theorem due to W. Fulton and J. Hansen and to G. Faltings.

19.6.7 Corollary: the Connectivity Theorem of Fulton–Hansen and Faltings. (See [20] and [15].) *Let K be an algebraically closed field, let $r \in \mathbb{N}$, and let $V, W \subseteq \mathbb{P}^r(K)$ be projective varieties. Then*

$$c(V \cap W) \geq \dim V + \dim W - r - 1.$$

Proof. By 19.1.10(iii), we have $c(V) = \dim V$ and $c(W) = \dim W$, and so the claim follows from 19.6.6(iii). □

19.6.8 Corollary (W. Fulton and J. Hansen [20, Corollary 1]). *Let K be an algebraically closed field, let $r \in \mathbb{N}$, and let $V, W \subseteq \mathbb{P}^r(K)$ be projective varieties such that $\dim V + \dim W > r$. Then $V \cap W$ is connected.*

Proof. By Corollary 19.6.7, we have $c(V \cap W) \geq 0$, so that $V \cap W$ is connected by 19.1.10(i). □

19.6.9 Example. Interpret the polynomial ring $\mathbb{C}[X_1, X_2, X_3]$ as $\mathcal{O}(\mathbb{A}^3)$, as in 2.3.1. In $\mathbb{A}^3$, we consider the two surfaces

$$\overset{\circ}{V} := V_{\mathbb{A}^3}(X_1 - X_2 X_3) \quad \text{and} \quad \overset{\circ}{W} := V_{\mathbb{A}^3}(X_1 + X_2^2 - X_2 X_3 - 1).$$

Then

$$\begin{aligned}
\overset{\circ}{V} \cap \overset{\circ}{W} &= V_{\mathbb{A}^3}(X_1 - X_2 X_3, X_1 + X_2^2 - X_2 X_3 - 1) \\
&= V_{\mathbb{A}^3}(X_1 - X_2 X_3, X_2^2 - 1) \\
&= V_{\mathbb{A}^3}(X_1 - X_2 X_3, X_2 + 1) \cup V_{\mathbb{A}^3}(X_1 - X_2 X_3, X_2 - 1) \\
&= V_{\mathbb{A}^3}(X_1 + X_3, X_2 + 1) \cup V_{\mathbb{A}^3}(X_1 - X_3, X_2 - 1),
\end{aligned}$$

so that $\overset{\circ}{V} \cap \overset{\circ}{W}$ is the union of the two lines $\overset{\circ}{L_1} = V_{\mathbb{A}^3}(X_1 + X_3, X_2 + 1)$ and $\overset{\circ}{L_2} = V_{\mathbb{A}^3}(X_1 - X_3, X_2 - 1)$ and is disconnected. On the other hand,

$$\dim \overset{\circ}{V} + \dim \overset{\circ}{W} = 2 + 2 > 3$$

and $\overset{\circ}{V}$ and $\overset{\circ}{W}$ are both irreducible (as their defining polynomials are). Therefore, the analogue of 19.6.8 for affine varieties is not always true.

Now interpret the polynomial ring $\mathbb{C}[X_0, X_1, X_2, X_3]$ as $\mathcal{O}(\mathbb{A}^4)$ and consider the projective varieties $V, W \subseteq \mathbb{P}^3$ defined by

$$V := V_{\mathbb{P}^3}(X_0 X_1 - X_2 X_3) \quad \text{and} \quad W := V_{\mathbb{P}^3}(X_0 X_1 + X_2^2 - X_2 X_3 - X_0^2).$$

(Note that both $X_0 X_1 - X_2 X_3$ and $X_0 X_1 + X_2^2 - X_2 X_3 - X_0^2$ are irreducible polynomials. One can think of V and W as the projective closures of $\overset{\circ}{V}$ and $\overset{\circ}{W}$ respectively with respect to the isomorphism of varieties $\sigma_0 : \mathbb{A}^3 \overset{\cong}{\longrightarrow} U_0 \mathbb{P}^3$ of 19.3.7(i): see [30, Chapter I, Exercise 2.9].) Since $\dim V = \dim W = 2$, Corollary 19.6.8 tells us that $V \cap W$ is connected.

Indeed, we have

$$V \cap W$$
$$= V_{\mathbb{P}^3}(X_0 X_1 - X_2 X_3, X_0 X_1 + X_2^2 - X_2 X_3 - X_0^2)$$
$$= V_{\mathbb{P}^3}(X_1 + X_3, X_2 + X_0) \cup V_{\mathbb{P}^3}(X_1 - X_3, X_2 - X_0) \cup V_{\mathbb{P}^3}(X_0, X_2)$$
$$= L_1 \cup L_2 \cup L,$$

where $L_1 := V_{\mathbb{P}^3}(X_1 + X_3, X_2 + X_0)$ and $L_2 := V_{\mathbb{P}^3}(X_1 - X_3, X_2 - X_0)$ are the projective closures of $\overset{\circ}{L_1}$ and $\overset{\circ}{L_2}$ respectively with respect to σ_0, and L is the 'line at infinity' $V_{\mathbb{P}^3}(X_0, X_2)$ which intersects L_1 at $p := (0 : 1 : 0 : -1)$ and L_2 at $q := (0 : 1 : 0 : 1)$.

19.6.10 Exercise. Let $\overset{\circ}{V} := V_{\mathbb{A}^3}(X_1^3 + X_1^2 - X_2^2)$, $\overset{\circ}{W} := V_{\mathbb{A}^3}(X_2)$, and

$$V := V_{\mathbb{P}^3}(X_1^3 + X_0 X_1^2 - X_0 X_2^2), \quad W := V_{\mathbb{P}^3}(X_2).$$

Determine $\overset{\circ}{V} \cap \overset{\circ}{W}$ and $V \cap W$.

19.7 The projective spectrum and connectedness

Let us take stock. Among the results we have presented so far in this chapter are the Connectedness Bound for Complete Local Rings 19.2.10, Grothendieck's Connectedness Theorem 19.2.12, the Bertini–Grothendieck Connectivity Theorem 19.5.3, the Intersection Inequality for the Connectedness Dimensions of Projective Algebraic Sets 19.6.6, and the Fulton–Hansen Connectivity Theorem 19.6.7. All of these are important results about connectivity, but none of them mentions local cohomology in its statement. However, our proofs above of these results depend on one proposition, crucial for our approach, that does use local cohomology, namely Proposition 19.2.8. The key argument in our proof of that proposition concerned part of an (exact) Mayer–Vietoris sequence

$$H_{\mathfrak{a} \cap \mathfrak{b}}^{d-1}(R) \longrightarrow H_{\mathfrak{a} + \mathfrak{b}}^d(R) \longrightarrow H_{\mathfrak{a}}^d(R) \oplus H_{\mathfrak{b}}^d(R),$$

where $(R, \mathfrak{m})$ is a d-dimensional complete local domain and $\mathfrak{a}$ and $\mathfrak{b}$ are non-zero proper ideals of R whose sum is $\mathfrak{m}$-primary. An exact sequence like the one displayed above was considered in [5, p. 484], and so we shall refer to the above sequence as *Rung's display*. We used it in conjunction with the local Lichtenbaum–Hartshorne Vanishing Theorem 8.2.1 and the non-vanishing result of 6.1.4. From these few arguments from local cohomology, we have derived far-reaching geometric consequences which do not involve local cohomology in their statements!

Our final part of this chapter is concerned with a graded analogue of Rung's display, which we shall use in conjunction with the Graded Lichtenbaum–Hartshorne Vanishing Theorem 14.1.16 and the non-vanishing result of Exercise 16.1.10(ii) in order to prove a ring-theoretic version of Zariski's Main Theorem on the Connectivity of Fibres of Blowing-up. To prepare for this, we remind the reader, in 19.7.1 below, of some facts concerning the projective spectrum of a positively graded commutative Noetherian ring.

19.7.1 Reminder and ♯Exercise. Assume that $R = \bigoplus_{n \in \mathbb{N}_0} R_n$ is positively graded.

(i) Show that, for any two prime ideals $\mathfrak{p}, \mathfrak{q} \in {}^*\mathrm{Spec}(R)$ with $\mathfrak{p} \subset \mathfrak{q}$ and $\mathrm{ht}\, \mathfrak{q}/\mathfrak{p} > 1$, there exists a prime ideal $\mathfrak{s} \in {}^*\mathrm{Spec}(R)$ with $\mathfrak{p} \subset \mathfrak{s} \subset \mathfrak{q}$. (Recall that $\subset$ denotes strict inclusion.)

Recall that the *projective spectrum* of R, denoted by $\mathrm{Proj}(R)$, is the set ${}^*\mathrm{Spec}(R) \setminus \mathrm{Var}(R_+)$ of all graded prime ideals of R which do not contain the irrelevant ideal R_+ (see 15.1.1). The *Zariski topology* on $\mathrm{Proj}(R)$ is defined as the topology induced by the Zariski topology on $\mathrm{Spec}(R)$. Note that, since, for an ideal $\mathfrak{b}$ of R, we have ${}^*\mathrm{Spec}(R) \cap \mathrm{Var}(\mathfrak{b}) = {}^*\mathrm{Spec}(R) \cap \mathrm{Var}(\overline{\mathfrak{b}})$ where $\overline{\mathfrak{b}}$ is the ideal of R generated by all the homogeneous components of all the elements of $\mathfrak{b}$, it follows that the set of closed sets for the Zariski topology on $\mathrm{Proj}(R)$ is $\{\mathrm{Proj}(R) \cap \mathrm{Var}(\mathfrak{c}) : \mathfrak{c} \text{ is a graded ideal of } R\}$.

Let $\pi : \mathrm{Proj}(R) \longrightarrow \mathrm{Spec}(R_0)$ be the natural map, defined by $\pi(\mathfrak{p}) = \mathfrak{p} \cap R_0$ for all $\mathfrak{p} \in \mathrm{Proj}(R)$. In the case when $(R_0, \mathfrak{m}_0)$ is local, we refer to $\pi^{-1}(\mathfrak{m}_0)$ as the *special fibre* of π.

(ii) Show that ${}^*\mathrm{Spec}(R)$, with the topology induced by the Zariski topology on $\mathrm{Spec}(R)$, is a Noetherian topological space.

(iii) Show that $\mathrm{Proj}(R)$ is a Noetherian topological space.

(iv) Show that the set of closed irreducible subsets of $\mathrm{Proj}(R)$ is

$$\{\mathrm{Proj}(R) \cap \mathrm{Var}(\mathfrak{p}) : \mathfrak{p} \in \mathrm{Proj}(R)\}.$$

(v) Let $\mathfrak{b}$ be a graded ideal of R. Show that

$$\mathrm{Proj}(R) \cap \mathrm{Var}(\mathfrak{b}) = \mathrm{Proj}(R) \cap \mathrm{Var}(\mathfrak{b} \cap R_+),$$

and deduce that each closed set in $\mathrm{Proj}(R)$ can be defined by finitely many homogeneous elements of positive degree in R.

(vi) Show that $\pi^{-1}(\mathrm{Var}(\mathfrak{b}_0)) = \mathrm{Proj}(R) \cap \mathrm{Var}(\mathfrak{b}_0 R)$ for each ideal $\mathfrak{b}_0$ of R_0, and deduce that $\pi : \mathrm{Proj}(R) \longrightarrow \mathrm{Spec}(R_0)$ is continuous.

(vii) Let $\mathfrak{b}$ be a graded ideal of R. Show that

$$\pi(\operatorname{Proj}(R) \cap \operatorname{Var}(\mathfrak{b})) = \operatorname{Var}\left(\bigcup_{n \in \mathbb{N}}(\mathfrak{b} :_R (R_+)^n) \cap R_0\right).$$

(Here are some hints: show that it is sufficient to prove that, for $\mathfrak{p} \in \operatorname{Proj}(R)$, we have $\pi(\operatorname{Proj}(R) \cap \operatorname{Var}(\mathfrak{p})) = \operatorname{Var}(\mathfrak{p} \cap R_0)$; then show that, in the special case in which $(R_0, \mathfrak{m}_0)$ is local, $R_+ \not\subseteq \sqrt{\mathfrak{m}_0 R + \mathfrak{p}}$ when $\mathfrak{p} \in \operatorname{Proj}(R)$.)

Deduce that π is closed and that $\pi(\operatorname{Proj}(R)) = \operatorname{Var}\left(R_0 \cap \Gamma_{R_+}(R)\right)$.

(viii) Assume that $(R_0, \mathfrak{m}_0)$ is local. Show that, for a graded ideal $\mathfrak{b}$ of R, we have $\mathfrak{m}_0 \in \pi(\operatorname{Proj}(R) \cap \operatorname{Var}(\mathfrak{b}))$ if and only if $R_+ \not\subseteq \sqrt{\mathfrak{m}_0 R + \mathfrak{b}}$. Show that $\pi^{-1}(\mathfrak{m}_0) \cap W \neq \emptyset$ for every non-empty closed subset W of $\operatorname{Proj}(R)$.

Deduce that $\pi^{-1}(\mathfrak{m}_0)$ is connected if and only if $\pi^{-1}(Z)$ is connected for each non-empty closed subset Z of $\operatorname{Spec}(R_0)$.

Our next lemma uses the promised graded analogue of Rung's display.

19.7.2 Lemma. *Assume that $R = \bigoplus_{n \in \mathbb{N}_0} R_n$ is positively graded and that $(R_0, \mathfrak{m}_0)$ is local and complete. Then the following statements are equivalent:*

(i) $\operatorname{Proj}(R)$ *is connected;*

(ii) *the special fibre $\pi^{-1}(\mathfrak{m}_0)$ (under the natural map $\pi : \operatorname{Proj}(R) \longrightarrow \operatorname{Spec}(R_0)$ of 19.7.1) is connected.*

Proof. (ii) $\Rightarrow$ (i) This is immediate from Exercise 19.7.1(viii).

(i) $\Rightarrow$ (ii) Since $\operatorname{Proj}(R) \neq \emptyset$, we have $\pi^{-1}(\mathfrak{m}_0) \neq \emptyset$, by 19.7.1(viii), and $R_+ \neq 0$. If $\dim R_0 = 0$, then $\pi^{-1}(\mathfrak{m}_0) = \operatorname{Proj}(R)$. Therefore we can, and do, assume that $\dim R_0 > 0$.

Suppose that $\pi^{-1}(\mathfrak{m}_0) = \operatorname{Proj}(R) \cap \operatorname{Var}(\mathfrak{m}_0 R)$ is disconnected; we shall obtain a contradiction. Then there exist two non-empty closed subsets Z_1, Z_2 of $\operatorname{Proj}(R)$ such that $Z_1 \cap Z_2 = \emptyset$ and $Z_1 \cup Z_2 = \pi^{-1}(\mathfrak{m}_0)$. Let $T_1, \ldots, T_r$ be the distinct irreducible components of $\operatorname{Proj}(R)$, and set

$$A := \{i \in \{1, \ldots, r\} : T_i \cap Z_1 \neq \emptyset\}, \quad B := \{j \in \{1, \ldots, r\} : T_j \cap Z_2 \neq \emptyset\}.$$

Clearly each of A and B is non-empty, since $Z_1 \neq \emptyset \neq Z_2$. By 19.7.1(viii), it follows that, for each $k \in \{1, \ldots, r\}$, we have $T_k \cap (Z_1 \cup Z_2) \neq \emptyset$. This shows that $A \cup B = \{1, \ldots, r\}$.

We show next that $A \cap B \neq \emptyset$. Suppose, on the contrary, that $A \cap B = \emptyset$, and seek a contradiction. Then $W := \left(\bigcup_{i \in A} T_i\right) \cap \left(\bigcup_{j \in B} T_j\right)$ would be non-empty because $\operatorname{Proj}(R)$ is connected; also, we would have

$$\pi^{-1}(\mathfrak{m}_0) \cap W = (Z_1 \cup Z_2) \cap W = (Z_1 \cap W) \cup (Z_2 \cap W) = \emptyset,$$

contrary to 19.7.1(viii). Hence there exists $k \in A \cap B$.

Then $Z_i \cap T_k \neq \emptyset$ for $i = 1, 2$ and so $\pi^{-1}(\mathfrak{m}_0) \cap T_k$ is disconnected. By 19.7.1(iv), there exists $\mathfrak{p} \in \operatorname{Proj}(R)$ for which $T_k = \operatorname{Proj}(R) \cap \operatorname{Var}(\mathfrak{p})$. Let $\overline{R} = \bigoplus_{n \in \mathbb{N}_0} \overline{R}_n = R/\mathfrak{p}$, graded in the natural way, let $\overline{\mathfrak{m}}_0$ be the maximal ideal of $\overline{R}_0$, and let $\overline{\pi} : \operatorname{Proj}(\overline{R}) \longrightarrow \operatorname{Spec}(\overline{R}_0)$ be the natural map. The natural homeomorphism between T_k and $\operatorname{Proj}(\overline{R})$ maps $\pi^{-1}(\mathfrak{m}_0) \cap T_k$ onto $\overline{\pi}^{-1}(\overline{\mathfrak{m}}_0)$, and so this latter special fibre is disconnected. Hence, in our search for a contradiction, we can, and do, assume that the graded ring R is a domain.

There exist graded ideals $\mathfrak{b}, \mathfrak{c}$ of R such that $\operatorname{Proj}(R) \cap \operatorname{Var}(\mathfrak{b}) \neq \emptyset$, $\operatorname{Proj}(R) \cap \operatorname{Var}(\mathfrak{c}) \neq \emptyset$, $\operatorname{Proj}(R) \cap \operatorname{Var}(\mathfrak{b}) \cap \operatorname{Var}(\mathfrak{c}) = \emptyset$, and

$$(\operatorname{Proj}(R) \cap \operatorname{Var}(\mathfrak{b})) \cup (\operatorname{Proj}(R) \cap \operatorname{Var}(\mathfrak{c})) = \pi^{-1}(\mathfrak{m}_0).$$

Since $\operatorname{Proj}(R) \cap \operatorname{Var}(\mathfrak{b}) \subseteq \pi^{-1}(\mathfrak{m}_0)$, we have

$$\operatorname{Proj}(R) \cap \operatorname{Var}(\mathfrak{b}) = \operatorname{Proj}(R) \cap \operatorname{Var}(\mathfrak{b} + \mathfrak{m}_0 R),$$

and a similar comment applies to $\mathfrak{c}$; we therefore assume that $\mathfrak{b} \supseteq \mathfrak{m}_0 R$ and $\mathfrak{c} \supseteq \mathfrak{m}_0 R$. We can now deduce that

$$\sqrt{\mathfrak{b} + \mathfrak{c}} = \mathfrak{m}_0 R + R_+ \quad \text{and} \quad \sqrt{\mathfrak{b} \cap \mathfrak{c}} = \sqrt{\mathfrak{m}_0 R}.$$

Set $d := \dim R$. We can now use 1.2.3 in conjunction with the Graded Mayer–Vietoris sequence 14.1.5 to see that there is an exact sequence of graded R-modules and homogeneous homomorphisms

$$H^{d-1}_{\mathfrak{m}_0 R}(R) \longrightarrow H^{d}_{\mathfrak{m}_0 R + R_+}(R) \longrightarrow H^{d}_{\mathfrak{b}}(R) \oplus H^{d}_{\mathfrak{c}}(R).$$

As $\operatorname{Proj}(R) \cap \operatorname{Var}(\mathfrak{b}) \neq \emptyset$ and $\operatorname{Proj}(R) \cap \operatorname{Var}(\mathfrak{c}) \neq \emptyset$, we have $\dim R/\mathfrak{b} > 0$ and $\dim R/\mathfrak{c} > 0$. Therefore, by the Graded Lichtenbaum–Hartshorne Vanishing Theorem 14.1.16, we have $H^{d}_{\mathfrak{b}}(R) = H^{d}_{\mathfrak{c}}(R) = 0$; also, Lemma 14.1.12 shows that $H^{d-1}_{\mathfrak{m}_0 R}(R)_n \cong H^{d-1}_{\overline{\mathfrak{m}}_0}(R_n)$ for all $n \in \mathbb{Z}$, and this is 0 for $n < 0$. Hence $H^{d}_{\mathfrak{m}_0 R + R_+}(R)_n = 0$ for all $n < 0$, contrary to 16.1.10(ii). $\square$

19.7.3 Theorem: the Connectedness Criterion for the Special Fibre. *Assume that $R = \bigoplus_{n \in \mathbb{N}_0} R_n$ is positively graded and that $(R_0, \mathfrak{m}_0)$ is local. Let $\widehat{R}_0$ denote the completion of R_0. As in 16.2.2, set $\widehat{R} := R \otimes_{R_0} \widehat{R}_0$; after an obvious identification, we can consider $\widehat{R}$ as a graded ring $\bigoplus_{n \in \mathbb{N}_0} (R_n \otimes_{R_0} \widehat{R}_0)$. Let $\pi : \operatorname{Proj}(R) \longrightarrow \operatorname{Spec}(R_0)$ be the natural map.*

The following statements are equivalent:

(i) $\operatorname{Proj}(\widehat{R})$ *is connected;*
(ii) *the special fibre $\pi^{-1}(\mathfrak{m}_0)$ is connected;*
(iii) *for each non-empty closed subset Z of $\operatorname{Spec}(R_0)$, the set $\pi^{-1}(Z)$ is connected.*

Proof. The equivalence of (ii) and (iii) is the subject of part of Exercise 19.7.1(viii).

(i) $\Leftrightarrow$ (ii) Let $\widehat{\pi} : \mathrm{Proj}(\widehat{R}) \longrightarrow \mathrm{Spec}(\widehat{R_0})$ be the natural map. Let $\widehat{\mathfrak{m}_0} = \mathfrak{m}_0 \widehat{R_0}$, the maximal ideal of $\widehat{R_0}$. Since there are natural homeomorphisms $\pi^{-1}(\mathfrak{m}_0) \approx \mathrm{Proj}(R/\mathfrak{m}_0 R)$ and $\widehat{\pi}^{-1}(\widehat{\mathfrak{m}_0}) \approx \mathrm{Proj}(\widehat{R}/\widehat{\mathfrak{m}_0}\widehat{R})$, and since there is a homogeneous ring isomorphism $R/\mathfrak{m}_0 R \xrightarrow{\cong} \widehat{R}/\widehat{\mathfrak{m}_0}\widehat{R}$, there is a homeomorphism $\pi^{-1}(\mathfrak{m}_0) \approx \widehat{\pi}^{-1}(\widehat{\mathfrak{m}_0})$. We can therefore apply Lemma 19.7.2 to complete the proof. $\square$

We are now going to apply the Connectedness Criterion for the Special Fibre 19.7.3 to the (ordinary) Rees ring of an ideal of a commutative Noetherian ring.

19.7.4 Notation. We use $\mathcal{R}(\mathfrak{a})$ to denote the ordinary Rees ring $\bigoplus_{n \in \mathbb{N}_0} \mathfrak{a}^n$ of $\mathfrak{a}$. If, as in 18.1.4, we let $\{a_1, \ldots, a_h\}$ be a generating set for $\mathfrak{a}$ and T be an indeterminate, then there is a homogeneous isomorphism of graded R-algebras $\mathcal{R}(\mathfrak{a}) \xrightarrow{\cong} R[a_1 T, \ldots, a_h T] =: R[\mathfrak{a}T]$.

We remark here that $\mathcal{R}(\mathfrak{a})$ is also called the *blowing-up ring* of $\mathfrak{a}$; this terminology has its roots in the fact that $\mathrm{Proj}\,(\mathcal{R}(\mathfrak{a}))$ is the topological space underlying the scheme obtained by blowing up $\mathrm{Spec}(R)$ with respect to $\mathfrak{a}$.

19.7.5 Corollary. *Let* $\mathfrak{p} \in \mathrm{Spec}(R)$ *be such that* $\widehat{R_\mathfrak{p}}$ *has only one minimal prime ideal and the ideal* $\mathfrak{a}R_\mathfrak{p}$ *of* $R_\mathfrak{p}$ *is not nilpotent. Let* $\mathcal{R}(\mathfrak{a})$ *denote the ordinary Rees ring of* $\mathfrak{a}$ *(see 19.7.4), and let* $\pi_\mathfrak{a} : \mathrm{Proj}(\mathcal{R}(\mathfrak{a})) \longrightarrow \mathrm{Spec}(R)$ *be the natural map. Then the fibre* $\pi_\mathfrak{a}^{-1}(\mathfrak{p})$ *of* $\mathfrak{p}$ *under* $\pi_\mathfrak{a}$ *is connected.*

Proof. We can use localization at $\mathfrak{p}$ to see that it is enough for us to prove the claim under the assumption that $(R, \mathfrak{m})$ is local and $\mathfrak{p} = \mathfrak{m}$; we make this assumption in what follows. Now $\widehat{R}$ is flat over R; also, $\mathcal{R}(\mathfrak{a})$ can be viewed as a subring of $\mathcal{R}(\mathfrak{a}\widehat{R})$ and there is a homogeneous isomorphism of $\mathcal{R}(\mathfrak{a})$-algebras $\mathcal{R}(\mathfrak{a}) \otimes_R \widehat{R} \xrightarrow{\cong} \mathcal{R}(\mathfrak{a}\widehat{R})$. Therefore $\mathrm{Proj}\,(\mathcal{R}(\mathfrak{a}) \otimes_R \widehat{R})$ and $\mathrm{Proj}\,(\mathcal{R}(\mathfrak{a}\widehat{R}))$ are homeomorphic. By Theorem 19.7.3, it is sufficient for us to prove that $\mathrm{Proj}(\mathcal{R}(\mathfrak{a}\widehat{R}))$ is connected.

In order to prove this, we let $\widehat{\mathfrak{p}}$ be the unique minimal prime ideal of $\widehat{R}$, and $\mathfrak{P}$ be the ideal of $\mathcal{R}(\mathfrak{a}\widehat{R})$ given by $\mathfrak{P} := \bigoplus_{n \in \mathbb{N}_0} (\mathfrak{a}^n \widehat{R} \cap \widehat{\mathfrak{p}})$; in fact $\mathfrak{P} \in {}^*\mathrm{Spec}(\mathcal{R}(\mathfrak{a}\widehat{R}))$. Since $\widehat{\mathfrak{p}} = \sqrt{0}$, there exists $t \in \mathbb{N}$ such that $\widehat{\mathfrak{p}}^t = 0$; hence $\mathfrak{P}^t = 0$, so that $\mathfrak{P}$ is the unique minimal prime of $\mathcal{R}(\mathfrak{a}\widehat{R})$.

It follows from the faithful flatness of $R \to \widehat{R}$ that $\mathfrak{a}\widehat{R} \not\subseteq \sqrt{0} = \widehat{\mathfrak{p}}$, and so $\mathfrak{a}^n \widehat{R} \cap \widehat{\mathfrak{p}} \subset \mathfrak{a}^n \widehat{R}$ for all $n \in \mathbb{N}$. Hence

$$\mathfrak{P} \in {}^*\mathrm{Spec}(\mathcal{R}(\mathfrak{a}\widehat{R})) \setminus \mathrm{Var}(\mathcal{R}(\mathfrak{a}\widehat{R})_+) = \mathrm{Proj}(\mathcal{R}(\mathfrak{a}\widehat{R})).$$

As $\mathfrak{P}$ is the unique minimal prime of $\mathcal{R}(\mathfrak{a}\widehat{R})$, we also have $\mathrm{Proj}(\mathcal{R}(\mathfrak{a}\widehat{R})) =$

$\operatorname{Proj}(\mathcal{R}(\mathfrak{a}\widehat{R})) \cap \operatorname{Var}(\mathfrak{P})$. Hence, by 19.7.1(iv), $\operatorname{Proj}(\mathcal{R}(\mathfrak{a}\widehat{R}))$ is irreducible, and so it is connected by 19.1.2(i). □

We are now able to deduce from 19.7.5 the last main connectedness result of this chapter.

19.7.6 Corollary: ring-theoretic version of Zariski's Main Theorem on the Connectivity of Fibres of Blowing-up. *Assume that R is a domain and that $\mathfrak{a} \neq 0$. Let $\mathfrak{p} \in \operatorname{Spec}(R)$ be such that $\widehat{R_{\mathfrak{p}}}$ is also a domain. Let $\pi_{\mathfrak{a}} :$ $\operatorname{Proj}(\mathcal{R}(\mathfrak{a})) \longrightarrow \operatorname{Spec}(R)$ be the natural map. Then the fibre $\pi_{\mathfrak{a}}^{-1}(\mathfrak{p})$ is connected.* □

19.7.7 Exercise. Assume that R is a domain and that $\mathfrak{a} \neq 0$. Define $\pi_{\mathfrak{a}} :$ $\operatorname{Proj}(\mathcal{R}(\mathfrak{a})) \longrightarrow \operatorname{Spec}(R)$ as in 19.7.5. Let $Z \subseteq \operatorname{Spec}(R)$ be a connected closed subset of $\operatorname{Spec}(R)$ such that $\widehat{R_{\mathfrak{m}}}$ is an integral domain for each maximal ideal $\mathfrak{m}$ of R which belongs to Z. Prove that $\pi_{\mathfrak{a}}^{-1}(Z)$ is connected.

19.7.8 Exercise. Assume that $(R, \mathfrak{m})$ is a local domain and that $\mathfrak{a}$ is non-zero and proper. We use $\mathcal{G}(\mathfrak{a})$ to denote the associated graded ring $\bigoplus_{i \in \mathbb{N}_0} \mathfrak{a}^i/\mathfrak{a}^{i+1}$: see 18.2.1.

(i) Show that, if $\widehat{R}$ is an integral domain, then $\operatorname{Proj}(\mathcal{G}(\mathfrak{a}))$ is connected.
(ii) Show that $\operatorname{Proj}(\mathcal{G}(\mathfrak{m}))$ is connected if and only if $\operatorname{Proj}(\mathcal{R}(\mathfrak{m}\widehat{R}))$ is connected.

19.7.9 Exercise. Let V denote the *Cartesian curve* $V_{\mathbb{A}^2}(X^3 + X^2 - Y^2)$, and W denote the *cuspidal curve* $V_{\mathbb{A}^2}(X^3 - Y^2)$.

(i) Let $(R, \mathfrak{m})$ be the local ring $\mathcal{O}_{V,0}$ of the origin 0 on V. Show that $\widehat{R}$ has two minimal primes, and that $\operatorname{Proj}(\mathcal{G}(\mathfrak{m}))$ is a discrete topological space with just two points.
(ii) Let $(R', \mathfrak{m}')$ be the local ring $\mathcal{O}_{W,0}$ of the origin 0 on W. Show that $\widehat{R'}$ is a domain and that $\operatorname{Proj}(\mathcal{G}(\mathfrak{m}'))$ is a singleton set.

20

Links with sheaf cohomology

In this last chapter we shall develop the links between local cohomology and the cohomology of quasi-coherent sheaves over certain Noetherian schemes. Here we shall assume for the first time that the reader has some basic knowledge about schemes and sheaves: our reference for these topics is Hartshorne's book [30]. The central idea in this chapter is to extend our earlier relations in 2.3.2 and 15.1.3 between ideal transforms and rings of regular functions on varieties to quasi-coherent sheaves over certain Noetherian schemes. We shall be very concerned with a generalization of the 'Deligne Isomorphism' (see [30, Chapter III, Exercise 3.7, p. 217]) which links the group of sections (over an open subset) of an induced sheaf on an affine scheme with an ideal transform. More precisely, let $\widetilde{M}$ denote the sheaf induced by an R-module M on the affine scheme $\mathrm{Spec}(R)$, and let $U = \mathrm{Spec}(R) \setminus \mathrm{Var}(\mathfrak{a})$, where $\mathfrak{a} \subset R$; then the group of sections $\Gamma(U, \widetilde{M})$ is isomorphic to the ideal transform $D_{\mathfrak{a}}(M)$. We shall use standard techniques involving negative strongly connected sequences of functors to extend this Deligne Isomorphism, and our generalization of it, to produce the Deligne Correspondence 20.3.11. This correspondence provides connections between higher cohomology groups of induced sheaves on the one hand, and local cohomology modules on the other.

We shall also examine the case when R is graded in some detail. Here the central result for us is the Serre–Grothendieck Correspondence 20.3.15, which we shall also derive from the Deligne Isomorphism by standard 'connected sequence' arguments. In this introduction, we mention only some consequences of the Serre–Grothendieck Correspondence for projective schemes. Consider, therefore, the special case in which $R = \bigoplus_{n \in \mathbb{N}_0} R_n$ is positively $\mathbb{Z}$-graded and homogeneous; set $T := \mathrm{Proj}(R) = {}^*\mathrm{Spec}(R) \setminus \mathrm{Var}(R_+)$ and consider the projective scheme $(T, \mathcal{O}_T)$ defined by R (see [30, Chapter II, §2, p. 76]). If $\mathcal{F}$ is a coherent sheaf of $\mathcal{O}_T$-modules, there is a finitely generated graded R-module N such that $\mathcal{F}$ is isomorphic to the sheaf of $\mathcal{O}_T$-modules associated to N on $\mathrm{Proj}(R)$. The classical form of the Serre–Grothendieck Correspondence yields, for each $i \in \mathbb{N}_0$ and $n \in \mathbb{Z}$, an R_0-isomorphism between the i-th cohomology group $H^i(T, \mathcal{F}(n))$ of the twisted sheaf $\mathcal{F}(n)$ and $\mathcal{R}^i D_{R_+}(N)_n$, so that, for $i > 0$, we have $H^i(T, \mathcal{F}(n)) \cong H^{i+1}_{R_+}(N)_n$. These results enable us to deduce quickly, from algebraic results about local cohomology established earlier in the book, significant results about the cohomology of coherent sheaves of $\mathcal{O}_T$-modules.

In the final two sections, we use this approach to present proofs of some fundamental theorems, and extensions thereof, from projective algebraic geometry, including Serre's Finiteness Theorem for the cohomology of coherent sheaves over projective schemes (see 20.4.8), Serre's Criterion for the global generation of coherent sheaves over projective schemes (see 20.4.13), the existence of a Hilbert polynomial for a coherent sheaf over a projective scheme over an Artinian base ring (see 20.4.16), Mumford's Regularity Bound for coherent sheaves of ideals over a projective space (see 20.4.18), the Severi–Enriques–Zariski–Serre Vanishing Theorem for the cohomology of coherent sheaves over projective schemes (see 20.4.23), Serre's Cohomological Criterion for local freeness of coherent sheaves over regular projective schemes (see 20.5.6), Horrocks' Splitting Criterion for coherent sheaves over projective spaces (see 20.5.8), and Grothendieck's Splitting Theorem for coherent locally free sheaves over the projective line (see 20.5.9).

20.1 The Deligne Isomorphism

The basic result in this chapter is Theorem 20.1.14, a generalized version of the Deligne Isomorphism. This result is not formulated explicitly in sheaf-theoretic terms, but rather in terms of certain local families of fractions. We now start to develop the notions of 'S-topology' and 'S-local family of fractions' which we shall use in our formulation and proof of Theorem 20.1.14.

20.1.1 Notation and Terminology. Throughout this chapter, we shall use S to denote a non-empty subset of R which is closed under multiplication. It should be noted that we do *not* assume that $1 \in S$.

(i) We denote by $\mathfrak{A}_S$ the set $\{(S') : \emptyset \neq S' \subseteq S\}$ of all ideals of R generated by elements of S. As R is Noetherian, each ideal in $\mathfrak{A}_S$ can be generated by finitely many elements of S. Note that, if $(\mathfrak{a}_i)_{i \in I}$ is a family of ideals in $\mathfrak{A}_S$, then $\sum_{i \in I} \mathfrak{a}_i \in \mathfrak{A}_S$, and if $\mathfrak{b}_1, \ldots, \mathfrak{b}_r \in \mathfrak{A}_S$, then $\prod_{j=1}^r \mathfrak{b}_j \in \mathfrak{A}_S$ too.

(ii) We shall say that a subset T of $\mathrm{Spec}(R)$ is *essential with respect to* S precisely when

  (a) $T \subseteq \mathrm{Spec}(R) \setminus \mathrm{Var}((S))$;
  (b) there exists $\mathfrak{b} \in \mathfrak{A}_S$ such that $T \subseteq \mathrm{Var}(\mathfrak{b})$; and
  (c) $\sqrt{\mathfrak{c}} = \bigcap_{\mathfrak{p}\in\mathrm{Var}((S))\cup(T\cap\mathrm{Var}(\mathfrak{c}))} \mathfrak{p}$ for all $\mathfrak{c} \in \mathfrak{A}_S$.

Observe that condition (a) is automatically satisfied if S contains a unit, that condition (b) is automatically satisfied if S contains 0, and that, for every subset T of $\mathrm{Spec}(R)$, we certainly have

$$\sqrt{\mathfrak{c}} \subseteq \bigcap_{\mathfrak{p}\in\mathrm{Var}((S))\cup(T\cap\mathrm{Var}(\mathfrak{c}))} \mathfrak{p} \quad \text{for all } \mathfrak{c} \in \mathfrak{A}_S.$$

Note also that, if $S = \{0\}$, then the empty set is the unique subset of $\mathrm{Spec}(R)$ which is essential with respect to S. In general, if the empty set is essential with respect to S, then $\sqrt{\mathfrak{c}} = \sqrt{(S)}$ for all $\mathfrak{c} \in \mathfrak{A}_S$, so that, in particular, $\mathrm{ht}(S) \leq 1$ if S contains a non-unit; if, in addition, $0 \in S$, then $\sqrt{(S)} = \sqrt{0}$.

Note also that condition (a) implies that, for each $\mathfrak{p} \in T$, there exists $s_{\mathfrak{p}} \in S \setminus \mathfrak{p}$.

(iii) Let T be a subset of $\mathrm{Spec}(R)$ which is essential with respect to S. It is an easy consequence of the last sentence in part (i) that

$$\{T \cap \mathrm{Var}(\mathfrak{b}) : \mathfrak{b} \in \mathfrak{A}_S\}$$

is the set of closed sets in a topology on T: we refer to this topology as the *S-topology on* T. We denote by $\mathcal{U}_T^{(S)}$ the set of open sets in the S-topology on T: thus $\mathcal{U}_T^{(S)} = \{T \setminus \mathrm{Var}(\mathfrak{b}) : \mathfrak{b} \in \mathfrak{A}_S\}$. For $\mathfrak{p} \in T$, we denote by $\mathcal{U}_{T,\mathfrak{p}}^{(S)}$ the set $\{U \in \mathcal{U}_T^{(S)} : \mathfrak{p} \in U\}$ of all open neighbourhoods of $\mathfrak{p}$ in the S-topology on T.

Since every non-empty set of closed subsets of T (in the S-topology) has a minimal member with respect to inclusion, the S-topology makes T into a Noetherian topological space, and hence (see 19.1.6(i)) every open subset of T is quasi-compact.

We now state the assumptions that will be in force throughout this chapter.

20.1.2 Standard hypotheses. Throughout this chapter, S will denote a non-empty subset of R which is closed under multiplication, $\mathfrak{A}_S$ will denote the set $\{(S') : \emptyset \neq S' \subseteq S\}$ of all ideals of R generated by non-empty subsets of S, T will denote a subset of $\mathrm{Spec}(R)$ which is essential with respect to S (see 20.1.1(ii)), $\mathfrak{a}$ will denote an ideal in $\mathfrak{A}_S$, and M will denote an R-module.

20.1.3 Examples. (The hypotheses of 20.1.2 apply.)

(i) If $S = R$, then $\mathfrak{A}_S$ is the set of all ideals of R, and $T := \mathrm{Spec}(R)$ is essential with respect to S. The S-topology on T is just the ordinary Zariski topology.

(ii) Assume that G is a finitely generated torsion-free Abelian group and $R = \bigoplus_{g \in G} R_g$ is G-graded, and take $S := \bigcup_{g \in G} R_g$ to be the set of homogeneous elements of R. Then $\mathfrak{A}_S$ is the set of all graded ideals of R, and $T := {}^* \mathrm{Spec}(R)$ is essential with respect to S. It is easy to see that the S-topology on T is again the ordinary Zariski topology.

20.1.4 ♯Exercise. Assume that $R = \bigoplus_{n \in \mathbb{N}_0} R_n$ is positively $\mathbb{Z}$-graded, and take $S := \bigcup_{n \in \mathbb{N}} R_n$, the set of homogeneous elements of R of positive degrees, together with the zero element of R; then $\mathfrak{A}_S$ is the set of all graded ideals of R which are contained in R_+.

Show that $T := \mathrm{Proj}(R) = {}^* \mathrm{Spec}(R) \setminus \mathrm{Var}(R_+)$ is essential with respect to S, and that the S-topology on T is the Zariski topology.

The next exercise suggests how the notion of S-topology on T can be regarded as a generalization of the Zariski topology on an affine variety.

20.1.5 Exercise. Let V be an affine variety over the algebraically closed field K. Take $R := \mathcal{O}(V)$, and $S := R$; by 20.1.3(i), we know that $T := \mathrm{Spec}(R)$ is essential with respect to S. Let $\max T$ denote the set of all maximal ideals of R; note that $\max T$ is also essential with respect to S (since R is a Hilbert ring), and that the S-topology on $\max T$ is the topology induced from the Zariski topology on T.

Use the Nullstellensatz to show that there is a homeomorphism $j : V \xrightarrow{\approx} \max T$ given by $j(p) = I_V(p)$ for all $p \in V$.

Show that the assignment $U \mapsto U \cap \max T$ defines a bijection between $\mathcal{U}_T^{(S)}$ and $\mathcal{U}_{\max T}^{(S)}$.

We next introduce the notion of 'S-local family of fractions': this generalizes the concept of regular function on a quasi-affine variety.

20.1.6 Notation and Terminology. (The hypotheses of 20.1.2 apply.) Note that, if $T \neq \emptyset$, then $\prod_{\mathfrak{p} \in T}(S \setminus \mathfrak{p}) \neq \emptyset$: a member of this set is called a *family of denominators in S*.

(i) Let $\mathfrak{p} \in T$. Since $S \backslash \mathfrak{p} \neq \emptyset$, we can form the commutative ring $(S \backslash \mathfrak{p})^{-1} R$ and the $(S \setminus \mathfrak{p})^{-1} R$-module $(S \setminus \mathfrak{p})^{-1} M$.

(ii) Let $\emptyset \neq U \in \mathcal{U}_T^{(S)}$. The elements of $\prod_{\mathfrak{p} \in U}(S \setminus \mathfrak{p})^{-1} M$ are called *families of fractions over U with numerators in M and denominators in S*. Such a family $\gamma = (\gamma_{\mathfrak{p}})_{\mathfrak{p} \in U} \in \prod_{\mathfrak{p} \in U}(S \setminus \mathfrak{p})^{-1} M$ is called *local*, or an *S-local family of fractions over U*, if and only if, for each $\mathfrak{q} \in U$, there exist $W \in \mathcal{U}_{T,\mathfrak{q}}^{(S)}$ and $(s, m) \in S \times M$ such that, for each $\mathfrak{p} \in U \cap W$, it is the case that $s \in S \setminus \mathfrak{p}$ and $\gamma_{\mathfrak{p}} = m/s$ in $(S \setminus \mathfrak{p})^{-1} M$.

The set of all S-local families of fractions over U is denoted by $\widetilde{M}(U)$. We set $\widetilde{M}(\emptyset) = 0$.

Note that $\prod_{\mathfrak{p} \in U}(S \setminus \mathfrak{p})^{-1} M$ has a natural structure as a module over the commutative ring $\prod_{\mathfrak{p} \in U}(S \setminus \mathfrak{p})^{-1} R$. It is easy to check that $\widetilde{R}(U)$ is a subring of $\prod_{\mathfrak{p} \in U}(S \setminus \mathfrak{p})^{-1} R$ and that $\widetilde{M}(U)$ is an $\widetilde{R}(U)$-submodule of $\prod_{\mathfrak{p} \in U}(S \setminus \mathfrak{p})^{-1} M$.

Note that $\widetilde{R}(\emptyset)$ is a trivial ring; of course, $\widetilde{M}(\emptyset) = 0$ is an $\widetilde{R}(\emptyset)$-module.

(iii) Now let $m \in M$ and let U be as in part (ii). Choose a family of denominators $(s_{\mathfrak{p}})_{\mathfrak{p} \in U} \in \prod_{\mathfrak{p} \in U}(S \setminus \mathfrak{p})$. For each $\mathfrak{p} \in U$, the fraction $s_{\mathfrak{p}} m / s_{\mathfrak{p}} \in (S \setminus \mathfrak{p})^{-1} M$ is independent of the choice of denominator $s_{\mathfrak{p}} \in S \setminus \mathfrak{p}$. We can therefore define a family of fractions

$$\widetilde{m} := \left(\frac{s_{\mathfrak{p}} m}{s_{\mathfrak{p}}} \right)_{\mathfrak{p} \in U} \in \prod_{\mathfrak{p} \in U}(S \setminus \mathfrak{p})^{-1} M,$$

which does not depend on the choice of family of denominators $(s_{\mathfrak{p}})_{\mathfrak{p} \in U}$. Moreover, for $\mathfrak{q} \in U$, we have, for each $\mathfrak{p} \in U \setminus \mathrm{Var}((s_{\mathfrak{q}}))$, the relation $s_{\mathfrak{p}} m / s_{\mathfrak{p}} = s_{\mathfrak{p}} s_{\mathfrak{q}} m / s_{\mathfrak{p}} s_{\mathfrak{q}} = s_{\mathfrak{q}} m / s_{\mathfrak{q}}$ in $(S \setminus \mathfrak{p})^{-1} M$. As $U \setminus \mathrm{Var}((s_{\mathfrak{q}})) \in \mathcal{U}_{T,\mathfrak{q}}^{(S)}$, we thus see that the family of fractions $\widetilde{m}$ is local. We can therefore define a map $\varepsilon_M^U : M \longrightarrow \widetilde{M}(U)$ by $\varepsilon_M^U(m) = \widetilde{m} = (s_{\mathfrak{p}} m / s_{\mathfrak{p}})_{\mathfrak{p} \in U}$ for all $m \in M$. Of course, we define $\varepsilon_M^{\emptyset} : M \longrightarrow \widetilde{M}(\emptyset) = 0$ to be the zero homomorphism.

Note that $\varepsilon_R^U : R \longrightarrow \widetilde{R}(U)$ is a ring homomorphism, and so turns $\widetilde{R}(U)$ into an R-algebra; furthermore, ε_M^U is an R-homomorphism. Similar comments apply to $\varepsilon_R^{\emptyset}$ and $\varepsilon_M^{\emptyset}$.

(iv) Now let $U, V \in \mathcal{U}_T^{(S)}$ with $V \subseteq U$. Suppose that $V \neq \emptyset$, and let $\gamma = (\gamma_{\mathfrak{p}})_{\mathfrak{p} \in U} \in \widetilde{M}(U)$. Then it is clear that the restriction $\gamma \lceil_V := (\gamma_{\mathfrak{p}})_{\mathfrak{p} \in V}$ belongs to $\widetilde{M}(V)$. We therefore have a *restriction map* $\rho_{UV} (= \rho_{UV,M}) : \widetilde{M}(U) \longrightarrow \widetilde{M}(V)$ for which $\rho_{UV}(\gamma) = \gamma \lceil_V$ for all $\gamma \in \widetilde{M}(U)$. Of course, we define $\rho_{U\emptyset} : \widetilde{M}(U) \longrightarrow \widetilde{M}(\emptyset) = 0$ to be the zero map.

It is easy to see that $\rho_{UV,R} : \widetilde{R}(U) \longrightarrow \widetilde{R}(V)$ is a homomorphism of R-algebras and that $\rho_{UV,M} : \widetilde{M}(U) \longrightarrow \widetilde{M}(V)$ is an $\widetilde{R}(U)$-homomorphism. Note also that the diagram

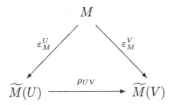

commutes. Statements similar to those in this paragraph hold if $V = \emptyset$.

The next exercise shows that the notion of local family of fractions, as introduced in 20.1.6, can be viewed as a generalization of the notion of regular function on a quasi-affine variety.

20.1.7 Exercise. Let V be an affine variety over the algebraically closed field K, and take $R := \mathcal{O}(V)$, $S := R$ and $T := \operatorname{Spec}(R)$, as in Exercise 20.1.5; let $j : V \xrightarrow{\approx} \max T$ be the homeomorphism introduced in that exercise.

Let $\mathfrak{b}$ be a non-zero ideal of $\mathcal{O}(V)$, and let U be the non-empty open subset $\{p \in V : f(p) \neq 0 \text{ for some } f \in \mathfrak{b}\}$ of V. We have $j(U) = \max T \setminus \operatorname{Var}(\mathfrak{b})$.

(i) Recall that, for $p \in U$, we have $\mathcal{O}_{V,p} = \mathcal{O}(V)_{I_V(p)} = \mathcal{O}(V)_{j(p)} = (S \setminus j(p))^{-1}R$, and that each $f \in \mathcal{O}(U)$ can be viewed as an element of the ring $\mathcal{O}_{V,p}$. Thus we can define a ring homomorphism $\max \iota_U : \mathcal{O}(U) \longrightarrow \tilde{R}(j(U))$ by the assignment $f \mapsto (f)_{\mathfrak{m} \in j(U)}$. Show that $\max \iota_U$ is an isomorphism.

(ii) Now let $\tilde{U}$ be the open subset $T \setminus \operatorname{Var}(\mathfrak{b})$ of $T = \operatorname{Spec}(R)$. Given $\mathfrak{p} \in \tilde{U}$, there exists $p \in U$ such that $\mathfrak{p} \subseteq j(p)$, so that $(S \setminus j(p))^{-1}R = R_{j(p)}$ can be viewed as a subring of $R_{\mathfrak{p}} = (S \setminus \mathfrak{p})^{-1}R$. Thus we can define a ring homomorphism $\iota_U : \mathcal{O}(U) \longrightarrow \tilde{R}(\tilde{U})$ by the assignment $f \mapsto (f)_{\mathfrak{p} \in \tilde{U}}$. Show that ι_U is an isomorphism.

In 2.3.2, we described, in terms of ideal transforms, the ring of regular functions on a non-empty open subset of an affine algebraic variety over an algebraically closed field. The last exercise therefore shows that, in one special case at least, a ring of local families of fractions can be described in terms of ideal transforms. The next major aim for this chapter is a description of general modules of local families of fractions in terms of ideal transforms, and we now embark on the preparations for this result.

20.1.8 Lemma. *(The standard hypotheses of 20.1.2 apply.)* Set $T \setminus \operatorname{Var}(\mathfrak{a}) =: U$; assume that $U \neq \emptyset$. Then

(i) *for each* $\gamma = (\gamma_\mathfrak{p})_{\mathfrak{p} \in U} \in \widetilde{M}(U)$, *there exist* $r \in \mathbb{N}$, $s_1, \ldots, s_r \in S \cap \mathfrak{a}$ *and* $m_1, \ldots, m_r \in M$ *such that* $U = \bigcup_{i=1}^r \left(T \setminus \mathrm{Var}((s_i)) \right)$ *and* $\gamma_\mathfrak{p} = m_i/s_i$ *for all* $\mathfrak{p} \in T \setminus \mathrm{Var}((s_i))$ *(for* $i = 1, \ldots, r$*);*

(ii) $\Gamma_\mathfrak{a}(\widetilde{M}(U)) = 0$*; and*

(iii) $\mathrm{Ker}(\varepsilon_M^U) = \Gamma_\mathfrak{a}(M)$, *where* $\varepsilon_M^U : M \longrightarrow \widetilde{M}(U)$ *is the homomorphism defined in* 20.1.6(iii).

Proof. (i) As U is quasi-compact (see 20.1.1(iii)), there exist $n \in \mathbb{N}$, $W_1, \ldots,$ $W_n \in \mathcal{U}_T^{(S)}$, $t_1, \ldots, t_n \in S$ and $l_1, \ldots, l_n \in M$ such that $U = \bigcup_{j=1}^n W_j$ and, for each $j = 1, \ldots, n$ and each $\mathfrak{p} \in W_j$, we have $t_j \notin \mathfrak{p}$ and $\gamma_\mathfrak{p} = l_j/t_j$.

Let $j \in \{1, \ldots, n\}$. Since $W_j \in \mathcal{U}_T^{(S)}$, there exists $\mathfrak{b}_j \in \mathfrak{A}_S$ such that $W_j = T \setminus \mathrm{Var}(\mathfrak{b}_j)$. Now $W_j \subseteq U = T \setminus \mathrm{Var}(\mathfrak{a})$, and so

$$W_j = W_j \cap U = (T \setminus \mathrm{Var}(\mathfrak{b}_j)) \cap (T \setminus \mathrm{Var}(\mathfrak{a})) = T \setminus \mathrm{Var}(\mathfrak{a}\mathfrak{b}_j).$$

We therefore can, and do, assume that $\mathfrak{b}_j \subseteq \mathfrak{a}$. Suppose that $\mathfrak{b}_j$ is generated by $w_{j1}, \ldots, w_{jn_j} \in S \cap \mathfrak{a}$. Set $W_{jk} := T \setminus \mathrm{Var}((w_{jk}))$ for all $k = 1, \ldots, n_j$; then $W_j = \bigcup_{k=1}^{n_j} W_{jk}$, and, for $k \in \{1, \ldots, n_j\}$ and all $\mathfrak{p} \in W_{jk}$, we have $t_j w_{jk} \in S \setminus \mathfrak{p}$ (so that $W_{jk} = T \setminus \mathrm{Var}((w_{jk})) = T \setminus \mathrm{Var}((t_j w_{jk}))$) and we can write

$$\gamma_\mathfrak{p} = \frac{l_j}{t_j} = \frac{w_{jk}l_j}{w_{jk}t_j} \quad \text{in } (S \setminus \mathfrak{p})^{-1}M.$$

We have therefore only to relabel the pairs

$$(w_{jk}t_j, w_{jk}l_j) \in (S \cap \mathfrak{a}) \times M \quad (k = 1, \ldots, n_j, j = 1, \ldots, n)$$

in order to complete the proof of part (i).

(ii) Let $\gamma = (\gamma_\mathfrak{p})_{\mathfrak{p} \in U} \in \Gamma_\mathfrak{a}(\widetilde{M}(U))$. There exists $n \in \mathbb{N}$ such that $\mathfrak{a}^n \gamma = 0$. Consider a $\mathfrak{p} \in U = T \setminus \mathrm{Var}(\mathfrak{a}^n)$. Since $\mathfrak{a}^n \not\subseteq \mathfrak{p}$ and $\mathfrak{a}^n$ can be generated by elements of S, there exists $s_\mathfrak{p} \in (S \cap \mathfrak{a}^n) \setminus \mathfrak{p}$. Since $s_\mathfrak{p} \gamma = 0$, we have $s_\mathfrak{p} \gamma_\mathfrak{p} = 0$ and we see that, in $(S \setminus \mathfrak{p})^{-1}M$,

$$\gamma_\mathfrak{p} = \frac{1}{s_\mathfrak{p}}(s_\mathfrak{p} \gamma_\mathfrak{p}) = 0.$$

Hence $\gamma = 0$.

(iii) As $\varepsilon_M^U : M \longrightarrow \widetilde{M}(U)$ is an R-homomorphism, we have

$$\varepsilon_M^U(\Gamma_\mathfrak{a}(M)) \subseteq \Gamma_\mathfrak{a}(\widetilde{M}(U)) = 0$$

(we have used part (ii) here), and so $\mathrm{Ker}(\varepsilon_M^U) \supseteq \Gamma_\mathfrak{a}(M)$.

Now let $m \in \mathrm{Ker}(\varepsilon_M^U)$. Choose a family of denominators

$$(s_\mathfrak{p})_{\mathfrak{p} \in U} \in \prod_{\mathfrak{p} \in U}(S \setminus \mathfrak{p}).$$

Then $0 = \varepsilon_M^U(m) = (s_{\mathfrak{p}}m/s_{\mathfrak{p}})_{\mathfrak{p} \in U}$, and so, for each $\mathfrak{p} \in U$, there exists $t_{\mathfrak{p}} \in S \setminus \mathfrak{p}$ such that $t_{\mathfrak{p}} s_{\mathfrak{p}} m = 0$.

Let $\mathfrak{b} = \sum_{\mathfrak{p} \in U} R t_{\mathfrak{p}} s_{\mathfrak{p}}$; then $\mathfrak{b}$ is an ideal belonging to $\mathfrak{A}_S$ and $\mathfrak{b}m = 0$. As, for each $\mathfrak{p} \in U$, we have $t_{\mathfrak{p}} s_{\mathfrak{p}} \notin \mathfrak{p}$, it follows that $U \cap \mathrm{Var}(\mathfrak{b}) = \emptyset$, so that $T \cap \mathrm{Var}(\mathfrak{b}) \subseteq \mathrm{Var}(\mathfrak{a})$ and $\mathrm{Var}((S)) \cup (T \cap \mathrm{Var}(\mathfrak{b})) \subseteq \mathrm{Var}(\mathfrak{a})$. Therefore, since T is essential with respect to S,

$$\sqrt{\mathfrak{b}} = \bigcap_{\mathfrak{p} \in \mathrm{Var}((S)) \cup (T \cap \mathrm{Var}(\mathfrak{b}))} \mathfrak{p} \supseteq \bigcap_{\mathfrak{p} \in \mathrm{Var}(\mathfrak{a})} \mathfrak{p} = \sqrt{\mathfrak{a}} \supseteq \mathfrak{a}.$$

Thus there exists $h \in \mathbb{N}$ such that $\mathfrak{a}^h \subseteq \mathfrak{b}$; therefore $\mathfrak{a}^h m \subseteq \mathfrak{b}m = 0$ and $m \in \Gamma_{\mathfrak{a}}(M)$. □

20.1.9 Lemma. *(The standard hypotheses of* 20.1.2 *apply.) Set* $T \setminus \mathrm{Var}(\mathfrak{a}) =:$ U, *assume that* $U \neq \emptyset$, *and let* $W \in \mathcal{U}_T^{(S)}$ *be such that* $U \subseteq W$. *Consider the restriction homomorphism* $\rho_{WU} : \widetilde{M}(W) \longrightarrow \widetilde{M}(U)$ *of* 20.1.6(iv). *Then* $\mathrm{Ker}(\rho_{WU}) = \Gamma_{\mathfrak{a}}(\widetilde{M}(W))$.

Proof. By 20.1.8(ii), we have $\rho_{WU}(\Gamma_{\mathfrak{a}}(\widetilde{M}(W))) \subseteq \Gamma_{\mathfrak{a}}(\widetilde{M}(U)) = 0$, so that $\mathrm{Ker}(\rho_{WU}) \supseteq \Gamma_{\mathfrak{a}}(\widetilde{M}(W))$.

There exists $\mathfrak{c} \in \mathfrak{A}_S$ such that $W = T \setminus \mathrm{Var}(\mathfrak{c})$. Let $\gamma = (\gamma_{\mathfrak{p}})_{\mathfrak{p} \in W} \in \mathrm{Ker}(\rho_{WU})$. By Lemma 20.1.8(i), there exist $r \in \mathbb{N}$, $s_1, \ldots, s_r \in S \cap \mathfrak{c}$ and $m_1, \ldots, m_r \in M$ such that $W = \bigcup_{i=1}^r (T \setminus \mathrm{Var}((s_i)))$ and, for each $i = 1, \ldots, r$ and each $\mathfrak{p} \in T \setminus \mathrm{Var}((s_i))$, we have $\gamma_{\mathfrak{p}} = m_i/s_i$. Since $\gamma \in \mathrm{Ker}(\rho_{WU})$, we have $\gamma_{\mathfrak{p}} = 0$ for all $\mathfrak{p} \in U = T \setminus \mathrm{Var}(\mathfrak{a})$.

Let $i \in \{1, \ldots, r\}$. Set

$$U' := (T \setminus \mathrm{Var}(\mathfrak{a})) \cap (T \setminus \mathrm{Var}((s_i))) = T \setminus \mathrm{Var}(s_i \mathfrak{a}).$$

Then $m_i/s_i = 0$ in $(S \setminus \mathfrak{p})^{-1}M$ for all $\mathfrak{p} \in U'$. This means that $\varepsilon_M^{U'}(m_i) = 0$, so that there exists $h_i \in \mathbb{N}$ such that $(s_i \mathfrak{a})^{h_i} m_i = 0$ by 20.1.8(iii). Let $h := \max\{h_i : i = 1, \ldots, r\}$.

Now let $\mathfrak{p} \in W$. There exists $i \in \{1, \ldots, r\}$ with $\mathfrak{p} \in T \setminus \mathrm{Var}((s_i))$, and then, for all $d \in \mathfrak{a}^h$, we have, in $(S \setminus \mathfrak{p})^{-1}M$,

$$d\gamma_{\mathfrak{p}} = \frac{dm_i}{s_i} = \frac{ds_i^h m_i}{s_i^{h+1}} = 0.$$

Hence $\mathfrak{a}^h \gamma = 0$, and $\gamma \in \Gamma_{\mathfrak{a}}(\widetilde{M}(W))$. □

20.1.10 Lemma. *(The standard hypotheses of* 20.1.2 *are in force.) Set* $U :=$ $T \setminus \mathrm{Var}(\mathfrak{a})$. *The homomorphism* $\varepsilon_M^U : M \longrightarrow \widetilde{M}(U)$ *of* 20.1.6(iii) *has* $\mathfrak{a}$-*torsion cokernel.*

Proof. We can assume that $U \neq \emptyset$.

Let $\gamma = (\gamma_{\mathfrak{p}})_{\mathfrak{p} \in U} \in \widetilde{M}(U)$. By 20.1.8(i), there exist $r \in \mathbb{N}$, $s_1, \ldots, s_r \in S \cap \mathfrak{a}$ and $m_1, \ldots, m_r \in M$ such that $\bigcup_{i=1}^{r} (T \setminus \text{Var}((s_i))) = U$ and, for each $i = 1, \ldots, r$ and each $\mathfrak{p} \in T \setminus \text{Var}((s_i))$, we have $\gamma_{\mathfrak{p}} = m_i/s_i$.

Let $i \in \{1, \ldots, r\}$, and let $U_i := T \setminus \text{Var}((s_i))$. Then, for each $\mathfrak{p} \in U_i$, we have $s_i \gamma_{\mathfrak{p}} = s_i m_i/s_i$ in $(S \setminus \mathfrak{p})^{-1} M$. But this means that

$$\rho_{UU_i}(s_i\gamma) = s_i \rho_{UU_i}(\gamma) = \varepsilon_M^{U_i}(m_i) = \rho_{UU_i}(\varepsilon_M^U(m_i)),$$

so that $s_i\gamma - \varepsilon_M^U(m_i) \in \text{Ker}(\rho_{UU_i}) = \Gamma_{(s_i)}(\widetilde{M}(U))$ by Lemma 20.1.9. Hence there exists $n_i \in \mathbb{N}$ such that $s_i^{n_i}(s_i\gamma - \varepsilon_M^U(m_i)) = 0$. Define

$$n := \max\{n_i + 1 : i = 1, \ldots, r\}.$$

Then, for all $i = 1, \ldots, r$, we have

$$s_i^n \gamma = s_i^{n-1-n_i} s_i^{n_i} s_i \gamma = s_i^{n-1-n_i} s_i^{n_i} \varepsilon_M^U(m_i) = \varepsilon_M^U(s_i^{n-1} m_i) \in \varepsilon_M^U(M).$$

As

$$T \setminus \text{Var}\left(\sum_{i=1}^{r} Rs_i^n\right) = T \setminus \text{Var}\left(\sum_{i=1}^{r} Rs_i\right) = \bigcup_{i=1}^{r} U_i = U = T \setminus \text{Var}(\mathfrak{a}),$$

it follows that $T \cap \text{Var}\left(\sum_{i=1}^{r} Rs_i^n\right) \subseteq \text{Var}(\mathfrak{a})$. Therefore, since T is essential with respect to S,

$$\sqrt{\sum_{i=1}^{r} Rs_i^n} = \bigcap_{\mathfrak{p} \in \text{Var}((S)) \cup (T \cap \text{Var}(\sum_{i=1}^{r} Rs_i^n))} \mathfrak{p} \supseteq \bigcap_{\mathfrak{p} \in \text{Var}(\mathfrak{a})} \mathfrak{p} = \sqrt{\mathfrak{a}} \supseteq \mathfrak{a}.$$

Thus there exists $h \in \mathbb{N}$ such that $\mathfrak{a}^h \subseteq \sum_{i=1}^{r} Rs_i^n$, and

$$\mathfrak{a}^h \gamma \subseteq \left(\sum_{i=1}^{r} Rs_i^n\right)\gamma = \sum_{i=1}^{r} Rs_i^n \gamma \subseteq \varepsilon_M^U(M). \qquad \square$$

20.1.11 Lemma. *(The standard hypotheses of* 20.1.2 *apply.) Set* $T \setminus \text{Var}(\mathfrak{a})$ $=: U$, *and let* $W \in \mathcal{U}_T^{(S)}$ *be such that* $U \subseteq W$. *Consider the restriction homomorphism* $\rho_{WU} : \widetilde{M}(W) \longrightarrow \widetilde{M}(U)$ *of* 20.1.6(iv). *The* R-*module* $\text{Coker} \, \rho_{WU}$ *is* $\mathfrak{a}$-*torsion.*

Proof. Since $\varepsilon_M^U = \rho_{WU} \circ \varepsilon_M^W$ by 20.1.6(iv), we see that $\text{Im} \, \varepsilon_M^U \subseteq \text{Im} \, \rho_{WU}$ and $\text{Coker} \, \rho_{WU}$ is a homomorphic image of $\text{Coker} \, \varepsilon_M^U$. The claim therefore follows from Lemma 20.1.10. $\qquad \square$

20.1.12 Remark and Notation. With the hypotheses of 20.1.2, let $h : M \to N$ be a homomorphism of R-modules, and let $\emptyset \neq U \in \mathcal{U}_T^{(S)}$.

(i) Now h induces, for each $\mathfrak{p} \in U$, an $(S \setminus \mathfrak{p})^{-1}R$-homomorphism

$$(S \setminus \mathfrak{p})^{-1}h : (S \setminus \mathfrak{p})^{-1}M \longrightarrow (S \setminus \mathfrak{p})^{-1}N.$$

There is therefore induced a $\prod_{\mathfrak{p} \in U}(S \setminus \mathfrak{p})^{-1}R$-homomorphism

$$\prod_{\mathfrak{p} \in U}(S \setminus \mathfrak{p})^{-1}h : \prod_{\mathfrak{p} \in U}(S \setminus \mathfrak{p})^{-1}M \longrightarrow \prod_{\mathfrak{p} \in U}(S \setminus \mathfrak{p})^{-1}N.$$

It is easy to see that the image under this map of an S-local family of fractions over U with numerators in M is an S-local family of fractions over U with numerators in N. Thus h induces an $\widetilde{R}(U)$-homomorphism $\widetilde{h}(U) : \widetilde{M}(U) \longrightarrow \widetilde{N}(U)$ for which

$$\widetilde{h}(U)\left((\gamma_{\mathfrak{p}})_{\mathfrak{p} \in U}\right) = \left((S \setminus \mathfrak{p})^{-1}h(\gamma_{\mathfrak{p}})\right)_{\mathfrak{p} \in U} \quad \text{for all } (\gamma_{\mathfrak{p}})_{\mathfrak{p} \in U} \in \widetilde{M}(U).$$

Of course, we define $\widetilde{h}(\emptyset) : \widetilde{M}(\emptyset) \longrightarrow \widetilde{N}(\emptyset)$ to be the zero homomorphism.

(ii) It is clear that $\widetilde{\mathrm{Id}_M}(U) = \mathrm{Id}_{\widetilde{M}(U)}$, and that, if $g : N \longrightarrow L$ is another homomorphism of R-modules, then $\widetilde{g \circ h}(U) = \widetilde{g}(U) \circ \widetilde{h}(U)$. Thus $\widetilde{\bullet}(U)$ is a covariant functor from $\mathcal{C}(R)$ to $\mathcal{C}(\widetilde{R}(U))$. It is straightforward to check that this functor is R-linear (note that an $\widetilde{R}(U)$-module can be regarded as an R-module by means of ε_R^U).

Clearly, similar comments can be made when U is replaced by $\emptyset$.

(iii) Now consider a second open set $V \in \mathcal{U}_T^{(S)}$ such that $V \subseteq U$. It is clear that (even if $V = \emptyset$) the diagram

$$
\begin{array}{ccc}
\widetilde{M}(U) & \xrightarrow{\;\widetilde{h}(U)\;} & \widetilde{N}(U) \\
{\scriptstyle \rho_{UV,M}}\big\downarrow & & \big\downarrow{\scriptstyle \rho_{UV,N}} \\
\widetilde{M}(V) & \xrightarrow{\;\widetilde{h}(V)\;} & \widetilde{N}(V)
\end{array}
$$

commutes, and so $\rho_{UV} : \widetilde{\bullet}(U) \longrightarrow \widetilde{\bullet}(V)$ is a natural transformation of functors (from $\mathcal{C}(R)$ to $\mathcal{C}(\widetilde{R}(U))$). Also, $\rho_{\emptyset\emptyset}$ is a natural transformation.

(iv) Finally, the diagram

$$
\begin{array}{ccc}
M & \xrightarrow{\;h\;} & N \\
{\scriptstyle \varepsilon_M^U}\big\downarrow & & \big\downarrow{\scriptstyle \varepsilon_N^U} \\
\widetilde{M}(U) & \xrightarrow{\;\widetilde{h}(U)\;} & \widetilde{N}(U)
\end{array}
$$

also commutes, and so $\varepsilon^U : \mathrm{Id} \longrightarrow \widetilde{\bullet}(U)$ is also a natural transformation of functors (from $\mathcal{C}(R)$ to itself). In addition, $\varepsilon^{\emptyset} : \mathrm{Id} \longrightarrow \widetilde{\bullet}(\emptyset)$ is a natural transformation.

20.1.13 ♯Exercise. Show that, in the situation of 20.1.12, the functor $\widetilde{\bullet}(U) : \mathcal{C}(R) \longrightarrow \mathcal{C}(\widetilde{R}(U))$ is left exact.

We are now ready to establish the central result of this chapter; this result is a version of the Deligne Isomorphism.

20.1.14 The Deligne Isomorphism Theorem. *(The standard hypotheses of 20.1.2 apply.) Set $U := T \setminus \mathrm{Var}(\mathfrak{a})$; assume that $U \neq \emptyset$.*

(i) *There is a unique R-isomorphism $\nu_{\mathfrak{a},M} : \widetilde{M}(U) \xrightarrow{\cong} D_{\mathfrak{a}}(M)$ such that the diagram*

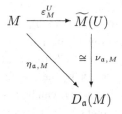

commutes. (Occasionally, $\nu_{\mathfrak{a},M}$ will be written as $\nu_{\mathfrak{a},M}^{U}$ when it is important to stress the dependence on the open set U.) Moreover, if $h : M \longrightarrow N$ is a homomorphism of R-modules, then the diagram

$$
\begin{array}{ccc}
\widetilde{M}(U) & \xrightarrow{\widetilde{h}(U)} & \widetilde{N}(U) \\
\cong \downarrow \nu_{\mathfrak{a},M} & & \cong \downarrow \nu_{\mathfrak{a},N} \\
D_{\mathfrak{a}}(M) & \xrightarrow{D_{\mathfrak{a}}(h)} & D_{\mathfrak{a}}(N)
\end{array}
$$

commutes, and so $\nu_{\mathfrak{a}} : \widetilde{\bullet}(U) \longrightarrow D_{\mathfrak{a}}$ is a natural equivalence of functors (from $\mathcal{C}(R)$ to itself).

(ii) *The map $\nu_{\mathfrak{a},R} : \widetilde{R}(U) \xrightarrow{\cong} D_{\mathfrak{a}}(R)$ is an isomorphism of R-algebras.*

Proof. By 20.1.8(iii) and 20.1.10, both the kernel and cokernel of ε_{M}^{U} are $\mathfrak{a}$-torsion. It is therefore immediate from 2.2.15(ii) that there is a unique

R-homomorphism $\nu_{\mathfrak{a},M} : \widetilde{M}(U) \longrightarrow D_{\mathfrak{a}}(M)$ such that the diagram

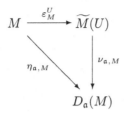

commutes, and, in fact, it follows from the formula for $\nu_{\mathfrak{a},M}$ provided by 2.2.15(ii) that this map is monomorphic, since we know from 20.1.8(ii) that $\Gamma_{\mathfrak{a}}(\widetilde{M}(U)) = 0$. Note also that it follows from 2.2.17 that $\nu_{\mathfrak{a},R}$ is a homomorphism of R-algebras, and so part (ii) will follow from part (i).

To show that $\nu_{\mathfrak{a},M}$ is surjective, let $y \in D_{\mathfrak{a}}(M)$. Then there exists $n \in \mathbb{N}$ and $h \in \operatorname{Hom}_R(\mathfrak{a}^n, M)$ such that y is the natural image of h in $D_{\mathfrak{a}}(M)$. As $\mathfrak{a}$ is generated by elements of S, there exists, for each $\mathfrak{p} \in U$, an element $s_{\mathfrak{p}} \in (\mathfrak{a} \cap S) \setminus \mathfrak{p}$. Note that

$$\delta := \left(\frac{h(s_{\mathfrak{p}}^n)}{s_{\mathfrak{p}}^n} \right)_{\mathfrak{p} \in U} \in \prod_{\mathfrak{p} \in U} (S \setminus \mathfrak{p})^{-1} M.$$

Now, for $\mathfrak{q} \in U$, we have, for each $\mathfrak{p} \in U \setminus \operatorname{Var}((s_{\mathfrak{q}}))$, that, in $(S \setminus \mathfrak{p})^{-1} M$,

$$\frac{h(s_{\mathfrak{p}}^n)}{s_{\mathfrak{p}}^n} = \frac{s_{\mathfrak{q}}^n h(s_{\mathfrak{p}}^n)}{s_{\mathfrak{q}}^n s_{\mathfrak{p}}^n} = \frac{h(s_{\mathfrak{q}}^n s_{\mathfrak{p}}^n)}{s_{\mathfrak{q}}^n s_{\mathfrak{p}}^n} = \frac{s_{\mathfrak{p}}^n h(s_{\mathfrak{q}}^n)}{s_{\mathfrak{p}}^n s_{\mathfrak{q}}^n} = \frac{h(s_{\mathfrak{q}}^n)}{s_{\mathfrak{q}}^n},$$

so that δ is an S-local family of fractions over U, that is, $\delta \in \widetilde{M}(U)$. Our immediate aim is to show that $\nu_{\mathfrak{a},M}(\delta) = y$.

For each $r \in \mathfrak{a}^n$, we have

$$r\delta = \left(r \frac{h(s_{\mathfrak{p}}^n)}{s_{\mathfrak{p}}^n} \right)_{\mathfrak{p} \in U} = \left(\frac{h(r s_{\mathfrak{p}}^n)}{s_{\mathfrak{p}}^n} \right)_{\mathfrak{p} \in U} = \left(\frac{s_{\mathfrak{p}}^n h(r)}{s_{\mathfrak{p}}^n} \right)_{\mathfrak{p} \in U} = \varepsilon_M^U(h(r)).$$

Hence $r\nu_{\mathfrak{a},M}(\delta) = \nu_{\mathfrak{a},M}(r\delta) = \nu_{\mathfrak{a},M}(\varepsilon_M^U(h(r))) = \eta_{\mathfrak{a},M}(h(r))$, and this is just the natural image in $D_{\mathfrak{a}}(M)$ of $h' \in \operatorname{Hom}_R(\mathfrak{a}^n, M)$, where $h'(r') = r'h(r) = rh(r')$ for all $r' \in \mathfrak{a}^n$. We thus see that $r\nu_{\mathfrak{a},M}(\delta) = ry$ for all $r \in \mathfrak{a}^n$. Hence $\nu_{\mathfrak{a},M}(\delta) - y \in \Gamma_{\mathfrak{a}}(D_{\mathfrak{a}}(M))$, which is zero by 2.2.10(iv). Thus $\nu_{\mathfrak{a},M}$ is surjective, and so is an isomorphism.

To prove the second part, we wish to show that, in the diagram

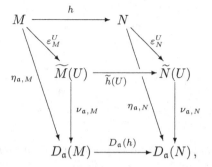

the front square commutes. We know from the first part of this proof that the two side triangles commute; furthermore, the top square commutes by 20.1.12(iv), while the sloping rectangle on the underside commutes because $\eta_\mathfrak{a}$ is a natural transformation. Therefore

$$\nu_{\mathfrak{a},N} \circ \widetilde{h}(U) \circ \varepsilon_M^U = \nu_{\mathfrak{a},N} \circ \varepsilon_N^U \circ h = \eta_{\mathfrak{a},N} \circ h$$
$$= D_\mathfrak{a}(h) \circ \eta_{\mathfrak{a},M} = D_\mathfrak{a}(h) \circ \nu_{\mathfrak{a},M} \circ \varepsilon_M^U.$$

However, by 2.2.13(ii), 20.1.8(iii) and 20.1.10, there is a unique R-homomorphism $h' : \widetilde{M}(U) \to D_\mathfrak{a}(N)$ such that the diagram

$$
\begin{array}{ccc}
M & \xrightarrow{\varepsilon_M^U} & \widetilde{M}(U) \\
{\scriptstyle h}\downarrow & & \downarrow{\scriptstyle h'} \\
N & \xrightarrow{\eta_{\mathfrak{a},N}} & D_\mathfrak{a}(N)
\end{array}
$$

commutes, and so $\nu_{\mathfrak{a},N} \circ \widetilde{h}(U) = D_\mathfrak{a}(h) \circ \nu_{\mathfrak{a},M}$, as required. $\qquad\square$

The observant reader might have noticed that, in the Deligne Isomorphism Theorem 20.1.14, we did not consider the case where $U = \emptyset$, that is, where $T \subseteq \mathrm{Var}(\mathfrak{a})$. The next exercise shows that, in some circumstances, there is a 'Deligne Isomorphism' in the case where $U = \emptyset$.

20.1.15 Definition and ♯Exercise. (The standard hypotheses of 20.1.2 apply.) We say that T is *large with respect to* S if and only if, whenever $\mathfrak{c} \in \mathfrak{A}_S$ is such that $T \subseteq \mathrm{Var}(\mathfrak{c})$, then $\mathrm{Var}(\mathfrak{c}) = \mathrm{Spec}(R)$, that is, $\mathfrak{c}$ is nilpotent. Suppose that this is the case and that $T \subseteq \mathrm{Var}(\mathfrak{a})$. Show that

(i) $\mathrm{Ker}(\varepsilon_M^\emptyset) = \Gamma_\mathfrak{a}(M)$ (compare 20.1.8(iii));
(ii) if $W \in \mathcal{U}_T^{(S)}$, then $\mathrm{Ker}(\rho_{W\emptyset}) = \Gamma_\mathfrak{a}(\widetilde{M}(W))$ (compare 20.1.9); and

(iii) there is a natural equivalence of functors $\nu_{\mathfrak{a}}^{\emptyset} : \widetilde{}(\emptyset) \longrightarrow D_{\mathfrak{a}}$ (compare 20.1.14(i)).

In several subsequent results that depend on the Deligne Isomorphism Theorem 20.1.14, the reader will find the hypothesis (about $U := T \setminus \mathrm{Var}(\mathfrak{a})$) that '$U \neq \emptyset$ (or $U = \emptyset$ and T is large with respect to S)'. The facts that the results concerned are still valid under the alternative hypothesis in parentheses are in most cases easy consequences of Exercise 20.1.15.

20.1.16 Exercise. Consider the special case of the situation of 20.1.12 in which $R = K[X, Y]$ is the polynomial ring in two indeterminates over a field K, and take $S = R$ and $T = \mathrm{Spec}(R)$, so that T is essential with respect to S (see 20.1.3(i)). Let $U := T \setminus \mathrm{Var}((X, Y))$. Use the natural homomorphism $R \to R/XR$ to show that the functor $\widetilde{}(U) : \mathcal{C}(R) \longrightarrow \mathcal{C}(\widetilde{R}(U))$ is not exact.

The following exercise shows that Theorem 20.1.14 can indeed be viewed as a natural generalization of the isomorphism established in 2.3.2.

20.1.17 Exercise. Consider again the situation of 20.1.5 and 20.1.7, so that V is an affine variety over the algebraically closed field K, $R := \mathcal{O}(V)$, $S := R$, $T := \mathrm{Spec}(R)$, $\mathfrak{b}$ is a non-zero ideal of $\mathcal{O}(V)$, U is the open subset of V determined by $\mathfrak{b}$, and $\widetilde{U}$ is the open subset $T \setminus \mathrm{Var}(\mathfrak{b})$ of $T = \mathrm{Spec}(R)$.

Let $j : V \xrightarrow{\approx} \max T$ be the homeomorphism of 20.1.5, and consider the ring isomorphisms $\iota_U : \mathcal{O}(U) \longrightarrow \widetilde{R}(\widetilde{U})$ and $\max \iota_U : \mathcal{O}(U) \longrightarrow \widetilde{R}(j(U))$ of 20.1.7. Show that both $\nu_{\mathfrak{b},R}^{\widetilde{U}} \circ \iota_U : \mathcal{O}(U) \longrightarrow D_{\mathfrak{b}}(R)$ and $\nu_{\mathfrak{b},R}^{j(U)} \circ \max \iota_U : \mathcal{O}(U) \longrightarrow D_{\mathfrak{b}}(R)$ coincide with the isomorphism $\nu_{V,\mathfrak{b}}$ of 2.3.2.

20.1.18 ♯Exercise. Consider again the situation and notation of the Deligne Isomorphism Theorem 20.1.14.

(i) Show that, when $D_{\mathfrak{a}}(M)$ is regarded as an $\widetilde{R}(U)$-module via $\nu_{\mathfrak{a},R}$, the map $\nu_{\mathfrak{a},M} : \widetilde{M}(U) \xrightarrow{\cong} D_{\mathfrak{a}}(M)$ is an $\widetilde{R}(U)$-isomorphism.
(ii) Let $\mathfrak{b} \in \mathfrak{A}_S$ be such that $\mathfrak{a} \subseteq \mathfrak{b}$, and set $W := T \setminus \mathrm{Var}(\mathfrak{b})$. Show that the diagram

$$
\begin{array}{ccc}
\widetilde{M}(W) & \xrightarrow{\ \rho_{WU}\ } & \widetilde{M}(U) \\[2mm]
{\scriptstyle \cong}\Big\downarrow{\scriptstyle \nu_{\mathfrak{b},M}} & & {\scriptstyle \cong}\Big\downarrow{\scriptstyle \nu_{\mathfrak{a},M}} \\[2mm]
D_{\mathfrak{b}}(M) & \xrightarrow{\ \alpha_{\mathfrak{b},\mathfrak{a},M}\ } & D_{\mathfrak{a}}(M)\,,
\end{array}
$$

in which $\alpha_{\mathfrak{b},\mathfrak{a},M}$ is the natural map of 2.2.23, commutes.

20.1.19 ♯Exercise. Consider once more the situation and notation of the Deligne Isomorphism Theorem 20.1.14.

(i) Show that, if $\mathfrak{a}$ contains a poor M-sequence of length 2, then the homomorphism $\varepsilon_M^U : M \longrightarrow \widetilde{M}(U)$ of 20.1.6(iii) is an isomorphism.

(ii) Deduce that, if S contains a unit of R, then $\varepsilon_M^T : M \longrightarrow \widetilde{M}(T)$ is an isomorphism.

20.1.20 Exercise. Let V be an affine variety over the algebraically closed field K, take $R := \mathcal{O}(V)$, let (as in 20.1.2) S be a non-empty subset of R which is closed under multiplication, and let T be a subset of $\mathrm{Spec}(R)$ which is essential with respect to S. Let $\mathfrak{b}$ be a non-zero ideal of $\mathfrak{A}_S$, and let $U = T \setminus \mathrm{Var}(\mathfrak{b})$; assume that $U \neq \emptyset$.

Show that the functor $\widetilde{\bullet}(U) : \mathcal{C}(R) \longrightarrow \mathcal{C}(\widetilde{R}(U))$ is exact if and only if the open subset $V \setminus V(\mathfrak{b})$ of V determined by $\mathfrak{b}$ (the notation is as in 6.4.1) is affine.

20.2 The Graded Deligne Isomorphism

We now intend to 'add graded frills' to the Deligne Isomorphism Theorem 20.1.14: we shall call the refined version the 'Graded Deligne Isomorphism Theorem'. One can view this refinement process as analogous to the improvement, in the case of rings of regular functions on affine varieties, in Theorem 2.3.2 afforded, in the $\mathbb{Z}$-graded case, by Proposition 15.1.3. For this work, we introduce the concept of 'homogeneous S-local family of fractions'.

20.2.1 Hypotheses for the section. The standard hypotheses of 20.1.2 will be in force throughout this section, and, in addition, we shall assume that G is a finitely generated torsion-free Abelian group and that $R = \bigoplus_{g \in G} R_g$ is G-graded, that $M = \bigoplus_{g \in G} M_g$ is a graded R-module, and that S consists entirely of homogeneous elements (so that that all the ideals in $\mathfrak{A}_S$ are graded).

20.2.2 Remark and Notation. (The hypotheses of 20.2.1 apply.) Let $\emptyset \neq U \in \mathcal{U}_T^{(S)}$.

(i) For each $\mathfrak{p} \in U$, the non-empty set $S \setminus \mathfrak{p}$ consists entirely of homogeneous elements, so that, in the light of 13.1.1, the ring $(S \setminus \mathfrak{p})^{-1}R$ carries a natural G-grading, and the $(S \setminus \mathfrak{p})^{-1}R$-module $(S \setminus \mathfrak{p})^{-1}M$ is also naturally graded; note also that the natural ring homomorphism $R \longrightarrow (S \setminus \mathfrak{p})^{-1}R$ is homogeneous in the sense of 14.1.3. For $g \in G$, we use $((S \setminus \mathfrak{p})^{-1}R)_g$ and $((S \setminus \mathfrak{p})^{-1}M)_g$ to denote the components of degree g of $(S \setminus \mathfrak{p})^{-1}R$ and $(S \setminus \mathfrak{p})^{-1}M$ respectively.

(ii) Let $g \in G$. The elements of $\prod_{\mathfrak{p} \in U} ((S \setminus \mathfrak{p})^{-1}M)_g$ are called *families of homogeneous fractions of degree g over U with numerators in M and denominators in S.* The set $\widetilde{M}(U) \cap \prod_{\mathfrak{p} \in U} ((S \setminus \mathfrak{p})^{-1}M)_g$ of such families which are also local will be denoted by $\widetilde{M}(U)_g$; the members of $\widetilde{M}(U)_g$ are called *homogeneous S-local families of fractions of degree g over U.* Of course, we set $\widetilde{M}(\emptyset)_g = 0$.

(iii) Let $g, h \in G$. It is easy to check that $\widetilde{R}(U)_0$ is an R_0-subalgebra of $\widetilde{R}(U)$, that $\widetilde{M}(U)_g$ is an $\widetilde{R}(U)_0$-submodule of $\widetilde{M}(U)$, and that $\sigma\gamma \in \widetilde{M}(U)_{g+h}$ for all $\sigma \in \widetilde{R}(U)_g$ and all $\gamma \in \widetilde{M}(U)_h$. An important aim for us is to establish that the $\widetilde{R}(U)_g$ $(g \in G)$ provide a G-grading on $\widetilde{R}(U)$, and that the $\widetilde{M}(U)_g$ $(g \in G)$ provide a grading on the $\widetilde{R}(U)$-module $\widetilde{M}(U)$. Lemma 20.2.3 below provides a key for the establishment of this aim.

(iv) Let $\gamma = (\gamma_{\mathfrak{p}})_{\mathfrak{p} \in U} \in \prod_{\mathfrak{p} \in U} (S \setminus \mathfrak{p})^{-1}M$, and let $g \in G$. Then, for each $\mathfrak{p} \in U$, the element $\gamma_{\mathfrak{p}} \in (S \setminus \mathfrak{p})^{-1}M$ has g-th component, denoted by $(\gamma_{\mathfrak{p}})_g$, in $((S \setminus \mathfrak{p})^{-1}M)_g$. We shall refer to the family $\gamma_g := ((\gamma_{\mathfrak{p}})_g)_{\mathfrak{p} \in U} \in \prod_{\mathfrak{p} \in U} ((S \setminus \mathfrak{p})^{-1}M)_g$ as the *g-th homogeneous part of the family γ.*

20.2.3 Lemma. *(The hypotheses of 20.2.1 apply.) Let $\emptyset \neq U \in \mathcal{U}_T^{(S)}$ and let $\gamma = (\gamma_{\mathfrak{p}})_{\mathfrak{p} \in U} \in \widetilde{M}(U)$. Then*

(i) $\gamma_g := ((\gamma_{\mathfrak{p}})_g)_{\mathfrak{p} \in U} \in \widetilde{M}(U)_g$ *for each $g \in G$;*
(ii) *the set $\{g \in G : \gamma_g \neq 0\}$ is finite;*
(iii) $\gamma = \sum_{g \in G} \gamma_g$; *and*
(iv) $\gamma = 0$ *if and only if $\gamma_g = 0$ for all $g \in G$.*

Proof. Because U is quasi-compact (see 20.1.1(iii)), there exist $r \in \mathbb{N}$, W_1, $\ldots, W_r \in \mathcal{U}_T^{(S)}$, $t_1, \ldots, t_r \in S$ and $l_1, \ldots, l_r \in M$ such that $U = \bigcup_{j=1}^r W_i$ and, for each $j = 1, \ldots, r$ and each $\mathfrak{p} \in W_j$, we have $t_j \notin \mathfrak{p}$ and $\gamma_{\mathfrak{p}} = l_j/t_j$. For each $j = 1, \ldots, r$, let $h_j \in G$ be such that $t_j \in R_{h_j}$ (recall that S consists of homogeneous elements of R), and let $(l_j)_g$ denote the g-th homogeneous component of l_j (for all $g \in G$).

(i) It is obvious that, for each $j = 1, \ldots, r$,

$$(\gamma_{\mathfrak{p}})_g = \left(\frac{l_j}{t_j} \right)_g = \frac{(l_j)_{g+h_j}}{t_j} \quad \text{for all } \mathfrak{p} \in W_j.$$

Hence the family $\gamma_g = ((\gamma_{\mathfrak{p}})_g)_{\mathfrak{p} \in U}$ is local, and so lies in $\widetilde{M}(U)_g$.

(ii) For each $j = 1, \ldots, r$, let H_j denote $\{g \in G : (l_j)_{g+h_j} \neq 0\}$, a finite set, and observe that $(\gamma_{\mathfrak{p}})_g = 0$ for all $g \in G \setminus H_j$ and all $\mathfrak{p} \in W_j$. Hence $\gamma_g = 0$ for all $g \in G \setminus (H_1 \cup \cdots \cup H_r)$.

(iii) As $\gamma_{\mathfrak{p}} = \sum_{g \in G} (\gamma_{\mathfrak{p}})_g$ for each $\mathfrak{p} \in U$, this is immediate from part (ii).

(iv) Just note that $\gamma = 0$ if and only if $\gamma_{\mathfrak{p}} = 0$ for all $\mathfrak{p} \in U$, and that this is the case if and only if $(\gamma_{\mathfrak{p}})_g = 0$ for all $\mathfrak{p} \in U$ and all $g \in G$. □

20.2.4 Remark and Definition. (The hypotheses of 20.2.1 apply.) Let $\emptyset \neq U \in \mathcal{U}_T^{(S)}$. It is now immediate from 20.2.2(iii) and 20.2.3 that $\left(\widetilde{R}(U)_g \right)_{g \in G}$ provides a grading on the ring $\widetilde{R}(U)$, and that $\left(\widetilde{M}(U)_g \right)_{g \in G}$ provides $\widetilde{M}(U)$ with the structure of a graded $\widetilde{R}(U)$-module: we refer to these gradings as the *natural gradings*, and any unexplained references to gradings on $\widetilde{R}(U)$ and $\widetilde{M}(U)$ should always be interpreted as references to these natural gradings. Trivially, similar conclusions apply to $\widetilde{R}(\emptyset)$ and $\widetilde{M}(\emptyset)$.

20.2.5 Remarks. (The hypotheses of 20.2.1 apply.) Let $U \in \mathcal{U}_T^{(S)}$.

(i) It is clear that the map $\varepsilon_M^U : M \longrightarrow \widetilde{M}(U)$ of 20.1.6(iii) is homogeneous.

(ii) Let $V \in \mathcal{U}_T^{(S)}$ with $V \subseteq U$. It is also clear that the restriction map $\rho_{UV} (= \rho_{UV,M}) : \widetilde{M}(U) \longrightarrow \widetilde{M}(V)$ of 20.1.6(iv) is homogeneous.

(iii) Now let N be a second graded R-module and let $h : M \longrightarrow N$ be a homogeneous R-homomorphism. It is easy to check that the $\widetilde{R}(U)$-homomorphism $\widetilde{h}(U) : \widetilde{M}(U) \longrightarrow \widetilde{N}(U)$ of 20.1.12(i) is homogeneous.

20.2.6 ♯Exercise. (The hypotheses of 20.2.1 apply.) Let $U \in \mathcal{U}_T^{(S)}$ and $g \in G$. This exercise involves the g-th shift functor described in 13.1.1.

(i) Use the fact that $(S \setminus \mathfrak{p})^{-1}(M(g)) = ((S \setminus \mathfrak{p})^{-1}M)(g)$ for all $\mathfrak{p} \in U$ to show that $\widetilde{M(g)}(U) = \widetilde{M}(U)(g)$.

(ii) Let $V \in \mathcal{U}_T^{(S)}$ with $V \subseteq U$. Show that $\rho_{UV,M(g)} = \widetilde{\rho_{UV,M}(g)}$, that is, that the restriction homomorphism $\rho_{UV,M(g)} : \widetilde{M(g)}(U) \longrightarrow \widetilde{M(g)}(V)$ is the g-th shift of the restriction homomorphism $\rho_{UV,M} : \widetilde{M}(U) \longrightarrow \widetilde{M}(V)$.

(iii) Let N be a second graded R-module and let $h : M \longrightarrow N$ be a homogeneous R-homomorphism. Show that $\widetilde{h(g)}(U) = \widetilde{h}(U)(g)$.

Our promised Graded Deligne Isomorphism Theorem can now be obtained very quickly.

20.2.7 The Graded Deligne Isomorphism Theorem. *(The hypotheses of 20.2.1 apply.) Define $U := T \setminus \mathrm{Var}(\mathfrak{a})$; assume that $U \neq \emptyset$.*

Then the Deligne Isomorphism $\nu_{\mathfrak{a},M} (= \nu_{\mathfrak{a},M}^U) : \widetilde{M}(U) \overset{\cong}{\longrightarrow} D_{\mathfrak{a}}(M)$ of

20.1.14 *is a homogeneous R-homomorphism (with respect to the natural grading of* 20.2.4 *on* $\widetilde{M}(U)$ *and the grading of* 13.3.14 *on* $D_{\mathfrak{a}}(M)$*)*.

Proof. Since $\mathfrak{a}$ is graded, $\widetilde{M}(U)$ is graded by 20.2.4, and $\varepsilon_M^U : M \longrightarrow \widetilde{M}(U)$ is homogeneous by 20.2.5(i), this result is now immediate from 13.5.4(ii) and the Deligne Isomorphism Theorem 20.1.14. $\square$

20.2.8 Remarks. (The hypotheses of 20.2.1 apply.) Set $T \setminus \mathrm{Var}(\mathfrak{a}) =: U$.

(i) It is clear from 20.2.4 and 20.2.5(iii) that the functor $\widetilde{\bullet}(U) : \mathcal{C}(R) \longrightarrow \mathcal{C}(\widetilde{R}(U))$ has the *restriction property of 13.3.6, and (by 20.2.5(i)) that the natural transformation $\varepsilon^U : \mathrm{Id} \longrightarrow \widetilde{\bullet}(U)$ has the *restriction property of 13.3.7.

(ii) Let $V \in \mathcal{U}_T^{(S)}$ with $V \subseteq U$. By 20.2.5(ii), the natural transformation of functors $\rho_{UV} : \widetilde{\bullet}(U) \longrightarrow \widetilde{\bullet}(V)$ (from $\mathcal{C}(R)$ to $\mathcal{C}(\widetilde{R}(U))$) of 20.1.12(iii) has the *restriction property.

(iii) Assume now that $U \neq \emptyset$ (or $U = \emptyset$ and T is large with respect to S). We see from 20.2.7 (or 20.1.15) that $\nu_{\mathfrak{a},M}$ is a homogeneous R-isomorphism; by 20.1.18(i), it is also an $\widetilde{R}(U)$-isomorphism; since the natural grading on $\widetilde{M}(U)$ is a grading of this as an $\widetilde{R}(U)$-module, it is therefore automatic that the grading of 13.3.14 on $D_{\mathfrak{a}}(M)$ is a grading of $D_{\mathfrak{a}}(M)$ as an $\widetilde{R}(U)$-module. Hence $D_{\mathfrak{a}} : \mathcal{C}(R) \longrightarrow \mathcal{C}(\widetilde{R}(U))$ has the *restriction property, and $\nu_{\mathfrak{a}} : \widetilde{\bullet}(U) \longrightarrow D_{\mathfrak{a}}$, when viewed as a natural equivalence of functors from $\mathcal{C}(R)$ to $\mathcal{C}(\widetilde{R}(U))$ (or, for that matter, from $\mathcal{C}(R)$ to itself), has the *restriction property.

20.3 Links with sheaf theory

Most serious readers of this chapter will by now have realised that, in view of the results we have obtained about local families of fractions, we have essentially started to discuss sheaves, even if only implicitly. We now intend to make the connection more explicit, and to formulate 20.1.14 and 20.2.7 in sheaf-theoretic terms. Henceforth, we assume that the reader has some familiarity with basic knowledge about schemes and sheaves, although we provide numerous references to Hartshorne's book [30].

20.3.1 Remarks and ♯Exercise: sheaf-theoretic interpretations. (The standard hypotheses of 20.1.2 apply.)

(i) It is not difficult to check that the family $\left(\widetilde{R}(U) \right)_{U \in \mathcal{U}_T^{(S)}}$, together with the restriction maps $\rho_{UV,R} : \widetilde{R}(U) \longrightarrow \widetilde{R}(V)$ $(U, V \in \mathcal{U}_T^{(S)}$ with $V \subseteq U)$ defines a sheaf $\widetilde{R}$ of R-algebras on the topological space T, so that $(T, \widetilde{R})$ is a *ringed space* (see [30, Chapter II, §2, p. 72]).

(ii) Similarly, the family $\left(\widetilde{M}(U)\right)_{U \in \mathcal{U}_T^{(S)}}$, together with the restriction maps $\rho_{UV,M} : \widetilde{M}(U) \longrightarrow \widetilde{M}(V)$ $(U, V \in \mathcal{U}_T^{(S)}$ with $V \subseteq U)$, defines a *sheaf $\widetilde{M}$ of $\widetilde{R}$-modules on T* (see [30, Chapter II, §5, p. 109]). We shall call this sheaf $\widetilde{M}$ the *S-sheaf induced by M*. We shall sometimes use standard notation from sheaf theory (see [30, Chapter II, §1, p. 61]) and denote $\widetilde{M}(U)$ by $\Gamma(U, \widetilde{M})$: the elements of $\Gamma(U, \widetilde{M})$ are the *sections* of the sheaf $\widetilde{M}$ over U.

(iii) Now let $h : M \longrightarrow N$ be a homomorphism of R-modules. It follows from 20.1.12(i),(iii) that the homomorphisms $\widetilde{h}(U) : \widetilde{M}(U) \longrightarrow \widetilde{N}(U)$ $(U \in \mathcal{U}_T^{(S)})$ define a *morphism* $\widetilde{h} : \widetilde{M} \longrightarrow \widetilde{N}$ *of sheaves of $\widetilde{R}$-modules* (see [30, Chapter II, §5, p. 109]), which we call the *morphism induced by h*.

(iv) It is now clear that $\widetilde{\bullet}$ is a functor from $\mathcal{C}(R)$ to the category $\mathcal{S}(\widetilde{R})$ of sheaves of $\widetilde{R}$-modules.

(v) Let $\mathfrak{p} \in T$. The *stalk* $\widetilde{R}_{\mathfrak{p}} := \varinjlim_{U \in \mathcal{U}_{T,\mathfrak{p}}^{(S)}} \widetilde{R}(U)$ of the sheaf $\widetilde{R}$ at $\mathfrak{p}$ (see [30, Chapter II, §1, p. 62]) has a natural structure as an R-algebra. Also, the stalk $\widetilde{M}_{\mathfrak{p}} := \varinjlim_{U \in \mathcal{U}_{T,\mathfrak{p}}^{(S)}} \widetilde{M}(U)$ of $\widetilde{M}$ at $\mathfrak{p}$ has a natural structure as an $\widetilde{R}_{\mathfrak{p}}$-module. For $U \in \mathcal{U}_{T,\mathfrak{p}}^{(S)}$, we shall use $\rho_{U,\mathfrak{p}}$ $(= \rho_{U,\mathfrak{p},M}) : \widetilde{M}(U) \longrightarrow \widetilde{M}_{\mathfrak{p}}$ to denote the natural map; thus $\rho_{U,\mathfrak{p},R}$ is a homomorphism of R-algebras and $\rho_{U,\mathfrak{p},M}$ is an $\widetilde{R}(U)$-homomorphism.

The universal property of direct limits leads to a map $\psi_M^{\mathfrak{p}} : \widetilde{M}_{\mathfrak{p}} \longrightarrow (S \setminus \mathfrak{p})^{-1}M$ for which $\psi_M^{\mathfrak{p}}(\rho_{U,\mathfrak{p},M}(\gamma)) = \gamma_{\mathfrak{p}}$ for all $U \in \mathcal{U}_{T,\mathfrak{p}}^{(S)}$ and $\gamma = (\gamma_{\mathfrak{q}})_{\mathfrak{q} \in U} \in \widetilde{M}(U)$. Show that $\psi_R^{\mathfrak{p}}$ is an isomorphism of R-algebras, and that, as M varies through $\mathcal{C}(R)$, the $\psi_M^{\mathfrak{p}}$ constitute a natural equivalence $\psi^{\mathfrak{p}} : \widetilde{\bullet}_{\mathfrak{p}} \longrightarrow (S \setminus \mathfrak{p})^{-1}$ of functors from $\mathcal{C}(R)$ to itself.

Deduce that the functor $\widetilde{\bullet} : \mathcal{C}(R) \longrightarrow \mathcal{S}(\widetilde{R})$ of part (iv) is exact (see [30, Chapter II, Exercise 1.2(c), p. 66]).

20.3.2 ♯Exercise. (The standard hypotheses of 20.1.2 apply.) Let $h : L \longrightarrow N$ be a homomorphism of R-modules. Assume that $T \neq \emptyset$.

(i) Show that the induced sheaf $\widetilde{M}$ is zero if and only if the R-module M is (S)-torsion.

(ii) Show that the induced morphism $\widetilde{h} : \widetilde{L} \longrightarrow \widetilde{N}$ of sheaves of $\widetilde{R}$-modules is injective (respectively surjective) if and only if $\operatorname{Ker} h$ (respectively $\operatorname{Coker} h$) is (S)-torsion.

20.3.3 Remarks. (The standard hypotheses of 20.1.2 are in force.) Let $U :=$ $T \setminus \operatorname{Var}(\mathfrak{a})$; assume that $U \neq \emptyset$ (or $U = \emptyset$ and T is large with respect to S). Here, among other things, we reformulate the Deligne Isomorphism Theorem in the language of sheaves.

(i) The functor $\tilde{\bullet}(U) : \mathcal{C}(R) \longrightarrow \mathcal{C}(\tilde{R}(U))$ of 20.1.12(ii) can be regarded as the composition of the functor $\tilde{\bullet} : \mathcal{C}(R) \longrightarrow \mathcal{S}(\tilde{R})$ of 20.3.1(iv) and the section functor $\Gamma(U, \bullet) : \mathcal{S}(\tilde{R}) \longrightarrow \mathcal{C}(\tilde{R}(U))$. Therefore, the Deligne Isomorphism Theorem 20.1.14, with the refinement afforded by Exercise 20.1.18(i), provides a natural equivalence $\nu_{\mathfrak{a}} : \Gamma(U, \tilde{\bullet}) \xrightarrow{\cong} D_{\mathfrak{a}}$ of functors from $\mathcal{C}(R)$ to $\mathcal{C}(\tilde{R}(U))$. Of course, we can interpret both $D_{\mathfrak{a}}$ and $\Gamma(U, \tilde{\bullet})$ as functors from $\mathcal{C}(R)$ to itself (strictly, we should then write $\Gamma(U, \tilde{\bullet}) \lceil_R$ in the latter case, but we shall often omit the '$\lceil_R$' in the interests of notational simplicity), so that $\nu_{\mathfrak{a}}$ can also be interpreted as a natural equivalence of functors from $\mathcal{C}(R)$ to itself.

(ii) The above interpretation can be refined in the case of the Graded Deligne Isomorphism Theorem 20.2.7, for which we assume, in addition, that R is G-graded, where G is a finitely generated torsion-free Abelian group, and that S consists entirely of homogeneous elements. Then, in view of 20.2.4, the sheaf $\tilde{R}$ becomes a sheaf of G-graded R-algebras, and the functor $\tilde{\bullet} : \mathcal{C}(R) \longrightarrow \mathcal{S}(\tilde{R})$ of 20.3.1(iv) restricts to a functor from $*\mathcal{C}(R)$ to the category $*\mathcal{S}(\tilde{R})$ of sheaves of graded $\tilde{R}$-modules. We shall again denote this functor by $\tilde{\bullet}$, as we do not expect this to cause confusion. It follows from 20.2.8 that $\Gamma(U, \tilde{\bullet})$, considered as a functor from $\mathcal{C}(R)$ to either $\mathcal{C}(R)$ or $\mathcal{C}(\tilde{R}(U))$, has the *restriction property, and that, also in the two cases, the natural equivalence $\nu_{\mathfrak{a}} : \Gamma(U, \tilde{\bullet}) \xrightarrow{\cong} D_{\mathfrak{a}}$ of part (i) has the *restriction property.

(iii) Note that, with the notation of part (ii), $(\tilde{R}(U)_0)_{U \in \mathcal{U}_T^{(S)}}$ defines a sheaf $\tilde{R}_0 = (\tilde{R})_0$ of R_0-algebras, so that $(T, \tilde{R}_0)$ is also a ringed space. Similarly, for a graded R-module M and $g \in G$, $(\tilde{M}(U)_g)_{U \in \mathcal{U}_T^{(S)}}$ defines a sheaf $\tilde{M}_g = (\tilde{M})_g$ of $\tilde{R}_0$-modules.

(iv) Again with the notation of parts (ii) and (iii), for each $\mathfrak{p} \in U$, the stalk $\tilde{M}_{\mathfrak{p}}$ inherits a structure as a G-graded R-module for which the natural homomorphism $\rho_{U,\mathfrak{p},M} : \tilde{M}(U) \longrightarrow \tilde{M}_{\mathfrak{p}}$ is homogeneous. It is straightforward to check that there is a natural isomorphism of R_0-modules $(\tilde{M}_g)_{\mathfrak{p}} \cong (\tilde{M}_{\mathfrak{p}})_g$ for all $g \in G$. We shall use these isomorphisms as identifications.

A special case of 20.3.3(i) yields the 'classical' form of Deligne's Isomorphism.

20.3.4 Example: Deligne's Isomorphism for Affine Schemes. Take $S = R$ and $T := \operatorname{Spec}(R)$ (which is essential with respect to S by 20.1.3(i)). Then the induced sheaf $\widetilde{R}$ of 20.3.1(i) is just the *structure sheaf* $\mathcal{O}_T$ *of the affine scheme* $(T, \mathcal{O}_T)$ defined by R (see [30, Chapter II, §2, p. 70]). Moreover, if M is an R-module, then the induced sheaf $\widetilde{M}$ of 20.3.1(ii) is just the *sheaf of $\mathcal{O}_T$-modules associated to M* (see [30, Chapter II, §5, p. 110]). It follows from [30, Chapter II, §5, p. 111, and Proposition 5.4, p. 113] that the induced sheaves in the sense of 20.3.1(ii) are, up to isomorphism, precisely the quasi-coherent sheaves of $\mathcal{O}_T$-modules, and that the sheaves which are induced by finitely generated R-modules are, up to isomorphism, precisely the coherent sheaves of $\mathcal{O}_T$-modules.

An arbitrary ideal $\mathfrak{b}$ of R automatically belongs to $\mathfrak{A}_S$ in this case. Let $U := T \setminus \operatorname{Var}(\mathfrak{b}) = \operatorname{Spec}(R) \setminus \operatorname{Var}(\mathfrak{b})$. In this special case, T is large with respect to S, and the isomorphism $\nu_{\mathfrak{b},M} : \Gamma(U, \widetilde{M}) \xrightarrow{\cong} D_{\mathfrak{b}}(M)$ of 20.1.14 (or 20.1.15) and 20.3.3(i) is just the classical isomorphism of Deligne (see [30, Chapter III, Exercise 3.7, p. 217]).

20.3.5 ♯Exercise.   Let $(T, \mathcal{O}_T)$ be the affine scheme defined by R, and let qcoh_T denote the category of all quasi-coherent sheaves of $\mathcal{O}_T$-modules. Let M be an R-module. Use Exercise 20.1.19 to show that $\varepsilon_M^T : M \longrightarrow \Gamma(T, \widetilde{M})$ is an isomorphism, and deduce that the functor $\widetilde{\bullet} : \mathcal{C}(R) \longrightarrow \operatorname{qcoh}_T$ is an equivalence of categories. (See [30, Chapter II, Corollary 5.5, p. 113].)

20.3.6 ♯Exercise.   (The standard hypotheses of 20.1.2 apply.) Let $\mathfrak{b}$ be a second ideal of $\mathfrak{A}_S$. Set $U := T \setminus \operatorname{Var}(\mathfrak{a})$ and $Z := T \cap \operatorname{Var}(\mathfrak{b})$. Set

$$\Gamma_Z(U, \widetilde{M}) := \{\gamma \in \Gamma(U, \widetilde{M}) : \rho_{U,\mathfrak{p},M}(\gamma) = 0 \text{ for all } \mathfrak{p} \in U \setminus Z\}.$$

(See [30, Chapter II, Exercise 1.14, p. 67, and Exercise 1.20, p. 68].) Note that $\Gamma_Z(U, \widetilde{\bullet})$ is a functor from $\mathcal{C}(R)$ to $\mathcal{C}(R)$, and also a functor from $\mathcal{C}(R)$ to $\mathcal{C}(\widetilde{R}(U))$.

Now assume that $U \neq \emptyset$ (or $U = \emptyset$ and T is large with respect to S). Use Lemma 20.1.10 to show that the restriction of the isomorphism $\nu_{\mathfrak{a},M} : \Gamma(U, \widetilde{M}) \xrightarrow{\cong} D_{\mathfrak{a}}(M)$ of 20.1.14 and 20.3.3(i) provides an isomorphism

$$\nu_{\mathfrak{a},\mathfrak{b},M} : \Gamma_Z(U, \widetilde{M}) \xrightarrow{\cong} \Gamma_{\mathfrak{b}}(D_{\mathfrak{a}}(M)),$$

and deduce that the functors $\Gamma_Z(U, \widetilde{\bullet})$ and $\Gamma_{\mathfrak{b}}(D_{\mathfrak{a}}(\bullet))$, whether considered as functors from $\mathcal{C}(R)$ to itself or from $\mathcal{C}(R)$ to $\mathcal{C}(\widetilde{R}(U))$, are naturally equivalent.

20.3.7 Example. Let $(T, \mathcal{O}_T)$ be the affine scheme defined by R. (We interpret S as R here and T as $\operatorname{Spec}(R)$, as in 20.3.4, so that R as well as $\mathfrak{a}$ belongs to $\mathfrak{A}_S$.) Set $Z := \operatorname{Var}(\mathfrak{a})$ and $U := T \setminus \operatorname{Var}(\mathfrak{a}) = \operatorname{Spec}(R) \setminus Z$. If we apply 20.3.6 to the open set T and the closed set Z, we obtain an isomorphism $\nu_{R,\mathfrak{a},M} : \Gamma_Z(T, \widetilde{M}) \xrightarrow{\cong} \Gamma_{\mathfrak{a}}(D_R(M))$. Bear in mind that $\eta_{R,M} : M \longrightarrow D_R(M)$ is an isomorphism. Also bear in mind 20.1.15. The diagram

$$
\begin{array}{ccccccc}
0 & \longrightarrow & \Gamma_Z(T, \widetilde{M}) & \longrightarrow & \Gamma(T, \widetilde{M}) & \xrightarrow{\rho_{TU}} & \Gamma(U, \widetilde{M}) \\
& & \downarrow{\mu_{\mathfrak{a},M}} \cong & & \downarrow{(\varepsilon_M^T)^{-1}} \cong & & \downarrow{\nu_{\mathfrak{a},M}} \cong \\
0 & \longrightarrow & \Gamma_{\mathfrak{a}}(M) & \longrightarrow & M & \xrightarrow{\eta_{\mathfrak{a},M}} & D_{\mathfrak{a}}(M) & \longrightarrow H_{\mathfrak{a}}^1(M) \longrightarrow 0,
\end{array}
$$

in which $\mu_{\mathfrak{a},M} = \Gamma_{\mathfrak{a}}((\eta_{R,M})^{-1}) \circ \nu_{R,\mathfrak{a},M}$, the second map in the top row is the inclusion map, and the lower row comes from 2.2.6(i)(c), has exact rows and commutes.

20.3.8 Exercise. Let $(T, \mathcal{O}_T)$ be the affine scheme defined by R, let Z be a closed subset of $T = \operatorname{Spec}(R)$, and let $U := \operatorname{Spec}(R) \setminus Z$. Let $\mathcal{F}$ be a coherent sheaf of $\mathcal{O}_T$-modules (see Example 20.3.4). Show that the restriction map $\rho_{TU} : \Gamma(T, \mathcal{F}) \longrightarrow \Gamma(U, \mathcal{F})$ associated with the sheaf $\mathcal{F}$ is injective if and only if, for all $\mathfrak{p} \in Z$, the stalk $\mathcal{F}_{\mathfrak{p}}$ of $\mathcal{F}$ at $\mathfrak{p}$ has positive depth. (Remember that the depth of a zero module over a local ring is interpreted as ∞.) Show further that ρ_{TU} is bijective if and only if $\operatorname{depth} \mathcal{F}_{\mathfrak{p}} > 1$ for all $\mathfrak{p} \in Z$.

The following exercise and reminders are in preparation for an extension of the sheaf-theoretic version of Deligne's Isomorphism, as presented in 20.3.3, to higher cohomology. This work will lead to what we shall call the Deligne Correspondence.

20.3.9 ♯Exercise. (The standard hypotheses of 20.1.2 apply.)

(i) Let I be an injective R-module. Show that the S-sheaf induced by I is flasque, that is (see [30, Chapter II, Exercise 1.16, p. 67]), for every pair of open sets $U, V \in \mathcal{U}_T^{(S)}$ with $V \subseteq U$, the restriction map $\rho_{UV,I} : \Gamma(U, \widetilde{I}) \longrightarrow \Gamma(V, \widetilde{I})$ is surjective. (You might find 20.1.18(ii) helpful.)

(ii) Now assume in addition that $R = \bigoplus_{g \in G} R_g$ is G-graded, where G is a finitely generated torsion-free Abelian group, and that S consists entirely of homogeneous elements. Let J be a *injective graded R-module. Show that the S-sheaf induced by J is flasque.

(iii) Let R, S and J be as in part (ii), and let $g \in G$. Thus $\widetilde{J}$ is a sheaf of graded $\widetilde{R}$-modules. Now $(\widetilde{R})_0$ is a sheaf of R_0-algebras and the g-th homogeneous component $\left(\widetilde{J}\right)_g$ of this sheaf carries a natural structure as a sheaf of $\left(\widetilde{R}\right)_0$-modules: show that $\left(\widetilde{J}\right)_g$ is flasque.

20.3.10 Reminders. (The standard hypotheses of 20.1.2 apply.) Set $U :=$ $T \setminus \mathrm{Var}(\mathfrak{a})$. Recall [30, Chapter II, Exercise 1.8, p. 66] that the section functor $\Gamma(U, \bullet)$ (from the category of sheaves of Abelian groups on T to the category $\mathcal{C}(\mathbb{Z})$ of Abelian groups) is left exact: for each $i \in \mathbb{N}_0$, the i-th right derived functor of $\Gamma(U, \bullet)$ is denoted by $H^i(U, \bullet)$ and is referred to as the *i-th sheaf cohomology functor on U.* See [30, Chapter III, §2, p. 207]. Thus $H^i(U, \bullet)$ is again a functor from the category of sheaves of Abelian groups on T to $\mathcal{C}(\mathbb{Z})$; however, if $\mathcal{F}$ is a sheaf of $\widetilde{R}$-modules on T, then, for all $i \in \mathbb{N}_0$, the Abelian group $H^i(U, \mathcal{F})$ carries a natural structure as $\widetilde{R}(U)$-module which is such that, for all $\lambda \in \widetilde{R}(U)$ and all $\sigma \in H^i(U, \mathcal{F})$, we have $\lambda\sigma = H^i(U, \lambda\,\mathrm{Id}_{\mathcal{F}})(\sigma)$. It then follows that $\left(H^i(U, \bullet)\right)_{i \in \mathbb{N}_0}$ is a negative strongly connected sequence of covariant functors from $\mathcal{S}(\widetilde{R})$ to $\mathcal{C}(\widetilde{R}(U))$.

We noted in 20.3.1(v) that the functor $\widetilde{} : \mathcal{C}(R) \longrightarrow \mathcal{S}(\widetilde{R})$ of 20.3.1(iv) is exact. If we follow this with the above sheaf cohomology functors, we find that $\left(H^i(U, \widetilde{\bullet})\right)_{i \in \mathbb{N}_0}$ is a negative strongly connected sequence of covariant functors from $\mathcal{C}(R)$ to $\mathcal{C}(\widetilde{R}(U))$. Of course, for an R-module M, we can regard the $\widetilde{R}(U)$-module $H^i(U, \widetilde{M})$ as an R-module by restriction of scalars: this device leads to a negative strongly connected sequence of covariant functors $\left(H^i(U, \widetilde{\bullet})\lceil_R\right)_{i \in \mathbb{N}_0}$ from $\mathcal{C}(R)$ to itself (although we shall normally drop the '$\lceil_R$' from the notation).

We are now ready to present the promised Deligne Correspondence. This correspondence presents a fundamental connection between sheaf cohomology and local cohomology.

20.3.11 The Deligne Correspondence Theorem. *(The standard hypotheses of 20.1.2 apply.) Set $U := T \setminus \mathrm{Var}(\mathfrak{a})$; assume that $U \neq \emptyset$ (or $U = \emptyset$ and T is large with respect to S).*

There is a unique isomorphism

$$\Theta = \left(\theta^i\right)_{i \in \mathbb{N}_0} : \left(H^i(U, \widetilde{\bullet})\right)_{i \in \mathbb{N}_0} \overset{\cong}{\longrightarrow} \left(\mathcal{R}^i D_\mathfrak{a}\right)_{i \in \mathbb{N}_0}$$

of negative strongly connected sequences of covariant functors from $\mathcal{C}(R)$ to itself for which θ^0 is the natural equivalence $\nu_\mathfrak{a} : \Gamma(U, \widetilde{\bullet}) \overset{\cong}{\longrightarrow} D_\mathfrak{a}$ of 20.3.3(i).

Consequently, by 2.2.6(iii), for each $i \in \mathbb{N}$, the functors $H^i(U, \widetilde{\bullet})$ and $H_\mathfrak{a}^{i+1}$ from $\mathcal{C}(R)$ to itself are naturally equivalent.

Proof. Let I be an injective R-module. Of course, $\mathcal{R}^i D_\mathfrak{a}(I) = 0$ for all $i \in \mathbb{N}$. By 20.3.9(i), the induced S-sheaf $\widetilde{I}$ is flasque, so that $H^i(U, \widetilde{I}) = 0$ for all $i \in \mathbb{N}$, by [30, Chapter III, Proposition 2.5, p. 208]. The claim is therefore immediate from 1.3.4(ii). □

20.3.12 Remark. In the situation of 20.3.11, we can, on account of 20.1.18, consider $D_\mathfrak{a}$ as a functor from $\mathcal{C}(R)$ to $\mathcal{C}(\widetilde{R}(U))$, so that $\left(\mathcal{R}^i D_\mathfrak{a}\right)_{i \in \mathbb{N}_0}$ is actually a negative strongly connected sequence of covariant functors from $\mathcal{C}(R)$ to $\mathcal{C}(\widetilde{R}(U))$; furthermore, $\nu_\mathfrak{a} : \Gamma(U, \widetilde{\bullet}) \overset{\cong}{\longrightarrow} D_\mathfrak{a}$ is actually a natural equivalence of functors from $\mathcal{C}(R)$ to $\mathcal{C}(\widetilde{R}(U))$ (see 20.3.3(i)).

One can therefore argue as in the above proof of 20.3.11 to see that there is a unique isomorphism $\left(\delta^i\right)_{i \in \mathbb{N}_0} : \left(H^i(U, \widetilde{\bullet})\right)_{i \in \mathbb{N}_0} \overset{\cong}{\longrightarrow} \left(\mathcal{R}^i D_\mathfrak{a}\right)_{i \in \mathbb{N}_0}$ of negative strongly connected sequences of covariant functors from $\mathcal{C}(R)$ to $\mathcal{C}(\widetilde{R}(U))$ for which δ^0 is $\nu_\mathfrak{a}$; moreover, the uniqueness aspect of Theorem 20.3.11 shows that $\delta^i \lceil_R = \theta^i$ for all $i \in \mathbb{N}_0$, so that, for each such i and each R-module N, the maps δ^i_N and θ^i_N coincide and θ^i_N is an $\widetilde{R}(U)$-isomorphism.

20.3.13 A Graded Version of the Deligne Correspondence. Here, we present refinements of 20.3.11 which are available in the graded case, and so we assume, in addition to the hypotheses of 20.3.11, that $R = \bigoplus_{g \in G} R_g$ is G-graded, where G is a finitely generated torsion-free Abelian group, and that S consists entirely of homogeneous elements. In particular, this means that $\mathfrak{a}$ is graded. We consider the isomorphism

$$\Theta = \left(\theta^i\right)_{i \in \mathbb{N}_0} : \left(H^i(U, \widetilde{\bullet})\right)_{i \in \mathbb{N}_0} \overset{\cong}{\longrightarrow} \left(\mathcal{R}^i D_\mathfrak{a}\right)_{i \in \mathbb{N}_0}$$

of negative strongly connected sequences of covariant functors from $\mathcal{C}(R)$ to $\mathcal{C}(\widetilde{R}(U))$ of 20.3.11 and 20.3.12: recall that θ^0 is the natural equivalence $\nu_\mathfrak{a} : \Gamma(U, \widetilde{\bullet}) \overset{\cong}{\longrightarrow} D_\mathfrak{a}$ of 20.3.3(i), which, by 20.3.3(ii), has the *restriction property.

(i) Whenever I is a *injective graded R-module, we have

$$H^i(U, \widetilde{I}) \cong \mathcal{R}^i D_\mathfrak{a}(I) \cong H^{i+1}_\mathfrak{a}(I) = 0 \quad \text{for all } i \in \mathbb{N},$$

by 20.3.11, 2.2.6(iii) and 13.2.6. Since $\Gamma(U, \widetilde{\bullet})$ has the *restriction property (see 20.2.8(i)), we can use Theorem 13.3.15 to deduce that there is exactly one choice of gradings on the $\widetilde{R}(U)$-modules $H^i(U, \widetilde{N})$ ($i \in \mathbb{N}$, N a graded R-module) with respect to which $\left(H^i(U, \widetilde{\bullet})\right)_{i \in \mathbb{N}_0}$ has the *restriction property of 13.3.9. We shall refer to these gradings as the *natural gradings*, and any unexplained gradings on these modules are to be interpreted as these natural ones. Notice that they provide gradings over R, and so must be the unique choice of gradings with respect to which $\left(H^i(U, \widetilde{\bullet})\right)_{i \in \mathbb{N}_0}$, considered as a negative strongly connected sequence of functors from $\mathcal{C}(R)$ to itself, has the *restriction property.

(ii) Similarly, $D_\mathfrak{a} : \mathcal{C}(R) \longrightarrow \mathcal{C}(\widetilde{R}(U))$ has the *restriction property (see 20.2.8(iii)), and there is exactly one choice of gradings on the $\widetilde{R}(U)$-modules $\mathcal{R}^i D_\mathfrak{a}(N)$ ($i \in \mathbb{N}$, N a graded R-module) with respect to which $\left(\mathcal{R}^i D_\mathfrak{a}\right)_{i \in \mathbb{N}_0}$ has the *restriction property. We again refer to these gradings as the *natural gradings*; the uniqueness aspect of Theorem 13.3.15 means that these natural gradings, which work over both $\widetilde{R}(U)$ and R, must be the ones found in 13.5.7(ii).

Now θ^0 has the *restriction property. We can define gradings on the $\mathcal{R}^i D_\mathfrak{a}(N)$ ($i \in \mathbb{N}$, N a graded R-module) in such a way that all the θ_N^i ($i \in \mathbb{N}$, N a graded R-module) are homogeneous. Since Θ is an isomorphism of connected sequences, it follows that, with respect to *these* gradings, $\left(\mathcal{R}^i D_\mathfrak{a}\right)_{i \in \mathbb{N}_0}$ has the *restriction property, and so these gradings must be the natural gradings.

Thus, with respect to the natural gradings, for all $i \in \mathbb{N}_0$, the natural equivalence $\theta^i : H^i(U, \widetilde{\bullet}) \xrightarrow{\cong} \mathcal{R}^i D_\mathfrak{a}$ of 20.3.11 has the *restriction property.

(iii) Since, by 13.5.7(iii), for $i \in \mathbb{N}$, the natural equivalence $\mathcal{R}^i D_\mathfrak{a} \xrightarrow{\cong} H_\mathfrak{a}^{i+1}$ of 2.2.6(ii) has the *restriction property, it follows that the natural equivalence $H^i(U, \widetilde{\bullet}) \xrightarrow{\cong} H_\mathfrak{a}^{i+1}$ of the final paragraph of 20.3.11 also has the *restriction property. This gives a satisfactory extension, to the graded case, of our fundamental connection between sheaf cohomology and local cohomology.

Next, we are going to study the cohomology of homogeneous components of induced S-sheaves in the graded situation. This theme has great importance in the study of sheaf cohomology over projective varieties and, more generally, over projective schemes. Our aim is to produce a link between cohomology of homogeneous components of induced S-sheaves and homogeneous components of graded ideal transforms and local cohomology modules. We begin with some preparations.

20.3.14 Remarks and Notation.   Here, the hypotheses of 20.2.1 apply. Let $g \in G$, and set $U := T \setminus \mathrm{Var}(\mathfrak{a})$; assume that $U \neq \emptyset$ (or $U = \emptyset$ and T is large with respect to S).

(i) In 20.3.3(ii), we noted that $\widetilde{R}$ is a sheaf of G-graded R-algebras, and we introduced the functor $\widetilde{\bullet}$ from $*\mathcal{C}(R)$ to the category $*\mathcal{S}(\widetilde{R})$ of sheaves of graded $\widetilde{R}$-modules; this functor is exact, by 20.3.1(v).

We shall use $(\bullet)_g : *\mathcal{S}(\widetilde{R}) \longrightarrow \mathcal{S}((\widetilde{R})_0)$ to denote the functor which associates to each sheaf $\mathcal{F}$ of graded $\widetilde{R}$-modules its g-th component $\mathcal{F}_g$, which is a sheaf of $(\widetilde{R})_0$-modules.

It is a consequence of Exercise 20.2.6 that the functors $(\widetilde{\bullet})_g$ and $(\widetilde{\bullet(g)})_0$ (from $*\mathcal{C}(R)$ to $\mathcal{S}((\widetilde{R})_0)$) are equal. (Here, $(\bullet)(g) : *\mathcal{C}(R) \longrightarrow *\mathcal{C}(R)$ denotes the g-th shift functor.)

(ii) It follows from the final paragraph of part (i) that, in $\mathcal{C}((\widetilde{R}(U))_0)$, we have $\Gamma(U, (\widetilde{M(g)})_0) = \Gamma(U, (\widetilde{M})_g) = \Gamma(U, \widetilde{M})_g$ (for all $M \in *\mathcal{C}(R)$). In alternative notation, $(\widetilde{M(g)})_0(U) = (\widetilde{M})_g(U) = \widetilde{M}(U)_g$ for all $M \in *\mathcal{C}(R)$. The natural equivalence $\nu_{\mathfrak{a}} : \Gamma(U, \widetilde{\bullet}) \xrightarrow{\cong} D_{\mathfrak{a}}$ of 20.3.3 has the *restriction property, by 20.3.3(ii). Hence, on taking g-th components, we see that

$$\nu_{\mathfrak{a},g} : \Gamma(U, \widetilde{\bullet})_g \xrightarrow{\cong} D_{\mathfrak{a}}(\bullet)_g$$

is a natural equivalence of functors from $*\mathcal{C}(R)$ to $\mathcal{C}((\widetilde{R}(U))_0)$, and also of functors from $*\mathcal{C}(R)$ to $\mathcal{C}(R_0)$.

(iii) Since the functor $(\widetilde{\bullet(g)})_0$ from $*\mathcal{C}(R)$ to $\mathcal{S}((\widetilde{R})_0)$ is exact, it follows that $\left(H^i(U, (\widetilde{\bullet(g)})_0) \right)_{i \in \mathbb{N}_0}$ can be considered as a negative strongly connected sequence of covariant functors from $*\mathcal{C}(R)$ to either $\mathcal{C}((\widetilde{R}(U))_0)$ or $\mathcal{C}(R_0)$. We propose, in the Serre–Grothendieck Correspondence Theorem 20.3.15 below, to compare this with $\left(\mathcal{R}^i D_{\mathfrak{a}}(\bullet)_g\right)_{i \in \mathbb{N}_0}$, which also can be considered as a negative strongly connected sequence of covariant functors from $*\mathcal{C}(R)$ to either $\mathcal{C}((\widetilde{R}(U))_0)$ or $\mathcal{C}(R_0)$.

20.3.15 The Serre–Grothendieck Correspondence Theorem. *For this, we adopt the hypotheses of 20.2.1. Let $g \in G$, and set $U := T \setminus \mathrm{Var}(\mathfrak{a})$; assume that $U \neq \emptyset$ (or $U = \emptyset$ and T is large with respect to S).*

There is a unique isomorphism

$$\Omega_g = \left(\omega_g^i\right)_{i \in \mathbb{N}_0} : \left(H^i\left(U, (\widetilde{\bullet(g)})_0\right) \right)_{i \in \mathbb{N}_0} \xrightarrow{\cong} \left(\mathcal{R}^i D_{\mathfrak{a}}(\bullet)_g\right)_{i \in \mathbb{N}_0}$$

of negative strongly connected sequences of covariant functors from $*\mathcal{C}(R)$ to $\mathcal{C}((\widetilde{R}(U))_0)$ for which ω_g^0 is the natural equivalence

$$\nu_{\mathfrak{a},g} : \Gamma\left(U, (\widetilde{\bullet(g)})_0\right) \xrightarrow{\cong} D_{\mathfrak{a}}(\bullet)_g$$

of functors of 20.3.14(ii).

Proof. Let I be a *injective graded R-module. Let $i \in \mathbb{N}$. Then

$$\mathcal{R}^i D_{\mathfrak{a}}(I) \cong H_{\mathfrak{a}}^{i+1}(I) = 0,$$

by 2.2.6(ii) and 13.2.6. Also, by 20.3.14(i), we have $(\widetilde{I(g)})_0 = (\widetilde{I})_g$, and, by 20.3.9(iii), this is a flasque sheaf of $(\widetilde{R})_0$-modules. Hence $H^i\left(U, (\widetilde{I(g)})_0\right) = 0$ by [30, Chapter III, Proposition 2.5, p. 208].

If we now endow $\left(\widetilde{R}(U)\right)_0$ with the trivial grading, we can use 13.3.5(ii) to complete the proof. □

20.3.16 Remarks. Here we consider further the situation of, and use the notation of, the Serre–Grothendieck Correspondence 20.3.15.

(i) Observe that Ω_g can be regarded as an isomorphism of negative strongly connected sequences of covariant functors from $*\mathcal{C}(R)$ to $\mathcal{C}(R_0)$.

(ii) Let $d \in G$ and $r_d \in R_d$. For each graded R-module N, let $\mu_{r_d} = \mu_{r_d,N} : N \longrightarrow N(d)$ be the homogeneous R-homomorphism given by multiplication by r_d. Set $u = g + d$. It is straightforward to check that

$$\left(H^i\left(U, \left(\widetilde{\mu_{r_d,\,\bullet}(g)}\right)_0\right)\right)_{i\in\mathbb{N}_0} :$$

$$\left(H^i\left(U, \left(\widetilde{\bullet(g)}\right)_0\right)\right)_{i\in\mathbb{N}_0} \longrightarrow \left(H^i\left(U, \left(\widetilde{\bullet(u)}\right)_0\right)\right)_{i\in\mathbb{N}_0}$$

is a homomorphism of connected sequences of functors from $*\mathcal{C}(R)$ to $\mathcal{C}(R_0)$.

(iii) Let $i \in \mathbb{N}_0$. One can check that $\bigoplus_{j\in G} H^i\left(U, (\widetilde{M(j)})_0\right)$ has a structure as a graded R-module such that $r_d z_g = H^i\left(U, \left(\widetilde{\mu_{r_d}(g)}\right)_0\right)(z_g)$ for $(d \in G$ and) $r_d \in R_d$ and $z_g \in H^i\left(U, \left(\widetilde{M(g)}\right)_0\right)$. It is also straightforward to use the uniqueness aspect of 13.3.5(i) to show that, with respect to this graded R-module structure, the R_0-isomorphism

$$\bigoplus_{j\in G} \omega^i_{j,M} : \bigoplus_{j\in G} H^i\left(U, \left(\widetilde{M(j)}\right)_0\right) \xrightarrow{\cong} \bigoplus_{j\in G} \mathcal{R}^i D_\mathfrak{a}(M)_j = \mathcal{R}^i D_\mathfrak{a}(M)$$

given by the Serre–Grothendieck Correspondence 20.3.15 (and part (i)) is actually a homogeneous isomorphism of graded R-modules. For $i = 0$, this is just the homogeneous R-isomorphism

$$\bigoplus_{j\in G} \omega^0_{j,M} = \nu_{\mathfrak{a},M} : \bigoplus_{j\in G} \Gamma\left(U, \left(\widetilde{M(j)}\right)_0\right) = \Gamma(U, \widetilde{M}) = \widetilde{M}(U)$$

$$\xrightarrow{\cong} \bigoplus_{j\in G} D_\mathfrak{a}(M)_j = D_\mathfrak{a}(M)$$

of 20.3.3(ii).

(iv) Let $i \in \mathbb{N}$. It follows from part (iii) and 13.5.7(iii) that there is a homogeneous isomorphism $\bigoplus_{j\in G} H^i\left(U, \left(\widetilde{M(j)}\right)_0\right) \xrightarrow{\cong} H^{i+1}_\mathfrak{a}(M)$ of graded R-modules.

20.3.17 ‡Exercise. Here the hypotheses of 20.1.2 and 20.2.1 apply. Consider a $\mathfrak{p} \in T$.

(i) Show that the isomorphism of R-algebras $\psi_R^{\mathfrak{p}} : \widetilde{R}_{\mathfrak{p}} \longrightarrow (S \setminus \mathfrak{p})^{-1} R$ of 20.3.1(v) is homogeneous, and that the natural equivalence

$$\psi^{\mathfrak{p}} : \widetilde{\bullet}_{\mathfrak{p}} \longrightarrow (S \setminus \mathfrak{p})^{-1}$$

of functors from $\mathcal{C}(R)$ to itself (again of 20.3.1(v)) has the *restriction property.

(ii) Let $U := T \setminus \mathrm{Var}(\mathfrak{a})$ and assume that $\mathfrak{p} \in U$. Show that there is a commutative diagram with homogeneous maps

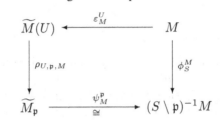

in which ϕ_S^M is the canonical map and the homogeneous homomorphisms ε_M^U and $\rho_{U,\mathfrak{p},M}$ are defined by 20.1.6(iii) and 20.3.1(v) respectively.

20.4 Applications to projective schemes

The Serre–Grothendieck Correspondence 20.3.15, as exploited in 20.3.16, can be used in a very effective manner, in various situations, to derive results about the cohomology of induced sheaves from purely algebraic results obtained earlier in the book. We have neither the space nor the intention to present a comprehensive approach to sheaf cohomology in this book, and so we shall content ourselves, in this section, with some applications of the Serre–Grothendieck Correspondence to the particular case of projective schemes induced by homogeneous, positively $\mathbb{Z}$-graded (Noetherian) rings. This situation is particularly fertile in this context because the quasi-coherent sheaves are, up to isomorphism, just the sheaves induced by graded modules.

We hope that the illustrations which we present in this section will convince the reader of the value of the Serre–Grothendieck Correspondence, and will whet her or his appetite for exploration of applications of the correspondence in other situations.

20.4.1 Hypotheses for the section. We shall assume throughout this section that $R = \bigoplus_{n \in \mathbb{N}_0} R_n$ is positively $\mathbb{Z}$-graded and homogeneous; we shall take

$S := \bigcup_{n \in \mathbb{N}} R_n$, the set of homogeneous elements of R of positive degrees together with 0 (so that $\mathfrak{A}_S$ is the set of all graded ideals of R which are contained in R_+ (see 20.1.4)); we shall take $T := \mathrm{Proj}(R) = {}^*\mathrm{Spec}(R) \setminus \mathrm{Var}(R_+)$ and we shall assume that $T \neq \emptyset$; also, we shall assume that the R-module M is graded. Note that our assumptions imply that T is large (see 20.1.15) with respect to S.

Recall from 20.1.4 that the S-topology on $\mathrm{Proj}(R)$ is the Zariski topology. For $\mathfrak{p} \in \mathrm{Proj}(R)$, the natural ring homomorphism $(S \setminus \mathfrak{p})^{-1}R \longrightarrow R_{(\mathfrak{p})}$ is an isomorphism.

20.4.2 Remarks.   (The hypotheses of 20.4.1 apply.)

(i) The sheaf $(\widetilde{R})_0$ of R_0-algebras of 20.3.9(iii) is just the structure sheaf $\mathcal{O}_T$ of the *projective scheme* $(T, \mathcal{O}_T)$ defined by R: see [30, Chapter II, §2, p. 76]. Note that, for each open subset U of T, we have $\mathcal{O}_T(U) = (\widetilde{R}(U))_0$, and, for $\mathfrak{p} \in T$, the stalk of $\mathcal{O}_T$ at $\mathfrak{p}$ is $(\widetilde{R}_0)_\mathfrak{p} = (\widetilde{R}_\mathfrak{p})_0$ (we are using 20.3.3(iv) here), and so 20.3.17(ii) yields an isomorphism of R_0-algebras $(\psi_R^\mathfrak{p})_0 : \mathcal{O}_{T,\mathfrak{p}} \xrightarrow{\cong} (R_{(\mathfrak{p})})_0$. The reader should notice that our use of the notation $R_{(\mathfrak{p})}$ is different from Hartshorne's in [30, p. 18]. From now on, we shall identify $\mathcal{O}_{T,\mathfrak{p}}$ with $(R_{(\mathfrak{p})})_0$ by means of the above isomorphism.

(ii) The sheaf $(\widetilde{M})_0$ of $(\widetilde{R})_0$-modules of 20.3.14(i) is just the *sheaf* of $\mathcal{O}_T$-modules *associated to* M on $\mathrm{Proj}(R)$, as defined by Hartshorne in [30, Chapter II, §5, p. 116]. Although Hartshorne's notation for this sheaf is commonly used, we shall continue to use the notation $(\widetilde{M})_0$, in an attempt to avoid confusing readers.

Note that, for each open subset U of T and $t \in \mathbb{Z}$, we have $(\widetilde{M})_t(U) = \widetilde{M}(U)_t$. Moreover, by 20.3.3(iv) and 20.3.17(i), for each $\mathfrak{p} \in T$ and $t \in \mathbb{Z}$, there is an isomorphism of R_0-modules $(\psi_M^\mathfrak{p})_t : ((\widetilde{M})_t)_\mathfrak{p} \xrightarrow{\cong} (M_{(\mathfrak{p})})_t$. We may consider $((\widetilde{M})_t)_\mathfrak{p}$ as an $(R_{(\mathfrak{p})})_0$-module via the isomorphism $((\psi_R^\mathfrak{p})_0)^{-1}$ of part (i). It then follows easily from 20.3.17(ii) that the above isomorphism $(\psi_M^\mathfrak{p})_t$ is an isomorphism of modules over $(R_{(\mathfrak{p})})_0 = \mathcal{O}_{T,\mathfrak{p}}$. Observe also, that by 20.3.14(i) we have $(\widetilde{M(t)})_0 = (\widetilde{M})_t$ for all $t \in \mathbb{Z}$. Consequently, $(\psi_{M(t)}^\mathfrak{p})_0 = (\psi_M^\mathfrak{p})_t$ for all $t \in \mathbb{Z}$.

20.4.3 ♯Exercise.   Let the situation be as in 20.4.2(ii), and let $\mathfrak{p} \in T$ and $t \in \mathbb{Z}$. Note that $(M_{(\mathfrak{p})})_t$ denotes the t-th component of $M_{(\mathfrak{p})}$, the homogeneous localization of M at $\mathfrak{p}$. Use the diagram of 20.3.17(ii), the fact that the maps in the diagrams in 20.1.14(i) are homogeneous, and the identifications $(\widetilde{M(t)})_0 = (\widetilde{M})_t$ and $(\psi_{M(t)}^\mathfrak{p})_0 = (\psi_M^\mathfrak{p})_t$ of 20.4.2(ii), to show that there is a commutative diagram

$$
\begin{array}{ccc}
\left(\widetilde{M(t)}\right)_0(T) & \xrightarrow[\cong]{(\nu^T_{R_+,M})_t} & D_{R_+}(M)_t \\
\Big\downarrow{\scriptstyle \rho_{T,\mathfrak{p},(\widetilde{M(t)})_0}} & & \Big\downarrow{\scriptstyle (\beta_{(\mathfrak{p})})_t} \\
\left(\left(\widetilde{M(t)}\right)_0\right)_{\mathfrak{p}} & \xrightarrow[\cong]{(\psi^{\mathfrak{p}}_M)_t} & \left(M_{(\mathfrak{p})}\right)_t
\end{array}
$$

in which $\beta_{(\mathfrak{p})} : D_{R_+}(M) \longrightarrow M_{(\mathfrak{p})}$ is defined as in 17.2.2.

20.4.4 Theorem: the classical form of the Serre–Grothendieck Correspondence. *(The hypotheses of 20.4.1 apply.) It follows from the Serre–Grothendieck Correspondence 20.3.15 and 20.3.16(iii) (with $\mathfrak{a} = R_+$) that there are homogeneous R-isomorphisms*

$$\bigoplus_{j\in\mathbb{Z}} H^i(T,(\widetilde{M(j)})_0) \xrightarrow{\cong} \mathcal{R}^i D_{R_+}(M) \xrightarrow{\cong} H^{i+1}_{R_+}(M) \quad \text{for all } i \in \mathbb{N}$$

and $\bigoplus_{j\in\mathbb{Z}} \Gamma(T,(\widetilde{M(j)})_0) = \bigoplus_{j\in\mathbb{Z}} \widetilde{M}(T)_j = \widetilde{M}(T) \xrightarrow{\cong} D_{R_+}(M).$ $\qquad\square$

We recall now some facts about sheaves over projective schemes which mean that the classical form of the Serre–Grothendieck Correspondence 20.4.4 is a powerful tool for translation of algebraic results about local cohomology into geometric results about sheaf cohomology.

20.4.5 Reminders. (The hypotheses of 20.4.1 apply.) Let $t \in \mathbb{Z}$.

(i) By [30, Chapter II, Proposition 5.11(c), p. 116], the sheaf $(\widetilde{M})_0$ is a quasi-coherent sheaf of $\mathcal{O}_T$-modules; furthermore, if M is finitely generated, then $(\widetilde{M})_0$ is a coherent sheaf of $\mathcal{O}_T$-modules.

(ii) The twisted sheaf $(\widetilde{M})_0(t)$, as defined by Hartshorne [30, Chapter II, §5, p. 117] is naturally isomorphic (as a sheaf of $\mathcal{O}_T$-modules) to $(\widetilde{M(t)})_0$, the sheaf associated to the shifted module $M(t)$: see [30, Chapter II, Proposition 5.12(b), p. 117]

(iii) Let $\mathcal{F}$ be a sheaf of $\mathcal{O}_T$-modules. Recall from [30, Chapter II, §5, p. 118] that $\Gamma_*(T,\mathcal{F}) := \Gamma_*(\mathcal{F}) := \bigoplus_{j\in\mathbb{Z}} \Gamma(T,\mathcal{F}(j))$ carries a natural structure as a graded R-module and is called the *graded R-module associated to* $\mathcal{F}$. We shall use the notation $\Gamma_*(T,\mathcal{F})$ instead of Hartshorne's $\Gamma_*(\mathcal{F})$.

(iv) Let $\mathcal{F}$ be a quasi-coherent sheaf of $\mathcal{O}_T$-modules. Then, by [30, Chapter II, Proposition 5.15, p. 119], there is a natural isomorphism

$$\left(\widetilde{\Gamma_*(T,\mathcal{F})}\right)_0 \xrightarrow{\cong} \mathcal{F}$$

of sheaves of $\mathcal{O}_T$-modules. This shows, in particular, that each quasi-coherent sheaf $\mathcal{F}$ of $\mathcal{O}_T$-modules is isomorphic to $(\widetilde{N})_0$ for some graded R-module N. Also, in this situation, the induced R_0-isomorphism

$$\Gamma_*(T,\mathcal{F}) := \bigoplus_{j\in\mathbb{Z}} \Gamma(T,\mathcal{F}(j)) \cong \bigoplus_{j\in\mathbb{Z}} \Gamma(T,(\widetilde{N(j)})_0)$$

(which arises in view of (ii) above) is a homogeneous R-isomorphism. This is easily seen by comparing, by means of the natural isomorphism of part (ii), the action of a homogeneous element $r_d \in R_d$ (where $d \in \mathbb{N}_0$) on $\Gamma_*(T,\mathcal{F})$ as described in [30, Chapter II, §5, p. 118] with the action of r_d on $\bigoplus_{j\in\mathbb{Z}} \Gamma(T,(\widetilde{N(j)})_0)$ as described in 20.3.16(iii).

20.4.6 ♯Exercise. (The hypotheses of 20.4.1 apply.) Let $\mathcal{F}$ be a coherent sheaf of $\mathcal{O}_T$-modules. Then by 20.4.5(iv) we can write $\mathcal{F} \cong (\widetilde{M})_0$, where M is a graded R-module. Our aim here is to show that we can choose M to be finitely generated. We do this in several steps.

 (i) Show that $(\widetilde{N})_0 = 0$ for each R_+-torsion graded R-module N.
 (ii) Show that we can replace M by $M_{\geq 0}$ (see 16.1.1) and hence assume that $\mathrm{beg}(M) \geq 0$. (You may find 20.3.14(i) helpful.)
 (iii) Write $R = R_0[f_1,\ldots,f_r]$ with $f_i \in R_1 \setminus \{0\}$. Let $i \in \{1,\ldots,r\}$. Observe that, by [30, Chapter II, Proposition 2.5, pp. 76,77], the open set $U_i := T \setminus \mathrm{Var}(f_iR)$ is (empty or) affine with $\mathcal{O}(U_i) = (R_{f_i})_0$, a Noetherian ring. Use 20.3.14(ii) to show that $\Gamma(U_i,\mathcal{F}) \cong D_{f_iR}(M)_0 \cong (M_{f_i})_0$. Conclude by [30, Chapter II, Corollary 5.5 p. 113] that the $(R_{f_i})_0$-module $(M_{f_i})_0$ is finitely generated. This is true for each $i = 1,\ldots,r$.
 (iv) Deduce that there is a finitely generated graded R-submodule P of M such that M/P is R_+-torsion.
 (v) Now use part (i) to show that $\mathcal{F} \cong (\widetilde{P})_0$.

20.4.7 ♯Exercise. (The hypotheses of 20.4.1 apply.)

 (i) Let $\mathfrak{b} \subseteq R$ be a graded ideal. Show (by use of 20.3.14(i)) that the inclusion map $\mathfrak{b} \longrightarrow R$ yields a monomorphism of sheaves of $\mathcal{O}_T$-modules $\mathcal{J} := (\widetilde{\mathfrak{b}})_0 \longrightarrow (\widetilde{R})_0 = \mathcal{O}_T$. This means that, by definition, $\mathcal{J}$ is a sheaf of ideals on T (see [30, p. 109]). Conclude that this sheaf is coherent.
 (ii) Let $\mathcal{J}$ be a coherent sheaf of $\mathcal{O}_T$-modules which is a sheaf of ideals on T, so that, by definition, there is a monomorphism of sheaves $\mathcal{J} \longrightarrow \mathcal{O}_T$. Use the exactness of the twisting functors on the category of sheaves of $\mathcal{O}_T$-modules, the left exactness of the section functor and the Serre–Grothendieck Correspondence (see 20.3.16(iii)) to show that

$\mathfrak{d} := \Gamma_*(T, \mathcal{J})$ is homogeneously R-isomorphic to a graded ideal $\mathfrak{c}$ of the graded R-algebra $D_{R_+}(R)$. Let $\mathfrak{b} \subseteq R$ be the inverse image of $\mathfrak{c}$ under the natural homomorphism of graded rings $\eta_R : R \longrightarrow D_{R_+}(R)$. Show that there are isomorphisms of sheaves of $\mathcal{O}_T$-modules $(\widetilde{\mathfrak{b}})_0 \cong (\widetilde{\mathfrak{c}})_0 \cong (\widetilde{\mathfrak{d}})_0 \cong \mathcal{J}$, so that $\mathcal{J}$ is induced by a graded ideal $\mathfrak{b}$ of R.

We can now prove the fundamental Finiteness Theorem of Serre concerning the cohomology of coherent sheaves over projective schemes.

20.4.8 Serre's Finiteness Theorem.   (See [77, §66, théorème 1 and théorème 2(b)] and [30, Chapter III, Theorem 5.2, p. 228].) *(The hypotheses of 20.4.1 apply.) Let $\mathcal{F}$ be a coherent sheaf of $\mathcal{O}_T$-modules. Then*

(i) *$H^i(T, \mathcal{F}(j))$ is a finitely generated R_0-module, for all $i \in \mathbb{N}_0$ and all $j \in \mathbb{Z}$; and*

(ii) *there exists $r \in \mathbb{Z}$ such that $H^i(T, \mathcal{F}(j)) = 0$ for all $i \in \mathbb{N}$ and all $j \geq r$.*

Proof.   By 20.4.6, there exists a finitely generated graded R-module N such that $\mathcal{F} \cong (\widetilde{N})_0$; furthermore, by 20.4.5(iv), there is a homogeneous isomorphism of graded R-modules

$$\Gamma_*(T, \mathcal{F}) \cong \bigoplus_{j \in \mathbb{Z}} \Gamma(T, (\widetilde{N(j)})_0).$$

Also, by 20.4.5(ii), for each $i \in \mathbb{N}_0$ and $j \in \mathbb{Z}$, there is an isomorphism of R_0-modules $H^i(T, \mathcal{F}(j)) \cong H^i(T, (\widetilde{N(j)})_0)$.

Let $j \in \mathbb{Z}$ and $i \in \mathbb{N}$. We can now use the classical form of the Serre–Grothendieck Correspondence 20.4.4 to see that there are R_0-isomorphisms $\Gamma(T, \mathcal{F}(j)) \cong D_{R_+}(N)_j$ and $H^i(T, \mathcal{F}(j)) \cong H_{R_+}^{i+1}(N)_j$. Since $D_{R_+}(N)_j$ is a finitely generated R_0-module by 16.1.6(ii), and $H_{R_+}^{i+1}(N)_j$ is a finitely generated R_0-module by 16.1.5(i), part (i) is now proved. Part (ii) follows from 16.1.5(ii). □

20.4.9 Exercise.   Consider the situation and use the notation of Serre's Finiteness Theorem 20.4.8. Show that, for all $r \in \mathbb{Z}$, the graded R-module

$$\Gamma_*(T, \mathcal{F})_{\geq r} := \bigoplus_{j \in \mathbb{Z}, \, j \geq r} \Gamma(T, \mathcal{F}(j))$$

is finitely generated.

The philosophy of the above proof of Theorem 20.4.8 suggests that a similar approach to Castelnuovo regularity of coherent sheaves of $\mathcal{O}_T$-modules might be profitable. This is indeed the case.

20.4.10 Definition. (The hypotheses of 20.4.1 apply.) Let $\mathcal{F}$ be a coherent sheaf of $\mathcal{O}_T$-modules, and let $r \in \mathbb{Z}$.

We say that $\mathcal{F}$ is *r-regular in the sense of Castelnuovo–Mumford* if and only if $H^i(T, \mathcal{F}(s - i)) = 0$ for all $i \in \mathbb{N}$ and all $s \in \mathbb{Z}$ with $s \geq r$. In practice, the phrase 'in the sense of Castelnuovo–Mumford' is usually omitted.

We define the *(Castelnuovo–Mumford) regularity* $\mathrm{reg}(\mathcal{F})$ *of* $\mathcal{F}$ by

$$\mathrm{reg}(\mathcal{F}) = \inf \left\{ r \in \mathbb{Z} : \mathcal{F} \text{ is } r\text{-regular} \right\}.$$

20.4.11 Remarks. In the situation, and with the notation, of 20.4.10, there exists a finitely generated graded R-module N such that $\mathcal{F} \cong (\tilde{N})_0$, by 20.4.6. Also, by 20.4.5(ii) and the classical form of the Serre–Grothendieck Correspondence 20.4.4, for each $i \in \mathbb{N}$ and each $j \in \mathbb{Z}$, there exists an R_0-isomorphism $H^i(T, \mathcal{F}(j)) \cong H^{i+1}_{R_+}(N)_j$; in particular,

$$H^i(T, \mathcal{F}(s - i)) \cong H^{i+1}_{R_+}(N)_{s+1-(i+1)} \quad \text{for all } i \in \mathbb{N} \text{ and } s \in \mathbb{Z}.$$

This enables us to use results from Chapter 16 to make deductions about the regularity of coherent sheaves of $\mathcal{O}_T$-modules.

  (i) By Theorem 16.2.5, if $H^i(T, \mathcal{F}(r - i)) = 0$ for all $i \in \mathbb{N}$, then $\mathcal{F}$ is r-regular.
 (ii) In view of Definition 16.2.1(ii), we can say that $\mathcal{F}$ is r-regular if and only if N is r-regular at and above level 2.
(iii) Hence $\mathrm{reg}(\mathcal{F}) = \mathrm{reg}^2(N)$ (see Definition 16.2.9).

An important property that a sheaf of $\mathcal{O}_T$-modules might have is that of being generated by global sections. We shall show that this property is closely related to the concept of regularity.

20.4.12 Definition. Let the notation and hypotheses be as in 20.4.10. We say that the coherent sheaf $\mathcal{F}$ is *generated by its global sections* precisely when, for each $\mathfrak{p} \in T$, the stalk $\mathcal{F}_\mathfrak{p}$ of $\mathcal{F}$ at $\mathfrak{p}$ is generated 'by germs of global sections of $\mathcal{F}$', that is, if and only if

$$\mathcal{F}_\mathfrak{p} = \sum_{\gamma \in \Gamma(T, \mathcal{F})} \mathcal{O}_{T,\mathfrak{p}} \rho_{T,\mathfrak{p},\mathcal{F}}(\gamma) \quad \text{for all } \mathfrak{p} \in T,$$

where $\rho_{T,\mathfrak{p},\mathcal{F}} : \Gamma(T, \mathcal{F}) \longrightarrow \mathcal{F}_\mathfrak{p}$ is the natural map.

20.4.13 Serre's Criterion for Generation by Global Sections. (See [77, §66, Théorème 2].) *The hypotheses of* 20.4.1 *apply. Let* $\mathcal{F}$ *be a coherent sheaf of* $\mathcal{O}_T$-modules and let $t \in \mathbb{Z}$ with $t \geq \mathrm{reg}(\mathcal{F})$. *Then the twisted sheaf* $\mathcal{F}(t)$ *is generated by its global sections.*

Proof. By 20.4.6, we may write $\mathcal{F} = \big(\widetilde{M}\big)_0$ for some finitely generated graded R-module M. By 20.4.5(ii), there is an isomorphism $\mathcal{F}(t) \cong \big(\widetilde{M(t)}\big)_0$. Then, by 17.2.2 and with the notation of that exercise, and in view of the identification of 20.4.2(i), we have

$$(M_{(\mathfrak{p})})_t = \sum_{m \in \mathcal{S}} (R_{(\mathfrak{p})})_0 (\beta_{(\mathfrak{p})})_t(m) = \sum_{m \in \mathcal{S}} \mathcal{O}_{T,\mathfrak{p}} (\beta_{(\mathfrak{p})})_t(m)$$

for some set $\mathcal{S} \subseteq D_{R_+}(M)_t$. Moreover, by 20.4.3 and 20.4.5(ii), we have a commutative diagram

$$
\begin{array}{ccccc}
\Gamma(T, (\mathcal{F}(t))) & = & \big(\widetilde{M(t)}\big)_0(T) & \xrightarrow[\cong]{(\nu^T_{R_+, M})_t} & D_{R_+}(M)_t \\
\downarrow{\scriptstyle \rho_{T,\mathfrak{p},\mathcal{F}(t)}} & & & & \downarrow{\scriptstyle (\beta_{(\mathfrak{p})})_t} \\
(\mathcal{F}(t))_\mathfrak{p} & = & \big(\big(\widetilde{M(t)}\big)_0\big)_\mathfrak{p} & \xrightarrow[\cong]{(\psi^\mathfrak{p}_M)_t} & (M_{(\mathfrak{p})})_t
\end{array}
$$

Set $\mathcal{T} := \big((\nu^T_{R_+,M})_t\big)^{-1}(\mathcal{S}) \subseteq \Gamma(T, (\mathcal{F}(t)))$. Then, since $(\psi^\mathfrak{p}_{M(t)})_0 = (\psi^\mathfrak{p}_M)_t$ is an $\mathcal{O}_{T,\mathfrak{p}}$-isomorphism (by 20.4.2(ii)), it follows that

$$(\mathcal{F}(t))_\mathfrak{p} = \sum_{\gamma \in \mathcal{T}} \mathcal{O}_{T,\mathfrak{p}} \rho_{T,\mathfrak{p},\mathcal{F}(t)}(\gamma),$$

and this proves the claim. $\square$

The above theorem extends a result of Serre [77, §66, Théorème 2] which states that, if R is the homogeneous coordinate ring of a projective variety and $\mathcal{F}$ is a coherent sheaf of $\mathcal{O}_T$-modules, then $\mathcal{F}(t)$ is generated by its global sections for all $t \gg 0$.

20.4.14 Definition and ♯Exercise. (The hypotheses of 20.4.1 apply.) Let $\mathcal{F}$ be a non-zero coherent sheaf of $\mathcal{O}_T$-modules. Recall that the *support of* $\mathcal{F}$, denoted by $\mathrm{Supp}\,\mathcal{F}$, is the set $\{\mathfrak{p} \in T : \mathcal{F}_\mathfrak{p} \neq 0\}$ (see [30, Chapter II, Exercise 1.14, p. 67]); the *dimension of* $\mathcal{F}$, denoted by $\dim \mathcal{F}$, is defined to be the dimension of $\mathrm{Supp}\,\mathcal{F}$ (see 19.1.8).

By 20.4.6, there exists a finitely generated graded R-module N such that $\mathcal{F} \cong (\widetilde{N})_0$. Show that

(i) $\mathrm{Supp}\,\mathcal{F} = \mathrm{Supp}\,N \cap \mathrm{Proj}(R)$ (a closed subset of $\mathrm{Proj}(R)$), and
(ii) when R_0 is Artinian, $\dim \mathcal{F} = \dim N - 1$ (so that $\dim \mathcal{F}$ is finite) (you might find 19.7.1(i),(iv) helpful here).

20.4.15 Notation. (The hypotheses of 20.4.1 apply.) Assume that R_0 is Artinian and let $\mathcal{F}$ be a coherent sheaf of $\mathcal{O}_T$-modules.

By Serre's Finiteness Theorem 20.4.8, for each $i \in \mathbb{N}_0$ and $n \in \mathbb{Z}$, the R_0-module $H^i(T, \mathcal{F}(j))$ is finitely generated, and so has finite length: we denote this length by $h^i(T, \mathcal{F}(j))$.

We show next how the Serre–Grothendieck Correspondence enables us to produce quickly, for a coherent sheaf $\mathcal{F}$ of $\mathcal{O}_T$-modules as in 20.4.15, a Hilbert polynomial of $\mathcal{F}$ and, for each $i \in \mathbb{N}_0$, an i-th cohomological Hilbert polynomial of $\mathcal{F}$ which play similar rôles to the corresponding polynomials in 17.1.8 and 17.1.11.

20.4.16 Theorem and Definitions. *(The hypotheses of 20.4.1 are in force.) Assume that R_0 is Artinian and let $\mathcal{F}$ be a non-zero coherent sheaf of $\mathcal{O}_T$-modules. Set $d := \dim \mathcal{F}$.*

Now $h^i(T, \mathcal{F}) = 0$ for all $i \gg 0$, and the Euler characteristic $\chi(\mathcal{F})$ *of $\mathcal{F}$ is defined by $\chi(\mathcal{F}) := \sum_{i \in \mathbb{N}_0} (-1)^i h^i(T, \mathcal{F})$ (see [30, Chapter III, Exercise 5.1, p. 230]).*

There is a (necessarily uniquely determined) polynomial $P_{\mathcal{F}} \in \mathbb{Q}[X]$ of degree $d = \dim \mathcal{F}$ such that $P_{\mathcal{F}}(n) = \chi(\mathcal{F}(n))$ for all $n \in \mathbb{Z}$. This polynomial $P_{\mathcal{F}}$ is called the Hilbert polynomial *of $\mathcal{F}$.*

With the notation of 17.1.1(v), for each $j = 0, \ldots, d$, we set $e_j(\mathcal{F}) := e_j(P_{\mathcal{F}})$ and refer to this as the j-th Hilbert coefficient *of $\mathcal{F}$. Thus*

$$P_{\mathcal{F}}(X) = \sum_{i=0}^{d} (-1)^i e_i(\mathcal{F}) \binom{X + d - i}{d - i}.$$

Moreover, if N is a finitely generated graded R-module with $\mathcal{F} = (\widetilde{N})_0$ (see 20.4.6), then we have $\dim N = d + 1$, $P_{\mathcal{F}} = P_N$, $\chi_{\mathcal{F}} = \chi_N$ and $e_i(\mathcal{F}) = e_i(N)$ for all $i \in \{0, \ldots, d\}$.

Proof. By 20.4.6, there exists a finitely generated graded R-module N such that $\mathcal{F} = (\widetilde{N})_0$; by 20.4.14, $\dim \mathcal{F} = \dim N - 1$, so that $\dim N = d + 1$. Let $n \in \mathbb{Z}$ and $i \in \mathbb{N}$. We can now use 20.4.5(ii) and 20.3.15 to see that there are R_0-isomorphisms $\Gamma(T, \mathcal{F}(n)) \cong D_{R_+}(N)_n$ and $H^i(T, \mathcal{F}(n)) \cong \mathcal{R}^i D_{R_+}(N)_n \cong H^{i+1}_{R_+}(N)_n$.

Hence $h^i(T, \mathcal{F}(n)) = 0$ for all $i > d$, and, with the notation of 17.1.4, we have $\chi(\mathcal{F}(n)) = \chi_N(n)$. The result therefore follows from 17.1.7. $\square$

20.4.17 Definitions and Exercise. (The hypotheses of 20.4.1 are in force.) Assume that R_0 is Artinian and let $\mathcal{F}$ be a non-zero coherent sheaf of $\mathcal{O}_T$-modules. Let $i \in \mathbb{N}_0$. The function $h^i_{\mathcal{F}} : \mathbb{Z} \longrightarrow \mathbb{N}_0$ defined by $h^i_{\mathcal{F}}(n) = h^i(T, \mathcal{F}(n))$ for all $n \in \mathbb{Z}$ is referred to as the *i-th cohomological Hilbert function of $\mathcal{F}$.*

Show that there is a polynomial $p_{\mathcal{F}}^i \in \mathbb{Q}[X]$ of degree at most i such that $h_{\mathcal{F}}^i(n) = h^i(T, \mathcal{F}(n)) = p_{\mathcal{F}}^i(n)$ for all $n \ll 0$. The (uniquely determined) polynomial $p_{\mathcal{F}}^i$ is called the *i-th cohomological Hilbert polynomial of* $\mathcal{F}$.

Let $d := \dim \mathcal{F}$. Show that the leading term of $p_{\mathcal{F}}^d$ is

$$\frac{(-1)^d e_0(\mathcal{F})}{d!} X^d,$$

where $e_0(\mathcal{F})$ is as defined in 20.4.16.

We come now to a basic bounding result, due to Mumford, for the regularity of coherent sheaves of ideals.

20.4.18 Theorem: Mumford's Regularity Bound [54, Theorem, p. 101].
Suppose that $R = R_0[X_0, X_1, \ldots, X_d]$ *is a polynomial ring in* $d + 1$ *indeterminates* $X_0, X_1, \ldots, X_d$ *over an Artinian local ring* R_0, *where* $d \in \mathbb{N}$. *Regard* $R = \bigoplus_{n \in \mathbb{N}_0} R_n$ *as* $\mathbb{N}_0$-*graded with* $\deg X_i = 1$ *for all* $i = 0, \ldots, d$. *As in* 20.4.1, *we set* $T = \mathrm{Proj}(R)$. (*Thus* T *is just* $\mathbb{P}^d_{R_0}$, *projective d-space over* R_0: *see* [30, Chapter II, Example 2.5.1, p. 77].)

Then there is a function $F : \mathbb{Z}^{d+1} \longrightarrow \mathbb{Z}$ *such that, for each non-zero coherent sheaf* $\mathcal{I}$ *of ideals on* T, *we have* $\mathrm{reg}(\mathcal{I}) \leq F(e_0(\mathcal{I}), \ldots, e_d(\mathcal{I}))$.

Proof. By 20.4.7, the coherent sheaves of ideals on T are exactly the sheaves of the form $\mathcal{I} = \left(\widetilde{\mathfrak{b}}\right)_0$, where $\mathfrak{b}$ is a graded ideal of R. If $\mathfrak{a} \neq 0$ is such a graded ideal, corresponding to the coherent sheaf of ideals $\mathcal{J}$, then $\dim \mathfrak{a} = \dim \left(\left(\widetilde{\mathfrak{a}}\right)_0\right) + 1 = \dim \mathcal{J} + 1 = d + 1$ and $e_i(\mathfrak{a}) = e_i(\mathcal{J})$ for all $i \in \{0, \ldots, d\}$ (see 20.4.16). Also, $\mathrm{reg}(\mathcal{J}) = \mathrm{reg}\left(\left(\widetilde{\mathfrak{a}}\right)_0\right) = \mathrm{reg}^2(\mathfrak{a})$ by 20.4.11(iii). The result therefore follows from 17.3.6. $\square$

So far in this chapter, the results which we have obtained about sheaf cohomology have not been obviously related to local properties of the underlying sheaves. In our next sequence of results and exercises, we aim for the so-called Severi–Enriques–Zariski–Serre Vanishing Theorem (see F. Severi [78], F. Enriques [13], O. Zariski [88], and J.-P. Serre [77, §76, Théorème 4]), which establishes a fundamental link between the local structure of a coherent sheaf of $\mathcal{O}_T$-modules (conveyed by information about the depths of its stalks, for example) and global properties of the sheaf (described by the vanishing of certain cohomology groups, for example). Our approach to this Vanishing Theorem again makes use of the Serre–Grothendieck Correspondence, this time in conjunction with the Graded Finiteness Theorem 14.3.10.

20.4.19 ♯Exercise. (The hypotheses of 20.4.1 apply.) Assume that the graded R-module M is finitely generated, and set $\mathcal{F} := (\widetilde{M})_0$. Fix $\mathfrak{p} = x \in T$, and let $\mathfrak{m}_{T,x}$ denote the maximal ideal $(\mathfrak{p}R_{(\mathfrak{p})})_0$ of the local ring $\mathcal{O}_{T,x}$, which we identify with $(R_{(\mathfrak{p})})_0$ by means of the isomorphism of 20.4.2(i).

(i) Use 19.7.1(vii) to show that, if x is a closed point of T (so that $\mathfrak{p}$ is a maximal member of $\mathrm{Proj}(R)$ with respect to inclusion), then $\mathfrak{p} \cap R_0$ is a maximal ideal of R_0 and $\dim R/\mathfrak{p} = 1$.

(ii) Since R is homogeneous, there exists $t \in R_1 \setminus \mathfrak{p}$. Denote the ring of fractions $\mathcal{O}_{T,x}[X]_X$ of the polynomial ring $\mathcal{O}_{T,x}[X]$ by $\mathcal{O}_{T,x}[X, X^{-1}]$. Show that there is a unique homogeneous isomorphism of $\mathcal{O}_{T,x}$-algebras
$$\phi_t : \mathcal{O}_{T,x}[X, X^{-1}] \xrightarrow{\cong} R_{(\mathfrak{p})} \text{ for which } \phi_t(X) = t.$$

(iii) Identify the stalk $\mathcal{F}_x$ with $(M_{(\mathfrak{p})})_0$ by means of the isomorphism of 20.4.3. Show that, when $M_{(\mathfrak{p})}$ is considered as an $\mathcal{O}_{T,x}[X, X^{-1}]$-module by means of the isomorphism ϕ_t of part (ii), there is a homogeneous isomorphism of graded $\mathcal{O}_{T,x}[X, X^{-1}]$-modules
$$\psi_{t,M} : \mathcal{F}_x \otimes_{\mathcal{O}_{T,x}} \mathcal{O}_{T,x}[X, X^{-1}] \xrightarrow{\cong} M_{(\mathfrak{p})}$$
for which
$$\psi_{t,M}(z \otimes f) = \phi_t(f)z \quad \text{for all } z \in (M_{(\mathfrak{p})})_0 \text{ and } f \in \mathcal{O}_{T,x}[X, X^{-1}].$$

(iv) Show that the isomorphism ϕ_t of part (ii) leads to a ring isomorphism $\mathcal{O}_{T,x}[X]_{\mathfrak{m}_{T,x}\mathcal{O}_{T,x}[X]} \xrightarrow{\cong} R_{\mathfrak{p}}$. Conclude that $\dim \mathcal{O}_{T,x} = \mathrm{ht}\,\mathfrak{p}$, that $\mathrm{depth}\,\mathcal{O}_{T,x} = \mathrm{depth}\,R_{\mathfrak{p}}$, and that $\mathcal{O}_{T,x}$ is a domain, respectively normal, Cohen–Macaulay, Gorenstein, regular, if and only if $R_{\mathfrak{p}}$ has the same property.

(v) Show that the isomorphism $\psi_{t,M}$ of part (iii) gives rise to an isomorphism $\mathcal{F}_x \otimes_{\mathcal{O}_{T,x}} \mathcal{O}_{T,x}[X]_{\mathfrak{m}_{T,x}\mathcal{O}_{T,x}[X]} \xrightarrow{\cong} M_{\mathfrak{p}}$ of $\mathcal{O}_{T,x}[X]_{\mathfrak{m}_{T,x}\mathcal{O}_{T,x}[X]}$-modules. Conclude that $\mathrm{depth}_{\mathcal{O}_{T,x}}\mathcal{F}_x = \mathrm{depth}_{R_{\mathfrak{p}}} M_{\mathfrak{p}}$ and that $\mathcal{F}_x$ is free of rank r over $\mathcal{O}_{T,x}$ if and only if $M_{\mathfrak{p}}$ is free of rank r over $R_{\mathfrak{p}}$.

(vi) Let $\mathfrak{q} \in \mathrm{Spec}(R)$ be such that $\mathfrak{q}^* \subseteq \mathfrak{p}$. Show that $R_{\mathfrak{q}}$ is a domain, respectively normal, Cohen–Macaulay, Gorenstein, regular, if $\mathcal{O}_{T,x}$ has the same property.

20.4.20 ♯Exercise. (The hypotheses of 20.4.1 are in force.) Let $\pi : T = \mathrm{Proj}(R) \longrightarrow \mathrm{Spec}(R_0)$ be the natural map, defined by $\pi(\mathfrak{q}) = \mathfrak{q} \cap R_0$ for all $\mathfrak{q} \in T$. Fix $\mathfrak{p} = x \in T$. In the graded ring $R/\mathfrak{p}$, let $\Sigma(\mathfrak{p})$ denote the set of non-zero homogeneous elements of degree 0. Note that $\Sigma(\mathfrak{p})^{-1}(R/\mathfrak{p})$ is positively graded and homogeneous: set $T_{\langle x \rangle} := \mathrm{Proj}(\Sigma(\mathfrak{p})^{-1}(R/\mathfrak{p}))$.

(i) Show that there is a homeomorphism $T_{\langle x \rangle} \xrightarrow{\approx} \pi^{-1}(\{\pi(x)\}) \cap \overline{\{x\}}$, where the 'overline' is used to indicate closure in T.

(ii) Show that $\dim T_{\langle x \rangle} = \mathrm{ht}(R_+ + \mathfrak{p})/\mathfrak{p} - 1$.

20.4.21 Notation and Remark. (The hypotheses of 20.4.1 apply.) Let $\mathcal{F}$ be a coherent sheaf of $\mathcal{O}_T$-modules. We set

$$\delta(\mathcal{F}) := \inf \left\{ \mathrm{depth}_{\mathcal{O}_{T,x}} \mathcal{F}_x + \dim T_{\langle x \rangle} : x \in T \right\},$$

where $T_{\langle x \rangle}$, for $x \in T$, is as defined in 20.4.20.

By 20.4.6, there exists a finitely generated graded R-module N such that $\mathcal{F} \cong (\widetilde{N})_0$; by 20.4.19(v), for $x = \mathfrak{p} \in T$, we have

$$\mathrm{depth}_{\mathcal{O}_{T,x}} \mathcal{F}_x = \mathrm{depth}_{R_\mathfrak{p}} N_\mathfrak{p}.$$

Therefore, by 20.4.20, with the notation of 9.2.2,

$$\delta(\mathcal{F}) = \inf \left\{ \mathrm{depth}\, N_\mathfrak{p} + \mathrm{ht}(R_+ + \mathfrak{p})/\mathfrak{p} - 1 : \mathfrak{p} \in {}^*\mathrm{Spec}(R) \setminus \mathrm{Var}(R_+) \right\}$$

$$= \inf \left\{ \mathrm{adj}_{R_+} \mathrm{depth}\, N_\mathfrak{p} - 1 : \mathfrak{p} \in {}^*\mathrm{Spec}(R) \setminus \mathrm{Var}(R_+) \right\}.$$

In the case when R is a homomorphic image of a regular (commutative Noetherian) ring, we can deduce from the Graded Finiteness Theorem 14.3.10 that $\delta(\mathcal{F}) = f_{R_+}(N) - 1$. This observation is the key to our proof of the Severi–Enriques–Zariski–Serre Vanishing Theorem 20.4.23 below.

20.4.22 Exercise. (The hypotheses of 20.4.1 apply.) Let $\pi : T \to \mathrm{Spec}(R_0)$ be the natural map. Assume that R_0 is Artinian, and let $\mathcal{F}$ be a coherent sheaf of $\mathcal{O}_T$-modules.

(i) Show that $\pi^{-1}(\{\pi(x)\}) \cap \overline{\{x\}} = \overline{\{x\}}$ for all $x \in T$.
(ii) Use 9.3.5 to show that

$$\delta(\mathcal{F}) = \inf \left\{ \mathrm{depth}_{\mathcal{O}_{T,x}} \mathcal{F}_x : x \text{ is a closed point of } T \right\}.$$

20.4.23 The Severi–Enriques–Zariski–Serre Vanishing Theorem. (See Severi [78], Enriques [13], Zariski [88], and Serre [77, §76, Théorème 4]). *(The hypotheses of* 20.4.1 *apply.) Assume that R_0 is a homomorphic image of a regular (commutative Noetherian) ring. Let $\mathcal{F}$ be a coherent sheaf of $\mathcal{O}_T$-modules, and let $r \in \mathbb{N}_0$. Then $\delta(\mathcal{F}) > r$ if and only if $H^i(T, \mathcal{F}(j)) = 0$ for all $i \leq r$ and all $j \ll 0$.*

Proof. By 20.4.6, there is a finitely generated graded R-module N such that $\mathcal{F} \cong (\widetilde{N})_0$. By 20.4.4, there are homogeneous R-isomorphisms

$$\bigoplus_{j \in \mathbb{Z}} H^i(T, (\widetilde{N(j)})_0) \xrightarrow{\cong} H^{i+1}_{R_+}(N) \quad \text{for all } i \in \mathbb{N}$$

and $\bigoplus_{j \in \mathbb{Z}} \Gamma(T, (\widetilde{N(j)})_0) \xrightarrow{\cong} D_{R_+}(N)$. Also, it follows from 2.2.6(i)(c) that $D_{R_+}(N)$ is finitely generated if and only if $H^1_{R_+}(N)$ is.

In view of [7, Proposition 1.5.4], the hypothesis on R_0 ensures that R is a homomorphic image of a regular (commutative Noetherian) ring. It therefore follows from 20.4.21 that $\delta(\mathcal{F}) = f_{R_+}(N) - 1$, so that $\delta(\mathcal{F}) > r$ if and only if the graded R-module $\bigoplus_{j \in \mathbb{Z}} H^i(T, (\widetilde{N(j)})_0)$ is finitely generated for all $i \leq r$. The desired conclusions follow from these observations and Serre's Finiteness Theorem 20.4.8. $\square$

20.4.24 Exercise.   (The hypotheses of 20.4.1 apply.) Assume that R_0 is Artinian, and let $\mathcal{F}$ be a coherent sheaf of $\mathcal{O}_T$-modules. Recall that $\operatorname{Ass} \mathcal{F} := \{x \in T : \operatorname{depth}_{\mathcal{O}_{T,x}} \mathcal{F}_x = 0\}$. Show that the graded R-module $\Gamma_*(T, \mathcal{F}) = \bigoplus_{j \in \mathbb{Z}} \Gamma(T, \mathcal{F}(j))$ is finitely generated if and only if $\operatorname{Ass} \mathcal{F}$ contains no closed point of T.

20.5 Locally free sheaves

In this final section, we give some applications to locally free sheaves. These correspond to vector bundles in algebraic geometry, and form an important class of sheaves.

20.5.1 Hypotheses for the section.   Throughout this section, the hypotheses of 20.4.1 will be in force.

20.5.2 Definition.   A coherent sheaf $\mathcal{F}$ of $\mathcal{O}_T$-modules is called *locally free* if the stalk $\mathcal{F}_x$ is a free $\mathcal{O}_{T,x}$-module (of finite rank, as $\mathcal{F}$ is coherent) for all $x \in T$. (See [30, Chapter II, §5, p. 109, and Exercise 5.7, p. 124].)

20.5.3 ♯Exercise.   Suppose that T is connected and $\mathcal{O}_{T,x}$ is an integral domain for all $x \in T$. Show that T has a unique minimal member (with respect to inclusion of prime ideals) and conclude that T is irreducible.

20.5.4 Reminder.   Recall from [30, Chapter II, Exercise 3.8, p. 91] that T is said to be *normal* if and only if $\mathcal{O}_{T,x}$ is a normal integral domain for all $x \in T$. Similarly, T is said to be *regular* if and only if the local ring $\mathcal{O}_{T,x}$ is regular for all $x \in T$. (See [30, Chapter II, Remark 6.11.1A, p. 142].)

Observe that, if T is regular, then it is normal. Note also that, by 20.4.19(iv), $\operatorname{Proj}(K[X_0, \ldots, X_r])$ is regular (where K is a field, $r \in \mathbb{N}$, $X_0, \ldots, X_r$ are indeterminates, and the polynomial ring is graded so that $\deg X_i = 1$ for all $i = 0, \ldots, r$ and K is the component of degree 0).

20.5.5 Exercise.   Assume that R_0 is a field and that $T = \operatorname{Proj}(R)$ is connected. Set $d := \dim T$.

Suppose that T is normal (see 20.5.4), and assume that $d \geq 2$. Let $\mathcal{F}$ be a locally free coherent sheaf of $\mathcal{O}_T$-modules. Show that $H^1(T, \mathcal{F}(j)) = 0$ for all $j \ll 0$. (Compare this with Hartshorne's presentation of the Enriques–Severi–Zariski Lemma in [30, Chapter III, Corollary 7.8, p. 244].)

20.5.6 Theorem: Serre's Cohomological Criterion for Local Freeness [77, §75, Théorème 3]. *Assume that R_0 is a field, and that $T = \operatorname{Proj}(R)$ is connected and regular (see 20.5.4). Let $d := \dim T$. Let $\mathcal{F}$ be a coherent sheaf of $\mathcal{O}_T$-modules. Then $\mathcal{F}$ is locally free if and only if $H^i(T, \mathcal{F}(n)) = 0$ for all $i < d$ and all $n \ll 0$.*

Proof. By 20.4.23, it is enough for us to show that $\mathcal{F}$ is locally free if and only if $\delta(\mathcal{F}) \geq d$. (See 20.4.21 for the definition of $\delta(\mathcal{F})$.) First of all observe that T has a unique minimal member $\mathfrak{q}$ (by 20.5.3). Since R_0 is a field, $\operatorname{Var}(R_+) = \{R_+\}$. Since $\mathfrak{q} \in T$, it follows that $\mathfrak{q}$ is the unique minimal prime of R. Let $x = \mathfrak{p} \in T$. Then we have $\mathfrak{q} \subseteq \mathfrak{p} \subseteq \mathfrak{p} + R_+ \in {}^*\operatorname{Spec}(R)$. So, on use of 20.4.19(iv) and 20.4.20, we obtain

$$\dim \mathcal{O}_{T,x} + \dim T_{\langle x \rangle} = \operatorname{ht} \mathfrak{p} + \operatorname{ht}(R_+ + \mathfrak{p})/\mathfrak{p} - 1 = \operatorname{ht}(R_+ + \mathfrak{p}) - 1$$
$$= \operatorname{ht} R_+ - 1 = \dim R - 1 = d$$

because R is a finitely generated R_0-algebra (and therefore catenary) and has a unique minimal prime $\mathfrak{q}$. Thus $\dim \mathcal{O}_{T,x} + \dim T_{\langle x \rangle} = d$ for all $x \in T$.

Now assume that $\mathcal{F}$ is locally free. Then, for each $x \in T$,

$$\operatorname{depth}_{\mathcal{O}_{T,x}} \mathcal{F}_x = \operatorname{depth} \mathcal{O}_{T,x} = \dim \mathcal{O}_{T,x},$$

as $\mathcal{O}_{T,x}$ is regular, and therefore Cohen–Macaulay. It follows that

$$\operatorname{depth}_{\mathcal{O}_{T,x}} \mathcal{F}_x + \dim T_{\langle x \rangle} = \dim \mathcal{O}_{T,x} + \dim T_{\langle x \rangle} = d \quad \text{for all } x \in T,$$

so that $\delta(\mathcal{F}) = d$.

Conversely, suppose that $\delta(\mathcal{F}) \geq d$. Then, for each $x \in T$, we have

$$\operatorname{depth}_{\mathcal{O}_{T,x}} \mathcal{F}_x + \dim T_{\langle x \rangle} \geq d = \dim \mathcal{O}_{T,x} + \dim T_{\langle x \rangle},$$

so that $\operatorname{depth}_{\mathcal{O}_{T,x}} \mathcal{F}_x \geq \dim \mathcal{O}_{T,x}$. As $\mathcal{O}_{T,x}$ is regular, it follows from the Auslander–Buchsbaum Theorem (see [50, Theorem 19.1], for example) that the $\mathcal{O}_{T,x}$-module $\mathcal{F}_x$ is projective, and therefore free (by [50, Theorem 2.5]). $\square$

Finally, we consider locally free sheaves over projective d-space $\mathbb{P}^d_K = \operatorname{Proj}(K[X_0, \ldots, X_d])$ over K, where K is an algebraically closed field.

20.5.7 Reminder and ♯Exercise. (The hypotheses of 20.4.1 apply.)

(i) In the category $\mathcal{S}(\mathcal{O}_T)$ of sheaves of $\mathcal{O}_T$-modules one may form direct sums, and the functors $(\widetilde{\,\bullet\,})_0$ (from $*\mathcal{C}(R)$ to $\mathcal{S}(\mathcal{O}_T)$) and $H^i(T, \bullet)$ ($i \in \mathbb{N}_0$) (from $\mathcal{S}(\mathcal{O}_T)$ to $*\mathcal{C}(R)$) are additive. So, if $r \in \mathbb{N}$, $N_1, \ldots, N_r$ are graded R-modules and $\mathcal{F}_1, \ldots, \mathcal{F}_r$ are sheaves of $\mathcal{O}_T$-modules, we have $\left(\bigoplus_{j=1}^r N_j\right)_0 \cong \bigoplus_{j=1}^r \left(\widetilde{N_j}\right)_0$ and

$$H^i\left(T, \bigoplus_{j=1}^r \mathcal{F}_j\right) \cong \bigoplus_{j=1}^r H^i(T, \mathcal{F}_j) \quad \text{for all } i \in \mathbb{N}_0.$$

(ii) Assume now that $d \in \mathbb{N}$, that K is a field and $R = K[X_0, \ldots, X_d]$ is a polynomial ring, so that $T = \operatorname{Proj}(R)$ becomes projective d-space $\mathbb{P}_K^d$ over K. A sheaf $\mathcal{F}$ of $\mathcal{O}_T$-modules is said to *split (completely)* (the word 'completely' is often omitted) if and only if $\mathcal{F} \cong \bigoplus_{j=1}^r \mathcal{O}_T(a_j)$ for some $r \in \mathbb{N}$ and some integers $a_1, \ldots, a_r$. Show that, if this is the case, then $\mathcal{F}$ is coherent and locally free, and if also the a_i are numbered so that $a_1 \geq a_2 \geq \cdots \geq a_r$, then $a := (a_1, \ldots, a_r) \in \mathbb{Z}^r$ is uniquely determined by $\mathcal{F}$. In this situation, $a = (a_1, \ldots, a_r)$ is called the *splitting type of $\mathcal{F}$*. (You might find it helpful to consider the cohomological Hilbert functions $n \mapsto h^0(T, \mathcal{F}(n))$ and $n \mapsto h^0(T, \mathcal{O}_T(a_j)(n))$, in conjunction with the Serre–Grothendieck Correspondence 20.4.4).

20.5.8 Theorem: Horrocks' Splitting Criterion [40]. *Let $d \in \mathbb{N}$, let K be a field, let $T = \mathbb{P}_K^d$ and let $\mathcal{F}$ be a non-zero coherent sheaf of $\mathcal{O}_T$-modules. Then the following statements are equivalent:*

(i) *$\mathcal{F}$ splits;*
(ii) *$H^0(T, \mathcal{F}(n)) = 0$ for all $n \ll 0$ and $H^i(T, \mathcal{F}(n)) = 0$ for all $i \in \{1, \ldots, d-1\}$ and all $n \in \mathbb{Z}$.*

Proof. Here, $T = \operatorname{Proj}(R)$ where R is the polynomial ring $K[X_0, \ldots, X_d]$ with $\deg X_i = 1$ for all $i = 0, \ldots, d$.

(i) $\Rightarrow$ (ii) Assume that $\mathcal{F}$ splits, so that $\mathcal{F} \cong \bigoplus_{j=1}^r \mathcal{O}_T(a_j)$ for some $r \in \mathbb{N}$ and some integers $a_1, \ldots, a_r$. On use of 20.5.7(i), we obtain

$$H^i(T, \mathcal{F}(n)) \cong \bigoplus_{j=1}^r H^i(T, \mathcal{O}_T(a_j + n)) \cong \bigoplus_{j=1}^r H^i\left(T, \widetilde{R}_0(a_j + n)\right)$$

$$\cong \bigoplus_{j=1}^r H^i\left(T, \widetilde{R(a_j + n)_0}\right) \quad \text{for all } i \in \mathbb{N}_0 \text{ and all } n \in \mathbb{Z}.$$

By the Serre–Grothendieck Correspondence 20.4.4,

$$H^0\left(T, \widetilde{R(a_j + n)_0}\right) \cong D_{R_+}(R)_{a_j+n},$$

and $D_{R_+}(R)_{a_j+n} \cong R_{a_j+n}$ as $H^0_{R_+}(R) = H^1_{R_+}(R) = 0$. Therefore

$$H^0\left(T, \widetilde{R(a_j + n)_0}\right) = 0 \quad \text{for all } n < -a_j,$$

and so $H^0(T, \mathcal{F}(n)) = 0$ for all $n \ll 0$. Also, by Theorem 20.4.4 once again, for all $i > 0$ and all $n \in \mathbb{Z}$, we have $H^i\big(T, \widetilde{R(a_j + n)}_0\big) \cong H^{i+1}_{R_+}(R)_{a_j+n}$, and this vanishes for all $i \in \{1, \ldots, d-1\}$ as grade $R_+ = d+1$. It follows that $H^i(T, \mathcal{F}(n)) = 0$ for all $i \in \{1, \ldots, d - 1\}$ and all $n \in \mathbb{Z}$. Hence statement (ii) is true.

(ii) $\Rightarrow$ (i) Assume that statement (ii) is true. By 20.4.6, we can write $\mathcal{F} = \big(\widetilde{M}\big)_0$ for a finitely generated graded R-module M. Then it follows from the Serre–Grothendieck Correspondence 20.4.4 that $H^i_{R_+}(M) = 0$ for all $i \in \{2, \ldots, d\}$ and $D_{R_+}(M)_n = 0$ for all $n \ll 0$. Consider the exact sequence

$$0 \longrightarrow \Gamma_{R_+}(M) \longrightarrow M \xrightarrow{\eta_M} D_{R_+}(M) \longrightarrow H^1_{R_+}(M) \longrightarrow 0$$

of 13.5.4(i), in which all the homomorphisms are homogeneous. As the K-vector space $H^1_{R_+}(M)_n$ has finite dimension for all $n \in \mathbb{Z}$ and vanishes for all large n (by 16.1.5), it follows that the graded R-module $D_{R_+}(M)$ is finitely generated. Moreover, if we apply the exact functor $(\widetilde{\bullet})_0$ to the above exact sequence and observe that both $\Gamma_{R_+}(M)$ and $H^1_{R_+}(M)$ are R_+-torsion (so that $\big(\widetilde{\Gamma_{R_+}(M)}\big)_0 = \big(\widetilde{H^1_{R_+}(M)}\big)_0 = 0$ (by 20.3.2(i))), we get an isomorphism of sheaves $\mathcal{F} = \big(\widetilde{M}\big)_0 \cong \big(\widetilde{D_{R_+}(M)}\big)_0$. Thus $D_{R_+}(M)$ is finitely generated and $\mathcal{F} \cong (\widetilde{D_{R_+}(M)})_0$. Now by 2.2.10(iv),(v), we have $H^i_{R_+}(D_{R_+}(M)) = 0$ for $i = 0, 1$ and $H^i_{R_+}(D_{R_+}(M)) \cong H^i_{R_+}(M)$ for all $i > 1$, As $H^i_{R_+}(M) = 0$ for all $i \in \{2, \ldots, d\}$, we get that $H^i_{R_+}(D_{R_+}(M)) = 0$ for all $i \in \{0, \ldots, d\}$. Therefore $\mathrm{grade}_{D_{R_+}(M)} R_+ \geq d + 1$. As $0 \neq \mathcal{F} \cong \big(\widetilde{D_{R_+}(M)}\big)_0$, we have $D_{R_+}(M) \neq 0$, so that $\mathrm{grade}_{D_{R_+}(M)} R_+ = d + 1$. So, by Hilbert's Syzygy Theorem (see [7, Corollary 2.2.15]), the finitely generated graded R-module $D_{R_+}(M)$ is free, and so $D_{R_+}(M) \cong \bigoplus_{j=1}^r R(a_j)$ (in $*\mathcal{C}(R)$) for some $r \in \mathbb{N}$ and some $a_1, \ldots, a_r \in \mathbb{Z}$. Therefore, in view of 20.4.5(ii),

$$\mathcal{F} \cong \Big(\widetilde{\bigoplus_{j=1}^r R(a_j)}\Big)_0 \cong \bigoplus_{j=1}^r \big(\widetilde{R(a_j)}\big)_0 \cong \bigoplus_{j=1}^r \big(\widetilde{R}\big)_0(a_j)$$
$$= \bigoplus_{j=1}^r \mathcal{O}_T(a_j).$$

Therefore $\mathcal{F}$ splits. $\qquad\square$

20.5.9 Corollary: Grothendieck's Splitting Theorem [23, Théorème 2.1]. *Let K be a field and let $T = \mathbb{P}^1_K$ be the projective line over K. Then each non-zero coherent locally free sheaf $\mathcal{F}$ of $\mathcal{O}_T$-modules splits.*

Proof. Here $T = \mathrm{Proj}(R)$ where R is the polynomial ring $K[X_0, X_1]$ with $\deg X_0 = \deg X_1 = 1$.

As T is regular (see 20.5.4) of dimension 1 and connected, Serre's Criterion for Local Freeness 20.5.6 implies that $H^0(T, \mathcal{F}(n)) = 0$ for all $n \ll 0$. We can therefore apply 20.5.8 to complete the proof. $\qquad\square$

References

[1] Y. AOYAMA, 'On the depth and the projective dimension of the canonical module', *Japanese J. Math.* 6 (1980) 61–66.

[2] Y. AOYAMA, 'Some basic results on canonical modules', *J. Math. Kyoto Univ.* 23 (1983) 85–94.

[3] H. BASS, 'On the ubiquity of Gorenstein rings', *Math. Zeit.* 82 (1963) 8–28.

[4] M. BRODMANN and C. HUNEKE, 'A quick proof of the Hartshorne–Lichtenbaum Vanishing Theorem', *Algebraic geometry and its applications* (Springer, New York, 1994), pp. 305–308.

[5] M. BRODMANN and J. RUNG, 'Local cohomology and the connectedness dimension in algebraic varieties', *Comment. Math. Helvetici* 61 (1986) 481–490.

[6] M. BRODMANN and R. Y. SHARP, 'Supporting degrees of multi-graded local cohomology modules', *J. Algebra* 321 (2009) 450–482.

[7] W. BRUNS and J. HERZOG, *Cohen–Macaulay rings*, Cambridge Studies in Advanced Mathematics 39, Revised Edition (Cambridge University Press, Cambridge, 1998).

[8] F. W. CALL and R. Y. SHARP, 'A short proof of the local Lichtenbaum–Hartshorne Theorem on the vanishing of local cohomology', *Bull. London Math. Soc.* 18 (1986) 261–264.

[9] C. D'CRUZ, V. KODIYALAM and J. K. VERMA, 'Bounds on the a-invariant and reduction numbers of ideals', *J. Algebra* 274 (2004) 594–601.

[10] D. EISENBUD, *Commutative algebra with a view toward algebraic geometry*, Graduate Texts in Mathematics 150 (Springer, New York, 1994).

[11] D. EISENBUD and S. GOTO, 'Linear free resolutions and minimal multiplicity', *J. Algebra* 88 (1984) 89–133.

[12] J. ELIAS, 'Depth of higher associated graded rings', *J. London Math. Soc.* (2) 70 (2004) 41–58.

[13] F. ENRIQUES, *Le superficie algebriche* (Zanichelli, Bologna, 1949).

[14] G. FALTINGS, 'Über die Annulatoren lokaler Kohomologiegruppen', *Archiv der Math.* 30 (1978) 473–476.

[15] G. FALTINGS, 'Algebraisation of some formal vector bundles', *Annals of Math.* 110 (1979) 501–514.

[16] G. FALTINGS, 'Der Endlichkeitssatz in der lokalen Kohomologie', *Math. Annalen* 255 (1981) 45–56.

[17] R. FEDDER and K. WATANABE, 'A characterization of F-regularity in terms of F-purity', *Commutative algebra: proceedings of a microprogram held June 15 – July 2, 1987*, Mathematical Sciences Research Institute Publications 15 (Springer, New York, 1989), pp. 227–245.

[18] D. FERRAND and M. RAYNAUD, 'Fibres formelles d'un anneau local Noethérien', *Ann. Sci. École Norm. Sup.* 3 (1970) 295–311.

[19] H.-B. FOXBY, 'Gorenstein modules and related modules', *Math. Scand.* 31 (1972) 267–284.

[20] W. FULTON and J. HANSEN, 'A connectedness theorem for projective varieties, with applications to intersections and singularities of mappings', *Annals of Math.* 110 (1979) 159–166.

[21] P. GABRIEL, 'Des catégories abéliennes', *Bull. Soc. Math. France* 90 (1962) 323–448.

[22] S. GOTO and K. WATANABE, 'On graded rings, II ($\mathbb{Z}^n$-graded rings)', *Tokyo J. Math.* 1 (1978) 237–261.

[23] A. GROTHENDIECK, 'Sur la classification des fibrés holomorphes sur la sphère de Riemann', *American J. Math.* 79 (1957) 121–138.

[24] A. GROTHENDIECK, 'Éléments de géométrie algébrique IV: étude locale des schémas et des morphismes de schémas', *Institut des Hautes Études Scientifiques Publications Mathématiques* 24 (1965) 5–231.

[25] A. GROTHENDIECK, *Local cohomology*, Lecture Notes in Mathematics 41 (Springer, Berlin, 1967).

[26] A. GROTHENDIECK, *Cohomologie locale des faisceaux cohérents et théorèmes de Lefschetz locaux et globaux (SGA 2)*, Séminaire de Géométrie Algébrique du Bois-Marie 1962 (North-Holland, Amsterdam, 1968).

[27] J. HARRIS, *Algebraic geometry: a first course*, Graduate Texts in Mathematics 133 (Springer, New York, 1992).

[28] R. HARTSHORNE, 'Complete intersections and connectedness', *American J. Math.* 84 (1962) 497–508.

[29] R. HARTSHORNE, 'Cohomological dimension of algebraic varieties', *Annals of Math.* 88 (1968) 403–450.

[30] R. HARTSHORNE, *Algebraic geometry*, Graduate Texts in Mathematics 52 (Springer, New York, 1977).

[31] M. HERRMANN, E. HYRY and T. KORB, 'On a-invariant formulas', *J. Algebra* 227 (2000) 254–267.

[32] M. HERRMANN, S. IKEDA and U. ORBANZ, *Equimultiplicity and blowing up* (Springer, Berlin, 1988).

[33] J. HERZOG and E. KUNZ, *Der kanonische Modul eines Cohen–Macaulay-Rings*, Lecture Notes in Mathematics 238 (Springer, Berlin, 1971).

[34] J. HERZOG and E. KUNZ, *Die Wertehalbgruppe eines lokalen Rings der Dimension 1*, Sitzungsberichte der Heidelberger Akademie der Wissenschaften Mathematisch-naturwissenschaftliche Klasse, Jahrgang 1971 (Springer, Berlin, 1971).

[35] L. T. HOA, 'Reduction numbers and Rees algebras of powers of an ideal', *Proc. American Math. Soc.* 119 (1993) 415–422.

[36] L. T. HOA and C. MIYAZAKI, 'Bounds on Castelnuovo–Mumford regularity for generalized Cohen–Macaulay graded rings', *Math. Annalen* 301 (1995) 587–598.

[37] M. HOCHSTER, 'Contracted ideals from integral extensions of regular rings', *Nagoya Math. J.* 51 (1973) 25–43.

[38] M. HOCHSTER and C. HUNEKE, 'Tight closure, invariant theory and the Briançon–Skoda Theorem', *J. American Math. Soc.* 3 (1990) 31–116.

[39] M. HOCHSTER and C. HUNEKE, 'Indecomposable canonical modules and connectedness', *Commutative algebra: syzygies, multiplicities, and birational algebra (South Hadley, MA, 1992)*, Contemporary Mathematics 159 (American Mathematical Society, Providence, RI, 1994), pp. 197–208.

[40] G. HORROCKS, 'Vector bundles on the punctured spectrum of a local ring', *Proc. London Math. Soc.* (3) 14 (1964) 689–713.

[41] C. HUNEKE, *Tight closure and its applications*, Conference Board of the Mathematical Sciences Regional Conference Series in Mathematics 88 (American Mathematical Society, Providence, RI, 1996).

[42] C. HUNEKE, 'Tight closure, parameter ideals and geometry', *Six lectures on commutative algebra (Bellaterra, 1996)*, Progress in Mathematics 166 (Birkhäuser, Basel, 1998), pp. 187–239.

[43] C. L. HUNEKE and R. Y. SHARP, 'Bass numbers of local cohomology modules', *Transactions American Math. Soc.* 339 (1993) 765–779.

[44] D. KIRBY, 'Coprimary decomposition of Artinian modules', *J. London Math. Soc.* (2) 6 (1973) 571–576.

[45] I. G. MACDONALD, 'Secondary representation of modules over a commutative ring', Symposia Matematica 11 (Istituto Nazionale di alta Matematica, Roma, 1973) 23–43.

[46] I. G. MACDONALD, 'A note on local cohomology', *J. London Math. Soc.* (2) 10 (1975) 263–264.

[47] I. G. MACDONALD and R. Y. SHARP, 'An elementary proof of the non-vanishing of certain local cohomology modules', *Quart. J. Math. Oxford* (2) 23 (1972) 197–204.

[48] T. MARLEY, 'The reduction number of an ideal and the local cohomology of the associated graded ring', *Proc. American Math. Soc.* 117 (1993) 335–341.

[49] E. MATLIS, 'Injective modules over Noetherian rings', *Pacific J. Math.* 8 (1958) 511–528.

[50] H. MATSUMURA, *Commutative ring theory*, Cambridge Studies in Advanced Mathematics 8 (Cambridge University Press, Cambridge, 1986).

[51] L. MELKERSSON, 'On asymptotic stability for sets of prime ideals connected with the powers of an ideal', *Math. Proc. Cambridge Philos. Soc.* 107 (1990) 267–271.

[52] L. MELKERSSON, 'Some applications of a criterion for artinianness of a module', *J. Pure and Applied Algebra* 101 (1995) 291–303.

[53] E. MILLER and B. STURMFELS, *Combinatorial commutative algebra*, Graduate Texts in Mathematics 227 (Springer, New York, 2005).

[54] D. MUMFORD, *Lectures on curves on an algebraic surface*, Annals of Mathematics Studies 59 (Princeton University Press, Princeton, NJ, 1966).

[55] M. P. MURTHY, 'A note on factorial rings', *Arch. Math. (Basel)* 15 (1964) 418–420.

[56] M. NAGATA, *Local rings* (Interscience, New York, 1962).

[57] U. NAGEL, 'On Castelnuovo's regularity and Hilbert functions', *Compositio Math.* 76 (1990) 265–275.

[58] U. NAGEL and P. SCHENZEL, 'Cohomological annihilators and Castelnuovo–Mumford regularity', *Commutative algebra: syzygies, multiplicities, and birational algebra (South Hadley, MA, 1992)*, Contemporary Mathematics 159 (American Mathematical Society, Providence, RI, 1994), pp. 307–328.

[59] D. G. NORTHCOTT, *Ideal theory*, Cambridge Tracts in Mathematics and Mathematical Physics 42 (Cambridge University Press, Cambridge, 1953).

[60] D. G. NORTHCOTT, *An introduction to homological algebra* (Cambridge University Press, Cambridge, 1960).

[61] D. G. NORTHCOTT, *Lessons on rings, modules and multiplicities* (Cambridge University Press, Cambridge, 1968).

[62] D. G. NORTHCOTT, 'Generalized Koszul complexes and Artinian modules', *Quart. J. Math. Oxford* (2) 23 (1972) 289–297.

[63] D. G. NORTHCOTT and D. REES, 'Reductions of ideals in local rings', *Proc. Cambridge Philos. Soc.* 50 (1954) 145–158.

[64] L. O'CARROLL, 'On the generalized fractions of Sharp and Zakeri', *J. London Math. Soc.* (2) 28 (1983) 417–427.

[65] A. OOISHI, 'Castelnuovo's regularity of graded rings and modules', *Hiroshima Math. J.* 12 (1982) 627–644.

[66] C. PESKINE and L. SZPIRO, 'Dimension projective finie et cohomologie locale', *Institut des Hautes Études Scientifiques Publications Mathématiques* 42 (1973) 323–395.

[67] L. J. RATLIFF, JR., 'Characterizations of catenary rings', *American J. Math.* 93 (1971) 1070–1108.

[68] D. REES, 'The grade of an ideal or module', *Proc. Cambridge Philos. Soc.* 53 (1957) 28–42.

[69] I. REITEN, 'The converse to a theorem of Sharp on Gorenstein modules', *Proc. American Math. Soc.* 32 (1972) 417–420.

[70] P. ROBERTS, *Homological invariants of modules over commutative rings*, Séminaire de Mathématiques Supérieures (Les Presses de l'Université de Montréal, Montréal, 1980).

[71] J. J. ROTMAN, *An introduction to homological algebra* (Academic Press, Orlando, FL, 1979).

[72] P. SCHENZEL, 'Einige Anwendungen der lokalen Dualität und verallgemeinerte Cohen–Macaulay-Moduln', *Math. Nachr.* 69 (1975) 227–242.

[73] P. SCHENZEL, 'Flatness and ideal-transforms of finite type', *Commutative algebra, Proceedings, Salvador 1988*, Lecture Notes in Mathematics 1430 (Springer, Berlin, 1990), pp. 88–97.

[74] P. SCHENZEL, 'On the use of local cohomology in algebra and geometry', *Six lectures on commutative algebra (Bellaterra, 1996)*, Progress in Mathematics 166 (Birkhäuser, Basel, 1998), pp. 241–292.

[75] P. SCHENZEL, 'On birational Macaulayfications and Cohen–Macaulay canonical modules', *J. Algebra* 275 (2004) 751–770.

[76] P. SCHENZEL, N. V. TRUNG and N. T. CUONG, 'Verallgemeinerte Cohen–Macaulay-Moduln', *Math. Nachr.* 85 (1978) 57–73.

[77] J.-P. SERRE, 'Faisceaux algébriques cohérents', *Annals of Math.* 61 (1955) 197–278.

[78] F. SEVERI, *Serie, sistemi d'equivalenza e corrispondenze algebriche sulle varietà algebriche* (a cura di F. Conforto e di E. Martinelli, Roma, 1942).

[79] R. Y. SHARP, 'Finitely generated modules of finite injective dimension over certain Cohen–Macaulay rings', *Proc. London Math. Soc.* (3) 25 (1972) 303–328.

[80] R. Y. SHARP, 'On the attached prime ideals of certain Artinian local cohomology modules', *Proc. Edinburgh Math. Soc.* (2) 24 (1981) 9–14.

[81] R. Y. SHARP, *Steps in commutative algebra: Second edition*, London Mathematical Society Student Texts 51 (Cambridge University Press, Cambridge, 2000).

[82] R. Y. SHARP and M. TOUSI, 'A characterization of generalized Hughes complexes', *Math. Proc. Cambridge Philos. Soc.* 120 (1996) 71–85.

[83] J. R. STROOKER, *Homological questions in local algebra*, London Mathematical Society Lecture Notes 145 (Cambridge University Press, Cambridge, 1990).

[84] J. STÜCKRAD and W. VOGEL, *Buchsbaum rings and applications* (Springer, Berlin, 1986).

[85] K. SUOMINEN, 'Localization of sheaves and Cousin complexes', *Acta Mathematica* 131 (1973) 27–41.

[86] N. V. TRUNG, 'Reduction exponent and degree bound for the defining equations of graded rings', *Proc. American Math. Soc.* 101 (1987) 229–236.

[87] N. V. TRUNG, 'The largest non-vanishing degree of graded local cohomology modules', *J. Algebra* 215 (1999) 481–499.

[88] O. ZARISKI, 'Complete linear systems on normal varieties and a generalization of a lemma of Enriques–Severi', *Annals of Math.* 55 (1952) 552–592.

[89] O. ZARISKI and P. SAMUEL, *Commutative algebra, Vol. II*, Graduate Texts in Mathematics 29 (Springer, Berlin, 1975).

Index

Printed in the United States
By Bookmasters